Jacly

Elementary Differential Equations

with Boundary Value Problems

David L. Powers

Clarkson University

 Prindle, Weber & Schmidt

Boston

PWS PUBLISHERS

Prindle, Weber & Schmidt • ♣ • Duxbury Press • ♠ • PWS Engineering • △
Statler Office Building • 20 Park Plaza • Boston, Massachusetts 02116

PWS Publishers is a division of Wadsworth, Inc.

89 88 87 86 85 — 10 9 8 7 6 5 4 3 2 1

Library of Congress Cataloging in Publication Data
Powers, David L.
    Elementary differential equations with
boundary value problems.

    Bibliography: p.
    Includes index.
    1. Differential equations. 2. Boundary value
problems.   I. Title.
QA371.P67   1985   515.3′5   85-3412
ISBN 0-87150-431-6

Sponsoring editor: David Pallai
Text design: Trisha Hanlon
Production coordinator: Helen Walden
Text composition: Universities Press
Printing and binding: Halliday Lithograph
Cover design: Helen Walden
Cover printing: New England Book Components
Cover photo: Tom Pantages

Preface

This book was written for second-year engineering and science students. The intended audience and my own experience as an engineering student have influenced the writing in three ways. First, applications motivate and illustrate the mathematics throughout. Second, methods are presented before theory wherever possible, so that the student approaches generalizations with a body of examples in mind. Third, most theorems are not proved, although they are explained, illustrated, and interpreted.

Two semesters of calculus is the required background. Infinite series, for Chapters 4, 9, and 10, and partial derivatives, for Chapter 10 and a few scattered sections, can be studied concurrently. Except for the bare facts about determinants, no linear algebra is needed in the bulk of the book. However most of Chapter 7, Systems of Linear Differential Equations, assumes a working knowledge of matrices. Appendix A is a self-contained text on matrix algebra and systems of algebraic equations for review or class presentation.

To accommodate different courses, curricula, teachers, and students, I have made this book as flexible as possible. Chapters 1–3 are to be taken in order; 4–8 are independent of each other; 9 and 10 form a sequence, and both draw on 4. Within each chapter, the material is arranged to allow different stopping points. The *Instructor's Manual* contains the particulars of chapter and section dependence as well as other useful information. Topics were chosen with a view to the current and probable future demands of the various engineering and science disciplines. There is enough material to design a two-semester course with any one of several different biases— toward classical applications, for instance, or systems and control, or computation.

The book has a number of special features that enhance its value as a text and reference.

* Over 225 examples illustrate definitions and theorems (in both the positive and negative senses) and guide the student in the use of new methods.

* More than 1500 exercises are provided, ranging from drill to novel applications, extensions of methods and theory, and previews of future material.

*   Solutions of odd-numbered exercises are in the back of the book. Answers to even-numbered exercises are available. Some 300 exercises are worked in detail in the *Student's Partial Solutions Manual* written to accompany this text.

*   Miscellaneous exercises conclude each chapter. Some of these are drill exercises for test preparation. Other problems require several sections' results, develop new methods, or take old methods in new directions.

*   Notes and references at the end of each chapter comment on the subject from a broader viewpoint, telling why and to whom it is important, how it is related to others and where to find out more about it. A bibliography is at the end of the book.

*   Each chapter has at least one section on an advanced or unusual topic. For example, Section 3.6 states and proves some theorems on boundedness and oscillation; Section 8.6 is an introduction to Jacobian elliptic functions.

*   Appendix B, Mathematical References, lists some useful formulas and theorems from trigonometry, algebra, and calculus.

It is my pleasure to acknowledge the many contributions, through conversations and helpful comments, of friends and colleagues including Mark Ablowitz, Heino Ainso, Bill Briggs, Axel Brinck, Susan Conry, George Davis, Larry Glasser, Charles Haines, Abdul Jerri, Victor Lovass-Nagy, Robert Meyer, Richard Miller, Gustave Rabson, Harvey Segur, Eric Thacher and the late R.G. Bradshaw. I also wish to thank the following reviewers:

Gregory F. Lawlor, Duke University
Theodore Burton, Southern Illinois University
B.J. Harris, Northern Illinois University
Wayne Dickey, University of Wisconsin
Richard Brown, Kent State University
Herbert Synder, Southern Illinois University
Robert E. Turner, University of Wisconsin
Beverly West, Cornell University
Erol Barbut, University of Washington
Euel W. Kennedy, California Polytechnical State University
Edward Scott, University of Illinois
Wlodek Proskuwoski, University of Southern California
Dennis R. Dunninger, Michigan State University
William Gilpin, Old Dominion University
C.H. Cook, University of Maryland
Gene M. Ortner, Michigan Technological University
Allan M. Krall, The Pennsylvania State University

Contents

1   First-Order Equations

2   Second-Order Linear Equations: Basic Methods and Applications

3   Theory of Linear Equations

4   Power Series Methods

5   Laplace Transform

6   Numerical Methods

7 Systems of Linear Differential Equations

8 Nonlinear Second-Order Equations

9 Boundary Value Problems

10 Partial Differential Equations

Appendix A   Matrix Algebra

Appendix B   Mathematical References

1 First-Order Equations

1.1

Introduction

In many important physical problems the development in time of a particular quantity is controlled by a fundamental physical law. A good example is provided by a chemical solution in a "stirred-tank chemical reactor." This is a tank containing a solution, initially at a particular concentration. When the period of observation begins, solution flows in continuously at a given rate and concentration, and the contents of the tank are drawn off continuously at a given rate. There is supposed to be a stirring device in the tank to ensure that the concentration of the solution is uniform throughout the tank at any time. It is usually required to find the amount (mass) of solute in the tank as a function of time. The equation governing this quantity is found by applying the law of conservation of mass in this form:

$$\text{accumulation rate} = \text{rate in} - \text{rate out.} \tag{1.1}$$

Example 1

A 200-liter tank is initially filled with brine (a solution of salt in water) at a concentration of 2 grams per liter. Then brine flows in at a rate of 8 liters per minute with a concentration of 4 grams per liter. The well-stirred contents of the tank are drawn off at a rate of 8 liters per minute. Express the law of conservation of mass for the salt in the tank.

  Let $u(t)$ be the mass of salt in the tank, measured in grams. The rate at which salt enters is

$$\text{rate in} = \frac{8 \text{ liter}}{\text{min}} \times \frac{4 \text{ g}}{\text{liter}} = \frac{32 \text{ g}}{\text{min}}.$$

The rate at which salt leaves is

$$\text{rate out} = \frac{8 \text{ liter}}{\text{min}} \times \frac{u(t) \text{ g}}{200 \text{ liter}} = \frac{0.04 \text{ g}}{\text{min}} u(t).$$

The rate at which salt accumulates in the tank is just du/dt, measured in

1

grams per minute. Thus the mass balance is

accumulation rate = rate in − rate out

$$\frac{du}{dt} = \frac{32 \text{ g}}{\text{min}} - \frac{0.04u(t) \text{ g}}{\text{min}} .$$

The units of measurement are included as a check on consistency. They are usually dropped at this stage, and the mass balance equation is written

$$\frac{du}{dt} = -0.04u + 32, \qquad 0 < t. \tag{1.2}$$

The inequality, $0 < t$, reminds us that the equation is valid after the experiment starts.

Many variants are possible in problems of this type: the inflow and outflow rates might be different or nonconstant, a chemical reaction might take place in the tank, the solution might become saturated, and so on. But in any event the accumulation rate term will cause the derivative of the unknown quantity u to appear in the mass balance equation. This brings us to the subject of our study.

Definition 1.1   A relationship between a function and its derivatives is called a *differential equation*. The highest-order derivative that appears is called the *order* of the differential equation.

The mass balance equation of Example 1, Eq. (1.2), is a first-order differential equation. We shall see many more examples of first-order equations in this chapter. In later chapters we shall see that certain simple mechanical or electrical systems can be described by second-order equations such as

$$\frac{d^2u}{dt^2} + 6\frac{du}{dt} + 10u = 2\cos t.$$

More complex systems may require differential equations of yet higher order.

Our objective, wherever possible, is to solve differential equations. Let us symbolize a general first-order equation as

$$\frac{du}{dt} = F(t, u).$$

Then a *solution* of this differential equation on an interval $\alpha < t < \beta$ is a function $u(t)$ that has a first derivative and satisfies the differential equation for all t in the interval $\alpha < t < \beta$. That is, substitution of $u(t)$ into the differential equation leads to an identity,

$$\frac{d}{dt} u(t) = F(t, u(t)), \qquad \alpha < t < \beta.$$

Example 2

The differential equation of Example 1,

$$\frac{du}{dt} = -0.04u + 32,$$

has for one solution the function

$$u(t) = 800 + 70e^{-0.04t} \tag{1.3}$$

over the interval $-\infty < t < \infty$. To confirm this claim, first note that the given function has a first derivative, which is

$$\frac{d}{dt} u(t) = 70(-0.04)e^{-0.04t} = -2.8e^{-0.04t}.$$

Substitution of $u(t)$ and its derivative into the given differential equation leads to the identity

$$-2.8e^{-0.04t} = -0.04(800 + 70e^{-0.04t}) + 32, \qquad -\infty < t < \infty.$$

It is also correct to say that the more general expression

$$u(t) = 800 + ce^{-0.04t}, \tag{1.4}$$

in which c is an arbitrary constant, is a solution of the differential equation. Indeed, substitution of this function into the differential equation again gives

$$-0.04ce^{-0.04t} = -0.04(800 + ce^{-0.04t}) + 32,$$

which is true for all t and any choice of the constant c.

Returning now to the chemical reactor problem of Example 1, we seem to have an unexpected problem: too many answers. Since Eq. (1.4) is a solution of our differential equation for any value of c, and each different value of c corresponds to a different function, we have an infinite family of solutions of Eq. (1.2). Yet the physical problem seemed perfectly definite, and we expect a single, definite solution.

This difficulty disappears, however, when we note that there is information given in Example 1 that we have not used. The initial concentration in the tank was given to be 2 grams per liter, which translates to an initial amount of 400 grams. In terms of the function u, we would state this condition as

$$u(0) = 400. \tag{1.5}$$

Now if we set $t = 0$ in the function of Eq. (1.4), we get

$$u(0) = 800 + ce^0 = 800 + c.$$

This quantity should equal 400. Thus $c = -400$, and the function we seek is

$$u(t) = 800 - 400e^{-0.04t}. \tag{1.6}$$

This function satisfies both the differential equation (1.2) and the auxiliary condition (1.5).

Definition 1.2

A first-order differential equation, together with a condition on the value of the solution at some point (an *initial condition*) is called an *initial value problem*. A solution of the differential equation that also satisfies the initial condition is a solution of the initial value problem. A general first-order initial value problem is denoted by

$$\frac{du}{dt} = F(t, u), \qquad u(t_0) = q.$$

A substantial part of any course in calculus is actually spent in dealing with the problem of solving first-order differential equations in which the right-hand side is a known function of t alone:

$$\frac{du}{dt} = f(t). \tag{1.7}$$

In words: the derivative of an unknown function is given, and the function is to be found. A solution of this problem is any antiderivative or indefinite integral of $f(t)$. The theorems of elementary calculus assure us that the most general solution is obtained by adding a constant to any solution. If an initial condition is imposed, the constant can be chosen to make the solution satisfy it.

Example 3

Solve the initial value problem

$$\frac{du}{dt} = e^{-2t}, \qquad t > 0,$$

$$u(0) = 5.$$

The right-hand side of the differential equation is a known function of t. By "integrating both sides" of the differential equation we find

$$u(t) = -\frac{e^{-2t}}{2} + c$$

as a solution of the differential equation. In order to fulfill the initial condition we must have

$$u(0) = 5,$$

$$\frac{-e^0}{2} + c = -\frac{1}{2} + c = 5.$$

Thus $c = \frac{11}{2}$, and the solution of the initial value problem is

$$u(t) = \frac{11}{2} - \frac{e^{-2t}}{2}.$$

Some functions $f(t)$ do not have an antiderivative that can be written down in closed form. In this case we must leave the integration of $f(t)$ to be done. To make our solution of the differential equation

$$\frac{du}{dt} = f(t)$$

perfectly definite, we write it as

$$u(t) = c + \int_a^t f(z)\, dz. \tag{1.8}$$

The lower limit, a, is any convenient fixed value (usually the initial value of t in initial value problems). We have used z as the dummy variable of integration; any other letter that is not busy elsewhere could be used instead. Elementary theorems of calculus assure us that Eq. (1.8) is a continuous function whose derivative is $f(t)$ at any t where f is continuous. We also use the form (1.8) to represent the solution of the differential equation when we do not want to specify the function f. However, this should not be used as a "formula for solving" the differential equation (1.4). It is much easier and more natural to think of integrating both sides.

Example 4

We attempt to solve the initial value problem

$$\frac{du}{dt} = e^{-t^2}, \qquad u(0) = \tfrac{1}{2}.$$

There is no function expressible in terms of polynomials, exponentials, etc., whose derivative is e^{-t^2}; therefore we must leave the integration to be done. We express the solution of the differential equation as

$$u(t) = \int_0^t e^{-z^2} \, dz + c.$$

The initial condition can be satisfied by setting $t = 0$ in the expression above and equating $u(0)$ to $\frac{1}{2}$:

$$u(0) = \int_0^0 e^{-z^2} \, dz + c = \tfrac{1}{2}.$$

We see that $c = \frac{1}{2}$ and that the solution of the initial value problem is

$$u(t) = \int_0^t e^{-z^2} \, dz + \tfrac{1}{2}.$$

Exercises

In Exercises 1–10, you are to solve the given differential equation. If there is an initial condition, choose the constant of integration to satisfy it.

1. $\dfrac{du}{dt} = 4,\ 0 < t;\ u(0) = 1$

2. $\dfrac{du}{dt} = e^{-5t}$

3. $\dfrac{du}{dt} = \sin 2t,\ u(0) = 0$

4. $\dfrac{du}{dt} = \cos 3t$

5. $\dfrac{du}{dt} = \dfrac{1}{t+1},\ t > 0$

6. $\dfrac{du}{dt} = \dfrac{t}{1+t^2}$

7. $\dfrac{du}{dt} = \dfrac{t}{\sqrt{1+t^2}}$

8. $\dfrac{du}{dt} = \dfrac{1}{t(t+1)},\ t > 1$

9. $\dfrac{du}{dt} = \dfrac{t-1}{t^2+3t+2},\ t > 0$

10. $\dfrac{du}{dt} = 2t + 3$

11. A tank is being filled with water at a rate of $q(t)$ liters per minute. If the tank starts empty, find an initial value problem describing the volume of water in the tank. (Assume that no water leaves the tank.)

12. Solve the initial value problem of Exercise 11 if $q(t) = 8$ liters per minute. For how long is your solution valid if the tank has a capacity of 100 liters?

13. Solve the initial value problem of Exercise 11 if the flow rate in liters per minute is

$$q(t) = \begin{cases} 4 - \dfrac{t}{10}, & 0 < t \le 40 \text{ min} \\ 0, & 40 < t. \end{cases}$$

14. Solve the initial value problem of Exercise 11 if $q(t) = e^{-t/20}$, $0 < t$.

15. Find an initial value problem for the amount of salt in a 50-liter tank if pure water enters at a rate of 5 liters per minute, solution is drawn off at a rate of 5 liters per minute, and the tank is initially filled with brine at a concentration of 10 grams per liter.

16. Suppose that a tank has a capacity of V liters; brine enters at a rate of Q liters per minute and concentration k; the well-stirred contents of the tank are drawn off at a rate of Q liters per minute; the tank is initially filled with brine at a concentration k_0. Show that an initial value problem for the amount of salt in the tank, $u(t)$, is

$$\frac{du}{dt} = -\frac{Q}{V}u + Qk, \qquad 0 < t,$$

$$u(0) = Vk_0.$$

17. Suppose a solid object (like a salt block) is to be dissolved in a liquid. The rate at which the solid dissolves is $-dV/dt$, where V is the volume of the solid. It is reasonable to assume that this rate depends on the area A of the solid that is in contact with the liquid and on the difference $(c_s - c)$ between saturation concentration of the solution and its current concentration. In mathematical terms we have said (k is a constant of proportionality)

$$\frac{dV}{dt} = -kA(c_s - c).$$

(a) Suppose the solid is in the form of a sphere of radius R. Rephrase the equation above as a differential equation for R. (Recall $V = \frac{4}{3}\pi R^3$, $A = 4\pi R^2$.)

(b) Solve the equation obtained in (a), assuming that $c_s - c$ (approximately) constant. Designate $R(0) = R_0$.

18. Suppose now that the solid has a "characteristic dimension" L (the radius of a sphere or the side of a cube) and that its shape retains the same proportions as it shrinks. Then $V = vL^3$, $A = aL^2$, where v and a are constants. Derive a differential equation for L and solve it.

1.2

Linear Equations

In this section and the next we will study methods for solving a very important kind of first-order differential equation. A *linear equation* is one that can be expressed as

$$\frac{du}{dt} = a(t)u + f(t). \tag{1.9}$$

The key feature is that the unknown function appears in just one place, as the multiplier of a given function, $a(t)$. For the time being, we assume that both $a(t)$ and $f(t)$ are continuous over some interval $\alpha < t < \beta$ that we are interested in.

A linear equation is further classified as being *homogeneous* if $f(t)$ is identically 0. The general linear homogeneous equation of first order is thus

$$\frac{du}{dt} = a(t)u. \tag{1.10}$$

If $f(t)$ is not identically 0, the equation is *nonhomogeneous*, and $f(t)$ is the *inhomogeneity*. The constant function $u(t) \equiv 0$ is always a solution of a homogeneous linear equation and is never a solution of a nonhomogeneous one.

Example 1

In Section 1.1 we derived an equation to describe the amount of salt in a mixing tank:

$$\frac{du}{dt} = -0.04u + 32.$$

This equation is linear, since the right-hand side has the form prescribed by Eq. (1.9). We can identify $a(t) = -0.04$ and $f(t) = 32$, both constant functions. Since $f(t)$ is nonzero, the equation is nonhomogeneous.

A homogeneous linear first-order equation can be solved with one integration. The thought process of someone solving

$$\frac{du}{dt} = a(t)u$$

runs as follows:

First divide through the equation by u to obtain

$$\frac{1}{u}\frac{du}{dt} = a(t). \tag{1.11}$$

Now the left-hand side is the derivative of a familiar function, $\ln|u|$. If we integrate both sides, we get

$$\ln|u| = A(t) + C, \tag{1.12}$$

where $A(t)$ represents in indefinite integral of $a(t)$. To recover u itself, first exponentiate both sides:

$$|u| = e^{A(t)+C} = e^{A(t)}e^C \tag{1.13}$$

Note that $|u|$ is either u or $-u$. Thus

$$u = \pm e^C e^{A(t)}. \tag{1.14}$$

Finally, rename the constant $\pm e^C$ as c:

$$u = ce^{A(t)}. \tag{1.15}$$

Example 2

Solve the differential equation

$$\frac{du}{dt} = -0.04u.$$

First divide through the equation by u, then integrate both sides:

$$\frac{1}{u}\frac{du}{dt} = -0.04,$$

$$\ln|u| = -0.04t + C.$$

Now exponentiate both sides of the last equation and solve for u:

$$|u| = e^{-0.04t+C} = e^{-0.04t}e^C$$

$$u = \pm e^C e^{-0.04t}.$$

Finally, replace the nonzero constant $\pm e^C$ by a general constant c to obtain

$$u(t) = ce^{-0.04t}.$$

This function is differentiable for all values of t, and its derivative is

$$\frac{du}{dt} = c(-0.04)e^{-0.04t}.$$

Since $du/dt = -0.04u$ identically, the function determined is indeed a solution of this differential equation. See Fig. 1.1.

There are several defects in the line of thought followed above. First, for the division by u to be legitimate, we must assume that u is nonzero. We subsequently obtain a function in Eq. (1.14) that indeed cannot be zero, and then we alter it in Eq. (1.15) so that it can be zero. (That is, $\pm e^C$ is nonzero, but c may be zero.) Surprisingly, these faults compensate for each other. Equation (1.15) represents the *general solution* of the differential equation (1.10) in the sense that every solution can be obtained from the expression in Eq. (1.15).

Figure 1.1

$u(t) = ce^{-0.04t}$
$(c > 0 \text{ shown})$

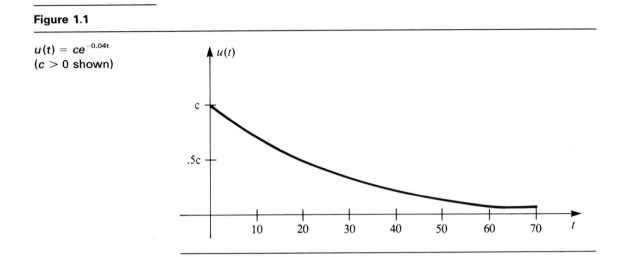

Example 3

Solve the initial value problem

$$(1 + t)\frac{du}{dt} = -2u, \qquad 0 < t,$$

$u(0) = 3.$

First, we must put the differential equation into standard form, with du/dt alone on one side of the equation:

$$\frac{du}{dt} = -\frac{2u}{1 + t}.$$

Next we solve the differential equation by the process developed above. The successive steps are as follows:

$$\frac{1}{u}\frac{du}{dt} = -\frac{2}{1 + t}$$

$$\ln|u| = -2\ln|1 + t| + C$$

$$|u| = e^{-2\ln|1+t|} \cdot e^C$$

$$u(t) = \frac{c}{(1 + t)^2}.$$

The function on the last line is continuous and differentiable for all $t > 0$, and substitution into the differential equation shows that it is a solution—indeed, the general solution.

Now we may choose c to satisfy the initial condition. We must have

$u(0) = 3$. Evaluating the last function above at $t = 0$ gives

$$u(0) = \frac{c}{1^2} = c.$$

Therefore we must choose $c = 3$ to satisfy the initial condition. The final result is (see Fig. 1.2)

$$u(t) = \frac{3}{(1 + t)^2}.$$

Figure 1.2

$u(t) = 3/(1 + t)^2$

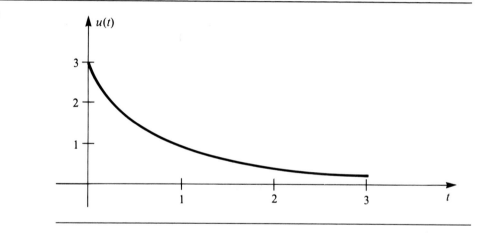

Exercises

In Exercises 1–14, find the general solution of the differential equation. If an initial condition is given, find the solution of the differential equation that satisfies it. Note any values of t where the coefficient of the differential equation is discontinuous. Some of the equations are not in standard form.

1. $\dfrac{du}{dt} = 2u$

2. $\dfrac{du}{dt} = -2u$

3. $5\dfrac{du}{dt} = 2u$

4. $\dfrac{du}{dt} = bu$ (b constant)

5. $t\dfrac{du}{dt} = u$

6. $\dfrac{du}{dt} + (\tan t)u = 0$

7. $\dfrac{du}{dt} = (\sec t)u$

8. $\dfrac{du}{dt} + (a + b \sin t)u = 0$

9. $\dfrac{du}{dt} + e^{-t}u = 0$

10. $\dfrac{du}{dt} = \dfrac{2t}{1 + t^2}u, \qquad u(0) = -2$

11. $\dfrac{du}{dt} + tu = 0, \qquad u(0) = 3$

12. $\dfrac{du}{dt} + t^k u = 0 \qquad u(1) = 1, (k \neq -1)$

13. $(\sin t)\dfrac{du}{dt} = (\cos t)u$

14. $t\dfrac{du}{dt} + u = 0$

15. Solve $t^k \dfrac{du}{dt} + kt^{k-1}u = 0$ if k is a real number

16. Let $u(t)$ represent the solution of an initial value problem of the form

$$\frac{du}{dt} = bu, \qquad u(0) = u_0.$$

If b and u_0 are unknown constants, show how to determine them from two "readings," $u(t_1) = U_1, u(t_2) = U_2$ (t_1 and t_2 are known).

17. Using the equation in Exercise 12, find b if $t_2 - t_1$ is known but not the t's individually.

18. Prove that if $u(t)$ and $v(t)$ are solutions of the differential equations

$$\frac{du}{dt} = a(t)u, \qquad \frac{dv}{dt} = b(t)v,$$

then their product $w(t) = u(t)v(t)$ is a solution of

$$\frac{dw}{dt} = (a(t) + b(t))w.$$

19. Solve the differential equation and sketch a solution:

$$\frac{du}{dt} = te^{-t}u.$$

20. Suppose that u is a solution of the initial value problem

$$\frac{du}{dt} = a(t)u, \qquad u(0) = 1,$$

where $a(t)$ is nonnegative and

$$\int_0^\infty a(t)\, dt = \alpha$$

is finite. Show that $u(t)$ increases, as t tends to infinity, to a limiting value. Find the limit. (See Exercise 19 for an example.)

21. Solve the differential equation

$$\frac{du}{dt} = (1 + \sin t)u.$$

22. Solve $\dfrac{du}{dt} = (\cos t)u.$

23. Suppose that u is the solution of the differential equation

$$\frac{du}{dt} + a(t)u = 0,$$

where $a(t)$ is periodic with period p: $a(t + p) = a(t)$ for all t. Show that the function $v(t) = e^{\alpha t}u(t)$ is periodic with period p, where α is the average value of $a(t)$ over a period

$$\alpha = \frac{1}{p}\int_{t_0}^{t_0+p} a(t)\, dt.$$

24. Figure 1.3 shows a cable wrapped around a fixed post; T_1 and T_0 are the tension forces on the two ends of the cable, and α is the angular

Figure 1.3

Cable Wrapped Around a Post

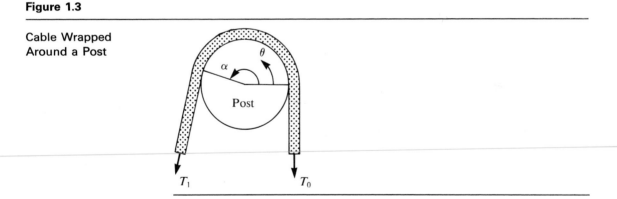

measure of the cable wrapped around the post. It can be shown that when the cable is on the point of slipping, the tension in the cable obeys the initial value problem

$$\frac{dT}{d\theta} = \mu T, \qquad T(0) = T_0,$$

in which θ is measured (in radians) as shown and μ is the coefficient of friction between cable and post. Solve this problem for $T(\theta)$.

25. If $\mu = 0.5$, find the maximum value of the ratio T_1/T_0; if $\alpha = \pi$; if $\alpha = 3\pi$.

1.3

Nonhomogeneous Linear Equations

Our objective in this section is to develop a method of solution for the nonhomogeneous linear equation

$$\frac{du}{dt} = a(t)u + f(t). \tag{NH}$$

We shall make use of a principle that applies quite broadly in linear equations: the nonhomogeneous equation (NH) can be solved if the general solution of the corresponding homogeneous equation

$$\frac{du}{dt} = a(t)u \tag{H}$$

is known.

Here is another way to look at the homogeneous equation. Recall that we can write our solution of (H) as

$$u(t) = ce^{A(t)}$$

where c is constant and $A(t)$ is an antiderivative (or indefinite integral) of $a(t)$. That is,

$$\frac{dA}{dt} = a(t).$$

The homogeneous equation (H) is algebraically equivalent to

$$\frac{du}{dt} - a(t)u = 0. \tag{1.16}$$

If this equation is multiplied through by $e^{-A(t)}$ (which is nonzero), we obtain another equivalent equation,

$$e^{-A(t)}\frac{du}{dt} - a(t)e^{-A(t)}u = 0. \tag{1.17}$$

The essential insight is to recognize that the left-hand side of this equation is the derivative of a product:

$$e^{-A(t)}\frac{du}{dt} - a(t)e^{-A(t)}u = \frac{d}{dt}(e^{-A(t)}u). \tag{1.18}$$

Thus Eq. (1.16) is equivalent to the very simple differential equation

$$\frac{d}{dt}(e^{-A(t)}u) = 0, \tag{1.19}$$

which may be solved to get

$$e^{-A(t)}u(t) = c \quad \text{or} \quad u(t) = ce^{A(t)}. \tag{1.20}$$

The function $e^{-A(t)}$ is called an *integrating factor* for the equation (1.16): after multiplication by the integrating factor the left-hand side becomes the derivative of a function expressible in terms of u.

This alternative approach to solving the homogeneous equation has the advantage that it generalizes to the nonhomogeneous equation. If we write (NH) in the form

$$\frac{du}{dt} - a(t)u = f(t),$$

then multiplication by $e^{-A(t)}$ again makes the left-hand side into the derivative of a recognizable function:

$$e^{-A(t)}\frac{du}{dt} - e^{-A(t)}a(t)u = e^{-A(t)}f(t)$$

$$\frac{d}{dt}(e^{-A(t)}u) = e^{-A(t)}f(t).$$

Now we have essentially an equation of the simplest kind, which can be solved with one integration for $e^{-A(t)}u$ and then for u.

Example 1

Use the method above to solve the nonhomogeneous equation

$$\frac{du}{dt} = 2u + t.$$

First we solve the corresponding homogeneous equation, which is

$$\frac{du}{dt} = 2u.$$

By the method of Section 1.2 we find that the general solution is $u(t) = ce^{2t}$. Equivalently, we can write $e^{-2t}u(t) = c$. The integrating factor is now identified as e^{-2t}.

Next we put both terms containing u in the nonhomogeneous equation on the same side,

$$\frac{du}{dt} - 2u = t,$$

and multiply through by the integrating factor to get

$$e^{-2t}\frac{du}{dt} - 2e^{-2t}u = e^{-2t}t.$$

The left-hand side is supposed to be the derivative of $e^{-2t}u$. This should be checked:

$$\frac{d}{dt}(e^{-2t}u) = e^{-2t}\frac{du}{dt} - 2e^{-2t}u.$$

Now our equation is reduced to

$$\frac{d}{dt}(e^{-2t}u) = e^{-2t}t.$$

Both sides are integrated (by parts) to obtain the general solution

$$e^{-2t}u = \int te^{-2t}\,dt$$

$$= t\frac{e^{-2t}}{-2} - \int \frac{e^{-2t}}{-2}\,dt$$

$$= -\tfrac{1}{2}te^{-2t} + \frac{1}{2}\cdot\frac{e^{-2t}}{-2} + c.$$

Finally, we recover $u(t)$, the general solution of the original nonhomogeneous equation

$$u(t) = -\tfrac{1}{2}t + \tfrac{1}{4} + ce^{2t}.$$

The outline of the method for solving a nonhomogeneous linear equation is as follows:

1. Solve the corresponding homogeneous linear equation to find the integrating factor. When the solution is written as $\phi(t)u(t) = c$, $\phi(t)$ is the integrating factor.

2. Multiply the nonhomogeneous equation by the integrating factor. The terms containing u and du/dt become the derivative of the product $\phi(t)u(t)$.

3. Integrate to find $\phi(t)u(t)$.

4. Solve for $u(t)$.

If the first step can be remembered, the rest of the calculation flows along naturally. It is good practice to check that the terms containing u and du/dt do indeed form the derivative of the product $\phi(t)u(t)$.

Example 2

Solve the initial value problem (from Section 1.1)

$$\frac{du}{dt} = -0.04u + 32, \qquad u(0) = 400.$$

First, we solve the corresponding homogeneous equation,

$$\frac{du}{dt} = -0.04u$$

to find that $u(t) = ce^{-0.04t}$ or $e^{0.04t}u(t) = c$. The integrating factor is identified as $e^{0.04t}$.

Now multiply the nonhomogeneous equation by the integrating factor and put all terms containing u or du/dt on the left:

$$e^{0.04t}\left(\frac{du}{dt} + 0.04u\right) = 32e^{0.04t}.$$

The left-hand member of this equation is indeed the derivative of a product:

$$e^{0.04t}\left(\frac{du}{dt} + 0.04u\right) = \frac{d}{dt}(e^{0.04t}u).$$

Thus the equation to be solved is

$$\frac{d}{dt}(e^{0.04t}u) = 32e^{0.04t}$$

$$e^{0.04t}u = \int 32e^{0.04t}\, dt$$

$$= \frac{32}{0.04}e^{0.04t} + c.$$

Finally, we obtain the general solution of the nonhomogeneous equation by isolating u:

$$u(t) = \frac{32}{0.04} + ce^{-0.04t} = 800 + ce^{-0.04t}.$$

The initial condition is now to be satisfied. Since $u(0)$ is to be 400, we must choose c to satisfy

$$800 + c = 400$$

or $c = -400$. Thus the solution of the initial value problem is

$$u(t) = 800 - 400e^{-0.04t}.$$

Example 3

Solve the initial value problem

$$\frac{du}{dt} + 2tu = 1, \qquad u(0) = 0.$$

We proceed in three steps: (1) solve the homogeneous equation; (2) solve the nonhomogeneous equation; (3) apply the initial condition.

1. The corresponding homogeneous equation is solved as in Section 1.2:

$$\frac{du}{dt} = -2tu$$

$$\frac{1}{u}\frac{du}{dt} = -2t$$

$$\ln|u| = -t^2 + C$$

$$u(t) = \pm e^C e^{-t^2} \quad \text{or} \quad u(t) = ce^{-t^2} \quad \text{or} \quad e^{t^2}u(t) = c.$$

2. The integrating factor is e^{t^2}. We multiply the nonhomogeneous equation by it to find

$$e^{t^2}\left(\frac{du}{dt} + 2tu\right) = e^{t^2}.$$

The left-hand side is the derivative of $e^{t^2}u$ (check this!), so the equation becomes

$$\frac{d}{dt}(e^{t^2}u) = e^{t^2}.$$

We wish to integrate both sides of this equation. However, the integration of the right-hand side cannot be completed in terms of elementary functions; we must leave the integration indicated:

$$e^{t^2}u(t) = \int_0^t e^{z^2}\,dz + c.$$

From here we obtain the general solution of the nonhomogeneous differential equation:

$$u(t) = e^{-t^2}\int_0^t e^{z^2}\,dz + ce^{-t^2}.$$

3. Now we can satisfy the initial condition. Evaluating u at 0, we get

$$u(0) = 0 + c \cdot 1.$$

Since the initial condition is $u(0) = 0$, we must take $c = 0$. Then the solution of the initial value problem is

$$u(t) = e^{-t^2}\int_0^t e^{x^2}\,dx.$$

This function is called Dawson's integral and is well known. (See Abramowitz and Stegun, 1964.)

Example 4

Find the general solution of the differential equation

$$(t^2 + 1)\frac{du}{dt} = 2tu + t^3.$$

As a preliminary step, it is advisable to write the equation in standard form:

$$\frac{du}{dt} = \frac{2t}{t^2 + 1}u + \frac{t^3}{t^2 + 1}.$$

Now we proceed with the solution.

1. The corresponding homogeneous equation is solved by the usual method:

$$\frac{du}{dt} = \frac{2t}{t^2 + 1}u$$

$$\frac{1}{u}\frac{du}{dt} = \frac{2t}{t^2 + 1}$$

$$\ln|u| = \ln|t^2 + 1| + C$$

$$u(t) = c(t^2 + 1) \qquad \text{or} \qquad \frac{u(t)}{t^2 + 1} = c.$$

2. The integrating factor has been identified as $1/(t^2 + 1)$. It is to multiply the equation *in standard form*, yielding

$$\frac{1}{t^2 + 1}\frac{du}{dt} - \frac{2t}{(t^2 + 1)^2}u = \frac{t^3}{(t^2 + 1)^2}.$$

The left-hand side is the derivative of the product $u/(t^2 + 1)$:

$$\frac{u'}{t^2 + 1} - \frac{2tu}{(t^2 + 1)^2} = \frac{d}{dt}\left(\frac{u}{t^2 + 1}\right),$$

so the differential equation becomes

$$\frac{d}{dt}\left(\frac{u}{t^2 + 1}\right) = \frac{t^3}{(t^2 + 1)^2}.$$

This can be solved by one integration (the substitution $t^2 + 1 = v$ helps) to get

$$\frac{u}{t^2 + 1} = \frac{1}{2}\left(\ln(t^2 + 1) + \frac{1}{t^2 + 1}\right) + c.$$

Finally, the general solution of the nonhomogeneous equation is

$u(t) = \frac{1}{2}(t^2 + 1)\ln(t^2 + 1) + \frac{1}{2} + c(t^2 + 1)$.

Exercises

In Exercises 1–16, find the general solution of the differential equation. If an initial condition is given, find the solution that satisfies it.

1. $\dfrac{du}{dt} = 3u - t$

2. $\dfrac{du}{dt} = -3u + t$

3. $\dfrac{du}{dt} + 2u = e^{-t}$

4. $\dfrac{du}{dt} = u + e^{t}$

5. $\dfrac{du}{dt} = t(u - 1)$

6. $(t - 1)\dfrac{du}{dt} = \dfrac{2t}{1 + t}u + 1, \ t > 1$

7. $\dfrac{du}{dt} + \dfrac{1}{t}u = t, \ t > 0$

8. $t\dfrac{du}{dt} = u + 1, \ t > 0$

9. $\dfrac{du}{dt} + (\cos t)u = \cos t, \ u(0) = 0$

10. $\dfrac{du}{dt} + \dfrac{k}{t}u = 1, \ t > 0$

11. $(\sin t)\dfrac{du}{dt} = u \cos t + 1, \ -\dfrac{\pi}{2} < t < \dfrac{\pi}{2}$

12. $\dfrac{du}{dt} + bu = e^{-at} \ (a \neq b, \text{ constants})$

13. $\dfrac{du}{dt} + bu = e^{-bt} \ (b \text{ constant})$

14. $\dfrac{du}{dt} + bu = \sin t, \ u(0) = 0 \ (b \text{ constant})$

15. $\dfrac{du}{dt} + tu = 1$

16. $\dfrac{du}{dt} = 2tu - t$

17. In Section 1.1 we derived this initial value problem for the amount of salt in a tank:

$$\frac{du}{dt} = -0.04u + 32, \qquad u(0) = 400.$$

A more general expression for the situation is

$$\frac{du}{dt} = -\frac{Q}{V}u + Qk, \qquad u(0) = Vk_0,$$

where Q is the rate of flow of solution into and out of the tank, V is the volume of the tank, k is the concentration of incoming solution, and k_0 is the initial concentration in the tank. Solve this initial value problem in terms of the parameters Q, V, k, and k_0, assuming them to be constants.

18. If $u(t)$ is the solution of the initial value problem in Exercise 17, describe what happens to $u(t)$ as $t \to \infty$.

19. A differential equation having the form

$$\frac{dy}{dx} = a(x)y + b(x)y^n \qquad (n \neq 0, 1)$$

is called a *Bernoulli* equation. If it is divided through by y^n, it becomes

$$y^{-n}\frac{dy}{dx} = a(x)y^{1-n} + b(x).$$

Show that the substitution $z = y^{1-n}$ turns a Bernoulli equation into a linear equation for z.

In Exercises 20–25 you are to solve the Bernoulli equation. If an initial condition is given, find the solution that satisfies it. Note any value(s) of x for which the solution becomes discontinuous.

20. $\dfrac{dy}{dx} = y - y^2,\ y(0) = \frac{1}{2}$

21. $\dfrac{dy}{dx} = y - \dfrac{1}{y}$

22. $\dfrac{dy}{dx} = \dfrac{y}{x} - y^2$

23. $\dfrac{dy}{dx} = y - x\sqrt{y},\ y(0) = 1$

24. $\dfrac{dy}{dx} = ay + by^2$ $(a, b$ constants$)$

25. In the next section we will see an application of the logistic equation

$$\frac{du}{dt} = bu - mu^2,$$

in which b and m are positive constants. Solve it as a Bernoulli equation.

1.4

Further Applications of Linear Equations

The first-order linear differential equations whose solutions were discussed in Sections 1.2 and 1.3 have such diverse applications in so many scientific disciplines that we can give only a brief sampling of some of the most common or interesting.

Radiocarbon Dating

Cosmic radiation of the earth's atmosphere is constantly producing carbon 14, a radioactive isotope of ordinary carbon 12. Both kinds of carbon appear in carbon dioxide, which is incorporated into the tissues of all plants and eventually into all animals. Because of the constant intake of carbon, all living matter may be assumed to have the same proportion of radioactive to ordinary carbon atoms as the atmosphere (about 1 to 10^{12}). In dead organic matter, however, the radioactive carbon decays and is not replaced from surroundings.

Radioactive decay proceeds at a constant percentage rate characteristic of the decaying substance. Thus the concentration c of carbon 14 in dead organic material obeys the initial value problem

$$\frac{dc}{dt} = -\lambda c, \qquad t > 0, \tag{1.21}$$

$$c(0) = c_0, \tag{1.22}$$

where $t = 0$ at the time of death, c_0 is the concentration of the isotope that gives equilibrium with the atmosphere, and λ is the characteristic constant ($\lambda = 1.24 \times 10^{-4}$ per year for carbon 14). The application is to predict the time of death of organic material from an examination of it now. The technique is to measure the present concentration $c(t)$ in some object and to determine t. We know that $c(t)$ is given by

$$c(t) = c_0 e^{-\lambda t}, \tag{1.23}$$

since it is a solution of Eqs. (1.21) and (1.22). If we assume that c_0 does not change, it can be found by examining living or recently dead matter. Thus we can determine

$$\frac{c(t)}{c_0} = e^{-\lambda t}, \tag{1.24}$$

and from this point we get

$$t = \frac{1}{\lambda} \ln (c_0/c(t)), \tag{1.25}$$

which is the time from the death of the organic material to the present.

Electric Circuits

Figure 1.4 shows an electric circuit consisting of a resistor R and a capacitor C in series. A differential equation relating current i in the

Figure 1.4

RC Circuit

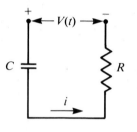

circuit, charge q on the capacitor, and voltage V measured at the points shown can be derived by applying Kirchhoff's law, which says that the voltage V must equal the sum of the voltage changes across resistor and capacitor. In addition it is known that these voltage changes are

$$\Delta V_R = Ri \qquad \text{for the resistor,} \tag{1.26}$$

$$\Delta V_C = q/C \qquad \text{for the capacitor.} \tag{1.27}$$

Therefore the voltage V is

$$V = Ri + q/C. \tag{1.28}$$

Furthermore, current is defined to be the flow rate of charge:

$$i = \frac{dq}{dt}. \tag{1.29}$$

By combining Eq. (1.29) with Eq. (1.28) we obtain a differential equation relating q and V:

$$R\frac{dq}{dt} + \frac{1}{C}q = V. \tag{1.30}$$

If R and C are constants, then q may be found as the solution of Eq. (1.30), and via Eq. (1.29) we may find the current.

In Fig. 1.5 is shown another simple electrical circuit, this time consisting of a resistor R and an inductor (or coil) L. The voltage change

Figure 1.5

RL Circuit

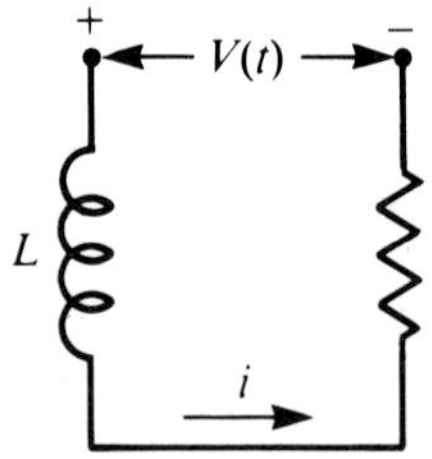

across the inductor is

$$\Delta V_L = L \frac{di}{dt}. \tag{1.31}$$

Using this together with Ohm's law, Eq. (1.26), and Kirchhoff's law, we obtain this differential equation for the current i in the circuit:

$$L \frac{di}{dt} + Ri = V. \tag{1.32}$$

Example 1

Suppose that the circuit of Fig. 1.4 has parameters $c = 10^{-6}$ farad and $R = 10^5$ ohms. Find $q(t)$ as a function of time if the capacitor starts with no charge and V is a constant 6 volts.

The initial value problem to be solved is

$$10^5 \frac{dq}{dt} + 10^6 q = 6, \qquad q(0) = 0,$$

or

$$\frac{dq}{dt} + 10q = 6 \times 10^{-6}, \qquad q(0) = 0.$$

1. The solution of the homogeneous differential equation, $dq/dt + 10q = 0$, is $q(t) = ce^{-10t}$, and the integrating factor is e^{10t}.

2. To solve the nonhomogeneous equation, multiply it by the integrating factor to obtain

$$e^{10t}\left(\frac{dq}{dt} + 10q\right) = 6 \times 10^{-5}e^{10t}.$$

The left-hand side is the derivative of the product $e^{10t}q$. Integration of both sides of the equation gives

$$e^{10t}q = \frac{6 \times 10^{-5}e^{10t}}{10} + c$$

or

$$q(t) = 6 \times 10^{-6} + ce^{-10t}.$$

3. The constant c is found by imposing the initial condition

$$q(0) = 6 \times 10^{-6} + c = 0.$$

Finally, the charge as a function of t is found to be

$$q(t) = 6 \times 10^{-6}(1 - e^{-10t}).$$

Note that t must be measured in seconds and charge in coulombs. (See Table 1.1.) Figure 1.6 shows $q(t)$.

Table 1.1	Quantity	Symbol	Unit
Electrical Quantities	Voltage	V	Volt
	Current	i	Ampere
	Charge	q	Coulomb
	Capacitance	C	Farad
	Resistance	R	Ohm
	Inductance	L	Henry

Relations among units:

1 coulomb = 1 amp sec	Eq. 1.29
1 ohm = 1 volt/amp	Eq. 1.26
1 farad = 1 amp sec/volt	Eq. 1.27
1 henry = 1 volt sec/amp	Eq. 1.31

Population Growth

A simple hypothesis for the prediction of the size of a population is that the rates of increase (by birth) and of decrease (by death) of a population are proportional to the population size. That is, if $n(t)$ is the number of members in a population, then the birth and death rates are

$$\beta n(t) \qquad \text{and} \qquad \mu n(t),$$

where β and μ are the specific birth and mortality rates (for example, β might be 11 births per thousand individuals per year or 0.011 per year).

Figure 1.6

$q(t) =$
$\quad 6 \times 10^{-6}(1 - e^{-10t})$

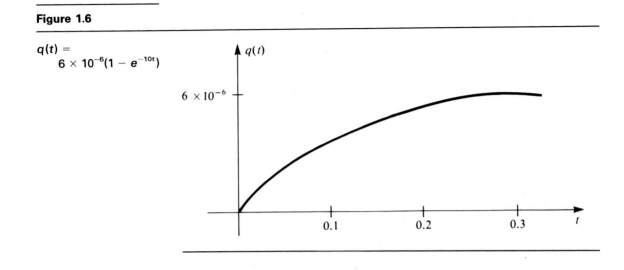

Then if the net immigration rate into the population is $f(t)$, we may apply a "conservation of mass" argument to obtain

$$\frac{dn}{dt} = (\beta - \mu)n + f(t) \tag{1.33}$$

as a differential equation for the population size. Note that we have approximated the number of individuals by a continuous function $n(t)$, which is permissible if n is large.

The long-term behavior of population size depends heavily on the net reproduction rate $\beta - \mu$. While this quantity is often taken to be constant, it may actually change as a function of time or, in a sentient population, as a function of n itself—thus making Eq. (1.33) nonlinear.

Table 1.2 contains data on the population of the United States, together with immigration figures. These show that the population increased at about a 3% rate from 1790 to 1860.

Example 2

Test the statement that the U.S. population grew at a 3% rate between 1790 and 1860, ignoring immigration.

Equation (1.33) can be used as a model of population size, with $\beta - \mu = 0.03$ and $f(t) = 0$ (since immigration is to be ignored). If t, measured in years, was 0 in 1790, an initial value problem for population (in millions) is

$$\frac{dn}{dt} = 0.03n, \quad n(0) = 3.93.$$

Table 1.2

	Year	Population (millions)	Immigration in Decade (millions)
U.S. Population	1790	3.93	—
	1800	5.31	—
	1810	7.24	—
	1820	9.64	—
	1830	12.9	0.1
	1840	17.1	0.6
	1850	23.2	1.7
	1860	31.4	2.6
	1870	38.6	2.3
	1880	50.2	2.8
	1890	63.0	5.2
	1900	76.0	3.7
	1910	92.0	8.8
	1920	106	6
	1930	123	4
	1940	132	1
	1950	151	1
	1960	179	3
	1970	203	3
	1980	227	4

The solution of this problem is given by

$$n(t) = 3.93e^{0.03t}.$$

This function gives, for the population in 1860 ($t = 70$),

$$n(70) = 3.93e^{0.03 \cdot 70} = 32.1.$$

This value is a bit too high.

Example 3

Between 1830 and 1860 the immigration rate increased steadily, being approximately $0.008t$ millions per year with $t = 0$ in 1830. Test this initial value problem as a model of population in those 30 years:

$$\frac{dn}{dt} = 0.025n + 0.008t, \qquad n(0) = 12.9.$$

The solution of the corresponding homogeneous equation is $n = ce^{0.025t}$. Multiplying through the nonhomogeneous equation by the integrating factor, $e^{-0.025t}$, we obtain

$$e^{-0.025t}\left(\frac{dn}{dt} - 0.025n\right) = 0.008te^{-0.025t}$$

or

$$\frac{d}{dt}(e^{-0.025t}n) = 0.008te^{-0.025t}.$$

The right-hand side can be integrated by parts to yield

$$e^{-0.025t}n = 0.008\left(\frac{t}{-0.025} - \frac{1}{(0.025)^2}\right)e^{-0.025t} + c.$$

Finally, we get the general solution as

$$n(t) = -12.8 - 0.32t + ce^{0.025t}.$$

The initial condition requires that c satisfy the equation

$$-12.8 + c = 12.9.$$

Thus $c = 25.7$, and the population size is given by

$$n(t) = -12.8 - 0.32t + 25.7e^{0.025t}.$$

The prediction of this function for 1850 is

$$n(20) = -12.8 - 6.4 + 25.7e^{0.025t} = 23.17,$$

in good agreement with Table 1.2.

Compound Interest and Inflation

Interest rates paid by various institutions are quoted as so many percent per year, with various provisions for compounding. If the amount of money on deposit is $A(t)$ (with t measured in years), then the addition of uncompounded interest after a year is represented by

$$A(1) = (1 + i)A(0), \tag{1.34}$$

where i is the annual interest rate. However, if the interest is compounded quarterly, the amounts on deposit on the quarter days are

$$A(\tfrac{1}{4}) = \left(1 + \frac{i}{4}\right)A(0),$$

$$A(\tfrac{1}{2}) = \left(1 + \frac{i}{4}\right)A(\tfrac{1}{4}) = \left(1 + \frac{i}{4}\right)^2 A(0),$$

$$A(\tfrac{3}{4}) = \left(1 + \frac{i}{4}\right)^3 A(0),$$

$$A(1) = \left(1 + \frac{i}{4}\right)^4 A(0).$$

It is easy to see that compounding means more interest, since

$$\left(1 + \frac{i}{4}\right)^4 = 1 + 4 \cdot \frac{i}{4} + 6\left(\frac{i}{4}\right)^2 + 4\left(\frac{i}{4}\right)^3 + \left(\frac{i}{4}\right)^4$$

is larger than $1 + i$. More compounding is more beneficial. Some banks offer monthly compounding, which amounts to

$$A(1) = \left(1 + \frac{i}{12}\right)^{12} A(0), \tag{1.35}$$

and also daily compounding, which yields

$$A(1) = \left(1 + \frac{i}{365}\right)^{365} A(0). \tag{1.36}$$

There is a limit to the frequency of compounding, namely, continuous compounding. In this extreme case the amount after one year is

$$A(1) = \lim_{n \to \infty} \left(1 + \frac{i}{n}\right)^n A(0), \tag{1.37}$$

which is $e^i A(0)$, because of the well-known limit

$$\lim_{n \to \infty} \left(1 + \frac{i}{n}\right)^n = e^i. \tag{1.38}$$

Table 1.3 contains the "effective annual interest rate," which has the

Table 1.3	n	5%	10%	15%
Effective Annual Interest Rates	1	5.00	10.00	15.00
	4	5.09	10.38	15.86
	12	5.12	10.47	16.08
	∞	5.13	10.52	16.18

functional form

$$i_{\text{eff}} = \left(1 + \frac{i}{n}\right)^n - 1. \tag{1.39}$$

In essence, then, continuous compounding means that the instantaneous rate of change of the amount A is proportional to A; that is,

$$\frac{dA}{dt} = iA. \tag{1.40}$$

The effect of inflation is also exponential, in the sense that the price $P(t)$ paid for a specific package of goods and services increases at a specific rate; that is,

$$\frac{dP}{dt} = rP, \qquad (1.41)$$

where r (usually variable) is the inflation rate. If $P(0)$ is taken to be the constant *value* (not price) of the package, then $P(0)/P(t) = V(t)$ measures the value of money. It is not hard to show that

$$\frac{dV}{dt} = -rV. \qquad (1.42)$$

Exercises

1. Instead of measuring the concentration of radiocarbon for dating purposes, one normally measures the decay rate, $|dc/dt|$. How is the age of an object calculated with this information?

2. If a sample of ancient charcoal has a carbon 14 decay rate of 100 disintegrations per minute per ounce, while modern charcoal has a rate of 190, what age can be attributed to the ancient sample? (Use Exercise 1.)

3. The half-life of a radioactive substance is the time required for the amount in a sample to decay to one-half the amount initially present. How is the half-life related to the constant λ in Eq. (1.21)?

4. Find the current i in an RC circuit if the impressed voltage is $V(t) = V_0 \sin \omega t$. (See Eq. (1.30).) Assume the initial charge to be 0. (Leave R, C, and V_0 as parameters.)

5. Solve the initial value problem for the charge on the capacitor in the circuit of Fig. 1.4 if $V = 10 \sin 20t$, $R = 10^5$ ohms, $C = 10^{-6}$ farad, and $q(0) = 0$. Also find the voltage drop across the capacitor and compare it to the impressed voltage V.

6. Find the current in the LR circuit of Fig. 1.5 if the impressed voltage is $V = 5$ volts, $L = 10$ henries, $R = 10^3$ ohms, and $i(0) = 0$.

7. Find the current i in an RL circuit if the impressed voltage is $V(t) = V_0 \sin \omega t$. (See Eq. (1.32).) Assume that the current is initially 0. (R, L, V_0, and ω are parameters.)

8. In Exercise 7 the solution can be written as a sum of an exponential term and terms containing $\sin \omega t$ and $\cos \omega t$. Sketch the sum of these latter terms, assuming that $R = L\omega$.

9. An RC circuit such as the one shown in Fig. 1.7 is used as a time-delay device for arming a burglar alarm. The capacitor is first charged by a 9-volt d.c. source. When the switch is thrown, the capacitor discharges through a short circuit. The alarm is armed (by a

Figure 1.7

Time Delay Circuit

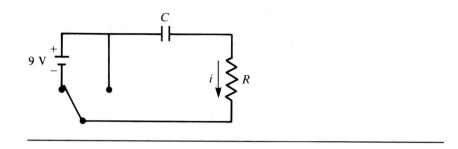

circuit not shown) when the voltage across the capacitor drops to 4 volts. Suppose that when the switch is thrown, the voltage across the capacitor is 9 volts. What value of the product RC will cause the time delay to be 2 minutes?

10. Show that during the charging phase the voltage across the capacitor approaches 9 volts, and the current in the circuit approaches 0.

11. Use the data in Table 1.2 to find the approximate growth rate in the U.S. population from 1790 to 1860, ignoring immigration. (Hint: find the percentage change in population size during each decade and divide by 10 for an average rate in the decade.)

12. In Exercise 11 it would be more correct to calculate the natural log of two successive populations and divide their differences by 10. Why is this "better"? Is there much difference between methods?

13. Follow the directions for Exercise 11, but use the period 1860–1930 and take immigration into account. Example: between 1850 and 1860 the population grew by 8.2 ($=31.4 - 23.2$) millions, of which 2.6 million were immigrants. The natural increase was 5.6 million, and the rate was

$$\frac{5.6}{23.2 \times 10} = 0.024 \quad \text{or} \quad 2.4\% \text{ per year.}$$

14. Bankers use this "rule of 72" to find the number of years required to double the principal left at an interest rate of r (percent):

$$T = \frac{72}{r}.$$

Show that the correct expression is

$$T = \frac{\ln 2}{\ln\left(1 + \dfrac{r}{100}\right)}.$$

15. Compare the results of the two formulas in Exercise 14 for these rates: $r = 1, 3, 5, 10, 20, 100$.

16. Suppose that $P(t)$ represents the principal of a loan after t years. If interest is charged at rate r (per year) and payments are made at the constant and continuous rate M (dollars per year), show that $P(t)$ obeys the differential equation

$$\frac{dP}{dt} = rP - M.$$

17. What value should M have if the loan is to be paid in full at the end of T years—that is, if $P(T) = 0$?

18. Derive Eq. (1.42) from Eq. (1.41) using the change of variables $P(t) = P(0)/V(t)$.

19. *Time* magazine (Nov. 14, 1983, p. 96) reported a serious pollution problem in the U.S.S.R.* About $4.5 \times 10^6 \, \text{m}^3$ of brine, containing 300 g/l of salt, were released into the Dniester river by the collapse of a dam at a fertilizer plant. Most of the salt was trapped behind another dam, the Novodnestrovsk. Treating the Novodnestrovsk reservoir as a stirred-tank chemical reactor (see Section 1.1), find an initial value problem for the amount of salt in the reservoir. Assume pure water inflow at 250 m³/sec and a reservoir size of $10^9 \, \text{m}^3$.

20. Solve the initial value problem found in Exercise 19.

21. Using the solution of Exercise 20, find the concentration of salt in the reservoir. How long will it take for the concentration to decrease by 50%? (Give time in days.)

1.5

Separable Equations

In the rest of this chapter we study first-order equations without regard to their being linear or not. First, we shall see how some of these equations arise.

Example 1

A branch of biology that has flourished in the last 50 years, called population dynamics, concerns itself with predicting the growth and change of population. In Section 1.4 we derived the simplest model of population growth, which assumes that the rate of change of the population size is proportional to the population size, $u(t)$, itself; that is,

$$\frac{du}{dt} = ru, \tag{1.43}$$

---

* The figures in the following are the author's responsibility. The reservoir size is an utter guess.

where the *specific growth rate r* is constant. Of course we recognize this as a linear equation whose solution is an exponential function, $u(t) = ce^{rt}$, where c is an arbitrary constant.

A defect of the simple model represented by Eq. (1.43) is the fact that exponential growth cannot be sustained indefinitely. Indeed, this is the well-known observation of the economist Malthus. A more subtle proposal, and one that corresponds more closely to observed facts, is that the growth rate, which is the difference between reproduction and death rates, decreases with population size and may even become negative if population size is too large. A differential equation that expresses this assumption is

$$\frac{du}{dt} = bu - mu^2 = (b - mu)u. \tag{1.44}$$

In the right-hand side the factor $b - mu$ is the specific growth rate that decreases with size. The differential equation (1.44) is called a logistic equation. The presence of the u^2 term shows that Eq. (1.44) is nonlinear. Nevertheless, we will see shortly that it can be solved in closed form as

$$u(t) = \frac{b}{m + ce^{-bt}}, \tag{1.45}$$

where c is an arbitrary constant.

Example 2

Two different populations that interact can lead to more complicated models. A well-known example is the predator–prey system associated with the names of Lotka and Volterra. Let $x(t)$ represent the size of a population of predators and let $y(t)$ be the size of a population of their prey. The prey without predators would tend to increase in numbers according to the simplest law of proportional increase, $dy/dt = by$. On the other hand, the predators without prey would tend to die out according to a similar law, $dx/dt = -mx$.

When both predators and prey are present in the same area, the frequency of encounters between them may be taken to be proportional to the product xy of population sizes. These encounters tend to diminish the number of prey but are beneficial for the predators. Thus we are led to propose this pair of equations for the two populations:

$$\frac{dx}{dt} = -mx + bxy \qquad \text{(predators)}$$

$$\frac{dy}{dt} = ay - cxy \qquad \text{(prey)}$$

where a, b, c, and m are constants.

This system of equations cannot be solved for x and y as functions of t in closed form. As a compromise, we may try to discover how x and y are related to each other, without regard to time. If we think of $x(t)$ and $y(t)$ as the coordinates of a point moving in the xy-plane, then the curve traced by that point has slope

$$\frac{dy}{dx} = \frac{dy/dt}{dx/dt} = \frac{ay - cxy}{-mx + bxy}. \tag{1.46}$$

Now we have a differential equation between x and y in which t is not mentioned.

Having seen how some nonlinear differential equations can arise, we now turn our attention to methods of solution. Our most general first-order differential equation may be written as

$$\frac{dy}{dx} = F(x, y). \tag{1.47}$$

(It makes no real difference whether we use t and u or x and y as the independent and dependent variables.)

To begin, remember that we can solve the differential equation (1.47) if $F(x, y)$ does not depend on y at all. That is, $F(x, y)$ is a function of x alone, and Eq. (1.47) is better written as

$$\frac{dy}{dx} = f(x). \tag{1.48}$$

As we have already seen, an equation of this kind is solved by integrating both sides.

One of our most fruitful techniques is to reduce a given equation to one like Eq. (1.48). In solving the homogeneous linear equation

$$\frac{dy}{dx} = a(x)y \tag{1.49}$$

we first divide by y to obtain

$$\frac{1}{y}\frac{dy}{dx} = a(x).$$

The left-hand side must be recognized as the derivative of $\ln |y|$. Thus the equation can be rewritten as

$$\frac{d}{dx} \ln |y| = a(x).$$

Now we have an equation of the simplest type that can be solved for $\ln|y|$ and then for y.

The solution technique for the linear homogeneous equation is a special case of the method used for a differential equation in which $F(x, y)$ is a product of a function of x by a function of y,

$$\frac{dy}{dx} = a(x)b(y). \tag{1.50}$$

An equation in the form of Eq. (1.50) is said to be *separable*, since we can divide through by $b(y)$ to separate the variables:

$$\frac{1}{b(y)}\frac{dy}{dx} = a(x) \quad \text{or} \quad \frac{dy}{b(y)} = a(x)\,dx. \tag{1.51}$$

(Note that if k is a number for which $b(k) = 0$, then $y(x) = k$ is a solution of the original equation (1.50) and Eq. (1.51) must be restricted by $y \neq k$.) Integration of both sides of Eq. (1.51) will lead to a relation of the form

$$H(y) = A(x) + c \tag{1.52}$$

with $A(x)$ an integral of $a(x)$ and $H(y)$ an integral of $1/b(y)$.

In the second version of Eq. (1.51) we used the differentials of y and x instead of the derivative of y with respect to x. Such an equation is said to be in *differential form*; an equation such as Eq. (1.47) in which the derivative is used is said to be in *normal form*.

Example 3

Solve the logistic equation (1.44) with $b = 3$, $m = 2$:

$$\frac{du}{dt} = (3 - 2u)u.$$

This equation is separable. In separated form it is

$$\frac{du}{(3 - 2u)u} = dt.$$

The left-hand side can be integrated by the method of partial fractions:

$$\frac{du}{(3 - 2u)u} = \frac{1}{3}\left(\frac{1}{u} + \frac{2}{3 - 2u}\right)du.$$

Now integrating both sides gives

$$\tfrac{1}{3}(\ln|u| - \ln|3 - 2u|) = t + C$$

or

$$\ln\left|\frac{u}{3 - 2u}\right| = 3(t + C).$$

This equation can be solved for u by exponentiating, eliminating the absolute value signs, and then solving algebraically for u. The steps are

$$\left| \frac{u}{3 - 2u} \right| = e^{3t+3C} = e^{3t} \cdot e^{3C}$$

$$\frac{u}{3 - 2u} = \pm e^{3C} e^{3t} = ce^{3t}.$$

Call the right-hand side E for short. Then

$$\frac{u}{3 - 2u} = E$$

$$u = 3E - 2uE$$

$$u + 2uE = 3E$$

$$u = \frac{3E}{1 + 2E}$$

or

$$u(t) = \frac{3ce^{3t}}{1 + 2ce^{3t}}.$$

In the exercises we shall study this solution further.

Example 4

Solve Eq. (1.46) from the predator–prey system with all coefficients equal to 1:

$$\frac{dy}{dx} = \frac{y - xy}{xy - x}. \tag{1.53}$$

This equation is separable; y is a factor of the numerator and x of the denominator. Thus we have

$$\frac{dy}{dx} = \frac{y}{y - 1} \cdot \frac{1 - x}{x}$$

or

$$\frac{y - 1}{y} dy = \frac{1 - x}{x} dx$$

in separated form. The integration of both sides can be done easily:

$$\int \left(1 - \frac{1}{y} \right) dy = \int \left(\frac{1}{x} - 1 \right) dx$$

$$y - \ln |y| = \ln |x| - x + c. \tag{1.54}$$

It is not possible to solve this equation for either x or y as a function of the other in closed form.

The result of our calculations in Example 4 is not wholly satisfying, but the difficulty is unavoidable and the solution fully correct. A relation between x and y—usually symbolized by writing $Q(x, y) = c$—is called an *implicit solution* of a differential equation if it defines implicitly at least one solution. The curves in the xy-plane defined by an implicit solution are called *solution curves*. To verify an implicit solution, just differentiate the relation and work backwards to the differential equation.

Example 4 continued

Verify that Eq. (1.54) is an implicit solution of Eq. (1.53).

Applying implicit differentiation (and the chain rule) to Eq. (1.54), we get

$$\frac{dy}{dx} - \frac{1}{y}\frac{dy}{dx} = \frac{1}{x} - 1.$$

This equation can be solved for the derivative, yielding

$$\frac{dy}{dx} = \frac{\dfrac{1}{x} - 1}{1 - \dfrac{1}{y}}.$$

This is algebraically equivalent to the original differential equation (1.53) as long as xy is not 0. (Multiply the numerator and denominator of the fraction above by xy to get Eq. (1.53).)

The distinction between the implicit and explicit solutions shows up most clearly in this example, in which both can be found.

Example 5

Solve the separable differential equation

$$\frac{dy}{dx} = -\frac{x}{y}.$$

On separating the variables we get the equation

$$y\frac{dy}{dx} = -x \qquad \text{or} \qquad y\,dy = -x\,dx.$$

Integrating both sides leads to the relation

$\frac{1}{2}y^2 = -\frac{1}{2}x^2 + C.$

This is an implicit solution of the original differential equation, as we can readily verify by differentiating. However, in this case we can also obtain the explicit solution, either

$y(x) = \sqrt{2C - x^2}$

or

$y(x) = -\sqrt{2C - x^2}.$

The implicit solution is the family of circles centered at the origin. The explicit solutions are the families of the upper and lower semicircles. Since multiplication by y was part of the process of separation of variables, it is no surprise that the correspondence between the implicit solution obtained and the explicit solutions of the original equation breaks down when $y = 0$. Indeed, this is the dividing line between the upper and lower semicircles.

Exercises

In Exercises 1–10, solve the differential equation by separating variables. If an initial condition is given, find the solution of the differential equation that satisfies it.

1. $\dfrac{dy}{dx} = \dfrac{x}{y}$

2. $\dfrac{dy}{dx} = -\dfrac{y}{x}$

3. $\dfrac{dy}{dx} = \dfrac{x}{y + xy}$

4. $\dfrac{dy}{dx} = e^{x-y}$

5. $\dfrac{dy}{dx} = 1 - y^2,\ y(0) = 0$

6. $\dfrac{dy}{dx} = 1 - y^2,\ y(0) = 1$

7. $\dfrac{dy}{dx} = \dfrac{1 + y^2}{1 + x^2},\ y(0) = 1$

8. $\dfrac{dy}{dx} = \sin x \cos y$

9. $\dfrac{dy}{dx} = \sqrt{xy},\ x > 0,\ y > 0$

10. $\dfrac{dy}{dx} = \dfrac{xy + x - y - x^2}{xy - y^2}$

11. Solve these three initial value problems and compare how rapidly the solutions approach 0.

    (a) $\dfrac{du}{dt} = -u,\ u(0) = 1$

    (b) $\dfrac{du}{dt} = -u^2,\ u(0) = 1$

    (c) $\dfrac{du}{dt} = -\sqrt{u},\ u(0) = 1$

12. Solve the three initial value problems composed of the logistic equation and the three different initial conditions:

$$\frac{du}{dt} = 5u - 0.01u^2,$$

(a) $u(0) = 100$, (b) $u(0) = 1000$, (c) $u(0) = -100$. Discuss the behavior of the solution as $t \to \infty$.

13. Solve the logistic equation below, in which b and m are positive parameters:

$$\frac{du}{dt} = bu - mu^2.$$

14. Show that $u(t) = b/m$ is a solution of the differential equation. Can it be obtained from your formula for the solution? Can the solution $u(t) \equiv 0$ be obtained?

15. Discuss the behavior as $t \to \infty$ of the solution of the equation in Exercise 13 for initial values in the indicated ranges: (a) $0 < u(0) < b/m$, (b) $u(0) > b/m$, (c) $u(0) < 0$.

16. Find the solution of the predator–prey equation with all parameters equal,

$$\frac{dy}{dx} = \frac{y - xy}{xy - x},$$

that satisfies the condition $y = 5$ when $x = 5$.

17. Show that the point with coordinates $x = y = 0.0348$ is also (approximately) on the solution curve found in Exercise 16. (That is, these values of x and y also satisfy approximately the relation that is the implicit solution.)

18. If we could find explicit solutions for the equation in Exercise 16, those lying in the first quadrant would be divided into "upper" and "lower" families. What is the dividing line between the families?

19. In a basic oxygen steel furnace, carbon is removed from iron by blowing oxygen through it. The carbon forms CO and CO_2. The carbon content $u(t)$ is governed by the empirically determined equation*

$$\frac{du}{dt} = \frac{-1}{b + (a/u)^2}.$$

where a and b are positive constants and $u > 0$. Solve this equation for $u(t)$.

---

* From Yoan D. Landau, *Adaptive Control*, Dekker, New York and Basel, 1979, p. 39.

20. If du/dt can be measured (from the gases coming out of the furnace), show how to find u from measurements of du/dt.

21. Show that the solution found in Exercise 19 approaches 0 as $t \to \infty$.

22. A soliton is a special kind of traveling wave. Lamb (1932, p. 424) cites this equation for the shape of a soliton on water:

$$\left(\frac{dz}{dx}\right)^2 = \frac{z^2}{b^2}\left(1 - \frac{z}{a}\right).$$

In this equation, x is horizontal distance, $z + h$ is the height of the water surface above the bottom, and h is a reference level chosen so that $z(x) \to 0$ as $x \to \infty$. The parameter a is the maximum value of z and $b^2 = h^2(h + a)/3a$. Find $z(x)$, assuming that $z(0) = a$. Take the negative root in solving for dz/dx.

1.6

Equations Reducible to Separable

Many differential equations can be solved by means of a substitution for the independent variable that reduces the equation to a separable one. We have already used this idea to solve the nonhomogeneous linear equation

$$\frac{dy}{dx} - a(x)y = b(x).$$

Multiplying through by the integrating factor $e^{-A(x)}$ (where $dA/dx = a(x)$) turns the left-hand side into

$$e^{-A(x)}\left(\frac{dy}{dx} - a(x)y\right) = \frac{d}{dx}(e^{-A(x)}y).$$

Thus the substitution $w = e^{-A(x)}y$ reduces the nonhomogeneous equation to a separable one—in fact, one of the simplest form:

$$\frac{dw}{dx} = e^{-A(x)}b(x).$$

With linear equations we rarely make the substitution explicitly.

Below we detail several cases in which a substitution can convert a first-order equation

$$\frac{dy}{dx} = F(x, y) \tag{1.55}$$

into a separable one. Of course, the function F must satisfy certain special conditions for each class of substitution.

Homogeneous-Polar Equations

If the function $F(x, y)$ in Eq. (1.55) can be expressed in terms of the ratio y/x, the differential equation

$$\frac{dy}{dx} = f\left(\frac{y}{x}\right) \tag{1.56}$$

is usually called a homogeneous equation. Because the ratio y/x is the tangent of the polar coordinate θ, we shall use the term *homogeneous-polar* to distinguish this type of equation from the unrelated homogeneous linear equation.

To solve this equation, we introduce the new independent variable

$$v = \frac{y}{x}. \tag{1.57}$$

Then since $y = vx$ and $y' = v + xv'$, the differential equation (1.56) becomes

$$v + x\frac{dv}{dx} = f(v). \tag{1.58}$$

This is a separable equation; the form with variables separated is

$$\frac{dv}{f(v) - v} = \frac{dx}{x}. \tag{1.59}$$

The integration of both sides will produce a logarithm of x, so x can be found as a function of $v = y/x$. But it often happens that neither x nor y can be cleared as a function of the other.

Example 1

Solve the homogeneous-polar equation

$$\frac{dy}{dx} = \frac{y^2 + x^2}{xy}.$$

The substitution $y = vx$ leads to the new equation in v:

$$v + x\frac{dv}{dx} = \frac{v^2x^2 + x^2}{xvx} = \frac{v^2 + 1}{v}.$$

Now we clear the equation for dv/dx:

$$x\frac{dv}{dx} = \frac{v^2 + 1}{v} - v = \frac{1}{v}$$

$$\frac{dv}{dx} = \frac{1}{xv}.$$

Next this separable equation can be solved easily:

$$v \, dv = \frac{dx}{x},$$

$$\tfrac{1}{2}v^2 = \ln|x| + c.$$

The last step is to restore the original variables, by reversing the substitution, $v = y/x$:

$$\frac{1}{2}\frac{y^2}{x^2} = \ln|x| + c.$$

In this case we can get explicit solutions for our equation:

$$y = \pm x\sqrt{2(\ln|x| + c)}.$$

Example 2

Solve the homogeneous-polar equation

$$\frac{dy}{dx} = \frac{3x - y}{x + y}.$$

Making the substitution $y = vx$ and developing the resulting equation, we obtain successively:

$$v + x\frac{dv}{dx} = \frac{3x - vx}{x + vx} = \frac{3 - v}{1 + v}$$

$$x\frac{dv}{dx} = \frac{3 - v}{1 + v} - v = \frac{3 - 2v - v^2}{1 + v} \qquad (*)$$

$$\frac{1 + v}{3 - 2v - v^2}\, dv = -\frac{1 + v}{(v + 3)(v - 1)}\, dv = \frac{dx}{x}.$$

Now the variables v and x are separated, and both sides may be integrated (using partial fractions on the left) to get

$$-\tfrac{1}{2}(\ln|v - 1| + \ln|v + 3|) = \ln|x| + C.$$

To simplify, multiply through by -2 and exponentiate both sides:

$$|(v - 1)(v + 3)| = e^{-2C}x^{-2}$$

$$(v - 1)(v + 3) = \pm e^{-2C}x^{-2} = cx^{-2}.$$

The general constant c has been introduced to replace $\pm e^{-2C}$. Finally, replace v by y/x and simplify:

$$\left(\frac{y}{x} - 1\right)\left(\frac{y}{x} + 3\right) = cx^{-2}$$

$$(y - x)(y + 3x) = c. \qquad\qquad (1.60)$$

The solution curves are a family of hyperbolas centered at the origin.

In the course of the algebraic manipulations—at the point marked (*)—we multiplied through by $v + 1$ and divided by $3 - 2v - v^2$. Values of v for which any of these are zero must be checked to see whether solutions of the original equation are affected.

1. $v + 1 = 0$ or $y = -x$ is not a solution;
2. $v = -2$ (one root of $3 - 2v - v^2 = 0$) yields $y = -3x$, which is a solution of the original equation;
3. $v = 1$ (the other root of $3 - 2v - v^2 = 0$) yields $y = x$, also a solution of the original equation.

Example 3

Solve the initial value problem

$$\frac{dy}{dx} = \frac{2xy + y^2}{x^2 + xy}, \qquad y(1) = 1.$$

The differential equation can be seen to be homogeneous-polar because when the substitution $y = vx$ is made, x disappears from the right-hand side:

$$v + x\frac{dv}{dx} = \frac{2xvx + (vx)^2}{x^2 + xvx} = \frac{2v + v^2}{1 + v}.$$

Now we may separate the variables, obtaining the equations

$$x\frac{dv}{dx} = \frac{2v + v^2}{1 + v} - v = \frac{2v + v^2 - v(1 + v)}{1 + v} = \frac{v}{1 + v}$$

$$\frac{1 + v}{v}\,dv = \frac{dx}{x}.$$

Upon integrating we obtain

$$\ln|v| + v = \ln|x| + C \tag{1.61}$$

or

$$\ln\left|\frac{y}{x}\right| + \frac{y}{x} = \ln|x| + C$$

as the implicit solution of our equation.

The initial values are $y = 1$, $x = 1$. When these are substituted into the solution, C is determined as being 1. The implicit solution of the initial value problem is thus

$$\ln\left|\frac{y}{x}\right| + \frac{y}{x} = \ln|x| + 1.$$

It is not possible to solve this equation for either y or x as a function of the other. However, we can return to Eq. (1.61) and obtain a parametric solution for our equation. In Eq. (1.61), set $c = 1$, as required by the initial condition, and solve for x:

$$\ln |x| = \ln |v| + v - 1$$

$$x = e^{\ln |v| + v - 1} = v e^{v-1}. \qquad \textbf{(1.62)}$$

Now, $y = vx$ gives an equation for y when x is replaced from Eq. (1.62):

$$y = v^2 e^{v-1}. \qquad \textbf{(1.63)}$$

Equations (1.62) and (1.63) are now a pair of parametric equations for the solution, starting from $v = 1$.

If we are trying to decide whether a function $F(x, y)$ can be expressed as $f(y/x)$, what should we look for? The answer is, roughly speaking:

   1. ratios of homogeneous polynomials (a homogeneous polynomial is a polynomial in x and y where each term has the same total degree, such as $x^4 + 2x^2 y^2 - xy^3$);
   2. transcendental functions with arguments that can be expressed in terms of y/x alone;
   3. deceptive combinations, such as $\ln x - \ln y = -\ln (y/x)$.
Of course, one can just replace each y by vx and see whether all the x's disappear. This is the most effective procedure.

Example 4

The function

$$4 \ln x + \frac{x^2 + 2xy}{y^2} \tan \left(\frac{x + y}{y} \right) - \ln (x^4 + xy^3 \sin (y/x))$$

is expressible as a function of y/x, as one can see by looking for functions of the types mentioned above. If we try $y = vx$, we get

$$F(x, y) = 4 \ln x + \frac{x^2 + 2x^2 v}{x^2 v^2} \tan \left(\frac{x + xv}{xv} \right) - \ln (x^4 + x^4 v \sin v).$$

Notice how $4 \ln x = \ln x^4$ cancels the x^4 in the other logarithm.

A more formal way for identifying a function of the kind we are considering is given in this theorem (due to Euler).

Theorem 1.1

If $F(x, y)$ has partial derivatives with respect to x and y in some region, then $F(x, y) = f(y/x)$ if and only if

$$x\frac{\partial F}{\partial x} + y\frac{\partial F}{\partial y} = 0.$$

This theorem is interesting but not very practical. If F is not obviously a function of y/x alone, the differentiation will probably be difficult.

Linear Fractional Equation

It is clear that a differential equation of the form

$$\frac{dy}{dx} = \frac{ax + by}{Ax + By} \tag{1.64}$$

is homogeneous-polar (the coefficients are constants). Such an equation was solved in Example 2. A slightly more general equation is the *linear fractional equation*

$$\frac{dy}{dx} = \frac{ax + by + c}{Ax + By + C}. \tag{1.65}$$

The numerator and denominator of the ratio in Eq. (1.64) call to mind the equations of two lines intersecting at the origin. Similarly, the numerator and denominator of the ratio in Eq. (1.65) call to mind the equations of lines intersecting at another point. If we identify the coordinates of the point of intersection as (h, k), then these numbers satisfy the equations

$$ah + bk + c = 0$$
$$Ah + bk + C = 0 \tag{1.66}$$

simultaneously. (We assume that this system has a unique solution: that is, $aB - Ab \neq 0$.)

Now we may shift the origin to the point with coordinates (h, k) by changing both the dependent and independent variables:

$$x = X + h, \qquad y = Y + k. \tag{1.67}$$

The derivative dy/dx is the same as dY/dX; using this fact and substituting for x and y in Eq. (1.65), we get

$$\frac{dY}{dX} = \frac{a(X + h) + b(Y + k) + c}{A(X + h) + B(Y + k) + C}$$

$$= \frac{aX + bY + ah + bk + c}{AX + BY + Ah + Bk + C}$$

$$= \frac{aX + bY}{AX + BY}.$$

We now have a homogeneous-polar equation that can be solved by routine methods. The final solution should be stated in terms of the original variables x and y.

Example 5

Solve the linear fractional equation

$$\frac{dy}{dx} = \frac{x + y - 3}{x - y - 1}.$$

First we must find the coordinates (h, k) of the point for which both numerator and denominator are 0. These satisfy the equations

$$h + k - 3 = 0$$

$$h - k - 1 = 0.$$

The solution is easily found to be $h = 2$, $k = 1$. Now making the substitution

$$x = X + 2, \qquad y = Y + 1,$$

we obtain the differential equation

$$\frac{dY}{dX} = \frac{X + 2 + Y + 1 - 3}{X + 2 - (Y + 1) - 1} = \frac{X + Y}{X - Y},$$

which is homogeneous-polar. The further substitution $Y = Xv$ reduces this to a separable equation. The steps in the solution follow:

$$v + X\frac{dv}{dX} = \frac{X + Xv}{X - Xv} = \frac{1 + v}{1 - v}$$

$$X\frac{dv}{dX} = \frac{1 + v}{1 - v} - v = \frac{1 + v^2}{1 - v}$$

$$\frac{1 - v}{1 + v^2}\, dv = \frac{dX}{X}$$

$$\tan^{-1} v - \tfrac{1}{2}\ln (1 + v^2) = \ln |X| + C$$

$$\tan^{-1}\left(\frac{Y}{X}\right) - \frac{1}{2}\ln\left(1 + \left(\frac{Y}{X}\right)^2\right) = \ln |X| + C$$

$$\tan^{-1}\left(\frac{y-1}{x-2}\right) - \frac{1}{2}\ln\left(1 + \left(\frac{y-1}{x-2}\right)^2\right) = \ln|x-2| + C.$$

This equation cannot be solved for either y or x.

Exercises

In Exercises 1–15, solve the differential equation. If an auxiliary condition is given, find the solution that satisfies it.

1. $\dfrac{dy}{dx} = \dfrac{y-x}{y+x}$

2. $\dfrac{dy}{dx} = \dfrac{2y+x}{2x+y}$

3. $\dfrac{dy}{dx} = \dfrac{y+2x}{x+2y}$

4. $\dfrac{dy}{dx} = \dfrac{x}{2x-y}$

5. $\dfrac{dy}{dx} = \dfrac{3x^2 - y^2}{x^2 + y^2}$

6. $\dfrac{dy}{dx} = \dfrac{2x^2 + xy}{xy}$

7. $\dfrac{dy}{dx} = \dfrac{2x^2 + 3xy + 2y^2}{x^2 + xy}$, $y = 1$ at $x = 1$

8. $\dfrac{dy}{dx} = \dfrac{x^2 - 2xy + y^2}{x^2 - y^2}$, $y = 1$ at $x = 0$

9. $\dfrac{dy}{dx} = (e^{-y})^{1/x} + \dfrac{y}{x}$

10. $x\dfrac{dy}{dx} = y - \dfrac{y^2}{x}$, $y(1) = 0$

11. $\dfrac{dy}{dx} = \dfrac{2y + x + 4}{2x + y + 5}$, $y(0) = 0$

12. $\dfrac{dy}{dx} = \dfrac{x+1}{y+2}$

13. $\dfrac{dy}{dx} = \dfrac{y}{x + y + 2}$

14. $x^2 \dfrac{dy}{dx} = (x^2 + y^2)\tan^{-1}(y/x) + xy$

15. $x\dfrac{dy}{dx} = y(\ln(ey) - \ln x)$

16. $x\dfrac{dy}{dx} = y - \sqrt{x^2 + y^2}$, $y(0) = 1$

17. $x^2 \dfrac{dy}{dx} = y(\sqrt{y^2 - \frac{1}{4}x^2} + x)$

18. If in the linear fractional equation

$$\frac{dy}{dx} = \frac{ax + by}{Ax + By}$$

it happens that $aB = bA$, show that the equation is actually of the simplest type.

19. If in the linear fractional equation

$$\frac{dy}{dx} = \frac{ax + by + c}{Ax + By + C}$$

it happens that $aB = bA \neq 0$, show (a) that the simultaneous equations (1.66) for h and k may have no solution and (b) that the substitution $z = Ax + By + C$ reduces the equation to a separable one.

20. Use the results of Exercise 19 to solve

$$\frac{dy}{dx} = \frac{2x - y + 2}{2x - y + 1}.$$

21. Express the solution of

$$\frac{dy}{dx} = \frac{x + y}{x - y}$$

in terms of the polar coordinates $r = \sqrt{x^2 + y^2}$ and $\theta = \tan^{-1}(y/x)$. Show that the solution curve is a spiral.

22. A mathematical model of an arms race (the Richardson model) is given by the equations

$$\frac{dx}{dt} = -c_1 x + d_1 y + g_1,$$

$$\frac{dy}{dt} = d_2 x - c_2 y + g_2,$$

where x and y are the sizes of the arms stocks of two competing countries. In the first equation, $-c_1 x$ shows the tendency to allow arms to go out of service because of costs, $d_1 y$ represents the tendency to build up arms as defense against the competitor's arsenal, and g_1 represents a grievance term. The other equation is interpreted similarly. Assuming that the c's, d's, and g's are positive constants, find a first-order differential equation relating x and y.

23. Find the solution of the equation found in Exercise 12 if $c_1 = c_2 = 3$, $d_1 = d_2 = 1$, $g_1 = 5$, and $g_2 = 1$.

24. In general, after solving a homogeneous-polar equation we have $x = c\phi(v)$, where

$$\phi(v) = \exp\left\{ \int \frac{dv}{f(v) - v} \right\}.$$

If this relation cannot be solved for y as a function of x, show how to find y as a function of v and thus obtain a parametric representation for the solution.

25. Apply the ideas of Exercise 16 to solving the homogeneous-polar equation

$$\frac{dy}{dx} = \frac{y^2 + xy + x^2}{x^2}.$$

1.7
Graphical Methods

Having now sampled some of the most important solution methods for first-order differential equations, we might well suspect that there are equations that will not yield to any sort of manipulation. It is indeed true that many equations cannot be solved, even implicitly, in terms of familiar functions. Nevertheless, there is a broad class of equations for which solutions are guaranteed to exist.

Theorem 1.2

Let $F(x, y)$ and its partial derivative with respect to y be continuous throughout some rectangular region of the xy-plane containing the point (x_0, y_0). Then on some interval $x_0 - h < x < x_0 + h$ there is one, and only one, solution of the initial value problem

$$\frac{dy}{dx} = F(x, y), \qquad y(x_0) = y_0.$$

A proof of this theorem is discussed in Section 1.7A.

Example 1

Find a region in the xy-plane in which the function $F(x, y) = 3xy^{1/3}$ satisfies the hypotheses of the theorem. For what x_0, y_0 might

$$\frac{dy}{dx} = 3xy^{1/3}, \qquad y(x_0) = y_0,$$

fail to have a solution or have more than one solution?

The function $F(x, y) = 3xy^{1/3}$ is continuous in the entire xy-plane. However its partial derivative with respect to y, $\partial F/\partial y = xy^{-2/3}$, fails to exist along the line $y = 0$. Therefore the function satisfies the hypotheses of the theorem in any rectangle that does not contain any part of the x-axis.

Any initial condition with $y_0 = 0$ can be satisfied by many solutions of the differential equation. For example, both $y(x) \equiv 0$ and $y(x) = x^3$ satisfy the differential equation and the condition $y(0) = 0$.

Naturally, $F(x, y)$ fails to be continuous at any point of the xy-plane at which $F(x, y)$ becomes infinite (for example, if division of a finite nonzero quantity by zero is required, the argument of a logarithm is 0, etc.). It may happen, however, that the related differential equation

$$\frac{dx}{dy} = \frac{1}{F(x, y)}$$

does have solutions there.

Example 2

Investigate the behavior of solutions of

$$\frac{dy}{dx} = -\frac{x}{y}$$

along the line $y = 0$.

The function $F(x, y) = -x/y$ is certainly discontinuous along this line. However, the differential equation

$$\frac{dx}{dy} = -\frac{y}{x}$$

has a unique solution passing through each point (except the origin) on the line $y = 0$. In fact, we have the same implicit solution

$$x^2 + y^2 = c$$

to either of these equations. To obtain $y(x)$ explicitly, we have to cut this family of circles along the line $y = 0$.

It sometimes happens that the function $F(x, y)$ has the indeterminate form 0/0 at some point of the plane. Such a point is called a *critical point*. Of course, this is usually a point of discontinuity of F. There may be no solution, a finite number of solutions, or an infinite number of solutions through such a point. (These various possibilities are explored in the exercises.) We note, however, that the behavior of solutions near a critical point is often an important problem.

Some of the surprising behavior touched on above—and much more information about solutions—can be obtained by using graphical methods of analyzing a differential equation. The general idea is this: a differential equation

$$\frac{dy}{dx} = F(x, y)$$

tells us that the *slope of the solution curve through the point (x, y) is*

$F(x, y)$. To convert this information to graphic form, we just draw at the point (x, y) a short line segment with slope $F(x, y)$; the line segment is a tangent to the solution curve there. When enough line segments are drawn in the plane, one can often see trends in the solution curves. Such a drawing is called a *direction field*.

Example 3

First consider the simple linear equation

$$\frac{dy}{dx} = x - y.$$

At the point $(1, 0)$ the slope of the solution curve is $1 - 0 = 1$; at $(1, 1)$ the slope is $1 - 1 = 0$; and at $(1, 2)$ the slope is $1 - 2 = -1$. Drawing in the line segments with the appropriate slope could well be a slow process, but we can expedite the work in this case by noting that $F(x, y) = 0$ at every point on the straight line $y = x$; $F(x, y) = 1$ at every point on the line $y = x - 1$; $F(x, y) = -1$ wherever $y = x + 1$; and so on.

Figure 1.8 shows a direction field for this differential equation. It is easy to see that every solution (at least in the part of the plane shown) is heading toward one special solution, the function $y(x) = x - 1$.

If we have a fairly simple function for $F(x, y)$, the labor of drawing a direction field is reduced by applying what is called the *method of isoclines*. The locus of all points (x, y) satisfying the equation

$$F(x, y) = m \tag{1.68}$$

is a curve in the plane, at each point of which we want to draw line segments with slope m. Such a curve is called an *isocline* of the differential equation; in Example 3 the isoclines were the straight lines $x - y = m$. Incidentally, it is very rare that any isocline is a solution of the differential equation (although that occurred in Example 3). The isoclines should be thought of as "construction lines" for the preparation of the direction field. They are not usually left on the drawing of the direction field.

Example 4

The isoclines of the differential equation

$$\frac{dy}{dx} = y - x^2$$

are the curves

$$y - x^2 = m \quad \text{or} \quad y = x^2 + m.$$

Figure 1.8

Direction Field for

$$\frac{dy}{dx} = x - y$$

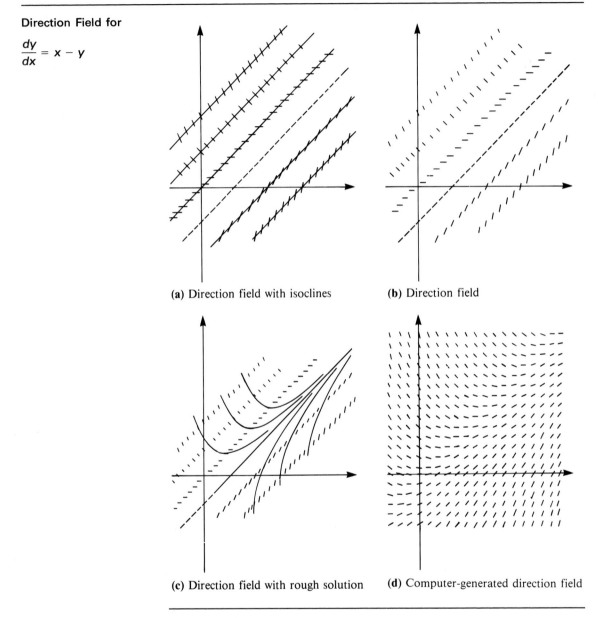

(**a**) Direction field with isoclines

(**b**) Direction field

(**c**) Direction field with rough solution

(**d**) Computer-generated direction field

In Fig. 1.9(a) are shown the isoclines for $m = -1$, 0, 1, 2, and in Fig. 1.9(b) is a direction field with the isoclines removed.

Theorem 1.2 has a graphical interpretation. It can be restated roughly to say that the differential equation $dy/dx = F(x, y)$ has one and only one solution curve through every point, except perhaps where F or one of its first partial derivatives is discontinuous. A corollary of the theorem is that *solution curves cannot cross* except where F or a derivative is discontinuous. As we have already mentioned, almost anything can happen at a critical point.

Example 5

Locate the critical points and find the isoclines for the equation (from the predator–prey system)

$$\frac{dy}{dx} = -\frac{y(1 - x)}{x(1 - y)}.$$

To locate the critical points, equate both numerator and denominator to 0:

$$y(1 - x) = 0 \quad \text{and} \quad x(1 - y) = 0.$$

In order to satisfy the first equation, either $y = 0$ or $x = 1$; from the second, either $x = 0$ or $y = 1$. By weeding out the contradictions among these four possibilities, we find the critical points $(0, 0)$ and $(1, 1)$. We shall see that every isocline passes through these points.

The isoclines are the curves defined by

$$-\frac{y(1 - x)}{x(1 - y)} = m.$$

This equation can be solved for y in the form

$$y = \frac{mx}{(m + 1)x - 1}.$$

We can identify the isoclines as hyperbolas (or a straight line for $m = -1$).

In Fig. 1.10(b) is shown a direction field for this differential equation. It appears that solutions in the first quadrant form closed curves about the critical point $(1, 1)$ and that no solution can cross from one quadrant to another. These observations illuminate the properties of the implicit solution

$$y - \ln|y| + x - \ln|x| = c.$$

Figure 1.9

Direction Field for

$$\frac{dy}{dx} = y - x^2$$

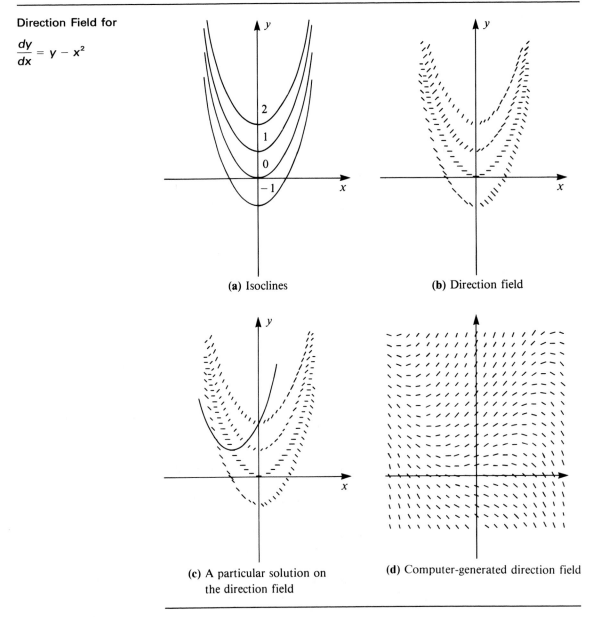

(a) Isoclines

(b) Direction field

(c) A particular solution on
the direction field

(d) Computer-generated direction field

Figure 1.10

Direction Field for

$$\frac{dy}{dx} = -\frac{y(1-x)}{x(1-y)}$$

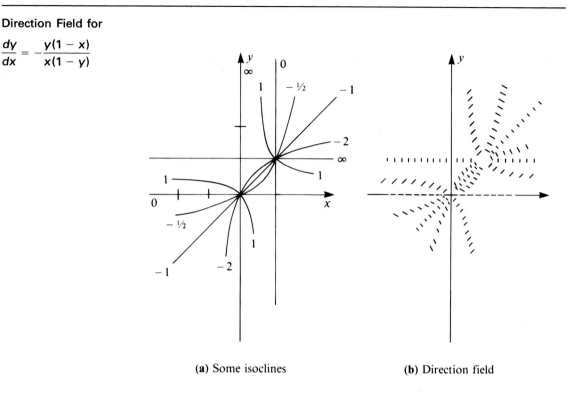

(a) Some isoclines **(b)** Direction field

In summary, to obtain graphical information about a differential equation, we carry out these steps:

1. Find all critical points (x, y) where $F(x, y) = 0/0$. These are points at which solutions and/or isoclines might cross.

2. Find the isoclines, described by $F(x, y) = m$, for several values of m—usually including $m = 0$ and $m = \infty$.

3. Construct the direction field.

What kind of information do we obtain about the solutions of a first-order differential equation by our graphical techniques?

1. *Trends in the solution.* In Example 3, for instance, it appears that all solutions tend toward the special solution $y = x - 1$ as x increases.

2. *Behavior near critical points.* In Example 5 we were able to see that solutions formed either closed curves or spirals about $(1, 1)$.

3. *Special regions.* In Example 1.29, one can see that no solution curve can leave a quadrant.

Exercises

In Exercises 1–5, sketch the direction field and sketch some solutions near the critical point $(0, 0)$.

1. $\dfrac{dy}{dx} = -\dfrac{x}{y}$

2. $\dfrac{dy}{dx} = \dfrac{x}{y}$

3. $\dfrac{dy}{dx} = \dfrac{2y - x}{y}$

4. $\dfrac{dy}{dx} = \dfrac{y + x}{y - x}$

5. $\dfrac{dy}{dx} = \dfrac{y - x}{y + x}$

In Exercises 6–9, locate the critical points and sketch the direction field.

6. $\dfrac{dy}{dx} = \dfrac{x + y - 2}{x - y}$

7. $\dfrac{dy}{dx} = \dfrac{x^2 - y}{x}$

8. $\dfrac{dy}{dx} = \dfrac{2y}{x^2 + y^2}$

9. $\dfrac{dy}{dx} = x^2 + y^2$

10. Analyze the behavior of the solutions of the logistic equation

$$\frac{du}{dt} = au - bu^2$$

(a, b positive constants) as t tends to infinity by making a direction field.

11. Find critical points and analyze the behavior of the solutions near them by making a direction field.

$$\frac{dy}{dx} = \frac{y(1 - y - x)}{x(2y - 1)}.$$

This equation arises from a predator–prey system in which the growth of prey without predators would follow a logistic equation.

12. Show that the point with coordinates (h, k), which is found in connection with the linear fractional equation (Section 1.6, Eq. (1.66)), is a critical point for that differential equation.

13. The origin is the only critical point for these four differential equations:

(a) $\dfrac{dy}{dx} = \dfrac{y}{x}$,   (b) $\dfrac{dy}{dx} = -\dfrac{x}{y}$,   (c) $\dfrac{dy}{dx} = \dfrac{x}{y}$,   (d) $\dfrac{dy}{dx} = -\dfrac{y}{x}$.

Determine how many solution curves pass through the origin for each of the equations.

14. In the Richardson model of an arms race (Section 1.6, Exercise 20) we obtained the linear fractional equation

$$\frac{dy}{dx} = \frac{d_2 x - c_2 y + g_2}{-c_1 x + d_1 y + g_1}.$$

Find the critical point of this differential equation, if there is one. Under what condition on the coefficients is it located in the first quadrant?

15. Show that the isoclines of a homogeneous-polar equation are lines passing through the origin.

16. For a homogeneous-polar equation, $dy/dx = f(y/x)$, what is the significance of the numbers α that satisfy $\alpha = f(\alpha)$ (if any exist)?

1.7A

Theoretical Matters

Here we look into the proof of the theorem stated in Section 7, which we restate in a slightly more precise form.

Theorem 1.3

Let $F(x, y)$ and its first partial derivative with respect to y be continuous in some rectangular region,

$$x_0 - a \leq x \leq x_0 + a, \qquad y_0 - b \leq y \leq y_0 + b.$$

Then on some interval $x_0 - h < x < x_0 + h$ there is one and only one solution of the initial value problem

$$\frac{dy}{dx} = F(x, y), \qquad y(x_0) = y_0. \tag{1.69}$$

We have already noted that if $\partial F/\partial y$ fails to be continuous, an initial value problem may have more than one solution. Specifically, the initial value problem

$$\frac{dy}{dx} = 3xy^{1/3}, \qquad y(0) = 0,$$

has among its solutions both $y(x) \equiv 0$ and $y(x) = x^3$. Note also that the theorem speaks of a solution on an interval $x_0 - h < x < x_0 + h$, which may be shorter than the interval $x_0 - a \leq x \leq x_0 + a$ in which F and $\partial F/\partial y$ are to be continuous. This is a necessity; the initial value problem

$$\frac{dy}{dx} = 1 + y^2, \qquad y(0) = 0,$$

has solution $y(x) = \tan x$, which is a solution only for $-\pi/2 < x < \pi/2$, although F and $\partial F/\partial y$ are continuous for all values of x and y (in this case, a and b may be taken as large as we wish).

The proof of the theorem hinges on two important ideas: the conversion of the initial value problem into an integral equation and the generation of a sequence of approximate solutions.

The conversion of Eq. (1.69) is accomplished by "integrating both sides" of the equation, between x_0 and x. The integration of the derivative returns the function, giving

$$y(x) - y(x_0) = \int_{x_0}^{x} F(z, y(z))\, dz$$

or, using the initial condition and a little algebra,

$$y(x) = y_0 + \int_{x_0}^{x} F(z, y(z))\, dz. \tag{1.70}$$

(Here z is a dummy variable of integration.) At first glance this appears to be a solution formula for Eq. (1.69). But the unknown function $y(x)$ appears in the integrand as well as on the left-hand side. Equation (1.70) is called an integral equation. It is equivalent to Eq. (1.69) in this sense: if $F(x, y)$ is continuous, then any continuous function $y(x)$ that satisfies the integral equation (1.70) is also differentiable and satisfies the differential equation (1.69). The continuity of F and that of y make the integrand in Eq. (1.70) continuous; then elementary calculus theorems guarantee that the right-hand side of Eq. (1.70) is a differentiable function whose derivative is $F(x, y(x))$. That is, differentiation of Eq. (1.70) is legitimate and gives back the differential equation (1.69). Of course the function in Eq. (1.70) automatically satisfies the initial condition.

Example 1

Find the integral equation that is equivalent to the initial value problem

$$\frac{dy}{dx} = y - xy^2, \qquad y(0) = 1.$$

The result of integrating both sides from 0 to x is

$$y(x) = 1 + \int_{0}^{x} [y(z) - zy^2(z)]\, dz.$$

It is left as an exercise to verify that the solution of both problems is

$$y(x) = \frac{1}{x - 1 + 2e^{-x}}.$$

The first advantage of the conversion to an integral equation is a technical one: the solution only has to be proved continuous; differentiability takes care of itself. The second advantage is quite practical. The integral equation suggests a way to generate a sequence of approximate solutions. This is the second main idea, developed by the French mathematician Emile Picard.

The sequence of approximate solutions of the integral equation (1.70) is created this way. The initial approximation is the constant function $\phi_0(x) = y_0$. Then we define

$$\phi_1(x) = y_0 + \int_{x_0}^{x} F(z, \phi_0(z))\, dz$$

$$= y_0 + \int_{x_0}^{x} F(z, y_0)\, dz.$$

Subsequent approximations are generated similarly, according to the formula

$$\phi_n(x) = y_0 + \int_{x_0}^{x} F(z, \phi_{n-1}(z))\, dz. \tag{1.71}$$

If the function F is simple enough, it will be possible to obtain at least some of the approximate solutions in closed form.

Example 2

Find approximate solutions through ϕ_2 of the initial value problem

$$\frac{dy}{dx} = x - y, \qquad y(0) = 1.$$

The integral equation equivalent to the initial value problem is

$$y(x) = 1 + \int_{0}^{x} (z - y(z))\, dz.$$

The initial approximation is $\phi_0(x) = 1$. Then

$$\phi_1(x) = 1 + \int_{0}^{x} (z - 1)\, dz = 1 + \frac{x^2}{2} - x,$$

$$\phi_2(x) = 1 + \int_{0}^{x} \left(z - \left(1 - z + \frac{z^2}{2} \right) \right) dz$$

$$= 1 + \frac{x^2}{2} - x + \frac{x^2}{2} - \frac{x^3}{6} = 1 - x + x^2 - \frac{x^3}{6}.$$

The rest of the proof of the existence of one or more solutions of the integral equation (1.70) requires some familiarity with the ideas of convergence—especially uniform convergence—usually studied in advanced calculus. We limit ourselves to a list of the main steps.

1. Proof that each approximate solution $\phi_n(x)$ remains within the bounds $y_0 - b \le \phi_n(x) \le y_0 + b$. At this stage it may be necessary to shorten the x-interval in order to guarantee the bounds.

2. Proof that the sequence of approximate solutions converges to a limit function: $\lim_{n \to \infty} \phi_n(x) = y(x)$.

3. Proof that the limit function $y(x)$ is continuous.

4. Proof that the limit function $y(x)$ actually satisfies the integral equation (1.70).

When these steps are complete, we have produced a solution of the integral equation (1.70) and also of the initial value problem (1.69). The last step in proving the theorem is:

5. Proof that any other solution of the integral equation (1.70) is identical to the one produced by the sequence of approximate solutions. This completes the proof.

The hypotheses of Theorem 1.3 are somewhat stronger than they need to be. As a result, somewhat more extensive conclusions can be drawn about the solution of the initial value problem (1.69), as explained in the following two corollaries. Both of them assume that the hypotheses of Theorem 1.3 are met, and both apply over a finite interval.

Corollary 1.1

The solution of the initial value problem (1.69) is a continuous function of the initial value y_0.

Corollary 1.2

If the function F depends continuously on some parameter τ, then the solution of the initial value problem (1.69) is also a continuous function of τ.

Example 3

Show that the solution of the initial value problem

$$\frac{dy}{dx} = by - my^2, \qquad y(0) = y_0,$$

depends continuously on b, m, and y_0.

The hypotheses of the corollaries are met on any finite interval $\alpha < x < \beta$. Therefore the continuous dependence follows. We may also write out the solution in closed form:

$$y(x) = \frac{by_0 e^{bx}}{b + my_0(e^{bx} - 1)},$$ (**1.72**)

from which the continuous dependence on the parameters is obvious.

These last two corollaries may be of greater interest than the theorem itself in many applications. Values of parameters are rarely known exactly, so it is important to know that a small approximation in the parameters does not cause a drastic change in the solution. Moreover, Corollary 1.2 often is (or should be) invoked in changing the form of a differential equation.

Example 4

If m is a small parameter, the solution of the initial value problem in Example 3 should be approximately the same as the solution of

$$\frac{dy}{dx} = by, \qquad y(0) = y_0.$$

Since the solution of this problem is

$$y(x) = y_0 e^{bx},$$ (**1.73**)

we can see that the two solutions are indeed nearly the same—in value, not form—provided that $e^{bx} - 1$ does not get too large, that is, for x restricted to a suitably small interval. Table 1.4 contains values of the functions from Eqs. (1.72) and (1.73) over the range $0 \le x \le 10$, with $b = 1$, $y_0 = 1$, and $m = 0.0001$ in Eq. (1.72).

Table 1.4

x	$y(x)$ (Eq. 1.73)	$y(x)$ (Eq. 1.72)
1	2.7183	2.7178
2	7.389	7.384
3	20.09	20.05
4	54.60	54.31
5	148.4	146.3
6	403.4	387.8
7	1097	988
8	2981	2296
9	8103	4476
10	22026	6878

Exercises

In Exercises 1–4, solve the integral equation by converting it to an initial value problem. Check that your solution does satisfy the integral equation.

1. $y(x) = 1 + \int_0^x y(z)\,dz$

2. $y(x) = \int_0^x (1 + y^2(z))\,dz$

3. $y(x) = \int_1^x e^{-y(z)}\,dz$

4. $y(x) = 1 + \int_0^x y^2(z)\,dz$

In Exercises 5–8, (a) write the equivalent integral equation and (b) find approximate solutions through the one indicated.

5. $\dfrac{dy}{dx} = x - y^2,\ y(0) = 0;\ \phi_3(x)$

6. $\dfrac{dy}{dx} = 1 + y^2,\ y(0) = 0;\ \phi_3(x)$

7. $\dfrac{dy}{dx} = \sqrt{y},\ y(0) = 1;\ \phi_2(x)$

8. $\dfrac{dy}{dx} = y - xy^2,\ y(0) = 1;\ \phi_2(x)$

9. (a) Solve the initial value problem

$$\frac{dy}{dx} = y^2, \qquad y(0) = 1.$$

(b) For what interval of x is the solution valid? (Find the largest interval containing $x_0 = 0$ in which the solution is continuous.)

(c) In what rectangle of the xy-plane is $F(x, y) = y^2$ continuous?

10. (a) Use the iterative procedure to find approximate solutions of

$$\frac{dy}{dx} = xy, \qquad y(0) = 1.$$

(b) Solve the equation and confirm that each approximate solution ϕ_k consists of the first $k + 1$ nonzero terms of the Taylor series of $y(x)$.

(c) Is this true also in Exercises 6 and 7 and in Example 2?

1.8

Exact Equations

In the previous sections there were several examples of separable or homogeneous equations whose solution arose in the form of a relation between x and y. That is, we had

$$Q(x, y) = c, \tag{1.74}$$

where Q was a function we found and c was an arbitrary constant. Now suppose that a relation like Eq. (1.74) is given. It represents a family of curves in the xy-plane, a different curve for each different admissible value of c. For example, the relation

$$x^2 + y^2 = c \qquad\qquad \textbf{(1.75)}$$

represents the family of circles centered on the origin, for $c > 0$.

We ask whether such a relation corresponds to a differential equation. The answer is yes, if Q has continuous partial derivatives. At all the points (x, y) on a curve defined by Eq. (1.74) the function Q has the same, constant value, so its *differential* is 0. In symbols we have $dQ = 0$ or—by the rule for the differential of a function of two variables—

$$\frac{\partial Q}{\partial x}\, dx + \frac{\partial Q}{\partial y}\, dy = 0 \qquad\qquad \textbf{(1.76)}$$

on each of the curves defined by Eq. (1.74). Thus the family of circles defined explicitly by Eq. (1.75) is also represented by the equation

$$2x\, dx + 2y\, dy = 0. \qquad\qquad \textbf{(1.77)}$$

Now if we are given a differential equation in what we called differential form,

$$M(x, y)\, dx + N(x, y)\, dy = 0, \qquad\qquad \textbf{(1.78)}$$

we say that the equation is *exact* if there is a differentiable function $Q(x, y)$ for which

$$\frac{\partial Q}{\partial x} = M(x, y), \qquad \frac{\partial Q}{\partial y} = N(x, y). \qquad\qquad \textbf{(1.79)}$$

If this is the case, the differential equation (1.78) may be restated as

$$\frac{\partial Q}{\partial x}\, dx + \frac{\partial Q}{\partial y}\, dy = 0$$

or, in other words, $dQ = 0$. Therefore the solution of Eq. (1.78) is $Q(x, y) = c$.

Equation (1.77) is easily seen to be exact—especially since we have the function Q in hand. In general, however, using the definition to test a differential equation for exactness is tantamount to solving it. However, there is another, easier test. A familiar theorem on partial differentiation says that if $Q(x, y)$ and the partial derivatives

$$\frac{\partial Q}{\partial x}, \ \frac{\partial Q}{\partial y}, \ \frac{\partial^2 Q}{\partial x\, \partial y}, \ \frac{\partial^2 Q}{\partial y\, \partial x}$$

are all continuous in a rectangle of the xy-plane, then the mixed second partial derivatives are equal:

$$\frac{\partial^2 Q}{\partial y \, \partial x} = \frac{\partial^2 Q}{\partial x \, \partial y}.$$

In an exact equation we can identify M and N as the first derivatives of some function

$$M = \frac{\partial Q}{\partial x}, \qquad N = \frac{\partial Q}{\partial y}.$$

Thus we expect to find

$$\frac{\partial M}{\partial y} = \frac{\partial N}{\partial x},$$

since these are the second mixed partials of Q. This observation gives part of the following theorem.

Theorem 1.4

Let M, N, $\partial M/\partial y$, and $\partial N/\partial x$ be continuous functions of x and y in a rectangle R of the xy-plane. Then the equation

$$M(x, y) \, dx + N(x, y) \, dy = 0$$

is exact in R if and only if

$$\frac{\partial M}{\partial y} = \frac{\partial N}{\partial x}$$

at every point of R.

Example 1

The differential equation

$$(2x + y^2) \, dx + (2xy + 1) \, dy = 0$$

has $M = 2x + y^2$, $N = 2xy + 1$. These functions and all their derivatives are continuous in the whole xy-plane. Furthermore, we find that

$$\frac{\partial M}{\partial y} = 2y, \qquad \frac{\partial N}{\partial x} = 2y.$$

These being equal, the original equation was exact.

    On the other hand, for another equation that looks almost the same,

$$(2x + y^2) \, dx + (xy + 1) \, dy = 0,$$

we find that the derivatives are

$$\frac{\partial M}{\partial y} = 2y, \qquad \frac{\partial N}{\partial x} = y.$$

Thus the latter equation is not exact.

Example 2

The differential equation

$$\frac{-y}{x^2 + y^2}\, dx + \frac{x}{x^2 + y^2}\, dy = 0$$

passes the test on the derivatives:

$$\frac{\partial M}{\partial y} = \frac{(x^2 + y^2)(-1) + y \cdot 2y}{(x^2 + y^2)^2} = \frac{y^2 - x^2}{(x^2 + y^2)^2},$$

$$\frac{\partial N}{\partial x} = \frac{(x^2 + y^2) - x \cdot 2x}{(x^2 + y^2)^2} = \frac{y^2 - x^2}{(x^2 + y^2)^2}.$$

However, both M and N and the derivatives are discontinuous at the origin. (They all have $x^2 + y^2$ in the denominator.) Thus the theorem guarantees exactness only in a rectangle that excludes the origin. The solution curves for this equation are all the straight lines through the origin. The fact that they all pass through the origin is an indication that the origin is a critical point of the differential equation. (See Section 1.6.)

Solving an exact differential equation is about as easy as solving a separable equation. The very fact that the equation

$$M(x, y)\, dx + N(x, y)\, dy = 0$$

is exact implies that there is a function $Q(x, y)$ for which

$$\frac{\partial Q}{\partial x} = M(x, y), \qquad \frac{\partial Q}{\partial y} = N(x, y). \qquad (1.80)$$

Since a partial derivative with respect to x is computed by treating y as constant and differentiating with respect to x, we may reverse the process and reconstruct Q from $\partial Q/\partial x$ by integrating with respect to x, treating y as a constant:

$$Q(x, y) = \int M(x, y)\, dx + k(y) = f(x, y) + k(y). \qquad (1.81)$$

In this equation, $f(x, y)$ is the result of the integration, and $k(y)$ is a function of y alone that plays the role of a "constant of integration." At this stage, $k(y)$ is unknown. To find it, we use the fact that $\partial Q/\partial y = N(x, y)$. That is, we differentiate, with respect to y, our candidate for $Q(x, y)$ in Eq. (1.81) and equate it to $N(x, y)$:

$$\frac{\partial}{\partial y}(f(x, y) + k(y)) = N(x, y). \tag{1.82}$$

The result is a differential equation for $k(y)$:

$$\frac{\partial f}{\partial y} + \frac{dk}{dy} = N(x, y)$$

or

$$\frac{dk}{dy} = N(x, y) - \frac{\partial f}{\partial y}. \tag{1.83}$$

It is important to check that the right-hand side of Eq. (1.83) is a function of y alone. (If not, the original equation was not exact, or a mistake was made.) Then Eq. (1.83) is an equation of simplest type, from which $k(y)$ can be found. Finally, the function $Q(x, y)$ can be constructed from Eq. (1.81) and $k(y)$, and the solution of our exact equation is

$$Q(x, y) = c.$$

Example 3

Solve the differential equation

$$(2x + y^2) \, dx + (2xy + 1) \, dy = 0.$$

We determined in Example 1 that the given equation was exact. Identifying $M(x, y) = 2x + y^2$, we find (cf. Eq. (1.81))

$$\frac{\partial Q}{\partial x} = 2x + y^2$$

$$Q(x, y) = \int (2x + y^2) \, dx + k(y)$$

$$= x^2 + xy^2 + k(y).$$

Now we require that $\partial Q/\partial y = N$, that is (cf. Eq. (1.82)),

$$\frac{\partial}{\partial y}(x^2 + xy^2 + k(y)) = 2xy + 1.$$

After carrying out the differentiation, we obtain

$$2xy + k'(y) = 2xy + 1,$$

which can be simplified in the obvious way to give (cf. Eq. (1.83))

$$k'(y) = 1.$$

From this we find that $k(y) = y$ and, finally,

$$Q(x, y) = x^2 + xy^2 + y.$$

The solution of the given exact equation is

$$x^2 + xy^2 + y = c.$$

Example 4

Solve the exact differential equation

$$(y \cos x + y^2) \, dx + (\sin x + 2xy + 3y^2) \, dy = 0.$$

First we find the candidate for Q:

$$Q(x, y) = \int M(x, y) \, dx + k(y)$$

$$= \int (y \cos x + y^2) \, dx + k(y)$$

$$= y \sin x + y^2 x + k(y).$$

Now applying the requirement that $\partial Q / \partial y = N$, we obtain

$$\frac{\partial Q}{\partial y} = \sin x + 2yx + k'(y) = \sin x + 2xy + 3y^2 = N.$$

All mention of x disappears from this equation by cancellation, leaving the easily solved equation

$$k'(y) = 3y^2.$$

Thus the solution of the original equation is

$$Q(x, y) = y \sin x + y^2 x + y^3 = c.$$

In the examples it can be seen that the equation for dk/dy is completely free of x. The general expression for dk/dy was given by Eq. (1.83):

$$\frac{dk}{dy} = N(x, y) - \frac{\partial f}{\partial y}$$

or, if we replace f by the integral it represents,

$$\frac{dk}{dy} = N(x, y) - \frac{\partial}{\partial y} \int M(x, y) \, dx. \tag{1.84}$$

To see why this equation is free of x, first note that the differentiation with respect to y of the integral with respect to x can be accomplished inside the integral (assuming $M(x, y)$ continuous):

$$\frac{\partial}{\partial y} \int M(x, y) \, dx = \int \frac{\partial M}{\partial y}(x, y) \, dx.$$

Now Eq. (1.84) can be written as

$$\frac{dk}{dy} = N(x, y) - \int \frac{\partial M}{\partial y}(x, y) \, dx.$$

But the exactness condition tells us that $\partial M/\partial y$ is the same as $\partial N/\partial x$. This allows the transformation of the last equation above to

$$\frac{dk}{dy} = N(x, y) - \int \frac{\partial N}{\partial x}(x, y) \, dx. \tag{1.85}$$

Finally, in this equation we see that the integration of $\partial N/\partial x$ with respect to x will restore to its original form every term of N that contained x. Thus the right-hand side of Eq. (1.85) must be free of x.

Example 5

The equation

$$(2x + y^2) \, dx + (xy + 1) \, dy = 0$$

was found in Example 2 to fail the test for exactness. If we try to carry out the solution procedure anyway, we find

$$\int M \, dx + k(y) = x^2 + xy^2 + k(y).$$

The derivative of this, with respect to y, should equal $N(x, y)$:

$$2xy + k'(y) = xy + 1.$$

However, this equation leads to

$$k'(y) = 1 - xy,$$

which is impossible because $k(y)$ *must not* depend on x. Thus k cannot be found, and the procedure cannot be finished.

It should not be assumed that every exact equation can be solved in explicit terms. In many (perhaps most) cases the solution will have to stay in the form of an integral.

Example 6

Solve this equation, which is exact in the entire xy-plane.

$$\frac{e^{xy}\sin x}{x}\,dx + \frac{e^{xy}(y\sin x - \cos x)}{y^2 + 1}\,dy = 0.$$

The first integration cannot be carried out explicitly:

$$\int M(x, y)\,dx + k(y) = \int \frac{e^{xy}\sin x}{x}\,dx + k(y).$$

However, its partial derivative with respect to y can be found:

$$\frac{\partial}{\partial y}\left(\int \frac{e^{xy}\sin x}{x}\,dx + k(y)\right) = \int \frac{\partial}{\partial y}\left(\frac{e^{xy}\sin x}{x}\right)dx + k'(y)$$

$$= \int \frac{xe^{xy}\sin x}{x}\,dx + k'(y)$$

$$= \int e^{xy}\sin x\,dx + k'(y)$$

$$= e^{xy}\frac{(y\sin x - \cos x)}{y^2 + 1} + k'(y).$$

On comparing this expression to $N(x, y)$ we find that $k'(y) = 0$, so k may be taken to be equal to 0. The solution of the original equation is

$$\int \frac{e^{xy}\sin x}{x}\,dx = c.$$

Exercises

In Exercises 1–20, test each equation for exactness. If exact, solve by the methods of this section. If not, try to solve by another method.

1. $dx + dy = 0$
2. $x\,dx + y\,dy = 0$
3. $y\,dx + x\,dy = 0$
4. $x^2\,dx + y^2\,dy = 0$
5. $y^2\,dx + x^2\,dy = 0$
6. $2xy\,dx + x^2\,dy = 0$
7. $(x^2 + y)\,dx + (x + e^y)\,dy = 0$
8. $(y - y^2)\,dx + x\,dy = 0$
9. $(2xy + y^2)\,dx + (x^2 + 2xy)\,dy = 0$
10. $\sin y\,dx + x\cos y\,dy = 0$
11. $y\cos x\,dx + \sin x\,dy = 0$
12. $dx + \sqrt{xy}\,dy = 0$

13. $2xe^y \, dx + x^2 e^y \, dy = 0$ 14. $\dfrac{y \, dx - x \, dy}{x^2 + y^2} = 0$

15. $y \, dx - x \, dy = 0$

16. $\dfrac{x^2 - y^2}{(x^2 + y^2)^2} \, dx + \dfrac{2xy}{(x^2 + y^2)^2} \, dy = 0$

17. $(x + x^2) \, dx - (y + xy) \, dy = 0$ 18. $\left(\dfrac{y}{x} + 1\right) dx - dy = 0$

19. $\dfrac{y}{x} \, dx + \ln x \, dy = 0$

20. $x(1 - y) \, dx + y(1 - x) \, dy = 0$

21. Prove: If $M \, dx + N \, dy = 0$ is exact, so is

$$[M + f(x)] \, dx + [N + g(y)] \, dy = 0,$$
for any differentiable functions f and g.

22. If $M \, dx + N \, dy = 0$ has solution $Q(x, y) = c$, this relation describes a family of curves on which the slope is $dy/dx = -M/N$. Show that a solution of $N \, dx - M \, dy = 0$ is a curve that cuts the curves $Q(x, y) = c$ at right angles (orthogonal trajectories).

23. Find the orthogonal trajectories of the curves defined by $x \, dx + y \, dy = 0$.

24. Find the orthogonal trajectories of the curves defined by $x \, dx - y \, dy = 0$.

25. Suppose $M \, dx + N \, dy = 0$ is exact, with solution $Q(x, y) = c$. What condition on the derivatives of Q will guarantee that $N \, dx - M \, dy = 0$ is also exact?

26. Show that the differential equation
$$(ax + by + c) \, dx - (Ax + By + C) \, dy = 0,$$
which is derived from the linear fractional equation
$$\frac{dy}{dx} = \frac{ax + by + c}{Ax + By + C},$$
is exact if and only if $A + b = 0$.

27. Assuming that the equation in Exercise 26 is indeed exact, find its solution.

1.9

Integrating Factors

A differential equation arising from applications is not likely to be exact. However, in many cases we make it exact by using an integrating factor. Suppose we are given the equation

$$M(x, y) \, dx + N(x, y) \, dy = 0. \tag{1.86}$$

An *integrating factor* for this equation is a nonzero function $\phi(x, y)$ such that the equation

$$\phi M \, dx + \phi N \, dy = 0, \tag{1.87}$$

obtained by multiplying through Eq. (1.86) by ϕ, is exact.

Example 1

The differential equation

$$y \, dx - x \, dy = 0$$

becomes exact when multiplied through by $1/(x^2 + y^2)$. Then the equation is

$$\frac{y}{x^2 + y^2} \, dx + \frac{-x}{x^2 + y^2} \, dy = 0$$

$$\frac{\partial}{\partial y}\left(\frac{y}{x^2 + y^2}\right) = \frac{x^2 - y^2}{(x^2 + y^2)^2} = \frac{\partial}{\partial x}\left(\frac{-x}{x^2 + y^2}\right).$$

In order for ϕ to be an integrating factor for Eq. (1.86), then, it must satisfy the relation

$$\frac{\partial(\phi M)}{\partial y} = \frac{\partial(\phi N)}{\partial x}, \tag{1.88}$$

which becomes, after development,

$$N\frac{\partial \phi}{\partial x} - M\frac{\partial \phi}{\partial y} = \phi\left(\frac{\partial M}{\partial y} - \frac{\partial N}{\partial x}\right). \tag{1.89}$$

This is a first-order partial differential equation, which seems (and usually is) more complex than the original equation. But we need just one solution of the partial differential equation (1.89) to get the *general* solution of the ordinary differential equation (1.87) or (1.86). Note, for instance, that if the original equation was exact, then the right-hand side of Eq. (1.89) is 0, and $\phi(x, y) = 1$ (or any other nonzero constant) is an integrating factor.

Example 2

For the differential equation

$$-y \, dx + x \, dy = 0$$

we find that $M = -y$, $N = x$, and

$$\frac{\partial M}{\partial y} - \frac{\partial N}{\partial x} = -2;$$

thus the equation is not exact. The equation to be fulfilled by an integrating factor is (from Eq. (1.89) above)

$$x\frac{\partial\phi}{\partial x} + y\frac{\partial\phi}{\partial y} = -2\phi.$$

Several integrating factors could be found by simple guesses. For instance, if one tries to find a solution in the form

$$\phi(x, y) = (x^2 + y^2)^a,$$

(suggested by the symmetry between x and y), it is easy to determine that $a = -1$ does give an integrating factor.

On the other hand, one might look for an integrating factor that depends on x alone—that is, assume that $\partial\phi/\partial y = 0$. Then ϕ must satisfy

$$x\frac{d\phi}{dx} = -2\phi.$$

This equation is both separable and linear. A solution is $\phi(x) = 1/x^2$, again an integrating factor for the original equation.

The second half of Example 2 illustrates one device that may lead to an integrating factor: the assumption that ϕ depends on only one of the two variables x and y. If we assume that $\partial\phi/\partial y = 0$, then Eq. (1.89) becomes

$$\frac{1}{\phi}\frac{d\phi}{dx} = \frac{\dfrac{\partial M}{\partial y} - \dfrac{\partial N}{\partial x}}{N}. \tag{1.90}$$

Of course, the right-hand side must be free of y; otherwise, the assumption that ϕ was independent of y is contradictory. Similarly, if we assume that $\partial\phi/\partial x = 0$, Eq. (1.89) becomes

$$-\frac{1}{\phi}\frac{d\phi}{dy} = \frac{\dfrac{\partial M}{\partial y} - \dfrac{\partial N}{\partial x}}{M}, \tag{1.91}$$

and the right-hand side of this equation must be free of x.

Example 3

To find an integrating factor for the equation

$$(x + y^2)\, dx + (2xy - \tfrac{1}{2}x^2 y)\, dy = 0,$$

we first calculate

$$\frac{\partial M}{\partial y} - \frac{\partial N}{\partial x} = 2y - (2y - xy) = xy.$$

Then we see that the assumption $\partial \phi / \partial y = 0$ leads to the equation

$$\frac{1}{\phi} \frac{d\phi}{dx} = \frac{xy}{2xy - \frac{1}{2}x^2 y} = \frac{2}{4 - x},$$

which can be solved easily. On the other hand, the other possibility leads to the contradictory equations $\partial \phi / \partial x = 0$ and

$$\frac{1}{\phi} \frac{d\phi}{dy} = \frac{xy}{x + y^2}.$$

As an application of the idea of integrating factors, let us consider the general first-order linear equation

$$\frac{dy}{dx} = a(x)y + b(x), \tag{1.92}$$

which can be converted to differential form as

$$(a(x)y + b(x))\, dx - dy = 0. \tag{1.93}$$

It is easy to find that

$$\frac{\partial M}{\partial y} - \frac{\partial N}{\partial x} = a(x),$$

and, since $N(x, y) = -1$, we can find an integrating factor that is a function of x alone. Equation (1.90) becomes, in this case,

$$\frac{1}{\phi} \frac{d\phi}{dx} = -a(x).$$

Thus our integrating factor for the linear equation (1.92) is

$$\exp\left(-\int a(x)\, dx \right) = e^{-A(x)}. \tag{1.94}$$

If now we multiply through by the integrating factor, Eq. (1.94) becomes

$$e^{-A(x)}(a(x)y - b(x))\, dx - e^{-A(x)}\, dy = 0.$$

This is exact, since it is now true that

$$\frac{\partial M}{\partial y} = \frac{\partial N}{\partial x} = e^{-A(x)}a(x).$$

Another application of our ideas is to the separable equation

$$\frac{dy}{dx} = a(x)b(y), \tag{1.95}$$

which might be written in differential form as

$$a(x)b(y)\,dx - dy = 0. \tag{1.96}$$

In this case we find

$$\frac{\partial M}{\partial y} - \frac{\partial N}{\partial x} = a(x)b'(y),$$

and so it is possible to find an integrating factor free of x:

$$\frac{\partial \phi}{\partial x} = 0, \qquad -\frac{1}{\phi}\frac{d\phi}{dy} = \frac{a(x)b'(y)}{a(x)b(y)}. \tag{1.97}$$

From here we determine that the integrating factor is $\phi = 1/b(y)$. The exact version of Eq. (1.96) is

$$a(x)\,dx - \frac{1}{b(y)}\,dy = 0, \tag{1.98}$$

an equation with the variables separated.

Exercises

In Exercises 1–6, find an integrating factor and solve the equation.

1. $(x + y)\,dx - x\,dy = 0$ 2. $y\,dx + 2x\,dy = 0$
3. $(1 + 2xye^y)\,dx + x(xe^y - 1)\,dy = 0$
4. $3x^2y\,dx - 2x^3\,dy = 0$
5. $(xy - 1)\,dx - dy = 0$ 6. $y\,dx - 2x\,dy = 0$

7. Find an integrating factor for the equation given in Example 3 and solve the equation.
8. Carry out the differentiation and algebra that lead from Eq. (1.88) to Eq. (1.89).
9. Suppose an equation is given in differential form and $M(x, y) = m(y/x)$, $N(x, y) = n(y/x)$; that is, both coefficients are homogeneous functions of x and y. Show that there is an integrating factor that is homogeneous also: $\phi(x, y) = \theta(y/x)$.
10. Use the results of Exercise 9 on the equation of Exercise 4. Divide through by x^3 to get homogeneous coefficients.
11. Use the results of Exercise 9 to find an integrating factor for

$$(x^2 + 2y^2)\,dx - xy\,dy = 0.$$

Notes and References

A beginning student often feels that solving differential equations is a matter of knowing enough "tricks." Each equation has to be classified as linear, separable, exact, etc., and then an appropriate sequence of operations is applied that eventually leaves us with the solution. The truth is that there is a unifying theory, but at a very high mathematical level.

Some individual types of equations have a theoretical structure. Linear equations are understood so well that it is possible to write down the solution of the initial value problem

$$\frac{du}{dt} = a(t)u + f(t), \qquad u(t_0) = u_0,$$

in terms of some integrals. (And you are challenged to do so.) Thus all properties of the solution can be determined from properties of $a(t)$, $f(t)$, t_0, and u_0. Similarly, the theory of the exact equation is well known and is often discussed in texts on multivariate calculus. Since the prerequisite of this book is just two semesters of calculus, we have not investigated the exact equation thoroughly.

The relationship between explicit and implicit solutions of a first-order equation is well explained in Chapter 1 of Birkhoff and Rota (1978). In many cases a first-order differential equation can be understood better if it is replaced by a system of two equations:

$$\frac{dy}{dx} = F(x, y) \rightarrow \begin{cases} \dfrac{dy}{dt} = h(x, y) \\ \dfrac{dx}{dt} = g(x, y) \end{cases}$$

where g and h are chosen to be well behaved and $F = h/g$. Such systems are discussed in Chapter 8.

The proof of the fundamental existence and uniqueness theorem was just sketched in Section 1.7A. A sound elementary proof is given by Hagin (1975). All proofs of this theorem require the concept of uniform convergence of a sequence of functions.

Braun (1975) writes in a very entertaining way about some applications of first-order differential equations. The story of the Van Meegeren art forgeries is particularly interesting.

Miscellaneous Exercises

In Exercises 1–24, classify the given equation as linear, separable, etc., and solve it. If an initial condition is given, also find the solution that satisfies it.

1. $\dfrac{dy}{dx} = \sqrt{xy}$, $y(0) = 1$ 2. $\dfrac{dy}{dx} = \dfrac{y}{\sqrt{x}}$

3. $\dfrac{dy}{dx} = 1 + y^2$, $y(0) = 1$ 4. $2xy\,dx + (1 + x^2)\,dy = 0$

5. $xy\,dx + (x^2 + y^2)\,dy = 0$ 6. $\dfrac{dy}{dx} = \sqrt{1 - y^2}$, $y(0) = \frac{1}{2}$

7. $\dfrac{dy}{dx} = -2xy^2$ 8. $\dfrac{dy}{dx} = \sqrt{1 + y^2}$

9. $\dfrac{dy}{dx} = x^2 - y$

10. $x\dfrac{dy}{dx} = y + \sqrt{x^2 + y^2}$, $y(0) = 1$

11. $x(x^2 + y^2 - 1)\,dx + y(x^2 + y^2 + 1)\,dy = 0$

12. $x^2\,dx - y^2\,dy = 0$

13. $y^2\,dx - x^2\,dy = 0$, $y(1) = 1$ 14. $\dfrac{dy}{dx} = \dfrac{x - 2y + 1}{y - 2x + 4}$

15. $x^2\dfrac{dy}{dx} = y(x - \sqrt{y^2 + 4x^2})$ 16. $\dfrac{dy}{dx} = (\cos x)y$

17. $\dfrac{dy}{dx} = -y\sqrt{1 - y^2}$ 18. $\dfrac{dy}{dx} = \dfrac{1 + y^2}{y}$

19. $(2x + 3y)\,dx + (3x + 4y)\,dy = 0$

20. $\dfrac{dy}{dx} = y + \dfrac{1}{y}$

21. $\dfrac{dr}{d\theta} - r\tan\theta = 1$

22. $(1 - t^2)\dfrac{du}{dt} + 2tu = 2t(1 - t^2)$

23. $\dfrac{dy}{dx} = \dfrac{y}{x + y}$ (Hint: treat x as the dependent variable.)

24. $\dfrac{du}{dt} + 5u = \sin t$

25. Make a direction field for the differential equation
$$\frac{dy}{dx} = \frac{x - y^2}{1 + y^2}.$$
What happens to y as $x \to \infty$? (Hint: use isoclines for $m = 0, \pm 1, \pm 2$.)

26. Make a direction field for the differential equation

$$\frac{dy}{dx} = \frac{-2y}{x^2 + y^2}.$$

27. Water resists the passage of a ship with a force $F = av + bv^2$, where v is speed and a, b are constants. Suppose the ship is proceeding with speed v_0 and cuts engines at time $t = 0$. Find an initial value problem that describes v as a function of t. (Hint: apply Newton's second law, $F = m\, dv/dt$.)

28. Use a direction field in the quadrant $v > 0$, $t > 0$ to determine the behavior of the solution of the initial value problem found in Exercise 27.

29. Determine how far the ship of Exercise 27 coasts. It is not necessary to solve the initial value problem.

30. Solve the initial value problem found in Exercise 27.

31. Solve each of these three initial value problems. All three solutions increase without limit; compare how rapidly they increase.

   (a) $y' = \sqrt{y}$ (b) $y' = y$ (c) $y' = y^2$
       $y(0) = 1$ $y(0) = 1$ $y(0) = 1$

32. The differential equation below does not fit any of our classifications; solve it by making the substitution $y = z^2$:

$$\frac{dy}{dx} = \sqrt{y} + x.$$

33. A large open tank, in the form of a vertical cylinder, is draining through a hole in the bottom. The velocity, and thus the flow rate, at the outlet is proportional to $\sqrt{h}$, where h is the height of the free surface above the outlet. By equating the rate of outflow to the rate of decrease in the volume of liquid in the tank—and thus to the rate of decrease of h—find a differential equation for h.

34. Suppose that the tank described above has a cross-sectional area of 2 square meters, the liquid is initially 4 meters deep, and the flow rate at the outlet is $0.016\sqrt{h}$ cubic meters per second (if h is measured in meters). Set up and solve an initial value problem that describes $h(t)$.

35. How long does it take for the tank in Exercise 34 to drain?

36. Suppose that the tank referred to in Exercise 33 has the shape of a solid of revolution with a vertical axis and that its cross-sectional area is $A(h)$. What function must $A(h)$ be, and what is the shape of the tank, if dh/dt is to be constant?

37. Find the solutions of the initial value problems consisting of the

differential equation

$$\frac{dy}{dx} = 1 - |y|, \qquad x > 0,$$

and the initial conditions (a) $y(0) = 2$, (b) $y(0) = \frac{1}{2}$, (c) $y(0) = -\frac{1}{2}$, (d) $y(0) = -2$. A direction field may be helpful.

2 Second-Order Linear Equations: Basic Methods and Applications

2.1

Introduction

The first-order linear equations that we saw in Chapter 1 have an unusual combination of features: they are important in applications and they can be solved explicitly. The higher-order linear equations that we are about to take up have these features also, as long as we restrict ourselves to equations with constant coefficients.

Our first order of business is to see where such an equation might arise. The classic example is a mechanical system consisting of a mass, a spring, and a damper, shown in idealized form in Fig. 2.1. We wish to describe the motion of the center of the mass. Newton's second law, $F = Ma$, is the governing principle, so we must study the forces that act on the mass. We assume that the mass is resting on a frictionless table and can move in only one direction (left to right in Fig. 2.1). Thus we analyze only the horizontal forces, which are due to the spring, the damper, and external influences.

1. A spring that is not subjected to any external force assumes its natural length L (see Fig. 2.2). If its length is changed from its natural length to another, $L + \Delta L$, the spring reacts with a force that is directed so as to restore its natural length. In this section we assume that the spring obeys Hooke's Law, which says that the magnitude of the restoring force is proportional to the change in length,*

$$|F| = k\,|\Delta L|.$$

---

* A spring that obeys Hooke's Law is also called linear. Some nonlinear springs are considerered in Chapter 8.

Figure 2.1

Mass–Spring–Damper
System

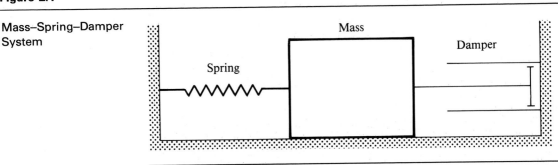

The constant of proportionality, k, is called the spring constant, measured in units of force/length. For example, if a spring stretches 2 inches when a 5-pound weight is hung from it, its spring constant is 2.5 lb/in.

2. A damper can stay at rest at any length; but if it is made to move, it reacts with a force that is directed opposite to the motion. We assume that the magnitude of the force is proportional to the speed,

$$|F| = p\,|v|.$$

The constant of proportionality, p, is called the damping constant, measured in units of force $\cdot$ time/length. For example, if a force of 7 pounds is required to maintain a speed of 5 inches per second, the damping constant is 1.4 lb sec/in.

A damper or shock absorber often has the form of a piston in a cylinder. A small hole in the piston allows the working fluid—air or oil—to move from one end of the cylinder to the other as the piston moves.

Figure 2.2

Force Exerted by
Spring

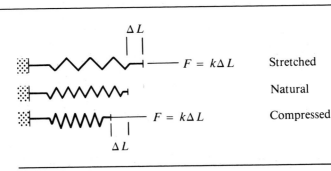

3. There may be other forces acting on the system (for example, gravitational force), and they may vary with time.

Now let us designate $u(t)$ as the position of the center of the mass. We choose the origin of the coordinate system so that the displacement is 0 when the spring has its natural length. Then we have these expressions for the forces exerted by the spring and damper on the mass:

$$F_s = -ku, \qquad F_d = -p\frac{du}{dt}.$$

The spring force has a negative sign because when u is positive, the spring is being stretched, and the force of the spring on the mass is directed so as to restore the spring's natural length. The sign for the damper force is explained similarly.

The sum of forces in the u-direction equals mass (M) times acceleration, according to Newton's second law. Thus we find the equation (see Fig. 2.3)

$$-ku - p\frac{du}{dt} + F(t) = M\frac{d^2u}{dt^2}. \tag{2.1}$$

Here $F(t)$ is the possible external force mentioned above.

Traditionally, we collect all terms containing u or its derivatives on one side of the equation and force a coefficient of 1 for the term containing the highest derivative. Thus we would divide Eq. (2.1) by M and rearrange terms to get

$$\frac{d^2u}{dt^2} + b\frac{du}{dt} + cu = f(t), \tag{2.2}$$

where the coefficients in this equation are

$$b = p/M, \qquad c = k/M, \qquad f(t) = F(t)/M.$$

The differential equation (2.2) contains the second derivative of the unknown function $u(t)$; for this reason it is a second-order equation. We

Figure 2.3

Free-Body Diagram

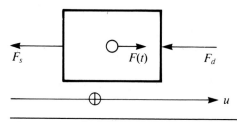

are going to assume that the coefficients b and c are *constants*. Constant or not, as long as b and c do not depend on the displacement u or the velocity du/dt, Eq. (2.2) is *linear*. The equation is called *homogeneous* if $u(t) \equiv 0$ is a solution; this can occur only if $f(t)$ is identically 0. As it stands, Eq. (2.2) is nonhomogeneous, and $f(t)$ is the inhomogeneity.

Our objective is to describe the position of the mass at any time, that is, to find $u(t)$. However, to do this, we need to know more than just the differential equation. As in the case of a falling body—whose position is also governed by a second-order differential equation obtained from Newton's second law—we must know the position of the mass and its speed at some instant. Usually, this is the time when the experiment is supposed to start, that is, time $t = 0$. Thus the complete problem to be solved is the initial value problem

$$\frac{d^2u}{dt^2} + b\frac{du}{dt} + cu = f(t), \qquad t > 0, \tag{2.3}$$

$$u(0) = u_0, \qquad \frac{du}{dt}(0) = v_0. \tag{2.4}$$

In the next several sections we shall deal with techniques for solving this problem. We are interested not only in producing a solution, but also in knowing how that solution depends on the coefficients b and c, on the initial conditions, and on the forcing function $f(t)$.

The mass–spring–damper system is not the only physical system that is described in the differential equation (2.2). Another one of importance is the simple electrical circuit shown in Fig. 2.4. Each of the devices shown causes a voltage change ΔV that is related in a simple way to the charge q or the current i:

Resistor: $\Delta V = Ri,$

Capacitor: $\Delta V = q/C,$

Inductor: $\Delta V = L\dfrac{di}{dt}.$

In these equations the resistance R, capacitance C, and inductance L are assumed to be known constants.

Kirchhoff's Law says that the sum of the voltage changes around a closed circuit must be 0. In other words, the applied voltage V must equal the sum of the voltage drops due to the devices in the circuit. For the case shown in Fig. 2.4 this law takes the form of the equation

$$V = Ri + \frac{1}{C}q + L\frac{di}{dt}. \tag{2.5}$$

This is not yet a differential equation. We need the additional

Figure 2.4

RCL Circuit

RCL Circuit

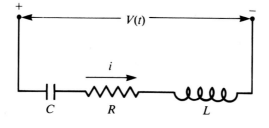

relation between current and charge:

$$i = \frac{dq}{dt},$$

(2.6)

that is, the current is the flow rate of the charge. Now we can convert Eq. (2.5) to a differential equation in q:

$$L\frac{d^2q}{dt^2} + R\frac{dq}{dt} + \frac{1}{C}q = V.$$

(2.7)

Charge, however, is not as convenient a concept as the current. We often restate Eq. (2.5) as a differential equation in i by differentiating it. In order for this to be helpful, however, the coefficients L, R, and C must be constants, independent of time, charge, and current. If this is true, we obtain

$$L\frac{d^2i}{dt^2} + R\frac{di}{dt} + \frac{1}{C}i = \frac{dV}{dt}.$$

(2.8)

The similarity between Eq. (2.7) and Eq. (2.1) suggests a well-known analog between mechanical and electrical systems, summarized in Table 2.1.

Table 2.1

Mechanical System	Electrical System
Displacement, u	Charge, q
Velocity, du/dt	Current, $i = dq/dt$
Mass, M	Inductor, L
Damper, p	Resistor, R
Spring, k	Capacitor, $1/C$
Applied force, $F(t)$	Applied voltage, $V(t)$

Exercises

1. Derive an initial value problem describing the position u of the mass in the system shown in Fig. 2.5. (Apply Newton's Law, using damping force and applied force $F(t)$.)

Figure 2.5

Mass–Damper
System

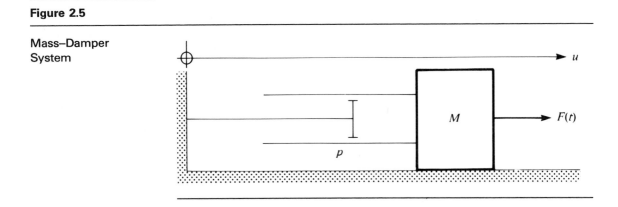

2. Suppose the coordinate system in Fig. 2.5 is moved so that the new and old coordinates are related by $U = u - h$. What is the differential equation for U?

3. What is the differential equation for the position of a mass in a mass–spring system as shown in Fig. 2.6? (Apply Newton's Law, using the spring force only. Assume that $u = 0$ when the spring has its natural length.)

Figure 2.6

Mass–Spring System

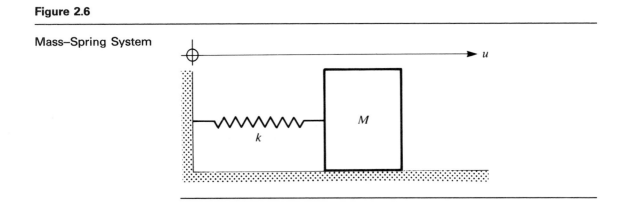

4. What differential equation results for the system in Fig. 2.6 if the

origin of the u-axis is moved so that $u = h$ when the spring is at its natural length?

5. Derive an initial value problem for displacement u of the mass in the series system shown in Fig. 2.7. Assume that y measures variation of spring length from its natural length and that the piston in the damper does not hit bottom. (Hints: (1) equate the spring and damper forces where they meet; (2) the damper force is proportional to the velocity of the piston relative to the cylinder.)

Figure 2.7

Mass, Damper, and Spring in Series

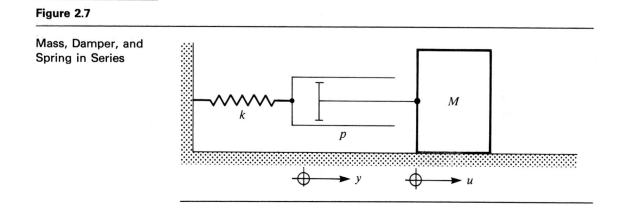

6. Derive an initial value problem for the current $i(t)$ in the circuit shown in Fig. 2.8. (Use Kirchhoff's Law.)

Figure 2.8

RL Circuit

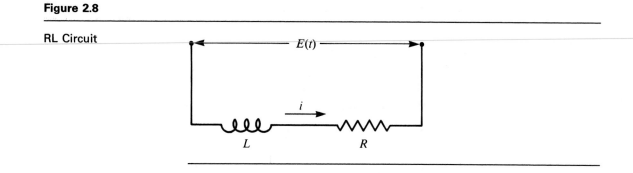

7. Derive an initial value problem for the charge $q(t)$ in the capacitor in the circuit shown in Fig. 2.9.

Figure 2.9

RC Circuit

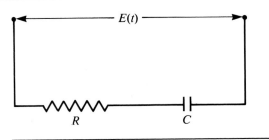

8. A one-pound weight hanging on a screen door spring causes the spring to lengthen 0.2 inches. What is the spring constant?

9. The coil spring in an automobile suspension deflects 12.5 cm under a weight of 350 kg. What is the spring constant?

In Exercises 10–14, verify that the given function is a solution of the differential equation. If initial conditions are given, determine values for the constants c_1 and c_2 so that they are satisfied.

10. $\dfrac{d^2u}{dx^2} + 5\dfrac{du}{dt} + 6u = 0$; $u(t) = c_1 e^{-2t} + c_2 e^{-3t}$

11. $\dfrac{d^2u}{dt^2} + 4u = 0$; $u(t) = c_1 \cos 2t + c_2 \sin 2t$; $u(0) = 1$, $\dfrac{du}{dt}(0) = -1$

12. $\dfrac{d^2u}{dt^2} + 4\dfrac{du}{dt} + 4u = 0$; $u(t) = (c_1 + c_2 t)e^{-2t}$

13. $\dfrac{d^2u}{dt^2} + 3\dfrac{du}{dt} + 2u = 0$; $u(t) = c_1 e^{-2t} + c_2 e^{-t}$; $u(0) = 2$, $\dfrac{du}{dt}(0) = 4$

14. $\dfrac{d^2u}{dt^2} + 2\dfrac{du}{dt} + 2u = 0$; $u(t) = e^{-t}(c_1 \cos t + c_2 \sin t)$

2.2

Solution of Homogeneous Equations with Constant Coefficients

For the time being, we are going to concentrate on solving the homogeneous, linear, second-order differential equation

$$\frac{d^2u}{dt^2} + b\frac{du}{dt} + cu = 0 \tag{2.9}$$

with constant coefficients b and c. We wish to find a *general solution* of the differential equation (2.9), that is, a solution, containing two constants, which can be made to satisfy any initial conditions

$$u(t_0) = u_0, \qquad \frac{du}{dt}(t_0) = v_0 \tag{2.10}$$

by an appropriate choice of those constants.

In Chapter 1 we found that a first-order linear homogeneous equation with constant coefficient

$$\frac{du}{dt} = au$$

has $u(t) = e^{at}$ for a solution. We are led to suspect that the second-order equation (2.9) may have one or more solutions of the same form. Indeed, we shall see that this is always true. Our method of solution is to assume that Eq. (2.9) has a solution in exponential form

$$u(t) = e^{mt} \tag{2.11}$$

and then find the constant m. To force this exponential to be a solution, we substitute it and its derivatives,

$$\frac{du}{dt} = me^{mt}, \qquad \frac{d^2u}{dt^2} = m^2 e^{mt}, \tag{2.12}$$

into the differential equation (2.9), obtaining

$$m^2 e^{mt} + bme^{mt} + ce^{mt} = 0. \tag{2.13}$$

Obviously, each term in Eq. (2.13) contains the factor e^{mt}, which is never zero and therefore can be divided out. The result is that the exponent multiplier m must satisfy the polynomial equation

$$m^2 + bm + c = 0. \tag{2.14}$$

This is called the *characteristic equation* of the differential equation (2.9), and the term $m^2 + bm + c$ is called its *characteristic polynomial*. The roots of this equation (there are usually two different roots) then provide two exponent multipliers, m_1 and m_2, for which

$$e^{m_1 t} \quad \text{and} \quad e^{m_2 t}$$

are both solutions of the differential equation (2.9).

Example 1

Solve the differential equation

$$\frac{d^2u}{dt^2} + 5\frac{du}{dt} + 6u = 0.$$

Assume the exponential form of the solution: $u = e^{mt}$. Substitution of this function and its derivatives into the differential equation leads to

$$m^2 e^{mt} + 5m e^{mt} + 6 e^{mt} = 0$$

and thence to the characteristic equation

$$m^2 + 5m + 6 = 0.$$

The roots of this equation are $m_1 = -2$ and $m_2 = -3$. Thus we obtain the two solutions

$$u_1(t) = e^{-2t}, \qquad u_2(t) = e^{-3t}.$$

Since the procedure developed above will usually produce two different solutions of a second-order differential equation (2.9), we must determine how they are to be used. First we make more precise the idea of "different" functions.

Definition 2.1

Two functions $u_1(t)$ and $u_2(t)$ are *linearly independent* on the interval $\alpha < t < \beta$ if the equality

$$c_1 u_1(t) + c_2 u_2(t) \equiv 0, \qquad \alpha < t < \beta,$$

is valid *only* for $c_1 = 0$ and $c_2 = 0$. If two functions are not linearly independent, they are linearly *dependent*. (Often the adjective "linearly" is dropped.)

If either $u_1(t)$ or $u_2(t)$ is identically 0, then the two functions are linearly dependent. For instance, with $u_1(t) \equiv 0$ and any $u_2(t)$ the linear combination $1 \cdot u_1(t) + 0 \cdot u_2(t)$ is identically 0, but u_1 has a nonzero coefficient.

If $u_1(t)$ and $u_2(t)$ are dependent over an interval, but neither function is 0 in that interval, then we can solve $c_1 u_1(t) + c_2 u_2(t) \equiv 0$ to find $u_2(t)/u_1(t) = -c_2/c_1$; the ratio is constant.

The ideas above can be refined to give this alternative test for dependence: if $u_1(t)$ and $u_2(t)$ are solutions of the same linear homogeneous differential equation, and neither is identically 0, then they are independent if and only if their ratio is not constant.[*]

---

[*] This alternative test can be used with two functions only. In the next chapter we consider independence of larger sets of functions.

Example 2

The functions e^{-2t} and e^{-3t} (see Example 1) are independent, $-\infty < t < \infty$, since their ratio is not constant. The same is true of $e^{m_1 t}$ and $e^{m_2 t}$ if m_1 and m_2 are distinct numbers.

Similarly, $\sin t$ and $\cos t$ have a nonconstant ratio and are independent. However, $\sin (t + \pi/4)$ and $\cos (t - \pi/4)$ are dependent—they are actually identical.

It is clear now that when the characteristic equation (2.14) has two distinct roots, m_1 and m_2, the two functions $e^{m_1 t}$ and $e^{m_2 t}$ are independent solutions of the differential equation (2.9). Both of these can be incorporated into the expression

$$u(t) = c_1 e^{m_1 t} + c_2 e^{m_2 t},$$

where c_1 and c_2 are constants. (A sum of constant multiples of some functions is called a *linear combination* of them.) In fact, this linear combination of solutions is again a solution. To check this, let us write $e^{m_1 t} = u_1(t)$ and $e^{m_2 t} = u_2(t)$. Then using $u(t)$ as given, we find

$$cu = cc_1 u_1(t) + cc_2 u_2(t)$$

$$b\frac{du}{dt} = bc_1 \frac{du_1}{dt} + bc_2 \frac{du_2}{dt}$$

$$\frac{d^2 u}{dt^2} = c_1 \frac{d^2 u_1}{dt^2} + c_2 \frac{d^2 u_2}{dt^2}.$$

Adding up both sides of these three equalities, we obtain

$$\frac{d^2 u}{dt^2} + b\frac{du}{dt} + cu = c_1 \left(\frac{d^2 u_1}{dt^2} + b\frac{du_1}{dt} + cu_1 \right)$$

$$+ c_2 \left(\frac{d^2 u_2}{dt^2} + b\frac{du_2}{dt} + cu_2 \right) = c_1 \cdot 0 + c_2 \cdot 0 = 0.$$

The last equalities follow from the fact that u_1 and u_2 are individually solutions of the differential equation.

Now we have shown that the linear combination

$$u(t) = c_1 e^{m_1 t} + c_2 e^{m_2 t} \qquad (2.15)$$

is a solution of the differential equation (2.9). In fact, this is a *general solution* of the differential equation (2.9) because the constants c_1 and c_2 can be chosen to make u satisfy any initial conditions. To see this, suppose that we have the initial conditions

$$u(0) = u_0, \qquad \frac{du}{dt}(0) = v_0.$$

Now u and its derivative are

$$u(t) = c_1 e^{m_1 t} + c_2 e^{m_2 t}, \qquad u(0) = c_1 + c_2,$$

$$\frac{du}{dt}(t) = c_1 m_1 e^{m_1 t} + c_2 m_2 e^{m_2 t}, \qquad \frac{du}{dt}(0) = m_1 c_1 + m_2 c_2.$$

In order for $u(t)$ to satisfy the initial conditions, we must have

$$c_1 + c_2 = u_0,$$

$$m_1 c_1 + m_2 c_2 = v_0.$$

The determinant of this system is $m_2 - m_1$. As long as this is not zero—that is, $m_2 \neq m_1$—the system can be solved for c_1 and c_2. Thus any initial conditions can be satisfied.

Example 3

Solve the initial value problem

$$\frac{d^2 u}{dt^2} + 5 \frac{du}{dt} + 6u = 0$$

$$u(0) = -4, \qquad \frac{du}{dt}(0) = 7.$$

In Example 2 we determined that e^{-2t} and e^{-3t} are independent solutions of this differential equation. Therefore the general solution is

$$u(t) = c_1 e^{-2t} + c_2 e^{-3t}.$$

In order to satisfy the initial conditions we need

$$\frac{du}{dt}(t) = -2c_1 e^{-2t} - 3c_2 e^{-3t}.$$

Now the initial conditions require that

$$c_1 + c_2 = -4,$$

$$-2c_1 - 3c_3 = 7.$$

These two equations are easily solved for $c_1 = -5$, $c_2 = 1$. Thus the solution of the given initial value problem is

$$u(t) = -5e^{-2t} + e^{-3t}.$$

Complex Roots

It usually happens that the constant coefficients of our differential equation (2.9) are real numbers. Even so, it is well known that the roots of the characteristic equation $m^2 + bm + c = 0$ need not be real. For example, the differential equation

$$\frac{d^2u}{dt^2} + 4\frac{du}{dt} + 5 = 0 \tag{2.16}$$

has a characteristic polynomial $m^2 + 4m + 5$, with roots

$$m = -2 \pm i.$$

These roots dictate the general solution

$$u(t) = c_1 e^{(-2+i)t} + c_2 e^{(-2-i)t}. \tag{2.17}$$

Is this really a solution? And how should it be interpreted?

The answers to both questions lie in the so-called Euler formulas (see the Appendix for more information)

$$\left.\begin{array}{l} e^{i\theta} = \cos\theta + i\sin\theta \\ e^{-i\theta} = \cos\theta - i\sin\theta \end{array}\right\} \tag{2.18}$$

which may be derived from the Taylor series representation for the exponential function. From these—or directly from the Taylor series—it can be shown that the following equations hold:

$$e^{(\alpha+i\beta)t} = e^{\alpha t}e^{i\beta t} = e^{\alpha t}(\cos\beta t + i\sin\beta t), \tag{2.19}$$

$$\frac{d}{dt}e^{(\alpha+i\beta)t} = (\alpha + i\beta)e^{(\alpha+i\beta)t}. \tag{2.20}$$

Equation (2.20) is the property that we need in order to justify the complex exponential as a solution of a differential equation with constant coefficients. Equation (2.19) provides us with a new form for our complex solutions. For instance, we may rewrite Eq. (2.17) in terms of real exponentials, sines, and cosines as follows.

$$u(t) = c_1 e^{-2t}(\cos t + i\sin t) + c_2 e^{-t}(\cos t - i\sin t)$$
$$= e^{-2t}(c_1 + c_2)\cos t + e^{-2t}(c_1 - c_2)i\sin t. \tag{2.21}$$

In the more general case, let us suppose that the coefficients b and c in the characteristic equation

$$m^2 + bm + c = 0$$

are real but the roots are complex. The two roots must be complex conjugates, $\alpha + i\beta$ and $\alpha - i\beta$. According to Eqs. (2.18) and (2.19), our independent solutions are

$$u_1(t) = e^{(\alpha+i\beta)t} = e^{\alpha t}(\cos\beta t + i\sin\beta t),$$

$$u_2(t) = e^{(\alpha-i\beta)t} = e^{\alpha t}(\cos\beta t - i\sin\beta t).$$

We may obtain from u_1 and u_2 these two real solutions:

$$u_3(t) = \frac{u_1(t) + u_2(t)}{2i} = e^{\alpha t} \cos \beta t,$$

$$u_4(t) = \frac{u_1(t) - u_2(t)}{2i} = e^{\alpha t} \sin \beta t.$$

These must be solutions of the original differential equation because they are linear combinations of the exponential solutions $u_1(t)$ and $u_2(t)$. Furthermore, the ratio of $u_3(t)$ and $u_4(t)$ is not constant, so they are independent.

Thus we conclude that if the differential equation (2.9) has real coefficients, but its characteristic equation has complex conjugate roots $\alpha + i\beta$ and $\alpha - i\beta$, then

$$e^{\alpha t} \cos \beta t, \qquad e^{\alpha t} \sin \beta t \hspace{3cm} \textbf{(2.22)}$$

are independent solutions of the differential equation (2.9). Note that our two real solutions are the real and imaginary parts of the complex exponential $e^{(\alpha + i\beta)t}$.

Example 4	Find the general solution of the differential equation

$$\frac{d^2u}{dx^2} + u = 0.$$

The exponential method leads us to the characteristic equation $m^2 + 1 = 0$, with roots $m = \pm i$. The two independent exponential solutions

$$e^{it} = \cos t + i \sin t, \qquad e^{-it} = \cos t - i \sin t$$

can be replaced by the two independent solutions $\cos t$ and $\sin t$ to give the general solution

$$u(t) = c_1 \cos t + c_2 \sin t.$$

Double Root

There is one case in which the exponential method does not give us two independent solutions of our differential equation. That occurs when the characteristic equation (2.14) has two equal roots. Of course, we must then have

$$m^2 + bm + c = \left(m + \frac{b}{2}\right)^2 = 0,$$

and $-b/2$ is the double root. In this case we obtain only

$$u_1 = e^{-bt/2}$$

as a solution of the differential equation (2.9), which is (because $c = b^2/4$)

$$\frac{d^2u}{dt^2} + b\frac{du}{dt} + \left(\frac{b}{2}\right)^2 u = 0. \tag{2.23}$$

However, we can find the complete solution by assuming

$$u(t) = v(t)e^{-bt/2}. \tag{2.24}$$

where $v(t)$ is a function to be found. Then

$$\frac{du}{dt} = \frac{dv}{dt}e^{-bt/2} - \frac{b}{2}ve^{-bt/2}$$

$$\frac{d^2u}{dt^2} = \frac{d^2v}{dt^2}e^{-bt/2} - b\frac{dv}{dt}e^{-bt/2} + \left(\frac{b}{2}\right)^2 ve^{-bt/2}.$$

When these expressions are substituted into the differential equation (2.23), there is a lot of cancellation, and only

$$\frac{d^2v}{dt^2} = 0 \tag{2.25}$$

survives. This trivial equation can be integrated twice to give

$$v(t) = c_1 + c_2t. \tag{2.26}$$

Since this is the general solution of Eq. (2.25), we obtain from it the general solution of Eq. (2.23),

$$u(t) = (c_1 + c_2t)e^{-bt/2}. \tag{2.27}$$

Example 5

Find the general solution of

$$\frac{d^2u}{dt^2} + 2\frac{du}{dt} + u = 0.$$

Clearly, the characteristic equation of this differential equation is

$$m^2 + 2m + 1 = 0$$

with double root $m = -1$. According to our observations above, the general solution of the differential equation is

$$u(t) = (c_1 + c_2t)e^{-t}.$$

We summarize our results concerning the solution of the differential equation (2.9) in the following theorem.

Theorem 2.1

For the differential equation

$$\frac{d^2u}{dt^2} + b\frac{du}{dt} + cu = 0$$

(where b and c are real constants) with characteristic equation

$$m^2 + bm + c = 0,$$

the general solution has the form:

1. $u(t) = c_1 e^{m_1 t} + c_2 e^{m_2 t}$

if the characteristic equation has distinct real roots m_1 and m_2;

2. $u(t) = (c_1 \cos \beta t + c_2 \sin \beta t)e^{\alpha t}$

if the characteristic equation has complex conjugate roots $\alpha + i\beta$ and $\alpha - i\beta$;

3. $u(t) = (c_1 + c_2 t)e^{mt}$

if the characteristic equation has m as a double root.

By way of illustration for the theorem, Table 2.2 shows the characteristic polynomials, their roots, and the general solution of the differential equation

$$\frac{d^2u}{dt^2} + 4\frac{du}{dt} + cu = 0$$

as c ranges from -5 to 8.

Exercises

In Exercises 1–16, find the characteristic polynomial and use it to find the general solution of the differential equation.

1. $\dfrac{d^2u}{dt^2} + 5\dfrac{du}{dt} + 6u = 0$

2. $\dfrac{d^2u}{dt^2} + 3\dfrac{du}{dt} + 2u = 0$

3. $\dfrac{d^2u}{dt^2} + 2\dfrac{du}{dt} = 0$

4. $\dfrac{d^2u}{dt^2} + 2\dfrac{du}{dt} + u = 0$

5. $\dfrac{d^2u}{dt^2} + 2\dfrac{du}{dt} + 2u = 0$

6. $\dfrac{d^2u}{dt^2} + 2\dfrac{du}{dt} + 5u = 0$

7. $\dfrac{d^2u}{dt^2} + 2\dfrac{du}{dt} - 3u = 0$

8. $\dfrac{d^2u}{dt^2} + 2\dfrac{du}{dt} + 10u = 0$

9. $\dfrac{d^2u}{dt^2} - 2\dfrac{du}{dt} + 10u = 0$

10. $4\dfrac{d^2u}{dt^2} + 4\dfrac{du}{dt} + u = 0$

Table 2.2

Differential Equation	Characteristic Equation and Roots	General Solution
$\dfrac{d^2u}{dt^2} + 4\dfrac{du}{dt} - 5u = 0$	$m^2 + 4m - 5 = 0$ $m = -5, 1$	$u(t) = c_1 e^{-5t} + c_2 e^{t}$
$\dfrac{d^2u}{dt^2} + 4\dfrac{du}{dt} = 0$	$m^2 + 4m = 0$ $m = -4, 0$	$u(t) = c_1 e^{-4t} + c_2$
$\dfrac{d^2u}{dt^2} + 4\dfrac{du}{dt} + 3u = 0$	$m^2 + 4m + 3 = 0$ $m = -3, -1$	$u(t) = c_1 e^{-3t} + c_2 e^{-t}$
$\dfrac{d^2u}{dt^2} + 4\dfrac{du}{dt} + 4u = 0$	$m^2 + 4m + 4 = 0$ $m = -2, -2$	$u(t) = (c_1 + c_2 t)e^{-2t}$
$\dfrac{d^2u}{dt^2} + 4\dfrac{du}{dt} + 5u = 0$	$m^2 + 4m + 5 = 0$ $m = -2 \pm i$	$u(t) = e^{-2t}(c_1 \cos t + c_2 \sin t)$
$\dfrac{d^2u}{dt^2} + 4\dfrac{du}{dt} + 8u = 0$	$m^2 + 4m + 8 = 0$ $m = -2 \pm 2i$	$u(t) = e^{-2t}(c_1 \cos 2t + c_2 \sin 2t)$

11. $\dfrac{d^2u}{dt^2} - u = 0$

12. $\dfrac{d^2u}{dt^2} + u = 0$

In Exercises 13–20, find the general solution of the differential equation, and then choose the coefficients so that the initial conditions are satisfied.

13. $\dfrac{d^2u}{dt^2} + 4u = 0,$

$u(0) = 1, \dfrac{du}{dt}(0) = -1$

14. $\dfrac{d^2u}{dt^2} + 4\dfrac{du}{dt} = 0,$

$u(0) = 10, \dfrac{du}{dt}(0) = 0$

15. $\dfrac{d^2u}{dt^2} + 4\dfrac{du}{dt} + 4u = 0,$

$u(0) = 2, \dfrac{du}{dt}(0) = 0$

16. $\dfrac{d^2u}{dt^2} + 2\dfrac{du}{dt} + 2u = 0$

$u(0) = 0, \dfrac{du}{dt}(0) = 0$

17. $\dfrac{d^2u}{dt^2} - 4u = 0,$

$u(0) = 1, \dfrac{du}{dt}(0) = 3$

18. $4\dfrac{d^2u}{dt^2} + 4\dfrac{du}{dt} + u = 0,$

$u(0) = 1, \dfrac{du}{dt}(0) = 1$

19. $\dfrac{d^2u}{dt^2} + 100u = 0,$

20. $\dfrac{d^2u}{dt^2} + 16\dfrac{du}{dt} + 100u = 0$

$u(0) = 0, \dfrac{du}{dt} = 1$

$u(0) = 1, \dfrac{du}{dt}(0) = 0$

21. Prove: If $u(t)$ is a solution of the differential equation (2.9) and b and c are *positive* constants, then $u(t)$ approaches 0 as t increases. (Hint: use Theorem 2.1 and consider the sign of the exponential multiplier.)

22. Let u be a solution of the equation

$$\frac{d^2u}{dt^2} + cu = 0,$$

where c is a positive constant. Show that the function

$$V(t) = \left(\frac{du}{dt}\right)^2 + cu^2$$

is constant. (Hint: differentiate $V(t)$ and use the differential equation.)

23. Let u be a solution of the differential equation

$$\frac{d^2u}{dt^2} + b\frac{du}{dt} + cu = 0.$$

Define the function

$$V(t) = \left(\frac{du}{dt}\right)^2 + bu\frac{du}{dt} + \left(c + \frac{b^2}{2}\right)u^2.$$

Assuming that b and c are positive constants, show the following.

(a) $V(t) = \left(\dfrac{du}{dt} + \dfrac{b}{2}u\right)^2 + \left(c + \dfrac{b^2}{4}\right)u^2$

and therefore $V = 0$ only if $u = 0$ and $du/dt = 0$, and $V > 0$ otherwise.

(b) $dV/dt \le 0$ (implication from the differential equation), and equality holds only if $u = 0$ and $du/dt = 0$.

(c) From the above, conclude that for any initial conditions, $V(t)$ decreases steadily toward 0 as t increases and also that $u(t)$ and du/dt approach 0 as t increases.

The functions $V(t)$ in Exercises 22 and 23 are called Lyapunov functions and are used to study the behavior of solutions of more complex equations.

24. Make a plot in the complex plane of the location of the roots of the polynomial $p(m) = m^2 + 4m + c$, as c varies from $-\infty$ to ∞. If this is

the characteristic polynomial of a differential equation, how does the nature of the solution of the differential equation reflect the position of the roots of $p(m)$? See Table 2.2.

25. Solve the differential equation $d^2u/dt^2 = 0$. Restate the equation in terms of curvature. Do the solutions have the predicted curvature?

26. (a) If $\alpha + i\beta$ and $\alpha - i\beta$ are complex conjugate roots of $m^2 + bm + c = 0$, find b and c in terms of α and β.
    (b) Find the differential equation of which this is the characteristic polynomial.
    (c) Verify that the functions in Eq. (2.22) are solutions of this differential equation.

27. (a) Verify that the function $u(t) = (c_1 + c_2t)e^{-bt/2}$ is a solution of

$$\frac{d^2u}{dt^2} + b\frac{du}{dt} + \frac{b^2}{4}u = 0$$

for any constants c_1 and c_2.
    (b) Show that $e^{-bt/2}$ and $te^{-bt/2}$ are independent
    (c) Show that c_1 and c_2 can be chosen so that u satisfies any initial conditions

$$u(0) = u_0, \qquad \frac{du}{dt}(0) = v_0.$$

28. (a) Express $\sinh at$ and $\cosh at$ as linear combinations of e^{at} and e^{-at}.
    (b) Show that $\sinh at$ and $\cosh at$ are independent.
    (c) Show that $\sinh t$ and $\cosh t$ are solutions of

$$\frac{d^2u}{dt^2} - u = 0.$$

    (d) What initial conditions does $\sinh t$ satisfy? $\cosh t$?

29. Suppose that $u_1(t)$ and $u_2(t)$ are two differentiable functions whose ratio $u_2(t)/u_1(t)$ is constant over some interval $\alpha < t < \beta$. Show that the relation

$$u_1(t)u_2'(t) - u_2(t)u_1'(t) \equiv 0, \qquad \alpha < t < \beta$$

must hold. (Hint: differentiate u_2/u_1.)

30. Show that the equation in Exercise 29 can be written as the determinantal relation

$$\begin{vmatrix} u_1(t) & u_2(t) \\ u_1'(t) & u_2'(t) \end{vmatrix} \equiv 0.$$

This determinant is called the Wronskian of u_1 and u_2. See Section 3.1.

2.3

Free Vibrations

Now we return to the mass–spring–damper system. In Section 2.1 we derived a second-order linear differential equation with constant coefficients to describe the motion of the mass. In the absence of an applied force the equation is homogeneous, so we are equipped to solve it. In this section we make a detailed study of the motion of that system. We imagine that the mass and the spring are the same in all cases but that the damping is variable.

No Damping

We first assume that the damper is not present; our system consists of just a mass and a spring. (See Fig. 2.10.) The position of the center of the mass is then described by this initial value problem

$$M\frac{d^2u}{dt^2} + ku = 0, \qquad 0 < t, \tag{2.28}$$

$$u(0) = u_0, \qquad \frac{du}{dt}(0) = v_0. \tag{2.29}$$

We divide through the differential equation by M and call $k/M = \omega^2$. The square is for later convenience; note, however, that k and M are both positive, so the implication that ω^2 is positive is true.

Our differential equation is now

$$\frac{d^2u}{dt^2} + \omega^2 u = 0 \tag{2.30}$$

Figure 2.10

Mass–Spring System

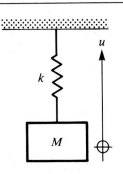

The assumed exponential solution, $u = e^{mt}$, leads to the characteristic equation

$$m^2 + \omega^2 = 0, \tag{2.31}$$

whose solutions are $m = \pm i\omega$. As we saw in Section 2.2, exponentials with imaginary exponents can be converted into sines and cosines. Thus we determine that the general solution of the differential equation (2.30) is

$$u(t) = c_1 \cos \omega t + c_2 \sin \omega t. \tag{2.32}$$

The two constants c_1 and c_2—which are arbitrary as far as satisfying the differential equation is concerned—are determined by the two initial conditions in the following way. When we evaluate the solution (Eq. 2.32) and its derivative at time $t = 0$, we obtain

$$u(0) = c_1 \quad \text{and} \quad \frac{du}{dt}(0) = \omega c_2. \tag{2.33}$$

Now the initial conditions, Eq. (2.29), tell us that

$$c_1 = u_0 \quad \text{and} \quad \omega c_2 = v_0. \tag{2.34}$$

Hence we find that the solution of the initial value problem, Eqs. (2.29) and (2.30), is

$$u(t) = u_0 \cos \omega t + \frac{v_0}{\omega} \sin \omega t. \tag{2.35}$$

In Fig. 2.11, graphs of the function $u(t)$ are shown for several different choices of u_0 and v_0.

Several observations could be made about the graphs of Fig. 2.11. One is that regardless of the initial conditions, the distance between successive maxima or minima is always the same. That distance is $2\pi/\omega$. Similarly, the successive *zeros* (values of t for which $u(t) = 0$) are exactly π/ω units apart for any initial conditions. Finally, we note that each solution curve $u(t)$ has the same sinusoidal shape; different initial conditions contribute different heights and different placements on the axes.

These observations can all be explained and clarified by collapsing the two terms of $u(t)$ in Eq. (2.35) into a single trigonometric function. To do this, we need the trigonometric identity

$$\cos (\omega t - \phi) = \cos \phi \cos \omega t + \sin \phi \sin \omega t \tag{2.36}$$

We can express $u(t)$ in terms of the single trigonometric function $\cos (\omega t - \phi)$ if we find an angle ϕ and a (positive) multiplier A for which

$$u_0 = A \cos \phi, \tag{2.37}$$

$$v_0/\omega = A \sin \phi. \tag{2.38}$$

Figure 2.11

Free Vibrations of a
Mass–Spring System
with no Damping

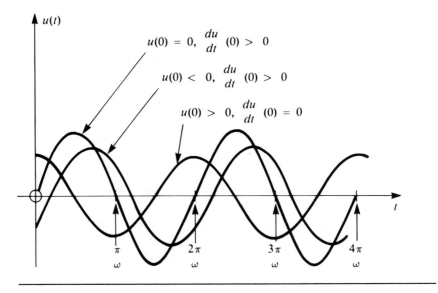

$$u(0) = 0, \quad \frac{du}{dt}(0) > 0$$

$$u(0) < 0, \quad \frac{du}{dt}(0) > 0$$

$$u(0) > 0, \quad \frac{du}{dt}(0) = 0$$

These are the same equations one uses to relate polar (A, ϕ) coordinates to rectangular $(u_0, v_0/\omega)$ coordinates. It is easy to find A:

$$A = \sqrt{u_0^2 + (v_0/\omega)^2}. \tag{2.39}$$

Then ϕ can be determined from Eqs. (2.37) and (2.38).

Now we can express the solution of the initial value problem in Eqs. (2.29) and (2.30) as

$$u(t) = A \cos(\omega t - \phi). \tag{2.40}$$

From here it is easy to see how the amplitude A (the maximum displacement) given by Eq. (2.39) and the *phase angle* ϕ determined by Eqs. (2.37) and (2.38) are affected by the initial conditions. Furthermore, it is obvious why the maxima, minima, and zeros of $u(t)$ are equally spaced— they are the maxima, minima, and zeros of a shifted cosine function. The *natural frequency* ω (measured in radians per unit time) comes directly from the differential equation (2.30), where

$$\omega^2 = k/M$$

is a coefficient.

The motion, or time-varying displacement, represented by Eq. (2.35) or Eq. (2.40) is called *simple harmonic motion,* and a device that carries out such a motion—or even the differential equation (2.28) or (2.30)—is called a harmonic oscillator.

Example 1

Find the motion, in terms of a single trigonometric function, of the center of the mass in a system with a mass that weighs 16 lb and a spring with a constant of 2 lb/in. if the initial displacement and velocity are

$$u_0 = 1 \text{ in.}, \qquad v_0 = -12 \text{ in./sec.}$$

The mass of the system is given by the equation $w = Mg$, where w is the weight and g is the acceleration due to gravity, about 32 ft/sec². Thus

$$M = \frac{w}{g} = \frac{16 \text{ lb}}{32 \text{ ft/sec}^2} = \frac{1}{2} \frac{\text{lb sec}^2}{\text{ft}}.$$

Next we need to find the coefficient ω^2 from

$$\omega^2 = \frac{k}{M} = \frac{2 \text{ lb/in.}}{\frac{1}{2} \text{ lb sec}^2/\text{ft}}$$

$$= \frac{4 \text{ ft}}{\text{in. sec}^2} \times \frac{12 \text{ in.}}{\text{ft}} = \frac{48}{\text{sec}^2}.$$

Note that feet had to be converted to inches (or vice versa) for consistency. The natural frequency of the system is $\sqrt{48} = 4\sqrt{3}$ radians per second. (Radian is actually a dimensionless measure.)

The initial value problem to be solved is

$$\frac{d^2u}{dt^2} + 48u = 0,$$

$$u(0) = 1, \qquad \frac{du}{dt}(0) = -12.$$

The general solution of the differential equation is

$$u(t) = c_1 \cos 4\sqrt{3}t + c_2 \sin 4\sqrt{3}t,$$

and c_1 and c_2 are determined from the initial conditions

$$c_1 = 1, \qquad 4\sqrt{3}c_2 = -12.$$

Thus the solution is

$$u(t) = \cos 4\sqrt{3}t - \sqrt{3} \sin 4\sqrt{3}t.$$

To put this solution in the form of a single trigonometric function, we need to find

$$A = \sqrt{1^2 + (-\sqrt{3})^2} = \sqrt{1 + 3} = 2$$

and then to solve Eqs. (2.37) and (2.38) for ϕ:

$$1 = 2 \cos \phi, \qquad -\sqrt{3} = 2 \sin \phi.$$

The phase angle ϕ is easily found to be $-\pi/3$ radians, and so the solution is

$$u(t) = 2 \cos (4\sqrt{3}t + \pi/3).$$

Slightly Damped Motion

We now assume that the damper is included in the system, so the initial value problem to be solved is

$$\frac{d^2u}{dt^2} + b\frac{du}{dt} + \omega^2 u = 0, \qquad 0 < t, \tag{2.41}$$

$$u(0) = u_0, \qquad \frac{du}{dt}(0) = v_0. \tag{2.42}$$

The characteristic equation is

$$m^2 + bm + \omega^2 = 0. \tag{2.43}$$

By "slightly damped" we mean that b is small enough that the solutions of the characteristic equation are complex conjugates,

$$m = -\frac{b}{2} \pm \sqrt{\left(\frac{b}{2}\right)^2 - \omega^2} = -\frac{b}{2} \pm i\sqrt{\omega^2 - \left(\frac{b}{2}\right)^2}. \tag{2.44}$$

Thus we wish $b < 2\omega$ for slight damping.

From our discussion of complex roots in Section 2.2 we know that the general solution of the differential equation (2.41) may be written as

$$u(t) = e^{-\alpha t}(c_1 \cos \beta t + c_2 \sin \beta t), \tag{2.45}$$

where we have used the shorthand

$$\beta = \sqrt{\omega^2 - \left(\frac{b}{2}\right)^2}, \qquad \alpha = b/2. \tag{2.46}$$

Now the initial conditions (2.42) are to be satisfied. Using $u(t)$ as given in Eq. (2.45), we get the initial conditions

$$u(0) = u_0: \qquad c_1 = u_0,$$

$$\frac{du}{dt}(0) = v_0: \qquad -\alpha c_1 + \beta c_2 = v_0.$$

These equations are easily solved for c_1 and c_2, to give us the solution of the initial value problem, Eqs. (2.41) and (2.42), as

$$u(t) = e^{-\alpha t}\left(u_0 \cos \beta t + \frac{v_0 + \alpha u_0}{\beta} \sin \beta t\right).$$ (2.47)

The combination of sine and cosine in this expression can be condensed in the same manner as above, to obtain

$$u(t) = Ae^{-\alpha t} \cos (\beta t - \phi)$$ (2.48)

The determination of A and ϕ is left as an exercise. (See Exercise 9 at the end of this section.)

Figure 2.12 shows graphs of $u(t)$ for various initial values. It should be noted that the zeros and the maxima and minima of $u(t)$ are still equally spaced. But of course the exponential factor $e^{-\alpha t}$ in Eq. (2.47) guarantees that $u(t)$ approaches 0 as t increases, as one can see in the figure.

Figure 2.12

Free Vibrations with Slight Damping

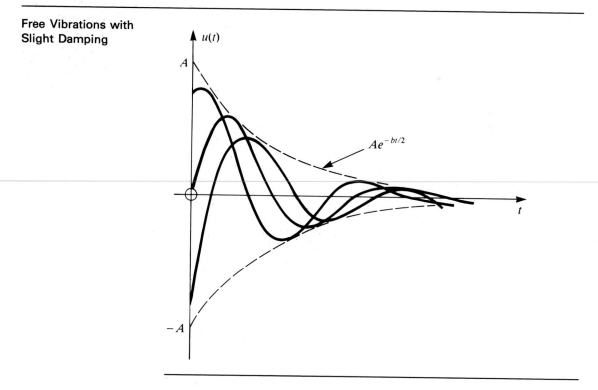

Example 2

Determine the motion of a system consisting of the mass and spring of Example 1 if a damper with constant $p = \frac{1}{3}$ lb sec/in. is added. Use the same initial conditions.

We have $M = \frac{1}{2}$ lb sec²/ft from Example 2. The coefficient b of the differential equation is

$$b = \frac{p}{M} = \frac{\frac{1}{3} \text{ lb sec/in.}}{\frac{1}{2} \text{ lb sec}^2/\text{ft}} = \frac{2 \text{ ft}}{3 \text{ in. sec}} = \frac{8}{\text{sec}}.$$

We can now write out the initial value problem to be solved:

$$\frac{d^2u}{dt^2} + 8\frac{du}{dt} + 48u = 0, \qquad 0 < t,$$

$$u(0) = 1, \qquad \frac{du}{dt}(0) = -12.$$

The characteristic equation and its roots are

$$m^2 + 8m + 48 = 0, \qquad m = -4 \pm i4\sqrt{2}.$$

Thus the general solution of the differential equation is

$$u(t) = e^{-4t}(c_1 \cos 4\sqrt{2}t + c_2 \sin 4\sqrt{2}t).$$

The initial conditions, which determine the coefficients in u, reduce to

$$c_1 = 1,$$
$$-4c_1 + 4\sqrt{2}c_2 = -12.$$

We easily find $c_2 = -\sqrt{2}$, and the solution is

$$u(t) = e^{-4t}(\cos 4\sqrt{2}t - \sqrt{2} \sin 4\sqrt{2}t).$$

The parenthetical term can be converted to a single trigonometric function, giving the alternative form for the solution,

$$u(t) = \sqrt{3}e^{-4t} \cos (4\sqrt{2}t - \phi)$$

with $\phi = -0.955$ radian, approximately.

Critical Damping

We now consider the very special case in which the damping constant, spring constant, and mass of our mechanical system are such that

$$\left(\frac{b}{2}\right)^2 = \omega^2. \tag{2.49}$$

The initial value problem governing our system is still

$$\frac{d^2u}{dt^2} + b\frac{du}{dt} + \omega^2 u = 0, \qquad 0 < t, \tag{2.50}$$

$$u(0) = u_0, \qquad \frac{du}{dt}(0) = v_0, \tag{2.51}$$

Now, however, the characteristic equation of the differential equation,

$$m^2 + bm + \omega^2 = 0,$$

has a double root,

$$m = -\frac{b}{2} \pm \sqrt{\left(\frac{b}{2}\right)^2 - \omega^2} = -\frac{b}{2} = -\omega. \tag{2.52}$$

because of our assumption in Eq. (2.49).

The developments of Section 2.2 told us that in this special case the general solution of the differential equation (2.50) is given by

$$u(t) = (c_1 + c_2 t)e^{-\omega t}. \tag{2.53}$$

In what features does the solution (2.53) of this "critically damped" system differ from the solution (2.45) of the "slightly damped" system? The most important and noticeable difference is that the solution of the critically damped system (2.53) has at most one zero, whereas the solution of the slightly damped system (2.45) has an infinite number of regularly recurring zeros. This is immediately obvious from the presence of the polynomial factor in (2.53) as compared with the sinusoidal factor in (2.45).

Thus the difference between slightly damped and critically damped is mathematically sharp. In the physical systems, however, the difference is not so noticeable. Indeed, if we make several experiments, each with the initial conditions, $u(0) = u_0$, $\dot{u}(0) = 0$, but with the damping constant increased from 0 toward the critical value $b_c = 2\omega$, we find that the time at which the system first crosses its equilibrium position (that is, the first $t > 0$ for which $u(t) = 0$) grows steadily with the damping constant b. However, physical phenomena not included in our analysis (sliding friction, internal friction in elastic members, turbulence in fluids, etc.), although small, have the effect of dissipating energy and extinguishing motion of the system in finite time. For this reason it may be impossible to distinguish physically between regimes that are mathematically quite different.

Example 3

Determine the motion of the system with the mass and spring of Example 1, if the damping is critical.

The damping must be such that $(b/2)^2 = \omega^2 = 48$. Thus $b = 8\sqrt{3}$/sec and $p = bM = \sqrt{3}/3$ lb sec/in. The initial value problem to be solved is

$$\frac{d^2u}{dt^2} + 8\sqrt{3}\frac{du}{dt} + 48u = 0,$$

$$u(0) = 1, \qquad \frac{du}{dt}(0) = -12.$$

The characteristic equation and its double root are

$$m^2 + 8\sqrt{3}m + 48 = 0, \qquad m = -4\sqrt{3}.$$

The general solution of the differential equation is thus

$$u(t) = (c_1 + c_2t)e^{-4\sqrt{3}t}.$$

The initial conditions determine the choice of c_1 and c_2. These equations are

$$c_1 = 1, \qquad -4\sqrt{3}c_1 + c_2 = -12.$$

Hence we find that the solution of the initial value problem is

$$u(t) = (1 + (4\sqrt{3} - 12)t)e^{-4\sqrt{3}t}.$$

Overdamping

Finally, we consider the overdamped or supercritical case, in which the damping constant exceeds the critical value

$$b > 2\omega. \tag{2.54}$$

Now the differential equation of our initial value problem,

$$\frac{d^2u}{dt^2} + b\frac{du}{dt} + \omega^2u = 0, \qquad 0 < t, \tag{2.55}$$

$$u(0) = u_0, \qquad \frac{du}{dt}(0) = v_0, \tag{2.56}$$

has the characteristic equation

$$m^2 + bm + \omega^2 = 0,$$

whose roots, on account of the assumption (2.54), are both real:

$$m_1 = -\frac{b}{2} - \sqrt{\left(\frac{b}{2}\right)^2 - \omega^2}, \qquad m_2 = -\frac{b}{2} + \sqrt{\left(\frac{b}{2}\right)^2 - \omega^2}. \tag{2.57}$$

It is easy to see, however, that both roots are negative, so in the solution

$$u(t) = c_1 e^{m_1 t} + c_2 e^{m_2 t}, \tag{2.58}$$

both terms—and therefore the solution itself, whatever the constants c_1 and c_2—tend to 0 as t increases.

It is true for the overdamped system, as for the critically damped system, that any solution $u(t)$ has at most one zero. Indeed, as b increases, increasingly large initial velocities are required to achieve conditions in which $u(t) = 0$ can ever be satisfied. And we might add that it is even more difficult to distinguish physically an overdamped system from a critically damped one than to distinguish the critically and subcritically damped cases.

Example 4

Determine the motion of the system having the spring and mass of Example 1 and a damper with constant 7 lb sec/ft. Use the same initial conditions.

The coefficient b of the differential equation is given by $b = p/M$ or

$$b = \frac{7 \text{ lb sec/ft}}{\frac{1}{2} \text{ lb sec}^2/\text{ft}} = 14/\text{sec}.$$

The initial value problem to be solved is

$$\frac{d^2 u}{dt^2} + 14 \frac{du}{dt} + 48u = 0,$$

$$u(0) = 1, \qquad \frac{du}{dt}(0) = -12.$$

The characteristic polynomial of the differential equation and its roots are

$$m^2 + 14m + 48 = 0, \qquad m = -6, -8.$$

Hence the general solution of the differential equation is

$$u(t) = c_1 e^{-6t} + c_2 e^{-8t},$$

and the initial conditions are satisfied if c_1 and c_2 satisfy

$$c_1 + c_2 = 1,$$
$$-6c_1 - 8c_2 = -12.$$

Finally, the solution is

$$u(t) = -2e^{-6t} + 3e^{-8t}.$$

The results of the four cases that we have considered are summarized in Table 2.3. Also see Fig. 2.13 for a comparison of the different degrees of damping.

Table 2.3

Case	Roots	Nature of Solution
Undamped $b = 0$	$m = \pm i\omega$	Oscillatory, undiminished
Underdamped (subcritical) $0 < b < 2\omega$	$m = -\dfrac{b}{2} \pm i\sqrt{\omega^2 - \left(\dfrac{b}{2}\right)^2}$	Oscillatory, decaying
Critical $b = 2\omega$	$m = -\dfrac{b}{2}$	Nonoscillatory, decaying
Overdamped (supercritical) $b > 2\omega$	$m = -\dfrac{b}{2} \pm \sqrt{\left(\dfrac{b}{2}\right)^2 - \omega^2}$	Nonoscillatory, decaying

Figure 2.13

Slight (I), Critical (II), and Supercritical (III) Damping

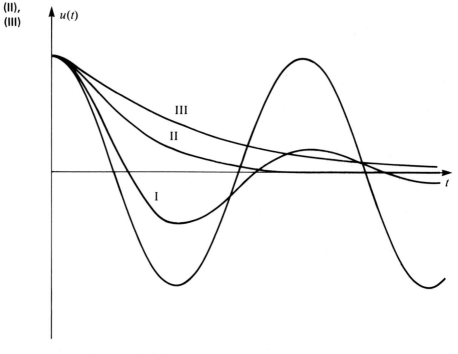

Exercises

Exercises 1–5 refer to the examples worked in this section.

1. Sketch the location in the complex plane of the roots of $m^2 + bm + 48 = 0$ as b varies from 0 to ∞. Designate the portions of the locus that correspond to the four cases of damping.

2. Sketch the roots m_1 and m_2 of the polynomial $m^2 + bm + 48 = 0$ as functions of b as b ranges from $8\sqrt{3}$ to ∞.

3. Solve the initial value problem

$$\frac{d^2u}{dt^2} + b\frac{du}{dt} + 48u = 0,$$

$$u(0) = 1, \quad \frac{du}{dt}(0) = -12,$$

for $b = 16, 19, 26, 49$.

4. Examples 1–4 and Exercise 3 all have the same ω^2 and the same initial conditions. Sketch the solutions, on the same axes, for $b = 0$, 8, $8\sqrt{3}$, and 16, for $0 \le t < 0.5$ (approximately).

5. Using the solutions in Examples 1–4 and the first case of Exercise 3, find for each the first $t > 0$ when $u(t) = 0$.

In Exercises 6–11, (a) identify the damping as subcritical, critical, or supercritical; (b) find the general solution of the differential equation; and (c) find the solution of the initial value problem.

6. $\dfrac{d^2u}{dt^2} + 6\dfrac{du}{dt} + 8u = 0,$

$u(0) = 0, \quad \dfrac{du}{dt}(0) = 1$

7. $\dfrac{d^2u}{dt^2} + 6\dfrac{du}{dt} + 9u = 0,$

$u(0) = 0, \quad \dfrac{du}{dt}(0) = 1$

8. $\dfrac{d^2u}{dt^2} + 6\dfrac{du}{dt} + 10u = 0,$

$u(0) = 0, \quad \dfrac{du}{dt}(0) = 1$

9. $\dfrac{d^2u}{dt^2} + 10\dfrac{du}{dt} + 9u = 0,$

$u(0) = 10, \quad \dfrac{du}{dt} = 10$

10. $\dfrac{d^2u}{dt^2} + 2\dfrac{du}{dt} + 10u = 0,$

$u(0) = 1, \quad \dfrac{du}{dt}(0) = 3$

11. $4\dfrac{d^2u}{dt^2} + 5\dfrac{du}{dt} + u = 0$

$u(0) = 1, \quad \dfrac{du}{dt}(0) = 0$

12. Show that any two consecutive maxima of the solution of

$$\frac{d^2u}{dt^2} + b\frac{du}{dt} + \omega^2u = 0,$$

$$u(0) = u_0, \quad \frac{du}{dt}(0) = v_0,$$

are spaced $2\pi/\beta$ apart, where $\beta = \sqrt{\omega^2 - (b/2)^2} > 0$.

13. Find an initial value problem satisfied by the derivative of the solution of the problem in Exercise 4. (Hint: differentiate the differential equation.)

14. Suppose $u(t)$ is the solution of Eqs. (2.41) and (2.42). If you know the values of u at two successive maxima, what can be found out about the parameters b, ω, u_0, and v_0?

15. Let $u(t)$ be the solution of Eqs. (2.41) and (2.42). Find A and ϕ as used in Eq. (2.48).

16. Let $u(t)$ be the function of Eq. (2.48). If t_1 and t_2 are two consecutive times when u is 0, what happens to the difference $t_2 - t_1$ as a function of b when b approaches the critical value 2ω?

17. Find c_1 and c_2 in Eq. (2.53) so that the initial conditions Eq. (2.51) are satisfied.

18. At what time t is the solution in Eq. (2.53) equal to 0?

19. Follow the directions of Exercise 17 but for u as given in Eq. (2.58) and initial conditions in Eq. (2.56).

20. At what time t (if any) is the solution (2.58) for the overdamped case equal to 0?

21. Express the answer to Exercise 20 in terms of initial values and system parameters. Is it true that large initial velocities are needed to get $u(t) = 0$ at some positive time?

22. An automobile weighing 2000 lb is observed to bounce with a frequency of about 1 cycle/sec (without shock absorbers). If the weight of the car is shared equally by four springs, what is the spring constant for each?

23. Design of a recoil mechanism. A gun barrel recoils against a spring until it stops; then it engages a damper and returns to its original position without overshoot. Data: barrel weight, 700 kg; initial velocity of barrel, 25 m/sec.
    (a) In the initial phase of the recoil the barrel and spring act like an undamped mass–spring system with initial velocity as given and initial displacement 0. What value should k have so that the barrel travels 1.5 m before stopping?
    (b) In the second phase of recoil the barrel, spring, and damper act like a critically damped or overdamped system. Supposing critical damping, what should be the value of the damping constant?

2.4
Nonhomogeneous Equations—Undetermined Coefficients

Before we can continue the study of vibrational systems, we need to obtain methods for the solution of differential equations that are not homogeneous ($u(t) \equiv 0$ is not a solution). We still accept the restriction that the coefficients of the differential equation be constant, so our model equation is

$$\frac{d^2u}{dt^2} + b\frac{du}{dt} + cu = f(t), \qquad\qquad (2.59)$$

where b and c are constants. Our solution of the nonhomogeneous first-order linear equation suggests a form for expressing the general solution of Eq. (2.59), as presented in Theorem 2.2.

Theorem 2.2

The general solution of the differential equation

$$\frac{d^2u}{dt^2} + b\frac{du}{dt} + cu = f(t) \qquad\qquad \text{(NH)}$$

has the form

$$u(t) = u_p(t) + u_c(t)$$

where u_p is any solution of the nonhomogeneous equation and u_c is the general solution of the corresponding homogeneous equation

$$\frac{d^2u}{dt^2} + b\frac{du}{dt} + cu = 0. \qquad\qquad \text{(H)}$$

Note: The two parts of the solution have traditional names: u_p is called a *particular solution* of (NH), and u_c is called the *complementary solution*.

By way of proof we will show that the sum of particular solution and complementary solution does really satisfy the equation (NH). Using $u = u_p + u_c$, we find that

$$\frac{d^2u}{dt^2} + b\frac{du}{dt} + cu = \left(\frac{d^2u_p}{dt^2} + b\frac{du_p}{dt} + cu_p\right) + \left(\frac{d^2u_c}{dt^2} + b\frac{du_c}{dt} + cu_c\right)$$

$$= f(t) + 0.$$

Thus the sum $u(t) = u_p(t) + u_c(t)$ is a solution of the nonhomogeneous equation (NH). Furthermore, the arbitrary coefficients in the complementary solution are available to satisfy any initial conditions.

Example 1

Find the general solution of the nonhomogeneous equation

$$\frac{d^2u}{dt^2} + 3\frac{du}{dt} + 2u = t. \qquad\qquad \text{(NH)}$$

1. A particular solution of this nonhomogeneous equation is

$$u_p(t) = \tfrac{1}{2}t - \tfrac{3}{4}.$$

(Later in this section we will see how to find this particular solution. For the time being, you are asked to treat it as given and to check that it is, indeed, a solution of (NH).)

2. The homogeneous equation that corresponds to (NH) is

$$\frac{d^2u}{dt^2} + 3\frac{du}{dt} + 2u = 0. \tag{H}$$

Using the methods of Section 2.2, we find that the characteristic polynomial is $m^2 + 3m + 2 = 0$, with roots $m = -1, -2$. Thus the general solution of (H), which is the complementary solution of (NH), is

$$u_c(t) = c_1e^{-t} + c_2e^{-2t}.$$

3. According to the theorem, the general solution of (NH) is

$$u(t) = u_p(t) + u_c(t) = \tfrac{1}{2}t - \tfrac{3}{4} + c_1e^{-t} + c_2e^{-2t}.$$

Another property that will help us in solving linear nonhomogeneous equations is Theorem 2.3.

Theorem 2.3

If u_1 and u_2 are particular solutions of the equations

$$\frac{d^2u}{dt^2} + b\frac{du}{dt} + cu = f_1(t),$$

$$\frac{d^2u}{dt^2} + b\frac{du}{dt} + cu = f_2(t),$$

respectively, then $k_1u_1 + k_2u_2$ is a particular solution of

$$\frac{d^2u}{dt^2} + b\frac{du}{dt} + cu = k_1f_1(t) + k_2f_2(t).$$

In other words, we can put together a particular solution from solutions of several other problems. The proof follows the same lines as the proof of Theorem 2.2.

Example 2

Find the general solution of the nonhomogeneous equation

$$\frac{d^2u}{dt^2} + 3\frac{du}{dt} + 2u = 5e^{2t} + 7t.$$

We break down the inhomogeneity as a linear combination of $f_1(t) = e^{2t}$ and $f_2(t) = t$. A particular solution of

$$\frac{d^2u}{dt^2} + 3\frac{du}{dt} + 2u = e^{2t}$$

is $u_1 = e^{2t}/12$. We already saw in Example 2.10 that

$$u_2 = \tfrac{1}{2}t - \tfrac{3}{4}$$

is a particular solution of the equation

$$\frac{d^2u}{dt^2} + 3\frac{du}{dt} + 2u = t.$$

Therefore we can conclude that a particular solution of

$$\frac{d^2u}{dt^2} + 3\frac{du}{dt} + 2u = 5e^{2t} + 7t$$

is

$$u_p(t) = 5\left(\frac{e^{2t}}{12}\right) + 7(\tfrac{1}{2}t - \tfrac{3}{4}).$$

Now we turn our attention to a method for finding particular solutions. Of the many available, we will first take up the method of *undetermined coefficients*, which is the easiest method when it works. Roughly speaking, we guess that the particular solution of

$$\frac{d^2u}{dt^2} + b\frac{du}{dt} + cu = f(t)$$

has the form of a linear combination of f and its derivatives

$$u_p(t) = Af(t) + Bf'(t) + Cf''(t) + \cdots. \tag{2.60}$$

When the appropriate form has been chosen, the proposed function is inserted into the nonhomogeneous equation, and the "undetermined coefficients" $A, B, C \ldots$ are determined to make the equation hold. This usually involves equating coefficients of like functions.

This method works satisfactorily if f is a function for which the series (2.60) can be cut off after a few terms. Such functions are:
1. polynomials—$f^{(n)}(t) \equiv 0$ if n is high enough;
2. exponentials—$f'(t)$ is proportional to $f(t)$;
3. sines and cosines—$f''(t)$ is proportional to $f(t)$;
4. sums and products of the above.

Example 3

Find a particular solution of the differential equation

$$\frac{d^2u}{dt^2} + 3\frac{du}{dt} + 2u = t.$$

Since $f(t) = t$, $f'(t) = 1$, and all higher derivatives are identically 0, we should guess that u_p has the form

$$u_p = At + B.$$

To determine A and B, we substitute the guessed function into the differential equation:

$$0 + 3(A) + 2(At + B) = t.$$

Now we equate coefficients of like terms to obtain this pair of equations for A and B:

$$3A + 2B = 0 \qquad \text{(equate constants)},$$
$$2A \qquad\quad = 1 \qquad \text{(equate coefficients of } t\text{)}.$$

These two equations are easily solved, giving us

$$u_p = \tfrac{1}{2}t - \tfrac{3}{4}.$$

Example 4

Solve the initial value problem

$$\frac{d^2u}{dt^2} + 3\frac{du}{dt} + 2u = e^{2t}, \qquad u(0) = \frac{1}{3}, \qquad \frac{du}{dt}(0) = \frac{1}{6}.$$

1. First find the complementary solution. The homogeneous equation associated with the given equation is

$$\frac{d^2u}{dt^2} + 3\frac{du}{dt} + 2u = 0.$$

Its characteristic polynomial is $m^2 + 3m + 2$, with roots -1 and -2. Thus the complementary solution is

$$u_c(t) = c_1 e^{-t} + c_2 e^{-2t}.$$

2. Next find the particular solution. Since all derivatives of e^{2t} are multiples of e^{2t}, we guess a solution in the form

$$u_p(t) = Ae^{2t}.$$

Inserting this function and its derivatives into the differential equation gives

$$A(4 + 3 \cdot 2 + 2)e^{2t} = e^{2t}.$$

This is satisfied by taking $A = \frac{1}{12}$. Thus we have found a particular solution,

$$u_p(t) = e^{2t}/12.$$

3. The general solution of the given differential equation is

$$u(t) = c_1 e^{-t} + c_2 e^{-2t} + \frac{1}{12} e^{2t}.$$

To this solution we apply the two initial conditions, which become equations for c_1 and c_2:

$$u(0) = \frac{1}{3}: \qquad c_1 + c_2 + \frac{1}{12} = \frac{1}{3},$$

$$\frac{du}{dt}(0) = \frac{1}{6}: \qquad -c_1 - 2c_2 + \frac{2}{12} = \frac{1}{6}.$$

The coefficients are easily found to be $c_1 = \frac{1}{2}$, $c_2 = -\frac{1}{4}$. Therefore the solution of the initial value problem is

$$u(t) = \frac{1}{2} e^{-t} - \frac{1}{4} e^{-2t} + \frac{1}{12} e^{2t}.$$

To see more clearly the workings of our method, let us try to find a particular solution of

$$\frac{d^2 u}{dt^2} + 3 \frac{du}{dt} + 2u = e^{rt},$$

where r is a parameter. We would guess that a particular solution has the form

$$u_p(t) = A e^{rt}.$$

Then by inserting u_p in this form into the differential equation we find that the condition to be satisfied is

$$A(r^2 + 3r + 2)e^{rt} = e^{rt}.$$

Obviously, we will have to choose

$$A = 1/(r^2 + 3r + 2),$$

provided that the denominator is not 0. Since the denominator is just the characteristic polynomial of the homogeneous equation, we see that our guess is correct (A can be found to make $A e^{rt}$ a particular solution of the nonhomogeneous equation) if and only if e^{rt} is *not* a solution of the homogeneous equation.

Whenever the inhomogeneity $f(t)$ is itself a solution of the homogeneous equation, our first guess for the form of the particular solution will fail. It is necessary in this case to revise the assumed form for

u_p: multiply the original form by the lowest power of t such that no solution of the homogeneous equation is a term of u_p.

Example 5

Find a particular solution of

$$\frac{d^2u}{dt^2} + 3\frac{du}{dt} + 2u = e^{-2t}.$$

Our first guess would be $u_p(t) = Ae^{-2t}$. However, e^{-2t} is a solution of the homogeneous equation, so we must revise the guess to

$$u_p(t) = Ate^{-2t}.$$

Inserting this into the differential equation yields the equation

$$A(4t - 2)e^{-2t} + 3A(-2t - 1)e^{-2t} + 2Ate^{-2t} = e^{-2t}.$$

From here it is easy to find $A = -1$. Notice that the terms containing t cancel among themselves.

A similar revision of the assumed form for $u_p(t)$ is necessary any time that form contains a term that is a solution of the homogeneous equation. Thus to find a particular solution of

$$\frac{d^2u}{dt^2} + 3\frac{du}{dt} + 2u = te^{-2t},$$

the first guess would be $u_p(t) = Ate^{-2t} + Be^{-2t}$. But since the last term is a solution of the homogeneous equation, the guess is revised to

$$u_p = t(Ate^{-2t} + Be^{-2t}),$$

in which no term is a solution of the homogeneous equation.

Now we are ready to summarize the steps in the method of undetermined coefficients.

1. Solve the corresponding homogeneous equation to find the complementary solution $u_c(t)$.

2. Break down the inhomogeneity into simple functions, each one being (a) a polynomial, (b) an exponential, (c) a sine and/or cosine (of the same frequency), or (d) a product of these.

3. For each simple nonhomogeneity f, determine the form of f', f'', etc. Guess the particular solution $u_p(t)$ to have the form of a linear combination of these functions. (Instead of f, f', f'', etc., we can use any set of functions from which f and its derivatives can be formed as linear combinations.) See Table 2.4.

Table 2.4	Inhomogeneity $f(t)$	Assumed Form for $u_p(t)$*
	$a_0 t^n + a_1 t^{n-1} + \cdots + a_n$	$A_0 t^n + A_1 t^{n-1} + \cdots + A_n$
	$e^{\alpha t}$	$A e^{\alpha t}$
	$a \cos \omega t + b \sin \omega t$	$A \cos \omega t + B \sin \omega t$
	$e^{\alpha t}(a_0 t^n + a_1 t^{n-1} + \cdots + a_n)$	$e^{\alpha t}(A_0 t^n + A_1 t^{n-1} + \cdots + A_n)$
	$e^{\alpha t}(a \cos \omega t + b \sin \omega t)$	$e^{\alpha t}(A \cos \omega t + B \sin \omega t)$
	$(a_0 t^n + a_1 t^{n-1} + \cdots + a_n) \cos \omega t$	$(A_0 t^n + A_1 t^{n-1} + \cdots + A_n) \cos \omega t$
	$\quad + (b_0 t^n + b_1 t^{n-1} + \cdots + b_n) \sin \omega t$	$\quad + (B_0 t^n + B_1 t^{n-1} + \cdots + B_n) \sin \omega t$

* If any term in the form listed is a solution of the homogeneous equation, revise the assumed form.

4. Test the guessed form to see whether any term or linear combination of terms is a solution of the corresponding homogeneous equation. If necessary, revise the guess: multiply it by the lowest power of t for which the revised guess passes this test.

5. Insert the guessed $u_p(t)$ into the differential equation and equate the coefficients of like terms. (If a function is missing, its coefficient is 0.) Solve for the coefficients.

Trouble shooting: If the equations from step 5 cannot be solved, check carefully for errors in algebra and double-check step 4.

Example 6

Find a particular solution of the nonhomogeneous equation

$$\frac{d^2 u}{dt^2} + 3 \frac{du}{dt} + 2u = (6t^2 - 2)e^{-2t}.$$

1. The solution of the corresponding homogeneous equation is

$$u_c(t) = c_1 e^{-t} + c_2 e^{-2t}.$$

2. The inhomogeneity is already a "simple function"—a product of a polynomial and an exponential.

3. Every derivative of f has the form of a second-degree polynomial times e^{-2t}. We may guess

$$u_p(t) = (At^2 + Bt + C)e^{-2t}.$$

4. Since e^{-2t} is a solution of the corresponding homogeneous equation, we must revise the guess. Multiplying by the first power of t is adequate to ensure that no term of the guess is a solution of the corresponding homogeneous equation. Thus our revised guess is:

$$u_p(t) = t(At^2 + Bt + C)e^{-2t}.$$

5. Insert the last form for $u_p(t)$ into the differential equation, finding

$$A[6t - 12t^2 + 4t^3 + 3(3t^2 - 2t^3) + 2t^3]e^{-2t}$$
$$+ B[2 - 8t + 4t^2 + 3(2t - 2t^2) + 2t^2]e^{-2t}$$
$$+ C[-4 + 4t + 3(1 - 2t) + 2t]e^{-2t} = (6t^2 - 2)e^{-2t}.$$

Now equate coefficients of like terms:

$2B - C = -2$	(coeffs. of e^{-2t}),
$6A - 2B + 0C = 0$	(coeffs. of te^{-2t}),
$-3A + 0B = 6$	(coeffs. of t^2e^{-2t}),
$0A = 0$	(coeffs. of t^3e^{-2t}).

These simultaneous equations are easily solved for $A = -2$, $B = -6$, $C = -10$. Thus we have found

$$u_p(t) = t(-2t^2 - 6t - 10)e^{-2t}.$$

Exercises

In Exercises 1–10, find (a) the complementary solution, (b) a particular solution, and (c) the general solution.

1. $\dfrac{d^2u}{dt^2} + \dfrac{du}{dt} = 1$
2. $\dfrac{d^2u}{dt^2} + 3\dfrac{du}{dt} = e^{-3t}$

3. $\dfrac{d^2u}{dt^2} + 4\dfrac{du}{dt} + 5u = 1$
4. $\dfrac{d^2u}{dt^2} + 2\dfrac{du}{dt} + 5u = e^{-t}$

5. $\dfrac{d^2u}{dt^2} + b\dfrac{du}{dt} + \omega^2 u = \cos \omega t$
6. $\dfrac{d^2u}{dt^2} + 9u = \cos 2t$

7. $\dfrac{d^2u}{dt^2} + \omega^2 u = \cos \omega t$
8. $\dfrac{d^2u}{dt^2} + \omega^2 u = e^{-\alpha t}\cos \omega t$

9. $\dfrac{d^2u}{dt^2} + 3\dfrac{du}{dt} + 2u = \cos t$
10. $\dfrac{d^2u}{dt^2} - u = e^{-t}$

In Exercises 11–20, find (a) the complementary solution and (b) the correct *form* for a particular solution.

11. $\dfrac{d^2u}{dt^2} + u = 1 + 2t + 3e^t$
12. $\dfrac{d^2u}{dt^2} + 2\dfrac{du}{dt} + 2u = \cos t$

13. $\dfrac{d^2u}{dt^2} + 2\dfrac{du}{dt} + u = 1 - e^{-t}$
14. $\dfrac{d^2u}{dt^2} + u = t \cos t$

15. $\dfrac{d^2u}{dt^2} = (1 + t)^2$
16. $\dfrac{d^2u}{dt^2} + u = e^{-t}\sin t$

17. $\dfrac{d^2u}{dt^2} - u = \sinh t$

18. $\dfrac{d^2u}{dt^2} - u = 1 + \cos t + \frac{1}{2}\cos 2t$

19. $\dfrac{d^2u}{dt^2} + u = 1 + \cos t + \frac{1}{2}\cos 2t$

20. $\dfrac{du}{dt} + u = 1 - e^{-t}$

In Exercises 21–30, find the general solution of the differential equation and the specific solution that satisfies the initial conditions.

21. $\dfrac{d^2u}{dt^2} + \dfrac{du}{dt} = -1,\ u(0) = 10,\ \dfrac{du}{dt}(0) = 0$

22. $\dfrac{d^2u}{dt^2} + 4u = \sin t,\ u(0) = 0,\ \dfrac{du}{dt}(0) = 0$

23. $\dfrac{d^2u}{dt^2} + 4u = \sin 2t,\ u(0) = 0,\ \dfrac{du}{dt}(0) = 0$

24. $\dfrac{d^2u}{dt^2} + 4\dfrac{du}{dt} = 1,\ u(0) = 1,\ \dfrac{du}{dt}(0) = 0$

25. $\dfrac{d^2u}{dt^2} + 2\dfrac{du}{dt} + u = e^{-t},\ u(0) = 1,\ \dfrac{du}{dt}(0) = 0$

26. $\dfrac{d^2u}{dt^2} + 2\dfrac{du}{dt} + 2u = \cos t,\ u(0) = 1,\ \dfrac{du}{dt}(0) = 2$

27. $4\dfrac{d^2u}{dt^2} + 4\dfrac{du}{dt} + u = 1,\ u(0) = 1,\ \dfrac{du}{dt}(0) = 0$

28. $\dfrac{d^2u}{dt^2} - u = \cos t,\ u(0) = 0,\ \dfrac{du}{dt}(0) = 1$

29. $\dfrac{d^2u}{dt^2} + 4\dfrac{du}{dt} + 8u = t,\ u(0) = 0,\ \dfrac{du}{dt}(0) = 0$

30. $\dfrac{d^2u}{dt^2} + \dfrac{du}{dt} - 2u = 6\cosh t,\ u(0) = -1,\ \dfrac{du}{dt}(0) = 2$

2.5

Forced Vibrations

We now continue with the study of the mass–spring–damper system. In Section 2.3 we considered the kinds of motion that would result if the system vibrated under the influence of nonzero initial conditions. Now we are prepared to take into account the action of an impressed force also.

The governing differential equation was derived originally (Section 2.1) as

$$M\frac{d^2u}{dt^2} + p\frac{du}{dt} + ku = F(t), \qquad 0 < t. \tag{2.61}$$

We divide through this equation by M and define the resulting coefficients as $b = p/M$, $\omega_0^2 = k/M$, and $f(t) = F(t)/M$. The equation has become

$$\frac{d^2u}{dt^2} + b\frac{du}{dt} + \omega_0^2 u = f(t), \qquad 0 < t. \tag{2.62}$$

The general initial conditions that could accompany the differential equation are

$$u(0) = u_0, \qquad \frac{du}{dt}(0) = v_0. \tag{2.63}$$

However, we shall simplify matters by taking both u_0 and v_0 to be 0. In words, our system starts from rest. Thus we are really interested in solving the initial value problem consisting of the differential equation (2.62) and the initial conditions

$$u(0) = 0, \qquad \frac{du}{dt}(0) = 0. \tag{2.64}$$

As a further simplification, we will suppose that $b < 2\omega_0$—damping is slight or absent.

There are so many possibilities for the inhomogeneity in Eq. (2.62) that we cannot consider even a significant sample. In this section we will restrict ourselves to $f(t) = f_0 \cos \omega t$ and study the effects of assumptions about damping and the relationship between the forcing frequency ω and the natural frequency ω_0.

Slight Damping

First we take the case of a slightly damped system: $0 < b < 2\omega_0$. The initial value problem to be solved is

$$\frac{d^2u}{dt^2} + b\frac{du}{dt} + \omega_0^2 u = f_0 \cos \omega t, \qquad 0 < t, \tag{2.65}$$

$$u(0) = 0, \qquad \frac{du}{dt}(0) = 0. \tag{2.66}$$

It is important to distinguish between ω_0, the natural frequency of the undamped system, and ω, the frequency of the impressed force.

Following our procedure for finding the general solution of Eq.

(2.65), we first find the complementary function, the solution of

$$\frac{d^2u}{dt^2} + b\frac{du}{dt} + \omega_0^2 u = 0,$$

the homogeneous equation corresponding to Eq. (2.65). By the methods of Section 2.2 we find

$$u_c(t) = e^{-bt/2}(c_1 \cos \beta t + c_2 \sin \beta t), \tag{2.67}$$

where $\beta = \sqrt{\omega_0^2 - (b/2)^2}$.

Next, since $\cos \omega t$ is not a solution of the homogeneous equation (not even if $\omega = \beta$, as long as $b \neq 0$!), we assume that the particular solution has the form

$$u_p(t) = A \cos \omega t + B \sin \omega t.$$

After substituting this function into the differential equation (2.65) and equating coefficients of like terms, we find

$$(\omega_0^2 - \omega^2)A + \omega b B = f_0 \qquad \text{(from } \cos \omega t),$$

$$-\omega b A + (\omega_0^2 - \omega^2)B = 0 \qquad \text{(from } \sin \omega t).$$

The solution of this system of equations is

$$A = \frac{f_0}{\Delta}(\omega_0^2 - \omega^2), \qquad B = \frac{f_0}{\Delta}\omega b,$$

where $\Delta = (\omega_0^2 - \omega^2)^2 + \omega^2 b^2$. We now have found the particular solution,

$$u_p(t) = \frac{f_0}{\Delta}((\omega_0^2 - \omega^2) \cos \omega t + b\omega \sin \omega t). \tag{2.68}$$

We now may write out the general solution of the differential equation (2.65). It is the sum of u_c from Eq. (2.67) and u_p from Eq. (2.68):

$$u(t) = \frac{f_0}{\Delta}((\omega_0^2 - \omega^2) \cos \omega t + b\omega \sin \omega t) + e^{-bt/2}(c_1 \cos \beta t + c_2 \sin \beta t). \tag{2.69}$$

The next task is to satisfy the initial conditions, Eq. (2.64), which require that both u and its derivative be 0 at $t = 0$. We leave it as an exercise in algebra to carry out these computations, which lead to

$$u(t) = \frac{f_0}{\Delta}[(\omega_0^2 - \omega^2) \cos \omega t + \omega b \sin \omega t]$$

$$- \frac{f_0}{\Delta}e^{-bt/2}\left[(\omega_0^2 - \omega^2) \cos \beta t + \frac{b}{2\beta}(\omega_0^2 + \omega^2) \sin \beta t\right]. \tag{2.70}$$

This complicated solution conveys some useful information, which we can obtain quite easily. First, note that the apparently simple initial conditions in Eq. (2.66) have not really turned out to be simple. In the presence of the particular solution (2.69) they caused the complementary solution u_c to have nonzero coefficients. Nevertheless, as long as some damping is present ($b > 0$), the decaying exponential factor $e^{-bt/2}$ in the complementary solution will cause that function to become negligible after a sufficient time. The particular solution, on the contrary, persists forever. The name *transient part* or *transient solution* is generally applied to those terms in a solution that disappear as t increases indefinitely. We shall describe the terms that do not disappear as the *persistent part** of a solution.

The persistent part of the solution of the differential equation (2.65) (which we shall call $v(t)$) is given by Eq. (2.68) or, after trigonometric manipulation, by

$$v(t) = \frac{f_0}{\sqrt{\Delta}} \cos(\omega t - \phi) \qquad\qquad (2.71)$$

with $\phi = \tan^{-1}(\omega b/(\omega_0^2 - \omega^2))$. It is common to compare this part of the output with the function

$$\frac{f(t)}{\omega_0^2} = \frac{f_0}{\omega_0^2} \cos \omega t.$$

(This is what the displacement $u(t)$ would be if the mass followed the driving force perfectly.) Evidently, the frequency of the driving force is retained by the persistent part of the solution, but there is a phase shift, represented by ϕ, and a difference in amplitude measured by the factor $\omega_0^2/\sqrt{\Delta}$ (called the magnification factor). Both ϕ and the magnification factor are shown in Fig. 2.14 as functions of ω, for various values of b. It is noteworthy that the persistent part of the solution, as given in Eq. (2.71), is valid for all values of b—not just $0 < b < 2\omega_0$.

Example 1

Find the persistent part of the solution of

$$\frac{d^2u}{dt^2} + 2\frac{du}{dt} + 17u = f_0 \cos 3t,$$

$$u(0) = 0, \qquad \frac{du}{dt}(0) = 0.$$

---

* Also called the steady state solution.

Figure 2.14

(a) Phase Shift ϕ
(b) Magnification

Factor $\omega_0^2/\sqrt{\Delta}$

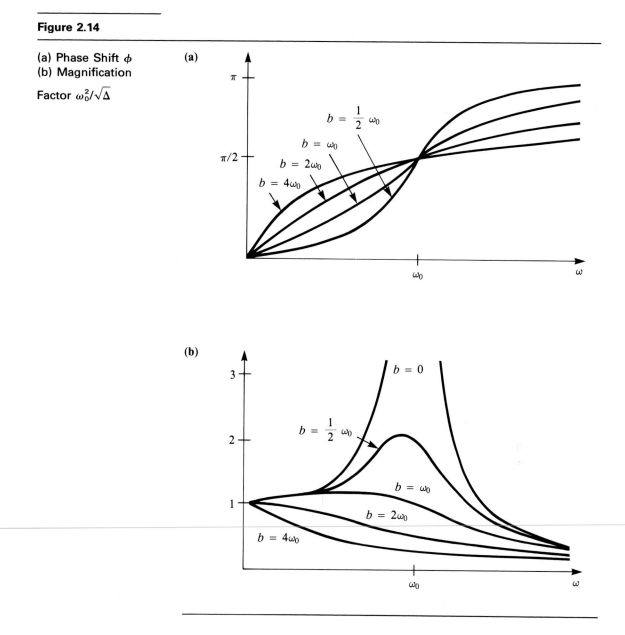

(a)

π

$\pi/2$

$b = \frac{1}{2}\,\omega_0$

$b = \omega_0$

$b = 2\omega_0$

$b = 4\omega_0$

ω_0

ω

(b)

3

$b = 0$

$b = \frac{1}{2}\,\omega_0$

2

$b = \omega_0$

1

$b = 2\omega_0$

$b = 4\omega_0$

ω_0

ω

We have seen that in this case the persistent part coincides with the particular solution found by the method of undetermined coefficients. We must guess that

$$u_p(t) = A \cos 3t + B \sin 3t.$$

After inserting u_p into the differential equation and equating coefficients of like terms, we find

$$8A + 6B = f_0,$$

$$-6A + 8B = 0.$$

From these we determine that $A = 0.08f_0$ and $B = 0.06f_0$, so

$$v(t) = u_p(t) = f_0(0.08 \cos 3t + 0.06 \sin 3t)$$

$$= \frac{f_0}{10} \cos (3t - 0.644).$$

For these parameters, $\sqrt{\Delta} = 10$ and $\omega_0^2 = 17$, so the magnification factor is 1.7

Beats

We now take up the case where the forcing function is sinusoidal but there is no damping. The initial value problem to be considered is

$$\frac{d^2u}{dt^2} + \omega_0^2 u = f_0 \cos \omega t, \qquad 0 < t, \tag{2.72}$$

$$u(0) = 0, \qquad \frac{du}{dt}(0) = 0. \tag{2.73}$$

Here we are assuming that the natural and forcing frequencies are different: $\omega \neq \omega_0$. We may simply set $b = 0$ in the solution of the preceding case (Eq. 2.70) to obtain the solution in this case:

$$u(t) = \frac{f_0}{\omega_0^2 - \omega^2} (\cos \omega t - \cos \omega_0 t). \tag{2.74}$$

If ω and ω_0 are very different, the graph of $u(t)$ is easy to draw: it appears to be a high-frequency wobble added to a low-frequency cosine curve. (See Fig. 2.15.) If ω and ω_0 are quite close in value, it is convenient to write the solution in this alternative form

$$u(t) = \frac{2f_0}{\omega_0^2 - \omega^2} \left[\sin \frac{\omega_0 - \omega}{2} t \right] \left[\sin \frac{\omega_0 + \omega}{2} t \right]. \tag{2.75}$$

Thus we can think of $u(t)$ as a sinusoid, represented by the factor with the larger frequency, $(\omega_0 + \omega)/2$, with a time-varying amplitude, represented by the factor with the slower frequency, $|\omega_0 - \omega|$. (See Fig. 2.16.) Indeed, in the case of acoustic vibrations, if $|\omega_0 - \omega|$ is much smaller than $\omega_0 + \omega$, the effect of the slow variation of $\sin (\omega_0 - \omega)t/2$ is quite audible as a fluttering sound. This is called the "beat" phenomenon, and $|\omega_0 - \omega|/2$ is

Figure 2.15

Forced Vibrations, ω and ω_0 very different: $u(t) = \cos t - \cos 8t$ (graph of $u/t) = \cos t$ shown for comparison)

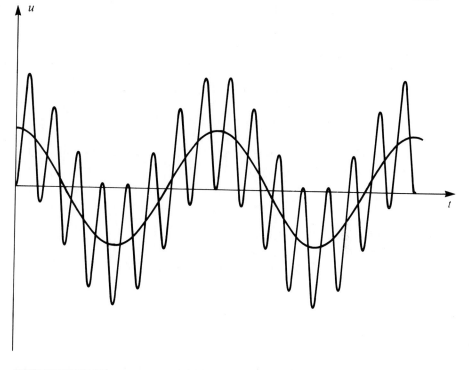

the beat frequency. Note that in this case there is no transient part of the solution, since there is no allowance for damping in our equation.

Example 2

Solve the initial value problem

$$\frac{d^2u}{dt^2} + u = \cos \omega t,$$

$$u(0) = 0, \qquad \frac{du}{dt}(0) = 0,$$

for $\omega = 0.1$, 0.9, 10.

It is convenient to use Eq. (2.74) for $\omega = 0.1$ and 10 and Eq. (2.75) for $\omega = 0.9$. The results are

$$(\omega = 0.1) \qquad u(t) = \frac{2}{0.99}(\cos 0.1t - \cos t),$$

Figure 2.16

Beats: $u(t) =$
$\cos t - \cos (1.2t)$ and
$u(t) = \pm 2 \sin (0.1t)$

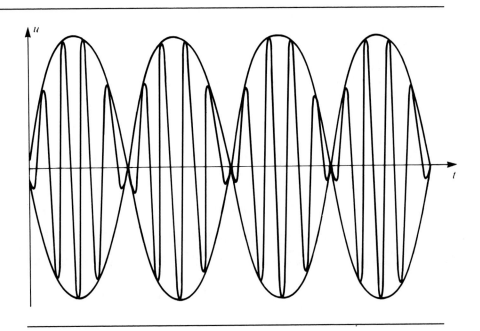

$$(\omega = 0.9) \qquad u(t) = \frac{2}{0.19} \sin (0.05t) \sin (0.95t),$$

$$(\omega = 10) \qquad u(t) = \frac{2}{-99} (\cos 10t - \cos t).$$

The maxima for the three functions are approximately 2.02, 10.53, 0.0202.

Resonance

Finally, we consider the case in which there is no damping and the forcing frequency matches the natural frequency of our system. The initial value problem to be solved is

$$\frac{d^2u}{dt^2} + \omega_0^2 u = f_0 \cos \omega_0 t, \qquad 0 < t, \tag{2.76}$$

$$u(0) = 0, \qquad \frac{du}{dt}(0) = 0. \tag{2.77}$$

Since the inhomogeneity here is itself a solution of the differential equation, the form to be guessed for the particular solution is

$$u_p(t) = t(A \cos \omega_0 t + B \sin \omega_0 t).$$

Substitution into the differential equation shows that the correct values for A and B are

$$A = 0, \qquad B = -\frac{f_0}{2\omega_0}.$$

Furthermore, in this case the particular solution so obtained also satisfies the initial conditions; thus the solution of the initial value problem of Eqs. (2.76) and (2.77) is

$$u(t) = -\frac{f_0}{2\omega_0} t \sin \omega_0 t. \tag{2.78}$$

It is obvious that we have again a sinusoidal function with a time-varying amplitude. In this instance, however, the amplitude increases steadily in time, without any limit. (See Fig. 2.17.)

There are many physical systems that can be modeled with fair accuracy by a differential equation like Eq. (2.76). Something like resonance happens when you start up a child's swing. However, no physical

Figure 2.17

Resonance: $u(t) = t \sin t$

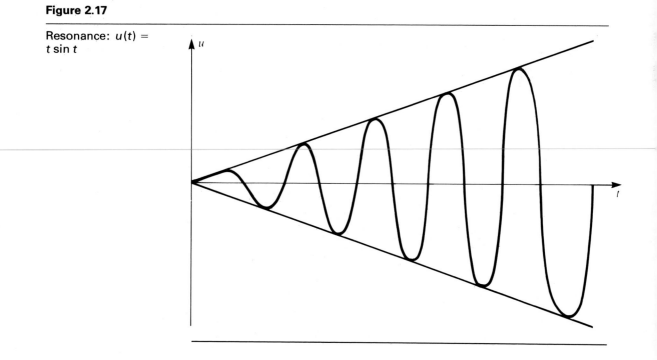

system can sustain indefinitely the growth of internal energy represented by the increasing amplitude in Eq. (2.78). Therefore there must come a time when the mathematical model represented by Eqs. (2.76) and (2.77) becomes divorced from physical reality. Many systems are self-limiting because of nonlinear behavior, so the linear differential equation (2.76) is simply not valid except for small values of u. This is certainly the case for a child's swing. Other systems do behave as predicted by our mathematical model, until something breaks. Many well-known bridge failures fall into this category.

Exercises

1. A constant inhomogeneity is called a "step input" because it is supposed to be zero for $t < 0$ and nonzero for $t > 0$. Find the response to a step input—that is, the solution of

$$\frac{d^2u}{dt^2} + b\frac{du}{dt} + \omega_0^2 u = f_0,$$

$$u(0) = 0, \qquad \frac{du}{dt}(0) = 0,$$

where f_0 is constant and $0 < b < 2\omega_0$.

2. What is the persistent part of the solution in Exercise 1?

3. Solve the problem of Exercise 1 in the following three cases: (a) $b = 0$; (b) $b = 2\omega_0$; (c) $b = 4\omega_0$.

4. Sketch the solutions found in Exercise 3.

5. Solve the nonhomogeneous initial value problem

$$\frac{d^2u}{dt^2} + b\frac{du}{dt} + \omega_0^2 u = kt,$$

$$u(0) = 0, \qquad \frac{du}{dt}(0) = 0.$$

(An inhomogeneity proportional to t is called a *ramp input*.)

6. What is the persistent part of the solution to Exercise 5? Sketch it and the input as functions of t.

7. Give an example of an inhomogeneity $f(t)$ in the differential equation

$$\frac{d^2u}{dt^2} + b\frac{du}{dt} + \omega_0^2 u = f(t)$$

for which the particular solution found by undetermined coefficients is *not* the same as the persistent part of the solution.

8. Suppose u_1 is the solution of the initial value problem composed of

Eqs. (2.62) and (2.64), while u_2 is the solution of the problem

$$\frac{d^2u}{dt^2} + b\frac{du}{dt} + \omega_0^2 u = 0,$$

$$u(0) = u_0, \qquad \frac{du}{dt}(0) = v_0.$$

Show that $u(t) = u_1(t) + u_2(t)$ is the solution of Eqs. (2.62) and (2.63).

9. Find the persistent part of the solution of Eqs. (2.65) and (2.66) when the impressed frequency ω equals ω_0.

10. What is the phase angle ϕ (see Eq. (2.71)) associated with the solution of Exercise 5?

11. Considering ω_0 and b fixed and ω variable, determine what value of ω produces a maximum in the magnification factor

$$\frac{\omega_0^2}{\sqrt{(\omega_0^2 - \omega^2)^2 + \omega^2 b^2}}.$$

12. What is the greatest value that the function $u(t)$ in Eq. (2.75) can assume? (Hint: do not differentiate; replace the variable factors by their greatest values.)

13. Show that the time required for $u(t)$ in Eq. (2.75) to get as large as possible is *about* $\pi/|\omega_0 - \omega|$.

14. Sketch the solution of the resonance problem, Eq. (2.78) as a function of t through several periods of the sine.

15. Let M be the maximum amplitude found in Exercise 12 and let T be the time found in Exercise 13. Find the limit of M/T as ω approaches ω_0.

2.6

Electrical Circuits

In Section 2.1 we observed that the charge q on the capacitor in the electrical circuit of Fig. 2.18 is described by the differential equation

$$L\frac{d^2q}{dt^2} + R\frac{dq}{dt} + \frac{1}{C}q = V(t). \tag{2.79}$$

Since this equation is the same as the one for the mass–spring–damper system, the behavior of the solution must be the same also.

Nevertheless, there is a certain difference in focus between electrical circuits and mechanical systems. For one thing, it is not the charge q, but rather its derivative, the current $i = dq/dt$, that is of interest. For

Figure 2.18

RCL Circuit

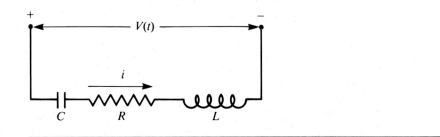

another, initial conditions may be unknown, and attention naturally centers on the persistent part of the solution of Eq. (2.79). This is often called "the steady state solution" of the differential equation, although it may be anything but steady. There are also some differences in methodology that we will survey briefly.

The most important special case of Eq. (2.79) corresponds to an impressed voltage that is sinusoidal:

$$V(t) = v_0 \cos \omega t$$

(v_0 is the peak voltage). There is a surprising gain in simplicity if, instead of using the cosine function itself, we replace it with the complex voltage

$$V(t) = v_0 e^{j\omega t}.$$

(The use of j for $\sqrt{-1}$ is common in electrical engineering, to free i for current.) The actual voltage is then the real part of $V(t)$.

$$\text{Re}\,(v_0 e^{j\omega t}) = v_0 \cos \omega t.$$

The charge or current corresponding to the actual voltage is just the real part of the complex charge or current. That is, if $q(t)$ is the solution of Eq. (2.79) with a complex $V(t)$, then

$$L\frac{d^2}{dt^2}\,\text{Re}\,(q) + R\frac{d}{dt}\,\text{Re}\,(q) + \frac{1}{C}\,\text{Re}\,(q) = \text{Re}\,(V(t)).$$

Similarly, the imaginary part of $q(t)$ corresponds to an impressed voltage that is the imaginary part of $V(t)$.

As an example of the use of complex exponentials, let us try to find the persistent part of the solution of

$$L\frac{d^2q}{dt^2} + R\frac{dq}{dt} + \frac{1}{C}q = v_0 e^{j\omega t}. \qquad (2.80)$$

The general solution of the corresponding homogeneous equation,

$$L\frac{d^2q}{dt^2} + R\frac{dq}{dt} + \frac{1}{C}q = 0, \tag{2.81}$$

can be found by using the exponential method. The characteristic polynomial is

$$Lr^2 + Rr + \frac{1}{C} = 0,$$

with roots

$$r = -\frac{R}{2L} \pm \sqrt{\left(\frac{R}{2L}\right)^2 - \frac{1}{LC}}. \tag{2.82}$$

Except in the critical case when the radicand above is 0, the general solution of Eq. (2.81) is

$$q_c(t) = c_1 e^{r_1 t} + c_2 e^{r_2 t}. \tag{2.83}$$

Since we are allowing complex arithmetic, there is no need to separate the cases of real and complex roots.

If the resistance R is not zero, it is clear that both roots of the characteristic equation have a negative real part. Thus $u_c(t)$ approaches 0 as t increases, and $u_c(t)$ does not enter into the persistent part of the solution.

Now we turn to the problem of finding a particular solution. Since the inhomogeneity in Eq. (2.80) is exponential, we may expect that a particular solution is exponential, too:

$$q_p(t) = Ae^{j\omega t}.$$

Substituting this function into the nonhomogeneous equation (2.80) leads to the algebraic equation

$$\left(-L\omega^2 + Rj\omega + \frac{1}{C}\right)A = v_0 \tag{2.84}$$

after the exponential has been cancelled from both sides.

It is an easy matter to solve Eq. (2.84) for A and thence for

$$q_p(t) = \frac{v_0}{\left(\frac{1}{C} - L\omega^2\right) + jR\omega} e^{j\omega t}.$$

This is the persistent part of the complex charge. The persistent part of the complex current is just its time derivative,

$$i_p(t) = \frac{j\omega v_0}{\left(\dfrac{1}{C} - L\omega^2\right) + jR\omega}\, e^{j\omega t}. \tag{2.85}$$

In order to study the relationship between current and voltage, it will be convenient to change the form of the complex current, Eq. (2.85). Let us designate

$$\Delta = \sqrt{\left(\frac{1}{C} - L\omega^2\right)^2 + R^2\omega^2}\,,$$

—that is, Δ is the magnitude of the complex number in the denominator of Eq. (2.85). Now multiply numerator and denominator of Eq. (2.85) by the conjugate of the denominator to find

$$i_p(t) = \frac{j\omega v_0}{\Delta^2}\left(\frac{1}{C} - L\omega^2 - jR\omega\right)e^{j\omega t}$$

$$= \frac{\omega v_0}{\Delta^2}\left(R\omega + j\left(\frac{1}{C} - L\omega^2\right)\right)e^{j\omega t} \tag{2.86}$$

$$= \frac{\omega v_0}{\Delta}\, e^{j(\omega t - \phi)} \tag{2.87}$$

In order to pass from Eq. (2.86) to Eq. (2.87) it is only necessary to solve

$$R\omega + j\left(\frac{1}{C} - L\omega^2\right) = \Delta e^{j\phi},$$

that is

$$\frac{R\omega}{\Delta} = \cos\phi, \qquad \frac{\dfrac{1}{C} - L\omega^2}{\Delta} = -\sin\phi. \tag{2.88}$$

With the current as shown in Eq. (2.86) or (2.87) we easily obtain several results.

1. The current corresponding to input voltage $v_0 \cos \omega t$ is the real part of Eq. (2.87), namely,

$$i(t) = \frac{\omega v_0}{\Delta}\cos(\omega t - \phi).$$

2. The current corresponding to input voltage $v_0 \sin \omega t$ is the imaginary part of Eq. (2.87):

$$i(t) = \frac{\omega v_0}{\Delta}\sin(\omega t - \phi).$$

3. The ratio of complex voltage to complex current is

$$\frac{v(t)}{i(t)} = \frac{\Delta}{\omega} e^{j\phi}. \tag{2.89}$$

Note that this is not a function of time, although the ratio of the actual voltage and current certainly is so. The ratio calculated above is called the *complex impedance*, an analog of resistance in being a ratio of voltage to current. It is given the symbol Z and can be expressed as

$$Z = \frac{\Delta}{\omega} e^{j\phi} = R + j\left(L\omega - \frac{1}{C\omega}\right),$$

from Eq. (2.85), or from Eq. (2.89), using the definition of Δ,

$$Z = \frac{\Delta}{\omega} e^{j\phi} = \sqrt{R^2 + \left(L\omega - \frac{1}{C\omega}\right)^2} (\cos \phi + j \sin \phi).$$

From either of the last two equations it can be seen that the impedance is minimized by choosing L and C so that

$$L\omega - \frac{1}{C\omega} = 0 \quad \text{or} \quad \frac{1}{LC} = \omega^2.$$

Examination of the original differential equation shows that this is equivalent to choosing the natural frequency of the circuit, $\omega_0 = 1/\sqrt{LC}$, to equal the impressed frequency. Varying L or C to accomplish this is called tuning the circuit.

Exercises

1. An *RLC* circuit contains a resistance of 10 ohms, an inductance of 10^{-4} henry, and a capacitance of 10^{-6} farad. If the impressed voltage is $v_0 \cos \omega t$, find the steady state complex charge.

2. Find the steady state complex current for the circuit of Exercise 1.

3. Express the solutions of Exercises 1 and 2 in the form of a positive multiple of $e^{j(\omega t - \phi)}$.

4. Find the complex impedance for this circuit and sketch its magnitude (absolute value) as a function of ω.

5. An *RLC* circuit has a 10-ohm resistor and a 100-microhenry ($= 10^{-4}$ h) inductance. What capacitance should be used if it is desired to have minimum impedance for a frequency of 10^4 hertz? (1 hertz $= 2\pi$ radians/sec.)

6. What is the minimum impedance in Exercise 5?

7. An *RC* circuit is shown in Fig. 2.19. Find an equation linking the input voltage $V(t)$ with the charge on the capacitor and the current

Figure 2.19

RC Circuit

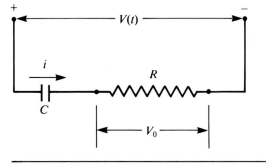

through the resistor. Find the steady state solution, the complex charge, if $V(t) = v_0 e^{j\omega t}$.

8. For the circuit of Exercise 7, find the steady state complex current and thus the output voltage, $V_0 = Ri$.

9. What is the relationship between output and input voltage as a function of frequency for Fig. 2.19?

10. Figure 2.20 shows a circuit similar to that in Fig. 2.19. Use the results of Exercises 7 to find the output voltage here, which is $V_0 = q/C$.

Figure 2.20

RC Circuit

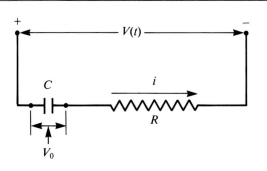

11. What is the relationship between input and output voltage as a function of frequency for Fig. 2.20?

Notes

There is a sharp contrast between solution methods for first-order equations and those used in this chapter. In solving first-order equations we make a series of transformations and operations—always including at least one integration—which finally lead us to the solution. In solving second-order linear equations with constant coefficients we know essentially what form the solution must have; we need only to determine appropriate values for parameters and coefficients. The mathematical tools used are mainly algebraic—solving polynomial equations and systems of simultaneous equations.

We shall see that this algebraic approach can be extended to higher-order linear equations with constant coefficients (Chapter 3) and to systems (Chapter 7). Both higher-order equations and systems are used in the analysis of mechanical vibrations and complex electrical circuits.

Miscellaneous Exercises

In Exercises 1–10, find the general solution of the differential equation.

1. $\dfrac{du}{dt} + 5u = \sin t$

2. $\dfrac{d^2u}{dt^2} + 9\dfrac{du}{dt} + 20u = 5t + 1$

3. $\dfrac{d^2u}{dt^2} + 5\dfrac{du}{dt} - 6u = e^t$

4. $4\dfrac{d^2u}{dt^2} + 20\dfrac{du}{dt} + 25u = 25\cos 5t$

5. $\dfrac{d^2u}{dt^2} - k^2u = 1 + 3t$

6. $\dfrac{d^2u}{dt^2} - u = \sin t$

7. $\dfrac{d^2u}{dt^2} + \dfrac{du}{dt} = \sin t$

8. $\dfrac{d^2u}{dt^2} + 2\dfrac{du}{dt} + 3u = e^{-t}$

9. $\dfrac{d^2u}{dt^2} + 4u = \sin 2t$

10. $\dfrac{d^2u}{dt^2} + 17\dfrac{du}{dt} + 72u = 0$

In Exercises 11–16, find the general solution of the differential equation and then solve the initial value problem.

11. $\dfrac{d^2u}{dt^2} + 4\dfrac{du}{dt} + 6u = 0$, $u(0) = 0$, $\dfrac{du}{dt}(0) = 1$

12. $\dfrac{d^2u}{dt^2} + 4\dfrac{du}{dt} + 16u = 0$, $u(0) = 1$, $\dfrac{du}{dt}(0) = 0$

13. $\dfrac{d^2u}{dt^2} + 8\dfrac{du}{dt} + 16u = 0$, $u(0) = -1$, $\dfrac{du}{dt}(0) = 0$

14. $\dfrac{d^2u}{dt^2} + \dfrac{du}{dt} + \tfrac{1}{2}u = 0$, $u(0) = 1$, $\dfrac{du}{dt}(0) = -1$

15. $\dfrac{d^2u}{dt^2} + 5\dfrac{du}{dt} + 4u = 0, \ u(0) = 0, \ \dfrac{du}{dt}(0) = 0$

16. $\dfrac{d^2u}{dt^2} + 7\dfrac{du}{dt} + 12u = 0, \ u(0) = 1, \ \dfrac{du}{dt}(0) = -1$

In Exercises 17–21, find (a) a particular solution, (b) the general solution, and (c) the solution of the initial value problem.

17. $\dfrac{du}{dt} + u = e^{-t}, \ u(0) = 0$

18. $\dfrac{d^2u}{dt^2} - u = \sin t, \ u(0) = 0, \ \dfrac{du}{dt}(0) = 0$

19. $\dfrac{d^2u}{dt^2} + u = e^{-t}, \ u(0) = 0, \ \dfrac{du}{dt}(0) = 1$

20. $\dfrac{d^2u}{dt^2} + 8\dfrac{du}{dt} + 16u = 16, \ u(0) = 0, \ \dfrac{du}{dt}(0) = 0$

21. $\dfrac{d^2u}{dt^2} + 4\dfrac{du}{dt} + 16u = 16, \ u(0) = 0, \ \dfrac{du}{dt}(0) = 0$

In Exercises 22–27, find the general solution of the differential equation.

22. $\dfrac{d^2u}{dt^2} + 4\dfrac{du}{dt} + 8u = 0$

23. $\dfrac{d^2u}{dt^2} + u = \cos t$

24. $\dfrac{d^2u}{dt^2} + \dfrac{du}{dt} = 1$

25. $\dfrac{d^2u}{dt^2} + 2\dfrac{du}{dt} + u = \cos t$

26. $\dfrac{d^2u}{dt^2} + 2\dfrac{du}{dt} + u = \sin t$

27. $\dfrac{d^2u}{dt^2} + u = t$

28. The position (positive up) of a parachutist is governed by the initial value problem

$$M\dfrac{d^2u}{dt^2} = -p\dfrac{du}{dt} - Mg, \quad 0 < t,$$

$$u(0) = u_0, \quad \dfrac{du}{dt}(0) = 0,$$

where M is the parachutist's mass, g ($\cong 10$ m/sec^2) is the acceleration of gravity, and p is a constant characteristic of the parachute. Solve the problem.

29. Show that there is a limiting speed for a parachutist, and find it in terms of the parameters in Exercise 28.

30. A mass–spring–damper system has unknown parameters. The following observations are made. (i) If subjected to a constant force of one pound, the mass oscillates and then settles down to a new equilibrium position one inch from the original equilibrium position. (ii) If subjected to a sinusoidal force of 10 cycles per second, the steady state (persistent) velocity is in phase with the force. (iii) Free oscillations die away like a damped sinusoid with a frequency of 5 cycles per second. Find the spring and damper constants and the mass.

3 Theory of Linear Equations

3.1

Theory of Linear Equations

In Chapter 2 we acquired a good deal of experience with second-order equations—finding the general solution of homogeneous and nonhomogeneous equations and satisfying initial conditions. We also got to know a number of properties of solutions; for instance, that the general solution of a homogeneous equation is a linear combination of independent solutions.

In this chapter our goal is to consolidate our experience and extend our knowledge to more general problems, particularly higher-order equations and equations with variable coefficients. In this section we are concerned with the theory of the nth-order linear equation

$$\frac{d^n u}{dt^n} + a_1(t)\frac{d^{n-1}u}{dt^{n-1}} + \cdots + a_{n-1}(t)\frac{du}{dt} + a_n(t)u = f(t). \tag{3.1}$$

If the highest derivative has a coefficient $a_0(t)$ not identically 1, divide through by it to get the normal form in Eq. (3.1).

There are two equivalent ways to define *linear*. One is to say that Eq. (3.1) is linear provided that the coefficients $a_1(t), a_2(t), \ldots, a_n(t)$ do not depend on the unknown function u or any of its derivatives. A more modern and useful approach is this. First, abbreviate the left-hand side of Eq. (3.1) as $L(u)$:

$$L(u) = \frac{d^n u}{dt^n} + a_1(t)\frac{d^{n-1}u}{dt^{n-1}} + \cdots + a_{n-1}(t)\frac{du}{dt} + a_n(t)u. \tag{3.2}$$

Now L is called an operator, a kind of superfunction whose argument or input, u, is itself a function and whose value or output, $L(u)$, is another function. (Of course, the input to L must be a function with at least n derivatives.)

An operator is said to be a *linear operator* if it has these two

properties:
1. $L(cu) = cL(u)$, for any constant c;
2. $L(u + v) = L(u) + L(v)$.

These are familiar and fundamental properties of the operator that we call differentiation,

$$D(u) = \frac{du}{dt},$$

which is therefore linear.

A differential equation $L(u) = f$ is called *linear* if the operator L is linear. It is easy to show that the operator L defined by Eq. (3.2) is linear if and only if the coefficients $a_1(t), a_2(t), \ldots, a_n(t)$ do not depend on u. Thus the old and new definitions are equivalent.

The most significant feature of a linear operator is the way in which it acts on a linear combination of functions. It is easy to show that

$$L(c_1 u_1 + c_2 u_2 + \cdots + c_k u_k) = c_1 L(u_1) + c_2 L(u_2) + \cdots + c_k L(u_k)$$
$$\tag{3.3}$$

if L is any linear operator (and $c_1, c_2, \ldots, c_k$ are constants, of course). This fact is simply an extension and combination of the defining properties 1 and 2 above.

We are interested in solutions of the differential equation (3.1), or $L(u) = f$. Since L can be applied only to functions with at least n derivatives, the definition of solution is more or less forced upon us.

Definition 3.1

A *solution* of the differential equation (3.1) is a function u, for which the derivatives

$$\frac{du}{dt}, \frac{d^2 u}{dt^2}, \ldots, \frac{d^n u}{dt^n}$$

all exist (and therefore u and all its derivatives up to $u^{(n-1)}$ are continuous) and which, when substituted in the differential equation (3.1) yields an identity. A *general solution* is a solution containing n constants such that any initial condition can be satisfied by appropriate choice of those constants.

We have seen that many differential equations appear as part of an initial value problem. In the case of first- and second-order equations with constant coefficients, we have been able to exhibit a solution in terms of elementary functions. In more general cases, however, this will not be true. For this reason we state the following existence and uniqueness theorem.

Theorem 3.1   If each of the coefficients $a_i(t)$, $i = 1, 2, \ldots, n$, and $f(t)$ are continuous on some interval $\alpha < t < \beta$ containing t_0, then there is one and only one function $u(t)$ defined on that interval that is a solution of the initial value problem

$$\frac{d^n u}{dt^n} + a_1(t)\frac{d^{n-1}u}{dt^{n-1}} + \cdots + a_{n-1}(t)\frac{du}{dt} + a_n(t)u = f(t), \quad \alpha < t < \beta,$$

(3.4)

$$u(t_0) = q_0, \quad \frac{du}{dt}(t_0) = q_1, \ldots, \quad \frac{d^{n-1}u}{dt^{n-1}}(t_0) = q_{n-1}.$$ (3.5)

The theorem tells us that there is one solution of the differential equation that also meets the initial condition. The conditions on the coefficients are much more severe than is necessary, as we will see in Exercises 4 and 5 at the end of this section.

Now let us consider the homogeneous equation

$$\frac{d^n u}{dt^n} + a_1(t)\frac{d^{n-1}u}{dt^{n-1}} + \cdots + a_n(t)u = 0,$$

which we know to be of special importance. Since the equation could be rephrased as $L(u) = 0$, we immediately obtain a key property of solutions.

Theorem 3.2   If $u_1(t), u_2(t), \ldots, u_k(t)$ are solutions of the homogeneous linear equation

$$\frac{d^n u}{dt^n} + a_1(t)\frac{d^{n-1}u}{dt^{n-1}} + \cdots + a_n(t)u = 0,$$

then any linear combination of them,

$$u(t) = c_1 u_1(t) + c_2 u_2(t) + \cdots + c_k u_k(t),$$

is also a solution.

Because the homogeneous linear equation is $L(u) = 0$, the proof of this theorem is just as direct application of Eq. (3.3).

As an example, we might take the operator

$$L(u) = \frac{d^3 u}{dt^3}.$$

Then we see that $1, t, t^2, 1 + t, (t - 1)^2$, and, in fact, any polynomial of degree 2 or less, is a solution of $L(u) = 0$. From this infinite collection of solutions, however, it is necessary to single out only $1, t,$ and t^2. Every solution—that is, any polynomial of degree 2 or less—is a linear conbination of these three.

Similarly, for the more general equation

$$\frac{d^n u}{dt^n} + a_1(t)\frac{d^{n-1}u}{dt^{n-1}} + \cdots + a_n(t)u = 0$$

it is always possible to select n solutions that are "different enough" to allow expressing any other solution as a linear combination of them. We clarify this statement with a definition and a theorem.

Definition 3.2

The n functions $u_1(t), \ldots, u_n(t)$ are *linearly independent* on the interval $\alpha < t < \beta$ if the only linear combination of them that is identically 0 is the combination with all coefficients equal to 0. They are linearly dependent otherwise.

Example 1

Show that the three functions $1, t,$ and t^2 are linearly independent, $-\infty < t < \infty$.

To prove this, we must show that the equality

$$c_1 + c_2 t + c_3 t^2 \equiv 0 \tag{3.6}$$

implies that $c_1 = 0, c_2 = 0,$ and $c_3 = 0$. Let us designate $p(t) = c_1 + c_2 t + c_3 t^2$. If $p(t) \equiv 0$, then $p(0) = c_1 = 0$ also. Now, $p(t) \equiv 0$ means that the derivative $p'(t) = c_2 + 2c_3 t \equiv 0$ also. In particular, $p'(0) = c_2 = 0$. Finally, we must have $p''(t) = 2c_2 \equiv 0$ also, so $c_3 = 0$. We have shown that Eq. (3.6) can be true only if $c_1 = 0, c_2 = 0,$ and $c_3 = 0$; that is, $1, t,$ and t^2 are linearly independent.

On the other hand, the three functions

$$t^2 - 1, \qquad t^2 - 2t + 1, \qquad t - 1$$

are not independent because some linear combination of them is identically zero. For instance, we have

$$(-1) \cdot (t^2 - 1) + 1 \cdot (t^2 - 2t + 1) + 2 \cdot (t - 1) \equiv 0,$$

as one can easily check.

It is worth noting that the inclusion of the constant function 0 in a set of functions automatically makes them dependent. Also note that the definition above coincides with the definition in Section 2.2. Testing by means of ratios only applies to two functions, however.

The definition of linear independence is not very convenient to work with. There is another way of detecting whether n given functions $u_1, u_2, \ldots, u_n$ are dependent. The question is whether or not it is possible to find coefficients $c_1, c_2, \ldots, c_n$—not all 0—such that

$$c_1 u_1(t) + c_2 u_2(t) + \cdots + c_n u_n(t) = 0, \qquad \alpha < t < \beta, \tag{3.7}$$

holds identically, But—assuming all the u's to be differentiable as many times as necessary—if Eq. (3.7) is true, it can be differentiated again and again to get n equations:

$$
\begin{aligned}
c_1 u_1(t) + \quad c_2 u_2(t) \quad + \cdots + c_n u_n(t) &= 0, \\
c_1 u_1'(t) + \quad c_2 u_2'(t) \quad + \cdots + c_n u_n'(t) &= 0, \\
&\vdots \\
c_1 u^{(n-1)}(t) + c_2 u^{(n-1)}(t) + \cdots + c_n u_n^{(n-1)}(t) &= 0.
\end{aligned}
\tag{3.8}
$$

Now this is a system of n homogeneous equations in the n "unknowns" $c_1, c_2, \ldots, c_n$. There can be a nonzero solution only if the determinant

$$
W(u_1, u_2, \ldots, u_n) =
\begin{vmatrix}
u_1 & u_2 & \cdots & u_n \\
u_1' & u_2' & \cdots & u_n' \\
\vdots & & & \\
u_1^{(n-1)} & u_2^{(n-1)} & \cdots & u_n^{(n-1)}
\end{vmatrix}
\tag{3.9}
$$

is identically 0, $\alpha < t < \beta$. This is called the Wronskian determinant, or just the *Wronskian*, of the functions $u_1, u_2, \ldots, u_n$.

Example 2

In Example 3.1 we showed that the three functions $t^2 - 1$, $t^2 - 2t + 1$, and $t - 1$ are linearly dependent. Thus their Wronskian should be identically 0, as indeed it is:

$$
\begin{vmatrix}
t^2 - 1 & t^2 - 2t + 1 & t - 1 \\
2t & 2t - 2 & 1 \\
2 & 2 & 0
\end{vmatrix}
\equiv 0.
$$

We showed above that dependence of some functions implies that their Wronskian is 0. If we restrict our functions to be solutions of a suitable linear differential equation, however, more is true.

Theorem 3.3

Let $a_1(t)$, $a_2(t)$, ..., $a_n(t)$ be continuous functions, $\alpha < t < \beta$ and let $u_1(t)$, $u_2(t)$, ..., $u_n(t)$ be n solutions of the nth-order homogeneous linear equation

$$\frac{d^n u}{dt^n} + a_1(t)\frac{d^{n-1}u}{dt^{n-1}} + \cdots + a_{n-1}(t)\frac{du}{dt} + a_n(t)u = 0, \qquad \alpha < t < \beta.$$

(3.10)

1. Then $u_1, u_2, \ldots, u_n$ are linearly independent if and only if their Wronskian is never 0 for any t, $\alpha < t < \beta$.

2. If $u_1, u_2, \ldots, u_n$ are linearly independent solutions of the homogeneous equation $L(u) = 0$, then the general solution is

$$c_1 u_1(t) + c_2 u_2(t) + \cdots + c_n u_n(t).$$

Proof. The first conclusion is stronger than what we showed above: we only proved the "only if" part there. We shall not prove the "if" part, but go on to the second conclusion. We wish to show that the linear combination is the general solution, that is, that the coefficients can be chosen to satisfy any arbitrarily specified initial conditions at a point t_0, $\alpha < t_0 < \beta$.

If the conditions are

$$u(t_0) = q_0, \qquad u'(t_0) = q_1, \ldots, u^{(n-1)}(t_0) = q_{n-1},$$

then these become n conditions on the c's, namely,

$$\begin{aligned}
c_1 u_1(t_0) + \quad c_2 u_2(t_0) \quad + \cdots + c_n u_n(t_0) &= q_0, \\
c_1 u_1'(t_0) + \quad c_2 u_2'(t_0) \quad + \cdots + c_n u_n'(t_0) &= q_1, \\
\vdots \\
c_1 u_1^{(n-1)}(t_0) + c_2 u_2^{(n-1)}(t_0) + \cdots + c_n u_n^{(n-1)}(t_0) &= q_n.
\end{aligned}$$

(3.11)

Now the determinant of this system of n algebraic equations in n unknowns is precisely the Wronskian of $u_1, u_2, \ldots, u_n$, evaluated at t_0. Since the solutions are supposed to be independent, the Wronskian is not 0, and the system (3.11) has a solution.

A set of n independent solutions of an nth-order homogeneous equation is called a *fundamental set* for that equation. Theorem 3.3 tells us that we construct a *general solution* by forming a linear combination of the solutions in a fundamental set. But does an nth-order linear homogeneous equation, such as Eq. (3.10), always have a fundamental set of solutions? The answer is yes, if the coefficients are continuous.

To "construct" a fundamental set, require each solution to satisfy some set of initial conditions at t_0 (that is, each solution satisfies an initial value problem). By Theorem 3.1, each of these solutions exists. The n sets of initial conditions are chosen to give the Wronskian a nonzero value at t_0. That the Wronskian stays away from 0 is a consequence of this theorem.

Theorem 3.4

If $u_1(t), u_2(t), \ldots, u_n(t)$ are solutions of the nth-order homogeneous equation (3.10) for $\alpha < t < \beta$ and if $W(t)$ is their Wronskian, then W itself satisfies the differential equation

$$\frac{dW}{dt} + a_1(t)W = 0, \qquad \alpha < t < \beta.$$

Now we see that the Wronskian in the theorem can be written down as

$$W(t) = W(t_0) \exp\left(-\int_{t_0}^{t} a_1(t')\, dt' \right).$$

We further see that the Wronskian of a set of solutions of an nth-order homogeneous equation (3.10) is *either identically* 0 *or never* 0, as long as the coefficients are continuous.

Example 3

The third-order equation

$$\frac{d^3 u}{dt^3} + 3\frac{d^2 u}{dt^2} + 2\frac{du}{dt} = 0$$

has among its solutions the functions 1, e^{-t}, e^{-2t}. Their Wronskian is

$$\begin{vmatrix} 1 & e^{-t} & e^{-2t} \\ 0 & -e^{-t} & -2e^{-2t} \\ 0 & e^{-t} & 4e^{-2t} \end{vmatrix} = -2e^{-3t},$$

which is indeed a solution of the first-order equation

$$\frac{dW}{dt} + 3W = 0.$$

Since the Wronskian is never 0, the three functions form a fundamental set.

Example 4

The functions t and t^2 are solutions of the second-order differential equation

$$t^2 \frac{d^2u}{dt^2} - 2t \frac{du}{dt} + 2u = 0.$$

Their Wronskian is

$$W(t, t^2) = \begin{vmatrix} t & t^2 \\ 1 & 2t \end{vmatrix} = t^2,$$

which is nonzero for $t > 0$ or $t < 0$.

Does the fact that the Wronskian is 0 at $t = 0$ conflict with the statement above in the text? No, because the equation, in normal form, is

$$\frac{d^2u}{dt^2} - \frac{2}{t}\frac{du}{dt} + \frac{2}{t^2}u = 0.$$

It is clear that the coefficients are discontinuous at $t = 0$.

The nonhomogeneous equation (3.1), or $L(u) = f$, also shares many properties with the second-order equations that we have studied. For one thing, the general solution is constructed in the same way.

Theorem 3.5

The general solution of the nonhomogeneous differential equation

$$\frac{d^nu}{dt^n} + a_1(t)\frac{d^{n-1}u}{dt^{n-1}} + \cdots + a_{n-1}(t)\frac{du}{dt} + a_n(t)u = f(t)$$

has the form

$$u(t) = u_p(t) + u_c(t),$$

where $u_p(t)$ is any solution of the nonhomogeneous equation and $u_c(t)$ is the general solution of the corresponding homogeneous equation.

The proof of this theorem follows the same lines as the proof of the analogous theorem in Section 2.5.

Another property that was helpful in solving second-order nonhomogeneous equations is true in the more general case. If we have particular solutions for each of several equations,

$$L(u_{p1}) = f_1, \qquad L(u_{p2}) = f_2, \qquad \ldots, \qquad L(u_{pk}) = f_k, \tag{3.12}$$

then we can combine them to get

$$L(c_1 u_{p1} + c_2 u_{p2} + \cdots + c_k u_{pk}) = c_1 f_1 + c_2 f_2 + \cdots + c_k f_k, \tag{3.13}$$

for any constants $c_1, c_2, \ldots, c_k$.

Example 5

Take the linear operator

$$L(u) = \frac{d^3 u}{dt^3} + \frac{du}{dt}.$$

These facts can easily be checked:
1. three independent solutions of $L(u) = 0$ are 1, $\cos t$, and $\sin t$;
2. $u_{p1}(t) = t$ is a solution of $L(u) = 1$;
3. $u_{p2}(t) = -\frac{1}{2} e^{-t}$ is a solution of $L(u) = e^{-t}$.

From these observations, and using various theorems above, we conclude that the general solution of

$$\frac{d^3 u}{dt^3} + \frac{du}{dt} = 1 - e^{-t}$$

is $u(t) = c_1 + c_2 \cos t + c_3 \sin t + t + \frac{1}{2} e^{-t}$.

Exercises

1. Let L_1 and L_2 be two linear operators—that is, operators that satisfy properties 1 and 2 at the beginning of this section. Show that the operator

$$L = a_1(t) L_1 + a_2(t) L_2$$

also satisfies properties 1 and 2.
2. Use the results of Exercise 1 to show that the operator in Eq. (3.2) satisfies the linearity properties 1 and 2.
3. Show that these are linear operators (i.e., they satisfy properties 1 and 2). In all cases, a is a fixed number and v is a given, fixed function.
   (a) $L(u) = u(a)v(t)$
   (b) $L(u) = \displaystyle\int_a^t u(x)v(x)\, dx$
   (c) $L(u) = \displaystyle\int_a^t u(x)v(t-x)\, dx$

4. Let $h(t)$ be the discontinuous function defined by

$$h(t) = \begin{cases} 0, & t < 0 \\ 1, & 0 \le t \end{cases}$$

Verify that the function u below is the general solution of the differential equation

$$\frac{du}{dt} + h(t)u = 0.$$

$$u(t) = \begin{cases} c, & t < 0 \\ ce^{-t}, & 0 \le t \end{cases}$$

(Check that u is continuous.) Sketch the function corresponding to $c = 1$. Is its derivative continuous?

5. Find the general solution of the differential equation

$$\frac{d^2u}{dt^2} + h(t)u = 0,$$

where $h(t)$ is as in Exercise 4. (Hint: solve first for $t < 0$, then for $t \ge 0$. Define the constants to make your solutions and their derivatives continuous at $t = 0$.)

6. Try to express the polynomial $t + 1$ as a linear combination of the three polynomials

$$t^2 - 1, \qquad t^2 - 2t + 1, \qquad t - 1.$$

(Write a linear combination, with unknown coefficients, equal to $t + 1$. Equate coefficients of like terms.) Where and why does the process break down?

7. The three functions in Exercise 6 all are equal to 0 at $t = 1$. What has this to do with their linear dependence?

8. Verify that the functions

$$u_1 = t, \qquad u_2 = t \ln t$$

are linearly independent solutions of the differential equation

$$t^2 \frac{d^2u}{dt^2} - t\frac{du}{dt} + u = 0 \qquad (t > 0).$$

9. Verify that the two functions

$$u_1 = t \cos(t^{-1}), \qquad u_2 = t \sin(t^{-1})$$

are independent for $t > 0$. They are solutions of

$$\frac{d^2u}{dt^2} + t^{-4}u = 0.$$

10. Verify that the two functions

$$u_1 = e^{-at}, \qquad u_2 = te^{-at}$$

are solutions of the differential equation

$$\frac{d^2u}{dt^2} + 2a\frac{du}{dt} + a^2u = 0$$

and that they are independent, $-\infty < t < \infty$.

11. Prove that any two exponentials

$$u_1 = e^{m_1t}, \qquad u_2 = e^{m_2t}$$

are independent, $-\infty < t < \infty$, if the constants m_1 and m_2 are different. (Find a differential equation of which both are solutions, and then use Theorem 3.3.)

12. Extend the results of Exercise 11 to any n exponentials

$$u_1 = e^{m_1t}, \qquad u_2 = e^{m_2t}, \qquad \ldots, \qquad u_n = e^{m_nt}.$$

13. Suppose that u_1 and u_2 are two independent functions with continuous first and second derivatives. The determinantal relation

$$\begin{vmatrix} u_1 & u_2 & u \\ u_1' & u_2' & u' \\ u_1'' & u_2'' & u'' \end{vmatrix} = 0$$

may be read as a linear second-order differential equation for u.
(a) Find the equation explicitly. (The coefficients are expressions in u_1, u_2 and their derivatives.)
(b) Explain why the general solution is

$$u(t) = c_1u_1(t) + c_2u_2(t).$$

14. Use the results of Exercise 13 to find a differential equation having $u_1 = t$ and $u_2 = t^2$ as solutions.

15. Follow the directions for Exercise 14, using $u_1 = t$, $u_2 = e^t$.

16. (a) Show that $u_1(t) = \cos t$ and $u_2(t) = e^{\cos t}$ are independent functions, without using the Wronskian.
(b) Show that their Wronskian is 0 at $t = 0$, $\pm\pi$, $\pm 2\pi, \ldots$.
(c) If u_1 and u_2 are solutions of a differential equation of the form $u'' + a_1(t)u' + a_2(t)u = 0$, what can you say about $a_1(t)$?

17. In the text it is shown that if n functions are dependent (and have $n - 1$ derivatives), $\alpha < t < \beta$, then their Wronskian is identically 0. Why is it that the converse is not always true? That is, why could two (or more) functions have a zero Wronskian and yet not be dependent? Can you supply an example?

3.2

The Solution of Equations with Constant Coefficients

In Section 2.2 we developed a way to solve second-order homogeneous equations with constant coefficients. Basically, the same method may be applied to higher-order equations. As a general form for such equations we take

$$\frac{d^n u}{dt^n} + b_1 \frac{d^{n-1} u}{dt^{n-1}} + \cdots + b_{n-1} \frac{du}{dt} + b_n u = 0, \tag{3.14}$$

where $b_1, b_2, \ldots, b_n$ are all constants. Again it will be convenient to abbreviate the left-hand side of Eq. (3.14) as $L(u)$.

The method of Section 2.2, and our method here, is to assume that the solution is of the form $u = e^{mt}$ and then determine the multiplier m so that this is true. Now the effect of the linear operator L on the exponential function is very simple:

$$L(e^{mt}) = (m^n + b_1 m^{n-1} + \cdots + b_{n-1} m + b_n) e^{mt}. \tag{3.15}$$

The polynomial in m is called the characteristic polynomial of L:

$$p(m) = m^n + b_1 m^{n-1} + \cdots + b_{n-1} m + b_n. \tag{3.16}$$

And, clearly, e^{mt} is a solution of $L(u) = 0$ if and only if $p(m) = 0$. Thus we can find an exponential solution of $L(u) = 0$ for each distinct root of the polynomial equation $p(m) = 0$.*

As in the case of the second-order equation, complex roots lead to complex exponentials, which can be replaced by an appropriate combination of sines and cosines:

$$e^{(\alpha+i\beta)t}, \; e^{(\alpha-i\beta)t} \leftrightarrow e^{\alpha t} \cos \beta t, \; e^{\alpha t} \sin \beta t.$$

Example 1

Find the general solution of

$$\frac{d^4 u}{dt^4} - u = 0. \tag{3.17}$$

The characteristic polynomial of this differential equation can be factored as

$$m^4 - 1 = (m^2 - 1)(m^2 + 1) = (m - 1)(m + 1)(m^2 + 1)$$

---

* See the Appendix for information on polynomials.

with roots, ± 1, $\pm i$. The four corresponding solutions are

$$u_1 = e^t, \qquad u_2 = e^{-t}, \qquad u_3 = \cos t, \qquad u_4 = \sin t.$$

The Wronskian test readily shows that these solutions are independent. Thus the general solution of Eq. (3.17) is

$$u(t) = c_1 e^t + c_2 e^{-t} + c_3 \cos t + c_4 \sin t.$$

Of course, we are interested in finding n independent solutions of Eq. (3.14) in order to form the general solution. The exponential method gives us one solution for each *different* root of the characteristic polynomial. Thus if $p(m) = 0$ has multiple roots, there will not be a sufficient number of independent exponential solutions. However, multiplying the exponential solution by powers of t will provide the right number of independent solutions. These observations are summarized in Theorem 3.6.

Theorem 3.6

The nth-order homogeneous equation with constant coefficients

$$\frac{d^n u}{dt^n} + b_1 \frac{d^{n-1} u}{dt^{n-1}} + \cdots + b_{n-1} \frac{du}{dt} + b_n u = 0$$

with characteristic polynomial

$$p(m) = m^n + b_1 m^{n-1} + \cdots + b_{n-1} m + b_n$$

has n independent solutions, of these forms:

1. e^{mt} if m is a real, simple root of $p(m) = 0$;
2. $e^{mt}, te^{mt}, \ldots, t^{k-1} e^{mt}$ if m is a real root of multiplicity k;
3. $e^{\alpha t} \cos \beta t$, $e^{\alpha t} \sin \beta t$ if $\alpha + i\beta$ and $\alpha - i\beta$ are a pair of simple, complex conjugate roots;
4. $e^{\alpha t} \cos \beta t$, $te^{\alpha t} \cos \beta t$, $\ldots$, $t^{k-1} e^{\alpha t} \cos \beta t$, $e^{\alpha t} \sin \beta t$, $te^{\alpha t} \sin \beta t$, $\ldots$, $t^{k-1} e^{\alpha t} \sin \beta t$ if $\alpha + i\beta$ and $\alpha - i\beta$ are conjugate complex roots of multiplicity k.

Example 2

The differential equation

$$\frac{d^4 u}{dt^4} + 2 \frac{d^2 u}{dt^2} + u = 0$$

has characteristic polynomial

$$m^4 + 2m^2 + 1 = (m^2 + 1)^2,$$

whose roots are i, i, $-i$, $-i$. Thus the four independent solutions are

$\cos t$, $t \cos t$, $\sin t$, $t \sin t$.

Example 3

Find the general solution of the sixth-order differential equation

$$\frac{d^6 u}{dt^6} - 3\frac{d^4 u}{dt^4} + \frac{d^2 u}{dt^2} = 0.$$

The characteristic polynomial of the differential equation is $m^6 - 3m^4 + m^2$. This is a bicubic polynomial that can be factored as

$$m^6 - 3m^4 + m^2 = (m^2 - 2)(m^2 - 1)m^2.$$

Thus there are four simple roots, ± 1, $\pm\sqrt{2}$, and the double root 0. The general solution is

$$u(t) = c_1 e^{\sqrt{2}t} + c_2 e^{-\sqrt{2}t} + c_3 e^t + c_4 e^{-t} + c_5 + c_6 t.$$

Thus we see that, in principle, solving a high-order homogeneous equation is no more difficult than solving one of second order. There are two practical complications however: (1) finding the roots of a high-degree polynomial equation is, in itself, a difficult numerical problem; (2) the higher order, the greater the possibility of a multiple root of the characteristic polynomial.

Now we turn our attention to the nth-order nonhomogeneous equation

$$\frac{d^n u}{dt^n} + b_1 \frac{d^{n-1} u}{dt^{n-1}} + \cdots + b_{n-1}\frac{du}{dt} + b_n u = f(t), \tag{3.18}$$

in which $b_1, b_2, \ldots, b_n$ are constants. Theorem 3.5 of Section 3.1 states that the general solution of Eq. (3.18) is $u(t) = u_c(t) + u_p(t)$, where $u_c(t)$ is the general solution of the corresponding homogeneous equation and $u_p(t)$ is any particular solution of the nonhomogeneous equation (3.15). Since we know already how to find $u_c(t)$, we focus on finding a particular solution—if possible, by the method of undetermined coefficients.

Again we refer to the theory developed in Section 3.1. Equations (3.12) and (3.13) allows us to break down the inhomogeneity $f(t)$ as a linear combination of simpler functions and treat these individually. By simple function we mean: (1) a polynomial; (2) a linear combination of $\sin \omega t$ and $\cos \omega t$ (same frequency for both functions); (3) an exponential $e^{\alpha t}$; (4) a product of functions of the preceding types.

On the basis of what we said above about solving homogeneous equations with constant coefficients, we can state succinctly just what a "simple function" is. It is a solution of a homogeneous equation with constant coefficients whose characteristic polynomial has just one root (or one pair of conjugate complex roots) with, possibly, a multiplicity greater than 1. (See Exercises 32–41 at the end of this section.)

The procedure for finding a particular solution of a differential equation with a simple inhomogeneity is the same as before.

1. Find $u_c(t)$, the general solution of the corresponding homogeneous equation.

2. Choose the appropriate form for the particular solution.

3. Test the chosen form to see whether it contains a solution of the homogeneous equation. If so, multiply the chosen form by the lowest power of t that eliminates this situation.

4. Determine the coefficients for the particular solution.

The test mentioned in part 3 can be rephrased in terms of roots of the characteristic polynomial. This is done in Table 3.1, in which inhomogeneities and the corresponding forms for $u_p(t)$ are listed.

Table 3.1

Inhomogeneity, $f(t)$	Characteristic Polynomial	Form of Particular Solution
$p_k(t)$	0 not a root	$P_k(t)$
	0 a root, multiplicity s	$t^s P_k(t)$
$p_k(t)e^{\alpha t}$	α not a root	$P_k(t)e^{\alpha t}$
	α a root, multiplicity s	$t^s P_k(t)e^{\alpha t}$
$p_k(t)\cos \beta t$ $+ q_k(t)\sin \beta t$	$\pm i\beta$ not roots	$P_k(t)\cos \beta t$ $+ Q_k(t)\sin \beta t$
	$\pm i\beta$ roots, multiplicity s	$t^s(P_k(t)\cos \beta t$ $+ Q_k(t)\sin \beta t)$
$e^{\alpha t}(p_k(t)\cos \beta t$ $+ q_k(t)\sin \beta t)$	$\alpha \pm i\beta$ not roots	$e^{\alpha t}(P_k(t)\cos \beta t$ $+ Q_k(t)\sin \beta t)$
	$\alpha \pm i\beta$ roots, multiplicity s	$t^s e^{\alpha t}(P_k(t)\cos \beta t$ $+ Q_k(t)\sin \beta t)$

$p_k(t)$ and $q_k(t)$ are polynomials of degree k at most; $P_k(t)$ and $Q_k(t)$ have the form of polynomials of degree k. Note that a constant is to be interpreted as a polynomial of degree 0.

Example 4

Obtain a particular solution of the differential equation

$$\frac{d^4 u}{dt^4} + 2 \frac{d^2 u}{dt^2} + u = \cos \omega t.$$

The general solution of the homogeneous equation, according to Example 3.7, is

$$u_c(t) = c_1 \cos t + c_2 t \cos t + c_3 \sin t + c_4 t \sin t.$$

Thus as long as $\omega \neq 1$, $\cos \omega t$ is not a solution of the homogeneous equation. In this case we may choose

$$u_p(t) = A \cos \omega t + B \sin \omega t$$

as the correct form for the particular solution. By substituting into the differential equation we find that A and B must satisfy

$$(\omega^4 - 2\omega^2 + 1)(A \cos \omega t + B \sin \omega t) = \cos \omega t.$$

Therefore $B = 0$, $A = 1/(\omega^2 - 1)^2$, and the particular solution is

$$u_p(t) = \frac{1}{(\omega^2 - 1)^2} \cos \omega t \qquad (\omega \neq 1).$$

However, if $\omega = 1$, then $\cos t$ and also $t \cos t$ are solutions of the homogeneous equation, and the correct form for the particular solution is

$$u_p(t) = t^2(A \cos t + B \sin t).$$

Substitution of this into the differential equation gives, after some algebra,

$$-8A \cos t - 8B \sin t = \cos t.$$

Therefore we find the particular solution to be

$$u_p(t) = -\tfrac{1}{8} t^2 \cos t \qquad (\omega = 1).$$

The reader may have some doubts about the usefulness of high-order differential equations. We shall derive one fourth-order equation, which will indicate how equations of quite high order might arise. The physical system considered is composed of two masses and three springs, linked as shown in Fig. 3.1.

The spring forces are shown in Fig. 3.2. The forces actually act in the direction shown—if $u_2 > u_1 > 0$ and the displacements u_1 and u_2 are measured from the equilibrium position. Now the application of Newton's

Figure 3.1

Two Masses and
Three Springs

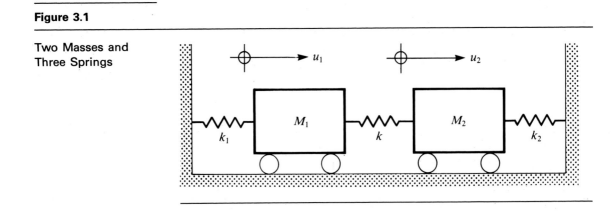

Figure 3.2

Free-Body Diagrams
of Masses in Fig. 3.1

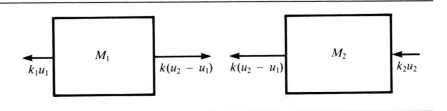

second law leads to these equations:

$$M_1 \frac{d^2u_1}{dt^2} = k(u_2 - u_1) - k_1u_1, \tag{3.19}$$

$$M_2 \frac{d^2u_2}{dt^2} = -k(u_2 - u_1) - k_2u_2. \tag{3.20}$$

Let us designate $(k_1 + k)/M_1 = \omega_1^2$ and $(k_2 + k)/M_2 = \omega_2^2$. (Either of these would be the natural frequency of oscillation of the corresponding mass if the other mass were held fixed.) Now, with these new parameters and some algebra, the equations above become

$$\frac{d^2u_1}{dt^2} + \omega_1^2u_1 = \frac{k}{M_1} u_2, \tag{3.21}$$

$$\frac{d^2u_2}{dt^2} + \omega_2^2u_2 = \frac{k}{M_2} u_1. \tag{3.22}$$

We shall make use again of the operator idea, designating the left-hand sides of Eqs. (3.21) and (3.22) as $L_1(u_1)$ and $L_2(u_2)$. Those two equations are now

$$L_1(u_1) = \frac{k}{M_1} u_2, \qquad L_2(u_2) = \frac{k}{M_2} u_1.$$

Now apply operator L_2 to both sides of the first equation to get

$$L_2(L_1(u_1)) = L_2\left(\frac{k}{M_1} u_2\right) = \frac{k^2}{M_1 M_2} u_1. \tag{3.23}$$

That is to say,

$$\frac{d^2}{dt^2}\left(\frac{d^2 u_1}{dt^2} + \omega_1^2 u_1\right) + \omega_2^2\left(\frac{d^2 u_1}{dt^2} + \omega_1^2 u_1\right) = \lambda^4 u_1, \tag{3.24}$$

where we have put $\lambda^4 = k^2/M_1 M_2$.

Now Eq. (3.24) is a fourth-order equation for u_1, obtained by eliminating u_2 between Eqs. (3.21) and (3.22). When the algebra has been done, Eq. (3.24) is

$$\frac{d^4 u}{dt^4} + (\omega_1^2 + \omega_2^2)\frac{d^2 u}{dt^2} + (\omega_1^2 \omega_2^2 - \lambda^4)u = 0, \tag{3.25}$$

and this equation is easily solved. Incidentally, the subscript has been left off u in Eq. (3.25) because it is valid for both u_1 and u_2.

Exercises

In Exercises 1–6, find (a) the characteristic polynomial and (b) the general solution of the differential equation.

1. $\dfrac{d^4 u}{dt^4} - k^4 u = 0$

2. $\dfrac{d^4 u}{dt^4} - 2\dfrac{d^2 u}{dt^2} + u = 0$

3. $\dfrac{d^4 u}{dt^4} + k^4 u = 0$

4. $\dfrac{d^4 u}{dt^4} + 4k\dfrac{d^3 u}{dt^3} + 6k^2\dfrac{d^2 u}{dt^2} + 4k^3\dfrac{du}{dt} + k^4 u = 0$

5. $\dfrac{d^3 u}{dt^3} - k^3 u = 0$

6. $\dfrac{d^3 u}{dt^3} + k^3 u = 0$

7. In Exercise 5 of Section 2.1 we found this equation for a spring–damper–mass system in series (Fig. 2.7 in Section 2.1)

$$M\frac{d^3 u}{dt^3} + \frac{Mk}{p}\frac{d^2 u}{dt^2} + k\frac{du}{dt} = 0.$$

Find the general solution of this equation.

8. Describe how the nature of the solution in Exercise 7 changes as a function of the damping constant p in the range from 0 to ∞.

9. Equation (3.25) describes the displacement of either of the two masses in Fig. 3.1. Is it possible for the coefficient of u to be negative?

10. Show that all four roots of the characteristic polynomial for Eq. (3.25) are imaginary. Can there be double roots?

11. Find the general solution of Eq. (3.25) in the special case that $k_1 = k_2 = k$ and $M_1 = M_2$. (Use $\lambda^2 = k/M$ as the main parameter.)

12. Consider the third-order equation

$$\frac{d^3u}{dt^3} + b\frac{du}{dt} + cu = 0.$$

Whatever values the constants b and $c \neq 0$ may have, some solution grows in magnitude as $t \to \infty$. Why?

13. What conditions on the constant coefficients b and c in the equation

$$\frac{d^4u}{dt^4} + 2b\frac{d^2u}{dt^2} + cu = 0$$

will guarantee that no solution grows without bound as $t \to \infty$?

14. Does Eq. (3.25) fulfill the requirements you found in Exercise 13?

In Exercises 15–23, find a particular solution of the differential equation.

15. $\dfrac{d^4u}{dt^4} + k^4u = \sin kt$

16. $\dfrac{d^4u}{dt^4} - k^4u = \sin \omega t \quad (\omega \neq k)$

17. $\dfrac{d^4u}{dt^4} - k^4u = \sin kt$

18. $\dfrac{d^4u}{dt^4} + 3\dfrac{d^2u}{dt^2} + 2u = \sin t$

19. $\dfrac{d^4u}{dt^4} - k^4u = e^{-t}$

20. $\dfrac{d^3u}{dt^3} = t$

21. $\dfrac{d^4u}{dt^4} + \dfrac{d^2u}{dt^2} = 1$

22. $\dfrac{d^3u}{dt^3} + u = t$

23. $\dfrac{d^6u}{dt^6} + 2\dfrac{d^3u}{dt^3} + u = 1$

In Exercises 24–30, give the correct *form* to use in finding a particular solution by the method of undetermined coefficients.

24. $\dfrac{d^3u}{dt^3} - u = \sin t + t$

25. $\dfrac{d^4u}{dt^4} + 4\dfrac{d^2u}{dt^2} + 4u = te^{-t}$

26. $\dfrac{d^4u}{dt^4} - u = e^t$

27. $\dfrac{d^3u}{dt^3} - 3\dfrac{d^2u}{dt^2} + 3\dfrac{du}{dt} - u = te^t$

28. $\dfrac{d^4 u}{dt^4} + u = t^2 + \cos t$ 29. $\dfrac{d^5 u}{dt^5} - \dfrac{d^3 u}{dt^3} = t + e^t$

30. $\dfrac{d^4 u}{dt^4} + 7 \dfrac{d^2 u}{dt^2} + 12u = e^{-2t} + \cos t$

31. Find the general solution of the nth-order differential equation

$$\dfrac{d^n u}{dt^n} = u.$$

In the text a simple inhomogeneity is identified as "a solution of a homogeneous equation whose characteristic polynomial has just one root (or one pair of conjugate complex roots) with, possibly, a multiplicity greater than 1. In Exercises 32–41 a simple inhomogeneity is given. (a) Find the homogeneous equation of lowest order with constant coefficients of which it is a solution. (b) Find the corresponding characteristic polynomial.

Example: $f(t) = e^{2t}$:
  (a) $u' - 2u = 0$; (b) $m - 2$.

32. $f(t) = t$
33. $f(t) = p_k(t)$, a polynomial of degree k
34. $f(t) = 3 \cos t + 2 \sin t$
35. $f(t) = e^{\alpha t}(a \cos \beta t + b \sin \beta t)$
36. $f(t) = e^{\alpha t}$
37. $f(t) = p_k(t) e^{\alpha t}$
38. $f(t) = t \cos t$
39. $f(t) = p_k(t) \cos \beta t + q_k(t) \sin \beta t$
40. $f(t) = t e^t \sin 2t$
41. $f(t) = e^{\alpha t}(p_k(t) \cos \beta t + q_k(t) \sin \beta t)$

3.3
The Euler-Cauchy Equation

Section 3.1 told us a great deal about how solutions of a linear equation should be used but nothing about how to find them. The unhappy fact is that very few significant equations with variable coefficients (except in the first-order case) can be solved in elementary terms. One of the important exceptions to this rule is the Euler-Cauchy (or equidimensional) equation,

$$t^n \dfrac{d^n u}{dt^n} + b_1 t^{n-1} \dfrac{d^{n-1} u}{dt^{n-1}} + \cdots + b_{n-1} t \dfrac{du}{dt} + b_n u = 0. \qquad (3.26)$$

(Here $b_1, b_2, \ldots, b_n$ are constants.) In this form the distinguishing feature of this type of equation is that the coefficient of the kth derivative is the kth power of t times a constant. Equation (3.26), however, is not in the standard form we have been using—the one in which the highest-order derivative has coefficient 1. In our standard form the equation becomes

$$\frac{d^n u}{dt^n} + \frac{b_1}{t}\frac{d^{n-1}u}{dt^{n-1}} + \cdots + \frac{b_{n-1}}{t^{n-1}}\frac{du}{dt} + \frac{b_n}{t^n}u = 0. \tag{3.27}$$

Of course, the coefficients become infinite at $t = 0$, so we must not admit this value. Usually, $t > 0$ is the accepted range for the independent variable, although $t < 0$ would be correct also.

The first-order version of this equation is

$$\frac{du}{dt} + \frac{b}{t}u = 0, \tag{3.28}$$

the solution of which is easily found to be

$$u(t) = ce^{-b\ln t} = ct^{-b}. \tag{3.29}$$

In fact, once we know the solution, we can see that the differential equation (3.28) is just the rule for differentiating $u = t^{-b}$:

$$\frac{du}{dt} = -b\frac{u}{t}.$$

As in the case of constant coefficients, we can use our knowledge of the solution of the first-order equation to obtain the solution of higher-order equations. The usual method of solution is to assume that the solution has the form of a power of t,

$$u = t^m,$$

and then find the correct value(s) for the exponent m.

Example 1

We solve the equation

$$\frac{d^2 u}{dt^2} + \frac{3}{t}\frac{du}{dt} - \frac{3}{t^2}u = 0$$

by assuming that the solution has the form $u = t^m$. Then the derivatives needed are

$$\frac{du}{dt} = mt^{m-1}, \qquad \frac{d^2 u}{dt^2} = m(m-1)t^{m-2}.$$

Substituting these expressions into the differential equation leads to the equation

$$m(m-1)t^{m-2} + \frac{3}{t}mt^{m-1} - \frac{3}{t^2}t^m = 0,$$

from which we eliminate the factor t^{m-2} that is common to all three terms. The result is

$$m(m - 1) + 3m - 3 = m^2 + 2m - 3 = 0,$$

a quadratic in m that is easily solved to find

$$m = -3, \qquad m = 1.$$

Thus the two functions t and t^{-3} are both solutions of the given equation, and the general solution is

$$u(t) = c_1 t^{-3} + c_2 t.$$

The generalization of the process in Example 1 is simple. In order to solve the nth-order equation (3.27) we would assume that a solution has the form $u = t^m$. The derivatives needed for the differential equation are

$$\frac{du}{dt} = m t^{m-1}, \qquad \frac{d^2 u}{dt^2} = m(m - 1) t^{m-2}, \qquad \cdots,$$

$$\frac{d^n u}{dt^n} = m(m - 1) \cdots (m - n + 1) t^{m-n}.$$

By substituting u and its derivatives into the differential equation (3.27) we are led first to an equation in which each term has the common factor t^{m-n}, and then to a polynomial equation of degree n:

$$q(m) = m(m - 1) \cdots (m - n + 1) + b_1 m(m - 1) \cdots (m - n + 2)$$
$$+ \cdots + b_{n-1} m + b_n = 0. \tag{3.30}$$

The polynomial $q(m)$ is called the characteristic polynomial of the equation (3.27). (Note how it differs from the characteristic polynomial of a differential equation with constant coefficients.) Each of the distinct roots of the characteristic polynomial gives an independent solution of the differential equation (3.27). Two matters remain to be cleared up: how to deal with complex roots of the characteristic polynomial and how to obtain a sufficient number of independent solutions in case $q(m)$ has multiple roots.

The first of these is not really difficult to understand. If we find a pair of complex conjugate roots of the characteristic equation, say, $\alpha + i\beta$ and $\alpha - i\beta$, then two independent solutions are

$$t^{\alpha + i\beta} = t^\alpha t^{i\beta}, \qquad t^{\alpha - i\beta} = t^\alpha t^{-i\beta}. \tag{3.31}$$

The imaginary exponents can be interpreted by the laws of exponents:

$$t^{i\beta} = (e^{\ln t})^{i\beta} = e^{i\beta \ln t} = \cos(\beta \ln t) + i \sin(\beta \ln t). \tag{3.32}$$

Thus we may replace the two complex solutions in Eq. (3.31) with the two real solutions:

$$t^\alpha \cos (\beta \ln t), \quad t^\alpha \sin (\beta \ln t). \tag{3.33}$$

Either pair of solutions may be said to correspond to the two complex roots $\alpha \pm i\beta$ of the characteristic polynomial.

Example 2

Find two independent solutions of the Euler-Cauchy equation

$$\frac{d^2u}{dt^2} + \frac{3}{t}\frac{du}{dt} + \frac{5}{t^2}u = 0.$$

Assume that $u = t^m$ and substitute into the differential equation. After cancelling the common factor we are left with the polynomial equation

$$m(m-1) + 3m + 5 = 0,$$

having roots $-1 \pm 2i$. According to the results above, two independent solutions of the differential equation are

$$t^{-1} \cos (2 \ln t), \quad t^{-1} \sin (2 \ln t).$$

By now it may begin to seem that $\ln t$ is playing the role that the independent variable t itself played in the equation with constant coefficients. In fact, it is true that the change of variable from t to

$$x = \ln t$$

would turn the Euler-Cauchy equation (3.27) into an equation with constant coefficients. The details of the general case are messy and not very interesting. However, we shall observe that if $m = a$ is a double root of the characteristic polynomial, then two independent solutions corresponding to this root are

$$t^a, \quad (\ln t)t^a.$$

Example 3

The Euler-Cauchy equation

$$\frac{d^2u}{dt^2} - \frac{1}{t}\frac{du}{dt} + \frac{1}{t^2}u = 0$$

has characteristic polynomial

$$m(m-1) - m + 1 = m^2 - 2m + 1$$

with $m = 1$ as a double root. We shall find, then, that two independent solutions of this equation are $u_1 = t$ and $u_2 = t \ln t$.

The considerations above give us some insight about finding an adequate number of independent solutions of an Euler-Cauchy equation. The theory of linear equations in Section 3.1 then tells us that the general solution of the homogeneous equation (3.27) is a linear combination of the n independent solutions that we find. Theorem 3.7 summarizes the forms of the independent solutions and their correspondence to the roots of the characteristic polynomial.

Theorem 3.7

The nth-order homogeneous Euler-Cauchy differential equation

$$\frac{d^n u}{dt^n} + \frac{b_1}{t}\frac{d^{n-1}u}{dt^{n-1}} + \cdots + \frac{b_{n-1}}{t^{n-1}}\frac{du}{dt} + \frac{b_n}{t^n}u = 0$$

with characteristic polynomial

$$q(m) = m(m-1)\cdots(m-n+1)$$
$$+ b_1 m(m-1)\cdots(m-n+2) + \cdots + b_{n-1}m + b_n$$

has n independent solutions, of these forms:
   1. t^m if m is a real, simple root of $q(m)$;
   2. $t^m, t^m \ln t, \ldots, t^m(\ln t)^{k-1}$ if m is a real root of multiplicity k;
   3. $t^\alpha \cos(\beta \ln t), t^\alpha \sin(\beta \ln t)$ if $\alpha + i\beta$ and $\alpha - i\beta$ are a pair of simple complex conjugate roots;
   4. the solutions in 3 multiplied by $1, \ln t, (\ln t)^2, \ldots, (\ln t)^{k-1}$, if $\alpha + i\beta$ and $\alpha - i\beta$ are conjugate complex roots, each of multiplicity k.

Example 4

Solve the Euler-Cauchy equation

$$\frac{d^4 u}{dt^4} + \frac{6}{t}\frac{d^3 u}{dt^3} + \frac{9}{t^2}\frac{d^2 u}{dt^2} + \frac{3}{t^3}\frac{du}{dt} + \frac{1}{t^4}u = 0.$$

We start by seeking solutions in the form $u = t^m$. The characteristic polynomial is found to be

$$m(m-1)(m-2)(m-3) + 6m(m-1)(m-2) + 9m(m-1)$$
$$+ 3m + 1 = m^4 + 2m^2 + 1 = (m^2 + 1)^2.$$

This biquadratic has $\pm i$ as double roots. By using case 4 of Theorem 3.7 we find the general solution to be

$$u(t) = (c_1 + c_2 \ln t) \cos (\ln t) + (c_3 + c_4 \ln t) \sin (\ln t).$$

Exercises

In Exercises 1–10, obtain the characteristic polynomial and the general solution of the differential equation.

1. $\dfrac{du}{dt} + \dfrac{2}{t} u = 0$

2. $\dfrac{du}{dt} - \dfrac{4}{t} u = 0$

3. $\dfrac{d^2u}{dt^2} - \dfrac{1}{t} \dfrac{du}{dt} = 0$

4. $\dfrac{d^2u}{dt^2} + \dfrac{1}{t} \dfrac{du}{dt} - \dfrac{1}{t^2} u = 0$

5. $\dfrac{d^2u}{dt^2} + \dfrac{1}{t} \dfrac{du}{dt} + \dfrac{1}{t^2} u = 0$

6. $\dfrac{d^2u}{dt^2} - \dfrac{1}{t} \dfrac{du}{dt} + \dfrac{2}{t^2} u = 0$

7. $\dfrac{d^2u}{dt^2} - \dfrac{2}{t^2} u = 0$

8. $\dfrac{d^2u}{dt^2} + \dfrac{2}{t^2} u = 0$

9. $\dfrac{d^2u}{dt^2} - \dfrac{2}{t} \dfrac{du}{dt} + \dfrac{2}{t^2} u = 0$

10. $\dfrac{d^2u}{dt^2} + \dfrac{4}{t} \dfrac{du}{dt} + \dfrac{2}{t^2} u = 0$

The Euler-Cauchy equation often arises in problems where the independent variable is a radial distance. In Exercises 11–17, obtain the general solution of the differential equation.

11. $\dfrac{1}{r} \dfrac{d}{dr} \left(r \dfrac{du}{dr} \right) = -1$ (Hint: multiply by r and integrate.)

12. $\dfrac{1}{r} \dfrac{d}{dr} \left(r \dfrac{du}{dr} \right) - \dfrac{k^2}{r^2} u = 0$ $(k = 1, 2, \ldots)$

13. $\dfrac{1}{r} \dfrac{d}{dr} \left(r \dfrac{du}{dr} \right) + \dfrac{k^2}{r^2} u = 0$ $(k = 1, 2, \ldots)$

14. $\dfrac{1}{\rho^2} \dfrac{d}{d\rho} \left(\rho^2 \dfrac{du}{d\rho} \right) = 0$

15. $\dfrac{1}{\rho^2} \dfrac{d}{d\rho} \left(\rho^2 \dfrac{du}{d\rho} \right) = -1$

16. $\dfrac{d}{d\rho} \left(\rho^2 \dfrac{du}{d\rho} \right) - (k^2 - \tfrac{1}{4})u = 0$ $(k = 0, 1, 2, \ldots)$

17. $\dfrac{d}{d\rho} \left(\rho^2 \dfrac{du}{d\rho} \right) + k(k + 1)u = 0$ $(k = 0, 1, 2, \ldots)$

18. Let $u(t)$ be a function of t and let $x = \ln t$, $v(x) = v(\ln t) = u(t)$. Find du/dt and d^2u/dt^2 in terms of dv/dx and d^2v/dx^2.

19. If u and v are as above and if u is a solution of

$$\frac{d^2u}{dt^2} + \frac{b}{t}\frac{du}{dt} + \frac{c}{t^2}u = 0,$$

find the equation satisfied by $v(x)$.

3.4

Reduction of Order

In Chapter 1 we saw how the knowledge of a solution of a homogeneous first-order equation could be exploited to solve a nonhomogeneous equation. Similarly, when we are dealing with a second-order equation, we can use knowledge of one solution of a homogeneous equation to find another or to solve a nonhomogeneous equation.

Second Solution of a Homogeneous Second-Order Equation

First, let us suppose that we know a solution $u_1(t)$ of a linear homogeneous second-order equation

$$\frac{d^2u}{dt^2} + a_1(t)\frac{du}{dt} + a_2(t)u = 0. \tag{3.34}$$

We seek a second, independent solution in the form

$$u_2(t) = v(t)u_1(t), \tag{3.35}$$

where $v(t)$ is a function to be found. The first and second derivatives of u_2 are then

$$\frac{du_2}{dt} = v'u_1 + vu_1',$$

$$\frac{d^2u_2}{dt^2} = v''u_1 + 2v_1'u_1' + vu_1''.$$

Substituting these two expressions into the differential equation (3.34) and collecting terms in v and its derivatives, we obtain

$$u_1\frac{d^2v}{dt^2} + (2u_1' + a_1u_1)\frac{dv}{dt} + (u_1'' + a_1u_1' + a_2u_1)v = 0. \tag{3.36}$$

Note, however, that the coefficient of v is exactly 0 because $u_1(t)$ is a solution of the differential equation (3.34). This fact provides a convenient check. If the coefficient of v is not 0, there has been a mistake in algebra or differentiation, or else u_1 was not really a solution of the equation.

Now our equation simplifies substantially. After dividing through Eq. (3.36) by u_1 we obtain

$$\frac{d^2v}{dt^2} + \left(2\frac{u_1'}{u_1} + a_1\right)\frac{dv}{dt} = 0. \tag{3.37}$$

Although Eq. (3.37) is a second-order equation with variable coefficients, the substitution $w = dv/dt$ makes it first order in w:

$$\frac{dw}{dt} + \left(2\frac{u_1'}{u_1} + a_1\right)w = 0. \tag{3.38}$$

This equation is easily solved for w:

$$w = \frac{dv}{dt} = c_1 \frac{e^{-A_1(t)}}{u_1^2(t)}, \tag{3.39}$$

where $A_1(t)$ is an integral of $a_1(t)$. Since we are looking for any solution—except a constant—of Eq. (3.39), we might choose c_1 to be any convenient nonzero number. We take $c_1 = 1$.

From this point we can get v by integrating Eq. (3.39) and, from there, u_2:

$$u_2(t) = u_1(t) \int_{t_0}^{t} \frac{e^{-A_1(z)}}{u_1^2(z)}\, dz. \tag{3.40}$$

Example 1

Find a second solution of the differential equation

$$\frac{d^2u}{dt^2} + 2\frac{du}{dt} + u = 0.$$

We know that $u_1(t) = e^{-t}$ is a solution. First, we assume that $u_2(t) = ve^{-t}$ and calculate

$$\frac{du_2}{dt} = v'e^{-t} - ve^{-t};$$

$$\frac{d^2u_2}{dt^2} = v''e^{-t} - 2v'e^{-t} + ve^{-t}.$$

Second, we substitute these expressions into the differential equation and simplify, obtaining

$$e^{-t}\frac{d^2v}{dt^2} + (-2e^{-t} + 2e^{-t})\frac{dv}{dt} + (e^{-t} - 2e^{-t} + e^{-t})v = 0$$

or

$$\frac{d^2v}{dt^2} = 0.$$

We may easily find the general solution of this equation for v. It is just $v = c_1 + c_2 t$. From here we determine that

$$u(t) = (c_1 + c_2 t)e^{-t}$$

is the general solution of the original equation.

If we need just any second independent solution of the equation for u, we take any *nonconstant* solution of the equation for v. The natural choice is $v(t) = t$, leading to

$$u_2(t) = te^{-t}$$

as the second independent solution of the original equation.

The effect of our assumption that $u_2 = vu_1$ is to produce a differential equation for v that is of second order but can be treated as if it were of first order. In principle, if we know a solution of an nth-order equation, we can go through the same process to get an equation of order $n - 1$ whose solution would give us the remaining solutions of the original equation. However, we have no real hope of solving a linear equation with variable coefficients of order greater than first.

In practice, it is not advisable to use the formula (3.40). Rather, one starts from the assumption $u_2 = vu_1$, substitutes into the differential equation, and then solves for v. Finding the general solution of the equation for v gives the general solution of the original equation. Finding any nonconstant solution of the equation for v gives a second independent solution of the original equation.

Solution of a Nonhomogeneous Equation

Now we try the same technique to solve a second-order nonhomogeneous equation,

$$\frac{d^2u}{dt^2} + a_1(t)\frac{du}{dt} + a_2(t)u = f(t). \tag{3.41}$$

Our objective is to find any particular solution, and we seek it in the form

$$u_p(t) = v(t)u_1(t), \tag{3.42}$$

where u_1 is supposed to be a solution of the homogeneous equation

$$\frac{d^2u_1}{dt^2} + a_1(t)\frac{du_1}{dt} + a_2(t)u_1 = 0. \tag{3.43}$$

After substituting $u_p(t)$ from Eq. (3.42), and its derivatives, into the differential equation (3.41) and collecting terms, we obtain

$$u_1 \frac{d^2v}{dt^2} + (2u_1' + a_1 u_1)\frac{dv}{dt} + (u_1'' + a_1 u_1' + a_2 u_1)v = f(t). \qquad (3.44)$$

Again, as in the homogeneous case, the coefficient of v is identically 0, by Eq. (3.43). Thus we are left with

$$\frac{d^2v}{dt^2} + \left(2\frac{u_1'}{u_1} + a_1\right)\frac{dv}{dt} = \frac{f(t)}{u_1(t)}. \qquad (3.45)$$

Again, we have an equation that is second order in v but first order in $w = dv/dt$. Although it appears rather complicated, it is not really difficult to solve. The corresponding homogeneous equation,

$$\frac{dw}{dt} + \left(2\frac{u_1'}{u_1} + a_1\right)w = 0, \qquad (3.46)$$

has as its solution

$$w(t) = c \exp\left(-\int \left(2\frac{u_1'}{u_1} + a_1\right) dt\right) = c\frac{e^{-A_1(t)}}{u_1^2(t)}. \qquad (3.47)$$

The nonhomogeneous equation (3.45) can now be solved easily. A formula for the solution is

$$\frac{dv}{dt} = \frac{e^{-A_1(t)}}{u_1^2(t)} \int e^{A_1(t)} u_1(t) f(t)\, dt. \qquad (3.48)$$

This function must be integrated one more time to finally find v, and then $u_p = vu_1$.

If we were to write down the formula for u_p, it would look awfully complicated—and would not be good for much. The calculations above are really intended to indicate how one is to go about the solution and to show that there is no real impediment to the solution. In practice, we start from the assumption $u_p = vu_1$, where u_1 is a known solution of the homogeneous equation, and work out the consequences.

It is worthwhile to note that this procedure is fairly general—certainly applicable in cases in which the method of undetermined coefficients is not. Notice, however, that the coefficient of dv/dt becomes infinite (therefore not continuous) wherever $u_1(t) = 0$. This may limit the usefulness of the technique. In the following section we introduce a still more general method.

Example 2

Find a particular solution of the nonhomogeneous differential equation

$$\frac{d^2u}{dt^2} - u = \frac{1}{e^t + e^{-t}}.$$

We seek a solution in the form $u_p(t) = v(t)e^t$, since $u_1 = e^t$ is a solution of the homogeneous equation. Below are the differential equations using u_p in its special form and equations occurring in the simplification and solution of the equation for v.

$$\frac{du_p}{dt} = v'e^t + ve^t \qquad \text{(Derivative of } u_p)$$

$$\frac{d^2u_p}{dt^2} = v''e^t + 2v'e^t + ve^t \qquad \text{(Second derivative)}$$

$$v''e^t + 2v'e^t + ve^t - ve^t = \frac{1}{e^t + e^{-t}} \qquad \text{(Differential equation)}$$

$$v'' + 2v' = \frac{e^{-t}}{e^t + e^{-t}} \qquad \text{(Simplified)}$$

$$v' = e^{-2t} \int e^{2t} \frac{e^{-t}}{e^t + e^{-t}} \, dt \qquad \text{(Solution of above)}$$

$$= e^{-2t} \int \frac{1}{1 + e^{-2t}} \, dt \qquad \text{(Integrand simplified)}$$

$$v' = e^{-2t}(t + \tfrac{1}{2}\ln(1 + e^{-2t})) \qquad \text{(Integration completed)}$$

$$v = -\left(\frac{t}{2} + \frac{1}{4}\right)e^{-2t} - \tfrac{1}{4}(1 + e^{-2t})(\ln(1 + e^{-2t}) - 1) \qquad \text{(Integral of } v')$$

$$u_p = ve^t = -\left(\frac{t}{2} + \frac{1}{4}\right)e^{-t} - \tfrac{1}{4}(e^t + e^{-t})(\ln(1 + e^{-2t}) - 1).$$

We know that any linear combination of e^t and e^{-t} is a solution of the homogeneous equation. Therefore any such terms may be struck out of the expression for u_p, and the result should still be a particular solution; for example,

$$u_p = -\frac{t}{2}e^{-t} - \tfrac{1}{4}(e^t + e^{-t})\ln(1 + e^{-2t}).$$

Note that the method of undetermined coefficients will not work for this problem.

Exercises

In Exercises 1–6, one solution of a homogeneous second-order differential equation is given. Use reduction of order to find a second independent solution

1. $(1 - t^2)\dfrac{d^2u}{dt^2} - 2t\dfrac{du}{dt} + 2u = 0, \qquad u_1(t) = t$

2. $t^2 \dfrac{d^2u}{dt^2} - t\dfrac{du}{dt} + u = 0, \qquad u_1(t) = t$

3. $t^2 \dfrac{d^2u}{dt^2} - 5t\dfrac{du}{dt} + 9u = 0, \qquad u_1(t) = t^3$

4. $\dfrac{d^2u}{dt^2} - \dfrac{t}{t-1}\dfrac{du}{dt} + \dfrac{1}{t-1}u = 0, \qquad u_1(t) = t$

5. $\dfrac{d^2u}{dt^2} + 4\dfrac{du}{dt} + 4u = 0, \qquad u_1(t) = e^{-2t}$

6. $\dfrac{d^2u}{dt^2} - \dfrac{t}{t-1}\dfrac{du}{dt} + \dfrac{1}{t-1}u = 0, \qquad u_1(t) = e^t$

In Exercises 7–11, use reduction of order to find a particular solution.

7. $\dfrac{d^2u}{dt^2} + 2\dfrac{du}{dt} + u = e^{-t}, \qquad u_1(t) = e^{-t}$

8. $\dfrac{d^2u}{dt^2} + u = e^{\sin t}, \qquad u_1(t) = \cos t$

9. $t^2 \dfrac{d^2u}{dt^2} - t\dfrac{du}{dt} + u = t, \qquad u_1(t) = t$

10. $\dfrac{d^2u}{dt^2} + 3\dfrac{du}{dt} + 2u = \sin(e^t), \qquad u_1(t) = e^{-t}$

11. $\dfrac{d^2u}{dt^2} + u = \tan t, \qquad u_1(t) = \cos t$

12. Verify that $w(t)$ as given in Eq. (3.39) in indeed a solution of Eq. (3.38).

13. Compute the Wronskian of $u_1(t)$ and $u_2(t)$ (from Eq. (3.40)).

3.5

Variation of Parameters

In Section 3.4 we saw how we might get a particular solution of a second-order nonhomogeneous equation,*

$$\frac{d^2u}{dt^2} + a_1(t)\frac{du}{dt} + a_2(t)u = f(t), \tag{3.49}$$

if we have in hand one solution of the corresponding homogeneous equation,

$$\frac{d^2u}{dt^2} + a_1(t)\frac{du}{dt} + a_2(t)u = 0. \tag{3.50}$$

---

* If the highest derivative has coefficient $a_0(t)$, divide through by it to obtain this normalized form.

From the theoretical point of view, the solution obtained has a defect: since division by the known solution is required, the process may fail if that solution has a zero. This defect can be remedied by assuming that the particular solution sought has the form

$$u_p(t) = v_1(t)u_1(t) + v_2(t)u_2(t), \tag{3.51}$$

where $u_1(t)$ and $u_2(t)$ are assumed to be *two* independent solutions of the homogeneous equation (3.50) and the v's are functions to be determined. The expression above for $u_p(t)$ resembles the general solution of the homogeneous equation (3.50), but v_1 and v_2 are functions (variables) instead of constants (parameters). For this reason the method we are developing is called *variation of parameters*.

The most straightforward attack is to substitute Eq. (3.51) and its derivatives into the nonhomogeneous equation. The result, however, is a (messy) second-order differential equation containing two unknown functions. This gives too much liberty—the unknown functions are not fully specified by a single equation. We can rectify the situation by assuming that

$$v_1'u_1 + v_2'u_2 = 0. \tag{3.52}$$

This requirement gives us an extra equation in v_1 and v_2 and has other salutary effects. One is that the derivative of u_p simplifies:

$$u_p' = v_1u_1' + v_2u_2' + v_1'u_1 + v_2'u_2 = v_1u_1' + v_2u_2'. \tag{3.53}$$

The benefit extends to the second derivative as well. It becomes

$$u_p'' = v_1u_1'' + v_2u_2'' + v_1'u_1' + v_2'u_2'. \tag{3.54}$$

Now when we substitute u_p and its derivatives (Eqs. 3.51, 3.53, and 3.54) into the nonhomogeneous differential equation (3.49), we obtain

$$(v_1u_1'' + v_2u_2'' + v_1'u_1' + v_2'u_2') + a_1(v_1u_1' + v_2u_2') + a_2(v_1u_1 + v_2u_2) = f. \tag{3.55}$$

This can be rearranged to get

$$v_1'u_1' + v_2'u_2' + v_1(u_1'' + a_1u_1' + a_2u_1) + v_2(u_2'' + a_1u_2' + a_2u_2) = f. \tag{3.56}$$

But here, both terms in parentheses are 0 because u_1 and u_2 are supposed to be solutions of the homogeneous equation (3.50) or

$$u'' + a_1u' + a_2u = 0.$$

Between Eq. (3.52) and Eq. (3.56) we get a system of two algebraic equations in the unknowns v_1', v_2':

$$\left. \begin{array}{l} v_1'u_1 + v_2'u_2 = 0 \\ v_1'u_1' + v_2'u_2' = f \end{array} \right\}. \tag{3.57}$$

Furthermore, the determinant of this system is precisely the Wronskian of u_1 and u_2! Since they were assumed to be independent solutions of the homogeneous equation (3.50), the Wronskian is nonzero, and we are assured of the existence of a solution of Eq. (3.57). In fact, we can write down explicitly

$$v_1'(t) = -\frac{u_2(t)f(t)}{W(t)}, \qquad v_2'(t) = \frac{u_1(t)f(t)}{W(t)}, \qquad (3.58)$$

where $W(t) = W(u_1, u_2)$ is the Wronskian,

$$W(t) = \begin{vmatrix} u_1(t) & u_2(t) \\ u_1'(t) & u_2'(t) \end{vmatrix}. \qquad (3.59)$$

Equations (3.58) have known right-hand sides. From them, $v_1(t)$ and $v_2(t)$ can be found by integration.

Example 1

Use variation of parameters to find a particular solution of

$$\frac{d^2u}{dt^2} + \frac{1}{t}\frac{du}{dt} - \frac{1}{t^2}u = \cos t.$$

The corresponding homogeneous equation is of the Euler-Cauchy type. We easily find

$$u_1(t) = t, \qquad u_2(t) = t^{-1}$$

as independent solutions of the homogeneous equation. The particular solution is to have the form

$$u_p(t) = v_1 t + v_2 t^{-1}.$$

The assumption of Eq. (3.52) is

$$v_1't + v_2't^{-1} = 0.$$

With this assumption we calculate

$$\frac{du_p}{dt} = v_1 - v_2 t^{-2}$$

$$\frac{d^2u_p}{dt^2} = v_1' - v_2't^{-2} + 2v_2 t^{-3}.$$

Now we substitute these expressions into the original differential equation to obtain

$$(v_1' - v_2't^{-2} + 2v_2t^{-3}) + t^{-1}(v_1 - v_2t^{-2}) - t^{-2}(v_1t + v_2t^{-1}) = \cos t$$

or

$$v_1' - v_2't^{-2} = \cos t.$$

(The terms that cancel additively do so because t and t^{-1} are solutions of the homogeneous equation.) Now the equations analogous to Eqs. (3.52) and (3.56) are

$$v_1't + v_2't^{-1} = 0,$$

$$v_1' - v_2't^{-2} = \cos t.$$

These can easily be solved, and the solutions integrated, to find

$$v_1(t) = \tfrac{1}{2} \sin t, \qquad v_2(t) = \left(1 - \frac{t^2}{2}\right) \sin t - t \cos t.$$

Then, finally, the particular solution is (after some algebraic cancellation)

$$u_p(t) = t^{-1} \sin t - \cos t.$$

Variation of parameters has virtues as a theoretical, as well as a practical, tool. To illustrate the power of what we have done, let us push on from Eq. (3.58) toward a formula for a particular solution of the nonhomogeneous equation (3.49).

To begin with, we can integrate Eq. (3.58) to get v_1 and v_2 explicitly. Let us suppose that t_0 is a special time (perhaps an initial time). Then

$$v_1(t) = \int_{t_0}^{t} -\frac{u_2(z)f(z)}{W(z)}\, dz, \qquad v_2(t) = \int_{t_0}^{t} \frac{u_1(z)f(z)}{W(z)}\, dz \tag{3.60}$$

are expressions for v_1 and v_2. We are not interested in these but rather in

$$u_p(t) = v_1(t)u_1(t) + v_2(t)u_2(t)$$

$$= u_1(t) \int_{t_0}^{t} -\frac{u_2(z)f(z)}{W(z)}\, dz + u_2(t) \int_{t_0}^{t} \frac{u_1(z)f(z)}{W(z)}\, dz. \tag{3.61}$$

Of course, $u_1(t)$ and $u_2(t)$ are constants relative to the integration with respect to z. Therefore they can pass inside the integrals to give

$$u_p(t) = \int_{t_0}^{t} \frac{u_2(t)u_1(z) - u_1(t)u_2(z)}{W(z)} f(z)\, dz. \tag{3.62}$$

This remarkable formula shows that a particular solution of Eq. (3.49) can be obtained in the form of an integral containing the inhomogeneity f as a factor in the integrand. The other factor in the integrand is called the *Green's function* for the initial value problem composed of the nonhomogeneous equation (3.49) and initial conditions at t_0. This Green's function is

$$G(t, z) = \frac{u_2(t)u_1(z) - u_1(t)u_2(z)}{W(z)}. \qquad (3.63)$$

It is left as an exercise to prove these properties of the Green's function:

   1. $u(t) = G(t, z)$ (considered as a function of t) is a solution of the homogeneous equation

$$\frac{d^2u}{dt^2} + a_1(t)\frac{du}{dt} + a_2(t)u = 0. \qquad (3.64)$$

   2. $u(t)$ satisfies these "initial conditions" at $t = z$:

$$u(z) = 0, \qquad \frac{du}{dt}(z) = 1. \qquad (3.65)$$

Example 2

The most noteworthy application of this idea is to the solution of the equation

$$\frac{d^2u}{dt^2} + \omega^2 u = f(t). \qquad (3.66)$$

Taking $u_1(t) = \cos \omega t$, $u_2(t) = \sin \omega t$ gives Wronskian (the denominators in Eq. (3.58)) identically equal to ω. Then we have from Eq. (3.58)

$$v_1' = -\frac{\sin \omega t}{\omega}f(t), \qquad v_2' = \frac{\cos \omega t}{\omega}f(t). \qquad (3.67)$$

These two equations are integrated to get

$$v_1(t) = \int_0^t -\frac{\sin \omega z}{\omega}f(z)\,dz, \qquad v_2(t) = \int_0^t \frac{\cos \omega z}{\omega}f(z)\,dz. \qquad (3.68)$$

The solution finally may be written as

$$u_p(t) = \cos \omega t \int_0^t -\frac{\sin \omega z}{\omega}f(z)\,dz + \sin \omega t \int_0^t \frac{\cos \omega z}{\omega}f(z)\,dz. \qquad (3.69)$$

But $\cos \omega t$ and $\sin \omega t$ are constants relative to the integration with respect to z. Thus we may put them inside the integrals and obtain

$$u_p(t) = \int_0^t \frac{\cos \omega z \sin \omega t - \sin \omega z \cos \omega t}{\omega}f(z)\,dz. \qquad (3.70)$$

Finally, application of a trigonometric identity allows the further simplification

$$u_p(t) = \int_0^t \frac{\sin \omega(t - z)}{\omega}f(z)\,dz, \qquad (3.71)$$

which coincides with Eq. (3.62). Thus we identify the Green's function

$$G(t, z) = \frac{\sin \omega(t - z)}{\omega}. \tag{3.72}$$

What we have done above for a second-order equation can be done for the nth-order equation

$$\frac{d^n u}{dt^n} + a_1(t)\frac{d^{n-1}u}{dt^{n-1}} + \cdots + a_{n-1}(t)\frac{du}{dt} + a_n(t)u = f(t). \tag{3.73}$$

We start by assuming that a particular solution has the form

$$u_p(t) = v_1(t)u_1(t) + v_2(t)u_2(t) + \cdots + v_n(t)u_n(t), \tag{3.74}$$

where $u_1, u_2, \ldots, u_n$ are n independent solutions of the homogeneous equation corresponding to Eq. (3.73):

$$u^{(n)} + a_1 u^{(n-1)} + \cdots + a_{n-1}u' + a_n u = 0. \tag{3.75}$$

In order to find the v's we follow the lead established previously and require that

$$v_1' u_1 + v_2' u_2 + \cdots + v_n' u_n = 0. \tag{3.76}$$

This effect is to leave

$$\frac{du_p}{dt} = v_1 u_1' + v_2 u_2' + \cdots + v_n u_n', \tag{3.77}$$

which looks like a linear combination of the derivatives of the u's. Similarly, we shall require that

$$v_1' u_1' + v_2' u_2' + \cdots + v_n' u_n' = 0, \tag{3.78}$$

which leaves the second derivative in the form

$$\frac{d^2 u_p}{dt^2} = v_1 u_1'' + v_2 u_2'' + \cdots + v_n u_n''. \tag{3.79}$$

In general, we shall assume that

$$v_1' u_1^{(k)} + v_2' u_2^{(k)} + \cdots + v_n' u_n^{(k)} = 0, \tag{3.80}$$

for $k = 0, 1, 2, \ldots, n-2$. (Equations (3.76) and (3.78) correspond to $k = 0$ and $k = 1$.) The effect is that each derivative up through order $n - 1$ has the simple form

$$\frac{d^k u_p}{dt^k} = v_1 u_1^{(k)} + v_2 u_2^{(k)} + \cdots + v_n u_n^{(k)}. \tag{3.81}$$

The last derivative to be calculated is

$$\frac{d^n u_p}{dt^n} = \frac{d}{dt}(v_1 u_1^{(n-1)} + v_2 u_2^{(n-1)} + \cdots + v_n u_n^{(n-1)})$$

$$= (v_1' u_1^{(n-1)} + v_2' u_2^{(n-1)} + \cdots + v_n' u_n^{(n-1)})$$

$$+ (v_1 u_1^{(n)} + v_2 u_2^{(n)} + \cdots + v_n u_n^{(n)}). \qquad \textbf{(3.82)}$$

Note that the term in the first set of parentheses is *not* assumed to be zero.

Now we substitute our calculated forms for the derivatives of u_p (Eq. 3.81) into the nonhomogeneous equation (3.73), which we abbreviate in the form $L(u) = f$. Obviously, L is the linear operator that makes up the left-hand side of Eq. (3.73). Because each derivative except the nth has the form of a linear combination of the u's (Eq. 3.81), we find that

$$L(u_p) = v_1 L(u_1) + \cdots + v_n L(u_n) + v_1' u_1^{(n-1)} + \cdots + v_n' u_n^{(n-1)}. \qquad \textbf{(3.83)}$$

Of course, each of the u_i's is a solution of the homogeneous equation, (3.74), or $L(u_i) = 0$. Thus the equation $L(u_p) = f$ boils down to

$$v_1' u_1^{(n-1)} + v_2' u_2^{(n-1)} + \cdots + v_n' u_n^{(n-1)} = f. \qquad \textbf{(3.84)}$$

Now consider what we know about the v's. Equation (3.80) is really $n - 1$ simultaneous linear equations in $v_1', v_2', \ldots, v_n'$, and Eq. (3.84) is another. Thus we have a system of n simultaneous equations:

$$\left. \begin{array}{l} v_1' u_1 + v_2' u_2 \quad + \cdots + v_n' u_n \quad = 0 \\ v_1' u_1' + v_2' u_2' \quad + \cdots + v_n' u_n' \quad = 0 \\ \quad \vdots \\ v_1' u_1^{(n-2)} + v_2' u_2^{(n-2)} + \cdots + v_n' u_n^{(n-2)} = 0 \\ v_1' u_1^{(n-1)} + v_2' u_2^{(n-1)} + \cdots + v_n' u_n^{(n-1)} = f(t) \end{array} \right\} \qquad \textbf{(3.85)}$$

Furthermore, the determinant of this system is precisely the Wronskian, $W(u_1, u_2, \ldots, u_n)$—which is nonzero, since the u's were assumed to be independent solutions of $L(u) = 0$.

The conclusion of all this calculation is that the system (3.68) "may" be solved for the derivatives $v_1', v_2', \ldots, v_n'$, and the v's may be determined by integration.

Theorem 3.8

If $u_1, u_2, \ldots, u_n$ are n independent solutions of the nth-order equation $L(u) = 0$, then a particular solution of the nonhomogeneous equation $L(u) = f$ has the form

$$u_p = v_1(t)u_1(t) + v_2(t)u_2(t) + \cdots + v_n(t)u_n(t).$$

The coefficient functions are found by integrating their derivatives, which are solutions of the system (3.85).

Exercises

1. Equation (3.63) gives the Green's function for the differential equation (3.49) in terms of u_1 and u_2. From this, show that $u(t) = G(t, z)$ has properties 1 and 2 (Eqs. 3.64 and 3.65).

2. Suppose that $G(t, z)$ is a Green's function for the second-order differential equation (3.49). Show that the solution of Eq. (3.49),

$$u_p(t) = \int_{t_0}^{t} G(t, z) f(z) \, dz,$$

satisfies the initial conditions

$$u_p(t_0) = 0, \qquad \frac{du_p}{dt}(t_0) = 0.$$

3. Verify by direct differentiation and substitution that the function $u_p(t)$ given in Exercise 2 really satisfies the differential equation (3.49). (Hint: assume the properties in Eqs. (3.64) and (3.65) or use the explicit formula (3.63).)

4. In Example 2 we found that the Green's function for the differential equation

$$\frac{d^2u}{dt^2} + \omega^2 u = f(t)$$

was $G(t, z) = \sin \omega(t - z)/\omega$. Verify properties 1 and 2 (Eqs. 3.64 and 3.65) for this function.

5. Use the Green's function in Eq. (3.72) to solve the differential equation

$$\frac{d^2u}{dt^2} + \omega^2 u = \cos \omega t.$$

6. Use the Green's function in Eq. (3.72) to solve

$$\frac{d^2u}{dt^2} + \omega^2 u = 1.$$

7. Find a formula for the solution of the nonhomogeneous initial value problem

$$\frac{d^2u}{dt^2} = f(t), \qquad u(0) = 0, \qquad \frac{du}{dt}(0) = 0,$$

by integrating the differential equation twice.

8. Find the Green's function for the initial value problem above. You can use Eq. (3.63) or the properties (3.64) and (3.65).

9. Use the Green's function of Exercise 8 to solve the initial value problem stated in Exercise 7.

10. Are the solutions found in Exericses 7 and 9 the same? Should they be?

11. Find the Green's function for the differential equation

$$\frac{d^2u}{dt^2} + \frac{1}{t}\frac{du}{dt} - \frac{k^2}{t^2}u = f(t).$$

12. Find two independent solutions of the corresponding homogeneous equation and apply variation of parameters to solve

$$\frac{d^2u}{dt^2} + \frac{1}{t}\frac{du}{dt} - \frac{k^2}{t^2}u = 1.$$

13. Use the Green's function from Exercise 11 to solve the given equation with $f(t) = 1$. Take $t_0 = 1$.

14. Find the Green's function for the differential equation

$$\frac{d^2u}{dt^2} - k^2u = f(t).$$

15. Use the result of Exercise 14 to get a formula for the solution of

$$\frac{d^2u}{dt^2} - k^2u = f(t), \qquad u(0) = 0, \qquad u'(0) = 0,$$

and use it to solve when $f(t) = 1$.

16. Obtain the solution of the initial value problem in Exercise 15 (with $f(t) = 1$) by other means.

17. Solve the nonhomogeneous equation

$$\frac{d^4u}{dt^4} = f(t)$$

by the method of variation of parameters, using these four independent solutions of the homogeneous equation:

$$u_1(t) = 1, \qquad u_2(t) = t, \qquad u_3(t) = t^2, \qquad u_4(t) = t^3.$$

18. Compare the result of Exercise 17 with the solution one would obtain by integration of the equation four times.

19. Use variation of parameters to find a particular solution of

$$2\frac{d^2u}{dt^2} + 2\frac{du}{dt} + u = \sqrt{t}.$$

20. Use variation of parameters to find a particular solution of

$$\frac{d^2u}{dt^2} - u = \frac{1}{e^t + e^{-t}}.$$

Compare with Example 2 in Section 3.4.

3.6
Second-Order Linear Equations with Variable Coefficients

In Chapter 2 we studied the solutions of linear equations with constant coefficients, and in this chapter we have learned some of the theory of linear equations in general. Our objective in this section will be to describe the qualitative features of the solutions of second-order equations with variable coefficients.

First, let us review some features of the solutions of the linear homogeneous equation

$$\frac{d^2u}{dt^2} + b_1\frac{du}{dt} + b_2u = 0 \tag{3.86}$$

with constant coefficients b_1, b_2. We know that the solution is usually

$$u(t) = c_1e^{m_1t} + c_2e^{m_2t}, \tag{3.87}$$

where c_1, c_2 are constants and m_1, m_2 are the roots of the characteristic polynomial:

$$m_1, m_2 = \tfrac{1}{2}(-b_1 \pm \sqrt{b_1^2 - 4b_2}). \tag{3.88}$$

If $b_1^2 - 4b_2 > 0$, no solution of Eq. (3.86) can have more than one zero. (A *zero* is a value of t for which $u(t) = 0$.) The same is true if $b_1^2 - 4b_2 = 0$, although Eq. (3.87) does not give the correct form for the solution in this case.

On the other hand, if $b_1^2 - 4b_2 < 0$, the roots (3.88) are complex, and our general solution is more conveniently written as

$$u(t) = e^{-b_1t/2}(c_1\cos\beta t + c_2\sin\beta t) \tag{3.89}$$

with $\beta = \tfrac{1}{2}\sqrt{4b_2 - b_1^2}$. In this case, every solution has an infinite number of zeros in the range $0 < t < \infty$. A function is called *oscillatory* for $t > t_0$ if it has an infinite number of zeros in that range.

Another important feature of a solution is whether its magnitude "stays finite" as t tends to infinity. More precisely, is there a constant M such that

$$-M \leqslant u(t) \leqslant M \tag{3.90}$$

for $t > t_0$? If so, we say that $u(t)$ is *bounded*. It is easy to check that all solutions of (3.86) are bounded if b_1 and b_2 are nonnegative (and at least one of them is positive). If both are positive, all solutions decay exponentially in time. In Table 3.2 and Fig. 3.3 the observations made above are summarized.

Table 3.2	Relation	Feature
Nature of the Solutions of $\dfrac{d^2u}{dt^2} + b_1\dfrac{du}{dt} + b_2u = 0$	$\begin{cases} b_1^2 - 4b_2 < 0 \\ b_1^2 - 4b_2 \geqslant 0 \end{cases}$ $\begin{cases} b_1 \geqslant 0,\, b_2 > 0 \\ b_1 < 0 \text{ or } b_2 < 0 \end{cases}$	$u(t)$ oscillatory $u(t) = 0$ at most once $u(t)$ bounded $u(t)$ unbounded

Figure 3.3

Nature of Solution of $u'' + b_1u' + b_2u = 0$ as a Function of b_1 and b_2

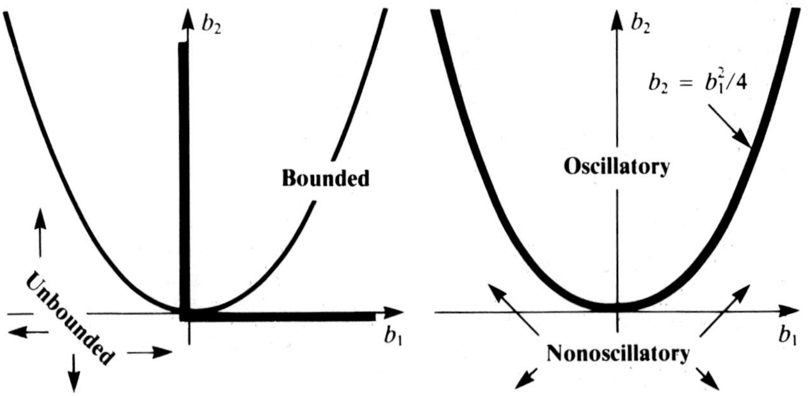

Now we want to explore the behavior of solutions of a more general equation

$$\frac{d^2u}{dt^2} + a_1(t)\frac{du}{dt} + a_2(t)u = 0. \tag{3.91}$$

It happens that a great deal can be found out about solutions even when they cannot be found explicitly. This is most fortunate, since only a few equations with variable coefficients can be solved in closed form. First we will quote without proof a theorem about the existence and uniqueness of solutions.

Theorem 3.9

Let $a_1(t)$ and $a_2(t)$ be continuous for $t \geq t_0$. Then there is one and only one solution of the initial value problem

$$\frac{d^2u}{dt^2} + a_1(t)\frac{du}{dt} + a_2(t)u = 0, \qquad t_0 < t,$$

$$u(t_0) = q_0, \qquad \frac{du}{dt}(t_0) = q_1.$$

One simple consequence of this theorem is that the constant function $u(t) \equiv 0$ is the *only* solution of a linear homogeneous differential equation that satisfies zero initial conditions. (This is often called the *trivial* solution of the differential equation.) Furthermore, if a solution and its derivative are ever simultaneously zero, we can conclude that the solution is identically zero.

Throughout this section we will assume not only that the coefficients $a_1(t)$ and $a_2(t)$ are continuous, but also that they have continuous derivatives. Our purpose is to show what kind of information can be gotten about the solutions of Eq. (3.91) by elementary methods. The next theorem uses a really elementary method.

Theorem 3.10

If $a_2(t) < 0$ for $t \geq t_0$, any nontrivial solution of Eq. (3.91) has at most one zero after t_0.

Proof: Suppose that $u(t)$ is a definite, nontrivial solution of Eq. (3.91). If $u'(t)$ is never zero, then $u(t)$ is steadily increasing or steadily decreasing, and its graph could not cut the t-axis twice.

On the other hand, if there is a time after t_0 (say, t_1) when $u'(t_1) = 0$, the differential equation tells us that the second derivative of u at that time is

$$u''(t_1) = -a_2(t_1)u(t_1).$$

Since we are dealing with a nontrivial solution, $u(t_1)$ cannot be 0. If $u(t_1)$ is positive, then so is $u''(t_1)$ (recall that $-a_2(t)$ is positive), and therefore $u(t_1)$ is a minimum. On the other hand, if $u(t_1)$ is negative, then so is $u''(t_1)$, and $u(t_1)$ is a maximum. Thus the graph of $u(t)$ turns away from the t-axis at a maximum or minimum, making it impossible for $u(t)$ to cross the axis at all.

Example 1

The differential equation

$$\frac{d^2u}{dt^2} + \frac{t}{1+t}\frac{du}{dt} - \frac{1}{1+t}u = 0$$

has $a_2(t) = -1/(1 + t) < 0$. We can verify from the general solution

$$u(t) = c_1 t + c_2 e^{-t}$$

that $u(t)$ does indeed have at most one zero.

If we reverse the reasoning of the proof and require that $a_2(t)$ be positive, it follows that $u(t)$ will have positive maxima and/or negative minima. While this is a necessary condition for $u(t)$ to be oscillatory, it is not sufficient. Just consider the equation

$$\frac{d^2u}{dt^2} + 2\frac{du}{dt} + u = 0,$$

where $a_2 = 1 > 0$. Nevertheless, no solution has more than one zero, essentially because of the damping term, $2\,du/dt$.

For the time being, let's limit ourselves to an equation without "damping":

$$\frac{d^2v}{dt^2} + p(t)v = 0. \tag{3.92}$$

Theorem 3.11

If $p(t)$ is continuous and there is some constant $k > 0$ such that $p(t) \geqslant k^2$ for all $t \geqslant 0$, then every solution v of Eq. (3.92) is oscillatory.

Proof: If $p(t)$ were equal to k^2, Eq. (3.92) and its solution would be

$$\frac{d^2v}{dt^2} + k^2v = 0, \qquad v(t) = A\cos\theta(t). \tag{3.93}$$

We know that $\theta(t) = kt - \phi$; but we could also say that $\theta(t) = -\tan^{-1}(v'/kv)$.

In the general case in which $p(t)$ is not constant, we may define analogously

$$\theta(t) = -\tan^{-1}\left(\frac{v'}{kv}\right). \tag{3.94}$$

Even though we may not be able to express $v(t)$ as a cosine of $\theta(t)$, the function $\theta(t)$ still indicates whether v is oscillatory; if so, the ratio v'/kv will vary from ∞ through 0 to $-\infty$ as $v(t)$ varies from 0 through a maximum or minimum and back to 0. This, in turn, causes $\theta(t)$ to increase by π from one odd multiple of $\pi/2$ to the next.

Now we need to show that $\theta(t)$ as defined by Eq. (3.94) increases steadily. The derivative is

$$\frac{d\theta}{dt} = -\frac{\dfrac{kvv'' - v'kv'}{(kv)^2}}{1 + (v'/kv)^2} = k\frac{(v')^2 - vv''}{(v')^2 + (kv)^2}. \tag{3.95}$$

We are assuming that v is a solution of Eq. (3.92). We may therefore replace v'' by $-p(t)v$ to get

$$\frac{d\theta}{dt} = k\frac{(v')^2 + pv^2}{(v')^2 + k^2v^2}. \tag{3.96}$$

But since $p(t) \geqslant k^2$, the numerator in Eq. (3.96) is at least as large as the denominator, and we have

$$\frac{d\theta}{dt} \geqslant k. \tag{3.97}$$

In other words, $\theta(t)$ increases at least as fast as kt, and therefore $v(t)$ has zeros at least as frequently as $\cos kt$—certainly an infinite number of them in the range $t_0 < t < \infty$.

The theorem above does indeed tell us conditions that make solutions of Eq. (3.92) oscillatory. But we wanted to know about solutions of Eq. (3.91), which might contain the first derivative. Actually, we can always turn Eq. (3.91) into Eq. (3.92) by a transformation. First, define

$$v(t) = F(t)u(t), \tag{3.98}$$

where $u(t)$ is to be a solution of Eq. (3.91),

$$\frac{d^2u}{dt^2} + a_1(t)\frac{du}{dt} + a_2(t)u = 0, \tag{3.99}$$

and the multiplier $F(t)$ is to be

$$F(t) = \exp\left\{\frac{1}{2}\int_{t_0}^{t} a_1(t')\, dt'\right\}. \tag{3.100}$$

It is easy to calculate that

$$\frac{dF}{dt} = \tfrac{1}{2}a_1(t)F(t), \qquad \frac{d^2F}{dt^2} = \left(\tfrac{1}{2}a_1'(t) + \tfrac{1}{4}a_1^2(t)\right)F(t). \tag{3.101}$$

From here we can find that

$$\frac{d^2v}{dt^2} = F\frac{d^2u}{dt^2} + 2F'\frac{du}{dt} + F''u = F\left[\frac{d^2u}{dt^2} + a_1\frac{du}{dt} + \left(\tfrac{1}{2}a_1' + \tfrac{1}{4}a_1^2\right)u\right]. \tag{3.102}$$

Now using the fact that $u(t)$ is to be a solution of Eq. (3.99), it is just a step to this equation for v:

$$\frac{d^2v}{dt^2} + (a_2(t) - \tfrac{1}{4}a_1^2(t) - \tfrac{1}{2}a_1'(t))v = 0. \qquad (3.103)$$

Of course, this is the equation referred to in the theorem, with

$$p(t) = a_2(t) - \tfrac{1}{4}a_1^2(t) - \tfrac{1}{2}a_1'(t). \qquad (3.104)$$

Thus the transformation $v = Fu$ replaces the differential equation (3.91) or (3.99) with the simpler one, Eq. (3.103). But since the multiplier $F(t)$ given by Eq. (3.100) is never zero, we have $u(t) = 0$ at exactly the same values of t where $v(t) = 0$. Therefore we can extend Theorem 3.11 as follows.

Theorem 3.12

If $a_1(t)$, $a_1'(t)$, and $a_2(t)$ are continuous and there is some constant $k > 0$ such that

$$a_2(t) - \tfrac{1}{4}a_1^2(t) - \tfrac{1}{2}a_1'(t) \geq k^2$$

for all $t \geq t_0$, then every solution u of Eq. (3.99) (or Eq. (3.91)) is oscillatory.

Note that Theorem 3.12 correctly predicts that the equation

$$\frac{d^2u}{dt^2} + b_1\frac{du}{dt} + b_2u = 0$$

with constant coefficients has oscillatory solutions whenever $4b_2 - b_1^2 > 0$.

Example 2

A very important example is the equation

$$\frac{d^2u}{dt^2} + \frac{1}{t+1}\frac{du}{dt} + u = 0,$$

which is a form of Bessel's equation. From $a_1 = 1/(t+1)$ and $a_2 = 1$ we find

$$a_2(t) - \tfrac{1}{4}a_1^2(t) - \tfrac{1}{2}a_1'(t) = 1 + \frac{1}{4(t+1)^2}.$$

Since this function is always greater than 1, the solutions of the given equation are oscillatory.

Finally, we are going to tackle this question: Under what conditions on $a_1(t)$ and $a_2(t)$ are the solutions of Eq. (3.91) bounded? Again we can find some nice results with elementary methods.

Theorem 3.13

Suppose that $a_1(t)$, $a_2(t)$, and $a_2'(t)$ are continuous, that

$$a_1(t) \geq 0, \qquad a_2'(t) \leq 0,$$

and that $a_2(t) \geq k^2$ (k is some constant), all for $t \leq t_0$. Then every solution of Eq. (3.91) is bounded.

Proof: First multiply through Eq. (3.91) by u' to obtain

$$u''u' + a_1(u')^2 + a_2uu' = 0. \tag{3.105}$$

The first and third terms remind one of the product rule for derivatives:

$$\frac{d}{dt}(u')^2 = 2u'u''; \qquad \frac{d}{dt}(a_2u^2) = 2a_2uu' + a_2'u^2.$$

In order to use these we have to make some algebraic adjustments, as shown in these three lines, which follow from Eq. (3.105).

$$u''u' + a_2uu' = -a_1(u')^2,$$

$$u''u' + a_2uu' + \tfrac{1}{2}a_2'u^2 = -a_1(u')^2 + \tfrac{1}{2}a_2'u^2,$$

$$\frac{1}{2}\frac{d}{dt}((u')^2 + a_2u^2) = -a_1(u')^2 + \tfrac{1}{2}a_2'u^2. \tag{3.106}$$

In the hypothesis we assumed that $a_1(t) \geq 0$, $a_2'(t) \leq 0$. Since u^2 and $(u')^2$ are positive (or zero), the right-hand side of Eq. (3.106) cannot be positive. Neither can the left-hand side:

$$\frac{d}{dt}((u')^2 + a_2u^2) \leq 0. \tag{3.107}$$

Now if the derivative of a function is negative or zero, that function cannot increase. It must be less than or equal to its initial value. Applied to the function above, this means

$$(u'(t))^2 + a_2(t)u^2(t) \leq (u'(t_0))^2 + a_2(t_0)u^2(t_0). \tag{3.108}$$

On the other hand, $a_2(t)$ is never less than the positive constant k^2, so the left-hand side of Eq. (3.108) actually limits the size of $u^2(t)$:

$$k^2u^2(t) \leq (u'(t))^2 + a_2(t)u^2(t). \tag{3.109}$$

Now putting together Eqs. (3.108) and (3.109), we find a bound for $u^2(t)$:

$$u^2(t) \leq \frac{1}{k^2}((u'(t_0))^2 + a_1(t_0)u^2(t_0)). \tag{3.110}$$

We are usually not concerned about the actual magnitude of the right-hand side, but only that it be finite.

Example 3

The differential equation

$$\frac{d^2u}{dt^2} + \frac{2}{t+1}\frac{du}{dt} + u = 0$$

has $a_2 = 1$, $a_1 = 2/(t+1)$. These functions fulfill the hypotheses of Theorem 3.13, and so any solution of this equation must be bounded. In fact, this is true, since the general solution is

$$u(t) = \frac{1}{t+1}(c_1\cos t + c_2\sin t).$$

Exercises

1. Find the solutions of the Euler-Cauchy equation

$$t^2\frac{d^2u}{dt^2} + t\frac{du}{dt} + \omega^2 u = 0.$$

Do Theorems 3.12 and 3.13 correctly predict the fact that the solutions are bounded and oscillatory for $t > 0$?

2. Two independent solutions of the equation

$$\frac{d^2u}{dt^2} + \frac{1}{t^4}u = 0$$

are $t\sin t^{-1}$ and $t\cos t^{-1}$. Are these solutions oscillatory for $t > 1$?

3. Two independent solutions of the equation

$$\frac{d^2u}{dt^2} - \frac{2}{t}\frac{du}{dt} + \frac{2+t^2}{t^2}u = 0$$

are $t\sin t$ and $t\cos t$. Does Theorem 3.12 correctly predict the oscillatory nature of these solutions for $t > 1$?

4. It sometimes happens that a function $u(t)$ is required to satisfy a relation of the form

$$\frac{1}{u}\frac{d}{dt}\left(\rho(t)\frac{du}{dt}\right) = \lambda\sigma(t),$$

where ρ and σ are positive functions and λ is constant. If $u(t)$ is supposed to be oscillatory, show that λ must be negative.

5. The proof of Theorem 3.11 really shows that every solution of

$$\frac{d^2u}{dt^2} + p(t)u = 0 \qquad (*)$$

has zeros at least as close as those of $\cos kt$ if $p(t) \geq k^2$. (See Eqs. (3.94)–(3.97).) Use this fact to show that if $p(t) \geq k^2$ for $t_0 \leq t \leq t_0 + T$, then every solution of $(*)$ has at least $(kT/\pi) - 1$ zeros in that interval.

6. Following the same lines as the proof of Theorem 3.13, show that the solutions of Eq. (3.91) are bounded if $a_2 > 0$ and $2a_1a_2 + a_2' \geq 0$. (Show that the function $V = u^2 + (u')^2/a_2$ has a nonpositive derivative.)

7. Apply the test developed in Exercise 6 to show that the solutions of

$$\frac{d^2u}{dt^2} + \frac{1}{t}\frac{du}{dt} + \frac{\omega^2}{t^2}u = 0$$

are bounded. (Compare Exercise 1.)

8. The equation of a mass–spring–damper system without forcing is

$$M\frac{d^2u}{dt^2} + p\frac{du}{dt} + ku = 0.$$

(See Section 2.1.) Show that the function used in the proof of Theorem 3.13 (Eqs. 3.106 and 3.107),

$$V = (u')^2 + a_2u^2,$$

is proportional to the energy stored in the system (potential plus kinetic).

9. Use Theorems 3.12 and 3.13 to get conditions on α that will guarantee that the solutions of

$$\frac{d^2u}{dt^2} - \frac{2\alpha}{t}\frac{du}{dt} + \left(\omega^2 + \frac{\alpha^2 + \alpha}{t^2}\right)u = 0, \qquad t \geq 1,$$

are (a) oscillatory, (b) bounded.

10. The general solution of the equation in Exercise 9 is

$$u(t) = t^\alpha(c_1 \cos \omega t + c_2 \sin \omega t).$$

What condition on α is necessary and sufficient for boundedness?

11. Use Theorem 3.12 and Exercise 6 to show that the solutions of

$$\frac{d^2u}{dt^2} - \frac{1}{t}\frac{du}{dt} + 4t^2u = 0, \qquad t > 1,$$

are oscillatory and bounded. Notice that the coefficient of du/dt is negative.

12. Follow the directions for Exercise 11 for the Bessel equation

$$\frac{d^2u}{dt^2} + \frac{1}{t}\frac{du}{dt} + \left(\lambda^2 - \frac{k^2}{t^2}\right)u = 0.$$

Notes and References

The theory of linear homogeneous equations presented in this chapter runs parallel to the theory of linear homogeneous systems presented in Chapter 7. Indeed, systems of first-order equations provide the most satisfactory way to prove the theorems of Section 3.1. The idea of the Wronskian fits in particularly well with the matrix notation of systems.

The Euler-Cauchy equation is important for many reasons. It does arise naturally in some problems in cylindrical and spherical coordinates. Furthermore, this equation is a model for more complex equations. We shall make use of this modeling idea in Chapter 4 when we develop a modified power series solution.

The properties of solutions that are developed in Section 3.6 are often of importance in the theory of boundary value problems. Two good references are Birkhoff and Rota (1978) and Leighton (1967).

Miscellaneous Exercises

In Exercises 1–5 you are given the characteristic polynomial of a homogeneous differential equation with constant coefficients. (a) Find the differential equation. (b) Find its general solution.

1. $r^3 - 2r^2 + r$
2. $r^4 - 2r^2 + 1$
3. $r^3 - 1$
4. $r^8 - 2r^4 + 1$
5. $r^4 + 2r^3 - 2r - 1$

In Exercises 6–10 a set of functions is given. (a) Find a linear differential equation (order equal to the number of functions) having them as solutions. (b) Check them for independence by finding their Wronskian.

6. $1, e^t, e^{-t}$
7. e^t, te^t, t^2e^t
8. $1, t, e^{2t}$
9. $t^2, t^2 + t, t^2 + t + 1$
10. $1, 1 + t, 1 + t + t^2$

11. (a) Find a fundamental set for the differential equation $u''' - u' = 0$.
    (b) Find the Wronskian of the functions in your fundamental set.

12. Follow the directions for Exercise 11 for the differential equation $u^{(4)} - u = 0$.

13. A third-order Cauchy-Euler equation has the three functions $u_1 = t, u_2 = t \ln t, u_3 = t(\ln t)^2$ as solutions. Prove that they are independent and find the differential equation.

In Exercises 14–19, find a linear homogeneous differential equation with real, constant coefficients, and of lowest possible order, that has the given function(s) among its solutions.

14. $1, \cos t$ 15. t

16. $e^{-t}, e^{t}, \cosh t$ 17. $\sin t, \sin 2t$

18. $t^2 - 2t + 1$ 19. $t \cosh t$

In Exercises 20–30, find the general solution of the differential equation. Determine the constants to satisfy the initial conditions, if any are given.

20. $\dfrac{d^3 u}{dt^3} + \dfrac{du}{dt} = 0, \quad u(0) = 0, \quad u'(0) = 1, \quad u''(0) = 0$

21. $\dfrac{d^4 u}{dt^4} - u = 0, \quad u(0) = 0, \quad u'(0) = 0, \quad u''(0) = 0, \quad u'''(0) = 1$

22. $\dfrac{d^4 u}{dt^4} + u = 1 + t$

23. $\dfrac{d^8 u}{dt^8} - 2\dfrac{d^4 u}{dt^4} + u = 0$

24. $\dfrac{d^3 u}{dt^3} - 6\dfrac{d^2 u}{dt^2} + 11\dfrac{du}{dt} - 6u = e^t$

25. $t^2 u''' + 3tu'' + 2u' = 0$

26. $t^2 u'' + u = 0$

27. $t^2 u'' + 2tu' + u = 0$

28. $t^3 u''' + tu' - u = 0$

29. $2t^3 u''' + 9t^2 u'' + 2tu' - 2u = 0$

30. $r\dfrac{d}{dr}\left(r\dfrac{du}{dr}\right) + \lambda^2 u = 0$

31. Find a solution of the differential equation below that does not become infinite as $r \to 0$:

$$r\dfrac{d}{dr}\left(r\dfrac{du}{dr}\right) - 4u = 0.$$

32. Follow the directions for Exercise 31 for the differential equation

$$\dfrac{d}{d\rho}\left(\rho^2 \dfrac{du}{d\rho}\right) - 6u = 0.$$

In Exercises 33–38, find the general solution of the given differential equation.

33. $t^2 u'' - tu' + u = t \ln t$

34. $u'' + tu' + b(t - b)u = 0$ (Hint: there is a solution of the form $u = e^{mt}$ where m is a constant.)

35. $x\dfrac{d^2y}{dx^2} - (2x + 1)\dfrac{dy}{dx} + (x + 1)y = 0$ (Hint: e^x is a solution.)

36. $x\dfrac{d^2y}{dx^2} - (2x + 1)\dfrac{dy}{dx} + (x + 1)y = x^2$

37. $\dfrac{d^2u}{dt^2} + u = \sec t$

38. $\dfrac{d^2u}{dt^2} + \dfrac{1}{4t^2}u = \dfrac{1}{\sqrt{t}}$

39. Verify that $y_1(x) = x$ is a solution and find a second independent solution:

$$\dfrac{d^2y}{dx^2} + 2x\dfrac{dy}{dx} - 2y = 0.$$

40. Show that $u_1(t) = (e^t + e^{-t})^2$ and $u_2(t) = (e^t - e^{-t})^2$ are both solutions of $u''' - 4u' = 0$ but neither e^t nor e^{-t} is a solution. Which of the following, together with u_1 and u_2, form a fundamental set for the differential equation: 1, e^{2t}, $\cosh 2t$, $\sinh 2t$?

41. Find the Green's function for

$$\dfrac{d^2u}{dt^2} - 2\dfrac{du}{dt} + u = f(t).$$

42. Use the Green's function form for the solution of

$$\dfrac{d^2u}{dt^2} + \omega_0^2 u = \cos \omega t$$

to explain resonance when $\omega = \omega_0$.

4 Power Series Methods

4.1
Power Series Solutions

Although the theory developed in Chapter 3 tells us that a linear differential equation has a solution (under rather broad conditions), there are few significant equations with nonconstant coefficients that can be solved in terms of familiar functions such as exponentials, trigonometric functions, logarithms, etc. A case in point is Bessel's equation

$$x^2 \frac{d^2y}{dx^2} + x \frac{dy}{dx} + (x^2 - k^2)y = 0,$$

which is probably the most frequently occurring equation in applied mathematics, after the constant coefficients and Euler-Cauchy equations. (See Section 3.3.) Indeed, it is common to take the viewpoint that a differential equation, together with suitable normalizing conditions, defines its solutions, usually as new transcendental functions. For instance, we might define the exponential function e^x as the solution of

$$\frac{dy}{dx} = y, \qquad y(0) = 1. \tag{4.1}$$

How shall we find out about these new functions? To some extent their properties can be deduced directly from the differential equation. (This was done in Section 3.6.) But developing them in power series reveals many analytic properties and also permits calculating numerical values. This is the standard treatment for the solutions of the differential equations that come up time after time in physics and engineering. Thus we might obtain the power series for the exponential function

$$e^x = 1 + x + \frac{x^2}{2} + \cdots + \frac{x^n}{n!} + \cdots = \sum_{n=0}^{\infty} \frac{x^n}{n!}$$

by applying the ideas of this section to the initial value problem (4.1).

From the series, one can derive many of the significant properties of this function and calculate its values.

In most of the following sections of this chapter we will be finding power series solutions for second-order linear equations of the form

$$a_0(x)\frac{d^2y}{dx^2} + a_1(x)\frac{dy}{dx} + a_2(x)y = 0, \tag{4.2}$$

where the coefficients a_0, a_1, a_2 are polynomials. (The independent variable is written as x because it is usually a space variable rather than a time variable.)

The method we pursue is simply to assume that the solution has the form of a power series with center x_0:

$$y = \sum_{n=0}^{\infty} c_n(x - x_0)^n. \tag{4.3}$$

We substitute y, in this form, into the differential equation and then choose the coefficients in such a way that the equation is satisfied. The first two coefficients are arbitrary, corresponding to two initial conditions that may be imposed on the solution of the differential equation. The choice of x_0 may be made for convenience or may be forced on us by the circumstances of the problem. The only restriction is that the lead coefficient of Eq. (4.2) be nonzero at x_0: $a_0(x_0) \neq 0$.

Example 1

The differential equation

$$\frac{d^2y}{dx^2} - xy = 0 \tag{4.4}$$

(called Airy's equation) is easily solved by the power series method. We first choose $x_0 = 0$, so the series for y is

$$y = c_0 + c_1x + c_2x^2 + \cdots + c_nx^n + \cdots.$$

The differential equation calls for the second derivative of y and the product xy. In Table 4.1 we have written out the series expressions for y and its first two derivatives. In order to make our calculations easy, we write down for each series the first two or three terms and then the term containing some general power, which we designate as x^n. Like powers of x are kept in columns to minimize errors:

$$y'' = 2c_2 + 6c_3x + 12c_4x^2 + \cdots + (n + 2)(n + 1)c_{n+2}x^n + \cdots$$
$$-xy = \quad - c_0x \quad - c_1x^2 - \cdots \quad - c_{n-1}x^n - \cdots.$$

Now, $y'' - xy = 0$, according to the differential equation. Thus the sum of the series on the right-hand side must be 0. In fact, the net

Table 4.1
Preliminary
Calculations

$$y = c_0 + c_1 x + c_2 x^2 + c_3 x^3 + \cdots + c_{n-1}x^{n-1} + c_n x^n + c_{n+1}x^{n+1} + c_{n+2}x^{n+2} + \cdots$$

$$y' = c_1 + 2c_2 x + 3c_3 x^2 + \cdots + (n-1)c_{n-1}x^{n-2} + nc_n x^{n-1} + (n+1)\, c_{n+1}x^n + (n+2)c_{n+2}x^{n+1} + \cdots$$

$$y'' = 2c_2 + 6c_3 x + \cdots + (n-1)(n-2)c_{n-1}x^{n-3} + n(n-1)c_n x^{n-2} + (n+1)nc_{n+1}x^{n-1} + (n+2)(n+1)c_{n+2}x^n + \cdots$$

coefficient of each power of x must be 0. Therefore we find

$$2c_2 = 0 \qquad \text{(constant term)}$$
$$6c_3 - c_0 = 0 \qquad \text{(terms in } x\text{)},$$
$$12c_4 - c_1 = 0 \qquad \text{(terms in } x^2\text{)},$$
$$(n+2)(n+1)c_{n+2} - c_{n-1} = 0 \qquad \text{(terms in } x^n\text{)}. \tag{4.5}$$

These equations are now solved. The first three give

$$c_2 = 0, \qquad c_3 = \frac{c_0}{6}, \qquad c_4 = \frac{c_1}{12}.$$

We can solve Eq. (4.5) for c_{n+2} in terms of c_{n-1}:

$$c_{n+2} = \frac{c_{n-1}}{(n+2)(n+1)}. \tag{4.6}$$

One can easily check that this relation has already been satisfied for $n = 1$ and 2 in determining c_3 and c_4. By substituting successive values of n in Eq. (4.6) we find

$$c_5 = \frac{c_2}{5 \cdot 4} = 0 \qquad \text{(using } n = 3\text{)},$$

$$c_6 = \frac{c_3}{6 \cdot 5} = \frac{c_0}{6 \cdot 5 \cdot 6} \qquad \text{(using } n = 4\text{)},$$

$$c_7 = \frac{c_4}{7 \cdot 6} = \frac{c_1}{7 \cdot 6 \cdot 12} \qquad \text{(using } n = 5\text{)}.$$

This process can be carried on as far as necessary. The coefficients already found give us the series for $y(x)$, through x^7:

$$y = c_0 + c_1 x + \frac{c_0}{6}x^3 + \frac{c_1}{12}x^4 + \frac{c_0}{6 \cdot 5 \cdot 6}x^6 + \frac{c_1}{7 \cdot 6 \cdot 12}x^7 + \cdots$$

$$= c_0\left(1 + \frac{x^3}{6} + \frac{x^6}{180} + \cdots\right) + c_1\left(x + \frac{x^4}{12} + \frac{x^7}{504} + \cdots\right).$$

We may or may not be able to find the general coefficient c_n as a function of n. Usually we cannot because this amounts to solving a *difference equation* (for example, Eq. (4.5) or (4.6)) in closed form, a job almost as hard as solving the original differential equation in closed form. However, any number of the c's can be found from their difference equation, either manually or by computer.

Example 2

Find a power series solution, with center $x_0 = 0$, of

$$(1 - x)\frac{d^2y}{dx^2} + y = 0.$$

We assume that y is a power series as in Table 4.1. The terms of the differential equation for y are

$$y'' = 2c_2 + 6c_3x + 12c_4x^2 + \cdots + (n + 2)(n + 1)c_{n+2}x^n + \cdots,$$
$$-xy'' = \qquad -2c_2x - 6c_3x^2 - \cdots - \qquad (n + 1)nc_{n+1}x^n - \cdots,$$
$$y = c_0 + c_1x + c_2x^2 + \cdots + \qquad c_nx^n + \cdots.$$

The sum of the terms on the left-hand side is 0. Therefore when we sum the terms on the right-hand side, we must obtain a power series whose coefficients are all 0. Thus we find

$$2c_2 + c_0 = 0 \qquad \text{(from the constant terms),}$$
$$6c_3 - 2c_2 + c_1 = 0 \qquad \text{(from the terms in } x\text{),}$$
$$12c_4 - 6c_3 + c_2 = 0 \qquad \text{(from the terms in } x^2\text{),}$$
$$(n + 2)(n + 1)c_{n+2} - (n + 1)nc_{n+1} + c_n = 0 \qquad \text{(from the terms in } x^n\text{).}$$
$$\tag{4.7}$$

From these equations we may determine the coefficients $c_2, c_3, c_4, \ldots$ in terms of the arbitrary coefficients c_0 and c_1:

$$c_2 = -\frac{c_0}{2},$$

$$c_3 = \frac{2c_2 - c_1}{6} = -\frac{c_0 + c_1}{6},$$

$$c_4 = \frac{6c_3 - c_2}{12} = \frac{-(c_0 + c_1) - \left(-\dfrac{c_0}{2}\right)}{12} = \frac{-c_0 - 2c_1}{24}.$$

And the subsequent coefficients can be calculated by using the difference equation

$$c_{n+2} = \frac{(n + 1)nc_{n+1} - c_n}{(n + 2)(n + 1)}. \tag{4.8}$$

The first few terms of the solution are

$$y(x) = c_0 + c_1 x - \frac{c_0}{2}x^2 - \frac{c_0 + c_1}{6}x^3 - \frac{c_0 + 2c_1}{24}x^4 + \cdots$$

$$= c_0\left(1 - \frac{x^2}{2} - \frac{x^3}{6} - \frac{x^4}{24} + \cdots\right) + c_1\left(x - \frac{x^3}{6} - \frac{x^4}{12} + \cdots\right).$$

We have rather brashly attacked a couple of differential equations, performing operations on power series without questioning their legitimacy. At the moment we must consider our results merely as formal solutions. However, a well-developed theory is available to justify our method. Before stating the main theorem, we recall some results about Taylor series.

1. A power series $\sum_{n=0}^{\infty} c_n(x - x_0)^n$ (with numerical coefficients) has a radius of convergence $\rho \geq 0$: the series converges for all x satisfying $|x - x_0| < \rho$ and, if ρ is finite, diverges for all x satisfying $|x - x_0| > \rho$. The set of x's that satisfy $|x - x_0| < \rho$ is called the interval of convergence. (If $\rho = 0$, the series is useless.)

2. For all x in the interval of convergence the function that is the sum of the series has derivatives of all orders, and the series for any derivative can be found by differentiating the original series term by term (as was done in Table 4.1).

3. A function having a convergent power series with center x_0 has just one power series with that center. This implies, in particular, that a power series converges to 0 throughout an interval if and only if all its coefficients are 0.

4. Convergent power series with the same center x_0 can be handled algebraically as if they were polynomials of infinite degree. They may be added, subtracted, and multiplied by powers of $x - x_0$ term by term, and they may multiply each other.

The key theorem about the power series method is as follows.

Theorem 4.1

Let a_0, a_1, and a_2 be polynomials in $(x - x_0)$ with no common polynomial factor. If $a_0(x_0) \neq 0$, the differential equation

$$a_0(x)\frac{d^2y}{dx^2} + a_1(x)\frac{dy}{dx} + a_2(x)y = 0$$

has two independent solutions of the form of a power series with center x_0:

$$y(x) = \sum_{n=0}^{\infty} c_n(x - x_0)^n.$$

The radius of convergence of such a series is not less than the distance from x_0 to the nearest root (in the complex plane) of the equation $a_0(x) = 0$.

Example 3

Airy's equation, solved in Example 1, has coefficients $a_0(x) = 1$, $a_1(x) = 0$, $a_2(x) = -x$. Since $a_0(x)$ has no roots, the radii of convergence of the solution series are infinite.

In Example 2 the coefficient $a_0(x) = 1 - x$ has one root, at $x = 1$. Since the center was $x_0 = 0$, both of the series we found have radii of convergence at least equal to 1.

The theorem above, and the procedure for solution, may be extended to equations in which the coefficients $a_0(x), a_1(x), a_2(x)$ have convergent power series with common center x_0. The radius of convergence of the solution series may be smaller than what is predicted above, however. Worse yet, the computations may become insufferably complicated. Fortunately, the case where coefficients are polynomials includes many differential equations of real practical interest.

In the examples we have given, the center of the series has been $x_0 = 0$. If it is necessary to use a different center, we generally change variables from x to $z = x - x_0$ so that, in terms of z, the series has center $z_0 = 0$.

Example 4

Find the power series solution with center $x_0 = 1$ for

$$x(x - 2)y'' + (x^2 - 1)y = 0.$$

First we change the variables from x to $z = x - 1$. If we define $y(x) = y(z + 1) = w(z)$, the derivatives of y are

$$\frac{dy}{dx} = \frac{dw}{dz}, \qquad \frac{d^2y}{dx^2} = \frac{d^2w}{dz^2},$$

and the coefficients of the differential equation are

$$x(x - 2) = (z + 1)(z + 1 - 2) = z^2 - 1,$$

and

$$x^2 - 1 = (z + 1)^2 - 1 = z^2 + 2z.$$

Now the differential equation for w is

$$(z^2 - 1)w'' + (z^2 + 2z)w = 0.$$

The determination of the series solution for w is left as an exercise. (See Exercise 13.)

Exercises

Use the power series method to solve Problems 1–12. In each case, find (a) the difference equation for the coefficients of the power series, (b) the first four nonzero coefficients, (c) the radius of convergence of the series, and (d) the solution of the problem by other means, if possible. Take $x_0 = 0$ unless otherwise noted.

1. $\dfrac{dy}{dx} = xy$

2. $\dfrac{dy}{dx} = y, \ y(0) = 1$

3. $\dfrac{dy}{dx} + xy = 1, \ y(0) = 0$

4. $(1 + x^2)\dfrac{dy}{dx} - y = 0$

5. $\dfrac{d^2y}{dx^2} + y = 0, \ y(0) = 1, y'(0) = 0$

6. $\dfrac{d^2y}{dx^2} + y = 0, \ y(0) = 0, \ y'(0) = 1$

7. $(x - 1)^2\dfrac{d^2y}{dx^2} + (x - 1)\dfrac{dy}{dx} - y = 0$

8. $\dfrac{d^2y}{dx^2} - xy = 0, \ x_0 = 1$

9. $\dfrac{d^2y}{dx^2} - (1 + x^2)y = 0$

10. $\dfrac{d^2y}{dx^2} - y = \dfrac{1}{1 + x^2} = 1 - x^2 + x^4 - + \cdots,$
    $y(0) = 0, \ y'(0) = 0$

11. $\dfrac{d^2y}{dx^2} + x\dfrac{dy}{dx} + y = 0$

12. $\dfrac{d^2y}{dx^2} + 2x\dfrac{dy}{dx} - 2y = 0$

13. Find the series solution, through terms in $(x - 1)^6$, of the differential equation in Example 4.

4.2
Legendre's Equation

In many problems in spherical coordinates the differential equation

$$\frac{d}{d\phi}\left(\sin\phi\,\frac{du}{d\phi}\right) + k(k + 1)\sin\phi u = 0 \tag{4.9}$$

appears. Here ϕ is the colatitude, varying from 0 at the north pole to π at the south (see Fig. 4.1). The change of variables

$$x = \cos \phi, \qquad u(\phi) = u(\cos^{-1} x) = y(x) \tag{4.10}$$

converts Eq. (4.9) into *Legendre's equation:*

$$\frac{d}{dx}\left((1 - x^2)\frac{dy}{dx}\right) + k(k + 1)y = 0 \tag{4.11a}$$

or, when the indicated differentiation is carried out,

$$(1 - x^2)\frac{d^2y}{dx^2} - 2x\frac{dy}{dx} + k(k + 1)y = 0 \tag{4.11b}$$

In all of these equations, k is a real parameter.

Since the coefficients of Eq. (4.11) are polynomials, we may apply the method of Section 4.1 to find the solutions. If we take the center of the power series to be $x_0 = 0$, our solutions will have a radius of convergence at least equal to 1, according to Theorem 4.1. (Since Eq. (4.11) comes from Eq. (4.9), $x = \cos \phi$ would naturally be limited to the range $-1 \le x \le 1$ anyway.)

Thus we are looking for solutions in the form of a power series

$$y(x) = c_0 + c_1 x + c_2 x^2 + \cdots + c_n x^n + \cdots . \tag{4.12}$$

The series corresponding to the four terms of the differential equation are

Figure 4.1

Spherical Coordinates

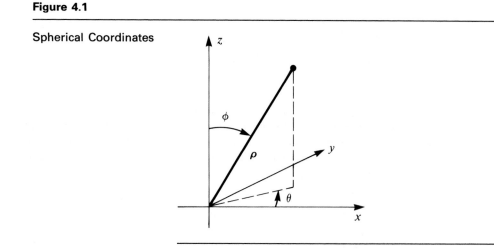

displayed below:

$$
\begin{array}{ll}
y'' = & 2c_2 + 6c_3x + \quad 12c_4x^2 + \cdots + (n+2)(n+1)c_{n+2}x^n + \cdots \\
-x^2y'' = & \quad\quad\quad\quad\quad -2c_2x^2 - \cdots \quad\quad -n(n-1)c_nx^n - \cdots \\
2xy' = & \quad\quad -2c_1x \quad -4c_2x^2 - \cdots \quad\quad\quad -2nc_nx^n - \cdots \\
k(k+1)y = & k(k+1)c_0 + k(k+1)c_1x + k(k+1)c_2x^2 + \cdots + \quad k(k+1)c_nx^n + \cdots
\end{array}
$$

Collecting the coefficients of the various powers of x and equating the results to 0 lead to these equations:

$$2c_2 + k(k+1)c_0 = 0,$$
$$6c_3 - (2 - k(k+1))c_1 = 0,$$
$$12c_4 - (6 - k(k+1))c_2 = 0,$$
$$(n+2)(n+1)c_{n+2} - (n(n-1) + 2n - k(k+1))c_n = 0. \tag{4.13}$$

The first three of these are actually special cases of the last one, which can be written more conveniently as

$$c_{n+2} = -\frac{(k-n)(k+n+1)}{(n+2)(n+1)}c_n. \tag{4.14}$$

(Notice that the coefficient of c_n in Eq. (4.13) has been simplified and factored.) A few of the coefficients are

$$c_2 = -\frac{k(k+1)}{2\cdot 1}c_0,$$

$$c_3 = -\frac{(k-1)(k+2)}{3\cdot 2}c_1,$$

$$c_4 = -\frac{(k-2)(k+3)}{4\cdot 3}c_2 = \frac{(k-2)k(k+1)(k+3)}{4\cdot 3\cdot 2\cdot 1}c_0,$$

$$c_5 = -\frac{(k-3)(k+4)}{5\cdot 4}c_3 = \frac{(k-3)(k-1)(k+2)(k+4)}{5\cdot 4\cdot 3\cdot 2}c_1.$$

Evidently, all of the coefficients with even indices will be multiples of c_0, and those with odd indices will be multiples of c_1. Thus the solution of the differential equation naturally separates into one that is a series of even powers of x and another that is a series of odd powers. We name these two functions y_0 and y_1:

$$y_0(x) = 1 - \frac{k(k+1)}{2!}x^2 + \frac{(k-2)k(k+1)(k+3)}{4!}x^4 - + \cdots$$

$$\tag{4.15}$$

$$y_1(x) = x - \frac{(k-1)(k+2)}{3!} x^3$$

$$+ \frac{(k-3)(k-1)(k+2)(k+4)}{5!} x^5 - + \cdots \qquad \textbf{(4.16)}$$

Since these two functions are independent, we have the general solution of Eq. (4.11) as

$$y(x) = c_0 y_0(x) + c_1 y_1(x). \qquad \textbf{(4.17)}$$

It is easy to see from Eq. (4.14) that if k is an integer, $c_{k+2} = 0$ and, consequently, $0 = c_{k+4} = c_{k+6} = \cdots$ also. In this most interesting case, one of the two functions $y_0(x)$ or $y_1(x)$ is a polynomial in x, while the other is a power series with radius of convergence equal to 1. The polynomial solution, appropriately normalized, is called the *Legendre polynomial* (or Legendre function of the first kind) *of order k* and is denoted by $P_k(x)$.

In order to study these polynomials, it is convenient to rewrite Eq. (4.14) as

$$c_n = -\frac{(n+1)(n+2)}{(k-n)(k+n+1)} c_{n+2}. \qquad \textbf{(4.18)}$$

Then, taking c_k as arbitrary (rather than c_0 or c_1), we work backwards from c_k to find

$$c_{k-2} = -\frac{(k-1)k}{2 \cdot (2k-1)} c_k \qquad (n = k-2),$$

$$c_{k-4} = -\frac{(k-3)(k-2)}{4(2k-3)} c_{k-2} \qquad (n = k-4),$$

$$= \frac{(k-3)(k-2)(k-1)k}{2 \cdot 4(2k-3)(2k-1)} c_k,$$

$$\vdots$$

$$c_{k-2r} = (-1)^r \frac{(k-2r+1)(k-2r+2)\cdots(k-1)k}{2 \cdot 4 \cdots (2r) \cdot (2k-2r+1)(2k-2r+3)\cdots(2k-1)} c_k.$$

The usual way to normalize these special solutions is by making $P_k(1) = 1$ for all k. In terms of the functions y_0 and y_1 in Eqs. (4.15) and (4.16),

$$P_k(x) = \begin{cases} y_0(x)/y_0(1) & \text{if } k \text{ is even,} \\ y_1(x)/y_1(1) & \text{if } k \text{ is odd.} \end{cases} \qquad \textbf{(4.19)}$$

It can then be proved that the last nonzero coefficient in P_k is c_k, with the formula

$$c_k = \frac{(2k)!}{2^k (k!)^2},$$ (4.20)

and that the general coefficient is

$$c_{k-2r} = (-1)^r \frac{(2k - 2r)!}{k!(k - r)!(k - 2r)!2^k}.$$ (4.21)

The first few of these polynomials are given in the Summary at the end of this section, and their graphs are shown in Fig. 4.2.

The second solution of Eq. (4.11) for integer values of k is called the Legendre function of the second kind of order k, written as $Q_k(x)$. The standard normalization of these solutions in terms of y_0 and y_1 in Eqs. (4.15) and (4.16) is

$$Q_k(x) = \begin{cases} y_0(1)y_1(x) & \text{if } k \text{ is even,} \\ -y_1(1)y_0(x) & \text{if } k \text{ is odd.} \end{cases}$$ (4.22)

Figure 4.2

Graphs of the First
Five Legendre
Polynomials

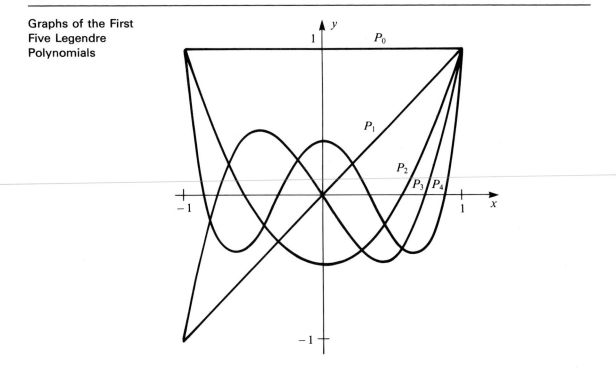

Summary

1. $(1 - x^2)\dfrac{d^2y}{dx^2} - 2x\dfrac{dy}{dx} + k(k + 1)y = 0$

is called Legendre's equation. Its general solution is

$y(x) = c_0 y_0(x) + c_1 y_1(x),$

where

$$y_0(x) = 1 - \frac{k(k + 1)}{2!}x^2 + \frac{(k - 2)k(k + 1)(k + 3)}{4!}x^4 - + \cdots$$

$$y_1(x) = x - \frac{(k - 1)(k + 2)}{3!}x^3$$

$$+ \frac{(k - 3)(k - 1)(k + 2)(k + 4)}{5!}x^5 - + \cdots.$$

These series converge for $|x| < 1$.

2. If k is a nonnegative integer, $k = 0, 1, 2, \ldots$, one of the series terminates, giving a polynomial solution.

3. The standard form for the polynomial solution is

$$P_k(x) = \begin{cases} y_0(x)/y_0(1) & \text{if } k \text{ is even,} \\ y_1(x)/y_1(1) & \text{if } k \text{ is odd.} \end{cases}$$

The first few polynomials are

$P_0(x) = 1$ $\qquad\qquad\qquad P_1(x) = x$

$P_2(x) = \frac{1}{2}(3x^2 - 1)$ $\qquad\qquad P_3(x) = \frac{1}{2}(5x^3 - 3x)$

$P_4(x) = \frac{1}{8}(35x^4 - 30x^2 + 3)$ $\qquad P_5(x) = \frac{1}{8}(63x^5 - 70x^3 + 15x)$

Exercises

1. Verify formula (4.20) for $k = 2, 3$. That is, compute the nonzero terms of the polynomial solution of Eq. (4.11b), normalize as in Eq. (4.19), and find the coefficient c_k.

2. Compute a polynomial solution of Eq. (4.11b) for $k = 4$. (That is, find $y_0(x)$.)

3. For $k = 1$, use $y = x$ as one solution and find a second independent solution by reduction of order. (See Section 3.4.)

4. Carry out the details of the change of variables mentioned in Eq. (4.10).

5. Write the functions $P_k(\cos \phi)$ for $k = 1, 2, 3, 4$. Expand $\cos^2 \phi$, etc. in terms of $\cos 2\phi$, $\sin 2\phi$, etc.

6. Show that $F_k(x) = (x^2 - 1)^k$ is a solution of the differential equation $(x^2 - 1)F' = 2xkF$.

7. Differentiate the equation above $k + 1$ times and show that $y = F_k^{(k)}(x)$ satisfies Legendre's equation (4.11b). (Hint: use the rule for the kth derivative of a product from the Appendix.)

8. Show that $F_k^{(k)}(x)$ is a polynomial of degree k and therefore is a multiple of $P_k(x)$. After the multiplier is determined, one obtains the *Rodrigues formula*:

$$P_k(x) = \frac{1}{k!2^k} \frac{d^k}{dx^k}[(x^2 - 1)^k].$$

9. The second-order differential equation

$$\frac{d^2y}{dx^2} - 2x\frac{dy}{dx} + 2ky = 0$$

is called *Hermite's equation*. Find the relation between the coefficients of the power series solution $(x_0 = 0)$. Show that the solution has the form

$$y = c_0y_0(x) + c_1y_1(x),$$

where $y_0(x)$ is a power series containing only even powers of x and y_1 is a series containing only odd powers of x.

10. Show that if k is an integer, one of the two solutions y_0 or y_1 is a polynomial of degree k. (When suitably normalized, these are called *Hermite polynomials*.)

11. Calculate the polynomial solution of Hermite's equation if $k = 2, 3$.

12. The differential equation

$$(1 - x^2)\frac{d^2y}{dx^2} - 2x\frac{dy}{dx} - y = 0$$

is not Legendre's equation (at least, not if k in Eq. (4.11) is to be real). Find the series solution with center $x_0 = 0$.

13. Prove that $y_0(x)$ and $y_1(x)$ in Eqs. (4.15) and (4.16) are independent by evaluating their Wronskian at $x = 0$. In what interval about 0 are they independent?

4.3
Frobenius' Method

In Section 4.1 we developed the power series method for solving a second-order linear equation,

$$a_0(x)\frac{d^2y}{dx^2} + a_1(x)\frac{dy}{dx} + a_2(x)y = 0. \tag{4.23}$$

One of the limitations imposed by Theorem 4.1 is that $a_0(x_0) \neq 0$: the center of the series solution must not coincide with a zero of $a_0(x)$. Unfortunately, such a point is often the most natural center for a series solution. This is the case for Bessel's equation,

$$x^2 \frac{d^2y}{dx^2} + x \frac{dy}{dx} + (x^2 - k^2)y = 0. \tag{4.24}$$

The choice of $x_0 = 0$ would seem to be the most convenient but is forbidden by Theorem 4.1.

In order to see why the power series method fails if $a_0(x_0) = 0$, let us recall the Euler-Cauchy equation,

$$x^2 \frac{d^2y}{dx^2} + ax \frac{dy}{dx} + by = 0, \tag{4.25}$$

in which a and b are constants. In Section 3.3 we found that at least one solution of this equation is a power of x, $y = x^r$. The exponent r must satisfy

$$r(r - 1) + ar + b = 0. \tag{4.26}$$

Now the only powers of x that have a power series development with center $x_0 = 0$ are $1, x, x^2, \ldots$ —the nonnegative integral powers. Thus it is quite likely that the solutions of Eq. (4.25) will *not* have a power series solution with center 0.

For example, the Euler-Cauchy equation

$$x^2 \frac{d^2y}{dx^2} + x \frac{dy}{dx} + \frac{y}{4} = 0 \tag{4.27}$$

has as its general solution

$$y(x) = c_0 x^{1/2} + c_1 x^{-1/2}.$$

No solution, other than $y(x) \equiv 0$, has a power series expansion about $x = 0$. Nevertheless, the solutions are not terribly complicated functions.

The example of the Euler-Cauchy equation leads us to hope for solutions of Eq. (4.23), in the case $a_0(x_0) = 0$, that act like powers of $x - x_0$, provided that the coefficients behave like those of the Euler-Cauchy equation near x_0. These considerations lead to the following definitions.

Definition 4.1

Let $a_0(x)$, $a_1(x)$, $a_2(x)$ have power series developments about x_0. Then the point x_0 is:

(a) a *regular point* of the differential equation (4.23) if both

$$\frac{a_1(x)}{a_0(x)} \quad \text{and} \quad \frac{a_2(x)}{a_0(x)}$$

have finite limits as $x \to x_0$;

(b) a *regular singular point* if it is not a regular point, but both

$$(x - x_0)\frac{a_1(x)}{a_0(x)} \quad \text{and} \quad (x - x_0)^2 \frac{a_2(x)}{a_0(x)}$$

have finite limits as $x \to x_0$;

(c) an *irregular singular point* otherwise.

Example 1

The equation of Legendre (see Section 4.2) is

$$(1 - x^2)y'' - 2xy' + k(k + 1)y = 0.$$

The ratios needed for testing points are

$$\frac{a_1(x)}{a_0(x)} = \frac{-2x}{1 - x^2}, \quad \frac{a_2(x)}{a_0(x)} = \frac{k(k + 1)}{1 - x^2}.$$

From these we can see that there are two regular singular points, at ± 1, and that all other points are regular points.

For Bessel's equation (4.24) the ratios to be studied are

$$\frac{a_1(x)}{a_0(x)} = \frac{x}{x^2}, \quad \frac{a_2(x)}{a_0(x)} = \frac{x^2 - k^2}{x^2}.$$

Evidently, the point $x_0 = 0$ is a regular singular point, while all others are regular points.

The following equation arises occasionally in the theory of beams:

$$x^4 y'' + y = 0, \tag{4.28}$$

It has an irregular singular point at $x_0 = 0$, but every other point is regular. From the general solution of this equation,

$$y(x) = x\left(c_0 \cos\left(\frac{1}{x}\right) + c_1 \sin\left(\frac{1}{x}\right)\right), \tag{4.29}$$

we can see that a power series expansion about the center $x_0 = 0$ does not exist.

From the definitions above, it is clear that Theorem 4.1 refers to a power series development about a regular point. Now we want to find a solution near a regular singlular point. The example of the Euler-Cauchy equation suggests the form of the solution as a power (not necessarily integral) times a series:

$$y(x) = (x - x_0)^r \sum_{n=0}^{\infty} c_n(x - x_0)^n. \tag{4.30}$$

The exponent r and the coefficients of the power series are to be chosen so that this is a solution of the differential equation. In order to make the exponent r quite definite, it is necessary to require that c_0 be nonzero. The use of the power-times-series form in Eq. (4.30) to solve differential equations near a regular singular point is called *Frobenius' method.*

Example 2

The differential equation

$$x^2 \frac{d^2 y}{dx^2} + x \frac{dy}{dx} - \frac{1 + 4x}{4} y = 0 \tag{4.31}$$

has a regular singular point at $x_0 = 0$. To study the behavior of the solutions near this point, we look for y in the form

$$y(x) = x^r \sum_0^\infty c_n x^n = \sum_0^\infty c_n x^{n+r}.$$

This series, and its derivatives, are shown in Table 4.2. With aid from the table we easily find the series corresponding to the terms of the differential equation, shown below:

$$x^2 y'' = r(r-1)c_0 x^r + (r+1)rc_1 x^{r+1} + \cdots + (r+n)(r+n-1)c_n x^{r+n} + \cdots$$
$$xy' = \qquad rc_0 x^r + (r+1)c_1 x^{r+1} + \cdots \qquad + (r+n)c_n x^{r+n} + \cdots$$
$$-xy = \qquad\qquad - c_0 x^{r+1} - \cdots \qquad\qquad - c_{n-1} x^{r+n} - \cdots$$
$$-\tfrac{1}{4}y = \qquad -\tfrac{1}{4}c_0 x^r \qquad -\tfrac{1}{4}c_1 x^{r+1} - \cdots \qquad\qquad -\tfrac{1}{4}c_n x^{r+n} - \cdots$$

As in the case of the simple power series solution, we maintain the terms in columns to expedite finding the net coefficients of each power of x. Since the four terms on the left must add up to 0, then so must the four series on the right, and the net coefficient of each power of x must be 0. Thus we obtain these equations:

$$(r(r-1) + r - \tfrac{1}{4})c_0 = 0, \tag{4.32}$$
$$((r+1)r + r + 1 - \tfrac{1}{4})c_1 - c_0 = 0, \tag{4.33}$$
$$\vdots$$
$$((r+n)(r+n-1) + r + n - \tfrac{1}{4})c_n - c_{n-1} = 0. \tag{4.34}$$

Table 4.2

$$
\begin{aligned}
y(x) &= \quad c_0 x^r \;+\; \qquad c_1 x^{r+1} + \qquad\qquad c_2 x^{r+2} + \cdots + \qquad\qquad c_n x^{r+n} \quad + \cdots \\
y'(x) &= \quad rc_0 x^{r-1} + \;(r+1)c_1 x^r \;+\; \qquad (r+2)c_2 x^{r+1} + \cdots + \qquad (r+n)c_n x^{r+n-1} + \cdots \\
y''(x) &= \; r(r-1)c_0 x^{r-2} + (r+1)rc_1 x^{r-1} + (r+2)(r+1)c_2 x^r \;+ \cdots + (r+n)(r+n-1)c_n x^{r+n-2} + \cdots
\end{aligned}
$$

The first of these equations says that c_0 times a polynomial in r must be 0. Since we have already agreed that c_0 must be nonzero, the polynomial in r must be zero. That is, r must be a root of the equation

$$r(r - 1) + r - \tfrac{1}{4} = 0, \tag{4.35}$$

called the *indicial* equation for the differential equation (4.31). Its roots are $\pm\tfrac{1}{2}$; we study the case $r = \tfrac{1}{2}$.

With this choice for r and some algebraic simplification, Eq. (4.34) becomes

$$c_n = \frac{1}{n(n + 1)} c_{n-1}. \tag{4.36}$$

(This includes Eq. (4.33) relating c_1 and c_0.) The first few coefficients are calculated as follows:

$$c_1 = \frac{1}{1 \cdot 2} c_0,$$

$$c_2 = \frac{1}{2 \cdot 3} c_1 = \frac{1}{2 \cdot 2 \cdot 3} c_0,$$

$$c_3 = \frac{1}{3 \cdot 4} c_2 = \frac{1}{2 \cdot 2 \cdot 3 \cdot 3 \cdot 4} c_0.$$

It is easy to see that a general formula for c_n is

$$c_n = \frac{c_0}{n!(n + 1)!} \tag{4.37}$$

and thus that one solution of the differential equation (4.31) is

$$y(x) = c_0 x^{1/2} \sum_{n=0}^{\infty} \frac{x^n}{n!(n + 1)!}. \tag{4.38}$$

Several useful observations can be made about this example. To begin with, requiring $c_0 \neq 0$ really was necessary for making r definite. Second, in contrast to the power series solutions of Section 4.1, c_1 was not found to be arbitrary. In fact, it was completely determined in terms of c_0. Finally, we should note that the choice $r = -\tfrac{1}{2}$ will not lead to a solution of the differential equation: for this value of r, the coefficient of c_1 in Eq. (4.33) is 0, so c_0 must be 0; this is a contradiction. Thus it may happen that we get only one solution by the present method, as suggested by the following theorem.

Theorem 4.2

Let x_0 be a regular singular point of the differential equation

$$a_0(x)\frac{d^2y}{dx^2} + a_1(x)\frac{dy}{dx} + a_2(x)y = 0,$$

in which $a_0(x)$, $a_1(x)$, $a_2(x)$ are polynomials without a common polynomial factor. Then there is at least one solution of the differential equation of the form

$$y(x) = (x - x_0)^r \sum_{n=0}^{\infty} c_n(x - x_0)^n,$$

where r is a root of the indicial equation,

$$r(r - 1) + p_0 r + q_0 = 0,$$

$$p_0 = \lim_{x \to x_0} (x - x_0)\frac{a_1(x)}{a_0(x)}, \qquad q_0 = \lim_{x \to x_0} (x - x_0)^2\frac{a_2(x)}{a_0(x)}.$$

The radius of convergence of the power series is not less than the distance from x_0 to the nearest (in the complex plane) root of $a_0(x) = 0$ (other than x_0).

Example 3

The radius of convergence of the series found in Example 2 is infinite. This confirms the statement in the theorem: $a_0(x) = 0$ has no root other than x_0 for the case in point.

On the other hand, the differential equation

$$x(2 - x)\frac{d^2y}{dx^2} + 2(1 - x)\frac{dy}{dx} + My = 0 \qquad (M = \text{constant}) \qquad \textbf{(4.39)}$$

has a regular singular point at $x = 0$. A solution centered at this point can be expected to have a series with radius of convergence 2.

Exercises

In Exercises 1–6, find at least one solution of the differential equation by the methods of this section. In all cases, $x_0 = 0$ is a regular singular point. (a) Find the indicial equation and its roots. (b) Using the larger of the two roots (if there are two real roots), find the first four nonzero coefficients of the series. (c) Find the general relation that determines the coefficients. (d) If possible, find a general formula for c_n.

1. $x\dfrac{d^2y}{dx^2} - y = 0$ 2. $x\dfrac{d^2y}{dx^2} + \dfrac{dy}{dx} - y = 0$

3. $4x^2 \dfrac{d^2y}{dx^2} + (1 + 4x)y = 0$ 4. $4x(1 - x)\dfrac{d^2y}{dx^2} - y = 0$

5. $x^2 \dfrac{d^2y}{dx^2} + x\dfrac{dy}{dx} + x^4 y = 0$ 6. $x(1 - x)\dfrac{d^2y}{dx^2} + y = 0$

7. Apply the method of this section to Eq. (4.39) with $x_0 = 0$. Find the indical equation, the first three coefficients of the power series, and the equation relating successive coefficients.

8. Show that Eq. (4.39) has a polynomial solution if $M = k(k + 1)$, k an integer.

9. Find the polynomial solution of Eq. (4.39) if $M = 12$ ($k = 3$).

10. Change the variable in Eq. (4.39) from x to $z = x - 1$. What is the resulting equation?

11. Find and classify (as regular or irregular) the singular points of these equations:

    (a) $(x^2 - 1)y'' + 4xy' + 2y = 0$

    (b) $(x - x^2)y'' + y = 0$

    (c) $(4 - x^2)^2 y'' + y' + \dfrac{y}{2x - x^2} = 0$

    (d) $xy'' + y' + y = 0$

12. Find the first few terms of the solution of Legendre's equation

    $$(1 - x^2)y'' - 2xy' + k(k + 1)y = 0$$

    near the point $x_0 = 1$. (Show that this is a regular singular point.)

4.4
Bessel's Equation I

Many problems of physics and engineering require the solution of *Bessel's equation*,

$$x^2 \frac{d^2y}{dx^2} + x\frac{dy}{dx} + (x^2 - k^2)y = 0. \tag{4.40}$$

In most applications, x is a multiple of the radial distance in a polar coordinate system, and k is a nonnegative parameter. We checked in Section 4.3 that the point $x_0 = 0$ is a regular singular point and, in fact, the only singular point.

As an illustration of the method of Frobenius we find a solution of Eq. (4.40) in the form

$$y(x) = x^r(c_0 + c_1 x + c_2 x^2 + \cdots). \tag{4.41}$$

The infinite series representing the various terms of the differential equation are displayed below.

$$x^2 y'' = r(r - 1)c_0 x^r + (r + 1)rc_1 x^{r+1} + (r + 2)(r + 1)c_2 x^{r+2} + \cdots + (r + n)(r + n - 1)c_n x^{r+n} + \cdots$$
$$xy' = \qquad rc_0 x^r + (r + 1)c_1 x^{r+1} + \qquad (r + 2)c_2 x^{r+2} + \cdots + \qquad (r + n)c_n x^{r+n} + \cdots$$
$$x^2 y = \qquad\qquad\qquad\qquad\qquad\qquad c_0 x^{r+2} + \cdots + \qquad c_{n-2} x^{r+n} + \cdots$$
$$-k^2 y = \quad -k^2 c_0 x^r \quad - k^2 c_1 x^{r+1} \qquad - k^2 c_2 x^{r+2} - \cdots - \qquad k^2 c_n x^{r+n} - \cdots.$$

Adding up the coefficients of x^r, the lowest power of x, and equating it to 0 give the equation

$$(r(r - 1) + r - k^2)c_0 = 0. \tag{4.42}$$

Since we have agreed that c_0 must be nonzero, the other factor in this equation must be 0:

$$r^2 - k^2 = 0. \tag{4.43}$$

This is the indicial equation for Bessel's equation. We choose the larger root $r = k \geq 0$ and proceed to find and equate to 0 the coefficients of the higher powers of x. The resulting equations are

$$((k + 1)^2 - k^2)c_1 = 0, \tag{4.44}$$

$$((k + 2)^2 - k^2)c_2 + c_0 = 0, \tag{4.45}$$

$$((k + n)^2 - k^2)c_n + c_{n-2} = 0, \qquad n \geq 2. \tag{4.46}$$

Because k is nonnegative, the factor in parentheses in Eq. (4.44) cannot be 0; therefore $c_1 = 0$. The next coefficient is

$$c_2 = -\frac{c_0}{(k + 2)^2 - k^2} = \frac{-c_0}{4 + 4k} \tag{4.47}$$

from Eq. (4.45). Similarly, Eq. (4.46) gives the relation

$$c_n = -\frac{c_{n-2}}{n^2 + 2nk} = -\frac{c_{n-2}}{n(n + 2k)}. \tag{4.48}$$

Evidently, all the coefficients with odd subscripts are 0 because they are all multiples of c_1. Those with even subscripts can be found from Eq. (4.48) as multiples of c_0. We have c_2 already. The next two are

$$c_4 = -\frac{c_2}{4(4 + 2k)} = \frac{c_0}{4(4 + 2k)2(2 + 2k)},$$

$$c_6 = -\frac{c_4}{6(6 + 2k)} = \frac{-c_0}{6(6 + 2k)4(4 + 2k)2(2 + 2k)}.$$

From this point it is easy to guess the general formula,

$$c_{2m} = \frac{(-1)^m c_0}{2 \cdot 4 \cdots 2m(2 + 2k)(4 + 2k) \cdots (2m + 2k)}$$

$$= \frac{(-1)^m c_0}{2^{2m} m!(1 + k)(2 + k) \cdots (m + k)}. \tag{4.49}$$

Naturally, we must verify that this really is a solution of the difference equation (4.48).

If k is an integer, it is customary to standardize the solution by choosing

$$c_0 = \frac{1}{2^k k!} \tag{4.50}$$

so that the general coefficient of Eq. (4.49) becomes

$$c_{2m} = \frac{(-1)^m}{m!(m + k)!2^{2m+k}}. \tag{4.51}$$

The resulting series defines a solution of Bessel's equation, called the *Bessel function of the first kind of* (integer) *order k*, and written as $J_k(x)$:

$$J_k(x) = \sum_{m=0}^{\infty} \frac{(-1)^m}{m!(m + k)!} \left(\frac{x}{2}\right)^{2m+k}. \tag{4.52}$$

The series converges for all x. Graphs of $J_0(x)$ and $J_1(x)$ are shown in Fig. 4.3.

If k is not an integer, the Bessel function is normalized by choosing

$$c_0 = \frac{1}{2^k \Gamma(1 + k)}. \tag{4.53}$$

The *gamma function* in the denominator is a function of a continuous variable that interpolates the factorial between the integers. (See Fig. 4.4.) It has the multiplicative property that

$$\Gamma(1 + k) = k\Gamma(k) \tag{4.54}$$

for all k. The definition $\Gamma(1) = 1$ then implies that

$$\Gamma(1 + n) = n!, \quad n = 0, 1, 2, \ldots. \tag{4.55}$$

On the other hand, $\Gamma(x)$ becomes infinite at $x = 0, -1, -2, \ldots$. As a consequence of Eq. (4.54), we can show that

$$\Gamma(1 + k) \cdot (1 + k)(2 + k) \cdots (m + k) = \Gamma(m + k + 1). \tag{4.56}$$

Thus the coefficient c_{2m} of the series for the Bessel function becomes

Figure 4.3

Graphs of the Bessel
Functions J_0, J_1

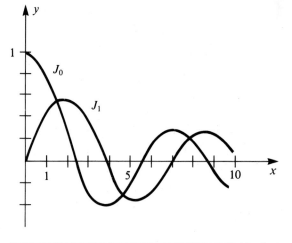

(from Eq. (4.49))

$$c_{2m} = \frac{(-1)^m}{2^{2m+k}m!\Gamma(m + k + 1)},$$ **(4.57)**

and the *Bessel function of the first kind of order k* is

$$J_k(x) = \sum_{m=0}^{\infty} \frac{(-1)^m}{m!\Gamma(m + k + 1)} \left(\frac{x}{2}\right)^{2m+k},$$ **(4.58)**

valid for all $k \geq 0$.

Now we try to find a second, independent solution of Bessel's equation for positive integer values of k. We return to the terms of the differential equation in series form. The second possible root of the indicial equation (4.43) is $r = -k$. If we use this value of r, the formula analogous to Eq. (4.46) is

$$((n - k)^2 - k^2)c_n + c_{n-2} = 0.$$ **(4.59)**

However, when $n = 2k$, this equation becomes $0 \cdot c_{2k} + c_{2k-2} = 0$. Since this cannot be solved for c_{2k}, we conclude that we cannot obtain a second solution by this method. (See Section 4.5.)

The situation is quite different when k is a nonintegral positive number; then the choice of $r = -k$ satisfies the indicial equation, while the rest of the calculations go through without difficulty. Choosing the

Figure 4.4

Graph of the Gamma
Function

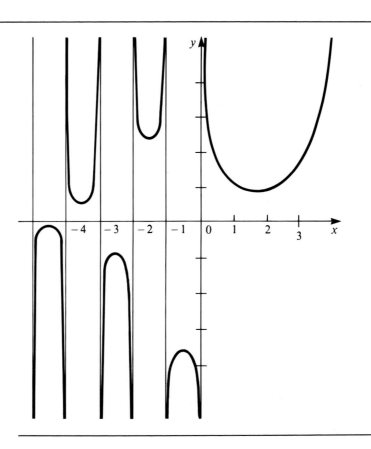

first coefficient as

$$c_0 = \frac{1}{2^{-k}\Gamma(1 - k)} \tag{4.60}$$

leads to the general coefficient

$$c_{2m} = \frac{(-1)^m}{2^{2m-k}m!\Gamma(m - k + 1)} \tag{4.61}$$

Because this is exactly analogous to Eq. (4.57) above, the power series solution found is called $J_{-k}(x)$:

$$J_{-k}(x) = \sum_{m=0}^{\infty} \frac{(-1)^m}{m!\Gamma(m - k + 1)}\left(\frac{x}{2}\right)^{2m-k} \qquad (k \neq 0, 1, 2, \ldots).$$

It remains to be seen that $J_k(x)$ and $J_{-k}(x)$ are independent solutions of Bessel's equation. In fact, this is easy to prove: the first terms of the two series are

$$J_k(x) = \frac{1}{\Gamma(k+1)} \left(\frac{x}{2}\right)^k + \cdots,$$

$$J_{-k}(x) = \frac{1}{\Gamma(-k+1)} \left(\frac{x}{2}\right)^{-k} + \cdots.$$

It is obvious that the ratio of these functions cannot be constant and therefore that they are indeed independent.

Because Bessel's equation is quite common in applications, it is important that we know something about the nature of the Bessel functions. Since the series (4.58) is now seen to be valid for all k except $-1, -2, \ldots$, we may say that $J_k(x)$ behaves approximately like the first term of its series, when x is very nearly 0. That is,

$$J_k(x) \cong \frac{1}{\Gamma(k+1)} \left(\frac{x}{2}\right)^k, \qquad |x| \ll 1. \tag{4.62}$$

Furthermore, we already determined in Section 3.6 that every solution of Bessel's equation is oscillatory—there are infinitely many solutions of $J_k(x) = 0$. These zeros of the Bessel function are important in themselves. The first few zeros of J_0 and J_1 are shown in the Summary at the end of this section.

We also showed in Section 3.6 that every solution of Bessel's equation is bounded. In fact, $J_k(x)$ approaches 0 as x increases, in such a way that $\sqrt{x} J_k(x)$ remains bounded.

Summary

1. $x^2 \dfrac{d^2 y}{dx^2} + x \dfrac{dy}{dx} + (x^2 - k^2)y = 0$

is called Bessel's equation of order k.

2. The Bessel function of the first kind of order k is defined by the power series

$$J_k(x) = \sum_{m=0}^{\infty} \frac{(-1)^m}{m!\,\Gamma(m+k+1)} \left(\frac{x}{2}\right)^{2m+k}.$$

The series converges for all x.

3. If k is not an integer, the general solution of Bessel's equation may be written

$$y(x) = AJ_k(x) + BJ_{-k}(x).$$

That is, J_k and J_{-k} are independent solutions.

If k is an integer, $J_k(x)$ is a solution of Bessel's equation, but $J_{-k}(x)$ is not an independent solution.

4. Every solution of Bessel's equation is bounded as $x \to \infty$ and oscillatory. Approximate values for the first few positive solutions of $J_0(\alpha_n) = 0$ and $J_1(\beta_n) = 0$ are given in Table 4.3.

Table 4.3	n	α_n	β_n
Zeros of J_0 and J_1	1	2.4048	3.8317
	2	5.5201	7.0156
	3	8.6537	10.1735
	4	11.7915	13.3237
	5	14.9309	16.4706

Exercises

1. The differential equation

$$x^2 \frac{d^2y}{dx^2} + x\frac{dy}{dx} - (x^2 + k^2)y = 0$$

is called the *modified Bessel equation*. Apply the Frobenius method to find a solution in the power-times-series form. Use center $x_0 = 0$.

2. Write out the first few terms of the series solutions of the modified Bessel equation for $k = 0$ and $k = 1$. Use the same c_0 that was used in the text for the Bessel functions, Eq. (4.50).

3. Verify that $y = J_k(\lambda x)$ is a solution of the equation

$$x^2 y'' + xy' + (\lambda^2 x^2 - k^2)y = 0.$$

4. If a circular plate of radius c is exposed to convection in a medium at temperature T, it can be shown that the temperature $u(r)$ is governed by the differential equation

$$\frac{1}{r}\frac{d}{dr}\left(r\frac{du}{dr}\right) = -\gamma^2(T - u), \qquad 0 < r < c,$$

where γ is a constant and r is a radial coordinate measured from the center of the plate. Show that (a) $u_p = T$ is a particular solution and (b) the corresponding homogeneous equation is equivalent to

$$r^2 \frac{d^2u}{dr^2} + r\frac{du}{dr} - \gamma^2 r^2 u = 0.$$

5. What is the relationship between the equation in Exercise 4 and the equation in Exercise 1?

6. Call the solution of the equation in Exercise 1 $I_k(x)$. Show that

$$u(r) = T + (T_1 - T)I_0(\gamma r)/I_0(\gamma c)$$

is a solution of the nonhomogeneous equation in Exercise 4 that satisfies the condition $u(c) = T_1$.

7. Use the method of reduction of order (Section 3.4) to find a second solution of Bessel's equation (4.40) with $k = 0$.

8. From the result of Exercise 7, show that the second solution of Bessel's equation with $k = 0$ behaves like a logarithm near $x = 0$.

9. Make the substitution $y = v/\sqrt{x}$ for Bessel's equation and find the differential equation for v.

10. Using the infinite series, Eq. (4.58), prove that

$$J_{1/2}(x) = \frac{1}{\sqrt{2}\Gamma(\frac{3}{2})}\frac{\sin x}{\sqrt{x}}.$$

11. Use the equation found for Exercise 9 to show that when $k = \frac{1}{2}$, the general solution of Bessel's equation is

$$y(x) = \frac{1}{\sqrt{x}}(A\cos x + B\sin x).$$

12. Using the series Eq. (4.58) for $J_0(x)$ and $J_1(x)$, show that

$$\frac{dJ_0}{dx} = -J_1.$$

4.5

The Second Solution

Theorem 4.2 at the end of Section 4.3 states that *one* solution of a second-order linear differential equation has the expected form

$$(x - x_0)^r \sum_{n=0}^{\infty} c_n(x - x_0)^n \tag{4.63}$$

near a regular singular point x_0 but promises nothing about a second solution. Our objectives in this section are to find the conditions under which the second solution also has the form (4.63) and to find the correct form for the second solution in the remaining, exceptional, cases.

In order to simplify matters we assume that the regular singular point is $x_0 = 0$ and rearrange the coefficients of the differential equation (4.23) of Section 4.3 to obtain

$$x^2\frac{d^2y}{dx^2} + xp(x)\frac{dy}{dx} + q(x)y = 0. \tag{4.64}$$

In terms of the former coefficient functions a_0, a_1, and a_2, we have

$$p(x) = x\frac{a_1(x)}{a_0(x)}, \qquad q(x) = x^2\frac{a_2(x)}{a_0(x)}. \tag{4.65}$$

It is part of the definition of regular singular point that $p(x)$ and $q(x)$ are finite at $x = 0$. Furthermore, since we have been assuming that the a's had power series developments, it follows that $p(x)$ and $q(x)$ have also:

$$p(x) = p_0 + p_1x + p_2x^2 + \cdots, \qquad q(x) = q_0 + q_1x + q_2x^2 + \cdots. \tag{4.66}$$

From Eqs. (4.65) and (4.66) and Theorem 4.2 of Section 4.3 it follows that the indicial equation for Eq. (4.64) is

$$r(r - 1) + p_0r + q_0 = 0. \tag{4.67}$$

In order to better understand the Frobenius method and the question of the second solution, we shall carry out the general computation for solving Eq. (4.64). We assume that the solution is

$$y(x) = x^r \sum_{n=0}^{\infty} c_nx^n = \sum_{n=0}^{\infty} c_nx^{n+r}. \tag{4.68}$$

Because we are working in general, it is convenient to use the summation notation instead of writing out terms of each series. We need the series

$$xy'(x) = \sum_{n=0}^{\infty} (n + r)c_nx^{n+r}, \qquad x^2y''(x) = \sum_{n=0}^{\infty} (n + r)(n + r - 1)c_nx^{n+r}. \tag{4.69}$$

Next we calculate the series for the products appearing in the differential equation. The first one is

$$\begin{aligned} q(x)y(x) &= (q_0 + q_1x + q_2x^2 + \cdots)(c_0 + c_1x + c_2x^2 + \cdots) \\ &= (q_0c_0 + (q_1c_0 + q_0c_1)x + (q_2c_0 + q_1c_1 + q_0c_2)x^2 + \cdots) \\ &= \sum_{n=0}^{\infty} \left(\sum_{k=0}^{n} q_{n-k}c_k \right)x^{n+r}. \end{aligned} \tag{4.70}$$

A similar calculation gives the series for the term in y':

$$p(x)xy'(x) = \sum_{n=0}^{\infty} \left(\sum_{k=0}^{n} p_{n-k}(k + r)c_k \right)x^{n+r}. \tag{4.71}$$

From Eqs. (4.69), (4.70), and (4.71) above, we gather the series for the three terms of the differential equation and set their sum to 0:

$$\sum_{n=0}^{\infty} (n + r)(n + r - 1)c_n x^{n+r} + \sum_{n=0}^{\infty} \left(\sum_{k=0}^{n} p_{n-k}(k + r)c_k \right) x^{n+r}$$

$$+ \sum_{n=0}^{\infty} \left(\sum_{k=0}^{n} q_{n-k}c_k \right) x^{n+r} = 0. \tag{4.72}$$

As usual, we argue that the net coefficient of each power of x must be 0 in order for the sum of the series to equal 0 for a whole interval of values of x. The resulting equation is

$$(n + r)(n + r - 1)c_n + \sum_{k=0}^{n} p_{n-k}(k + r)c_k + \sum_{k=0}^{n} q_{n-k}c_k = 0. \tag{4.73}$$

From this we collect on one side the terms containing c_n, obtaining

$$[(n + r)(n + r - 1) + p_0(n + r) + q_0]c_n = -\sum_{k=0}^{n-1} (p_{n-k}(k - r) + q_{n-k})c_k. \tag{4.74}$$

This equation is valid for $n = 1, 2, 3, \ldots$. For $n = 0$, Eq. (4.73) reads

$$(r(r - 1) + p_0 r + q_0)c_0 = 0. \tag{4.75}$$

Since we require c_0 to be nonzero, we conclude that r must satisfy the indicial equation (4.67). If we define the *indicial polynomial*

$$I(r) = r(r - 1) + p_0 r + q_0, \tag{4.76}$$

we can recognize that the indicial equation is $I(r) = 0$ and that Eq. (4.74) may be rewritten as

$$I(n + r)c_n = -\sum_{k=0}^{n-1} (p_{n-k}(k - r) + q_{n-k})c_k. \tag{4.77}$$

This is the equation that must be solved for the c's if we are to find a solution of the differential equation in the assumed form.

Let us designate the roots of the indicial equation (4.75) as r_1 and r_2:

$$r_1 = \frac{1 - p_0 + \sqrt{(p_0 - 1)^2 - 4q_0}}{2}, \qquad r_2 = \frac{1 - p_0 - \sqrt{(p_0 - 1)^2 - 4q_0}}{2}. \tag{4.78}$$

The different cases that arise depend on the relationship between these roots. We assume, of course, that $p(x)$ and $q(x)$ are real-valued functions.

Case 1: r_1 and r_2 are conjugate complex numbers. In this case, $I(r_1) = I(r_2) = 0$, but $I(r + n) \neq 0$ for all $n = 1, 2, 3, \ldots$. Thus the difference equation (4.77) can be solved for the coefficients corresponding to each of the roots. The resulting functions are clearly independent solutions of the differential equation. The complications arising from the complex quantities in this case are so great that it is never used in practice.

Case 2: $r_1 = r_2$; the indicial equation has a double root. Here the assumed form obviously gives one, but only one, solution. We must seek the second solution in another form.

Case 3: r_1 and r_2 are real and distinct. First, we consider the choice $r = r_1$. Because r_1 is the larger root of the indicial polynomial, $I(r_1 + n)$ is positive for all $n = 1, 2, 3, \ldots$, and Eq. (4.77) may be solved for the c's—as, in fact, we have done in several examples.

Now consider what happens with $r = r_2$, the smaller root of the indicial equation. If $I(r_2 + n)$ is nonzero for $n = 1, 2, 3, \ldots$, then—as before—the c's corresponding to this choice can be found, providing us with a second solution, necessarily independent. When does this process break down? It *may* break down if $I(r_2 + m) = 0$ for some positive integer m. Since $I(r)$ is a second-degree polynomial, this means that $r_2 + m = r_1$ or, in other words, $r_1 - r_2 = m$. This is the other instance in which the usual form may fail to provide a solution and one that we have seen (Example 2 in Section 4.3).

The findings above can be summarized this way: the second solution of the differential equation (4.64) *cannot* have the power-times-series form (4.68) if the indicial equation has a double root and *might not* have that form if the two roots differ by a positive integer.

The investigations above have identified the cases in which the usual form is not adequate for the second solution. Our experience with the Euler-Cauchy equation (Section 3.3) suggests that logarithms might play a role—and they do indeed, as Theorem 4.3 indicates.

Theorem 4.3

Let $r_1 \geq r_2$ be real roots of the indicial equation for the differential equation

$$x^2 \frac{d^2y}{dx^2} + xp(x)\frac{dy}{dx} + q(x)y = 0,$$

where p and q have power series expansions with center 0; and let $y_1(x)$ be a solution of the form

$$y_1(x) = x^{r_1} \sum_{n=0}^{\infty} c_n x^n, \qquad c_0 = 1.$$

Then a second, independent solution has the following form:
1. $y_2(x) = x^{r_2} \sum_{n=0}^{\infty} b_n x^n$, $b_0 = 1$, if $r_1 - r_2$ is not an integer;
2. $y_2(x) = y_1(x) \ln x + x^{r_1} \sum_{n=1}^{\infty} b_n x^n$ if $r_1 = r_2$ is a double root;

3. $y_2(x) = \beta y_1(x) \ln x + x^{r_2} \sum_{n=0}^{\infty} b_n x^n$ if $r_1 - r_2$ is a positive integer (β may be 0).
The radius of convergence of the series in y_1 and y_2 is not less than the smaller of the radii of convergence of the series for p and q.

A couple of examples will make clearer the effects of the logarithmic terms.

Example 1

The differential equation

$$x^2 \frac{d^2 y}{dx^2} + x \frac{dy}{dx} - xy = 0 \tag{4.79}$$

has 0 as a double root of its indicial equation. We easily find that the first solution is

$$y_1(x) = \sum_{n=0}^{\infty} \frac{x^n}{(n!)^2}. \tag{4.80}$$

The second solution has the form predicted in case (2) of Theorem 4.3. The three terms in the differential equation are

$$-xy = -xy_1 \ln x - \sum_{n=1}^{\infty} b_n x^{n+1},$$

$$xy' = xy_1' \ln x + x \cdot \frac{1}{x} y_1 + \sum_{n=1}^{\infty} n b_n x^n,$$

$$x^2 y'' = x^2 y_1'' \ln x + x^2 \cdot \frac{2}{x} y_1' + x^2 \left(-\frac{1}{x^2}\right) y_1 + \sum_{n=1}^{\infty} n(n-1) b_n x^n.$$

When these are added, we obtain

$$0 = \ln x (x^2 y_1'' + xy_1' - xy_1) + 2xy_1' + \sum_{n=1}^{\infty} n^2 b_n x^n - \sum_{n=1}^{\infty} b_n x^{n+1}.$$

The fact that y_1 is a solution of the differential equation means that the logarithm disappears from this equation. We are left with

$$0 = 2xy_1' + (b_1 x + 2^2 b_2 x^2 + \cdots + n^2 b_n x^n + \cdots)$$
$$- (b_1 x^2 + b_2 x^3 + \cdots + b_{n-1} x^n + \cdots).$$

Now we replace y_1 with its series expansion (recall that $r_1 = 0$) to get

$$0 = \sum_{n=1}^{\infty} \frac{2n x^n}{(n!)^2} + b_1 x + \sum_{n=2}^{\infty} (n^2 b_n - b_{n-1}) x^n. \tag{4.81}$$

We collect the net coefficient of each power of x and equate it to 0. The resulting equations are

$$2 + b_1 = 0 \qquad \text{(terms in } x\text{)},$$

$$\frac{4}{(2!)^2} + 4b_2 - b_1 = 0 \qquad \text{(terms in } x^2\text{)},$$

$$\frac{2n}{(n!)^2} + n^2 b_n - b_{n-1} = 0 \qquad \text{(terms in } x^n\text{)}. \tag{4.82}$$

Finally, the coefficients $b_1, b_2, \ldots,$ are determined from Eq. (4.82). The final result is that

$$b_n = -\frac{2h_n}{(n!)^2}, \qquad h_n = 1 + \frac{1}{2} + \cdots + \frac{1}{n}. \tag{4.83}$$

Thus the second independent solution of Eq. (4.79) is

$$y_2(x) = y_1(x) \ln x + \sum_{n=1}^{\infty} b_n x^n$$

with $y_1(x)$ and b_n given above.

Example 2

The differential equation

$$x^2 \frac{d^2 y}{dx^2} - xy = 0 \tag{4.84}$$

has indicial polynomial $r(r - 1)$, the roots of which are 1 and 0. Corresponding to the larger root, we find the solution

$$y_1(x) = x \sum_{n=0}^{\infty} c_n x^n, \qquad c_n = \frac{1}{n!(n+1)!}. \tag{4.85}$$

Thus we seek a second solution in the form

$$y_2(x) = \beta y_1(x) \ln x + \sum_{n=0}^{\infty} b_n x^n$$

as predicted by case 3 of Theorem 4.3. We insert y_2 in this form into the differential equation, finding here, as in Example 1, that the terms containing logarithms disappear, leaving

$$0 = \beta(2xy_1' + y_1) + \sum_{n=0}^{\infty} n(n-1)b_n x^n - x \sum_{n=0}^{\infty} b_n x^n. \tag{4.86}$$

Now we replace y_1 by its series, collect terms in the various powers of x, and equate the net coefficients to 0. The calculations are shown in Table 4.4.

Table 4.4

$$\beta \cdot 2xy_1' = \beta(2c_0x + 2 \cdot 2c_1x^2 + 2 \cdot 3c_2x^3 + \cdots + \qquad 2nc_{n-1}x^n + \cdots)$$
$$-\beta y_1 = \beta(-c_0x \qquad -c_1x^2 \qquad -c_2x^3 - \cdots \qquad -c_{n-1}x^n - \cdots)$$
$$\sum_{n=0}^{\infty} n(n-1)b_nx^n = \qquad 2 \cdot 1b_2x^2 + 3 \cdot 2b_3x^3 + \cdots + n(n-1)b_nx^n + \cdots$$
$$-x\sum_{n=0}^{\infty} b_nx^n = \qquad -b_0x \qquad -b_1x^2 \qquad -b_2x^3 - \cdots \qquad -b_{n-1}x^n - \cdots$$

$$\beta c_0 - b_0 = 0 \qquad \text{(terms in } x)$$
$$\beta \cdot 3c_1 + 2b_2 - b_1 = 0 \qquad \text{(terms in } x^2)$$
$$\beta \cdot 5c_2 + 3 \cdot 2b_3 - b_2 = 0 \qquad \text{(terms in } x^3)$$
$$\vdots$$
$$\beta(2n-1)c_{n-1} + n(n-1)b_n - b_{n-1} = 0 \qquad \text{(terms in } x^n)$$

We must first decide whether $\beta = 0$ or not. If β is chosen equal to 0, then $b_0 = 0$. The rest of the b's need not be 0, but if we take $b_1 \neq 0$, the solution we come up with actually starts with the first power of x and—it turns out—is just a multiple of $y_1(x)$. Thus β must be nonzero, and we take $\beta = 1$ for convenience. Now the equation arising from the coefficient of terms in x (see Table 4.4) requires that $b_0 = c_0$, and c_0 is 1. The next equation, arising from the coefficient of terms in x^2 is

$$3c_1 + 2b_2 - b_1 = 0.$$

The choice of b_1 is arbitrary; a nonzero choice simply incorporates into $y_2(x)$ a multiple of $y_1(x)$. This is perfectly acceptable, and in this case we take $b_1 = -1$ for convenience. Then the general equation

$$n(n-1)b_n = b_{n-1} - (2n-1)c_{n-1} \qquad (4.87)$$

can be solved in nearly closed form for $n = 1, 2, 3, \ldots$:

$$b_n = \frac{-d_n}{n!(n-1)!}, \qquad d_n = 2\left(1 + \frac{1}{2} + \cdots + \frac{1}{n}\right) - \frac{1}{n}. \qquad (4.88)$$

Example 3

To show that the logarithm is not always necessary for obtaining a second independent solution, we recall from Section 4.4 the Bessel equation with

$k = \frac{1}{2}$:

$$x^2 \frac{d^2y}{dx^2} + x\frac{dy}{dx} + (x^2 - \frac{1}{4})y = 0.$$

The roots of the indicial equation

$$r^2 - \frac{1}{4} = 0$$

are $r = \pm\frac{1}{2}$, with difference equal to 1. Thus we should apply case 3 of Theorem 4.3 to find the second solution. However, we have already found that $J_{1/2}(x)$ and $J_{-1/2}(x)$ are independent solutions of this equation, and neither of these functions has a logarithmic term. Hence in this case, $\beta = 0$. (See Exercise 1.)

Exercises

1. Show that the second solution of the Bessel equation with $k = \frac{1}{2}$,

$$x^2 \frac{d^2y}{dx^2} + x\frac{dy}{dx} + (x^2 - \frac{1}{4})y = 0,$$

has no logarithmic term: start the calculations for the second solution using case 3 of Theorem 4.3. Abbreviate $y_1 = J_{1/2}(x)$ as

$$\sum_0^\infty c_n x^{n+1/2}, \qquad c_0 \neq 0.$$

2. Suppose that a second solution of the differential equation (4.64) is to have the form $y_2(x) = \beta y_1(x) \ln x + v$. Show that v must satisfy the differential equation

$$x^2 \frac{d^2v}{dx^2} + xp(x)\frac{dv}{dx} + q(x)v + \beta\left(2x\frac{dy_1}{dx} + (p(x) - 1)y_1\right) = 0.$$

3. The Bessel equation with $k = 1$,

$$x^2\frac{d^2y}{dx^2} + x\frac{dy}{dx} + (x^2 - 1)y = 0,$$

has as one solution $y_1 = J_1(x)$, which we write as $y_1 = c_0 x + c_1 x^2 + c_2 x^3 + \cdots$. (Recall that $c_1 = c_3 = c_5 = \cdots = 0$.) Determine whether the second solution has a logarithmic term.

4. Does the second solution of Legendre's equation near $x_0 = 1$ have a logarithmic term?

5. Find the second solution of the equation

$$x\frac{d^2y}{dx^2} - y = 0$$

near $x_0 = 0$. See Exercise 1 of Section 4.3.

6. Same task as Exercise 5 for the differential equation

$$4x^2\frac{d^2y}{dx^2} + (1 + 4x)y = 0$$

from Exercise 3 of Section 4.3.

7. What form should be used for the second solution of the differential equation of Exercise 5 of Section 4.3?

4.6

Bessel's Equation II

We are now in a position to complete our study of Bessel's equation,

$$x^2\frac{d^2y}{dx^2} + x\frac{dy}{dx} + (x^2 - k^2)y = 0. \tag{4.89}$$

In Section 4.4 we applied the Frobenius method to find one solution

$$J_k(x) = \sum_{m=0}^{\infty} \frac{(-1)^m}{m!\Gamma(m + k + 1)}\left(\frac{x}{2}\right)^{2m+k}, \tag{4.90}$$

corresponding to the larger root of the indicial equation,

$$r^2 - k^2 = 0. \tag{4.91}$$

According to the theory in Section 4.5, the second solution cannot have the power-times-series form if $k = 0$, since then 0 is a double root of the indicial equation. Furthermore, the second solution might not have the power-times-series form if the two roots $\pm k$ differ by an integer—that is, if $2k$ is an integer. Nevertheless, we have seen that $J_k(x)$ and $J_{-k}(x)$ are indeed independent solutions when $2k$ is an *odd* integer: $k = \frac{1}{2}, \frac{3}{2}$, etc.

In the remaining cases, when $k = 1, 2, 3$, etc., the series (4.90) for $J_{-k}(x)$ gives a function dependent on $J_k(x)$. Usually, one takes $\Gamma(z) = \infty$ and its reciprocal equal to 0, if $z = 0, -1, -2, \ldots$. With this agreement the relation between J_k and J_{-k} for integer k is

$$J_{-k}(x) = (-1)^k J_k(x), \qquad k = 0, 1, 2, \ldots. \tag{4.92}$$

Now we set about the problem of finding a second solution of Bessel's equation in the cases for which the series (4.90) does not provide both solutions. First we assume $k = 0$, so that the differential equation (4.89) becomes

$$x^2\frac{d^2y}{dx^2} + x\frac{dy}{dx} + x^2y = 0. \tag{4.93}$$

Of course, a common factor of x could be cancelled from all terms.

Taking the first solution to be

$$y_1(x) = J_0(x),$$

we must seek the second solution in the form dictated by case 2 of Theorem 4.3 in Section 5:

$$y_2(x) = y_1(x) \ln x + \sum_{n=1}^{\infty} b_n x^n. \qquad (4.94)$$

After substituting y_2 in this form into the differential equation (4.93) we are left with

$$2xy_1' + \sum_{n=1}^{\infty} n^2 b_n x^n + \sum_{n=3}^{\infty} b_{n-2} x^n = 0. \qquad (4.95)$$

For simplicity we represent y_1 as the power series

$$y_1 = \sum_0^{\infty} c_n x^n.$$

Of course, the c's are known: $c_n = 0$ if n is odd, and $c_{2m} = (-1)^m/(2^m m!)^2$ if $n = 2m$ is even. Computing the series for $2xy_1'$ and then equating the net coefficient of each power of x in Eq. (4.95) to 0, we find

$$2c_1 + b_1 = 0 \qquad \text{(terms in } x\text{)}, \qquad (4.96)$$

$$2 \cdot 2c_2 + 2^2 b_2 = 0 \qquad \text{(terms in } x^2\text{)}, \qquad (4.97)$$

$$\vdots$$

$$2nc_n + n^2 b_n + b_{n-2} = 0, \qquad n \geq 3 \qquad \text{(terms in } x^n\text{)}. \qquad (4.98)$$

From Eq. (4.96) we see that $b_1 = -c_1$, which is 0. Recalling that all the c's with odd index are 0, we see from Eq. (4.98) that all the b's with odd index are 0 also. Thus we need only solve Eq. (4.97) for

$$b_2 = -c_2$$

and then solve Eq. (4.98) for even n. By means of various manipulations the coefficients b_{2m} can be found in nearly closed form as

$$b_{2m} = \frac{(-1)^{m-1} h_m}{2^{2m}(m!)^2}, \qquad h_m = 1 + \frac{1}{2} + \cdots + \frac{1}{m}. \qquad (4.99)$$

We now have a second solution of the Bessel equation of order 0, so the general solution of the differential equation (4.93) is

$$y(x) = Ay_1(x) + By_2(x),$$

where A and B are arbitrary constants. It is usual (in the English-language literature) to choose the standard second solution as this linear combination of our y_1 and y_2:

$$Y_0(x) = \frac{2}{\pi}[y_2(x) - (\ln 2 - \gamma)J_0] = \frac{2}{\pi}\left[J_0(x)\left(\ln\frac{x}{2} + \gamma\right) + \sum_{m=1}^{\infty} b_{2m}x^{2m}\right].$$

(4.100)

The resulting function is called the *Bessel function of the second kind of order 0.* (See Fig. 4.5.) The number γ is called Euler's constant, defined as

$$\gamma = \lim_{m\to\infty}\left(1 + \frac{1}{2} + \cdots + \frac{1}{m} - \ln m\right) \cong 0.57722.$$

Figure 4.5

Graphs of Y_0 and Y_1, Bessel Functions of the Second Kind

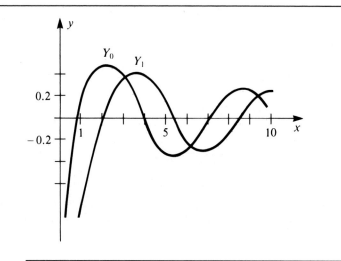

In the cases in which $k = 1, 2, 3, \ldots$, our first solution of Bessel's equation (4.90) is

$$J_k(x) = \sum_{m=0}^{\infty} \frac{(-1)^m}{m!(m + k)!}\left(\frac{x}{2}\right)^{2m+k}.$$

(4.101)

For brevity we shall write this as $y_1(x)$ and the series as

$$y_1(x) = \sum_{n=0}^{\infty} c_n x^{n+k},$$

(4.102)

remembering that the c's with odd index are 0. According to case 3 of

Theorem 4.3, the second solution is to have the form

$$y_2(x) = \beta y_1(x) \ln x + \sum_{N=0}^{\infty} b_N x^{N-k}. \tag{4.103}$$

When $y_2(x)$ is substituted into the differential equation (4.89), the result is

$$0 = \beta \cdot 2xy_1' + \sum_{0}^{\infty} (((N-k)^2 - k^2)b_N + b_{N-2})x^{N-k}. \tag{4.104}$$

The series for the first term is

$$\beta \cdot 2xy_1' = \sum_{n=0}^{\infty} \beta \cdot 2(n+k)c_n x^{n+k}.$$

Now we equate to 0 the net coefficient of each power of x, incorporating the simplification $(N-k)^2 - k^2 = N(N-2k)$. The resulting equations are shown in Table 4.5; the power of x to which each equation corresponds is shown in brackets.

First we note that $b_1 = 0$, from Eq. $[1-k]$ in Table 4.5. Then, recalling that the c's with odd index are 0, we see that all the b's with odd index must likewise be 0: they are related by an equation like Eq. $[N-k]$ in Table 4.5, which makes each one a multiple of b_1.

Next we must decide whether β is 0 or not. It first appears in Eq. $[k]$ in Table 4.5, which is

$$\beta \cdot 2kc_0 + b_{2k-2} = 0. \tag{4.105}$$

If we choose $\beta = 0$, then $b_{2k-2} = 0$, and the same is true of the b's with smaller, even indices. Now if b_{2k} is *chosen* different from 0, the function defined by the power series is actually a constant multiple of $J_k(x)$. Thus choosing $\beta = 0$ cannot produce a second independent solution.

Table 4.5

$[-k]$	$0 \cdot (-2k)b_0 = 0$
$[1-k]$	$1(1-2k)b_1 = 0$
$[2-k]$	$2(2-2k)b_2 + b_0 = 0$
	$\vdots$
$[N-k]$	$N(N-2k)b_N + b_{N-2} = 0, \qquad 2 \leq N < 2k$
$[k]$	$\beta \cdot 2kc_0 + 2k \cdot 0 \cdot b_{2k} + b_{2k-2} = 0$
$[1+k]$	$\beta \cdot 2(1+k)c_1 + (2k+1) \cdot 1 \cdot b_{2k+1} + b_{2k-1} = 0$
	$\vdots$
$[n+k]$	$\beta \cdot 2(n+k)c_n + (n+2k)nb_{n+2k} + b_{n+2k-2} = 0$

Suppose, then, that $\beta = 1$. Equation (4.105) defines b_{2k-2}, and then equations like Eq. $[N - k]$ in Table 4.5 define all the even-indexed b's back to b_0. Equations like $[n + k]$ in Table 4.5 with $n = 2, 4, \ldots$ define all the even-indexed b's forward from b_{2k-2}. In this way, all the b's are found.

As a specific example, let us suppose $k = 3$. Then Eqs. $[2 - 3]$, $[4 - 3]$, and $[3]$ are:

$$2(2 - 6)b_2 + b_0 = 0,$$
$$4(4 - 6)b_4 + b_2 = 0,$$
$$6c_0 + b_4 = 0.$$

These are easily solved, bottom to top, to find

$$b_4 = -6c_0,$$
$$b_2 = -2 \cdot 4 \cdot 6c_0,$$
$$b_0 = -2^2 \cdot 4^2 \cdot 6c_0.$$

Next look at Eq. $[2 + 3]$ and in general $[2m + 3]$:

$$2(2 + 3)c_2 + (2 \cdot 3 + 2) \cdot 2b_8 + b_6 = 0,$$
$$2(2m + 3)c_{2m} + (2m + 6)2mb_{2m+6} + b_{2m+4} = 0.$$

As we noted, $b_{2k} = b_6$ is arbitrary: if it is nonzero, a multiple of $J_3(x)$ will be incorporated as part of our second solution. But we shall take $b_6 = 0$, so that

$$b_8 = -2 \cdot 5 \cdot c_2$$

$$b_{2m+6} = -\frac{2(2m+3)c_{2m} + b_{2m+4}}{4(m + 3)m}, \qquad m = 2, 3, \ldots.$$

Instead of pursuing the rather depressing details of solving these equations, let us content ourselves with observing that the second solution of Bessel's equation *does* contain a logarithmic term whenever k is an integer. A standard definition for the second solution is this:

$$Y_k(x) = \frac{J_k(x) \cos k\pi - J_{-k}(x)}{\sin k\pi}, \qquad k \neq 0, 1, 2, \ldots, \tag{4.106}$$

$$Y_k(x) = \lim_{\kappa \to k} Y_\kappa(x), \qquad k = 0, 1, 2, \ldots. \tag{4.107}$$

These are the functions usually tabulated under the name of *Bessel functions of the second kind*. (See Fig. 4.5.)

Exercises

1. Let $y_2(x)$ represent the "second solution" of Bessel's equation. Prove that $|y_2(x)| \to \infty$ as $x \to 0$, for any $k \geq 0$.

2. Prove that the second solution of Bessel's equation is oscillatory. (Hint: see Section 3.6.)

3. What is the relation (from Eq. (4.106)) between $Y_k(x)$ and $J_{-k}(x)$ when $k = \frac{1}{2}$?

4. Use Eq. (4.106) to verify the relation

$$Y_0(x) = \frac{2}{\pi} \frac{\partial}{\partial \kappa} J_\kappa(x)\big|_{\kappa=0}.$$

5. Verify that b_{2m} as given in Eq. (4.99) satisfies Eq. (4.98) with $n = 2m$.

6. The solution Y_0 as given in Eq. (4.100) is supposed to be a linear combination of our y_1 and y_2. Find the coefficients in this combination.

4.7

The Point at Infinity; Asymptotic Series

It is often necessary to study the behavior of the solutions of a linear differential equation

$$\frac{d^2y}{dx^2} + p(x)\frac{dy}{dx} + q(x)y = 0 \tag{4.108}$$

for very large values of the independent variable. If the coefficients $p(x)$ and $q(x)$ are rational functions (ratios of polynomials) this is not difficult. We simply make a change of variables

$$z = \frac{1}{x}, \qquad v(z) = y\left(\frac{1}{z}\right) = y(x). \tag{4.109}$$

Then the study of the solutions $y(x)$ for large x reduces to the study of $v(z)$ for small z (for which the power series is an excellent tool).

To carry out this change of variable, we first use the chain rule to obtain these expressions for the derivatives of y:

$$\frac{dy}{dx} = \frac{dv}{dz}\frac{dz}{dx} = -z^2\frac{dv}{dz}, \tag{4.110}$$

$$\frac{d^2y}{dx^2} = \frac{d^2v}{dz^2}\left(\frac{dz}{dx}\right)^2 + \frac{dv}{dz}\frac{d^2z}{dx^2} = z^4\frac{d^2v}{dz^2} + 2z^3\frac{dv}{dz}. \tag{4.111}$$

We replace the derivatives in Eq. (4.108) with the appropriate expressions and find an equation for v:

$$z^4\frac{d^2v}{dz^2} + (2z^3 - z^2\bar{p}(z))\frac{dv}{dz} + \bar{q}(z)v = 0. \tag{4.112}$$

(Here we have used the notation $\bar{p}(z)$ for $p(x) = p(1/z)$, etc.) The behavior of the solution $v(z)$ near $z = 0$ will depend on the classification of this point. Restating the definitions of Section 4.3 for Eq. (4.112), we find that the point $z_0 = 0$ is

1. a regular point if these limits exist:

$$\lim_{z \to 0} \frac{2z^3 - z^2 \bar{p}(z)}{z^4}, \qquad \lim_{z \to 0} \frac{\bar{q}(z)}{z^4}; \tag{4.113}$$

2. a regular singular point if not a regular point but these limits exist:

$$\lim_{z \to 0} \frac{2z^3 - z^2 \bar{p}(z)}{z^3}, \qquad \lim_{z \to 0} \frac{\bar{q}(z)}{z^2}; \tag{4.114}$$

3. an irregular singular point otherwise.

It is really more convenient to work with the original differential equation. We can easily convert the above to criteria in terms of x. For the differential equation (4.108) with rational coefficients p and q the point at infinity is

1. a regular point if these limits exist:

$$\lim_{x \to \infty} (2x - x^2 p(x)), \qquad \lim_{x \to \infty} x^4 q(x); \tag{4.115}$$

2. a regular singular point if not a regular point but these limits exist:

$$\lim_{x \to \infty} (2 - xp(x)), \qquad \lim_{x \to \infty} x^2 q(x); \tag{4.116}$$

3. an irregular singular point otherwise.

We assume that the functions p and q themselves have limits as $x \to \infty$. This means for rational functions that the polynomial in the denominator has degree at least as high as the numerator.

Example 1

The point at infinity is never regular for the equation with constant coefficients

$$y'' + ay' + by = 0.$$

It is a regular singular point if $a = b = 0$ (with solution $c_0 + c_1 x$) and irregular otherwise.

Another important example is the Euler-Cauchy equation

$$y'' + \frac{a}{x} y' + \frac{b}{x^2} y = 0,$$

for which the point at infinity is a regular point if $a = 2$ and $b = 0$ or a regular singular point otherwise. It is interesting to note that under the change of variables from x to $z = 1/x$, one Euler-Cauchy equation is transformed into another.

The Airy equation, whose solution for small x was found in Section 4.1, is

$$y'' + xy = 0.$$

The point at infinity is not classified because $q(x)$ does not have a limit as x increases.

What should one expect of the solution y of Eq. (4.108) for large values of x? The transformation between x and z tells us about the two simpler cases:

1. If ∞ is a regular point of the differential equation (4.108) with rational coefficients, then the general solution will be found in the form of a series of descending powers of x,

$$y(x) = c_0 + \frac{c_1}{x} + \frac{c_2}{x^2} + \cdots. \tag{4.117}$$

Coefficients c_0 and c_1 will be arbitrary, and the series will converge for $|x| > R$, where R is the magnitude of the largest zero of the denominators of $p(x)$ and $q(x)$.

2. If ∞ is a regular singular point of the differential equation (4.108), then there will be at least one solution of the equation in the form

$$y(x) = x^r\left(c_0 + \frac{c_1}{x} + \frac{c_2}{x^2} + \cdots\right) \tag{4.118}$$

with $c_0 \neq 0$.

Example 2

The point at infinity is a regular point for the differential equation

$$x^4\frac{d^2y}{dx^2} + 2x^3\frac{dy}{dx} + y = 0.$$

We should look for solutions in the form of an infinite series of descending powers of x,

$$y(x) = c_0 + \frac{c_1}{x} + \frac{c_2}{x^2} + \cdots + \frac{c_n}{x^n} + \cdots.$$

Below are the series for the terms of the differential equation:

$$y = \qquad\qquad c_0 \qquad + c_1 x^{-1} + \cdots \qquad\qquad\qquad + c_n x^{-n} + \cdots,$$
$$2x^3 y' = - 2c_1 x - 2\cdot 2c_2 - 2\cdot 3c_3 x^{-1} - \cdots \qquad - 2\cdot(n+2)c_{n+2}x^{-n} - \cdots,$$
$$x^4 y'' = \quad 2c_1 x + 3\cdot 2c_2 + 4\cdot 3c_3 x^{-1} + \cdots + (n+3)(n+2)c_{n+2}x^{-n} + \cdots.$$

Equating to 0 the net coefficient of each power of x leads to the following equations:

$$2c_2 + c_0 = 0,$$
$$6c_3 + c_1 = 0,$$
$$\vdots$$
$$(n+1)(n+2)c_{n+2} + c_n = 0.$$

These equations can be solved easily for the coefficients; all the c's with even index are multiples of c_0, while those with odd index are multiples of c_1. The series that result give these solutions:

$$y(x) = c_0\left(1 - \frac{1}{2!x^2} + \frac{1}{4!x^4} - + \cdots\right) + c_1\left(\frac{1}{x} - \frac{1}{3!x^3} + \frac{1}{5!x^5} - + \cdots\right).$$

It is not hard to recognize these as power series and present the solution as

$$y(x) = c_0 \cos\left(\frac{1}{x}\right) + c_1 \sin\left(\frac{1}{x}\right).$$

The situation is quite different when the point at infinity is an irregular singular point. The most readily available example is the equation with constant coefficients. The exponential solutions, which we know to be valid for all finite values of the argument, do not have power series with center at infinity. That is to say, $e^{1/z}$ cannot be expressed as a series in powers of z. On the other hand, exponential functions (and the sines and cosines, which are their complex aspect) are well known and easy to calculate and work with. We might then expect the solution to have the form

$$y(x) = e^{mx} x^r \sum_{n=0}^{\infty} c_n x^{-n} \qquad\qquad (4.119)$$

because we know that the solution of an equation with constant coefficients behaves like an exponential, while the solution of an equation with a regular singular point might behave like a power-times-series (of descending powers). As in the case of just a regular singular point, it is necessary to require that $c_0 \neq 0$.

The procedure of solution follows the same lines as in the familiar cases: we insert y and its derivatives, as found from Eq. (4.119) above, into the differential equation. The net coefficient of each power of x is set equal to 0, and the resulting equations are solved to determine first m, then r, and then the c's recursively. An example will make the details of the process clearer.

Example 3

We consider the differential equation

$$\frac{d^2y}{dx^2} - \left(1 - \frac{2}{x}\right)y = 0, \tag{4.120}$$

which has an irregular singular point at infinity. We calculate the necessary terms for the differential equation:

$$\frac{d^2y}{dx^2} = m^2 e^{mx}(c_0 x^r + c_1 x^{r-1} + \cdots + c_n x^{r-n} + \cdots)$$
$$+ 2m e^{mx}(rc_0 x^{r-1} + (r-1)c_1 x^{r-2} + \cdots$$
$$+ (r - n + 1)c_{n-1} x^{r-n} + \cdots)$$
$$+ e^{mx}(r(r-1)c_0 x^{r-2} + (r-1)(r-2)c_1 x^{r-3} + \cdots$$
$$+ (r - n + 2)(r - n + 1)c_{n+2} x^{r-n} + \cdots)$$
$$-y = e^{mx}(-c_0 x^r - c_1 x^{r-1} - \cdots - c_n x^{r-n} - \cdots)$$
$$\frac{2}{x}y = 2e^{mx}(c_0 x^{r-1} + c_1 x^{r-2} + \cdots + c_{n-1} x^{r-n} + \cdots).$$

The *highest* power of x that appears in the series is x^r, with coefficient

$$(m^2 - 1)c_0. \tag{4.121}$$

Since c_0 is by agreement nonzero, we conclude that m must be either 1 or -1. First take the case of $m = 1$.

The next highest power of x is x^{r-1}. Equating its coefficient to zero gives us the equation

$$(m^2 - 1)c_1 + (2mr + 2)c_0 = 0. \tag{4.122}$$

Taking into account the fact that $m^2 - 1 = 0$ already, we see that r must be chosen to satisfy

$$2mr + 2 = 0. \tag{4.123}$$

That is, $r = -1$, since $m = 1$. The rest of the coefficients are determined from the difference equation

$$(m^2 - 1)c_n + (2m(r - n + 1) - 2)c_{n-1}$$
$$+ (r - n + 1)(r - n + 2)c_{n-2} = 0. \quad \textbf{(4.124)}$$

When the substitutions $m = 1$ and $r = -1$ are made, this simplifies to

$$2c_{n-1} + (-n)c_{n-2} = 0, \quad n \geq 2. \quad \textbf{(4.125)}$$

The first few coefficients, along with the formula for the general coefficient, are given below:

$$c_1 = c_0,$$
$$c_2 = \tfrac{3}{2}c_1 = \tfrac{3}{2}c_0,$$
$$c_3 = \frac{4}{2}c_2 = \frac{4 \cdot 3}{4}c_0,$$
$$\vdots$$
$$c_n = \frac{(n + 1)!}{2^n}c_0. \quad \textbf{(4.126)}$$

We have now arrived at the following as our candidate for the solution of the differential equation (4.120):

$$y(x) = c_0 e^x x^{-1} \sum_{n=0}^{\infty} \frac{(n + 1)!}{2^n x^n}$$
$$= c_0 e^x \left(\frac{1}{x} + \frac{1}{2x^2} + \frac{3}{2x^3} + \cdots \right). \quad \textbf{(4.127)}$$

This solution has just one defect—the series does not converge for any value of x! It sounds as though the formal solution above is of no use, but that is far from true.

The utility of an ordinary convergent series is that the sum of a finite number of terms provides an approximation to the value of the function it represents. The accuracy can always be improved by adding up more terms. With a series such as the one above, the value of the function it represents can still be approximated, but only to limited accuracy. It can be proved that the error introduced by adding up only a finite number of terms is smaller than the magnitude of the first term neglected. For fixed x the magnitude of the successive terms first decreases, then increases. Thus the best one can do is to add up the terms of decreasing size and then stop.

If x is quite large, the smallest term may be very small indeed. In fact, asymptotic or semiconvergent series such as the one above are often used in preference to other methods for the evaluation of special functions.

Example 4

The Bessel function $J_0(x)$ has this asymptotic representation, which is developed by the same method as above:

$$J_0(x) = \sqrt{\frac{2}{\pi x}} \cos\left(x - \frac{\pi}{4}\right)\left(1 - \frac{3^2}{2(8x)^2} + \frac{(3 \cdot 5 \cdot 7)^2}{4!(8x)^4} - + \cdots\right)$$
$$+ \sqrt{\frac{2}{\pi x}} \sin\left(x - \frac{\pi}{4}\right)\left(\frac{1}{8x} - \frac{(3 \cdot 5)^2}{3!(8x)^3} + \cdots\right). \tag{4.128}$$

In Table 4.6 are shown the beginning terms of the Taylor series and of the semiconvergent series, for several values of x. The superiority of the semiconvergent series for large values of x is clear, as is the superiority of the Taylor series for small values.

Table 4.6

	x = 1		x = 5		x = 10	
N	P_N	A_N	P_N	A_N	P_N	A_N
0	1	1	1	1	1	1
1	2.5 E-1	1.3 E-1	6.3 E0	2.5 E-2	2.5 E1	1.3 E-2
2	1.6 E-2	7.0 E-2	9.8 E0	2.8 E-3	1.6 E2	7.0 E-4
3	4.3 E-4	7.3 E-2	6.8 E0	5.8 E-4	4.3 E2	7.3 E-5
N_{opt}		7.0 E-2		1.0 E-5		4.0 E-10
		(2)		(10)		(20)

$$P_N = \frac{1}{(N!)^2}\left(\frac{x}{2}\right)^{2N}, \qquad A_N = \frac{(1 \cdot 3 \cdot 5 \cdots (2N-1))^2}{N! \,(8x)^N}$$

P_N is the Nth term of the power series for the Bessel function J_0; A_N is the Nth term of the asymptotic series—see Eq. (4.128). Opposite N_{opt} is the smallest A_N and in parentheses is the index, N_{opt}. Values shown are very rough.

Exercises

1. Find the solution of the Euler-Cauchy equation for which ∞ is a regular point.

2. Confirm the statement that the change of variables from x to $z = 1/x$ transforms one Euler-Cauchy equation into another.

3. Find and classify all the singular points of the *hypergeometric equation*

$$x(1 - x)\frac{d^2y}{dx^2} + (c - (a + b + 1)x)\frac{dy}{dx} - aby = 0.$$

The nature and location of the singular points depend on the parameters a, b, and c.

4. Verify that the substitution $y = e^{mx}v$ transforms Eq. (4.108) into

$$v'' + (2m + p(x))v' + (m^2 + mp(x) + q(x))v = 0.$$

5. Find an asymptotic series for the solution of Bessel's equation

$$x^2 \frac{d^2y}{dx^2} + x \frac{dy}{dx} + x^2 y = 0.$$

6. Find m and r (see Eq. (4.119)) for the asymptotic expansion for the solution of

$$x \frac{d^2y}{dx^2} - (1 - x)y = 0.$$

7. Follow the instructions for Exercise 6 for the equation

$$\frac{d^2y}{dx^2} + \left(a + \frac{b}{x}\right)y = 0.$$

4.8

A Convergence Proof

In Section 4.1 we observed that it is permissible to add, subtract, multiply, and differentiate infinite series within their interval of convergence as if they were polynomials, in the sense that the series that results from any of these operations also converges to the right function. The method developed in that section consists in assuming that the solution of the given differential equation has a convergent power series and then determining what its coefficients must be. Once they are found, we can test the series for convergence. If it has a nonzero interval of convergence, all the operations that had been carried out on the series were legitimate—the sum of the series is a solution of the differential equation. Now we are going to prove a theorem about that method. Basically, we follow the same method as in Section 4.1, down to the point of obtaining an expression that determines the coefficients. We cannot find them in closed form, of course, but we can prove that the series so obtained does converge and even estimate its radius of convergence.

Before starting the proof, we set the stage by choosing the form for the equation, namely,

$$\frac{d^2y}{dx^2} + p(x)\frac{dy}{dx} + q(x)y = 0, \tag{4.129}$$

in which we assume that the coefficients $p(x)$ and $q(x)$ both have power series:

$$p(x) = p_0 + p_1 x + p_2 x^2 + \cdots, \qquad q(x) = q_0 + q_1 x + q_2 x^2 + \cdots. \tag{4.130}$$

Let R_1 be the smaller of the radii of convergence of the two series. If p and q are both rational functions—this is the usual case—then R_1 will be the absolute value of the smallest zero of their common denominator.

In the proof we will make use of some facts about infinite series. First, recall that the series

$$p_0 + p_1 R + p_2 R^2 + \cdots, \qquad q_0 + q_1 R + q_2 R^2 + \cdots \qquad \text{(4.131)}$$

certainly converge if R satisfies $0 < R < R_1$. This implies that the terms of the series approach 0 as n increases:

$$p_n R^n \to 0, \qquad q_n R^n \to 0. \qquad \text{(4.132)}$$

Therefore it will always be possible to find positive constants M_p and M_q for which

$$|p_n| \le \frac{M_p}{R^n}, \qquad |q_n| \le \frac{M_q}{R^n}, \qquad n = 0, 1, 2, \ldots . \qquad \text{(4.133)}$$

(These are called Cauchy's inequalities.) It is not necessary that M_p and M_q be the best values that fulfill the inequalities. In fact, it turns out to be convenient to use a worse inequality for the q's, namely,

$$|q_n| \le \frac{M_q(n + 1)}{R^n}, \qquad n = 0, 1, 2, \ldots . \qquad \text{(4.134)}$$

Second, recall the comparison test for series. If $|c_n| \le C_n$ for $n = 0, 1, 2, \ldots$ and if the series

$$\sum_{n=0}^{\infty} C_n R^n = C_0 + C_1 R + C_2 R^2 + \cdots$$

converges, then the radius of convergence of the power series

$$\sum_{n=0}^{\infty} c_n x^n = c_0 + c_1 x + c_2 x^2 + \cdots \qquad \text{(4.135)}$$

is at least equal to R.

Finally, it will be convenient to recall these identities for the geometric series and its derivative:

$$\sum_{n=0}^{\infty} \left(\frac{x}{R}\right)^n = \frac{1}{1 - \dfrac{x}{R}}, \qquad \sum_{n=1}^{\infty} n\left(\frac{x}{R}\right)^{n-1} = \sum_{n=0}^{\infty} (n + 1)\left(\frac{x}{R}\right)^n = \frac{1}{\left(1 - \dfrac{x}{R}\right)^2}.$$

$$\text{(4.136)}$$

Both of these series have radius of convergence equal to R.

Now we are ready to state and prove the main theorem concerning the power series method for differential equations.

Theorem 4.4

Let the functions $p(x)$ and $q(x)$ have power series with center $x_0 = 0$, both convergent for $|x| < R_1$, and let c_0 and c_1 be any two numbers.

Then the initial value problem

$$\frac{d^2y}{dx^2} + p(x)\frac{dy}{dx} + q(x)y = 0,$$

$$y(0) = c_0, \qquad y'(0) = c_1,$$

has a unique solution,

$$y(x) = c_0 + c_1 x + c_2 x^2 + \cdots.$$

The coefficients $c_2, c_3, \ldots$ are found by formally substituting the series into the differential equation. The series converges for $|x| < R_1$.

Table 4.7

Series for terms of
$y'' + py' + qy = 0$

$$y'' = 2 \cdot 1 c_2 + 3 \cdot 2 c_3 x \qquad\qquad + 4 \cdot 3 c_4 x^2 \qquad\qquad + \cdots + (n + 2)(n + 1)c_{n+2} x^n + \cdots$$

$$py' = p_0 c_1 + (p_1 c_1 + p_0 \cdot 2 c_2)x + (p_2 c_1 + p_1 \cdot 2 c_2 + p_0 \cdot 3 c_3)x^2 + \cdots + \left(\sum_{k=0}^{n} p_{n-k}(k + 1)c_{k+1}\right)x^n + \cdots$$

$$qy = q_0 c_0 + (q_1 c_0 + q_0 c_1)x \qquad + (q_2 c_0 + q_1 c_1 + q_0 c_2)x^2 \qquad + \cdots + \left(\sum_{k=0}^{n} q_{n-k} c_k\right)x^n + \cdots$$

Proof: We start by substituting y in the form of a power series into the differential equation. The necessary expressions are shown in Table 4.7. By collecting the net coefficient of x^n and equating it to 0 we obtain this difference equation for the coefficients

$$(n + 2)(n + 1)c_{n+2} = -\sum_{k=0}^{n} p_{n-k}(k + 1)c_{k+1} - \sum_{k=0}^{n} q_{n-k} c_k. \qquad (4.137)$$

This equation is valid for $n = 0, 1, 2, \ldots$ and determines successively $c_2, c_3, c_4, \ldots$ in terms of c's with lower index. Recall that c_0 and c_1 are given in the initial conditions.

Now we establish an inequality valid for the absolute values of the c's:

$$(n + 2)(n + 1)|c_{n+2}| \le \sum_{k=0}^{n} |p_{n-k}|(k + 1)|c_{k+1}| + \sum_{k=0}^{n} |q_{n-k}| \cdot |c_k|. \quad (4.138)$$

We have simply replaced each quantity by its absolute value. Since this eliminates the possibility of additive cancellation in the sums, the right-hand side has become greater than (or equal to) the left-hand side. The right-hand side is further increased if we replace the p's and q's by the bounds in Eq. (4.133) (for p) and Eq. (4.134) (for q). We find now that

$$(n + 2)(n + 1) |c_{n+2}| \le \sum_{k=0}^{n} \frac{M_p}{R^{n-k}} (k + 1) |c_{k+1}| + \sum_{k=0}^{n} \frac{(n - k + 1)M_q}{R^{n-k}} |c_k|.$$
(4.139)

Here R can be any positive number less than R_1.

We cannot solve Eq. (4.137) in closed form; that is one reason for moving to an inequality. As it happens, we can solve a difference equation related to the last inequality above. First, define

$$C_0 = |c_0|, \qquad C_1 = |c_1|$$
(4.140)

and then let subsequent C's be determined by

$$(n + 2)(n + 1)C_{n+2} = \sum_{k=0}^{n} \frac{M_p}{R^{n-k}} (k + 1)C_{k+1} + \sum_{k=0}^{n} \frac{(n - k + 1)M_q}{R^{n-k}} C_k.$$
(4.141)

It is easy to see that all the C's are nonnegative and then to prove by induction that

$$|c_n| \le C_n, \qquad n = 0, 1, 2, \ldots.$$
(4.142)

Rather than write down the C's in closed form, let us note that Eq. (1.141) is exactly what we would get if we tried to solve the initial value problem

$$Y'' = \frac{M_p}{1 - \frac{x}{R}} Y' + \frac{M_q}{\left(1 - \frac{x}{R}\right)^2} Y, \qquad Y(0) = C_0, \qquad Y'(0) = C_1, \quad (4.143)$$

for the series

$$Y(x) = C_0 + C_1 x + C_2 x^2 + \cdots.$$
(4.144)

But Eq. (4.143) is an Euler-Cauchy equation, whose solutions are

$$Y(x) = A\left(1 - \frac{x}{R}\right)^{m_1} + B\left(1 - \frac{x}{R}\right)^{m_2},$$
(4.145)

where m_1 and m_w are the two roots of the indicial equation

$$m(m - 1) + mM_p R - M_q R^2 = 0.$$
(4.146)

If $M_q \ne 0$, both roots of Eq. (4.146) are real, and at least one is negative. Thus if $Y(x)$ from Eq. (4.145) is expanded in power series as in Eq. (4.144), the series will have radius of convergence R. Because of the inequalities (4.142) the series for $y(x)$ will have radius of convergence at least equal to R. Thus all the manipulations necessary to obtain the coefficients $c_2, c_3, \ldots$ in the power series for $y(x)$ were legitimate: that series is indeed the solution of the given initial value problem, at least for $|x| < R$.

Now if x is any number satisfying $|x| < R_1$, the series for $y(x)$ converges: $|x|$ is less than some R less than R_1, and that is all we ask of R. Thus we can say that the radius of convergence of the series for $y(x)$ is at least R_1, and the theorem is proved.

Exercises

1. Find the coefficients A and B in Eq. (4.145) in terms of C_0 and C_1.
2. Verify that $Y(x)$ as given in Eq. (4.145) is the general solution of the differential equation (4.143).
3. Confirm the statement that Eq. (4.146) has two real roots, at least one of which is negative, if $M_q \neq 0$.
4. What is the importance of one of the roots of Eq. (4.146) being negative?
5. Determine what radius of convergence (if any) is expected for the series solution (center 0) of this problem:

$$\frac{d^2 y}{dx^2} + \ln(1 - x)y = 0, \qquad y(0) = 1, \qquad y'(0) = 0.$$

Notes and References

Most of the special functions of mathematical physics are studied through their power-series developments. The Bessel functions and Legendre polynomials are the most frequently occurring of the special functions because they are related to certain simple partial differential equations in spherical and cylindrical coordinates.

Much specific information about special functions, including tables of values, can be found in Abramowitz and Stegun (1964). This book is an indispensible reference for the serious applied mathematician.

Miscellaneous Exercises

1. Find a solution in the form of a series, $y = \sum c_n x^n$, for the initial value problem

$$\frac{d^2 y}{dx^2} + y = \frac{1}{1 - x}, \qquad y(0) = 0, \qquad y'(0) = 0.$$

(This problem cannot be solved in closed form.) Find the relation that determines the c's and find the first four c's.

2. Investigate the series solution of the initial value problem

$$\frac{d^2y}{dx^2} - y = 2e^x, \qquad y(0) = 0, \qquad y'(0) = 1.$$

Also solve the problem by other means.

3. Obtain the first few terms of the series $y(x) = \sum c_n x^n$ that solves the nonlinear initial value problem

$$\frac{dy}{dx} = x^2 + y^2, \qquad y(0) = 1.$$

4. Assume a series solution $y = \sum c_n x^n$ and find the first few terms:

$$\frac{dy}{dx} = 1 + y^2, \qquad y(0) = 0.$$

Also solve by other means.

5. Find the first few terms of a series solution of the *Blasius equation*

$$\frac{d^3y}{dx^3} + 2y\frac{d^2y}{dx^2} = 0.$$

Auxiliary conditions are $y(0) = 0$, $y'(0) = 0$, $y''(0) = 1$. Assume that y has this special form:

$$y = c_0 x^2 + c_1 x^5 + c_2 x^8 + c_3 x^{11} + \cdots.$$

6. Find the first few terms of a series solution $y = \sum c_n x^n$ for the differential equation

$$y\frac{d^2y}{dx^2} + \left(\frac{dy}{dx}\right)^2 = 0.$$

7. Find a series solution (center 0) for the differential equation

$$\frac{d^2y}{dx^2} + \frac{y}{1-x} = 0.$$

Find the relation between the coefficients of the series and find the first few terms. What will be the radius of convergence of the series?

8. Find a series solution (center 0) for the differential equation of Exercise 7, in the form

$$(1 - x)\frac{d^2y}{dx^2} + y = 0.$$

9. Find the first few terms of the series solution of the differential equation

$$\frac{d^2y}{dx^2} = e^x y.$$

10. What should be the radius of convergence for the series in Exercise 9?

11. Express the polynomial $x^2 + 2x + 1$ as a linear combination of the Legendre polynomials $P_0(x)$, $P_1(x)$, $P_2(x)$.

12. Show that any polynomial of degree n can be expressed as a linear combination of the Legendre polynomials $P_0(x), P_1(x), \ldots, P_n(x)$.

13. Find a formal power series solution (center 0) for the nonstandard differential equation

$$\frac{dy(x)}{dx} + y(\beta x) = 0,$$

where β is a parameter between 0 and 1.

14. What is the radius of convergence of the series found in Exercise 13?

15. The point $x = 0$ is an irregular singular point of the differential equation

$$\frac{d^2 y}{dx^2} - \sqrt{x}\,y = 0$$

because $\sqrt{x}$ does not have a power series with center 0. Try to find a solution in the form of a generalized series

$$y = \sum_{n=0}^{\infty} c_n x^{r+n/2},$$

with the assumption that $c_0 = 1$.

16. Use the power series for the Bessel function to show that

$$y(x) = \sqrt{x}\,J_{1/3}(\tfrac{2}{3}x^{3/2})$$

is a solution of *Airy's equation* $y'' + xy = 0$.

17. Find a solution of the differential equation

$$x^2 \frac{d^2 y}{dx^2} + x \frac{dy}{dx} + x^k y = 0$$

centered at 0. Here k is a positive integer.

18. Express the solution found for Exercise 17 in terms of the Bessel function J_0.

19. The differential equation

$$\frac{d^2 y}{dx^2} + \left(1 - \frac{2\alpha}{x} - \frac{k(k+1)}{x^2}\right)y = 0$$

is called the *Coulomb wave equation*. Taking $\alpha = 0$ and k a nonnegative integer, find a solution in the power-times-series form.

20. Check the solution of Exercise 19 by taking $k = 0$.

21. Find a solution valid near $x = 0$ for the equation

$$x^2 \frac{d^2y}{dx^2} + x \frac{dy}{dx} - \frac{1 + 4x}{4} y = 0.$$

22. What is the correct form for the second solution of the equation in Exercise 21?

23. Find the second solution of the equation in Exercise 21.

24. Determine the nature of the point at infinity for Legendre's equation

$$(x^2 - 1) \frac{d^2y}{dx^2} + 2x \frac{dy}{dx} - k(k + 1)y = 0.$$

25. Find a solution of Legendre's equation valid "near infinity."

5 Laplace Transform

5.1

Introduction

We are about to start the study of the Laplace transform, a mathematical device that will allow us to convert an initial value problem into an algebraic problem. The usefulness of a transform (there are many kinds besides the Laplace transform) lies in the ability to change one kind of operation into another, simpler one. A familiar example is the logarithmic transform, which allows us to replace multiplication by addition and exponentiation by multiplication. The Laplace transform applies to functions and replaces the operations of calculus by those of algebra.

Definition 5.1

If $f(t)$ is a function defined for $t > 0$, its Laplace transform is

$$\mathcal{L}(f) = \int_0^\infty e^{-st} f(t)\, dt,$$

provided that the integral converges for at least some s.

Since the definition of the Laplace transform involves an improper integral, it will be convenient to recall how to handle them. Suppose $g(t)$ is a given function. Then the meaning of the improper integral is

$$\int_0^\infty g(t)\, dt = \lim_{T \to \infty} \int_0^T g(t)\, dt. \tag{5.1}$$

If we know an indefinite integral of g—that is, another function $G(t)$ whose derivative is $G'(t) = g(t)$—then we may also express the improper integral this way:

$$\int_0^\infty g(t)\, dt = \lim_{T \to \infty} G(T) - G(0). \tag{5.2}$$

Example 1

The Laplace transform of $f(t) = e^{at}$ is, by definition

$$\int_0^\infty e^{-st} e^{at}\, dt = \int_0^\infty e^{-(s-a)t}\, dt.$$

242

An indefinite integral of e^{kt} is e^{kt}/k, so the latter integral is

$$\int_0^\infty e^{-(s-a)t}\, dt = \lim_{T\to\infty} \frac{e^{-(s-a)T}}{-(s-a)} - \frac{1}{-(s-a)}.$$

If s is greater than a (that is, $s - a$ is positive), then $e^{-(s-a)T}$ approaches 0 as T increases. Thus we find that

$$\mathscr{L}(e^{at}) = \frac{1}{s-a} \qquad (s > a).$$

Example 2

If in the preceding example we set $a = 0$, our function is $f(t) = 1$, and its transform is $1/s$ $(s > 0)$.

Since the integral defining the Laplace transform is a definite (although improper) integral, the result no longer contains the variable t. However, s is present as a parameter in the exponential factor e^{-st}, so the Laplace transform is always a function of s. Indeed, we often use the symbolism

$$\mathscr{L}(f(t)) = F(s);$$

that is, we denote a function of t by a lowercase letter and its transform, a function of s, by the capital of the same letter.

We should establish the fact that the functions we are interested in do indeed possess Laplace transforms. Theorem 5.1 identifies a class of functions that can be transformed.

Theorem 5.1

Let $f(t)$ be a function that is continuous, $0 \le t < \infty$. Suppose that

$$\lim_{t\to\infty} e^{-\alpha t} f(t) = 0 \tag{5.3}$$

for some value of α. Then $f(t)$ has a Laplace transform.

The proof of this theorem amounts to showing that the improper integral defining the Laplace transform of f converges, at least for $s > \alpha$. Some details and examples will be found in Exercises 16 and 17 at the end of this section. Usually, the functions used in the study of linear differential equations have a Laplace transform, so existence is not a concern in our work.

The transforms of many simple functions—sines, cosines, polynomials, etc.—can be found by direct calculation, usually after some tedious integration by parts. For example, one could find this way that

$$\mathscr{L}(\cos \omega t) = \frac{s}{s^2 + \omega^2}.$$

However, it is much more satisfying, and ultimately more useful, to develop some properties of the Laplace transform that will allow us to avoid actual integration in many cases.

The most significant property of the Laplace transform is its linearity, as expressed in the equation

$$\mathscr{L}(c_1 f_1 + c_2 f_2) = c_1 \mathscr{L}(f_1) + c_2 \mathscr{L}(f_2). \tag{5.4}$$

Here c_1, c_2 are any constants, and f_1, f_2 are any functions having Laplace transforms. This property follows from the analogous property of definite integrals. In fact it remains true even when the constants or the functions are complex. If f is a complex-valued function with real and imaginary parts u and v, $f = u + iv$, then it follows directly from the definition that

$$\mathscr{L}(f) = \mathscr{L}(u) + i\mathscr{L}(v).$$

We may use the linearity to find the Laplace transforms of the sine and cosine. We know the real and imaginary parts of the exponential

$$e^{i\omega t} = \cos \omega t + i \sin \omega t.$$

We also know that the transform of the exponential is

$$\mathscr{L}(e^{i\omega t}) = \frac{1}{s - i\omega} = \frac{s + i\omega}{s^2 + \omega^2}.$$

Now the linearity of the Laplace transform guarantees that

$$\mathscr{L}(e^{i\omega t}) = \mathscr{L}(\cos \omega t) + i\mathscr{L}(\sin \omega t).$$

By equating real and imaginary parts in the two expressions for $\mathscr{L}(e^{i\omega t})$ we obtain

$$\mathscr{L}(\cos \omega t) = \frac{s}{s^2 + \omega^2}, \qquad \mathscr{L}(\sin \omega t) = \frac{\omega}{s^2 + \omega^2}. \tag{5.5}$$

While linearity is indispensible, the property that makes the Laplace transform important is this one:

$$\mathscr{L}(f'(t)) = sF(s) - f(0+). \tag{5.6}$$

Here $F(s)$ is supposed to be the transform of $f(t)$, which *must be a continuous function* for all $t > 0$. By $f(0+)$ we mean the limit of $f(t)$ as t approaches 0 through *positive* values. It often happens in applications that f has a discontinuity at 0, making $f(0)$ an ambiguous notation.

Example 3

Having obtained the transform of $f(t) = \cos \omega t$, we apply the property above to obtain

$$\mathcal{L}(f'(t)) = \mathcal{L}(-\omega \sin \omega t) = s \cdot \frac{s}{s^2 + \omega^2} - 1 = \frac{-\omega^2}{s^2 + \omega^2}.$$

Now the linearity of the Laplace transform lets us "divide out" the factor $-\omega$, giving again

$$\mathcal{L}(\sin \omega t) = \frac{\omega}{s^2 + \omega^2}. \tag{5.7}$$

We can give a justification of the formula (5.6) for the transform of a derivative. We assume that both f and f' satisfy the existence theorem stated earlier in this section. By definition, the transform of f' is

$$\mathcal{L}(f'(t)) = \int_0^\infty e^{-st} f'(t)\, dt.$$

Using integration by parts in the natural way, we get

$$\mathcal{L}(f'(t)) = e^{-st} f(t)\big|_0^\infty - \int_0^\infty (-s) e^{-st} f(t)\, dt.$$

Since f must satisfy condition (5.3), we may assume that the upper limit in the first term of the equation above is

$$\lim_{T \to \infty} e^{-sT} f(T) = 0,$$

at least for s large enough. The lower limit contributes $-f(0+)$. In the second term, two minus signs cancel; and s, which acts as a constant in the integration with respect to t, can come outside the integral, leaving $s\mathcal{L}(f(t))$ or $sF(s)$.

The formula (5.6) for the transform of a derivative has numerous extensions. In one direction we could apply it several times over to a function with enough continuous derivatives, to get formulas for the transforms of the second- and higher-order derivatives. The results are

$$\mathcal{L}(f''(t)) = s^2 F(s) - sf(0+) - f'(0+), \tag{5.8}$$
$$\mathcal{L}(f'''(t)) = s^3 F(s) - s^2 f(0+) - sf'(0+) - f''(0+), \tag{5.9}$$

and, in the general case,

$$\mathcal{L}(f^{(n)}(t)) = s^n F(s) - s^{n-1} f(0+) - s^{n-2} f'(0+) - \cdots - f^{(n-1)}(0+). \tag{5.10}$$

In the last formula, $F(s) = \mathcal{L}(f(t))$ and $f, f', \ldots, f^{(n-1)}$ are all assumed to be continuous.

Example 4

Calculate the Laplace transform of $f(t) = t^2$ using the derivative formula.
For this function we have the following derivatives:

$$f'(t) = 2t, \qquad f''(t) = 2, \qquad f'''(t) = 0.$$

Thus applying Eq. (5.9), we have

$$\mathscr{L}(f'''(t)) = 0 = s^3 F(s) - s^2 \cdot 0 - s \cdot 0 - 2.$$

This equation can be solved for $F(s)$, the transform of t^2, yielding

$$\mathscr{L}(t^2) = \frac{2}{s^3}.$$

Another application of the formula (5.6) for the transform of a derivative uses it in reverse. Suppose that $f(t)$ is a given function with Laplace transform $F(s)$. Then

$$g(t) = \int_0^t f(x)\, dx$$

is a continuous function for which

$$g'(t) = f(t), \qquad g(0) = 0.$$

According to the rule for the transform of a derivative,

$$\mathscr{L}(g'(t)) = F(s) = s\mathscr{L}(g(t)) - 0.$$

We now solve this last equation to get $\mathscr{L}(g(t))$ or

$$\mathscr{L}\left(\int_0^t f(x)\, dx \right) = \frac{1}{s} F(s). \qquad\qquad (5.11)$$

Equations (5.6) and (5.11) justify what we had said about the Laplace transform, that it replaces the operations of calculus with those of algebra. Equation (5.11) says that integration of a function can be replaced by dividing its transform by s, while Eq. (5.6) says that differentiation of a function can be replaced by multiplying its transform by s and subtracting $f(0+)$.

Finally, we mention another lesser, but still convenient, property, called the "shifting" theorem. It states that if

$$\mathscr{L}(f(t)) = F(s),$$

then

$$\mathscr{L}(e^{bt}f(t)) = F(s - b). \qquad\qquad (5.12)$$

Table 5.1

Basic Properties of
the Laplace Transform
$\mathscr{L}(f(t))$
$= \int_0^\infty e^{-st} f(t)\, dt = F(s)$

$\mathscr{L}(c_1 f_1(t) + c_2 f_2(t)) = c_1\mathscr{L}(f_1) + c_2\mathscr{L}(f_2)$

$\mathscr{L}(f'(t)) = sF(s) - f(0+)$

$\mathscr{L}(f''(t)) = s^2 F(s) - sf(0+) - f'(0+)$

$\mathscr{L}(f'''(t)) = s^3 F(s) - s^2 f(0+) - sf'(0+) - f''(0+)$

$\mathscr{L}(f^{(n)}(t)) = s^n F(s) - s^{n-1} f(0+) - s^{n-2} f'(0+) - \cdots - f^{(n-1)}(0+)$

$\mathscr{L}\left(\int_0^t f(x)\, dx\right) = \dfrac{1}{s} F(s)$

$\mathscr{L}(e^{bt} f(t)) = F(s - b)$

This formula is a result of simple manipulations with the definition, for

$$\mathscr{L}(e^{bt} f(t)) = \int_0^\infty e^{-st} e^{bt} f(t)\, dt$$

$$= \int_0^\infty e^{-(s-b)t} f(t)\, dt.$$

The last integral is the one defining $\mathscr{L}(f)$, except that $s - b$ has replaced s. Thus it is $F(s - b)$.

Example 5

Find the Laplace transform of $e^{bt} \cos \omega t$.

Let us designate $f(t) = \cos \omega t$. Then the transform of $f(t)$ is, according to Eq. (5.5),

$$F(s) = \mathscr{L}(\cos \omega t) = \frac{s}{s^2 + \omega^2}.$$

Now by Eq. (5.12) we find

$$\mathscr{L}(e^{bt} \cos \omega t) = \frac{s - b}{(s - b)^2 + \omega^2} = \frac{s - b}{s^2 - 2bs + b^2 + \omega^2}.$$

Exercises

In Exercises 1–10, use any of the basic properties (Table 5.1) to find the Laplace transform of the given function

1. $f(t) = e^{at} \sin \omega t$

2. $f(t) = e^{-2t} \sin 5t$

3. $f(t) = \cosh at$

4. $f(t) = \sinh 3t$

5. $f(t) = \dfrac{e^{mt} - e^{nt}}{m - n} \quad (m \neq n)$

6. $f(t) = \dfrac{me^{mt} - ne^{nt}}{m - n} \quad (m \neq n)$

7. $f(t) = te^{kt}$ 8. $f(t) = t^2 e^{4t}$

9. $f(t) = e^{-bt} \sinh at$ 10. $f(t) = e^{-bt} \cosh at$

11. Find the limit of the function f of Exercise 5 as $n \to m$, the transform of that function and compare to the limit of the transform.

12. Verify the integration rule (Eq. 5.11) for these functions: (a) $f(t) = \cos \omega t$, (b) $f(t) = e^{-at}$, (c) $f(t) = t$. (Find $\int_0^t f(x)\,dx$; find its transform; compare to $(1/s)\mathscr{L}(f)$.)

13. Find the transform of $f(t) = \cos^2 kt$ by using a trigonometric identity and linearity.

14. Use the trigonometric identity $\sin^2 A + \cos^2 A = 1$ and the results of Exercise 13 to find the transform of $\sin^2 kt$.

15. Find $\mathscr{L}(\sin kt \cos kt)$.

16. Suppose that $f(t)$ satisfies the hypotheses of Theorem 5.1. Why is it true that

$$\int_0^T e^{-st} f(t)\,dt$$

exists for all $T \geq 0$ and for all s?

17. Show that f has a Laplace transform by showing that the integral in Exercise 16 converges (i.e., the limit as $T \to \infty$ exists) if s is greater than the α in Eq. (5.3).

5.2

Applications to Initial Value Problems

In Section 5.1 we developed many properties of the Laplace transform; in this section we shall apply the linearity and the rule for the transform of derivatives,

$$\mathscr{L}(f^{(n)}(t)) = s^n F(s) - s^{n-1}f(0+) - s^{n-2}f'(0+) - \cdots - f^{(n-1)}(0+) \quad \textbf{(5.13)}$$

to the solution of initial value problems.

As a first example, consider the following problem, which we solved in Chapter 1:

$$\frac{du}{dt} + 2u = 0, \qquad 0 < t, \tag{5.14}$$

$$u(0) = 3. \tag{5.15}$$

The differential equation (5.14) expresses a relation between u and its derivative, which is valid for all $t > 0$. We may "take the transform" of both sides: formally, apply the operator $\mathscr{L}$; more precisely, multiply through by e^{-st} and integrate from $t = 0$ to ∞. Since $\mathscr{L}(0) = 0$, the result is

$$\mathscr{L}\left(\frac{du}{dt} + 2u\right) = 0.$$

But now we use the linearity of the Laplace transform to get

$$\mathscr{L}\left(\frac{du}{dt}\right) + 2\mathscr{L}(u) = 0.$$

Finally, we use Eq. (5.13) with $n = 1$, obtaining

$$s\mathscr{L}(u) - 3 + 2\mathscr{L}(u) = 0$$

or, if we designate $U(s) = \mathscr{L}(u(t))$,

$$sU - 3 + 2U = 0. \tag{5.16}$$

As we had foreseen in Section 5.1, the Laplace transform has changed the differential equation (5.14) into an algebraic equation (5.16). But note that the initial condition, Eq. (5.15), was incorporated by the relationship between $\mathscr{L}(u')$ and $\mathscr{L}(u)$. It is really the initial value problem, Eqs. (5.14) and (5.15), that has been converted into the algebraic equation (5.16).

Now Eq. (5.16) can be solved easily for the transform

$$U(s) = \frac{3}{s + 2}. \tag{5.17}$$

We readily recognize the right-hand side of Eq. (5.17) as the transform of $3e^{-2t}$ and therefore would like to conclude that the solution of the initial value problem, Eqs. (5.14) and (5.15), is

$$u(t) = 3e^{-2t}. \tag{5.18}$$

It is easy to verify that this is indeed the solution sought. But might it not be possible that two different functions have the same transform? If so, we would have to devise some process for deciding which one to use. Fortunately, this unpleasant possibility does not occur, at least in the cases that concern us, as the following theorem states.

Theorem 5.2

If $f(t)$ is continuous and $e^{-\alpha t}f(t) \to 0$ as $t \to \infty$ (at least for α large enough), then (1) $f(t)$ has a Laplace transform $F(s)$, and (2) no other function satisfying the hypotheses has $F(s)$ as its transform.

The utility of the theorem is this: if we find a function $U(s)$ that is the solution of a transformed problem, and if we can recognize $U(s)$ as the Laplace transform of some specific $u(t)$, then $u(t)$ must be the solution of the original problem. However well we justify this process, it is still best to verify a proposed solution—at least by spot-checking the initial conditions. Algebraic errors are easy to make.

Example 1

Solve the initial value problem

$$\frac{du}{dt} + 3u = 1, \quad 0 < t,$$

$$u(0) = -1.$$

Below is the sequence of steps (with explanation) carried out in obtaining the transform U of the solution:

$$\mathscr{L}\left(\frac{du}{dt} + 3u\right) = \mathscr{L}(1) \quad \text{(Transform both sides)}$$

$$\mathscr{L}\left(\frac{du}{dt}\right) + 3\mathscr{L}(u) = \frac{1}{s} \quad \text{(Linearity)}$$

$$sU - (-1) + 3U = \frac{1}{s} \quad \text{(Transform of derivative)}$$

$$U(s) = \frac{1}{s+3}\left(\frac{1}{s} - 1\right) \quad \text{(Solved for } U\text{)}$$

$$= \frac{1}{s(s+3)} - \frac{1}{s+3}.$$

To complete the solution, we must find the function $u(t)$ whose transform is $U(s)$. Certainly, we may use linearity to justify working on the two terms of $U(s)$ separately.

One way to find the function whose transform is $1/s(s+3)$ is to use partial fractions.* This technique breaks down a complicated rational function into simple ones, like $1/(s-a)^k$. These simple functions are transforms of familiar functions of t.

For the case at hand we find the partial fractions decomposition

$$\frac{1}{s(s+3)} = \frac{1}{3}\frac{1}{s} - \frac{1}{3}\frac{1}{s+3}.$$

From here we find that

$$U(s) = \frac{1}{s(s+3)} - \frac{1}{s+3} = \frac{1}{3}\frac{1}{s} - \frac{4}{3}\frac{1}{s+3},$$

* See the appendix to this chapter.

and this is recognized as the transform of

$$u(t) = \tfrac{1}{3} - \tfrac{4}{3}e^{-3t}.$$

Another way to find the function whose transform is $1/s(s + 3)$ is to use the integration rule (Eq. (5.11) of Section 5.1):

$$\mathcal{L}\left(\int_0^t f(x)\, dx\right) = \frac{1}{s}\mathcal{L}(f(t)).$$

Now we may identify the right-hand side with our function of s, $1/s(s + 3)$. We conclude that $\mathcal{L}(f(t)) = 1/(s + 3)$, so $f(t) = e^{-3t}$, and

$$\mathcal{L}\left(\int_0^t e^{-3x}\, dx\right) = \frac{1}{s(s + 3)}.$$

Finally, we find the function $u(t)$ whose transform is $U(s)$ above:

$$u(t) = \int_0^t e^{-3x}\, dx - e^{-3t}$$

$$= \tfrac{1}{3}(1 - e^{-3t}) - e^{-3t}$$

$$= \tfrac{1}{3} - \tfrac{4}{3}e^{-3t}.$$

Let us try another example, but this time with a second-order differential equation.

Example 2

A mass–spring–damper system, free of impressed forces, starts from given nonequilibrium position. Solve this initial value problem for the displacement of the mass:

$$\frac{d^2u}{dt^2} + 2\frac{du}{dt} + 5u = 0, \qquad 0 < t,$$

$$u(0) = 3, \qquad \frac{du}{dt}(0) = 0.$$

Here the steps used in obtaining the transform of the solution:

$$\mathcal{L}\left(\frac{d^2u}{dt^2} + 2\frac{du}{dt} + 5u\right) = \mathcal{L}(0) \qquad \text{(Transform both sides)}$$

$$\mathcal{L}\left(\frac{d^2u}{dt^2}\right) + 2\mathcal{L}\left(\frac{du}{dt}\right) + 5\mathcal{L}(u) = 0 \qquad \text{(Linearity)}$$

$$\left. \begin{array}{l} \mathcal{L}(u) = U \\[6pt] \mathcal{L}\!\left(\dfrac{du}{dt}\right) = sU - 3 \\[8pt] \mathcal{L}\!\left(\dfrac{d^2u}{dt^2}\right) = s^2U - 3s - 0 \end{array} \right\} \quad \text{(Rule for transform of derivatives)}$$

$$s^2U - 3s + 2(sU - 3) + 5U = 0 \qquad \text{(Substitution)}$$

$$U(s) = \frac{3s + 6}{s^2 + 2s + 5} \qquad \text{(Solved for } U\text{)}.$$

This function does not appear among the transforms we have calculated. Furthermore, since the polynomial in the denominator has complex roots, partial fractions may be inconvenient. However, the denominator can be simplified by completing the square:

$$s^2 + 2s + 5 = s^2 + 2s + 1 + 4 = (s + 1)^2 + 4.$$

This grouping now suggests that we express $U(s)$ as a function of $s + 1$:

$$U(s) = \frac{3s + 6}{s^2 + 2s + 5} = \frac{3(s + 1) + 3}{(s + 1)^2 + 4} = V(s + 1).$$

Then we see that $V(s)$ is just

$$V(s) = \frac{3s + 3}{s^2 + 4} = 3\,\frac{s}{s^2 + 4} + \frac{3}{2}\,\frac{2}{s^2 + 4}$$

$$= \mathcal{L}(3\cos 2t + \tfrac{3}{2}\sin 2t).$$

Thus by the shifting theorem, $U(s)$ is the transform of

$$u(t) = e^{-t}(3\cos 2t + \tfrac{3}{2}\sin 2t).$$

This is the solution of the given initial value problem.

It is becoming clear that transforming an initial value problem and then finding the transform of the solution ($U(s)$ in the examples) is quite easy. The hard part is to find the function $u(t)$ of which $U(s)$ is the transform. Incidentally, this relation is symbolized as

$$u(t) = \mathcal{L}^{-1}(U(s)) \qquad \qquad \textbf{(5.19)}$$

and is called the *inverse Laplace transform*. Like the direct transform, the inverse transform is linear:

$$\mathcal{L}^{-1}(c_1F_1(s) + c_2F_2(s)) = c_1\mathcal{L}^{-1}(F_1(s)) + c_2\mathcal{L}^{-1}(F_2(s)). \qquad \textbf{(5.20)}$$

For the purpose of inverting transforms, one usually constructs a table of functions of s in their simplest form and the corresponding functions of t. Table 5.2 is just such a listing, constructed from the solutions of the Exercises and Examples of Section 5.1. Far more extensive tables are available. But no table, it seems, is quite large enough to

Table 5.2	$F(s)$	$f(t)$
Laplace Transforms	$\dfrac{1}{s}$	1
	$\dfrac{1}{s^2}$	t
	$\dfrac{1}{s^3}$	$\dfrac{t^2}{2}$
	$\dfrac{1}{s^k}$	$\dfrac{t^{k-1}}{(k-1)!}$
	$\dfrac{1}{s-a}$	e^{at}
	$\dfrac{1}{(s-a)^2}$	te^{at}
	$\dfrac{1}{(s-a)^3}$	$\dfrac{t^2}{2}e^{at}$
	$\dfrac{1}{s^2+\omega^2}$	$\dfrac{1}{\omega}\sin \omega t$
	$\dfrac{s}{s^2+\omega^2}$	$\cos \omega t$
	$\dfrac{1}{(s-b)^2+\omega^2}$	$\dfrac{1}{\omega}e^{bt}\sin \omega t$
	$\dfrac{s}{(s-b)^2+\omega^2}$	$e^{bt}\left(\cos \omega t + \dfrac{b}{\omega}\sin \omega t\right)$
	$\dfrac{1}{(s^2+\omega^2)^2}$	$\dfrac{\sin \omega t}{2\omega^3} - \dfrac{t\cos \omega t}{2\omega^2}$
	$\dfrac{s}{(s^2+\omega^2)^2}$	$\dfrac{t\sin \omega t}{\omega}$

solve all the problems one encounters. One needs methods for breaking down the functions of s into simpler ones. If we can get $U(s)$ into a linear combination of simple functions then we can use the linearity of the inverse transform to find the inverse transform of U.

It is useful to know what kind of function to expect for $U(s)$.

Suppose $u(t)$ is the solution of an initial value problem

$$\frac{d^n u}{dt^n} + b_1 \frac{d^{n-1} u}{dt^{n-1}} + \cdots + b_{n-1} \frac{du}{dt} + b_n u = f(t), \tag{5.21}$$

$$u(0) = q_0, \qquad \frac{du}{dt}(0) = q_1, \qquad \ldots, \qquad \frac{d^{n-1} u}{dt^{n-1}}(0) = q_{n-1}. \tag{5.22}$$

We would carry out these steps in finding the transform U of the solution:

$$\mathcal{L}\left(\frac{d^n u}{dt^n} + b_1 \frac{d^{n-1} u}{dt^{n-1}} + \cdots + b_{n-1} \frac{du}{dt} + b_n u\right) = \mathcal{L}(f(t))$$

$$\mathcal{L}\left(\frac{d^n u}{dt^n}\right) + b_1 \mathcal{L}\left(\frac{d^{n-1} u}{dt^{n-1}}\right) + \cdots + b_{n-1} \mathcal{L}\left(\frac{du}{dt}\right) + b_n \mathcal{L}(u) = \mathcal{L}(f(t))$$

$$\mathcal{L}(f(t)) = F(s)$$

$$\mathcal{L}(u(t)) = U$$

$$\mathcal{L}\left(\frac{du}{dt}\right) = sU - q_0$$

$$\vdots$$

$$\mathcal{L}\left(\frac{d^{n-1} u}{dt^{n-1}}\right) = s^{n-1} U - s^{n-2} q_0 - \cdots$$

$$\mathcal{L}\left(\frac{d^n u}{dt^n}\right) = s^n U - s^{n-1} q_0 - \cdots.$$

Substituting these last expressions into the transformed equation gives

$$s^n U - s^{n-1} q_0 - \cdots - q_{n-1} + b_1(s^{n-1} U - s^{n-2} q_0 - \cdots) + \cdots$$
$$+ b_{n-1}(sU - q_0) + b_1 U = F(s). \tag{5.23}$$

(The most common error in applying Laplace transforms is to forget the parentheses in this kind of expression.) We can see that this equation can be solved, resulting in this form for U:

$$U(s) = \frac{F(s) + I(s)}{p(s)}, \tag{5.24}$$

where $F(s) = \mathcal{L}(f(t))$ is the transform of the inhomogeneity, $I(s)$ is a polynomial of degree $n - 1$ or less coming from the initial conditions, and $p(s)$ is the characteristic polynomial of the homogeneous equation. A quick glance at Table 5.2 shows that all the transforms there—which might be used as $F(s)$—are rational functions. (A rational function is a ratio of two polynomials.) Thus it seems that we should expect $U(s)$ in the form of a rational function.

The most satisfactory way of breaking down a rational function into simpler rational functions is the method of partial fractions, as seen in Example 1 and the following.

Example 3

Solve the initial value problem

$$\frac{d^2u}{dt^2} + 3\frac{du}{dt} + 2u = 1,$$

$$u(0) = 0, \qquad \frac{du}{dt}(0) = 0.$$

We first transform the initial value problem and then solve for $U(s)$:

$$s^2U + 3sU + 2U = \frac{1}{s}, \qquad U(s) = \frac{1}{s(s^2 + 3s + 2)}.$$

To use partial fractions, we must have the denominator in factored form:

$$s(s^2 + 3s + 2) = s(s + 1)(s + 2).$$

Then we seek coefficients A, B, and C such that

$$\frac{1}{s(s + 1)(s + 2)} = \frac{A}{s} + \frac{B}{s + 1} + \frac{C}{s + 2}.$$

By elementary algebra, or otherwise, we find the coefficients

$$A = \tfrac{1}{2}, \qquad B = -1, \qquad C = \tfrac{1}{2}.$$

Thus we have $U(s)$ in terms of simple functions of s:

$$U(s) = \tfrac{1}{2}\frac{1}{s} - \frac{1}{s + 1} + \tfrac{1}{2}\frac{1}{s + 2},$$

and we can easily find the inverse transform from tables:

$$u(t) = \tfrac{1}{2} - e^{-t} + \tfrac{1}{2}e^{-2t}.$$

Our procedure for solving initial value problems by means of the Laplace transform can be summarized as follows.
1. Transform the differential equation.
2. Solve for $U(s)$.
3. Find the corresponding $u(t)$.
As methods for carrying out the third step, we have used these in the examples. (1) Look up $U(s)$ in tables. (2) Express $U(s)$ as $V(s + b)$ and apply the shifting theorem. (3) Decompose $U(s)$ by partial fractions.

Of course, one is free to use any property of the Laplace transform—such as the integration property—to achieve a recognizable form for $U(s)$. The method of partial fractions is reviewed in detail in the appendix to this chapter. One consequence of the partial fractions decomposition is the Heaviside inversion formula, in the conclusion of Theorem 5.3.

Theorem 5.3

Suppose $N(s)$ and $D(s)$ are polynomials and $D(s)$ has higher degree than $N(s)$. Let $D(s)$ have only simple roots: $r_1, r_2, \ldots, r_k$. Then

$$\mathcal{L}^{-1}\left(\frac{N(s)}{D(s)}\right) = \frac{N(r_1)}{D'(r_1)} e^{r_1 t} + \frac{N(r_2)}{D'(r_2)} e^{r_2 t} + \cdots + \frac{N(r_k)}{D'(r_k)} e^{r_k t}.$$

Exercises

In Exercises 1–15, find the transform of the solution, $U(s)$, and then invert the transform to find $u(t)$.

1. $\dfrac{du}{dt} + 2u = e^{-t}$, $u(0) = 3$

2. $\dfrac{du}{dt} + 2u = t$, $u(0) = 0$

3. $\dfrac{du}{dt} + 2u = e^{-2t}$, $u(0) = 0$

4. $\dfrac{du}{dt} + u = \sin t$, $u(0) = u_0$

5. $\dfrac{d^2u}{dt^2} + 2\dfrac{du}{dt} + 4u = 0$, $u(0) = 1$, $\dfrac{du}{dt}(0) = 0$

6. $\dfrac{d^2u}{dt^2} + 2\dfrac{du}{dt} + 2u = 0$, $u(0) = 0$, $\dfrac{du}{dt}(0) = 1$

7. $\dfrac{d^2u}{dt^2} + 2\dfrac{du}{dt} + u = e^{-t}$, $u(0) = 0$, $\dfrac{du}{dt}(0) = 0$

8. $\dfrac{d^2u}{dt^2} + 2\dfrac{du}{dt} = 0$, $u(0) = 1$, $\dfrac{du}{dt}(0) = -1$

9. $\dfrac{d^2u}{dt^2} + 2\dfrac{du}{dt} = 1$, $u(0) = 0$, $\dfrac{du}{dt}(0) = 0$

10. $\dfrac{d^2u}{dt^2} + 4u = 0$, $u(0) = 0$, $\dfrac{du}{dt}(0) = 1$

11. $\dfrac{d^2u}{dt^2} + 4u = e^{-2t}$, $u(0) = 0$, $\dfrac{du}{dt}(0) = 0$

12. $\dfrac{d^2u}{dt^2} + 4u = \sin t, \ u(0) = 1, \ \dfrac{du}{dt}(0) = 0$

13. $\dfrac{d^2u}{dt^2} + 4u = \cos t, \ u(0) = 1, \ \dfrac{du}{dt}(0) = 0$

14. $\dfrac{d^2u}{dt^2} + 4u = \sin 2t, \ u(0) = 0, \ \dfrac{du}{dt}(0) = 0$

15. $\dfrac{d^2u}{dt^2} = 1, \ u(0) = 0, \ \dfrac{du}{dt} = 0$

16. Prove that the inverse transform Eq. (5.19) has the linearity property (5.20). (Hint: you only have to transform both sides of Eq. (5.20) and show them equal.)

5.3

Further Applications of the Laplace Transform

The Laplace transform can be used in a quite mechanical way to solve specific initial value problems with numerical initial values. In this section we shall point out other ways in which the Laplace transform can help to solve and understand differential equations.

Example 1

Obtain two independent solutions of the differential equation

$$\frac{d^2u}{dt^2} + 2\frac{du}{dt} + u = 0. \tag{5.25}$$

The Laplace transform requires initial data in addition to the differential equation. If we add two initial conditions in parametric form—say,

$$u(0) = u_0, \qquad \frac{du}{dt}(0) = v_0 \tag{5.26}$$

—then we may make specific choices for u_0 and v_0 to produce the required solutions.

The transformed problem is readily solved to find

$$U(s) = \frac{u_0 s + v_0 + 2u_0}{s^2 + 2s + 1} = \frac{u_0(s + 2) + v_0}{(s + 1)^2}. \tag{5.27}$$

We see immediately that the choice $u_0 = 1$, $v_0 = -1$ causes the numerator of Eq. (5.27) to simplify to $(s + 1)$. The transform $U(s)$ is just $1/(s + 1)$, corresponding to the solution $u_1(t) = e^{-t}$. A second independent solution can be found by choosing $u_0 = 0$, $v_0 = 1$. Then the numerator of Eq. (5.27) is 1, and $U(s) = 1/(s + 1)^2$. This is the transform of $u_2(t) = te^{-t}$. Other choices of the parameters u_0, v_0 could have been made. For example, the two choices

$$u_0 = 1, \qquad v_0 = 0: \qquad u(t) = (1 + t)e^{-t},$$
$$u_0 = 0, \qquad v_0 = 1: \qquad u(t) = te^{-t}$$

give us two solutions of the differential equation that satisfy especially simple initial conditions. In this simple case we could even leave u_0 and v_0 as parameters and compute the inverse of $U(s)$, getting

$$u(t) = u_0(1 + t)e^{-t} + v_0te^{-t}.$$

The Laplace transform tends to flood us with a quantity of detail that gives more information than we really want. Sometimes, one uses the Laplace transform just "part way," as in this example.

Example 2

Find the general solution of the fourth-order equation

$$\frac{d^4u}{dt^4} + 5\frac{d^2u}{dt^2} + 4u = \sin 2t. \tag{5.28}$$

We know that the general solution of the nonhomogeneous equation (5.28) has the form $u_c + u_p$, where u_p is a particular solution. It follows that the transform of the general solution will be

$$\mathscr{L}(u) = \mathscr{L}(u_c) + \mathscr{L}(u_p).$$

Now u_c is easily found by the usual method. We only need to factor the characteristic polynomial

$$m^4 + 5m^2 + 4 = (m^2 + 4)(m^2 + 1)$$

to see that the complementary function is

$$u_c = c_1 \cos 2t + c_2 \sin 2t + c_3 \cos t + c_4 \sin t.$$

A particular solution of the nonhomogeneous equation can be found by using the Laplace transform of Eq. (5.28) with zero initial conditions. The transform of this particular solution can be found with a minimum of algebra:

$$U(s) = \frac{2}{(s^2 + 4)(s^4 + 5s^2 + 4)} = \frac{2}{(s^2 + 4)^2(s^2 + 1)}. \tag{5.29}$$

We might now start the partial fractions expansion for Eq. (5.29). Since the transform is a function of s^2, we see that the correct form is

$$\frac{2}{(s^2 + 4)^2(s^2 + 1)} = \frac{A}{(s^2 + 4)^2} + \frac{B}{s^2 + 4} + \frac{C}{s^2 + 1}. \tag{5.30}$$

Clearly, the last two terms are transforms of solutions of the homogeneous equation. The significant term is thus the first one. We can quickly find $A = -\frac{2}{3}$ (replace s^2 by z and use the cover-up rule*) and the inverse transform

$$\mathscr{L}^{-1}\left(\frac{-\frac{2}{3}}{(s^2 + 4)^2}\right) = -\frac{2}{3}(\frac{1}{16})(\sin 2t - 2t \cos 2t).$$

We again recognize $\sin 2t$ as a solution of the homogeneous equation. After discarding that term we find the particular solution

$$u_p(t) = \frac{1}{12}t \cos 2t. \tag{5.31}$$

We might also see from inspection of Eq. (5.29) or (5.30) that a particular solution must have the form

$$u_p(t) = D_1 t \cos 2t + D_2 t \sin 2t$$

and then find D_1 and D_2 by the method of undetermined coefficients.

In the preceding example our objective was just to find a particular solution of a nonhomogeneous differential equation. We were therefore justified in adding or deleting solutions of the corresponding homogeneous equation (or their transforms). In many problems in applied mathematics we are interested in the long-term behavior of a solution. We are willing to ignore *transient* terms—functions that approach 0 as t increases. Any term that does not have limit 0 as t increases is called *persistent*.

The Laplace transform provides a handy method for distinguishing persistent and transient parts. If $U(s)$, the transform of the solution of an initial value problem, is a rational function, say,

$$U(s) = \frac{N(s)}{D(s)}, \tag{5.32}$$

where N and D are polynomials, then we know that U has a partial fractions expansion as a linear combination of terms like $1/(s - r)$ and its powers. The number r in such a term is a root of the denominator $D(s)$. Now a term like $1/(s - r)^k$ contributes a term like $t^{k-1}e^{rt}$ to the solution $u(t)$. If r is negative, $t^{k-1}e^{rt}$ vanishes as t increases; if r is 0 or positive, the function persists. Similarly, complex roots of $D(s)$, which occur as conjugate pairs $r = \alpha \pm i\beta$, produce terms in the solution like

$$t^m e^{\alpha t}(c_1 \cos \beta t + c_2 \sin \beta t)$$

* See the appendix to this chapter.

(where m is less than the multiplicity of r). Again, if α is negative, this function disappears in time; if α is 0 or positive, the term persists.

Thus to find the persistent parts of the inverse transform of $U(s)$ of Eq. (5.32), we should find the terms of the partial fractions expansion that correspond to roots of D with *nonnegative real part*. The remaining terms are representatives of parts of $u(t)$ that decay exponentially in time.

Example 3

A simple spring–mass–damper system (see Fig. 5.1) has the differential equation

$$\frac{d^2u}{dt^2} + 2\frac{du}{dt} + 5u = \cos t. \tag{5.33}$$

Find the persistent part of u for large t.

Figure 5.1

Mass–Spring–Damper System

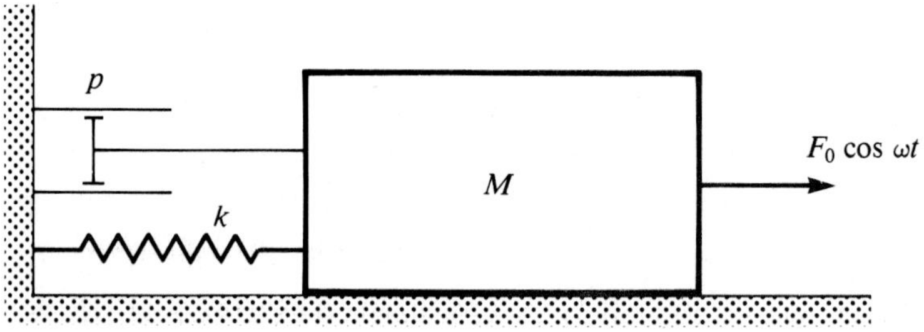

Assuming that p is positive, any solution of the homogeneous equation approaches 0 as t increases. Thus we need only examine some particular solution for persistent parts. For this purpose we impose the initial conditions

$$u(0) = 0, \qquad \frac{du}{dt}(0) = 0.$$

Then the transform of the solution is

$$U(s) = \frac{s}{(s^2 + 2s + 5)(s^2 + 1)}. \tag{5.34}$$

The denominator has two roots with real part -1 and two pure imaginary roots. Thus the partial fraction expansion of $U(s)$ has the form

$$\frac{s}{(s^2 + 2s + 5)(s^2 + 1)} = \frac{As + B}{s^2 + 1} + \frac{Cs + D}{s^2 + 2s + 5}. \tag{5.35}$$

Obviously, the second term is the transform of a transient part of the solution. Therefore it is necessary to find only A and B, which we may do by the complex cover-up:

$$A = \tfrac{1}{5}, \quad B = \tfrac{1}{10}. \tag{5.36}$$

From here we have the inverse transform of the first term of Eq. (5.35):

$$\tfrac{1}{5} \cos t + \tfrac{1}{10} \sin t$$

as the persistent part of the solution of Eq. (5.33)—whatever the initial conditions.

In general, the displacement of the mass in the mass–spring–damper system of Fig. 5.1 will have the differential equation

$$F_0 \cos \omega t - ku - p\frac{du}{dt} = M\frac{d^2u}{dt^2},$$

and the presence of the forcing function, $F_0 \cos \omega t$, guarantees that the solution of this problem has a persistent part in the form

$$u(t) = A \cos \omega t + B \sin \omega t.$$

In many actual systems this kind of response is undesirable. A vibration absorber can eliminate it, as the following example shows.

Example 4

The mechanical system of Example 3 is to be modified by the addition of another spring and mass, as shown in Fig. 5.2. If u and u_1 are the

Figure 5.2

Mass–Spring–Damper
System with
Vibration Absorber

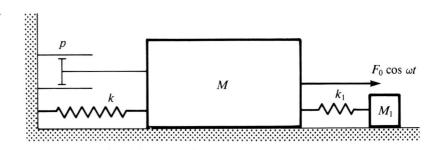

horizontal displacements of the centers of the masses M and M_1, the sum of horizontal forces on each mass leads to these equations as expressions of Newton's second law:

$$F_0 \cos \omega t - ku - p\frac{du}{dt} - k_1(u - u_1) = M\frac{d^2u}{dt^2} \, , \, k_1(u - u_1) = M_1\frac{d^2u_1}{dt^2}.$$

Note that the length of the added spring is $|u - u_1|$. By algebraic manipulation and renaming of coefficients as indicated, we arrive at these equations for the displacements u and u_1 of the masses M and M_1:

$$\frac{d^2u}{dt^2} + 2\frac{du}{dt} + (5 + \lambda^2)u - \lambda^2 u_1 = \cos t,$$

$$\frac{d^2u_1}{dt^2} + \omega^2 u_1 - \omega^2 u = 0, \tag{5.37}$$

$$\frac{p}{M} = 2, \quad \frac{k}{M} = 5, \quad \frac{F_0}{M} = 1, \quad \frac{k_1}{M_1} = \omega^2, \quad \frac{k_1}{M} = \lambda^2.$$

The transform of the system of equations (5.37), taking zero initial conditions, is

$$(s^2 + 2s + 5 + \lambda^2)U - \lambda^2 U_1 = \frac{s}{s^2 + 1}, \tag{5.38}$$

$$-\omega^2 U + (s^2 + \omega^2)U_1 = 0.$$

These two equations can be solved by elimination or otherwise to obtain the following expressions for the transforms:

$$U(s) = \frac{s(s^2 + \omega^2)}{(s^2 + 1)\,\Delta(s)}, \qquad U_1(s) = \frac{s\omega^2}{(s^2 + 1)\,\Delta(s)} \tag{5.39}$$

where

$$\Delta(s) = (s^2 + \omega^2)(s^2 + 2s + 5 + \lambda^2) - \lambda^2\omega^2. \tag{5.40}$$

Evidently, by choosing k_1 and M_1 to make $\omega^2 = 1$—tuning the system, as it were—we can eliminate the factor $s^2 + 1$ from the denominator of $U(s)$. This will free $u(t) = \mathcal{L}^{-1}(U(s))$ from persistent parts, provided that the polynomial $\Delta(s)$ has no roots with nonnegative real part. Nevertheless, the factor $s^2 + 1$ remains in the denominator of $U_1(s)$. Thus the device suggested passes on the persistent oscillation to the second mass but does not eliminate it from the system.

Now we must ask whether $\Delta(s)$ does indeed have roots with negative real parts. In general, it is very difficult to decide such a question. For this special case, however, we might observe that (1) $\Delta(s)$ cannot have a purely imaginary root as long as $\lambda^2 > 0$ (this is easy to prove); (2) the roots of $\Delta(s)$ are continuous functions of λ^2 and ω^2. Now if the roots of

$\Delta(s)$ have negative real parts for some specially selected positive values of λ^2 and ω^2, the foregoing observations guarantee that the roots of $\Delta(s)$ have negative real parts for all positive λ^2 and ω^2.

Exercises

In Exercises 1–10, use the Laplace transform to find a particular solution of the differential equation.

1. $\dfrac{du}{dt} + 4u = e^{-4t}$

2. $\dfrac{du}{dt} + ku = te^{-kt}$

3. $\dfrac{d^2u}{dt^2} + u = \sin 2t$

4. $\dfrac{d^2u}{dt^2} + 4\dfrac{du}{dt} + 3u = e^{-t}$

5. $\dfrac{d^2u}{dt^2} + 4\dfrac{du}{dt} + 5u = e^{-t}$

6. $\dfrac{d^2u}{dt^2} + 2b\dfrac{du}{dt} + cu = e^{-bt}$

7. $\dfrac{d^2u}{dt^2} + u = t\cos t$

8. $\dfrac{d^2u}{dt^2} - u = \cos t$

9. $\dfrac{d^2u}{dt^2} + 2\dfrac{du}{dt} + 2u = \sin t$

10. $\dfrac{d^2u}{dt^2} + 2\dfrac{du}{dt} + 2u = e^{-t}\cos t$

11. Find all solutions of the differential equation

$$\frac{d^4u}{dx^4} - \lambda^4 u = 0$$

satisfying the conditions

(a) $u(0) = 0, \ \dfrac{d^2u}{dx^2}(0) = 0$

(b) $u(0) = 0, \ \dfrac{du}{dx}(0) = 0$

(c) $\dfrac{d^2u}{dx^2}(0) = 0, \ \dfrac{d^3u}{dx^3}(0) = 0$

12. Find the persistent part of $u_1(t)$ from Example 4 (see Eq. (5.39)).

13. Show that $\Delta(s)$ in Eq. (5.40) cannot have a pure imaginary root if ω^2 and λ^2 are positive: set $s = i\theta$, find the real and imaginary parts of $\Delta(s) = \Delta(i\theta)$, and show that when one is 0 the other is not.

14. The delay differential equation

$$\frac{du}{dt}(t) = u(t - 1), \qquad t > 0,$$

$$u(t) = 1, \qquad -1 < t < 0$$

can be attacked by using the Laplace transform. If $U(s) = \mathcal{L}(u(t))$, then

$$\mathcal{L}(u(t-1)) = \int_0^\infty e^{-st}u(t-1)\,dt = \int_{-1}^\infty e^{-s(\tau+1)}u(\tau)\,d\tau$$

$$= e^{-s}\left[U(s) + \int_{-1}^0 e^{-s\tau}u(\tau)\,d\tau\right]$$

with the change of variables $\tau = t - 1$. Complete the integration and find the transformed equation.

15. Find the transform of the solution of the equation in Exercise 14.

16. Find the persistent part of the solution to the problem in Exercise 14. (Hint: assume that the equation $s = e^{-s}$ has just one solution with nonnegative real part.)

5.4

Discontinuity, Shift, and Impulse

One of the great virtues of the Laplace transform is its ability to handle discontinuous functions, which often occur as inhomogeneities in differential equations. The kind of discontinuity we can deal with is called a jump.

Definition 5.2

A function $f(t)$ has a *jump discontinuity* at $t = a$ if both

$$\lim_{h>0} f(a+h) = f(a+) \qquad \text{and} \qquad \lim_{h>0} f(a-h) = f(a-)$$

exist but are different.

The two limits—called the right-hand limit and left-hand limit—are symbolized by $f(a+)$ and $f(a-)$, although it may well happen that these are not values of the function $f(t)$. The simplest example of a function with a jump discontinuity is this one, called the *Heaviside step function,* defined by

$$h(t) = \begin{cases} 0, & t < 0, \\ 1, & t \geq 0. \end{cases} \tag{5.41}$$

For this function we have $h(0+) = 1$ and $h(0-) = 0$. (See Fig. 5.3.) The location of the jump can be changed by shifting the argument of h: the function $h(t - a)$ is 0 to the left of a and 1 to the right.

Other functions having a jump can be expressed easily by using the Heaviside step function. For instance, if we define

$$f(t) = \begin{cases} 0, & t < 1 \\ t - 1, & t \geq 1 \end{cases}$$

Figure 5.3

Heaviside Step
Function

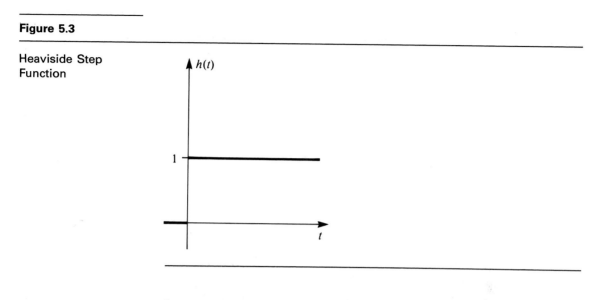

(see Fig. 5.4), we can more easily describe the same function with the formula

$$f(t) = (t - 1)h(t - 1).$$

The factor $h(t - 1)$ is "off" until $t = 1$ and "on" thereafter. By combining Heaviside step functions, one can create various effects such as a "window" (see Fig. 5.5):

$$h(t - a) - h(t - b) = \begin{cases} 1, & a \le t < b, \\ 0, & \text{otherwise.} \end{cases} \tag{5.42}$$

Figure 5.4

$f(t) = (t - 2)h(t - 1)$

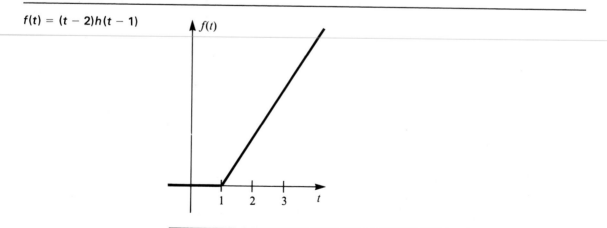

Figure 5.5

A "Window"

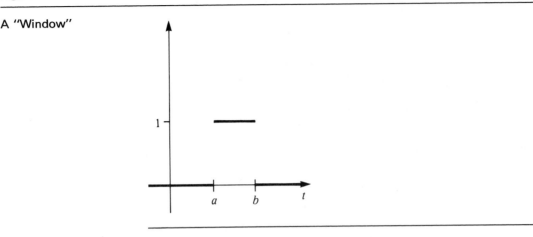

The theorem on existence of Laplace transforms in Section 5.1 refers to continuous functions only, but it can be broadened to include functions with jump discontinuities.

Theorem 5.4

Let $f(t)$ be a function that is continuous, except perhaps for a finite number of jump discontinuities, in every finite interval $0 \leq t \leq T$. Suppose that

$$\lim_{t \to \infty} e^{-\alpha t} f(t) = 0$$

for some α. Then f has a Laplace transform.

A function satisfying the continuity condition of this theorem is said to be *sectionally* (or piecewise) *continuous*, since it is made up of continuous segments between jumps. Note that an infinite number of discontinuities is allowed, as long as just a finite number occur on a finite interval.

As to the Laplace transform of $h(t)$, we quickly see that it is just $1/s$, since the transform depends only on values of the function for $t > 0$, where $h(t) = 1$. On the other hand, the shifted Heaviside function, $h(t - a)$, has a new transform:

$$\mathcal{L}(h(t - a)) = \int_0^\infty h(t - a)e^{-st}\, dt = \int_a^\infty e^{-st}\, dt = \frac{e^{-sa}}{s}, \qquad (5.43)$$

provided that $a \geq 0$. From this transform we can get the transform of the "window" of Eq. (5.42) as

$$\mathcal{L}(h(t - a) - h(t - b)) = \frac{e^{-sa} - e^{-sb}}{s}. \tag{5.44}$$

The Heaviside function often participates in the shifting of a function, say, $f(t)$. The "shifted function" is

$$h(t - a)f(t - a) = \begin{cases} 0, & t < a, \\ f(t - a), & t \geq a. \end{cases} \tag{5.45}$$

It is not difficult to find the transform of the shifted function:

$$\mathcal{L}(h(t - a)f(t - a)) = \int_0^\infty h(t - a)f(t - a)e^{-st} \, dt$$

$$= \int_a^\infty f(t - a)e^{-st} \, dt = \int_0^\infty f(t')e^{-s(t'+a)} \, dt'$$

$$= e^{-as} \int_0^\infty f(t')e^{-st'} \, dt'$$

(where $t' = t - a$). By identifying the integral in the last equation as the transform of f we obtain

$$\mathcal{L}(h(t - a)f(t - a)) = e^{-as}\mathcal{L}(f(t)), \tag{5.46}$$

known as the shifting theorem. In general, when e^{-as} appears in a transform, one should think of this theorem as a means of inverting the transform.

Example 1

Find the inverse transform of the function

$$U(s) = \frac{e^{-s}}{s^2 + 4}.$$

According to the shifting theorem, Eq. (5.46), this function is the transform of $h(t - 1)f(t - 1)$, where $f(t)$ has the transform

$$\mathcal{L}(f(t)) = \frac{1}{s^2 + 4}.$$

Hence we identify $f(t) = \frac{1}{2}\sin 2t$ and determine

$$\mathcal{L}^{-1}\left(\frac{e^{-s}}{s^2 + 4}\right) = \frac{1}{2}h(t - 1)\sin 2(t - 1).$$

Example 2

At time $t = 0$ the current in the circuit of Fig. 5.6 is i_0. The impressed voltage v is 0 until time T, when it jumps to E. Find the current $u(t)$ as a function of t.

Figure 5.6

An *RL* Circuit

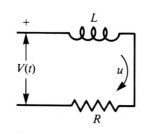

Kirchhoff's law (sum of voltage drops around a circuit equals the impressed voltage) gives the relation

$$L\frac{du}{dt} + Ru = v(t), \qquad t > 0.$$

The impressed voltage can be described as

$$v(t) = \begin{cases} 0, & 0 \le t < T \\ E, & T \le t \end{cases} = Eh(t - T).$$

Thus the transform of the differential equation is

$$L(sU - i_0) + RU = \frac{E}{s}e^{-sT}.$$

From here we can easily find $U(s)$:

$$U(s) = \frac{i_0 L}{Ls + r} + \frac{Ee^{-sT}}{s(Ls + R)}.$$

The inverse transform of the first term is

$$\mathscr{L}^{-1}\!\left(\frac{i_0 L}{Ls + R}\right) = \mathscr{L}^{-1}\!\left(\frac{i_0}{s + R/L}\right) = i_0 e^{-Rt/L}.$$

The second term contains the exponential that reminds us of the shifting theorem, Eq. (5.46). Thus

$$\frac{Ee^{-sT}}{s(Ls + R)} = \frac{E}{R}\!\left(\frac{1}{s} - \frac{1}{s + R/L}\right)\!e^{-sT}$$

is easily recognized as the transform of $h(t - T)f(t - T)$, where $f(t)$ is

$$f(t) = \mathscr{L}^{-1}\left(\frac{E}{s(Ls + R)}\right) = \frac{E}{R}(1 - e^{-Rt/L}).$$

Therefore we have

$$\mathscr{L}^{-1}(U(s)) = u(t) = i_0 e^{-Rt/L} + f(t - T)h(t - T).$$

This function is shown in Fig. 5.7.

Figure 5.7

Current in an *RL*
Circuit with
Impressed Voltage
That Jumps

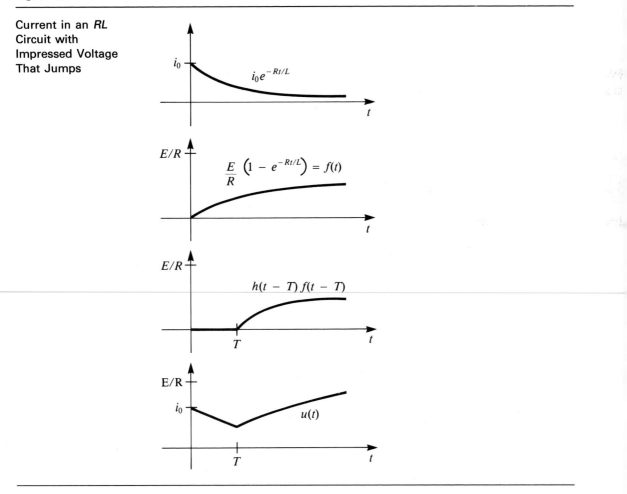

Example 3

A mass–spring system, starting from rest, is forced by a sinusoidal input at the natural frequency for one cycle. Find the displacement, which is the solution of

$$\frac{d^2u}{dt^2} + u = (h(t) - h(t - 2\pi)) \sin t, \qquad 0 < t,$$

$$u(0) = 0, \qquad \frac{du}{dt}(0) = 0.$$

The transform of this problem is

$$s^2 U - s \cdot 0 - 0 + U = \frac{(1 - e^{-2\pi s})}{s^2 + 1},$$

and from this we easily determine the transform of the solution:

$$U(s) = \frac{1 - e^{-2\pi s}}{(s^2 + 1)^2}.$$

Now from Table 5.2, the table of Laplace transforms in Section 5.2, we see that

$$\frac{1}{(s^2 + 1)^2} = \mathscr{L}\left(\frac{\sin t - t \cos t}{2}\right),$$

and then, using the shift theorem, Eq. (5.46), we find

$$u(t) = \frac{\sin t - t \cos t}{2}$$

$$- h(t - 2\pi)\frac{\sin(t - 2\pi) - (t - 2\pi)\cos(t - 2\pi)}{2}.$$

Another way to write this solution (taking advantage of the fact that $\sin(t - 2\pi) = \sin t$, etc.) is

$$u(t) = (1 - h(t - 2\pi))\frac{\sin t - t \cos t}{2} - h(t - 2\pi)\pi \cos t.$$

From the positions of the Heaviside functions we see that the first term is the solution for $0 < t < 2\pi$ and the second term is the solution for $2\pi \le t$. (See Fig. 5.8.)

Some physical problems lead to a nonhomogeneous differential equation whose inhomogeneity is a function having large value over a short duration. A mass–spring–damper system in which the mass is struck

Figure 5.8

Displacement of a Mass–Spring System Harmonically Excited for One Period

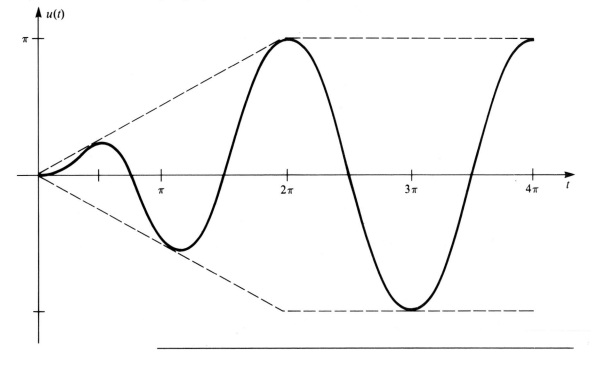

by a hammer provides a simple example. Typically, the values of the function are unknown, but its total effect, corresponding to the definite integral of the function, can be found. (In mechanics the time integral of a force is called impulse.)

The Heaviside unit step function can be used to construct a function having integral 1. For example,

$$\delta_\varepsilon(t) = \frac{1}{\varepsilon}\left(h(t) - h(t - \varepsilon)\right) = \begin{cases} 0, & t < 0, \\ \dfrac{1}{\varepsilon}, & 0 \le t < \varepsilon, \\ 0, & \varepsilon \le t, \end{cases} \tag{5.47}$$

has magnitude $1/\varepsilon$ for ε time units and 0 thereafter. Thus we can easily evaluate the integral

$$\int_0^\infty \delta_\varepsilon(t)\, dt = 1. \tag{5.48}$$

Example 4

Solve this initial value problem that models a mass–spring–damper system struck by a hammer.

$$\frac{d^2u}{dt^2} + 2\frac{du}{dt} + 5u = \delta_\varepsilon(t),$$

$$u(0) = 0, \qquad \frac{du}{dt}(0) = 0.$$

Apply the Laplace transform in the usual way to find the transform of the solution:

$$s^2U + 2sU + 5U = \frac{1}{\varepsilon}\left(\frac{1}{s} - \frac{e^{-\varepsilon s}}{s}\right),$$

$$U(s) = \frac{1}{\varepsilon}\frac{1}{s(s^2 + 2s + 5)}(1 - e^{-\varepsilon s}).$$

In order to determine $u(t)$, we find first the partial fractions expansion and then the inverse transform of the rational function

$$V(s) = \frac{1}{s(s^2 + 2s + 5)} = \frac{1}{5}\left(\frac{1}{s} - \frac{s + 2}{s^2 + 2s + 5}\right),$$

$$v(t) = \mathcal{L}^{-1}(V(s)) = \tfrac{1}{5}(1 - e^{-t}\cos 2t - \tfrac{1}{2}e^{-t}\sin 2t).$$

Now, using the shifting property, Eq. (5.46), we may find the inverse transform of $U(s)$:

$$u(t) = \mathcal{L}^{-1}(U(s)) = \frac{v(t) - h(t - \varepsilon)v(t - \varepsilon)}{\varepsilon}. \tag{5.49}$$

The parameter ε in Eq. (5.47), which is the "duration" of $\delta_\varepsilon(t)$, may be awkward, since it is usually known only to be small. One way out of this situation is to allow ε to approach 0. If we designate

$$\delta(t) = \lim_{\varepsilon \to 0} \delta_\varepsilon(t), \tag{5.50}$$

then we expect the integral of δ to remain equal to 1,

$$\int_0^\infty \delta(t) = 1, \tag{5.51}$$

and we expect its transform to be the limit of $\mathcal{L}(\delta_\varepsilon(t))$, that is,

$$\mathcal{L}(\delta(t)) = \lim_{\varepsilon \to 0}\frac{1 - e^{-\varepsilon s}}{\varepsilon s} = 1. \tag{5.52}$$

Unfortunately, the function defined by Eq. (5.50) has value 0, except at

$t = 0$, and thus cannot have the properties of Eqs. (5.51) and (5.52). Nevertheless, correct results can be obtained by treating the symbol $\delta(t)$ as if it were a function with properties (5.51) and (5.52). The usual names for $\delta(t)$ are *Dirac's delta* or the *unit impulse*.

Example 5

Use Laplace transforms to solve the initial value problem

$$\frac{d^2u}{dt^2} + 2\frac{du}{dt} + 5u = \delta(t),$$

$$u(0) = 0, \qquad \frac{du}{dt}(0) = 0.$$

The transformed equation is obtained by using the usual rules of Laplace transform and Eq. (5.45):

$$U(s) = \frac{1}{s^2 + 2s + 5}.$$

The inverse transform can be found by completing the squares in the denominator. It is

$$u(t) = \tfrac{1}{2}e^{-t}\sin 2t. \tag{5.53}$$

Another important property attributed to the unit impulse is this. If f is an integrable function defined at t, then

$$f(t) = \int_0^\infty \delta(t - \tau)f(\tau)\,d\tau. \tag{5.54}$$

This equation follows from Eq. (5.50) if the limit is moved outside the integral. Equation (5.54) is independent of the choice of the limits of integration, as long as t lies between them.

Exercises

In Exercises 1–12, find the transform of the solution of the problem and then find the solution. In each case, sketch the inhomogeneity.

1. $\dfrac{du}{dt} + au = h(t) - h(t - 1),\ u(0) = 0$

2. $\dfrac{du}{dt} + u = h(t - 1),\ u(0) = 1$

3. $\dfrac{du}{dt} + au = h(t) - 2h(t - 1) + h(t - 2),\ u(0) = 0$

4. $\dfrac{du}{dt} + u = (t-1)h(t-1),\ u(0) = 1$

5. $\dfrac{d^2u}{dt^2} + u = h(t) - h(t-1),\ u(0) = 0,\ \dfrac{du}{dt}(0) = 0$

6. $\dfrac{d^2u}{dt^2} + u = h(t) - h(t-\pi),\ u(0) = 0,\ \dfrac{du}{dt}(0) = 0$

7. $\dfrac{d^2u}{dt^2} + u = h(t) - 2h(t-\pi) + h(t-2\pi),\ u(0) = 0,\ \dfrac{du}{dt}(0) = 0$

8. $\dfrac{d^2u}{dt^2} + 2\dfrac{du}{dt} + u = \sin t + h(t-\pi)\sin(t-\pi),$

   $u(0) = 0,\ \dfrac{du}{dt}(0) = 0$

9. $\dfrac{d^2u}{dt^2} + 3\dfrac{du}{dt} + 2u = t - (t-1)h(t-1),$

   $u(0) = 0,\ \dfrac{du}{dt}(0) = 0$

10. $\dfrac{d^2u}{dt^2} + u = 0,\ u(0) = 0,\ \dfrac{du}{dt}(0) = 1$

11. $\dfrac{d^2u}{dt^2} + u = \delta(t),\ u(0) = 0,\ \dfrac{du}{dt}(0) = 0$

12. $\dfrac{d^2u}{dt^2} + 5\dfrac{du}{dt} + 6u = -\delta(t-1),\ u(0) = 0,\ \dfrac{du}{dt}(0) = 1$

13. The "staircase function" defined by $f(t) = n,\ n - 1 \le t < n$, can be expressed as a series of Heaviside functions:

$$f(t) = \sum_{k=0}^{\infty} h(t-k).$$

Find the transform of f in the form of an infinite series; then add up the series to obtain a closed form for $F(s)$.

14. Sketch the graph of the function $g(t) = f(t) - t$, where $f(t)$ is as in Exercise 13, and find its transform.

15. Is the unit impulse, $\delta(t)$, the derivative of the Heaviside step function?

16. Show that the function in Eq. (5.53) is the limit, as ε goes to 0, of the function in Eq. (5.49). Does that make sense, considering the meanings of the functions?

17. A machine gun mount is designed so that gun and recoil mechanism

are a critically damped system described by the differential equation

$$\frac{d^2u}{dt^2} + 2\frac{du}{dt} + u = f(t),$$

where u is the displacement of the muzzle from rest position. Assume that each bullet fired gives a unit impulse to the system, so that a burst of $K + 1$ bullets at intervals of τ time units gives

$$f(t) = \delta(t) + \delta(t - \tau) + \delta(t - 2\tau) + \cdots + \delta(t - K\tau)$$

$$= \sum_{k=0}^{K} \delta(t - k\tau).$$

Find $u(t)$ with this function for $f(t)$ in the differential equation and taking $u(0) = 0$, $du/dt(0) = 0$.

5.5

The Convolution

It frequently happens, in a very natural way, that we need to invert a transform that has the form of a product of two functions of s whose inverses are known. That is, we want to find $\mathcal{L}^{-1}(F(s)G(s))$ when $\mathcal{L}^{-1}(F(s))$ and $\mathcal{L}^{-1}(G(s))$ are both known. It turns out that one can calculate the inverse of such a product in terms of the known inverses, with the intervention of an integral.

Theorem 5.5

Let $f(t)$ and $g(t)$ be two functions that satisfy the hypotheses of Theorem 5.1 in Section 5.1 with transforms $F(s)$ and $G(s)$. Then the product $F(s)G(s)$ is the transform of

$$\mathcal{L}^{-1}(F(s)G(s)) = \int_0^t f(t - z)g(z)\, dz. \tag{5.55}$$

Note: The integral, which depends on both f and g, is called the *convolution* of f and g, written $f * g$. Notice that the defining integral is definite and that t appears both in the upper limit and in the integrand.

Proof: By the definition of Laplace transform we have

$$F(s)G(s) = \int_0^\infty e^{-su}f(u)\, du \int_0^\infty e^{-sv}g(v)\, dv. \tag{5.56}$$

(Variables u and v have been used in order to distinguish them. Since the integral defining the transform is definite, the name assigned to the variable of integration is immaterial.) Because the integrands $e^{-su}f(u)$ and

$e^{-sv}g(v)$ approach 0 exponentially as $u \to \infty$ or $v \to \infty$, we can treat these integrals as if the intervals of integration were finite. (Technically speaking, we have absolute and uniform convergence.) In particular, we can treat the product as an iterated double integral:

$$F(s)G(s) = \int_0^\infty \int_0^\infty e^{-s(u+v)} f(u)g(v) \, du \, dv \qquad (5.57)$$

over the quarter-plane $u > 0$, $v > 0$. Next we change the variables from u and v to $t = u + v$, $z = v$. The Jacobian of this transformation is 1, and the region of integration is described by the inequalities (see Fig. 5.9)

$$0 < t < \infty, \qquad 0 < z < t. \qquad (5.58)$$

Figure 5.9

Region of Integration
for Convolution

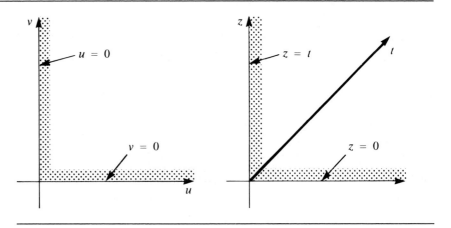

In terms of the new variables the integral for $F(s)G(s)$ is

$$\int_{t=0}^\infty \int_{z=0}^t e^{-st} f(t - z)g(z) \, dz \, dt = \int_0^\infty e^{-st} \int_0^t f(t - z)g(z) \, dz \, dt. \qquad (5.59)$$

This, of course, is just the definition of the transform of $f * g$.

    The convolution is an unfamiliar way of combining functions, but it follows many of the same algebraic rules as ordinary multiplication. Table 5.3 contains a summary of important theorems concerning convolution. In the following example we show how the convolution theorem can be used in the inversion of the Laplace transform.

Table 5.3

$$f * g = \int_0^t f(t - z)g(z)\, dz \qquad \text{Definition}$$

$$f * g = g * f \qquad \text{Commutative law}$$

$$f * (g * h) = (f * g) * h \qquad \text{Associative law}$$

$$f * (g + h) = f * g + f * h \qquad \text{Distributive law}$$

$$\mathcal{L}(f * g) = \mathcal{L}(f)\mathcal{L}(g) \qquad \text{Theorem 5.5}$$

Example 1

The solution of the initial value problem

$$\frac{d^2 u}{dt^2} + u = e^{-2t}, \qquad u(0) = 0, \qquad \frac{du}{dt}(0) = 0,$$

has as its transform

$$U(s) = \frac{1}{(s^2 + 1)(s + 2)}.$$

This can be recognized as the product of two well-known transforms

$$F(s) = \frac{1}{s^2 + 1} = \mathcal{L}(\sin t), \qquad G(s) = \frac{1}{s + 2} = \mathcal{L}(e^{-2t}).$$

Thus we may compute the inverse transform as

$$u(t) = \int_0^t \sin(t - z)e^{-2z}\, dz = f(t) * g(t)$$

$$= \int_0^t \sin z e^{-2(t-z)}\, dz = g(t) * f(t)$$

$$= e^{-2t} \int_0^t \sin z e^{2z}\, dz$$

$$= e^{-2t} \frac{e^{2t}(2 \sin t - \cos t) + 1}{1 + 2^2}$$

$$= \frac{2 \sin t - \cos t + e^{-2t}}{5}.$$

Example 1 could have been solved by numerous other methods, and the transform could have been inverted by partial fractions, for instance. But the convolution theorem is a convenient last resort when other methods fail. It also provides some useful general information.

Consider the nth-order differential equation with constant coefficients

$$\frac{d^n u}{dt^n} + b_1 \frac{d^{n-1} u}{dt^{n-1}} + \cdots + b_{n-1} \frac{du}{dt} + b_n u = f(t). \tag{5.60}$$

In order to find a formula for a solution, let us add the initial conditions

$$u(0) = 0, \qquad \frac{du}{dt}(0) = 0, \qquad \ldots, \qquad \frac{d^{n-1} u}{dt^{n-1}}(0) = 0. \tag{5.61}$$

The transform of the solution of the initial value problem expressed in Eqs. (5.55) and (5.56) is easily found to be

$$U(s) = \frac{F(s)}{p(s)} \tag{5.62}$$

where $F(s) = \mathscr{L}(f(t))$ and $p(s)$ is the characteristic polynomial of the homogeneous differential equation corresponding to Eq. (5.60):

$$p(s) = s^n + b_1 s^{n-1} + \cdots + b_{n-1} s + b_n. \tag{5.63}$$

We can apply the convolution theorem to inverting $U(s)$ in Eq. (5.57) provided that we can identify the function of t whose transform is

$$G(s) = \frac{1}{p(s)}. \tag{5.64}$$

This is not difficult, though, because the function G is the transform of the solution of the homogeneous differential equation corresponding to Eq. (5.60) with the initial conditions that will give us the right numerator. These conditions are just

$$u(0) = 0, \qquad \frac{du}{dt}(0) = 0, \qquad \ldots, \qquad \frac{d^{n-2} u}{dt^{n-2}}(0) = 0, \qquad \frac{d^{n-1} u}{dt^{n-1}}(0) = 1. \tag{5.65}$$

From these observations we can give an explicit formula for the solution of a nonhomogeneous differential equation.

Theorem 5.6

Let $u = g(t)$ be the solution of the initial value problem

$$\frac{d^n u}{dt^n} + b_1 \frac{d^{n-1} u}{dt^{n-1}} + \cdots + b_{n-1} \frac{du}{dt} + b_n u = 0,$$

$$u(0) = 0, \qquad \ldots, \qquad \frac{d^{n-2} u}{dt^{n-2}}(0) = 0, \qquad \frac{d^{n-1} u}{dt^{n-1}}(0) = 1.$$

where $b_1, b_2, \ldots, b_n$ are constants. Then the solution of the initial value problem

$$\frac{d^n u}{dt^{n-1}} + b_1 \frac{d^{n-1}u}{dt^{n-1}} + \cdots + b_{n-1}\frac{du}{dt} + b_n u = f(t)$$

$$u(0) = 0, \qquad \ldots, \qquad \frac{d^{n-1}u}{dt^{n-1}}(0) = 0,$$

is the convolution of f and g:

$$u(t) = f * g = \int_0^t g(t - z)f(z)\, dz.$$

This theorem is easier to use than the variation-of-parameters technique of Section 3.5 and is applicable even if $f(t)$ does not have a Laplace transform.

Example 2

A solution of the nonhomogeneous differential equation

$$\frac{d^2 u}{dt^2} + u = \frac{1}{1 + \cos t}$$

is provided by the theorem, as

$$u(t) = \int_0^t \frac{\sin (t - z)}{1 + \cos z}\, dz$$

$$= \sin t\left(t - \tan \frac{t}{2}\right) + \cos t\left(\ln \left|\frac{1 + \cos t}{2}\right|\right).$$

Because $1/(1 + \cos t)$ becomes infinite at $t = \pm\pi$, this function has no transform. Nevertheless the function $u(t)$ that we have determined is a solution of the given equation for $-\pi < t < \pi$.

A further application of the convolution theorem arises in solving certain kinds of integral equations which sometimes arise in applications, as illustrated by this example.

Suppose $n(t)$ is the number (approximated by a continuous function) of tools of a certain type needed by a large factory at time t. We may also assume that we know the *survival function* h for these tools: $h(t)$ is the proportion of a lot of tools that are still in service at age t. If the replacement rate for the tools is $u(t)$, then by accounting for all tools in service we come upon this equation:

$$n(t) = n(0)h(t) + \int_0^t u(z)h(t - z)\,dz. \tag{5.66}$$

The first term is the number of tools surviving from the original lot. The integral can be viewed as the limit of a sum

$$\sum u(z_k)h(t - z_k)\,\Delta z_k.$$

where $u(z_k)\,\Delta z_k$ is the number of tools bought in the kth time interval, $t - z_k$ time units prior to time t.

If we are trying to predict $u(t)$, knowing $n(t)$ and $h(t)$, we can easily find the transform $U(s)$, by using the convolution theorem:

$$N(s) = n(0)H(s) + U(s)H(s)$$

$$U(s) = \frac{N(s) - n(0)H(s)}{H(s)}. \tag{5.67}$$

In other circumstances it might be $n(t)$ that is the unknown function. For instance, we might be dealing with a population of animals in which the replacement rate (read: birth rate) is proportional to population size: $u(t) = \beta n(t)$. Then Eq. (5.62) becomes

$$n(t) = n(0)h(t) + \int_0^t \beta n(t')h(t - t')\,dt'. \tag{5.68}$$

Now we may find the transform of the unknown function n as

$$N(s) = n(0)H(s) + \beta N(s)H(s)$$

$$N(s) = \frac{n(0)H(s)}{1 - \beta H(s)}. \tag{5.69}$$

If $h(t)$ and hence $H(s)$ are known, we have a chance at finding the unknown function by inverting the transform in Eq. (5.67) or (5.69).

Example 3

Suppose that the survival function is $h(t) = e^{-\alpha t}$ in a biological population for which the replacement rate is $r(t) = \beta n(t)$. Then, from Eq. (5.69),

$$N(s) = \frac{n(0)}{s + \alpha} \frac{1}{1 - \dfrac{\beta}{s + \alpha}} = \frac{n(0)}{s + (\alpha - \beta)}$$

is the transform of the population size, which is

$$n(t) = n(0)e^{(\beta - \alpha)t}.$$

Consequently, if the specific birth rate β exceeds α, the population size grows exponentially.

Exercises

1. Compute the convolution $f * g$ of these functions:
   (a) $f(t) = t$, $g(t) = e^{-t}$
   (b) $f(t) = e^{-t}$, $g(t) = \sin t$
   (c) $f(t) = \cos t$, $g(t) = 1$

2. Prove these properties of the convolution:
   (a) $1 * f'(t) = f(t) - f(0)$
   (b) $(f * g)' = g(0)f(t) + g' * f = f(0)g(t) + g * f'$

3. Find a formula for the solution of the initial value problem below in terms of a convolution:

   $$\frac{d^2u}{dt^2} + \omega^2 u = f(t), \qquad u(0) = 0, \qquad \frac{du}{dt}(0) = 0.$$

4. Obtain the inverse transform of $U(s) = 1/s(s^2 + \omega^2)$ by (a) partial fractions; (b) integration; (c) convolution.

5. Compute the convolution $f * g$ of these functions by using Theorem 5.5 and a table of Laplace transforms:
   (a) $f(t) = g(t) = e^{at}$
   (b) $f(t) = \sin t$, $g(t) = e^{-at}$
   (c) $f(t) = g(t) = \sin \omega t$

6. How does the result of Exercise 5c help explain resonance?

7. Use Theorem 5.6 to solve the initial value problem

   $$\frac{d^2u}{dt^2} - u = \frac{1}{e^t + e^{-t}}, \qquad u(0) = 0, \qquad \frac{du}{dt}(0) = 0.$$

8. Does the inhomogeneity in Exercise 7 have a Laplace transform? Can you find it?

9. Use Laplace transform and the convolution theorem to solve

   $$\frac{d^2u}{dt^2} + \omega^2 u = \sin \omega t, \qquad u(0) = 0, \qquad \frac{du}{dt}(0) = 0.$$

10. Use Laplace transform and the convolution theorem to solve

    $$\frac{d^2u}{dt^2} - k^2 u = e^{kt}, \qquad u(0) = 0, \qquad \frac{du}{dt}(0) = 0.$$

11. Solve the integral equation (5.66) for the replacement rate $u(t)$ if $n(t)$ is to be always equal to $n(0)$ and the survival function is $h(t) = e^{-\alpha t}$:

    $$n(0) = n(0)e^{-\alpha t} + \int_0^t u(z)e^{-\alpha(t-z)} \, dz.$$

12. Solve the integral equation (5.68) for $n(t)$ if there is no loss—that is, $h(t) = 1$:

$$n(t) = n(0) + \int_0^t \beta n(z)\, dz.$$

13. Solve the integral equation (5.68) for $n(t)$ if the survival function is $h(t) = (1 + \cos t)/2$ and $\beta = \frac{4}{3}$:

$$n(t) = n(0)h(t) + \int_0^t \beta n(z)h(t - z)\, dz.$$

5.6

Applications to Control Theory

In electrical, mechanical, and chemical engineering a frequently occurring problem is to direct or control a device so as to make a specified variable behave in a certain way. A daily example is driving a car: the driver manipulates the steering wheel, accelerator, and brake pedals to make the car move prudently from the starting point to the destination. The ideas of control theory have penetrated beyond the traditional areas of applied mathematics into economics, biology, psychology, and so on.

The mechanical system pictured in Fig. 5.10 can be interpreted as a control system. It is composed of a disk mounted on a long shaft with

Figure 5.10

Disk and Dial System

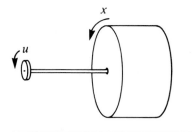

frictionless bearings. The output or controlled variable is the angular position of the disk, denoted by x, while the input or controlling variable is the angular displacement u of a dial at the end of the shaft. We assume that the shaft itself is flexible in torsion, acting like a spring with constant k (measured in units of torque per radian). Thus the torque acting on the disk is $k(u - x)$, and the motion of the disk is governed by the initial value problem

$$J\frac{d^2x}{dt^2} = k(u - x), \qquad 0 < t, \tag{5.70}$$

$$x(0) = x_0, \qquad \frac{dx}{dt}(0) = y_0. \tag{5.71}$$

Here J is the moment of inertia of the disk. The control obtained is called proportional: the controlling torque $k(u - x)$ is proportional to the *error* $u - x$, which is the difference between the desired and actual angular displacements.

One test of the effectiveness of a control system is its response to an abrupt change of input, starting from rest. That is to say, one examines the solution of the initial value problem

$$J\frac{d^2x}{dt^2} = k(u - x), \qquad 0 < t, \tag{5.72}$$

$$x(0) = 0, \qquad \frac{dx}{dt}(0) = 0, \tag{5.73}$$

with the assumption that $u(t) = u_1 h(t)$, where h is the Heaviside step function and u_1 is a constant.

We can easily find the transform of the solution of the problem above, using $\mathcal{L}(h(t)) = 1/s$. It is

$$X(s) = \frac{ku_1}{s(Js^2 + k)}. \tag{5.74}$$

The inverse transform can be found by partial fractions, convolution, or the integration theorem. Using the last alternative, we find

$$x(t) = u_1(1 - \cos \omega t), \tag{5.75}$$

where $\omega = \sqrt{(k/J)}$ is the natural frequency of the system.

This control system would generally be unacceptable, since the controlled variable $x(t)$ does not even approach the desired state. (See Fig. 5.11.) Of course, friction in a real system would eventually bring $x(t)$ to u_1. We might help things along by adding some damping—shown in Fig. 5.12 as a fan. This modified system would be governed by the differential equation

$$J\frac{d^2x}{dt^2} + p\frac{dx}{dt} = k(u - x). \tag{5.76}$$

We again test the system with a step function input, $u(t) = u_1 h(t)$.

Figure 5.11

Response of the Disk
and Dial System to a
Step Input

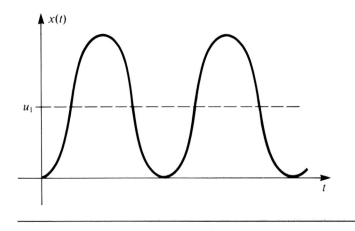

The transform of the solution (with initial conditions (5.73) again) is

$$X(s) = \frac{ku_1}{s(Js^2 + ps + k)}.$$ (5.77)

Usually, p is chosen to leave the system slightly underdamped. Under this assumption we may use partial fractions or tables to find

$$x(t) = u_1\left[1 - e^{-bt/2}\left(\cos \beta t + \frac{b}{2\beta}\sin \beta t\right)\right]$$ (5.78)

with

$$b = \frac{p}{J}, \qquad \beta = \sqrt{\frac{k}{J} - \left(\frac{p}{2J}\right)^2}.$$ (5.79)

Figure 5.12

Disk and Dial with
Fan as Damper

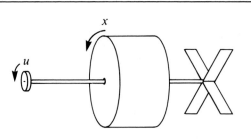

Inspection of $x(t)$ in Eq. (5.78) and Fig. 5.13 show that, with damping, $x(t)$ does indeed approach the set value u_1. Two numerical measures of the control system's performance for this "step-change" of input are (1) the overshoot

$$\frac{x_{\max} - u_1}{u_1} = e^{-b\pi/2\beta},\qquad\qquad\text{(5.80)}$$

and (2) the time lapse between the step change in input and the occurrence of $x_{\max}$ ("rise time"),

$$T = \pi/\beta.\qquad\qquad\text{(5.81)}$$

(See Exercises 1 and 2 at the end of this section.)

Another test of the effectiveness of a control system is its response to a "ramp function" input,

$$u(t) = \alpha t.\qquad\qquad\text{(5.82)}$$

In terms of the mechanical system of Fig. 5.12 the dial is moved with constant angular velocity α. We transform the initial value problem composed of the differential equation (5.76), with $u(t)$ from Eq. (5.82) and zero initial conditions

$$J\frac{d^2x}{dt^2} + p\frac{dx}{dt} + ku = k\alpha t,\qquad 0 < t,\qquad\qquad\text{(5.83)}$$

$$x(0) = 0,\qquad \frac{dx}{dt}(0) = 0.\qquad\qquad\text{(5.84)}$$

Figure 5.13

Response of the
Damped System to a
Step Input

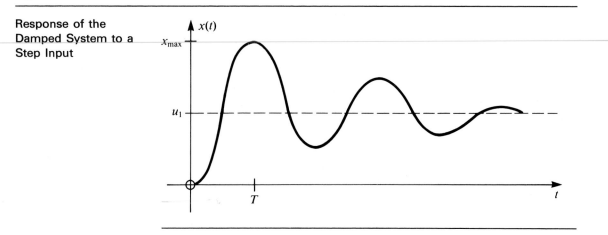

The transform of the solution is

$$X(s) = \frac{k\alpha}{s^2(Js^2 + ps + k)}.$$ (5.85)

The partial fractions expansion for $X(s)$ must have the form

$$X(s) = \frac{A}{s} + \frac{B}{s^2} + \frac{Cs + D}{Js^2 + ps + k}.$$ (5.86)

The presence of the characteristic polynomial of Eq. (5.83) in the denominator of the last term indicates that this is the transform of a solution of the corresponding homogeneous equation. Because of the damping, this solution decreases exponentially. The first two terms, however, represent the response to the input. By the cover-up rule and differentiation (or undetermined coefficients), we find

$$A = -\alpha p/k, \qquad B = \alpha,$$

whence the persistent part of the solution of Eq. (5.83) is

$$u_p(t) = \alpha t - \alpha p/k.$$ (5.87)

While the first term here is identical with the input, the second term is a constant error between input and output, called "velocity lag." (See Fig. 5.14.) Thus the damping, which seems necessary to suppress the oscilla-

Figure 5.14

Response to Ramp
Function Input

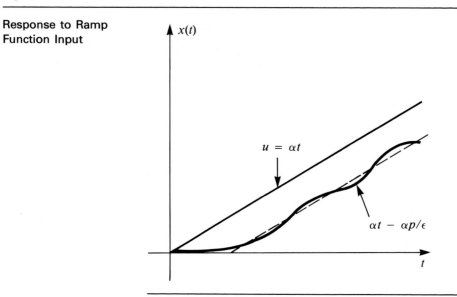

tions in the response, has the undesirable side effect of producing a lag in the response to a ramp input. In addition, of course, the damper consumes energy, and this may also be undesirable.

In the examples considered so far, we have limited ourselves to extremely simple mechanical devices, with the results that our differential equations are of order 2 and there is no possibility of instability. Now let us suppose that an ideal electric motor supplies the torque to our system and that the controller is an electronic device capable of performing arithmetic and calculus (differentiation and integration) on the input u and output x, individually or in combination. (See Fig. 5.15.)

Figure 5.15

Disk with Damper and Active Control System

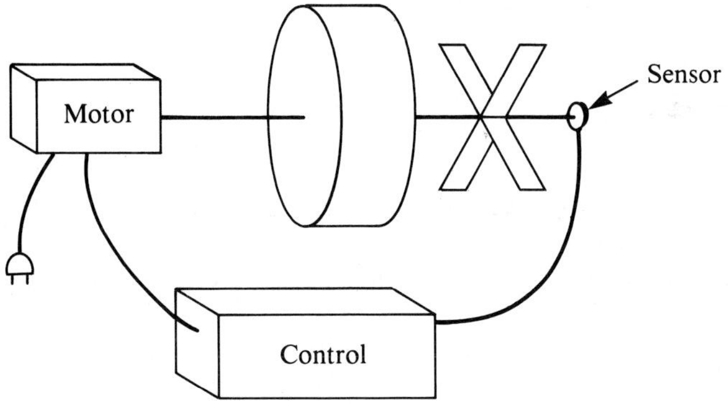

Then the torque on the shaft may be made to respond in many different ways to input and output. We shall suppose that the governing differential equation has the form

$$J\frac{d^2x}{dt^2} + p\frac{dx}{dt} = \phi(u, x) \tag{5.88}$$

where ϕ is a linear operator. We proceed to investigate a few of the choices for ϕ available to the control system designer.

Error Rate Damping

The controller may be designed so that the torque exerted is proportional to a combination of error and rate of change of error. In these circumstances the differential equation for x is

$$J\frac{d^2x}{dt^2} + p\frac{dx}{dt} = k(u - x) + a\left(\frac{du}{dt} - \frac{dx}{dt}\right). \tag{5.89}$$

Since the coefficient of dx/dt in this equation is now $a + p$ instead of p, we see that the beneficial effects of damping can be obtained without the expense of a dissipative device. Furthermore, if the input is a ramp function $u = \alpha t$, the persistent part of the output is

$$x = \alpha t - \frac{\alpha p}{k}. \tag{5.90}$$

Thus if viscous friction represented by p (not only the fan, but other friction as well) can be eliminated, then the velocity lag can be eliminated also.

Velocity Feed-Forward

A controller need not respond only to the error $u - x$; it may respond to the input or output or both separately. For instance, the torque supplied may be determined by

$$\phi(u, x) = k(u - x) + h\frac{du}{dt}. \tag{5.91}$$

In a sense, this controller is designed on the assumption that extra power will be needed when the input changes sharply. Comparison with the preceding case suggests that velocity feed-forward ought to have somewhat the same effect as error rate damping. The fact is that it goes even further. Even in the presence of viscous damping ($p > 0$) it can eliminate completely the velocity lag in the response to input $u = \alpha t$, and even turn it into a lead. (See Exercise 11 at the end of this section.)

Integral Control

Another often used option open to the designer of a control system is to use the integral of the error. Thus the torque supplied in our example might be

$$\phi(u, x) = k(u - x) + h\int_0^t (u(t') - x(t'))\, dt'. \tag{5.92}$$

Then the differential equation (5.88) becomes

$$J\frac{d^2x}{dt^2} + p\frac{dx}{dt} = k(u - x) + h\int_0^t (u(t') - x(t'))\, dt'. \tag{5.93}$$

This should more properly be called an integro-differential equation. But it can still be analyzed with the Laplace transform. The transform of Eq. (5.93), with zero initial conditions for $x(t)$, is

$$(Js^2 + ps)X = k(U - X) + \frac{h}{s}(U - X). \tag{5.94}$$

From here it is easy to find the transform of the solution

$$X(s) = \frac{k + h/s}{Js^2 + ps + k + h/s} U(s) = \frac{ks + h}{Js^3 + ps^2 + ks + h} U(s). \qquad (5.95)$$

The surprise here is that for some (positive) values of the parameters the equation is unstable; that is, some solutions of the homogeneous equation do not die out as t increases! To see how this may happen, first consider the denominator of the right-hand side of Eq. (5.95):

$$Js^3 + ps^2 + ks + h. \qquad (5.96)$$

If all three roots of this equation are negative real numbers, then the three independent solutions of the homogeneous equation are decaying exponentials which disappear in time. Similarly, if there is one negative real root and a pair of complex conjugate roots with negative real parts, then one of the three independent solutions of the homogeneous equation is a decaying exponential, and the other two have the form of a sine or cosine multiplied by a decaying exponential. Again, all solutions disappear.

On the other hand, if any root of the polynomial (5.96) is a real, positive number or is a complex number with positive real part, some solutions will grow exponentially in time. Thus the borderline case would be the one in which there is one negative real root and a pair of conjugate imaginary roots. (See Fig. 5.16.)

Now suppose that the polynomials in Eq. (5.96) does have conjugate imaginary roots $\pm i\omega$ and negative real root $-r$ $(r > 0)$. Then

$$Js^3 + ps^2 + ks + h = J(s^2 + \omega^2)(s + r) \qquad (5.97)$$

Figure 5.16

Two Imaginary Roots and One Real

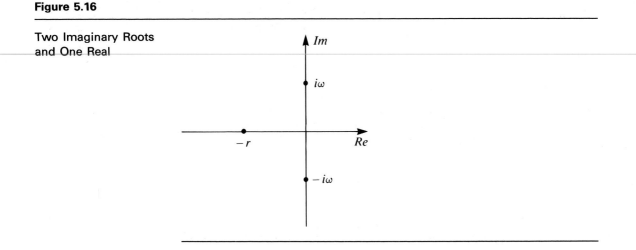

must hold, and we may match coefficients of like powers of s to find

$$p = Jr, \quad k = J\omega^2, \quad h = J\omega^2 r. \tag{5.98}$$

These relations then require that

$$\omega^2 = k/J \quad \text{and} \quad pk = Jh. \tag{5.99}$$

From here we can see that the condition characterizing this situation that divides the unstable from the stable is just $pk = Jh$. (We have assumed all along that p, k, and h are positive.) It turns out that the condition for stability is

$$pk > Jh. \tag{5.100}$$

(See Exercise 12 below.) Thus the relation between the coefficients k and h in the integral controller, Eq. (5.92), must satisfy Eq. (5.100) to avert disaster.

Exercises

1. Find the time when the function $x(t)$ given in Eq. (5.78) has its first maximum. (Hint: note the value of dx/dt at $t = 0$.)
2. Given Eq. (5.81)—or the result of Exercise 1—derive Eq. (5.80).
3. Show that the overshoot, Eq. (5.80), is a decreasing function of b and the rise time, Eq. (5.81), is an increasing function of b. (Limit p to the range $0 < p^2 < 4k/J$.) This means that small overshoot and fast rise time are conflicting goals.
4. What values should C and D have in Eq. (5.86) so that $x(t)$ satisfies the initial conditions (5.84)?
5. Two mechanical systems are shown in Fig. 5.17. Find the differential equations governing each, using x for vertical displacement of the

Figure 5.17

Two Mechanical
Systems

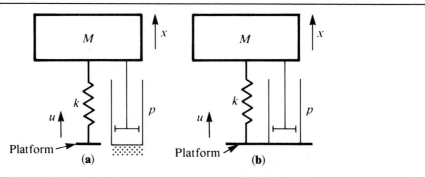

mass and u for displacement of the platform. (Assume that $x = 0$, $u = 0$ gives the natural length of the spring.)

6. Of the two systems of Exercise 5, one corresponds to a system with damping and the other to error rate damping. Which is which?

7. Let u and x be the displacements (measured from the same origin) of two automobiles traveling down a straight road. Suppose the driver of the second car (displacement x) accelerates to keep a constant distance d between the cars, in accordance with the differential equation

$$\frac{d^2x}{dt^2} = k(u - x - d),$$

where k is a positive constant. What is the solution $x(t)$ to the equation above with $u = \alpha t$, starting with zero initial conditions?

8. Safety organizations suggest that the distance in the equation of Exercise 7 should be not constant, but either (a) one carlength for each 10 mph of speed or (b) the distance that would be traveled in 2 sec at current speed. Show these are the same for carlength 29 feet 4 inches.

9. Combining the second suggestion of Exercise 8 with the equation in Exercise 7 gives

$$\frac{d^2x}{dt^2} = k\left(u - x - 2\frac{dx}{dt}\right).$$

Show that all solutions of the homogeneous equation tend to 0 as t increases.

10. Sketch the locus of roots of the characteristic polynomial for the equation above as a function of k $(0 < k < \infty)$.

11. Find the persistent part of the solution of Eq. (5.88) if the input is a ramp function, $u = \alpha t$ and the control law is given by Eq. (5.91).

12. We found that the condition $pk = Jh$ separates the stable from unstable cases of the polynomial in Eq. (5.96). Show that $(s + 1)^3$ is stable and that it satisfies Eq. (5.100).

13. Obtain the general solution of the homogeneous integro-differential equation obtained from Eq. (5.93) by setting $u(t) \equiv 0$ and using $J = 1$, $p = 2$, $k = 4$. Take $h = 3, 8$, and 21. (Hint: in each case the polynomial Eq. (5.96) has a negative integer root.)

In Exercises 14–17, find the particular solution containing no decaying exponential if $u = \sin \omega t$. Assume $a, b, c, h > 0$.

14. $\dfrac{d^2x}{dt^2} + b\dfrac{dx}{dt} = c(u - x)$

15. $\dfrac{d^2x}{dt^2} + b\dfrac{dx}{dt} = c(u - x) + a\left(\dfrac{du}{dt} - \dfrac{dx}{dt}\right)$

16. $\dfrac{d^2x}{dt^2} + b\dfrac{dx}{dt} = c(u - x) + h\dfrac{du}{dt}$

17. $\dfrac{d^3x}{dt^3} + b\dfrac{d^2x}{dt^2} = c\left(\dfrac{du}{dt} - \dfrac{dx}{dt}\right) + h(u - x)$

5.7

Periodic Functions

A function $f(t)$ is said to be periodic with period p if $f(t + p) = f(t)$ for all t. The familiar functions $\sin t$ and $\cos t$, which are periodic with period 2π, come immediately to mind as examples that are important in almost all fields of applied mathematics. Furthermore, periodic functions frequently appear as inhomogeneities and as solutions of differential equations.

The defining condition means that the graph of a periodic function is composed of an infinite sequence of copies of a portion that is one period long. (See Fig. 5.18.) This property also allows us to compute the Laplace transform in a special way.

Suppose that f is periodic with period p. Let us break up the interval of integration for the Laplace transform into segments of length p, so that

$$\int_0^\infty e^{-st}f(t)\,dt = \int_0^p e^{-st}f(t)\,dt + \int_p^{2p} e^{-st}f(t)\,dt + \cdots$$

$$+ \int_{kp}^{(k+1)p} e^{-st}f(t)\,dt + \cdots. \tag{5.101}$$

Figure 5.18

A Periodic
Function

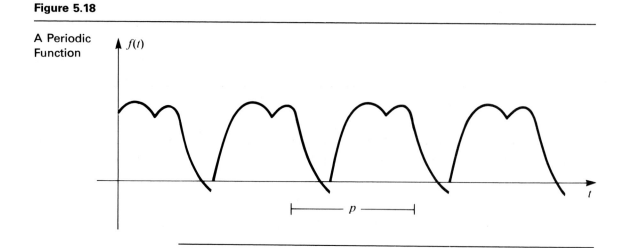

A typical one of these integrals, say, the one from kp to $(k + 1)p$, can be transformed by the substitution $t = t' + kp$, yielding

$$\int_{kp}^{(k+1)p} e^{-st}f(t)\, dt = \int_0^p e^{-s(t'+kp)}f(t' + kp)\, dt'. \tag{5.102}$$

By the periodicity condition,

$$f(t') = f(t' + p) = f(t' + 2p) = \cdots = f(t' + kp),$$

so the integral above becomes

$$\int_0^p e^{-s(t'+kp)}f(t')\, dt' = e^{-skp} \int_0^p e^{-st'}f(t')\, dt'. \tag{5.103}$$

Thus each of the integrals in Eq. (5.101) contains the factor

$$G(s) = \int_0^p e^{-st'}f(t')\, dt', \tag{5.104}$$

and the whole Laplace transform can be written

$$\mathscr{L}(f(t)) = G(s)(1 + e^{-sp} + \cdots + e^{-skp} + \cdots). \tag{5.105}$$

The infinite series above is just a geometric series in e^{-sp}, which converges when s is positive to $(1 - e^{-sp})^{-1}$. Thus we have done the computations for this theorem.

Theorem 5.7

Let $f(t)$ be periodic with period p. If f has a Laplace transform, it is

$$F(s) = \mathscr{L}(f(t)) = \frac{G(s)}{1 - e^{-sp}}, \tag{5.106}$$

where

$$G(s) = \int_0^p e^{-st}f(t)\, dt. \tag{5.107}$$

Example 1

A periodic function that is used frequently in electrical engineering is the "square wave" defined by

$$f(t) = \begin{cases} 1, & 0 < t < a \\ -1, & a < t < 2a \end{cases}$$

and the condition that f is periodic with period $2a$. (See Fig. 5.19.)

Figure 5.19

Square Wave

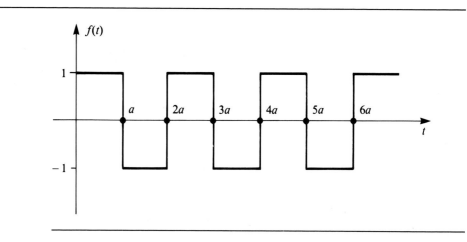

According to the theorem, it is necessary only to compute

$$G(s) = \int_0^{2a} e^{-st} f(t)\, dt$$

$$= \int_0^{a} e^{-st}\, dt + \int_a^{2a} e^{-st}(-1)\, dt$$

$$= \frac{e^{-st}}{-s}\Big|_0^a - \frac{e^{-st}}{-s}\Big|_a^{2a} = \frac{1 - 2e^{-as} + e^{-2as}}{s} = \frac{(1 - e^{-as})^2}{s}.$$

Thus the square wave function has the Laplace transform

$$\mathcal{L}(f) = \frac{(1 - e^{-as})^2}{s(1 - e^{-2as})} = \frac{1 - e^{-as}}{s(1 + e^{-as})}. \tag{5.108}$$

There is another way to interpret a periodic function and its transform. Let $f(t)$ be periodic with period p, and let $g(t)$ be another function that is identical with $f(t)$ for $0 < t < p$ but is 0 for $t > p$. (See Fig. 5.20.) Then $f(t)$ can be thought of as the sum of an infinite series of shifted copies of $g(t)$:

$$f(t) = g(t) + g(t - p)h(t - p) + g(t - 2p)h(t - 2p) + \cdots. \tag{5.109}$$

where h is the Heaviside step function. Thus the transform of $f(t)$ should

Figure 5.20

Decomposition of a
Periodic Function

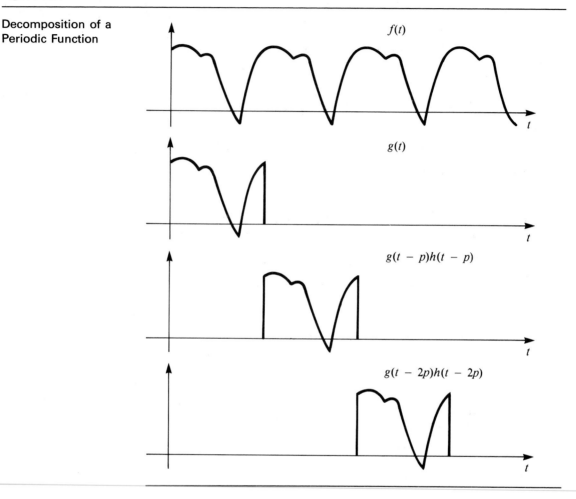

be

$$\mathcal{L}(f(t)) = G(s) + G(s)e^{-ps} + G(s)e^{-2ps} + \cdots$$

$$= G(s)(1 + e^{-ps} + e^{-2ps} + \cdots) = \frac{G(s)}{1 - e^{-ps}}. \qquad (5.110)$$

Note that $G(s) = \mathcal{L}(g(t))$ boils down to the integral in Eq. (5.107) because $g(t) = 0$ for $t > p$.

Example 2

Find the Laplace transform of the function $f(t)$ defined by

$$f(t) = e^{-\alpha t}, \qquad 0 < t < p, \tag{5.111}$$

and the condition that f is periodic with period p. (See Fig. 5.21.)

Figure 5.21

A Periodic Function

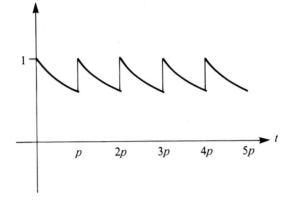

It is a simple matter to compute

$$G(s) = \int_0^p e^{-st} e^{-\alpha t}\, dt = \frac{1 - e^{-(s+\alpha)p}}{s + \alpha}.$$

Then, according to the theorem, we have for the transform of f

$$\mathcal{L}(f) = \frac{1 - e^{-(s+\alpha)p}}{(s + \alpha)(1 - e^{-ps})}. \tag{5.112}$$

As we have seen before, computing transforms is less of a challenge than computing inverse transforms. The first question we must ask is: How can one determine that a given function of s is the transform of a periodic function? Referring to Eqs. (5.106) and (5.107), note that if $G(s)$ exists, it remains finite for all values of s, real or complex. Thus the transform $F(s)$ given in Eq. (5.106) has poles* only where

$$1 - e^{-sp} = 0 \qquad \text{or} \qquad e^{-sp} = 1. \tag{5.113}$$

It can be proved that this condition is fulfilled for $s = 0$, $\pm 2\pi i/p$, $\pm 4\pi i/p, \dots$. Thus we may say:

* A pole of F is a zero of its denominator.

Theorem 5.8	A given function $F(s)$ is the transform of a periodic function only if its poles are all simple and all among the numbers

$$0, \pm \omega i, \pm 2\omega i, \ldots \qquad \textbf{(5.114)}$$

for some real ω.

Example 3	Recall that the transform of $\sin \alpha t$ is $F(s) = \alpha/(s^2 + \alpha^2)$. The only poles of $F(s)$ are at $s = \pm i\alpha$. This shows that F may have a finite number of poles.

Example 4	We have already found in Example 1 that a square wave has the transform

$$F(s) = \frac{1 - e^{-as}}{s(1 + e^{-as})}.$$

There is no pole at $s = 0$, since $F(s)$ has a finite limit there. However, the factor $1 + e^{-as}$ of the denominator is zero where

$$e^{-as} = -1.$$

To locate these poles, set $s = i\omega$ and use the identity

$$e^{iA} = \cos A + i \sin A. \qquad \textbf{(5.115)}$$

Hence we must solve the equation

$$\cos a\omega - i \sin a\omega = -1$$

or

$$\cos a\omega = -1, \qquad \sin a\omega = 0.$$

Thus $a\omega$ is an odd multiple of π, and the poles of $F(s)$ occur at

$$s = \pm i\pi/a, \pm 3i\pi/a, \pm 5i\pi/a, \ldots.$$

These are, as required, among the numbers listed in Eq. (5.114) if $\omega = \pi/a$.

Once a function is identified as being the transform of a periodic function, we may attempt to invert it by means of the Heaviside formula. This is in essence an extension of the partial-fractions technique.

Theorem 5.9

Suppose that

$$F(s) = \frac{N(s)}{D(s)},$$

where $N(s)$ has no poles and $D(s) = 0$ only at points among

$$s = 0, \pm \omega i, \pm 2\omega i, \ldots .$$

If $F(s)$ is the transform of some $f(t)$, it is

$$f(t) = \sum_{-\infty}^{\infty} \frac{N(n\omega i)}{D'(n\omega i)} e^{in\omega t}.$$ **(5.116)**

Remark: It is not immediately apparent that $f(t)$ as given in Eq. (5.116) is periodic. However, we have the identity

$$e^{in\omega t} = \cos{(n\omega t)} + i \sin{(n\omega t)},$$

and this function (although complex) is periodic with period $2\pi/n\omega$. This is the shortest period; any integer multiple of the shortest period is also a period. Thus all terms of the series have common period $2\pi/\omega$ and so will the sum of the series, if it exists.

Example 4

Let us try to invert the transform

$$F(s) = \frac{1 - e^{-as}}{s(1 + e^{-as})}$$

found in Example 1. In Example 3 we determined that

$$N(s) = \frac{1 - e^{-as}}{s}$$

has no poles and that

$$D(s) = 1 + e^{-as}$$

has simple zeros at points $s_n = in\omega$, $\omega = \pi/a$ for odd integers n (both positive and negative). Now we need to evaluate

$$D'(s_n) = -ae^{-as_n} = a,$$

$$N(s_n) = \frac{1 - e^{-as_n}}{in\omega} = \frac{-2i}{n\omega}.$$

Thus we determine that the candidate for $\mathcal{L}^{-1}(F(s))$ is

$$f_1(t) = \sum_{-\infty}^{\infty} \frac{-2i}{n\pi} e^{in\omega t} \qquad (n \text{ odd})$$

$$= \frac{-2i}{\pi} \left\{ \begin{array}{l} e^{i\omega t} + \frac{1}{3}e^{i3\omega t} + \cdots \\ -e^{-i\omega t} - \frac{1}{3}e^{-i3\omega t} - \cdots \end{array} \right.$$

Using the identity (equivalent to Eq. (5.115))

$$\sin A = \frac{1}{2i}(e^{iA} - e^{-iA}),$$

we can convert the series above to the form

$$f_1(t) = \frac{4}{\pi}(\sin \omega t + \frac{1}{3}\sin 3\omega t + \cdots). \tag{5.117}$$

This series actually converges to the square wave of Example 1. Figure 5.22 shows the sum of the terms through $\sin 7\omega t$.

Figure 5.22

Sum of a Few Terms
of a Series that
Converges to a
Square Wave

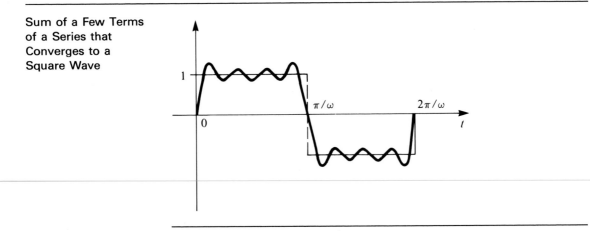

Example 5

Use Laplace transform techniques to find the persistent part of the solution of

$$R\frac{dq}{dt} + \frac{1}{C}q = E\,|\sin t|, \tag{5.118}$$

which describes the charge on the capacitor in the circuit shown in Fig. 5.23. The function $|\sin t|$ is called a *rectified sine wave*.

Figure 5.23

An *RC* Circuit

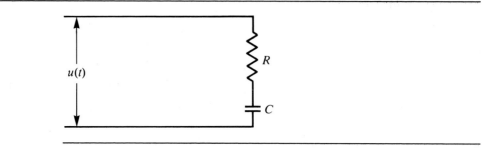

The transformed equation, with zero initial conditions, is

$$\left(Rs + \frac{1}{C}\right)Q = E\frac{1 + e^{-s\pi}}{(s^2 + 1)(1 - e^{-s\pi})}.$$

(It is left as an exercise to obtain the transform on the right.) The transform of the solution is

$$Q(s) = \frac{E}{R(s + \alpha)}\frac{1 + e^{-s\pi}}{(s^2 + 1)(1 - e^{-s\pi})}, \qquad \textbf{(5.119)}$$

where $\alpha = 1/RC$.

Although $s + \alpha$ has a zero, it is real and negative and corresponds to a decaying part of the solution. We shall therefore neglect it and concentrate on the transform corresponding to the poles of $Q(s)$ lying at $0, \pm 2i, \pm 4i, \dots$. Incidentally, $\pm i$ are not poles of $Q(s)$ in Eq. (5.119).

Now we specify

$$N(s) = \frac{E(1 + e^{-s\pi})}{R(s + \alpha)(s^2 + 1)},$$

$$D(s) = 1 - e^{-s\pi},$$

$$s_n = 2ni \qquad (n = 0, \pm i, \pm 2, \dots).$$

The choice of the numerator N seems a bit shocking at first, but N is required only to be a function that is finite at the zeros of the denominator. Now we find that

$$N(s_n) = \frac{E \cdot 2}{R(\alpha + 2ni)(-4n^2 + 1)}, \qquad D'(s_n) = \pi.$$

Thus the persistent part of the solution is given by the series

$$q(t) = \frac{2E}{R\pi}\sum_{-\infty}^{\infty}\frac{1}{(\alpha + 2ni)(1 - 4n^2)}e^{2nit}. \qquad \textbf{(5.120)}$$

Here it is convenient to pair the terms corresponding to n and $-n$:

$$\frac{e^{2nit}}{(\alpha + 2ni)(1 - 4n^2)} + \frac{e^{-2nit}}{(\alpha - 2ni)(1 - 4n^2)}$$

$$= \frac{1}{1 - 4n^2}\left(\frac{(\alpha - 2ni)e^{2nit} + (\alpha + 2ni)e^{-2nit}}{\alpha^2 + 4n^2}\right).$$

The two quantities in parentheses are complex conjugates; and $z + \bar{z} = 2\,\text{Re}\,(z)$ for any complex number z. Thus we need to calculate the real part of

$$(\alpha - 2ni)e^{2nit} = (\alpha - 2ni)(\cos 2nt + i \sin 2nt)$$

$$= (\alpha \cos 2nt + 2n \sin 2nt) + i(\alpha \sin 2nt - 2n \cos 2nt).$$

Finally, we have $q(t)$ from Eq. (5.120) in the real form

$$q(t) = \frac{2E}{R\pi}\left[\frac{1}{\alpha} - 2\sum_{n=1}^{\infty}\frac{\alpha \cos 2nt + 2n \sin 2nt}{(4n^2 - 1)(4n^2 + \alpha^2)}\right]$$

$$= EC\frac{2}{\pi}\left[1 - 2\alpha\sum_{n=1}^{\infty}\frac{\alpha \cos 2nt + 2n \sin 2nt}{(4n^2 - 1)(4n^2 + \alpha^2)}\right], \tag{5.121}$$

Figure 5.24 shows graphs of $q(t)$ for $\alpha = 0.1, 1, 10$. The series converges

Figure 5.24

Charge on the
Capacitor in an *RC*
Circuit with Rectified
Sine Input

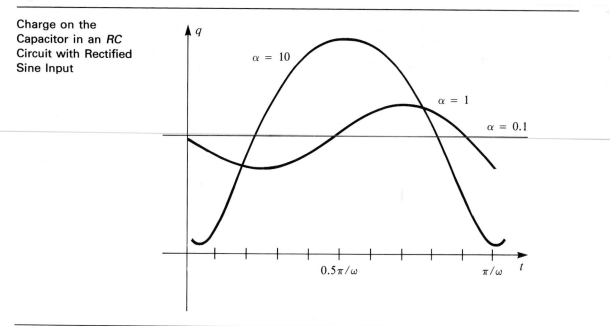

quickly—only about two terms are needed for accuracy sufficient for the figures.

Exercises

In Exercises 1–6, sketch the given periodic function and use Theorem 5.7 to find its Laplace transform.

1. $\sin \omega t$

2. $\cos \omega t$

3. $f(t) = \begin{cases} t, & 0 < t \le a \\ 2a - t, & a < t \le 2a \end{cases}$, $f(t + 2a) = f(t)$

4. $f(t) = t,\ 0 < t < p,\ f(t + p) = f(t)$

5. $f(t) = |\sin \omega t|$

6. $f(t) = \begin{cases} \sin \omega t, & 0 < t \le \pi/\omega \\ 0, & \pi/\omega < t \le 2\pi/\omega \end{cases}$, $f\left(t + \dfrac{2\pi}{\omega}\right) = f(t)$

7. Verify that $s = -\alpha$ is not a pole of the function in Eq. (5.112).

8. Verify that $s = \pm i$ are not poles of the function in Eq. (5.119).

9. Find the complete inverse transform of $Q(s)$ given in Eq. (5.119). (That is, $q(t)$ as given in Eq. (5.121) plus the contribution due to the pole of $Q(s)$ at $s = -\alpha$.)

Notes and References

It has been said that no problem can be solved with Laplace transform that cannot be solved otherwise. The statement may be true, but Laplace transform can often guide us to solutions of unfamiliar problems such as integral equations or delay-differential equations. (See the exercises in Sections 5.3 and 5.5.) Classical control theory was worked out in terms of the Laplace transform almost exclusively. More modern treatments use matrices and systems of differential equations, as in Chapter 7.

A standard text on Laplace transform is Churchill's (1972) book. Many Laplace transforms have been tabulated by Roberts and Kaufman (1966).

The Laplace transform is just one of a family of integral transforms. Some others that are important in applied mathematics are the Fourier transform,

$$\mathscr{F}(f) = \int_{-\infty}^{\infty} e^{-i\omega t} f(t)\, dt,$$

used especially by electrical engineers in analysis of signals, and the Hankel transform,

$$\mathscr{H}_n(f) = \int_0^{\infty} t J_n(xt) f(t)\, dt.$$

Here J_n is the Bessel function of the first kind and order n. (See Chapter 4.) A convenient general reference is Tranter (1966).

Miscellaneous Exercises

In Exercises 1–8, use Laplace transforms to find a particular solution of the differential equation or, if initial conditions are given, find the solution that satisfies them.

1. $\dfrac{du}{dt} - 3u = te^{2t}$, $u(0) = 0$

2. $2\dfrac{du}{dt} + 3u = e^{5t}$, $u(0) = 7$

3. $\dfrac{d^2u}{dt^2} + 2\dfrac{du}{dt} + 10u = \cos t$

4. $\dfrac{d^2u}{dt^2} + 4\dfrac{du}{dt} + 9u = \sin 3t$

5. $\dfrac{d^2u}{dt^2} - 2\dfrac{du}{dt} + u = te^t$

6. $\dfrac{d^2u}{dt^2} + 3\dfrac{du}{dt} + 2u = 4t^2$, $u(0) = 0$, $\dfrac{du}{dt}(0) = 0$

7. $\dfrac{d^2u}{dt^2} + u = t(1 - h(t - 1))$, $u(0) = 0$, $\dfrac{du}{dt}(0) = 0$

8. $\dfrac{d^2u}{dt^2} + 4\dfrac{du}{dt} + 5 = t$, $u(0) = 1$, $\dfrac{du}{dt}(0) = 0$

In Exercises 9–14, find an initial value problem whose solution has the given $U(s)$ as its transform. The differential equation is to be linear with constant coefficients and of the indicated order.

9. $U(s) = 1/(s + 1)$, first order

10. $U(s) = 1/s(s + 1)$, first order

11. $U(s) = s/(s^2 + 3s + 2)$, second order

12. $U(s) = 1/(s + 1)(s^2 + 1)$, second order

13. $U(s) = s^2/(s^2 + 1)(s^2 + 4)$, second order

14. $U(s) = 1/s(s + 1)$, second order

15. Let $F(s) = \mathscr{L}(f(t))$. Show that

$$\frac{dF}{ds} = \mathscr{L}(-tf(t)).$$

16. Use the formula of Exercise 15 to find the transform of
    (a) te^{at}; (b) $t \sin t$; (c) $t \cos t$.

17. In Exercise 17 of Section 5.4 the motion of a machine gun firing a short burst was described. If the burst is indefinitely long, the equation of motion is

$$\frac{d^2u}{dt^2} + 2\frac{du}{dt} + u = \sum_{k=0}^{\infty} \delta(t - k\tau).$$

Find $U(s)$, the transform of the solution of this problem. Transform the series term by term. (Use zero initial conditions.)

18. Find a closed form for $U(s)$ by identifying the sum of the series.

19. For $U(s)$ found in Exercise 17, find all real roots of the denominator and the corresponding terms in $u(t)$.

20. Find the persistent part of $U(s)$ from Exercise 17.

APPENDIX

Partial Fractions

The partial fractions decomposition of a function of the form

$$U(s) = \frac{N(s)}{D(s)},$$

where $N(s)$ and $D(s)$ are polynomials, with N of lower degree than D, allows us to represent U as a linear combination of simple rational functions of the form $1/(s - a)$, $1/(s - a)^2$, etc. Suppose $D(s)$ is a polynomial of degree n. Then $D(s)$ can be factored as

$$D(s) = a_0(s - r_1)^{m_1}(s - r_2)^{m_2} \cdots (s - r_k)^{m_k}. \tag{5.122}$$

where $r_1, r_2, \ldots, r_k$ are the distinct roots of $D(s) = 0$ and $m_1, m_2, \ldots, m_k$ are their respective multiplicities. Recall that a polynomial of degree n has n roots, counting multiplicities, so $m_1 + m_2 + \cdots + m_k = n$.

Form

The partial fractions decomposition of $U(s)$ has the form

$$\frac{N(s)}{D(s)} = R_1(s) + R_2(s) + \cdots + R_k(s), \tag{5.123}$$

where k is the number of distinct roots of $D(s)$ and each $R_i(s)$ is a simple rational function corresponding to a root of $D(s)$ as specified below.

    1. If r is a simple root of $D(s)$, the corresponding factor of $D(s)$ is $s - r$, and the corresponding R in the partial fractions decomposition has the form

$$R(s) = \frac{A}{s - r}, \tag{5.124}$$

where A is a constant to be found.

    2. If r is a root of $D(s)$ with multiplicity m, the corresponding factor of $D(s)$ is $(s - r)^m$, and the corresponding R in the partial fractions decomposition has the form

$$R(s) = \frac{B_1}{(s - r)^m} + \frac{B_2}{(s - r)^{m-1}} + \cdots + \frac{B_m}{s - r}. \tag{5.125}$$

    3. Complex roots of $D(s)$ can be treated just as in items 1 and 2 above. However, if the coefficients of $D(s)$ are real, complex roots occur in conjugate pairs. The corresponding factors of $D(s)$ can be paired to

give a quadratic factor, $s^2 + bs + c$, with real coefficients. The pair of conjugate roots of D corresponding to this factor gives rise to this term in the partial fractions decomposition

$$R(s) = \frac{Cs + D}{s^2 + as + b} \tag{5.126}$$

if the quadratic factor is simple or

$$R(s) = \frac{C_1 s + D_1}{(s^2 + as + b)^m} + \frac{C_2 s + D_2}{(s^2 + as + b)^{m-1}} + \cdots + \frac{C_m s + D_m}{s^2 + as + b} \tag{5.127}$$

if the quadratic factor has multiplicity m.

    Note that the number of coefficients A, B, C, D, etc. corresponding to one root of $D(s)$ is equal to the multiplicity of that root. The total number of coefficients is n, the degree of $D(s)$.

Determining Coefficients by Algebra

Once the correct *form* for the partial fractions decomposition is found, the coefficient must be determined. In the classical (algebraic) method the terms in the decomposition are "put over a common denominator"—which must be $D(s)$. The resulting numerator is equated to $N(s)$, yielding a system of n equations in the n unknown coefficients.

Example 1

Find the partial fractions decomposition of

$$U(s) = \frac{s^2 + 5s - 4}{s^3 + 3s^2 + 2s}.$$

The denominator is easily factored as

$$D(s) = s^3 + 3s^2 + 2s = s(s + 1)(s + 2).$$

According to 1 above, the partial fractions decomposition has the form

$$\frac{s^2 + 5s - 4}{s(s + 1)(s + 2)} = \frac{A_1}{s} + \frac{A_2}{s + 1} + \frac{A_3}{s + 2}.$$

When we put the right-hand side over the common denominator, we get

$$\frac{s^2 + 5s - 4}{s(s + 1)(s + 2)} = \frac{A_1(s + 1)(s + 2) + A_2 s(s + 2) + A_3 s(s + 1)}{s(s + 1)(s + 2)}.$$

Equating the coefficients of like powers of s in the numerator gives these equations:

$$A_1 + \;\; A_2 + A_3 = 1 \qquad \text{(coefficients of } s^2\text{)},$$
$$3A_1 + 2A_2 + A_3 = 5 \qquad \text{(coefficients of } s\text{)},$$
$$2A_1 \qquad\qquad = -4 \qquad \text{(coefficients of 1)},$$

which are solved to find $A_1 = -2$, $A_2 = 8$, and $A_3 = -5$. Thus our decomposition is

$$\frac{s^2 + 5s - 4}{s^3 + 3s^2 + 2s} = \frac{-2}{s} + \frac{8}{s+1} + \frac{-5}{s+2}.$$

It is easy to see that the classical method, although simple in concept, is nasty in practice. However, it is worth noting that the "equating of coefficients of like terms" only establishes n equations to be solved. As an alternative, we can evaluate both sides of the partial fractions decomposition at n different values of s—not roots of $D(s)$!—to get n equations. Also, one equation can be obtained by taking the limit of both sides as $s \to \infty$, after multiplying through both sides by the lowest power of s that will give a finite nonzero limit for the right-hand side.

Example 2

To find the coefficients in

$$\frac{s^3}{(s^2 + 1)^2} = \frac{As + B}{s^2 + 1} + \frac{Cs + D}{(s^2 + 1)^2},$$

we evaluate both sides at $s = 0, 1, -1$ to obtain these equations:

$$0 = B + D,$$
$$\frac{1}{4} = \frac{A + B}{2} + \frac{C + D}{4},$$
$$-\frac{1}{4} = \frac{-A + B}{2} + \frac{-C + D}{4}.$$

One more equation can be obtained by taking the limit of both sides as $s \to \infty$, after multiplying through both sides by s.

$$\frac{s^4}{(s^2 + 1)^2} = s\left(\frac{As + B}{s^2 + 1} + \frac{Cs + D}{(s^2 + 1)^2}\right),$$

$$\lim_{s \to \infty} \frac{s^4}{(s^2 + 1)^2} = 1 = A.$$

The solution of these equations is left as an exercise.

Determining Coefficients by Calculus

Application of calculus to the problem of finding a partial fractions decomposition simplifies calculations greatly. We start with the simplest case.

   Case 1: Simple root of $D(s)$.   Suppose that r is a simple root of $D(s)$, so $D(r) = 0$, but $D'(r) \neq 0$. Then we know that

$$\frac{N(s)}{D(s)} = \frac{A}{s - r} + R(s), \tag{5.128}$$

where A is a constant to be found and $R(s)$ is a rational function of s (actually, the rest of the partial fractions decomposition), which does not have a pole at r. That is, $R(s)$ is finite.

   The coefficient A can be found easily with a little calculus. Since Eq. (5.128) is to hold for all s, we can multiply through by $(s - r)$ to obtain

$$\frac{(s - r)N(s)}{D(s)} = A + (s - r)R(s). \tag{5.129}$$

Now, take the limit as s approaches r on both sides. The right-hand side has limit A, since $R(r)$ is finite. To obtain the limit for the left-hand side, we can proceed in either of two ways. If it is easy to divide $s - r$ out of D, do so and then set $s = r$ in the remaining expression. (This is sometimes called the "cover-up rule," since one covers up or cancels the factor $s - r$ in $D(s)$.) The result is a numerical value for A.

   If it is not convenient to divide the factor $s - r$ out of $D(s)$, we can apply L'Hopital's rule to find the limit of the left-hand side of Eq. (5.129):

$$\lim_{s \to r} \frac{(s - r)N(s)}{D(s)} = \lim_{s \to r} \frac{(s - r)N'(s) + N(s)}{D'(s)} = \frac{N(r)}{D'(r)}.$$

Finally, we have found a simple formula for the coefficient A in Eq. (5.128):

$$A = \frac{N(r)}{D'(r)}. \tag{5.130}$$

Example 3

Find the terms corresponding to the simple roots $s = 0$ and $s = -2$ in the partial fractions expansion of

$$\frac{N(s)}{D(s)} = \frac{2s + 3}{s^4 + 4s^3 + 5s^2 + 2s}.$$

   We know that the correct form is

$$\frac{N(s)}{D(s)} = \frac{A_1}{s} + \frac{A_2}{s + 2} + R(s).$$

Since it is easy to factor s out of the denominator, we use the cover-up rule to find

$$A_1 = \lim_{s \to 0} \frac{2s + 3}{s^3 + 4s^2 5s + 2} = \frac{3}{2}.$$

Dividing the factor $s + 2$ out of the denominator would be tedious, so we use the alternative rule to calculate

$$A_2 = \frac{N(-2)}{D'(-2)} = \frac{-4 + 3}{4(-8) + 12(4) + 10(-2) + 2} = \frac{1}{2}.$$

Case 2: Multiple root of $D(s)$.   Suppose now that $D(s)$ has a root of multiplicity m at r. This fact can be expressed by setting

$$D(s) = (s - r)^m H(s), \tag{5.131}$$

where we assume that $H(r) \neq 0$. The corresponding form for the partial fractions expansion is

$$\frac{N(s)}{D(s)} = \frac{B_1}{(s - r)^m} + \frac{B_2}{(s - r)^{m-1}} + \cdots + \frac{B_m}{s - r} + R(s). \tag{5.132}$$

The leftover piece, $R(s)$, would contain the terms corresponding to other roots of $D(s)$, but $R(s)$ is finite at $s = r$. There is no very easy way to find the coefficients $B_1, B_2, \ldots, B_m$. However, we may carry on in the spirit of the previous development. If we multiply both sides of Eq. (5.132) by $(s - r)^m$, we obtain

$$\frac{(s - r)^m N(s)}{D(s)} = B_1 + B_2(s - r) + \cdots + B_{m-1}(s - r)^{m-2}$$
$$+ B_m(s - r)^{m-1} + R(s)(s - r)^m. \tag{5.133}$$

The left-hand side is a function of s that can be expressed in closed form; the right-hand side is its Taylor polynomial, expanded about $s = r$. Therefore if we call

$$\frac{(s - r)^m N(s)}{D(s)} = G(s), \tag{5.134}$$

then the coefficients are

$$B_1 = G(r), \qquad B_2 = G'(r), \qquad \ldots, \qquad B_m = \frac{1}{m!} G^{(m)}(r). \tag{5.135}$$

These evaluations may be done by dividing $(s - r)^m$ out of $D(s)$ to get $G(s)$ or by using L'Hopital's rule on the left-hand side of Eq. (5.134) and its derivatives.

Example 4

Find the part of the partial fractions expansion corresponding to the double root $(m = 2)$ of the denominator at $s = -1$ for

$$\frac{N(s)}{D(s)} = \frac{2s + 3}{s^4 + 4s^3 + 5s^2 + 2s}.$$

We know that the correct form for the desired part of the expansion is

$$\frac{N(s)}{D(s)} = \frac{B_1}{(s + 1)^2} + \frac{B_2}{s + 1} + R(s).$$

Multiplying through by $(s + 1)^2$ gives

$$\frac{(s + 1)^2 N(s)}{D(s)} = \frac{2s + 3}{s^2 + 2s} = B_1 + B_2(s + 1) + (s + 1)^2 R(s).$$

Then, by evaluating both sides at $s = -1$, we find

$$\frac{1}{-1} = -1 = B_1.$$

Evaluating the derivatives of both sides at $s = -1$ gives

$$\frac{(s^2 + 2s) \cdot 2 - (2s + 3)(2s + 2)}{(s^2 + 2s)^2}\bigg|_{s=-1} = -2 = B_2.$$

Case 3: Simple conjugate complex roots. The methods above apply equally well to complex roots of $D(s)$. However, if all of the coefficients in $D(s)$ and $N(s)$ are real, the work can be simplified because the roots of $D(s)$ must then occur in complex conjugate pairs. Suppose that r and $\bar{r}$ are simple complex conjugate roots of $D(s)$, so that $D(s)$ can be factored as

$$D(s) = (s^2 + bs + c)H(s), \tag{5.136}$$

where r and $\bar{r}$ are the roots of the first factor. Then the partial fractions expansion has the form

$$\frac{N(s)}{D(s)} = \frac{As + B}{s^2 + bs + c} + R(s). \tag{5.137}$$

Following the pattern of our previous work, we multiply through by $s^2 + bs + c$ to obtain

$$\frac{(s^2 + bs + c)N(s)}{D(s)} = As + B + (s^2 + bs + c)R(s). \tag{5.138}$$

Now take the limit of both sides as s approaches r. The factor $s^2 + bs + c$

becomes 0, leaving

$$Ar + B = \lim_{s \to r} \frac{(s^2 + bs + c)N(s)}{D(s)} = \frac{N(r)}{H(r)}. \tag{5.139}$$

If $D(s)$ is available in factored form (so $H(s)$ is known explicitly), the last number of Eq. (5.139) can be found easily. Otherwise, one applies L'Hopital's rule to the limit.

Once established, Eq. (5.139) is an equation between complex quantities. However, A and B must be real; thus they are fully determined by this one equation.

Example 5

Find the partial fractions expansion of

$$\frac{s^2 + s - 1}{(s^2 + 4)(s^2 + 2s + 2)} = \frac{As + B}{s^2 + 4} + \frac{Cs + D}{s^2 + 2s + 2}.$$

(a) First we find A and B. According to our development, we have

$$\frac{s^2 + s - 1}{s^2 + 2s + 2} = As + B + (s^2 + 4)\frac{Cs + D}{s^2 + 2s + 2}.$$

Evaluating this expression at $s = 2i$, we find

$$2iA + B = \frac{(2i)^2 + 2i - 1}{(2i)^2 + 4i + 2} = \frac{2i - 5}{4i - 2} = \frac{9 + 8i}{10}.$$

Now using the fact that A and B are real, we get these equations:

$$2A = \tfrac{8}{10},$$
$$B = \tfrac{9}{10}.$$

(b) Now we find C and D. The sequence of steps is the same as in part (a). First, we have

$$\frac{s^2 + s - 1}{s^2 + 4} = Cs + D + (s^2 + 2s + 2)(As + B).$$

When we evaluate both sides at $s = -1 + i$, we find

$$\frac{-3 - 4i}{10} = C(-1 + i) + D = (D - C) + iC.$$

Now equating the real and imaginary parts of the two members of this equation gives these equations

$$D - C = -\tfrac{3}{10},$$
$$C = -\tfrac{4}{10},$$

from which we find $D = -\tfrac{7}{10}$.

Finally, we note that a judicious combination of methods may well prove to give the best results. Thus in a problem with many terms the best overall strategy would be to first find the coefficients that can be determined by the cover-up rule and then find the remaining coefficients by setting up a system of simultaneous equations.

Example 6

To evaluate the coefficients in the partial fractions expansion

$$\frac{1}{(s + 1)(s + 2)^2(s^2 + 1)} = \frac{A}{s + 1} + \frac{B_1}{(s + 2)^2} + \frac{B_2}{s + 2} + \frac{Cs + D}{s^2 + 1},$$

we first find A and B_1 by the cover-up rule:

$$A = \lim_{s \to -1} \frac{1}{(s + 2)^2(s^2 + 1)} = \frac{1}{2},$$

$$B_1 = \lim_{s \to -2} \frac{1}{(s + 1)(s^2 + 1)} = \frac{-1}{5}.$$

Now we have

$$\frac{1}{(s + 1)(s + 2)^2(s^2 + 1)} = \frac{\frac{1}{2}}{s + 1} - \frac{\frac{1}{5}}{(s + 2)^2} + \frac{B_2}{s + 2} + \frac{Cs + D}{s^2 + 1}.$$

First, multiply both sides by s and take the limit as $s \to \infty$:

$$0 = \tfrac{1}{2} - 0 + B_2 + C. \tag{5.140}$$

Next evaluate both sides at $s = 0$ and at $s = 1$:

$$\frac{1}{1 \cdot 4 \cdot 2} = \frac{\frac{1}{2}}{1} - \frac{\frac{1}{5}}{4} + \frac{B_2}{2} + D, \tag{5.141}$$

$$\frac{1}{2 \cdot 9 \cdot 2} = \frac{\frac{1}{2}}{2} - \frac{\frac{1}{5}}{9} + \frac{B_2}{3} + \frac{C + D}{2}. \tag{5.142}$$

Now Eqs. (5.140), (5.141), and (5.142) are solved simultaneously to get $B_2 = -\frac{9}{25}$, $C = -\frac{7}{50}$, $D = -\frac{1}{50}$.

Exercises

In Exercises 1–9, find the partial fractions decomposition of the given rational function.

1. $\dfrac{1}{(s - 1)^2(s^2 + 1)}$

2. $\dfrac{s - 1}{s^2(s + 1)}$

3. $\dfrac{s - 1}{s^2(s + 1)^2}$

4. $\dfrac{s^2 - s + 1}{(s + 1)(s + 2)(s + 3)}$

5. $\dfrac{s^2 + 1}{(s^2 - 1)(s^2 - 4)}$

6. $\dfrac{2s + 3}{s^2(s^2 - 1)}$

7. $\dfrac{s^2 - 1}{(s^2 + 1)(s^2 + 4)}$

8. $\dfrac{s^2 + 2}{s^4 - 1}$

9. $\dfrac{s^2}{(s^2 - 1)^2}$

10. $\dfrac{s^3}{(s^2 - 1)^2}$

11. $\dfrac{s^3}{(s^2 + 1)^2}$ (see Example 6)

6   Numerical Methods

6.1
Elementary Methods

It is obvious that many differential equations simply cannot be solved in closed form; even some that can be solved come out in such a complicated way that the solution formula provides no insight. In such cases we often turn to "numerical methods," techniques for obtaining an approximate solution for the problem at hand. To begin with, we shall think about a first-order initial value problem:

$$\frac{du}{dt} = f(t, u), \qquad u(t_0) = u_0. \tag{6.1}$$

Our objective is to obtain a sequence of numbers, $u_0, u_1, \ldots, u_n, \ldots$ that approximate the values of the solution $u(t)$ at selected times $t_0, t_1, \ldots, t_n, \ldots$:

$u_0 = u(t_0),$

$u_1 \cong u(t_1),$

$u_2 \cong u(t_2),$

$\vdots$

Usually, the times are chosen to be equally spaced, at least over some interval. Thus if we call the spacing, or step length, h, the times are

$$t_0, \qquad t_1 = t_0 + h, \qquad t_2 = t_0 + 2h, \qquad \ldots .$$

A numerical method for solution of Eq. (6.1) replaces the differential equation with a "difference equation":

$$u_{n+1} = F(u_n, t_n, h). \tag{6.2}$$

The choice of the function F depends on the choice of method and includes the function f from Eq. (6.1). The reason for using an equation like Eq. (6.2) is that we can carry out whatever calculation is necessary by arithmetic operations, avoiding derivatives, integrals, and, in general, all limit-related processes.

Most elementary methods can be understood in terms of approximating a derivative. If we know the values of the solution of Eq. (6.1) at two points, t_n and $t_{n+1} = t_n + h$, then the difference quotient

$$\frac{u(t_{n+1}) - u(t_n)}{t_{n+1} - t_n}$$

is, by the mean value theorem, a value assumed by the derivative u' at some point $\bar{t}$ between t_n and t_{n+1}:

$$\frac{u(t_{n+1}) - u(t_n)}{h} = u'(\bar{t}). \tag{6.3}$$

Now if f does not vary too quickly, the right-hand side can be decently approximated by $u'(t_n) = f(t_n, u(t_n))$. Thus we have the approximation

$$\frac{u(t_{n+1}) - u(t_n)}{h} \cong f(t_n, u(t_n)) \tag{6.4}$$

or, put more carefully,

$$u(t_{n+1}) = u(t_n) + hf(t_n, u(t_n)) + \varepsilon_n, \tag{6.5}$$

where ε_n is the error made in the approximation.

The simplest numerical method, called *Euler's method*, just drops the ε_n from Eq. (6.5) and requires that the numbers $u_1, u_2, \ldots$ satisfy the difference equation

$$u_{n+1} = u_n + hf(t_n, u_n). \tag{6.6}$$

We illustrate the use of Euler's method with the following example.

Example 1

The initial value problem to be solved is

$$\frac{du}{dt} = \sqrt{u} + t + 1, \qquad u(0) = 1. \tag{6.7}$$

We can identify u_0, t_0, and $f(t, u)$ by examining Eq. (6.7). The version of Eq. (6.6) to be used is

$$u_{n+1} = u_n + h(\sqrt{u_n} + t_n + 1). \tag{6.8}$$

The results of the calculations are summarized in Table 6.1. We have used $h = 0.1$.

The numbers in the columns headed t_n and u_n are used in the calculation of $f(t_n, u_n) = \sqrt{u_n} + t_n + 1$, which then contributes to the formation of u_{n+1}, as shown in Eq. (6.8). Eight digits were used in the calculations, although fewer are shown.

Table 6.1

n	t_n	u_n	$f(t_n, u_n)$
0	0.0	1.0	2.0
1	0.1	1.2	2.195
2	0.2	1.4195	2.491
3	0.3	1.6687	2.592
4	0.4	1.9279	2.788
5	0.5	2.2067	2.986
6	0.6	2.5053	3.183
7	0.7	2.8235	3.380
8	0.8	3.1616	3.578
9	0.9	3.5195	3.776
10	1.0	3.8970	

The initial value problem in Example 1 is unusual in that its solution can be written down simply:

$$u(t) = (1 + t)^2. \tag{6.9}$$

From here it is easy to see that the difference between u_{10} and $u(1) = 4$ is about 0.103. The approximation between derivative and difference quotient on which Euler's method is based gets better as h is made smaller. In Table 6.2 we show the results of applying Euler's method to Eq. (6.7) with $h = 0.1$ and $h = 0.05$, together with the exact solution.

Table 6.2

t	$u(t)$	$u_n[h = 0.1]$	Error	$u_n[h = 0.05]$	Error
0.0	1.0	1.0	0	1.0	0
0.1	1.21	1.2	0.01	1.2049	0.0051
0.2	1.44	1.4195	0.0205	1.4297	0.0103
0.3	1.69	1.6687	0.0213	1.6742	0.0158
0.4	1.96	1.9279	0.0321	1.9385	0.0215
0.5	2.25	2.2067	0.0433	2.2227	0.0273
0.6	2.56	2.5053	0.0547	2.5268	0.0332
0.7	2.89	2.8235	0.0665	2.8507	0.0393
0.8	3.24	3.1616	0.0784	3.1945	0.0455
0.9	3.61	3.5194	0.0906	3.5582	0.0518
1.0	4.0	3.8970	0.1030	3.9418	0.0582

It is clear that cutting the step length in half has reduced the error, in about the same proportion—but at the cost of twice as much calculation. We can do much better.

Geometrically, Euler's method amounts to following a tangent line, instead of an unknown solution curve, from u_n to the value we accept for u_{n+1}. (See Fig. 6.1.) Since the slope of the solution curve varies throughout the interval from t_n to t_{n+1}, the value calculated by Euler's method generally does not agree with the value on the solution curve.

Figure 6.1

Graphical
Representation of
One Step of Euler's
Method

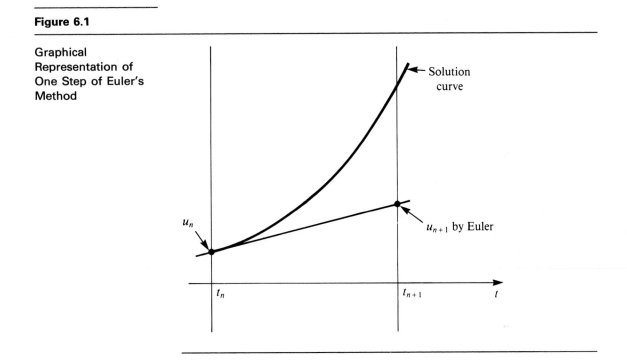

Another method, called the *modified Euler* or *Euler-Cauchy* method makes a half-step by Euler's method and then uses the slope at the point $(t_{n+1/2}, u_{n+1/2})$ to step across the interval. (See Fig. 6.2.) The formulas for this method are

$$u_{n+1/2} = u_n + \frac{h}{2} f(t_n, u_n), \tag{6.10}$$

$$t_{n+1/2} = t_n + \frac{h}{2}, \tag{6.11}$$

$$u_{n+1} = u_n + h f(t_{n+1/2}, u_{n+1/2}). \tag{6.12}$$

As Fig. 6.2 suggests, this method really is much better than Euler's method.

Figure 6.2

Graphical
Representation of
One Step of the
Modified Euler
Method

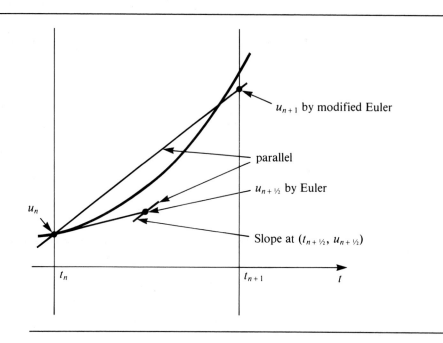

u_{n+1} by modified Euler

parallel

$u_{n+½}$ by Euler

Slope at $(t_{n+½}, u_{n+½})$

t_n t_{n+1} t

Example 2

The initial value problem to be solved is

$$\frac{du}{dt} = u + t, \qquad u(0) = 1. \tag{6.13}$$

The formulas for the calculation are taken from Eq. (6.10) and (6.12):

$$u_{n+1/2} = u_n + \frac{h}{2}(u_n + t_n), \tag{6.14}$$

$$u_{n+1} = u_n + h(u_{n+1/2} + t_{n+1/2}). \tag{6.15}$$

In Table 6.3 are listed values calculated by the modified Euler method with $h = 0.1$, along with values of the exact solution, $u(t) = 2e^t - t - 1$, and the values calculated by the original Euler method.

From the examples above, we can see that using the modified Euler instead of the original method (both with the same h) requires about twice as much work but improves the error by much more than a factor of two. If we have a fixed accuracy to achieve, we can think of the

Table 6.3	t_n	Euler	Modified	Exact
	0.0	1.0	1.0	1.0
	0.1	1.1	1.1100	1.1103
	0.2	1.22	1.2421	1.2428
	0.3	1.362	1.3985	1.3997
	0.4	1.5282	1.5818	1.5837
	0.5	1.7210	1.7949	1.7974
	0.6	1.9431	2.0409	2.0424
	0.7	2.1974	2.3231	2.3275
	0.8	2.4872	2.6456	2.6511
	0.9	2.8159	3.0124	3.0192
	1.0	3.1875	3.4282	3.4366

relationship another way. If one value of h provides the desired accuracy with Euler's method, we can use a much larger h and far fewer steps to get the same accuracy with the modified Euler method. To illustrate this last point, Table 6.4 shows the results of solving Eq. (6.13) by the modified Euler method with several different values of h. Even $h = 0.5$ gives results comparable with $h = 0.1$ in Euler's method.

Table 6.4	t_n	$u_n[h = 0.1]$	$u_n[h = 0.5]$	$u_n[h = 1]$
	0.0	1.0	1.0	1.0
	0.5	1.7949	1.75	—
	1.0	3.4282	3.2813	3.0

It is not hard to analyze why the modified method is so much better than the original. Let us restate the problem being considered.

$$\frac{du}{dt} = f(t, u), \qquad u(t_0) = u_0. \tag{6.16}$$

Now if $u(t)$ is the true solution, Taylor's theorem tells us that

$$u(t_1) = u(t_0 + h) = u(t_0) + hu'(t_0) + \frac{h^2}{2} u''(\bar{t}) \tag{6.17}$$

for some value of $\bar{t}$ between t_0 and t_1 (provided the second derivative of u is continuous). Euler's method gives

$$u_1 = u_0 + hf(t_0, u_0). \tag{6.18}$$

Now because $u(t)$ satisfies the differential equation (i.e., $u'(t_0) = f(t_0, u_0)$) and initial condition of Eq. (6.16), we see on comparing the two expressions in Eqs. (6.17) and (6.18) that

$$u(t_1) - u_1 = \frac{h^2}{2} u''(\bar{t}). \tag{6.19}$$

Thus the error in the first step of Euler's method is proportional to h^2. The method is said to be of *first order* because it accounts for terms through the first power of h in Taylor's series.

After the first step, our numerical solution has slipped off the solution curve $u(t)$ and onto another—call it $U_1(t)$—that represents the solution of

$$\frac{du}{dt} = f(t, u), \qquad u(t_1) = u_1.$$

(See Fig. 6.3.) On the second step, our calculated value u_2 will differ from

Figure 6.3

Numerical Solution by Euler's Method Slips Away from True Solution

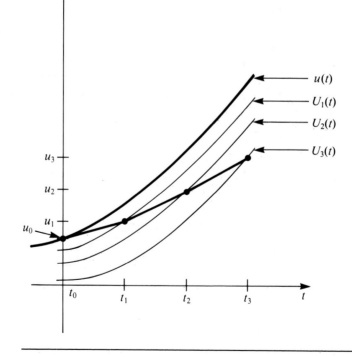

$U_1(t_2)$ by an error proportional to h^2 again. Similarly, at each step, Euler's method accepts a deviation proportional to h^2 between the calculated value u_{n+1} and the solution of the differential equation that satisfies

$$u(t_n) = u_n.$$

Each of these errors, called *local errors*, is proportional to h^2.

The local errors compound in a complicated way; only the first local error measures deviation from the desired solution curve. The total error at time $t = t_0 + T$ (fixed, independent of h), called the *global error*, represents an accumulation of $N = T/h$ errors, each of magnitude proportional to h^2. Thus the global error should be proportional to

$$Nh^2 = Th.$$

A somewhat more complicated analysis shows that the modified Euler method is of *second order*: it accounts for the terms through the one containing h^2 in the Taylor series of u. Thus the local error is of order h^3, and the global error is of order h^2.

Numerical Methods and Computing Equipment

Serious problems requiring accurate solutions demand the resources of some sort of computer able to evaluate complicated functions, remember past results, and produce storable copies of the final solutions. Some students do not have access to a computer but do have a calculator. Among the exercises in this chapter, those marked with a computer symbol require extensive calculation, while those marked with a calculator can be done with a programmable calculator. The remaining exercises are theoretical problems or numerical problems that can be done with a grocery store calculator (one memory, four arithmetic operations, and square root).

For example, Exercise 2 can be done on a grocery store calculator. The formula to be evaluated is

$$u_{n+1} = u_n + 0.1\sqrt{1 + u_n^2} \tag{E}$$

(Euler's method with $h = 0.1$), and the results are to be written down in tabular form. The "program" or keystroke sequence is

$$\left.\begin{array}{l} \text{CM} \\ 1 \\ \text{M+} \\ \text{RM} \end{array}\right\} \text{initialization}$$

$$\left. \begin{array}{l} \times \\ = \\ + \\ 1 \\ = \\ \sqrt{} \\ \times \\ .1 \\ = \\ M+ \\ RM \end{array} \right\} \text{computation}$$

Abbreviations: CM = clear memory; RM = recall memory; M+ = add to memory.

The initialization portion places the initial condition in the memory. The computation portion computes the right-hand side of Eq. (E), adds it to the former value of u in the memory, and displays the result, which is to be written down in the table. The computation portion is carried out repeatedly until the desired goal has been reached. (In Exercise 2, the goal is an approximation to $u(1)$.) As a control on mistakes, it is often helpful to record the quantity added to memory on each cycle. If this shows abrupt or irregular changes, either a mistake was made or the value of h is too large.

Exercises

In Exercises 1–5, use Euler's method with $h = 0.1$ to approximate $u(0.5)$ and $u(1)$. Record three decimals.

1. $\dfrac{du}{dt} = u,\ u(0) = 1$

2. $\dfrac{du}{dt} = \sqrt{1 + u^2},\ u(0) = 1$

3. $\dfrac{du}{dt} = 1 + u^2,\ u(0) = 0$

4. $\dfrac{du}{dt} = \dfrac{1 + u}{u},\ u(0) = 1$

5. $\dfrac{du}{dt} = -2u,\ u(0) = 1$

6. $\dfrac{du}{dt} = t^2 - u^2,\ u(0) = 0$

7. Obtain the exact solutions of the initial value problems in Exercises 1–5 and compute $u(1)$ for each.

In Exercises 8–10, use the modified Euler method with $h = 0.2$ and $h = 0.1$ to obtain approximate values for $u(1)$.

8. $\dfrac{du}{dt} = \sqrt{2 \cos u - 1},\ u(0) = 0$

9. $\dfrac{du}{dt} = u - u^2,\ u(0) = 0.1$

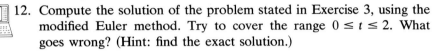

10. $\dfrac{du}{dt} = \ln(1 + u),\ u(0) = 1$

11. Try several initial conditions between 0 and 2 with the differential equation of Exercise 9 and graph the results for $0 \le t \le 5$. Compare with a graphical analysis.

12. Compute the solution of the problem stated in Exercise 3, using the modified Euler method. Try to cover the range $0 \le t \le 2$. What goes wrong? (Hint: find the exact solution.)

13. Solve the problem

$$\frac{du}{dt} = \frac{u}{1 + u^2}, \qquad u(0) = 1,$$

over the range $0 \le t \le 1$ with both Euler and modified Euler methods. Use $h = 0.2, 0.1, 0.05,$ and 0.025 with each method.

14. Let $E(h)$ be the error in the approximation to $u(1)$ found in Exercise 13 by Euler's method with step length h. (The correct value is $u(1) = 1.48591387$.) Is $E(h)/h$ approaching a limit as h decreases? Carry out a similar analysis with the modified Euler method and $E(h)/h^2$.

6.2

Analysis of Error

The starting point for analyzing error in numerical methods is the Taylor expansion

$$u(t + h) = u(t) + hu'(t) + \frac{h^2}{2}u''(t) + \cdots + \frac{h^n}{n!}u^{(n)}(t) + R_n. \qquad \textbf{(6.20)}$$

This expression is valid if u has derivatives through order $n + 1$ in the interval from t to $t + h$. The remainder R_n is

$$R_n = \frac{h^{n+1}}{(n + 1)!}u^{(n+1)}(\bar{t}), \qquad \textbf{(6.21)}$$

and $\bar{t}$ is somewhere between t and $t + h$. If we know the value of a function u and its derivatives at a point t, we can use the Taylor expansion to find $u(t + h)$ to within a tolerance estimated by R_n.

We are interested, of course, in the solution of an initial value problem,

$$\frac{du}{dt} = f(t, u), \qquad u(t_0) = u_0. \qquad \textbf{(6.22)}$$

By substituting t_0 and u_0 in f we can find $u'(t_0)$. As many more derivatives as are desired can be found by differentiating both sides of the differential equation; for example

$$\frac{d^2u}{dt^2} = f_t + f_u u',$$ (6.23)

$$\frac{d^3u}{dt^3} = f_{tt} + f_{tu}u' + f_{ut}u' + f_{uu}(u')^2 + f_u u''.$$ (6.24)

(Subscripts denote partial derivatives. These expressions were obtained by using the chain rule.) These formulas can be used to find values of the second and third derivatives of u at t_0. Then we could approximate $u(t_0 + h)$ by

$$u(t_0 + h) \cong u_1 = u_0 + h u'(t_0) + \frac{h^2}{2} u''(t_0) + \frac{h^3}{6} u'''(t_0).$$ (6.25)

According to Eq. (6.20), the approximation is in error by R_3, an amount proportional to h^4. Supposing this to be negligible for a suitably chosen value of h, we accept u_1 as $u(t_1)$ and repeat the process to find an approximation to $u(t_0 + 2h)$, etc.

Example 1

Consider the initial value problem

$$\frac{du}{dt} = t - u^2, \qquad u(0) = 1.$$

By differentiating directly we find

$$\frac{d^2u}{dt^2} = 1 - 2uu',$$

$$\frac{d^3u}{dt^3} = -2uu'' - 2(u')^2.$$

Now at $t = 0$ we have $u(0) = 1$ and, from the above expressions,

$$\frac{du}{dt}(0) = 0 - 1^2 = -1,$$

$$\frac{d^2u}{dt^2}(0) = 1 - 2 \cdot 1 \cdot (-1) = 3,$$

$$\frac{d^3u}{dt^3}(0) = -2(\cdot 1 \cdot 3 + (-1)^2) = -8.$$

The first terms of the Taylor expansion at $t = 0$ are

$$u(h) = 1 + h(-1) + \frac{h^2}{2}(3) + \frac{h^3}{6}(-8) + R_3$$

$$= 1 - h + \tfrac{3}{2}h^2 - \tfrac{4}{3}h^3 + R_3,$$

where R_3 contains the fourth power of h. We disregard the remainder to obtain an approximate value for u at h. For instance, if $h = 0.1$, $u(0.1) \cong u_1 = 1 - 0.1 + \frac{3}{2}(0.01) - \frac{4}{3}(0.001) = 0.91367$.

Now that we have u_1, an approximate value for $u(0.1)$, we can repeat the calculations:

$$\frac{du}{dt}(0.1) \cong u_1' = 0.1 - u_1^2 = -0.7348,$$

$$\frac{d^2u}{dt^2}(0.1) \cong u_1'' = 1 - 2u_1u_1' = 2.343,$$

$$\frac{d^3u}{dt^3}(0.1) \cong u_1''' = -2u_1u_1'' - 2(u_1')^2 = -5.361,$$

$$u(0.2) \cong u_2 = u_1 + 0.1u_1' + \frac{(0.1)^2}{2}u_1'' + \frac{(0.1)^3}{6}u_1''' = 0.85101.$$

Example 1 illustrates the first two steps of the "four-term Taylor series method." It is not hard to use but has the severe defect that one has to do explicit differentiation, just what we want to avoid. More sophisticated methods attempt to obtain higher accuracy by using more evaluations of the function $f(t, u)$ from the differential equation, rather than its derivatives. A whole family of such methods follows this pattern, which we now develop.

Starting from t_n, u_n, calculate two slopes

$$s_1 = f(t_n, u_n), \tag{6.26}$$

$$s_2 = f(t_n + \alpha h, u_n + \beta h s_1) \tag{6.27}$$

and then combine them to obtain the next solution value:

$$u_{n+1} = u_n + h(\gamma s_1 + \delta s_2). \tag{6.28}$$

The parameters α, β, γ, δ are to be chosen to make the local error "small," in the sense of being proportional to the highest possible power of h.

In order to compute the local error in Eq. (6.27) we will need to use the Taylor expansion for a function of two variables:

$$f(t + \Delta t, u + \Delta u) = f + \Delta t f_t + \Delta u f_u + \bar{R}. \tag{6.29}$$

Again we have used subscripts to indicate partial derivatives, and the functions on the right are evaluated at (t, u). The remainder $\bar{R}$ contains multiples of Δt^2, Δu^2, and $\Delta t \Delta u$.

Now by Eq. (6.29) the quantity s_2 of Eq. (6.27) is

$$s_2 = f(t_n + \alpha h, u_n + \beta h s_1)$$
$$= f + \alpha h f_t + \beta h s_1 f_u + \bar{R}$$
$$= f + \alpha h f_t + \beta h f f_u + \bar{R}.$$

Combining this with the rest of Eq. (6.28), we have a new expression for u_{n+1}:

$$u_{n+1} = u_n + h\gamma f + h\delta(f + \alpha h f_t + \beta h f f_u + \bar{R})$$
$$= u_n + h(\delta + \gamma)f + h^2\delta(\alpha f_t + \beta f_u f) + h\delta\bar{R}. \tag{6.30}$$

On the other hand, the first terms of the Taylor series for $u(t_{n+1})$ (starting at $u(t_n) = u_n$) are

$$u(t_{n+1}) = u(t_n + h) = u_n + hf + \frac{h^2}{2}(f_t + ff_u) + R_2. \tag{6.31}$$

Now we wish the right-hand side of Eq. (6.30) to match, as well as possible, the right-hand side of Eq. (6.31). In order for the terms containing h to match, we must have

$$\delta + \gamma = 1. \tag{6.32}$$

The terms containing h^2 will match if we make

$$\delta\alpha = \delta\beta = \tfrac{1}{2}. \tag{6.33}$$

The remainders cannot be matched by any choice for the parameters (although this is not obvious from Eqs. (6.30) and (6.31)); even though Eqs. (6.32) and (6.33) are satisfied, there will still be a difference between u_{n+1} and $u(t_{n+1})$ that is proportional to h^3.

Euler's method corresponds to the choice $\gamma = 1$, $\delta = 0$, α and β arbitrary. Evidently, Eq. (6.32) will be satisfied, but Eq. (6.33) will not. In this case the error is proportional to h^2. The modified Euler method corresponds to

$$\delta = 1, \qquad \gamma = 0, \qquad \alpha = \beta = \tfrac{1}{2}$$

and thus has an error proportional to h^3.

Equations (6.32) and (6.33) are three nonlinear equations in four unknowns. The general solution can be found by taking δ to be any nonzero quantity; then

$$\gamma = 1 - \delta, \qquad \alpha = \beta = \frac{1}{2\delta}. \tag{6.34}$$

Another popular elementary method, called the improved Euler, or Heun, method corresponds to $\delta = \gamma = \frac{1}{2}, \alpha = \beta = 1$. The formulas are

$$s_1 = f(t_n, u_n),$$
$$s_2 = f(t_n + h, u_n + hs_1),$$
$$u_{n+1} = u_n + \frac{h}{2}(s_1 + s_2).$$
(6.35)

All the methods found this way have local error compounded of R_2 and $h\bar{R}$ (see Eqs. (6.30) and (6.31)) and therefore proportional to h^3. Their global error is proportional to h^2, and they are all called second-order methods. The same idea of matching series may be used to generate methods of any desired order, usually called Runge-Kutta methods. (See Section 6.3.)

The purpose of error analysis is not simply to keep numerical analysts busy. Our objectives are to control or, even better, to eliminate error. While the total elimination of error is not possible, the process of *extrapolation* can do very nicely. Suppose we wish to calculate $u(T)$, where u is the solution of the initial value problem, Eq. (6.22). We use some method known to have a global error of order r, say. That means that $U(h)$, the approximation to $u(T)$ calculated with step length h, has an error proportional to h^r:

$$u(T) - U(h) = ch^r.$$
(6.36)

We might repeat the calculation with a different value of h (call it h_1) to get

$$u(T) - U(h_1) = c_1 h_1^r.$$
(6.37)

Actually, c and c_1 will differ somewhat. However, if they were equal (and they are nearly so), we could use the relation (6.36) and (6.37) to eliminate most of the error. Rewrite the equations assuming that $c = c_1$:

$$u(T) - ch^r = U(h)$$
$$u(T) - ch_1^r = U(h_1).$$

These may be thought of as two simultaneous linear equations in the unknowns $u(T)$ and c. We may solve them by elimination to get

$$u(T) = \frac{h^r U(h_1) - h_1^r U(h)}{h^r - h_1^r}.$$
(6.38)

This is not really $u(T)$, since c and c_1 are not really equal. But the right-hand side of Eq. (6.38) will generally be a better approximation to $u(T)$ than either $U(h)$ or $U(h_1)$.

Example 2

We apply the extrapolation idea to the data of Example 1 in Section 6.1. There Euler's method (order 1) was used to approximate the solution of

$$\frac{du}{dt} = \sqrt{u} + t + 1, \qquad u(0) = 1.$$

In Table 6.2 of Section 6.1 we have these approximations to $u(1)$:

$U(0.1) = 3.8970$,

$U(0.05) = 3.9418$.

Eq. (6.38), adapted to this case, is

$$u(1) \cong \frac{0.1 \times 3.9418 - 0.05 \times 3.8970}{0.1 - 0.05}$$

$$= 2 \times 3.9418 - 3.8970 = 3.9866.$$

This latter quantity is in error by 0.0134, while the original approximations showed errors of about 0.06 and 0.10.

The most frequent application of the extrapolation formula, Eq. (6.38), is to the case where $h_1 = h/2$. The formula can be simplified algebraically, then, to

$$u(T) \cong \frac{2^r U\left(\frac{h}{2}\right) - U(h)}{2^r - 1}. \tag{6.39}$$

Notice that it is indispensible to know r, the order of the method.

Example 3

Use extrapolation with the modified Euler method ($r = 2$) to improve the numerical solution of the initial value problem

$$\frac{du}{dt} = u^2 + t, \qquad u(0) = 0.$$

In Table 6.5 are given the calculated approximations to $u(1)$

Table 6.5	h	$U(h)$
	0.2	0.5518178
	0.1	0.5557065
	0.05	0.5567836
	0.025	0.5570655

The extrapolation formula, Eq. (6.39), becomes

$$u(1) \cong \frac{4U\left(\frac{h}{2}\right) - U(h)}{3}.$$

Applying this to the first two entries in the table ($h = 0.2$) results in the approximation

$$u(1) \cong 0.5570027.$$

In fact we can apply the extrapolation idea to each pair of rows and get successively

$$u(1) \cong 0.5571426$$

$$u(1) \cong 0.5571594.$$

The last result has five accurate digits, yet it is derived from information having only three or four accurate digits!

Exercises

1. Table 6.2 of Section 6.1 contains solutions by Euler's method (global error proportional to h) of

   $$\frac{du}{dt} = \sqrt{u} + t + 1, \qquad u(0) = 1,$$

   computed with $h = 0.1$ and 0.05. Use extrapolation (Eq. 6.39) to improve the solutions at $t = 0.2, 0.4, 0.6, 0.8, 1.0$. Compare the results with the exact solution $u(t) = (t + 1)^2$.

2. Extrapolate the data in this table using Eq. (6.38):

$h =$	0.5	0.1
$u(1) \cong$	3.28125	3.42816

   These approximations were obtained by using the modified Euler method on the initial value problem

   $$\frac{du}{dt} = u + t, \qquad u(0) = 1.$$

3. The same initial value problem mentioned in Exercise 2 is solved in Section 6.3 by a method with global error proportional to h^4. The results are

h	$u(1) \cong$
1.0	3.41666667
0.5	3.43469238
0.25	3.43641988

Using $h = 1$ and 0.5, then $h = 0.5$ and 0.25, extrapolate these results (Eq. 6.39). Compare with the best approximation, $u(1) \cong 3.43656366$.

4. The method used to produce Table 6.5 of this section has a global error proportional to h^2. Below is a table of the *extrapolated* values

h	$u(1) \cong$
0.1	0.5570027
0.05	0.5571426
0.025	0.5571594

Assuming these values to have an error proportional to h^3, carry out another extrapolation on them.

5. Carry out one more extrapolation on the two numbers obtained in solving Exercise 4. Assume their error proportional to h^4.

6. Set up the Taylor series, through the term containing the third derivative, for the initial value problem

$$\frac{du}{dt} = -2tu^2, \qquad u(0) = 1.$$

Use the series to approximate $u(1)$ in one step (i.e., use $h = 1$). Compare to the exact value.

7. Use the series prepared in Exercise 6 to compute an approximation for $u(1)$ in two steps ($h = 0.5$). Compare the results.

6.3

Runge-Kutta Methods

We observed in Section 6.2 that series matching can be used to derive methods with any specified order of accuracy. One of the most popular, because of its simplicity, is the fourth-order Runge-Kutta method (abbreviated RK4). As applied to the initial value problem

$$\frac{du}{dt} = f(t, u), \qquad u(t_0) = u_0, \tag{6.40}$$

the formulas for the step from u_n to u_{n+1} are

$$\left. \begin{aligned} s_1 &= f(t_n, u_n) \\ s_2 &= f(t_{n+1/2}, u_n + \tfrac{1}{2}hs_1) \\ s_3 &= f(t_{n+1/2}, u_n + \tfrac{1}{2}hs_2) \\ s_4 &= f(t_{n+1}, u_n + hs_3) \end{aligned} \right\} \tag{6.41}$$

$$u_{n+1} = u_n + \frac{h}{6}(s_1 + 2s_2 + 2s_3 + s_4). \tag{6.42}$$

Geometrically, we may think of the four s_i as slopes and the quantities

$$u_n + \tfrac{1}{2}hs_1, \qquad u_n + \tfrac{1}{2}hs_2, \qquad u_n + hs_3$$

as preliminary approximations to u at $t_{n+1/2}$ (the first two) and t_{n+1} (the last one). See Fig. 6.4.

Figure 6.4

Graphical Representation of One Step of RK4

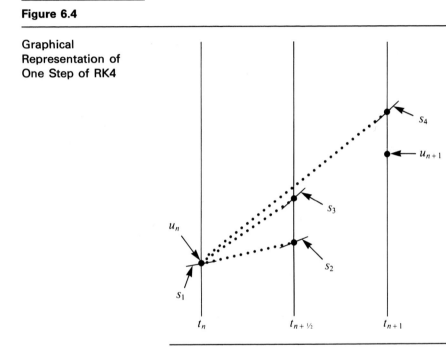

Example 1

Use the RK4 method with $h = 0.5$ to obtain a numerical approximation to the solution of

$$\frac{du}{dt} = u + t, \qquad u(0) = 1, \tag{6.43}$$

at $t = 1$.

Since $h = 0.5$, only two steps of RK4 are needed. The detailed calculations for the first step follow.

$$\left.\begin{aligned}
s_1 &= 1 + 0 = 1 \\
s_2 &= (1 + \tfrac{1}{2} \times 0.5 \times 1) + 0.25 = 1.5 \\
s_3 &= (1 + \tfrac{1}{2} \times 0.5 \times 1.5) + 0.25 = 1.625 \\
s_4 &= (1 + 0.5 \times 1.625) + 0.5 = 2.3125
\end{aligned}\right\}$$

$$u_1 = 1 + \frac{0.5}{6}(1 + 2 \times 1.5 + 2 \times 1.625 + 2.3125)$$

$$= 1.796875.$$

Another step gives $u(1) \cong u_2 = 3.4347$. The result compares favorably with the modified Euler with $h = 0.1$ but takes about half as much work. (See Example 2 of Section 6.1.)

A very serious practical problem is the choice of the step size h. If it is too large, the desired accuracy will not be achieved. Even the character of the solution may be falsified, in the sense that a decreasing solution may be computed as an increasing one, and so on. On the other hand, if h is too small, a great deal of computational effort may be wasted. This latter point is illustrated by the numbers in Table 6.6. In the column headed u_n are approximations to $u(1)$, where u is the solution of the initial value problem Eq. (6.43). The RK4 method was used for the various values of h shown. No further improvement can be obtained by reducing h, because the calculations were done on a microcomputer that has nine-digit precision.

Table 6.6	h	u_n	Error
	1.0	3.41666667	2×10^{-2}
	0.5	3.43469238	1.9×10^{-3}
	0.25	3.43641988	1.4×10^{-4}
	0.2	3.43650228	6.1×10^{-5}
	0.1	3.43655949	4.1×10^{-6}
	0.05	3.43656339	2.7×10^{-7}
	0.02	3.43656366	$< 10^{-8}$
	0.01	3.43656365	

Theoretical error analysis of a numerical method does not tell us how large the global error is, but only its "order." That is, a fourth-order method such as RK4 has an error that approaches 0 in proportion to the fourth power of h, as h decreases. (See Exercise 11 at the end of this section.) However, this theoretical fact has practical application. If it is known that the global error in a method is, say, of order 4,

$$\varepsilon = ch^4,$$

then a reduction in h by a factor of 2 should cause a reduction in the error by a factor of 16. In other words a quantity calculated with $h/2$ should have about one decimal digit more accuracy than a quantity calculated with h. The experimental results shown in Table 6.6 certainly confirm this prediction.

A reasonable way to estimate accuracy is to carry out a calculation twice, once with a value of h that is believed to be satisfactory and again with a step size twice as large. The first decimal place in the difference of the two results is the last decimal place that is accurate in the better result—the one calculated with h. This check on accuracy increases the cost of computation by 50%. (If we recalculate with $h/2$ instead of h, the check on accuracy costs twice as much as the original computation.) Programs for solving a differential equation over a long interval usually make such checks and automatically adjust the step size to maintain the desired accuracy without overdoing it.

Example 2

Suppose we want to obtain a numerical solution to the initial value problem

$$\frac{du}{dt} = \frac{u^2 + 1}{u}, \qquad u(0) = 1,$$

over the range $0 \le t \le 1$, with five-digit accuracy. We first choose $h = 0.1$, solve, and then solve again with $h = 0.2$. The approximations for $u(1)$ are

$$h = 0.1, \qquad u(1) \cong 3.7118901,$$
$$h = 0.2, \qquad u(1) \cong 3.7119277.$$

Since these two differ by about 4×10^{-5}, the value found with $h = 0.1$ can be taken to have six accurate digits. The exact value is

$$u(1) = \sqrt{2e^2 - 1} = 3.7118877,$$

so the approximation found with $h = 0.1$ really does have six good digits.

Second-Order Equations

Only a few of the problems of interest in engineering and science are of first order. More frequently, we have to solve a second-order initial value problem, say,

$$\frac{d^2u}{dt^2} = g(t, u, u'), \qquad u(t_0) = u_0, \qquad \frac{du}{dt}(t_0) = v_0. \tag{6.44}$$

Although special methods exist for second-order differential equations, the usual treatment is to convert the second-order equation into a pair of first-order equations. This may be done by defining du/dt to a new variable, say, v. Then the problem stated in Eq. (6.44) becomes

$$\left.\begin{aligned}\frac{du}{dt} &= v \\[2mm] \frac{dv}{dt} &= g(t, u, v)\end{aligned}\right\} \tag{6.45}$$

$$u(t_0) = u_0, \qquad v(t_0) = v_0. \tag{6.46}$$

It is no more complicated to consider the more general case of a system of two first-order equations represented by

$$\left.\begin{aligned}\frac{du}{dt} &= f(t, u, v) \\[2mm] \frac{dv}{dt} &= g(t, u, v)\end{aligned}\right\} \tag{6.47}$$

$$u(t_0) = u_0, \qquad v(t_0) = v_0. \tag{6.48}$$

Note that the initial data give us, via the differential equations, values for the derivatives of u and v at t_0. An analog of Euler's method is obvious. However, we prefer to set up the RK4 method, shown in the formulas below.

$$\left.\begin{aligned}r_1 &= f(t_n, u_n, v_n) & s_1 &= g(t_n, u_n, v_n) \\[2mm] r_2 &= f\left(t_{n+1/2}, u_n + \frac{h}{2}r_1, v_n + \frac{h}{2}s_1\right) & s_2 &= g\left(t_{n+1/2}, u_n + \frac{h}{2}r_1, v_n + \frac{h}{2}s_1\right) \\[2mm] r_3 &= f\left(t_{n+1/2}, u_n + \frac{h}{2}r_2, v_n + \frac{h}{2}s_2\right) & s_3 &= g\left(t_{n+1/2}, u_n + \frac{h}{2}r_2, v_n + \frac{h}{2}s_2\right) \\[2mm] r_4 &= f(t_{n+1}, u_n + hr_3, v_n + hs_3) & s_4 &= g(t_{n+1}, u_n + hr_3, v_n + hs_3)\end{aligned}\right\} \tag{6.49}$$

$$u_{n+1} = u_n + \frac{h}{6}(r_1 + 2r_2 + 2r_3 + r_4), \tag{6.50}$$

$$v_{n+1} = v_n + \frac{h}{6}(s_1 + 2s_2 + 2s_3 + s_4). \tag{6.51}$$

Example 3

Obtain a numerical solution of the initial value problem

$$\frac{d^2u}{dt^2} - tu = 0, \qquad u(0) = 1, \qquad u'(0) = 0.$$

over the range $0 \le t \le 2$ with five-digit accuracy.

First, we express the second-order differential equation as a system:

$$\frac{du}{dt} = v, \tag{6.52}$$

$$\frac{dv}{dt} = tu, \tag{6.53}$$

$$u(0) = 1, \qquad v(0) = 0. \tag{6.54}$$

Below are the formulas for the RK4 method, written out in detail for this specific example.

$$
\left.
\begin{aligned}
r_1 &= v_n & s_1 &= t_n u_n \\
r_2 &= v_n + \frac{h}{2} s_1 & s_2 &= t_{n+1/2}\left(u_n + \frac{h}{2} r_1\right) \\
r_3 &= v_n + \frac{h}{2} s_2 & s_3 &= t_{n+1/2}\left(u_n + \frac{h}{2} r_2\right) \\
r_4 &= v_n + h s_3 & s_4 &= t_n(u_n + h r_3)
\end{aligned}
\right\}
$$

$$u_{n+1} = u_n + \frac{h}{6}(r_1 + 2r_2 + 2r_3 + r_4),$$

$$v_{n+1} = v_n + \frac{h}{6}(s_1 + 2s_2 + 2s_3 + s_4).$$

We attempt the solution with $h = 0.1$ first and then, as a check on accuracy, with $h = 0.2$. (See Table 6.7.) A comparison of the approximations to $u(1)$ and $u(2)$ shows a difference in the fifth digit at $t = 2$ (and in

Table 6.7	h	$u(1) \cong$	$u(2) \cong$
	0.2	1.1722937	2.7307751
	0.1	1.1722996	2.7308757

the seventh at $t = 1$). This suggests that $h = 0.1$ does give five good digits. Thus we accept the values calculated with $h = 0.1$, rounded to five digits.

The information sought from a differential equation is not always the solution. In the next example we find the value of a parameter instead.

Example 4

Sizes of the populations of predators and prey are represented by the Volterra-Lotka system (see Section 1.5, Example 2)

$$\frac{du}{dt} = u(1 - v) \tag{6.55}$$

$$\frac{dv}{dt} = v(u - 1). \tag{6.56}$$

Theoretical investigations show that u and v are periodic, with a period (around 2π) that depends on the initial conditions. We attempt to find the period for

$$u(0) = 1, \qquad v(0) = 0.5. \tag{6.57}$$

The RK4 method is set up with an easily changed h, and the first run covers the interval $0 \le t \le 10$ with $h = 0.5$. The numbers generated are, of course, only a rude approximation to the solution, but show a return to the initial values somewhere around $t = 6.5$. Refinement of h to 0.25 shows the same, but provides no more exact information. In order to make the period more accessible, we scale the time in the system above to obtain

$$\frac{du}{dt} = pu(1 - v), \tag{6.58}$$

$$\frac{dv}{dt} = pv(u - 1) \tag{6.59}$$

and attempt to find p for which the solution of this system has period 1. Thus we solve Eqs. (6.58) and (6.59) numerically with different values of p, trying to make

$$u(1) = 1, \qquad v(1) = 0.5. \tag{6.60}$$

The first attempt, with $h = 0.1$ and $p = 6.5$, gives

$$u(1) \cong 1.004, \qquad v(1) \cong 0.502,$$

while the same h and $p = 6.49$ gives (see Fig. 6.5a)

$$u(1) \cong 0.999, \qquad v(1) \cong 0.502.$$

Apparently, the calculated value of $u(1)$ varies in the same direction as p; the situation for $v(1)$ is not clear.

Now we reduce h to 0.05, keeping $p = 6.49$. The RK4 method then calculates

$$u(1) \cong 1.0013, \qquad v(1) \cong 0.5001,$$

Figure 6.5

(a) Search for *p*, Part 1
Values are Deviations
of *u*(1) from 1

(a)

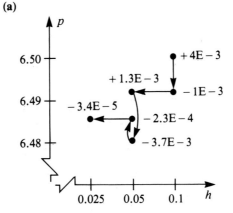

(b) Search for *p*, Part 2

(b)

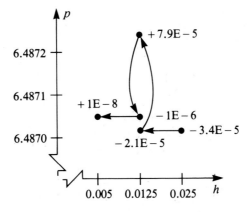

and with the same *h*, but slightly smaller *p* = 6.48, the results are

$$u(1) \cong 0.9963, \qquad v(1) \cong 0.5001.$$

The approximations to *u*(1) as a function of *p* have given this partial table of *u*(1) as a function of *p*:

p	u(1) ≅
6.48	0.9963
6.49	1.0013

By linear interpolation we determine *p* = 6.487 as the value to be tried, since we want $u(1) = 1$. Indeed, the RK4 solution with this p and h still equal to 0.05 gives

$$u(1) \cong 0.99977, \qquad v(1) \cong 0.50005.$$

Any further adjustment in p at the moment is probably futile, since the calculated value of $u(1)$ has only three reliable decimals. Thus we drop h to 0.025 and recalculate with $p = 6.487$, finding

$$u(1) \cong 0.999966, \qquad v(1) \cong 0.500001.$$

It seems that we have to decrease h further in order to get better accuracy in p. (See Fig. 6.5b.) With $h = 0.0125$ and $p = 6.487$ we get

$$u(1) \cong 0.999979, \qquad v(1) \cong 0.50000002.$$

Evidently, the calculated value of $u(1)$ has five-decimal-place accuracy. A modest increase in p, to 6.4872, produces

$$u(1) \cong 1.000079, \qquad v(1) = 0.50000002.$$

By using the idea of interpolation again, we determine that a better value of p is 6.48704, which leads to

$$u(1) \cong 0.999999, \qquad v(1) \cong 0.500000.$$

Reduction of h to 0.005 gives these values:

$$u(1) \cong 1.00000001, \qquad v(1) \cong 0.5.$$

Apparently, we are not going to get much better results; the calculated values have errors around 10^{-6} and are closer than that to the desired values. We can take the final value of $p = 6.48704$ as the period sought after.

Exercises

Solve Exercises 1–5, using the fourth-order Runge-Kutta method. The range and step length are given.

1. $\dfrac{du}{dt} = t - u^2$, $u(0) = 1$, $0 \le t \le 1$, $h = 0.1$

2. $\dfrac{du}{dt} = u$, $u(0) = 1$, $0 \le t \le 1$, $h = 1.0$

3. $\dfrac{du}{dt} = \dfrac{1 + u^2}{u}$, $u(0) = 1$, $0 \le t \le 1$, $h = 0.5$

4. $\dfrac{du}{dt} = \dfrac{1 + u^2}{u}$, $u(0) = 1$, $0 \le t \le 5$, $h = 0.5$

5. $\dfrac{du}{dt} = \sqrt{1 - u^2}$, $u(0) = 0$, $0 \le t < 1.6$, $h = 0.1$

6. Solve the initial value problem of Exercise 1 over the range $0 \leq t \leq 10$ using RK4 with $h = 0.2$. Confirm that $u(t)$ is approximately $\sqrt{t}$ for large t. Use graphical analysis to check this conclusion.

7. Solve the initial value problem

$$\frac{du}{dt} = -\frac{t}{u}, \qquad u(0) = 1,$$

using RK4 with $h = 0.1$, over the range $0 \leq t \leq 2$. What happens at $t = 1$? Obtain the exact solution.

8. How small must h be to obtain six-digit accuracy for the initial value problem of Example 3?

9. Solve the initial value problem

$$\frac{du}{dt} = t^2 - u^2, \qquad u(0) = 0,$$

on the interval $0 \leq t \leq 5$, using $h = 1, 0.5, 0.2, 0.1$. Comment on the results and their accuracy.

10. By comparing successive lines in Table 6.6 (as illustrated in Example 2), estimate the global error in the results for $h = 0.5, 0.25, 0.1, 0.05$. Do your estimates agree with the actual errors given in Table 6.6?

11. For each line in Table 6.6, compute the ratio of the error to h^4. Is this error apparently approaching a constant as $h \to 0$? That is, is the global error approximately proportional to h^4?

In Exercises 12–18, solve using RK4 over the indicated range with the given value(s) of h.

12. $\dfrac{d^2u}{dt^2} + u = 0$, $u(0) = 1$, $u'(0) = 0$, $0 \leq t \leq 1$, $h = 1.0$

13. $\dfrac{d^2u}{dt^2} + u = 0$, $u(0) = 1$, $u'(0) = 0$, $0 \leq t \leq 7$, $h = 1.0, 0.5$

14. $\dfrac{d^2u}{dt^2} + tu = 0$, $u(0) = 1$, $u'(0) = 0$, $0 \leq t \leq 1$, $h = 1.0$

15. $\dfrac{d^2u}{dt^2} + tu = 0$, $u(0) = 1$, $u'(0) = 0$, $0 \leq t \leq 2$, $h = 0.2, 0.1, 0.05$

16. $\dfrac{d^2u}{dt^2} + \sin u = 0$, $u(0) = 0$, $u'(0) = 1$, $0 \leq t \leq 1$, $h = 1.0$

17. $\dfrac{d^2u}{dt^2} + \sin u = 0$, $u(0) = 0$, $u'(0) = 1$, $0 \leq t \leq 7$, $h = 0.5, 0.25$.

18. $\dfrac{d^2u}{dt^2} + \dfrac{1}{t+1}\dfrac{du}{dt} + \dfrac{4}{(t+1)^2} u = 0$, $u(0) = 0$, $u'(0) = 2$, $0 \leq t \leq 4$, $h = 0.5, 0.25$

19. For the initial value problem of Exercise 18, find the first and second positive zeros of $u(t)$. Locate them approximately by solving with RK4 at $h = 0.25$.

20. It can be proved that the solution of the initial value problem of Exercise 17 is periodic. Find the period approximately, by solving with RK4 at $h = 0.25$ and finding the t where $u(t)$ and $u'(t)$ return to their initial conditions.

21. Find the first positive zero of the solution of the initial value problem in Exercise 15.

6.4
Predictor–Corrector Methods

The cost of calculating a numerical solution of an initial value problem

$$\frac{du}{dt} = f(t, u), \qquad u(t_0) = u_0, \tag{6.61}$$

is usually measured by the number of times it is necessary to evaluate the function f. An examination of the methods of Sections 6.1 and 6.3 shows that the number of function calls per step is the same as the order of the method: 1 for Euler, 2 for modified Euler, 4 for RK4. This observation is approximately true for all the *single-step* methods, those that calculate the new value of u using only the current value and information generated from it. If an error estimate is desired, it usually increases the cost by about 50%.

Single-step methods are so simple that their cost is often ignored. However, in a long calculation, cost may be significant. It is possible to reduce cost to two function calls per step and at the same time get an estimate of local error. The key to obtaining these benefits is the judicious use of history, in the form of previously computed values.

As an illustration of this idea, consider the possibility of calculating a numerical solution of Eq. (6.61) by

$$u_{n+1} = u_{n-1} + 2hf(t_n, u_n). \tag{6.62}$$

This formula is appealing because it is so intimately related to the definition of the derivative. Furthermore, it has a local error of order 3:

$$u(t_{n+1}) - u_{n+1} = -\frac{h^3}{3} u'''(t_1). \tag{6.63}$$

Unfortunately, however, Eq. (6.62) has a tendency to produce garbage over the long haul. To illustrate this point, the initial value problem

$$\frac{du}{dt} = -u, \qquad u(0) = 1,$$

was solved numerically in the range $0 \leq t \leq 10$ with $h = 0.1$. The numbers produced are within the expected accuracy for a while. But after $t = 3$, something peculiar happens—the numbers start to wobble around the true solution. (See Table 6.8.) Later, the u_n's even become negative! Making h smaller just postpones the onset of the difficulty.

Table 6.8	t_n	u_n	$u(t_n)$	Error
	3.0	0.0515	0.0498	−0.0017
	3.1	0.0436	0.0450	0.0014
	3.2	0.0428	0.0408	−0.0020
	3.3	0.0351	0.0369	0.0018
	3.4	0.0358	0.0334	−0.0024
	3.5	0.0279	0.0302	0.0023
	3.6	0.0302	0.0273	−0.0029
	3.7	0.0219	0.0247	0.0028
	3.8	0.0258	0.0224	−0.0034
	3.9	0.0167	0.0202	0.0035
	4.0	0.0225	0.0183	−0.0042

The method of Eq. (6.62) is good for short range but defective over a long range. Fortunately, it is possible to modify the simple method above. At each stage we use Eq. (6.62) to *predict* the next value of u:

$$u_{n+1}^{p} = u_{n-1} + 2hf(t_n, u_n), \tag{6.64}$$

and then we use a different formula to *correct* the prediction:

$$u_{n+1} = u_n + \frac{h}{2}[f(t_n, u_n) + f(t_{n+1}, u_{n+1}^{p})]. \tag{6.65}$$

The two formulas (6.64) and (6.65) give one step of this *predictor–corrector method.* Notice the similarity to the "improved Euler method" mentioned in Section 6.2.

It is important to know that both Eqs. (6.64) and (6.65) have local errors proportional to h^3. In particular, one can show that

$$u_{n+1}^{p} - u(t_{n+1}) = \frac{h^3}{3} u'''(\tau), \tag{6.66}$$

$$u_{n+1} - u(t_{n+1}) = -\frac{h^3}{12} u'''(\bar{\tau}) \tag{6.67}$$

(assuming continuous derivatives), where τ and $\bar{\tau}$ are points near t_n. If the third derivative of u does not change too rapidly, we might assume that $u'''(\bar{\tau}) = u'''(\tau)$ and then solve the above as a pair of simultaneous equations to determine the local error:

$$u_{n+1} - u(t_{n+1}) \cong \tfrac{1}{5}(u_{n+1} - u^p_{n+1}). \tag{6.68}$$

Example 1

To test the statements above, Eqs. (6.64) and (6.65) were used to solve the initial value problem

$$\frac{du}{dt} = u + t, \qquad u(0) = 0,$$

which has solution $u(t) = e^t - t - 1$.

In Table 6.9 we have listed the approximations to $u(1) = e - 2$ that were computed with various values of h, along with the sum of the magnitudes of the local error that were calculated by Eq. (6.68). Evidently, this latter is a decent estimate of the global error.

Table 6.9

h	$u(1) \cong$	Estimated Global Error	Exact Global Error
0.2	0.719339	4.4×10^{-3}	1.1×10^{-3}
0.1	0.719393	1.4×10^{-3}	1.1×10^{-3}
0.05	0.718692	3.8×10^{-4}	4.1×10^{-4}
0.025	0.718403	1.0×10^{-4}	1.2×10^{-4}

The reader may have noticed that Eq. (6.64) cannot be used to predict u_1, because it would require knowledge of a u_{-1}. It is a general failing of multistep methods that some other method must be used to get things started. The usual solution is to generate as many starting values as are needed by a single-step method having a local error of the same order as that of the predictor–corrector. Thus Eqs. (6.64) and (6.65) would normally be started by using a second-order method such as the modified Euler.

Now Eqs. (6.64) and (6.65) do not have much advantage over the modified Euler method used to start them up. Both methods require two function calls per step, and they have the same order of accuracy. Indeed, the only reason for bringing up this method is to have a simple example. In actual practice it is most common to use a fourth-order predictor–corrector. There are many such methods. One of the most commonly used is

$$u^p_{n+1} = u_n + \frac{h}{24}(55f_n - 59f_{n-1} + 37f_{n-2} - 9f_{n-3}), \tag{6.69}$$

$$f^p_{n+1} = f(t_{n+1}, u^p_{n+1}),$$

$$u_{n+1} = u_n + \frac{h}{24}(9f^p_{n+1} + 19f_n - 5f_{n-1} + f_{n-2}). \tag{6.70}$$

(The predictor, Eq. (6.69), is called an Adams-Bashforth formula, and the corrector an Adams-Moulton formula.) Note that we use past values of the slope:

$$f_n = f(t_n, u_n), \qquad f_{n-1} = f(t_{n-1}, u_{n-1}), \qquad \text{etc.}$$

instead of past values of u.

The method of Eqs. (6.69) and (6.70) require the calculation of f_n and f^p_{n+1} at each step—two function calls instead of the four needed for a comparable single-step method. It is important to remember that f_n is to be calculated once and then stored for future use. It should not be recalculated.

As for the simple method of Eqs. (6.64) and (6.65), a cheap error estimate is available. It can be shown that the local error for Eqs. (6.69) and (6.70) is approximately

$$u_{n+1} - u(t_{n+1}) \cong \tfrac{1}{14}(u_{n+1} - u^p_{n+1}). \tag{6.71}$$

Example 2

In order to compare the RK4 method and the predictor–corrector of Eqs. (6.69) and (6.70) we solve the initial value problem

$$\frac{du}{dt} = \frac{u^2 + 1}{u}, \qquad u(0) = 1,$$

over the range $0 \le t \le 10$. Table 6.10 shows the calculated approximation which may be compared to the true value, $u(10) = \sqrt{2e^{10} - 1} = 31150.127$. Clearly the accuracy of the two methods is about the same.

Table 6.10

h	Adams $u(10) \cong$	Runge-Kutta $u(10) \cong$
0.5	30963.285	31065.766
0.25	31147.351	31143.568
0.10	31150.615	31149.934

We must add the same comment here as in Section 6.3: second-order equations are more likely to come up than first-order. As before, we treat the first-order system

$$\frac{du}{dt} = f(t, u, v), \tag{6.72}$$

$$\frac{dv}{dt} = g(t, u, v), \tag{6.73}$$

$$u(t_0) = u_0, \qquad v(t_0) = v_0. \tag{6.74}$$

The Adams-Bashforth-Moulton method of Eqs. (6.69) and (6.70) becomes, for the system above,

$$u_{n+1}^p = u_n + \frac{h}{24}(55f_n - 59f_{n-1} + 37f_{n-2} - 9f_{n-3}),$$

$$v_{n+1}^p = v_n + \frac{h}{24}(55g_n - 59g_{n-1} + 37g_{n-2} - 9g_{n-3}),$$

$$f_{n+1}^p = f(t_{n+1}, u_{n+1}^p, v_{n+1}^p), \qquad g_{n+1}^p = g(t_{n+1}, u_{n+1}^p, v_{n+1}^p),$$

$$u_{n+1} = u_n + \frac{h}{24}(9f_{n+1}^p + 19f_n - 5f_{n-1} + f_{n-2}),$$

$$v_{n+1} = v_n + \frac{h}{24}(9g_{n+1}^p + 19g_n - 5g_{n-1} + g_{n-2}).$$

Exercises

1. Use the second-order predictor–corrector with $h = 0.2$ (Eqs. 6.64 and 6.65) to obtain approximate values for $u(1)$, if u is the solution of

$$\frac{du}{dt} = u^2 + t, \qquad u(0) = 0.$$

2. Follow the directions of Exercise 1 for

$$\frac{du}{dt} = u + t, \qquad u(0) = 0.$$

Find the solution in closed form.

3. Analyze the errors in Table 6.9 to confirm that the method of Eqs. (6.64) and (6.65) has global error of order h^2.

4. Derive Eq. (6.68) from Eqs. (6.66) and (6.67).

5. Rewrite Eqs. (6.66) and (6.67) as

$$u_{n+1}^p - u_{n+1} = 4E, \tag{6.66'}$$

$$u_{n+1}^c - u_{n+1} = -E, \tag{6.67'}$$

where $E = h^3 u'''(\tau)/12$ and τ and $\bar{\tau}$ are assumed to be equal. Now eliminate E between the two equations and solve for u_{n+1}. (This u_{n+1} is said to be "mopped up": in essence the local error estimated in

Eq. (6.68) is eliminated from the accepted value of u_{n+1}.)

6. What is wrong with using the mopped-up values obtained by using the results of Exercise 5?

7. Use the formula for mopped-up values found in Exercise 5 to solve the problem of Exercise 2 with $h = 0.1$.

8. In the table below are approximate values for $u(1)$, where

$$\frac{du}{dt} = u + t, \qquad u(0) = 0,$$

which were found by the mopped-up second-order method. Analyze the errors to confirm that the global error is of order h^3.

h	$u(1) \cong$	Error
0.2	0.71442199	3.2 E-3
0.1	0.71774329	9.6 E-4
0.05	0.71821059	7.1 E-5
0.025	0.71827266	9.2 E-6
0.0125	0.71828067	1.2 E-6

9. Extrapolate the last two lines of the table in Exercise 8 to obtain an approximate value for $u(1)$ with error of order h^4.

10. Use the fourth-order predictor–corrector, Eqs. (9) and (10), with $h = 0.2$, to solve

$$\frac{du}{dt} = u, \qquad u(0) = 1, \qquad 0 \le t \le 1.$$

Starting values are given in the table.

t_n	0	0.2	0.4	0.6
u_n	1	1.2214	1.4918	1.8221

11. Follow the directions in Exercise 10 for the problem

$$\frac{du}{dt} = -2tu^2, \qquad u(0) = 1, \qquad 0 \le t \le 1$$

t_n	0	0.2	0.4	0.6
u_n	1	0.9615	0.8621	0.7353
f_n	0	-0.3698	-0.5946	-0.6488

12. Use graphical analysis to show that the solution of

$$\frac{du}{dt} = t^2 - u^2, \qquad u(0) = 1,$$

approaches t as t increases, and so do all solutions of the differential equation in the first quadrant.

 13. Use Eqs. (6.69) and (6.70) to solve the initial value problem in Exercise 12, over the range $0 \le t \le 10$, with $h = 0.2, 0.1, 0.05$. Comment.

14. Follow the directions in Exercise 13 for the initial value problem

$$\frac{du}{dt} = 2t(t - u), \qquad u(0) = 1.$$

Should the solutions of these two problems be similar?

15. Use Eqs. (6.69) and (6.70) to solve the initial value problem

$$\frac{du}{dt} = -2tu^2, \qquad u(0) = 1.$$

Estimate $u(1)$, $u(2)$, and $u(10)$, using $h = 0.5, 0.2, 0.1, 0.05$.

16. Use Eqs. (6.69) and (6.70) to solve

$$\frac{du}{dt} = \ln(1 + u), \qquad u(0) = 1,$$

over the range $0 \le t \le 3$ with $h = 0.2, 0.1, 0.05$.

17. Use Eqs. (6.69) and (6.70) to solve

$$\frac{du}{dt} = t - u^2, \qquad u(0) = 1,$$

over the range $0 \le t \le 1$ with $h = 0.1$. See the solution of Exercise 1 in Section 6.3 for starting values.

18. Follow the directions in Exercise 17 for the problem

$$\frac{du}{dt} = \sqrt{1 - u^2}, \qquad u(0) = 0,$$

over the range $0 \le t \le 1$ with $h = 0.1$. See the solution of Exercise 5 in Section 6.3 for starting values.

6.5
Stability and Step Length

We can see how numerical methods work by applying them to the simple initial value problem

$$\frac{du}{dt} = Au, \qquad u(0) = 1, \tag{6.75}$$

with solution $u(t) = \exp(At)$. First consider Euler's method; the approximations to $u(t)$ are computed from

$$u_{n+1} = u_n + hAu_n = (1 + hA)u_n. \tag{6.76}$$

It is easy to see, after computing a few values, that

$$u_n = (1 + hA)^n. \tag{6.77}$$

Now when hA is small, the quantity $(1 + hA)^n$ is not a bad approximation to $u(nh) = \exp(Anh)$. If hA is positive and fairly large, the approximation is not good but has the right tendencies—that is, u_n is positive and growing geometrically with n. However, if hA is less than -1, u_n will alternate in sign, thus badly misrepresenting the nature of the solution of the initial value problem (6.75).

The modified Euler method applied to Eq. (6.75) gives these equations for the problem in Eq. (6.75):

$$\left.\begin{array}{l} u_{n+1/2} = u_n + \dfrac{h}{2} A u_n \\[2mm] u_{n+1} = u_n + hAu_{n+1/2}. \end{array}\right\} \tag{6.78}$$

When $u_{n+1/2}$ is eliminated between these two, we find that

$$u_{n+1} = (1 + hA + \tfrac{1}{2}h^2 A^2)u_n. \tag{6.79}$$

Again one can determine a formula for u_n directly:

$$u_n = (1 + hA + \tfrac{1}{2}h^2 A^2)^n. \tag{6.80}$$

For hA small or positive the modified Euler method gives better results than the original Euler method. If hA is negative, u_n is always positive, but increasing rather than decreasing as a function of n for values of hA less than -2.

The relative merits of these two methods show up nicely in this comparison. (See Fig. 6.6.)

$$\left.\begin{array}{ll} u_n = (1 + hA)^n & \text{(Euler)} \\[2mm] u_n = (1 + hA + \tfrac{1}{2}h^2 A^2)^n & \text{(modified)} \\[2mm] u_n = (e^{hA})^n = (1 + hA + \tfrac{1}{2}h^2 A^2 + \tfrac{1}{6}h^3 A^3 + \cdots)^n & \text{(exact)} \end{array}\right\} \tag{6.81}$$

Next we apply the second-order predictor–corrector of Section 6.4 to the initial value problem (6.75). The equations are

$$\left.\begin{array}{l} u^p_{n+1} = u_{n-1} + 2hAu_n \\[2mm] u_{n+1} = u_n + \dfrac{h}{2}[Au_n + Au^p_{n+1}]. \end{array}\right\} \tag{6.82}$$

When the predicted value is eliminated, we find that

$$u_{n+1} = (1 + \tfrac{1}{2}hA + h^2 A^2)u_n + \tfrac{1}{2}hAu_{n-1}. \tag{6.83}$$

It is no longer a trivial matter to solve this equation for the general term. Nevertheless, it can be shown that

$$u_n = c_1 \rho_1^n + c_2 \rho_2^n, \tag{6.84}$$

where ρ_1 and ρ_2 are the roots of the polynomial equation

$$\rho^2 = (1 + \tfrac{1}{2}hA + h^2 A^2)\rho + \tfrac{1}{2}hA, \tag{6.85}$$

called the characteristic equation of Eq. (6.83). The two roots are rather

Figure 6.6

(a) $1 + hA$ and e^{hA}

(b) $1 + hA + \frac{1}{2}h^2A^2$
and e^{hA}

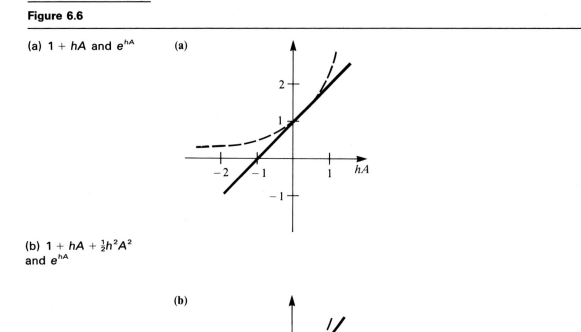

complicated functions of hA, shown in Fig. 6.7. The coefficients c_1 and c_2 depend on the choice of u_1, which must be supplied in order to get Eq. (6.83) started. Specifically, we can give the theoretical solution of Eq. (6.83) as

$$u_n = \frac{u_1 - \rho_2}{\rho_1 - \rho_2} \rho_1^n + \frac{\rho_1 - u_1}{\rho_1 - \rho_2} \rho_2^n. \tag{6.86}$$

Now if hA is positive and not too large, ρ_1 is nearly equal to e^{hA}, while ρ_2 is negative and not greater than 1 in magnitude. Thus in Eq. (6.86), ρ_1^n

Figure 6.7

The roots of $\rho^2 = (1 + \frac{1}{2}hA + h^2A^2)\rho + \frac{1}{2}hA$ and e^{hA}

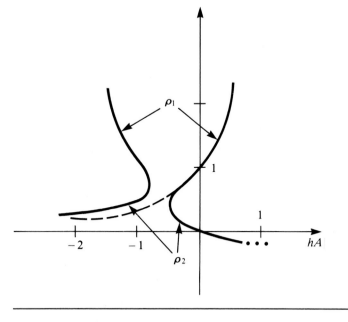

has a coefficient about equal to 1 ($u_1 \cong e^{hA} \cong \rho_1$), while ρ_2^n has a small coefficient—and goes to 0 anyway, as n increases.

On the other hand, if hA is negative, the relative values of the coefficients shift. For $hA < -1$, it is ρ_2 that is nearly equal to r^{hA}. Thus the coefficient of ρ_2^n is about equal to 1, while ρ_1^n gets a small coefficient. Nevertheless, for $hA < -1$, ρ_1 is greater than 1, and ρ_1^n grows with n. In spite of the small coefficient, this growing component of the solution will eventually dominate the other and produce invalid results.* This is an example of *numerical instability*.

Similar analyses on the fourth-order Runge-Kutta and predictor-corrector methods show similar results. Roughly speaking, $|hA|$ should be kept below 1 to avoid all the effects of numerical instability.

Now we must see how to apply our observations to the more general problem

$$\frac{du}{dt} = f(t, u), \qquad u(t_0) = u_0. \tag{6.87}$$

---

* The analysis is somewhat complicated by the fact that ρ_1 and ρ_2 are complex for hA between -0.5 and about -0.86. (Note the gap between solid curves in Fig. 6.7.)

Suppose we have arrived at time t_n and have accepted u_n as the approximate value for $u(t_n)$. If f is not changing too rapidly with t, we may expect the solution of Eq. (6.87) near (t_n, u_n) to behave about like the solution of the linear equation

$$\frac{du}{dt} = f(t_n, u_n) + \frac{\partial f}{\partial u}(t_n, u_n)(u - u_n). \tag{6.88}$$

Thus we should make sure that the product

$$\left| h \frac{\partial f}{\partial u}(t_n, u_n) \right| \tag{6.89}$$

is small enough to avoid numerical instability. Of course, no one wants to find and evaluate the partial derivative. But in some methods—especially the fourth-order Runge-Kutta (see Exercise 8 at the end of this section)—we can estimate it by using a formula such as

$$\frac{\partial f}{\partial u} \cong \frac{f(t, u + \Delta u) - f(t, u)}{\Delta u} \tag{6.90}$$

with values of f that have to be computed anyway. In many cases the stability problem we have been discussing does not arise because h is kept small to attain sufficient accuracy.

We have already noted that the predictor–corrector methods allow for an estimate of local error. By adding up the absolute values of these local errors we get an estimate for the accumulated error. This information can be used to select step length also. Suppose a differential equation is to be solved over an interval of length L within a global error of magnitude τ. It is reasonable to parcel out an allowance for local error of magnitude $h\tau/L$ in one step of length h. If the local error exceeds the allowance, the step length is reduced so that the local error on the next step will be within its allowance. It is also advisable to provide for increasing the step length in case a local error is much smaller than the targeted amount.

Since the predictor–corrector methods make available a good estimate of local error, one can use the estimate to try to eliminate the local error. This strategy, called "mopping up" (see Exercises 5–8 in Section 6.4) is seldom used by professionals in this field. The estimate on the error and its contribution to step length control are held to be more important than the potential for reducing the error.

Finally, let us recall that computers, as well as people, do arithmetic with only a finite number of significant digits, usually about eight. The result of any arithmetic operation is rounded to the correct number of digits. After a large number of operations, the accumulated roundoff error can seriously erode results. However, the effects are most noticeable when nearly equal quantities are subtracted. The most reliable digits cancel each other, and much (perhaps all) of what remains may represent trash. This effect has some influence in predictor–corrector methods (see Eqs. (6.69) and (6.70) in Section 6.4) where there is some subtraction.

Exercises

1. If $T = hn$ is fixed, show that the Euler approximation to $u(T)$ (see Eq. (6.77)) approaches $\exp(At)$ as n increases and h decreases.

2. Show that the fourth-order Runge-Kutta method (Section 6.3) applied to Eq. (6.75) gives the relation $u_{n+1} = R(hA) \cdot u_n$, where

   $$R(x) = 1 + x + \tfrac{1}{2}x^2 + \tfrac{1}{6}x^3 + \tfrac{1}{24}x^4.$$

   Sketch the graphs of $R(x)$ and e^x as in Fig. 6.6.

3. Verify that the formula in Eq. (6.84) gives a solution of Eq. (6.83) for any constants c_1 and c_2, as long as ρ_1 and ρ_2 are solutions of Eq. (6.85).

4. Check points on the graph of Fig. 6.7 by substituting these values of ρ and finding the values of hA that satisfy Eq. (6.85): $\rho = 1, \tfrac{1}{2}, 0, -1$.

5. Analyze the stability of the second-order predictor

   $$u_{n+1} = u_{n-1} + 2hf(t_n, u_n). \tag{P}$$

   (Apply it to the case $f(t, u) = Au$; solve the resulting relation among u_{n+1}, u_n, u_{n-1} by assuming $u_k = \rho^k$ for all k; study the roots of the characteristic polynomial.)

6. Considering the results of Exercise 5, explain the comment in Section 6.4 that the use of the predictor formula alone (Eq. (P) above) "has a tendency to produce garbage over the long haul."

7. In Exercise 9 of Section 6.3 and Exercises 12 and 13 of Section 6.4 we considered the differential equation

   $$\frac{du}{dt} = t^2 - u^2.$$

   Can you explain why both Runge-Kutta and predictor–corrector methods eventually produced bad results with this equation?

8. Show that the expression

   $$\frac{2(s_3 - s_2)}{h(s_2 - s_1)},$$

   with s_1, s_2, s_3 from the Runge-Kutta formulas of Section 6.3 (Eq. 6.41) gives an approximation to $\partial f/\partial u$ evaluated at $t_{n+1/2}$ and some u near u_n and u_{n+1}.

9. How can we estimate $\partial f/\partial u$ in the predictor–corrector method of Section 6.4 (Eqs. 6.69 and 6.70) without additional function evaluations?

10. Suppose we are using a fourth-order (local error $c_n h^5$ in u_n) and obtain an estimated local error ε_n that is different from the allowance $h\tau/L$. Show that the step length for the next step should be

$$\bar{h} = h(\tau h/L\varepsilon_n)^{1/4}.$$

Notes and References

Our objective in this chapter has been to see how numerical methods are derived, how they work, and what their weaknesses are. The methods presented are satisfactory for learning about differential equations, especially in conjunction with computer graphics, but they should not be used blindly for extensive calculations or where high accuracy is required.

Perhaps the most surprising aspect of this subject is that so many methods are available. The explanation for this phenomenon is that many criteria have to be balanced—accuracy, efficiency, simplicity, etc.—and each method has its drawbacks. An intelligent choice of methods for the serious calculation requires more detailed information, which may be found in the book of Burden, Faires and Reynolds (1978) or that of Dahlquist and Björk (1974). Many other books are available as are published programs in print and software.

Computer Programs

The BASIC computer programs that follow were used to do the numerical calculations for this chapter. The first four were written for the Commodore VIC-20 computer. Its BASIC has the peculiarity that it will not accept functions with more than one argument. Thus the function $f(t, u) = t - u^2$ has to be disguised as FNF(U) = T − U ∗ U; T is treated as a parameter. The program is written so as to overcome this oddity, and it should run on any machine using BASIC.

The last program, written in standard BASIC, solves a system of two first-order equations by the RK4 method. No effort has been made to maximize the speed of execution, but some attention was given to minimizing the programmer's time. The program can be extended easily to accommodate systems of more than two equations.

```
10 REM*****EULER*****
20 REM PROGRAM SOLVES     U'=FNF(T,U),U(0)=U0    BY EULER'S METHOD.
40 REM DEFINE FNF IN 50; U0,TF (FINAL T), AND H (STEP) IN 60.
50 DEF FNF(U)=
60 U0=   :TF=    :H=
70 U=U0
80 FOR T=0 TO TF STEP H
90 U=U+H*FNF(U)
100 T1=T+H
110 PRINT T1;U
120 NEXT T
```

```
10 REM***MODEULER*****
20 REM PROGRAM SOLVES     U'=FNF(T,U),U(0)=U0
30 REM SPECIFY FNF IN 40; U0, TF (FINAL T) AND H (STEP) IN 50.
40 DEF FNF(U)=
50 U0=    :TF=    :H=
60 U=U0
70 FOR T1=0 TO TF STEP   H
80 T=T1
90 U1=U+(H/2)*FNF(U)
100 T=T1+(H/2)
110 U=U+H*FNF(U1)
120 PRINT T1+H;U
130 NEXT T1
```

```
10 REM*******RK4********
20 REM PROGRAM SOLVES     U'=FNF(T,U),U(0)=U0    BY RUNGE-KUTTA.
30 REM SPECIFY FNF IN 50; U0,TF (FINAL T), AND H (STEP) IN 60.
40 DEF FNF(U)=
50 U0=    :TF=    :H=
60 U=U0
70 FOR T1=0 TO TF STEP H
80 T=T1
90 S1=FNF(U)
100 U1=U+(H/2)*S1
110 T=T1+(H/2)
120 S2=FNF(U1)
130 U2=U+(H/2)*S2
140 S3=FNF(U2)
150 U3=U+H*S3
160 T=T1+H
170 S4=FNF(U3)
180 U=U+(H/6)*(S1+2*S2+2*S3+S4)
190 PRINT T;U
200 NEXT T1
```

```
10 REM*******ADAMS*****
20 REM PROGRAM SOLVES     U'=FNF(T,U),U(0)=U0    BY ADAMS METHOD.
30 REM SPECIFY FNF IN 60; U0 ,TF(FINAL T) AND H (STEP) IN 70.
40 DIM D(4)
50 REM STARTUP
60 DEF FNF(U)=
70 U0=    :TF=    :H=
80 U=U0
90 GOSUB 500
100 D(1)=FNF(U0)
110 FOR N=0 TO 2
120 GOSUB 700
130 D(N+2)=FNF(U)
```

```
140 GOSUB 500
150 NEXT N
160 REM PRED-CORR LOOP
170 M=INT(TF/H)
180 FOR N=3 TO M
190 T=N*H
200 UP=U+(H/24)*(55*D(4)-59*D(3)+37*D(2)-9*D(1))
210 T=T+H
220 S=FNF(UP)
230 U=U+(H/24)*(9*S+19*D(4)-5*D(3)+D(2))
240 GOSUB 500
250 FOR I=1 TO 3
260 D(I)=D(I+1)
270 NEXT I
280 D(4)=FNF(U)
290 NEXT N
300 END
500 REM PRINTS T & U
510 PRINT T
520 PRINT TAB(4);U
530 RETURN
700 REM RK4 STARTER
710 T=T0+N*H
720 S1=FNF(U)
730 U1=U+(H/2)*S1
740 T=T+(H/2)
750 S2=FNF(U1)
760 U2=U+(H/2)*S2
770 S3=FNF(U2)
780 U3=U+H*S3
790 T=T+(H/2)
800 S4=FNF(U3)
810 U=U+(H/6)*(S1+2*S2+2*S3+S4)
820 RETURN


10 REM**********RK4SYSTEM***********
20 PRINT "   SOLVES U'=F(T,U,V), V'=G(T,U,V);  U(T0)=U0, V(T0)=V0.
30 REM   SUPPLY INITIAL DATA, ETC. IN LINE 50, FUNCTIONS IN LINES 60 & 70.
40 REM   TF IS FINAL TIME; H IS STEPLENGTH.
50 U0=         :V0=          :T0=        :TF=          :H=
60 DEF FNF(T,U,V)=
70 DEF FNG(T,U,V)=
80 REM PREPARATIONS
90 DIM M(4),W(4),U(2),S(2),K(2,4)
100 FOR J=1 TO 4: READ M(J): M(J)=M(J)*H: NEXT J
110 DATA 0,0.5,0.5,1
120 FOR J=1 TO 4: READ W(J): NEXT J
130 DATA 1,2,2,1
140 PRINT T0; U0; V0
150 U(1)=U0: U(2)=V0: T=T0
160 REM MAIN RK4 CALCULATION
170 FOR T=T0 TO TF STEP H
180 FOR J=1 TO 4
190 K(1,J)=FNF(T+M(J),U(1)+M(J)*K(1,J-1),U(2)+M(J)*K(2,J-1))
200 K(2,J)=FNG(T+M(J),U(1)+M(J)*K(1,J-1),U(2)+M(J)*K(2,J-1))
210 FOR I=1 TO 2
220 S(I)=S(I)+K(I,J)*W(J)
230 NEXT I
240 NEXT J
250 FOR I=1 TO 2
260 U(I)=U(I)+S(I)*H/6
270 S(I)=0
280 NEXT I
290 PRINT T+H; U(1); U(2)
300 NEXT T
310 END
```

Miscellaneous Exercises

In Exercises 1–3, solve the given initial value problem. (a) Use the modified Euler method with $h = 0.5$ for $0 \le t \le 1$. (b) Use the fourth-order Runge-Kutta method with $h = 0.5$ for $0 \le t \le 1$.

1. $\dfrac{du}{dt} = \dfrac{2u}{t + 1}, u(0) = 1$

2. $\dfrac{du}{dt} = 2(t + 1), u(0) = 1$

3. $\dfrac{du}{dt} = 2\sqrt{u}, u(0) = 1$

4. Solve analytically the three problems in Exercises 1–3. Explain why the modified Euler method gives exact results in Exercise 2 but not in Exercises 1 and 3.

5. Use the fourth-order Runge-Kutta method with $h = 0.5$ to solve

$$\left(\frac{du}{dt}\right)^2 = (1 - u^2)(3 - u^2), \qquad u(0) = 0,$$

over the range $0 \le t \le 2$. Assume that du/dt is positive at $t = 0$.

6. How is the solution of Exercise 5 altered if du/dt is made to change sign when $u = 1$?

 7. The differential equation in Exercise 5 comes from a second-order nonlinear equation. It can be proved that $u(t)$ is periodic and its period is $4T$, where T is the first solution of $u(T) = 1$. Use a fourth-order Runge-Kutta method to estimate T.

8. Solve $du/dt = -2tu^2$, $u(0) = 1$, by the modified Euler method with $h = 1$, over the interval $0 \le t \le 2$. Why is the result so bad?

9. Use the three-term Taylor-series method (include terms through $\frac{1}{2}h^2u''$) to solve $du/dt = tu^2$, $u(0) = 1$, with $h = 0.2$ over $0 \le t \le 1$.

10. What numerical method should give results comparable to those found in Exercise 9? Use that method over the same range and compare results.

11. Solve the initial value problem $du/dt = -t/u$, $u(0) = 1$, by the fourth-order Runge-Kutta method with $h = 0.2$ for $0 \le t \le 1.2$.

12. Why are the results of Exercise 11 poor? Why will every method give poor results near $t = 1$?

13. Use a numerical method to approximate the value of t where the solution of

$$\frac{du}{dt} = t + u^2, \qquad u(0) = 1,$$

becomes infinite. (Hint: let $v = 1/u$; find the initial value problem for v; find the t where $v(t) = 0$. It can be shown that $\pi/4 < t < 1$.)

14. Use a fourth-order method to study the behavior of the solution of

$$\frac{du}{dt} = u^t - 1, \qquad u(0) = \tfrac{1}{2}.$$

15. Many special numerical methods have been developed for equations of special form. The method of *Numerov* applies to the second-order linear equation

$$\frac{d^2 u}{dt^2} + p(t)u = f(t).$$

Successive values of the approximate solution are computed by the formula

$$u_{n+1} = \frac{1}{1 + \sigma p_{n+1}} [(2 - 10\sigma p_n)u_n - (1 + \sigma p_{n-1})u_{n-1}$$
$$+ \sigma(f_{n-1} + 10 f_n + f_{n+1})]$$

where $\sigma = h^2/12$, $p_n = p(t_n)$, etc. Apply Numerov's method to solve $u'' + u = 0$, $u(0) = 0$, $u'(0) = 1$, with $h = \pi/4$. Use $u_1 = 0.7071$.

16. Use Numerov's method to solve $u'' - u = 0$, $u(0) = 1$, $u'(0) = 1$, over the interval $0 \le t \le 2$ with several different values of h. Supply u_1 by solving the initial value problem exactly.

17. Use a fourth-order method to study the behavior of the solution of

$$\frac{d^2 u}{dt^2} + t\frac{du}{dt} + t^2 u = 0, \qquad u(0) = 1, \qquad u'(0) = 0,$$

over the interval $0 \le t \le 5$.

7 Systems of Linear Differential Equations

7.1

Introduction

In this chapter we study systems of differential equations in several dependent variables. In advanced mathematics and in many branches of engineering and science this is the most convenient way to treat differential equations, partly because of the generality of the systems approach and partly because of its naturalness. In this section we see how some systems of differential equations arise.

Every nth-order differential equation can be converted to a system of n first-order differential equations. A standard way to carry out the conversion is to designate the unknown function and its first $n - 1$ derivatives to be new variables. For instance, for the differential equation

$$\frac{d^2u}{dt^2} + 2\frac{du}{dt} + 3u = f(t) \tag{7.1}$$

we designate

$$x_1 = u, \qquad x_2 = \frac{du}{dt} \tag{7.2}$$

and then obtain two equations, one for the first derivative of each:

$$\frac{dx_1}{dt} = x_2 \tag{7.3}$$

$$\frac{dx_2}{dt} = -3x_1 - 2x_2 + f(t). \tag{7.4}$$

Equation (7.3) is really the definition of x_2, turned around; and Eq. (7.4) is just the original differential equation (7.1).

The same device works for equations of higher order. Consider the general linear fourth-order equation

$$\frac{d^4u}{dt^4} + a_1\frac{d^3u}{dt^3} + a_2\frac{d^2u}{dt^2} + a_3\frac{du}{dt} + a_4u = f(t). \tag{7.5}$$

The new variables are defined as

$$x_1 = u, \qquad x_2 = \frac{du}{dt}, \qquad x_3 = \frac{d^2u}{dt^2}, \qquad x_4 = \frac{d^3u}{dt^3}, \tag{7.6}$$

and then the first-order system that replaces Eq. (7.5) is

$$\frac{dx_1}{dt} = x_2,$$

$$\frac{dx_2}{dt} = x_3,$$

$$\frac{dx_3}{dt} = x_4, \tag{7.7}$$

$$\frac{dx_4}{dt} = -a_4x_1 - a_3x_2 - a_2x_3 - a_1x_4 + f(t).$$

Again the first equations are, in effect, the definitions of the new variables, and the last is the differential equation (7.5).

Note that if Eq. (7.5) is accompanied by initial conditions

$$u(t_0) = q_1, \qquad \frac{du}{dt}(t_0) = q_2. \qquad \frac{d^2u}{dt^2}(t_0) = q_3, \qquad \frac{d^3u}{dt^3}(t_0) = q_4,$$

these become four initial conditions on the four new unknown functions

$$x_1(t_0) = q_1, \qquad x_2(t_0) = q_2, \qquad x_3(t_0) = q_3, \qquad x_2(t_0) = q_4. \tag{7.8}$$

The artificial creation of a system of first-order equations illustrated above is extremely useful. For instance, we make use of it in numerical methods for second-order equations. And we shall see that it helps clarify such ideas as general solution and variation of parameters.

There are many settings in which a system of first-order equations is the natural way to model a physical phenomenon. As an example, consider the problem of describing the temperature in a laminated material. (See Fig. 7.1.) We suppose (1) that the rate of heat flow between

Figure 7.1

Laminated Material

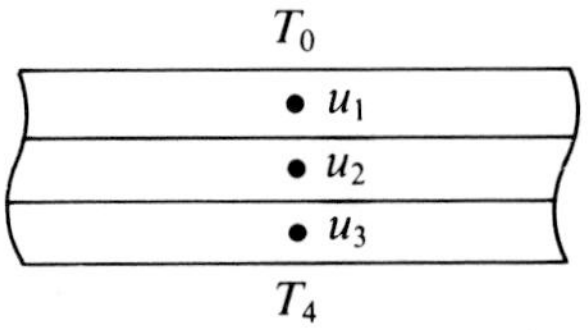

two objects in thermal contact is proportional to the difference in their temperatures and (2) that the rate at which heat is stored in an object is proportional to the rate of change of its temperature. Now the net heat flow rate into an object must balance the heat storage rate. This balance for each of the three layers gives three equations:

$$C_1 \frac{du_1}{dt} = k_{12}(u_2 - u_1) + k_{01}(T_0 - u_1),$$

$$C_2 \frac{du_2}{dt} = k_{12}(u_1 - u_2) + k_{23}(u_3 - u_2), \qquad \textbf{(7.9)}$$

$$C_3 \frac{du_3}{dt} = k_{23}(u_2 - u_3) + k_{34}(T_4 - u_3),$$

in which the C's and k's are constants of proportionality, and T_0 and T_4 are the temperatures at the two sides of the laminated material. It is easy to imagine how a larger or more complicated system might arise.

Another example is a system of equations devised to model an arms race between two countries. Let $x_i(t)$ represent the size of the arms stocks of country i ($i = 1, 2$). If a country were isolated, it might let its arms diminish according to a law like $dx_i/dt = -c_i x_i$, because of the costs of maintenance. The competition between countries, however, causes each one to increase its arms. The system suggested to describe the evolution of both countries' arsenals is

$$\frac{dx_1}{dt} = -c_1 x_1 + d_1 x_2 + g_1,$$

$$\frac{dx_2}{dt} = d_2 x_1 - c_2 x_2 + g_2, \qquad \textbf{(7.10)}$$

called Richardson's model. The c's are called cost factors, the d's defense factors, and the g's grievance terms. All these coefficients are assumed to be constant. Evidently, the participation of more countries would add more equations to the system.

All of the examples introduced above are linear systems. The most general form for such a system is

$$\frac{dx_1}{dt} = a_{11}(t)x_1 + a_{12}(t)x_2 + \cdots + a_{1n}(t)x_n + g_1(t),$$

$$\frac{dx_2}{dt} = a_{21}(t)x_1 + a_{22}(t)x_2 + \cdots + a_{2n}(t)x_n + g_2(t),$$

$$\vdots \qquad\qquad\qquad\qquad\qquad\qquad\qquad \textbf{(7.11)}$$

$$\frac{dx_n}{dt} = a_{n1}(t)x_1 + a_{n2}(t)x_2 + \cdots + a_{nn}(t)x_n + g_n(t).$$

(This is sometimes called the *normal form*, meaning that the derivatives of the unknown functions have been solved for.) If all of the g's are identically 0, the system is called homogeneous. Otherwise, it is nonhomogeneous, and the g's are the "inhomogeneities." The appropriate initial conditions would specify values of the x's at the initial time:

$$x_1(t_0) = q_1, \qquad x_2(t_0) = q_2, \qquad \ldots, \qquad x_n(t_0) = q_n. \tag{7.12}$$

Exercises

1. In Section 3.2 we derived the equations of motion of two masses and three springs (Fig. 7.2):

$$M_1 \frac{d^2 u_1}{dt^2} = -k_1 u_1 + k(u_2 - u_1),$$

$$M_2 \frac{d^2 u_2}{dt^2} = k(u_1 - u_2) - k_2 u_2.$$

(Here u_1 and u_2 are displacements.) Convert these two second-order

Figure 7.2

A System of Two Masses and Three Springs, and the Corresponding Free-Body Diagrams

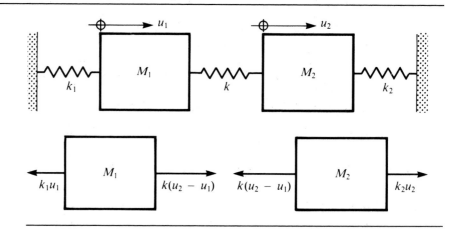

equations into four first-order equations, using these unknowns:

$$x_1 = u_1, \qquad x_2 = u_2, \qquad x_3 = \frac{du_1}{dt}, \qquad x_4 = \frac{du_2}{dt}.$$

2. Write the equation of motion of a simple mass–spring system as a system of two first-order equations.

3. In Fig. 7.3 an electric circuit is shown. The voltage drop across each of

Figure 7.3

An Electric Circuit
with Two Loops

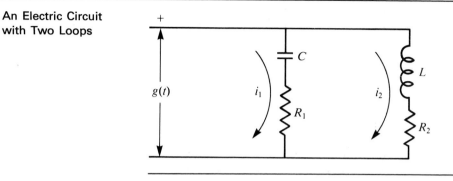

the four devices is

$$E_{R1} = R_1 i_1, \qquad E_{R2} = R_2 i_2,$$

$$E_C = \frac{1}{C} \int i_1 \, dt, \qquad E_L = L \frac{di_2}{dt}.$$

Kirchhoff's law, that the net voltage drop around any loop be zero, gives these equations:

$$u = E_{R1} + E_C, \qquad u = E_{R2} + E_L.$$

Using unknowns $x_1 = E_C$ and $x_2 = i_2$, convert the two equations arising from Kirchhoff's law into a system of two nonhomogeneous differential equations in normal form.

4. A simple dynamic model of a national economy is given by the equations

$$C = (1 - s)Y + A, \tag{i}$$

$$\frac{dK}{dt} = \mu(rY - K), \tag{ii}$$

$$\frac{dY}{dt} = \lambda\left(C + \frac{dK}{dt} - Y\right). \tag{iii}$$

The variables are Y, the real output; K, the fixed capital; and C, consumption. Also, A is the part of consumption that is independent of income, s is the marginal savings rate, r is the optimal ratio of capital to output, and λ and μ are positive constants. Explain the equations in words. Do they sound reasonable?

5. (a) Eliminate C to obtain two first-order equations in K and Y.
   (b) Obtain equations for K and Y in normal form.

7.2

Elimination Method

Because systems of equations do arise naturally in many contexts, it is convenient to have a method of solution that can be used without much theoretical background. The elimination method that we will discuss can be used on small systems (two or three equations) if a solution is all one needs. However, it gives practically no insight into the way systems work.

The objective of elimination is to convert a system of n linear equations into a single nth-order equation—just the reverse of what we did in Section 7.1. To start, let us consider a specific system of two homogeneous equations:

$$\frac{dx}{dt} = 5x + 2y, \tag{7.13}$$

$$\frac{dy}{dt} = 3x + 4y. \tag{7.14}$$

We can solve the first equation for y to get

$$y = \frac{1}{2}\left(\frac{dx}{dt} - 5x\right). \tag{7.15}$$

Substituting this for y in Eq. (7.14) gives us the differential equation

$$\frac{1}{2}\frac{d}{dt}\left(\frac{dx}{dt} - 5x\right) = 3x + \frac{4}{2}\left(\frac{dx}{dt} - 5x\right)$$

or

$$\frac{d^2x}{dt^2} - 9\frac{dx}{dt} + 14x = 0. \tag{7.16}$$

On the other hand, we might have solved Eq. (7.14) for x and have substituted into Eq. (7.13).

It is important to realize that the original system of equations has now been replaced by a new system consisting of Eq. (7.16), a second-order differential equation, and Eq. (7.15), which is to be thought of as an algebraic equation for y—not a differential equation.

We obtain the solution of Eq. (7.16) by routine means, finding

$$x(t) = c_1 e^{7t} + c_2 e^{2t}. \tag{7.17}$$

Now Eq. (7.15) gives us y by direct substitution:

$$y(t) = c_1 e^{7t} - \tfrac{3}{2}c_2 e^{2t}. \tag{7.18}$$

If initial conditions accompany Eqs. (7.13) and (7.14), say,

$$x(0) = q_1, \qquad y(0) = q_2, \tag{7.19}$$

then these are to be satisfied by an appropriate choice of c_1 and c_2. Namely, the initial conditions become

$$c_1 + c_2 = q_1, \tag{7.20}$$
$$c_1 - \tfrac{3}{2}c_2 = q_2,$$

which are to be solved for c_1 and c_2.

  The manipulations on systems are a little easier if we abbreviate the derivative by writing

$$\frac{dx}{dt} = Dx \qquad \text{or} \qquad \frac{dy}{dt} = Dy. \tag{7.21}$$

By using this notation, Eq. (7.15) might be written

$$y = \tfrac{1}{2}(D - 5)x, \tag{7.22}$$

and the substitution of this expression into Eq. (7.14) gives

$$\tfrac{1}{2}D(D - 5)x = 3x + \tfrac{4}{2}(D - 5)x \tag{7.23}$$

or

$$(D^2 - 9D + 14)x = 0. \tag{7.24}$$

This is the same as Eq. (7.16).

  This elimination method can also be used on more general systems. For example, the system below is not in normal form; it has not been solved for the derivatives:

$$\frac{dx}{dt} + \frac{dy}{dt} = 5x + 2y, \tag{7.25}$$

$$\frac{dx}{dt} + 2\frac{dy}{dt} = 9x + 3y. \tag{7.26}$$

Using the operator notation introduced above, we may write this system as

$$(D - 5)x + (D - 2)y = 0, \tag{7.27}$$

$$(D - 9)x + (2D - 3)y = 0. \tag{7.28}$$

It is not convenient to solve one of these equations for x or y. This would involve "dividing by" a polynomial in the D operator, an operation we are not prepared to discuss. Nevertheless, we can eliminate one or the other of the unknowns. For instance, we can eliminate y by (1) multiplying Eq. (7.27) on the left by $2D - 3$, (2) multiplying Eq. (7.28) on the left by $D - 2$, and (3) subtracting the resulting equations:

$$(2D + 3)(D - 5)x + (2D - 3)(D - 2)y = 0, \tag{7.29}$$

$$(D - 2)(D - 9)x + (D - 2)(2D - 3)y = 0, \tag{7.30}$$

$$[(2D - 3)(D - 5) - (D - 2)(D - 9)]x = 0. \tag{7.31}$$

In effect, we are handling the differentiation operator almost as if it were a number. When the calculations are completed, Eq. (7.31) becomes the second-order ordinary differential equation

$$\frac{d^2x}{dt^2} - 2\frac{dx}{dt} - 3x = 0, \tag{7.32}$$

which we can solve easily.

Now it is necessary to find y. We either eliminate the derivative of y between Eqs. (7.25) and (7.26), and solve for y in terms or x, or else just use one of these two equations as a differential equation for y. In the latter case we might take Eq. (7.25), in the form

$$\frac{dy}{dt} - 2y = 5x - \frac{dx}{dt}. \tag{7.33}$$

The solution of Eq. (7.32) contains two arbitrary constants, and the solution of Eq. (7.33) brings in one more. Of these three, only two are really free. One of them must be eliminated by requiring that Eq. (7.26)—which was not used to make Eq. (7.33)—be satisfied.

Example 1

The general solution of Eq. (7.32) is

$$x = c_1 e^{-t} + c_2 e^{3t}.$$

Then the differential equation (7.33) becomes

$$\frac{dy}{dt} + 2y = 6c_1 e^{-t} + 2c_2 e^{3t},$$

with solution

$$y = 6c_1 e^{-t} + \tfrac{2}{5}c_2 e^{3t} + c_3 e^{-2t}.$$

However, in order for x and y above to satisfy Eq. (7.26) it is necessary for c_3 to be 0.

The key to making y disappear from Eqs. (7.29) and (7.30) is the fact that

$$(2D - 3)(D - 2)y = (D - 2)(2D - 3)y.$$

We must point out that differential operators are not numbers, and, in contrast to numbers, the order in which they are applied does matter. Nevertheless, it can be proved that linear differential operators of any order, but with *constant coefficients*, always commute. On the other hand, it is unusual for differential operators with variable coefficients to commute.

If we have a general system of two linear differential equations, we might write it as

$$L_1x + L_2y = g_1, \tag{7.34}$$

$$L_3x + L_4y = g_2, \tag{7.35}$$

with the understanding that the L's represent linear differential operators—not necessarily of first order nor with constant coefficients. Following our former procedure, we attempt to eliminate y by applying L_4 to Eq. (7.34) and L_2 to Eq. (7.35):

$$L_4L_1x + L_4L_2y = L_4g_1,$$

$$L_2L_3x + L_2L_4y = L_2g_2.$$

Now, provided that L_2 and L_4 commute—that is, if

$$L_2L_4 = L_4L_2 \tag{7.36}$$

—we may subtract the two equations above to get

$$(L_4L_1 - L_2L_3)x = L_4g_1 - L_2g_2. \tag{7.37}$$

This equation would be solved for x, and then either Eq. (7.34) or Eq. (7.35) can be solved for y. The total number of arbitrary constants between x and y should be equal to the order of the operator $L_4L_1 - L_2L_3$ that appears in Eq. (7.37). If more are introduced in the course of finding y, then another relation among them must be established by requiring that x and y satisfy both of the original equations, Eqs. (7.34) and (7.35).

Example 2

Eliminating y from the system of equations

$$2\frac{dx}{dy} + \frac{dy}{dt} = x + 2y,$$

$$\frac{dx}{dt} + \frac{dy}{dt} = x + y$$

leads to the second-order equation

$$(D^2 - 1)x = 0 \quad \text{or} \quad \frac{d^2x}{dt^2} - x = 0,$$

having solution $x = c_1 e^t + c_2 e^{-t}$. Now we may take the second of the equations above to determine y:

$$\frac{dy}{dt} - y = x - \frac{dx}{dt} = 2c_2 e^{-t}.$$

The solution is $y = c_3 e^t - c_2 e^{-t}$. The three constants c_1, c_2, c_3 must be subjected to a relation implied by the first of the original equations:

$$2\frac{dx}{dt} + \frac{dy}{dt} = x + 2y,$$

$$2c_1 e^t - 2c_2 e^{-t} + c_3 e^t + c_2 e^{-t} = c_1 e^t + c_2 e^{-t} + 2c_3 e^t - 2c_2 e^{-t}.$$

By matching the coefficients of the independent functions e^t and e^{-t} we find that

$$2c_1 + c_3 = c_1 + 2c_3,$$

$$-c_2 = -c_2,$$

or that $c_3 = c_1$. Thus the solution of the original system is

$$x = c_1 e^t + c_2 e^{-t}, \qquad\qquad y = c_1 e^t - c_2 e^{-t}.$$

The elimination method is applicable to higher-order systems and to systems with more than two equations. However, it must be used with the greatest caution if coefficients are not constant.

Exercises

In Exercises 1–14, use the elimination method to find the general solution.

1. $\dfrac{dx}{dt} = 2x + y$

    $\dfrac{dy}{dt} = x + 2y$

2. $\dfrac{dx}{dt} = -2x + y + 1$

    $\dfrac{dy}{dt} = x - 2y$

3. $\dfrac{dx}{dt} = 2x + y$

    $\dfrac{dx}{dt} + \dfrac{dy}{dt} = y$

4. $\dfrac{dx}{dt} = 2x$

    $\dfrac{dy}{dt} = x - y$

5. $\dfrac{dx}{dt} = y$

    $\dfrac{dy}{dt} = -x$

6. $\dfrac{dx}{dt} = x + y$

    $\dfrac{dy}{dt} = y$

7. $\dfrac{dx}{dt} + \dfrac{dy}{dt} = x + 2y$

   $\dfrac{dx}{dt} - \dfrac{dy}{dt} = x$

8. $\dfrac{dx}{dt} + \dfrac{dy}{dt} = x + y$

   $-\dfrac{dx}{dt} + 2\dfrac{dy}{dt} = 2x - y$

9. $\dfrac{dx}{dt} = x + y$

   $\dfrac{dy}{dt} = y + 2z$

   $\dfrac{dz}{dt} = x$

10. $\dfrac{dx}{dt} = x + y$

    $\dfrac{dy}{dt} = y + z$

    $\dfrac{dz}{dt} = z$

11. $\dfrac{dx}{dt} = y$

    $\dfrac{dy}{dt} = 2x + 2z$

    $\dfrac{dz}{dt} = y$

12. $\dfrac{dx}{dt} = -2x + y$

    $\dfrac{dy}{dt} = x - 2y + z$

    $\dfrac{dz}{dt} = y - 2z$

13. $\dfrac{dx}{dt} = y + e^t$

    $\dfrac{dy}{dt} = x + e^{-t}$

14. $\dfrac{dx}{dt} = y + \cos t$

    $\dfrac{dy}{dt} = -x + \sin t$

15. In Eq. (7.9) of Section 7.1 we obtained a system of equations for the temperatures in a three-layer material. If all the C's and k's are taken equal to 1, the system becomes

$$\frac{du_1}{dt} = -2u_1 + u_2 + T_0,$$

$$\frac{du_2}{dt} = u_1 - 2u_2 + u_3,$$

$$\frac{du_3}{dt} = u_2 - 2u_3 + T_4.$$

Find the third-order equation satisfied by u_1.

16. Show that a particular solution of the system in Exercise 15 is

$$u_1 = \tfrac{1}{4}(3T_0 + T_4),$$
$$u_2 = \tfrac{1}{2}(T_0 + T_4),$$
$$u_3 = \tfrac{1}{4}(T_0 + 3T_4).$$

17. Solve this nonconstant system by first finding and solving an equation for $x + y$:

$$\frac{dx}{dt} = (1 + t)x + ty,$$

$$\frac{dy}{dt} = -tx + (1 - t)y.$$

18. In Exercises 4 and 5 of Section 7.1 we obtained this system of equations for output Y and fixed capital K in an economy:

$$\frac{dK}{dt} = -\mu K + \mu r Y,$$

$$\frac{dY}{dt} - \lambda \frac{dK}{dt} = -s\lambda Y + \lambda A.$$

Assuming A and the coefficients constant, eliminate Y to get a second-order equation for K.

19. Find a particular, constant solution of the system in Exercise 18. What conditions on the coefficients guarantee that $K(t)$ will approach the constant solution as t increases? What additional conditions will make the approach nonoscillatory?

20. Find the general solution of Richardson's arms race equations (Eq. (7.10) in Section 7.1), using $c_1 = c_2 = 1$, $d_1 = d_2 = 3$, $g_1 = 1$, $g_2 = 5$.

7.3
Eigenvalues and Eigenvectors

The most important special type of system is the one with constant coefficients. As in the case of nth-order equations, we can obtain closed-form solutions when coefficients are constant. The complications are purely algebraic. Our model system for this section will be a homogeneous matrix differential equation,

$$\frac{dx}{dt} = Ax, \tag{7.38}$$

in which A is an $n \times n$ matrix of constants. In analogy to the scalar case, we use an "exponential guess." We look for solutions in the form

$$x(t) = e^{\lambda t}z, \tag{7.39}$$

where the parameter λ and the constant column matrix z are to be found. To determine them, we substitute the proposed solution into the differential equation (7.38) to obtain

$$\lambda e^{\lambda t} z = A e^{\lambda t} z. \tag{7.40}$$

The scalar multiplier $e^{\lambda t}$ may be cancelled from both sides of this equation, leaving

$$\lambda z = Az \quad \text{or} \quad (\lambda I - A)z = 0. \tag{7.41}$$

The second form shows clearly that z is a solution of a homogeneous algebraic system. For most values of the parameter λ the coefficient matrix $\lambda I - A$ is nonsingular, and $z = 0$ is the only possibility. However, by Eq. (7.39) the solution of the original system of differential equations would then be $x(t) \equiv 0$. This is the *trivial* solution, of no interest because it was obvious from the beginning.

Our only hope of obtaining nontrivial solutions will come if the homogeneous algebraic system, Eq. (7.41), has a singular coefficient matrix. Thus we see that

$$x(t) = e^{\lambda t} z$$

is a nontrivial solution of the matrix differential equation

$$\frac{dx}{dt} = Ax$$

if and only if the parameter λ and constant column $z \neq 0$ satisfy

$$\det (\lambda I - A) = 0, \tag{7.42}$$

$$(\lambda I - A)z = 0. \tag{7.43}$$

Definition 7.1

A number λ (real or complex) that satisfies Eq. (7.42) is called an *eigenvalue* of A, and any nonzero column matrix that then satisfies Eq. (7.43) is called an *eigenvector* of A associated with λ.*

In the rest of this section we study some algebraic properties of eigenvalues and eigenvectors and see how they can be found.

---

* Other words for eigenvalue are characteristic value, proper value, secular value, and latent root. Other words for eigenvector are characteristic vector, proper vector, and modal vector. Also, one might distinguish between eigencolumns z ($Az = \lambda z$) and eigenrows w ($w^T A = \lambda w^T$).

Example 1

Find the eigenvalues and eigenvectors of the 2×2 matrix

$$A = \begin{bmatrix} 2 & 2 \\ 3 & 1 \end{bmatrix}.$$

To begin with, we should evaluate the determinant

$$\det(\lambda I - A) = \det \begin{bmatrix} \lambda - 2 & -2 \\ -3 & \lambda - 1 \end{bmatrix}$$
$$= (\lambda - 2)(\lambda - 1) - 6 = \lambda^2 - 3\lambda - 4.$$

The values of λ that make the determinant zero are $\lambda = 4$ and $\lambda = -1$. The eigenvector associated with the eigenvalue 4 is a nonzero solution of the system $(4I - A)z = 0$:

$$(4I - A)z = \left(4 \begin{bmatrix} 1 & 0 \\ 0 & 1 \end{bmatrix} - \begin{bmatrix} 2 & 2 \\ 3 & 1 \end{bmatrix} \right) z = \begin{bmatrix} 2 & -2 \\ -3 & 3 \end{bmatrix} z = 0.$$

The solution of this homogeneous system is $z = [c, c]^T$, where c is an arbitrary scalar. Choosing $c = 1$, we obtain the eigenvector $z = [1, 1]^T$.
For the eigenvalue $\lambda = -1$ we must solve $(-I - A)z = 0$:

$$\left(-\begin{bmatrix} 1 & 0 \\ 0 & 1 \end{bmatrix} - \begin{bmatrix} 2 & 2 \\ 3 & 1 \end{bmatrix} \right) z = \begin{bmatrix} -3 & -2 \\ -3 & -2 \end{bmatrix} z = 0.$$

Again there is an arbitrary scalar c in the solution of this homogeneous system, $z = c[-\frac{2}{3}, 1]^T$. Now taking $c = 3$ to get integer elements, we obtain the eigenvector $z = [-2, 3]^T$.
We may summarize the results as follows: the eigenvalues of A are 4 and -1. The eigenvectors are

$$z_1 = \begin{bmatrix} 1 \\ 1 \end{bmatrix} (\lambda = 4), \qquad z_2 = \begin{bmatrix} -2 \\ 3 \end{bmatrix} (\lambda = -1).$$

It should be verified that $Az_1 = 4z_1$ and $Az_2 = -z_2$.

The example brings out the fact that the general solution of the homogeneous equation $(\lambda I - A)z = 0$ contains an arbitrary scalar multiplier. That is, any (nonzero) scalar multiple of an eigenvector is still an eigenvector. The scalar is usually chosen for convenience: in the example we chose it to obtain integer components in z.
There are a number of simple properties of eigenvalues and eigencolumns that can be of help in finding them

About Eigenvalues

1. The expression that defines the eigenvalues of A, $\det(\lambda I - A)$, is a polynomial in λ of degree n. (Recall that A is an $n \times n$ matrix.) If this

polynomial—called the *characteristic polynomial* of A—is written out in the usual way, the coefficient of λ^n is 1:

$$\det (\lambda I - A) = \lambda^n + \alpha_1 \lambda^{n-1} + \cdots + \alpha_{n-1} \lambda + \alpha_n. \tag{7.44}$$

2. The eigenvalues of A are the roots (or zeros) of its characteristic polynomial. According to the fundamental theorem of algebra, then, *an $n \times n$ matrix has exactly n eigenvalues, counting multiplicities.*

3. The coefficients of this polynomial can be expressed in terms of the elements of the matrix $A = [a_{ij}]$. Two that are especially easy to exploit are

$$\alpha_1 = -(a_{11} + a_{22} + \cdots + a_{nn}),$$

$$\alpha_n = (-1)^n \det A.$$

(The sum of the diagonal entries of a matrix is called its *trace*: tr $A = a_{11} + a_{22} + \cdots + a_{nn}$.)

4. If A is a real matrix, the coefficients of the characteristic polynomial are real also. In this case the eigenvalues of A are real or complex conjugate pairs.

5. If λ and z are an eigenvalue and associated eigenvector of B and if $A = \alpha I + \beta B$, then $\alpha + \beta \lambda$ is an eigenvalue of A, and z is an associated eigenvector.

Example 2

Below are given a real 4×4 matrix A and $\lambda I - A$.

$$A = \begin{bmatrix} 1 & 1 & 0 & 0 \\ 0 & 1 & 0 & 0 \\ 0 & 0 & 0 & -2 \\ 1 & 0 & 1 & 2 \end{bmatrix},$$

$$\lambda I - A = \begin{bmatrix} \lambda - 1 & -1 & 0 & 0 \\ 0 & \lambda - 1 & 0 & 0 \\ 0 & 0 & \lambda & 2 \\ -1 & 0 & -1 & \lambda - 2 \end{bmatrix}.$$

Expanding the determinant of $\lambda I - A$ by the first row, we find

$$\det (\lambda I - A)$$

$$= (\lambda - 1) \det \begin{bmatrix} \lambda - 1 & 0 & 0 \\ 0 & \lambda & 2 \\ 0 & -1 & \lambda - 2 \end{bmatrix} - (-1) \det \begin{bmatrix} 0 & 0 & 0 \\ 0 & \lambda & 2 \\ -1 & -1 & \lambda - 2 \end{bmatrix}$$

$$= (\lambda - 1)^2 \det \begin{bmatrix} \lambda & 2 \\ -1 & \lambda - 2 \end{bmatrix} = (\lambda - 1)^2 (\lambda^2 - 2\lambda + 2)$$

$$= \lambda^4 - 4\lambda^3 + 7\lambda^2 - 6\lambda + 2.$$

This is the characteristic polynomial of A. It is easy to see that the coefficient of λ^3 is $-4 = -\mathrm{tr}\, A$ and that the "constant term" is $2 = (-1)^4 \det A$. Furthermore, the eigenvalues are

$$\lambda = 1, \quad 1, \quad 1 + i, \quad 1 - i.$$

The eigenvalue 1 has multiplicity 2, and the two complex eigenvalues are conjugates.

About Eigenvectors

1. To each different eigenvalue of A there must correspond at least one eigenvector. If an eigenvalue has multiplicity m as a root of the characteristic polynomial, there *may* be as many as m independent eigenvectors corresponding to it.

2. Eigenvectors corresponding to different eigenvalues are independent.

3. If A is real and λ is a real eigenvalue, the corresponding eigenvector(s) will naturally be real. If λ is a complex eigenvalue and z is a corresponding eigenvector, then $\bar{z}$ is an eigenvector corresponding to $\bar{\lambda}$.

4. It must always be borne in mind that an eigenvector must be nonzero but that any nonzero scalar multiple of an eigenvector is still an eigenvector. It is often convenient to "normalize" an eigenvector so that its elements have a simple form or that the largest element is a 1.

Example 3

We use the matrix A of Example 2. Any eigenvector z corresponding to the eigenvalue 1 must satisfy $(I - A)z = 0$ or

$$\begin{bmatrix} 0 & -1 & 0 & 0 \\ 0 & 0 & 0 & 0 \\ 0 & 0 & 1 & 2 \\ -1 & 0 & -1 & -1 \end{bmatrix} z = 0.$$

Since the coefficient matrix has rank 3, there is only one independent solution, which we normalize to

$$z = \begin{bmatrix} 1 \\ 0 \\ -2 \\ 1 \end{bmatrix}.$$

Note that 1 has multiplicity 2 as a root of the characteristic polynomial but has only one independent eigenvector.

The eigenvector corresponding to the eigenvalue $1 + i$ is a solution of $((1 + i)I - A)z = 0$, or

$$\begin{bmatrix} i & -1 & 0 & 0 \\ 0 & i & 0 & 0 \\ 0 & 0 & 1+i & 2 \\ -1 & 0 & -1 & -1+i \end{bmatrix} z = 0.$$

We find a normalized eigenvector to be as given below. Also, its conjugate $\bar{z}$ is an eigenvector corresponding to the eigenvalue $\bar{\lambda} = 1 - i$:

$$z = \begin{bmatrix} 0 \\ 0 \\ 1-i \\ 1 \end{bmatrix}, \quad \bar{z} = \begin{bmatrix} 0 \\ 0 \\ 1+i \\ 1 \end{bmatrix}.$$

We may summarize the procedure for finding the eigenvalues and eigenvectors of an $n \times n$ matrix A as follows:

1. Find the characteristic polynomial of A, $\det (\lambda I - A)$.
2. Find the eigenvalues of A, which are the roots of the characteristic polynomial.
3. For each eigenvalue λ, find as many independent eigenvectors as possible. That is, find as many independent solutions as possible of $(\lambda I - A)z = 0$.

We note that finding a characteristic polynomial is usually quite a tedious task, and finding its roots may require numerical approximation. On the other hand, finding eigenvectors is relatively easy—just a matter of solving a homogeneous system. As a practical matter, it is advisable to use $A - \lambda I$ instead of $\lambda I - A$ in steps 1 and 3 above; this saves many sign changes and potential errors.

Example 4

Find the eigenvalues and eigenvectors of the matrix

$$A = \begin{bmatrix} 3 & 2 & -2 \\ 2 & 0 & -1 \\ -2 & -1 & 0 \end{bmatrix}.$$

We first find the characteristic polynomial by calculating (expand by the last column)

$$\det (A - \lambda I) = \det \begin{bmatrix} 3-\lambda & 2 & -2 \\ 2 & -\lambda & -1 \\ -2 & -1 & -\lambda \end{bmatrix}$$

$$= -\lambda(-\lambda(3-\lambda)-4) - (-1)(-(3-\lambda)+2\cdot2)$$
$$+(-2)(2(-1)-2\lambda)$$
$$= -\lambda(\lambda^2 - 3\lambda - 4) + (1+\lambda) + 4(1+\lambda)$$
$$= -\lambda^3 + 3\lambda^2 + 9\lambda + 5.$$

By changing the signs in this polynomial we get the characteristic polynomial $\lambda^3 - 3\lambda^2 - 9\lambda - 5$. We check that tr $A = 3$ is indeed the negative of the coefficient of the next-to-highest term.

Now we set about finding eigenvalues. Since the polynomial has integer coefficients, the candidates for integer roots are the factors of the constant term; that is, ±1 and ±5. By trying these we find that -1 is a root. When the characteristic polynomial is divided by $\lambda + 1$, the quotient (without remainder, since $\lambda = -1$ is a root) is found to be $\lambda^2 - 4\lambda - 5$, with roots 5 and -1. Thus the eigenvalues of A are -1 (multiplicity 2) and 5.

Now we must find eigenvectors. First we solve $(A - 5I)z = 0$ to find an eigenvector corresponding to $\lambda = 5$:

$$\begin{bmatrix} -2 & 2 & -2 \\ 2 & -5 & -1 \\ -2 & -1 & -5 \end{bmatrix} z = 0, \qquad z = \begin{bmatrix} -2 \\ -1 \\ 1 \end{bmatrix}.$$

The arbitrary scalar was chosen to give integer components in z. Next we solve $(A - (-1)I)z = 0$:

$$\begin{bmatrix} 4 & 2 & -2 \\ 2 & 1 & -1 \\ -2 & -1 & 1 \end{bmatrix} z = 0.$$

The coefficient matrix is row-reduced to

$$A + I \rightarrow \begin{bmatrix} 1 & \frac{1}{2} & -\frac{1}{2} \\ 0 & 0 & 0 \\ 0 & 0 & 0 \end{bmatrix},$$

a matrix with rank 1. The general solution of $(A + I)z = 0$ thus contains two arbitrary scalars, designated a and b:

$$z = \begin{bmatrix} \frac{1}{2}(a-b) \\ b \\ a \end{bmatrix}.$$

By choosing first $a = 1$, $b = 0$, and then $a = 0$, $b = 1$, we obtain two independent solutions

$$\begin{bmatrix} \frac{1}{2} \\ 0 \\ 1 \end{bmatrix}, \quad \begin{bmatrix} -\frac{1}{2} \\ 1 \\ 0 \end{bmatrix}.$$

These are *both* eigenvectors of A corresponding to the eigenvalue -1.

Exercises

In Exercises 1–14, find the eigenvalues of the given matrix A and corresponding eigenvectors.

1. $\begin{bmatrix} 1 & 2 \\ 2 & 1 \end{bmatrix}$

2. $\begin{bmatrix} 2 & 1 \\ 1 & 2 \end{bmatrix}$

3. $\begin{bmatrix} a & b \\ b & a \end{bmatrix}$

4. $\begin{bmatrix} 1 & 2 \\ -2 & 1 \end{bmatrix}$

5. $\begin{bmatrix} a & b \\ -b & a \end{bmatrix}$

6. $\begin{bmatrix} 1 & 0 \\ 0 & 2 \end{bmatrix}$

7. $\begin{bmatrix} a & 0 \\ 0 & b \end{bmatrix}$

8. $\begin{bmatrix} -1 & 1 \\ 0 & -2 \end{bmatrix}$

9. $\begin{bmatrix} -1 & 0 \\ 1 & -2 \end{bmatrix}$

10. $\begin{bmatrix} 1 & 1 \\ 0 & 1 \end{bmatrix}$

11. $\begin{bmatrix} a & b \\ 0 & c \end{bmatrix} (c \neq a)$

12. $\begin{bmatrix} 0 & 0 & 1 \\ 1 & 0 & 0 \\ 0 & 1 & 0 \end{bmatrix}$

13. $\begin{bmatrix} a & b & b \\ b & a & b \\ b & b & a \end{bmatrix}$

14. $\begin{bmatrix} 1 & 2 & 1 \\ 0 & 1 & 0 \\ 0 & 2 & 1 \end{bmatrix}$

In Exercises 15–20 a matrix and its eigenvalues are given. Find all the eigenvectors corresponding to each eigenvalue.

15. $\begin{bmatrix} -2 & 2 & 0 & 0 \\ 1 & -2 & 1 & 0 \\ 0 & 1 & -2 & 1 \\ 0 & 0 & 2 & -2 \end{bmatrix}$
$(0, -1, -3, -4)$

16. $\begin{bmatrix} 0 & 1 & 1 & 0 \\ 1 & 0 & 0 & 1 \\ 1 & 0 & 0 & 1 \\ 0 & 1 & 1 & 0 \end{bmatrix}$
$(-2, 0, 0, 2)$

17. $\begin{bmatrix} 1 & 0 & -1 & 1 \\ 2 & 3 & 0 & 2 \\ 0 & 0 & 1 & 3 \\ 0 & 0 & 3 & 1 \end{bmatrix}$
$(-2, 1, 3, 4)$

18. $\begin{bmatrix} 5 & -1 & -1 & 1 \\ -1 & 5 & 1 & -1 \\ -1 & 1 & 5 & -1 \\ 1 & -1 & -1 & 5 \end{bmatrix}$
$(4, 4, 4, 8)$

19. $\begin{bmatrix} 3 & 1 & -1 & 1 \\ -1 & 2 & 1 & 2 \\ -1 & 3 & -1 & 3 \\ -1 & 2 & 1 & 2 \end{bmatrix}$
$(4, 4, 0, -2)$

20. $\begin{bmatrix} 0 & 1 & 0 & 0 \\ 0 & 0 & 1 & 0 \\ 0 & 0 & 0 & 1 \\ 1 & 0 & 0 & 0 \end{bmatrix}$
$(1, -1, i, -i)$

21. Suppose that z_1 and z_2 are eigenvectors of a matrix A corresponding to the same eigenvalue λ. Prove that any linear combination $z = c_1 z_1 + c_2 z_2$ is also an eigenvector of A corresponding to λ.

22. Check the statement in Exercise 21, using the matrix of Example 4.

23. Suppose that z_1 and z_2 are eigenvectors of a matrix A corresponding to different eigenvalues λ_1 and λ_2. Prove that z_1 and z_2 are linearly independent matrices.

24. Suppose that A is an upper triangular matrix (that is, all the elements of A below the diagonal are 0). Show that the eigenvalues of A are just the numbers on the diagonal.

25. Use property 5 of eigenvalues to show that the solution of the initial value problem

$$\frac{dx}{dt} = (aI + A)x, \qquad x(0) = q,$$

is $x = e^{at}y$, where y is the solution of

$$\frac{dy}{dt} = Ay, \qquad y(0) = q.$$

26. Suppose that A and B are $m \times m$ matrices. Show that the $2m \times 2m$ matrix

$$M = \begin{bmatrix} A & B \\ B & A \end{bmatrix}$$

has as eigenvectors (and eigenvalues)

$$\begin{bmatrix} u \\ u \end{bmatrix}(\gamma), \quad \begin{bmatrix} v \\ -v \end{bmatrix}(\mu),$$

where $(A + B)u = \gamma u$ and $(A - B)v = \mu v$.

27. Use the results of Exercise 26 to obtain the eigenvalues and eigenvectors of the matrix

$$M = \begin{bmatrix} 0 & 1 & 1 & 0 \\ 3 & 2 & 0 & 1 \\ 1 & 0 & 0 & 1 \\ 0 & 1 & 3 & 2 \end{bmatrix}$$

28. Use the results of Exercise 26 to obtain the eigenvalues of the matrix

$$A = \begin{bmatrix} 0 & 1 & 0 & 0 \\ -\omega^2 & 0 & \lambda^2 & 0 \\ 0 & 0 & 0 & 1 \\ \lambda^2 & 0 & -\omega^2 & 0 \end{bmatrix}$$

which comes from a system of two masses and three springs. (See Section 3.2, Eqs. (3.21) and (3.22): $\omega_1^2 = \omega_2^2 = \omega^2 > \lambda^2$.)

7.4

Homogeneous Systems with Constant Coefficients

We saw at the beginning of Section 7.3 that the matrix function

$$x(t) = e^{\lambda t}z \tag{7.45}$$

(λ a scalar and z a matrix of constants) is a solution of the linear homogeneous system

$$\frac{dx}{dt} = Ax \tag{7.46}$$

(A an $n \times n$ constant matrix) if and only if λ is an eigenvalue of A and z is a corresponding eigenvector. Now we must inquire how to solve an initial value problem

$$\frac{dx}{dt} = Ax, \qquad x(0) = q. \tag{7.47}$$

The answer lies in the fundamental property of solutions of the linear homogeneous system (7.46), the principle of superposition: if $x_1(t)$, $x_2(t), \ldots, x_k(t)$ are solutions of Eq. (7.46), then so is any linear combination of them,

$$x(t) = c_1x_1(t) + c_2x_2(t) + \cdots + c_kx_k(t). \tag{7.48}$$

(The coefficients $c_1, c_2, \ldots, c_k$ must be constants.) The truth of this principle rests on the distributive property of matrix multiplication and elementary properties of differentiation.

Now we see that our eigenvector solutions can be put together in a linear combination to form a more general solution. Next we ask whether a linear combination like Eq. (7.48) includes the solution of the initial value problem (7.49), whatever q may be. The initial condition is satisfied if

$$c_1x_1(0) + c_2x_2(0) + \cdots + c_kx_k(0) = q$$

is true for some choice of the constant coefficients $c_1, c_2, \ldots, c_k$. Such a choice is possible, for arbitrary q, if and only if (1) the number k of solutions is equal to the number n of equations in the system and (2) the n solutions, at the initial time, are independent vectors. A set of solutions of Eq. (7.46) with properties (1) and (2) above is called a *fundamental set of solutions*, and the general linear combination of them is called the *general solution* of Eq. (7.46).

Example 1

Find a fundamental set of solutions of the system

$$\frac{dx}{dt} = \begin{bmatrix} -1 & -1 & 1 \\ -1 & 1 & -1 \\ 1 & -1 & -1 \end{bmatrix} x.$$

The eigenvectors (and their corresponding eigenvalues) are found to be

$$\begin{bmatrix} 1 \\ 1 \\ 1 \end{bmatrix} (-1), \quad \begin{bmatrix} 1 \\ 0 \\ -1 \end{bmatrix} (-2), \quad \begin{bmatrix} 1 \\ -2 \\ 1 \end{bmatrix} (2).$$

From the three eigenvectors and eigenvalues we can form three solutions:

$$x_1(t) = e^{-t} \begin{bmatrix} 1 \\ 1 \\ 1 \end{bmatrix}, \quad x_2(t) = e^{-2t} \begin{bmatrix} 1 \\ 0 \\ -1 \end{bmatrix}, \quad x_3(t) = e^{2t} \begin{bmatrix} 1 \\ -2 \\ 1 \end{bmatrix}.$$

This is our candidate for a fundamental set. Property (2) requires that the three solutions at the initial time, $x_1(0)$, $x_2(0)$, $x_3(0)$, be linearly independent. One way to check is by forming the matrix with these three as columns,

$$\begin{bmatrix} 1 & 1 & 1 \\ 1 & 0 & -2 \\ 1 & -1 & 1 \end{bmatrix},$$

and calculating its determinant, which comes out to be $-6 \neq 0$. Thus the solutions are independent at time 0, and they form a fundamental set.

Example 2

Solve the initial value problem

$$\frac{dx}{dt} = \begin{bmatrix} -1 & -1 & 1 \\ -1 & 1 & -1 \\ 1 & -1 & -1 \end{bmatrix} x, \quad x(0) = \begin{bmatrix} 0 \\ 4 \\ 2 \end{bmatrix}.$$

In Example 1 we found a fundamental set for this system. A general solution is

$$x(t) = c_1 e^{-t} \begin{bmatrix} 1 \\ 1 \\ 1 \end{bmatrix} + c_2 e^{-2t} \begin{bmatrix} 1 \\ 0 \\ -1 \end{bmatrix} + c_3 e^{2t} \begin{bmatrix} 1 \\ -2 \\ 1 \end{bmatrix},$$

where c_1, c_2, c_3 are arbitrary constants.

To satisfy the initial conditions, the coefficients must be chosen so that

$$x(0) = c_1 \begin{bmatrix} 1 \\ 1 \\ 1 \end{bmatrix} + c_2 \begin{bmatrix} 1 \\ 0 \\ -1 \end{bmatrix} + c_3 \begin{bmatrix} 1 \\ -2 \\ 1 \end{bmatrix} = \begin{bmatrix} 0 \\ 4 \\ 2 \end{bmatrix}.$$

Thus the c's are the solution of the algebraic system

$$\begin{bmatrix} 1 & 1 & 1 \\ 1 & 0 & -2 \\ 1 & -1 & 1 \end{bmatrix} \begin{bmatrix} c_1 \\ c_2 \\ c_3 \end{bmatrix} = \begin{bmatrix} 0 \\ 4 \\ 2 \end{bmatrix},$$

which may be solved easily by elimination to yield $c_1 = 2$, $c_2 = -1$, $c_3 = -1$. Thus the solution of the initial value problem is

$$x(t) = 2e^{-t} \begin{bmatrix} 1 \\ 1 \\ 1 \end{bmatrix} - e^{-2t} \begin{bmatrix} 1 \\ 0 \\ -1 \end{bmatrix} - e^{2t} \begin{bmatrix} 1 \\ -2 \\ 1 \end{bmatrix}.$$

The examples make it clear that we can find a fundamental set of solutions of a homogeneous $n \times n$ system $dx/dt = Ax$, provided that A has n independent eigenvectors. The complete description of all such matrices would be too complicated to present here, but we may identify some easily recognized cases.

Theorem 7.1

An $n \times n$ matrix A has n independent eigenvectors if A is a real symmetric matrix or if the eigenvalues of A are all distinct.

The matrix in Example 1 was real symmetric and had three distinct eigenvalues. In the next example we see a matrix that is symmetric with a multiple eigenvalue.

Example 3

Find a fundamental set of solutions of

$$\frac{dx}{dt} = \begin{bmatrix} 1 & 2 & -1 \\ 2 & -2 & 2 \\ -1 & 2 & 1 \end{bmatrix} x.$$

The eigenvalues of the coefficient matrix A are -4, 2, 2. Corresponding to -4 we find eigenvector $[1, -2, 1]^T$. To find two independent eigenvectors corresponding to 2, we row-reduce

$$A - 2I = \begin{bmatrix} -1 & 2 & -1 \\ 2 & -4 & 2 \\ 1 & 2 & -1 \end{bmatrix} \rightarrow \begin{bmatrix} 1 & -2 & 1 \\ 0 & 0 & 0 \\ 0 & 0 & 0 \end{bmatrix}.$$

The latter matrix has rank 1, and the general solution of $(A - 2I)z = 0$ contains two arbitrary parameters: $z = [2b - a, b, a]^T$. We may obtain two independent eigenvectors for this eigenvalue by choosing first $a = 1$, $b = 0$ and then $a = 0$, $b = 1$. The resulting vectors are

$$\begin{bmatrix} -1 \\ 0 \\ 1 \end{bmatrix} \quad \text{and} \quad \begin{bmatrix} 2 \\ 1 \\ 0 \end{bmatrix}.$$

Now a fundamental set may be constructed using the eigenvalues and eigenvectors. It is

$$x_1(t) = e^{-4t} \begin{bmatrix} 1 \\ -2 \\ 1 \end{bmatrix}, \quad x_2(t) = e^{2t} \begin{bmatrix} -1 \\ 0 \\ 1 \end{bmatrix}, \quad x_3(t) = e^{2t} \begin{bmatrix} 2 \\ 1 \\ 0 \end{bmatrix}.$$

It is desirable to check the independence of these solutions at $t = 0$; that is left as an exercise.

In finding a fundamental set for $dx/dt = Ax$ our guiding principle is this: *for each eigenvalue λ, find as many independent solutions as λ's multiplicity as a root of the characteristic polynomial.* Now we examine some further items of interest.

Complex Eigenvalues

We know that a real matrix A may have complex eigenvalues. It is sometimes convenient to work out a real solution to the matrix differential equation, so as to avoid computations with complex numbers. Suppose then that A is a real matrix having the complex number $\lambda = \alpha + i\beta$ and its conjugate $\bar{\lambda} = \alpha - i\beta$ as simple eigenvalues. Then λ is associated with a complex eigenvector z and $\bar{\lambda}$ with $\bar{z}$. We may write

$$z = u + iv,$$

where u and v are real vectors.

The solution of the matrix differential equation

$$\frac{dx}{dt} = Ax$$

that corresponds to λ and z is

$$
\begin{aligned}
x(t) = e^{\lambda t}z &= e^{(\alpha + i\beta)t}(u + iv) \\
&= e^{\alpha t}(\cos \beta t + i \sin \beta t)(u + iv) \\
&= e^{\alpha t}(u \cos \beta t - v \sin \beta t) + ie^{\alpha t}(u \sin \beta t + v \cos \beta t).
\end{aligned} \tag{7.49}
$$

The real and imaginary parts of this solution must themselves be independent solutions of the differential equation. Thus we may replace the two complex solutions

$$ze^{\lambda t}, \quad \bar{z}e^{\bar{\lambda} t}$$

by these two real solutions in the fundamental set:

$$e^{\alpha t}(u \cos \beta t - v \sin \beta t), \quad e^{\alpha t}(u \sin \beta t + v \cos \beta t). \tag{7.50}$$

Example 4

Find a fundamental set for the system

$$
\frac{dx}{dt} = \begin{bmatrix} 0 & 0 & 4 \\ 1 & 0 & -2 \\ 0 & 1 & -1 \end{bmatrix} x.
$$

The coefficient matrix of this differential equation has distinct eigenvalues: $1, -1 + i\sqrt{3}, -1 - i\sqrt{3}$. The eigenvector corresponding to the eigenvalue $-1 + i\sqrt{3}$ is

$$
z = u + iv = \begin{bmatrix} -1 \\ 0 \\ 1 \end{bmatrix} + i\begin{bmatrix} -\sqrt{3} \\ \sqrt{3} \\ 0 \end{bmatrix}.
$$

Thus two independent real solutions of the original equation are

$$
e^{-t}\left(\cos \sqrt{3}t \begin{bmatrix} -1 \\ 0 \\ 1 \end{bmatrix} - \sin \sqrt{3}t \begin{bmatrix} -\sqrt{3} \\ \sqrt{3} \\ 0 \end{bmatrix} \right),
$$

$$
e^{-t}\left(\sin \sqrt{3}t \begin{bmatrix} -1 \\ 0 \\ 1 \end{bmatrix} + \cos \sqrt{3}t \begin{bmatrix} -\sqrt{3} \\ \sqrt{3} \\ 0 \end{bmatrix} \right).
$$

We note, however, that the complex eigenvector solutions are indeed solutions. That is, the real solutions above are merely replacing the complex solutions

$$
e^{(-1+i\sqrt{3})t}\begin{bmatrix} -1 - i\sqrt{3} \\ i\sqrt{3} \\ 1 \end{bmatrix}, \quad e^{(-1-i\sqrt{3})t}\begin{bmatrix} -1 + i\sqrt{3} \\ -i\sqrt{3} \\ 1 \end{bmatrix}.
$$

The completion of the given problem is left as an exericse.

If avoiding calculations with complex numbers is a serious objective, even the real and imaginary parts of the eigenvector $z = u + iv$ corresponding to the eigenvalue $\lambda = \alpha + i\beta$ can be found with real arithmetic. It is necessary only to separate the real and imaginary parts of the equation $(A - \lambda I)z = 0$. The result is two real matrix equations (remember that A is real):

$$(A - \alpha I)u + \beta v = 0,$$
$$-\beta u + (A - \alpha I)v = 0. \tag{7.51}$$

These are to be viewed as $2n$ equations in the $2n$ unknowns, which are the components of u and v.

Deficiency

We noted before that an eigenvalue of A with multiplicity m in the characteristic polynomial may have as many as m independent eigenvectors associated with it. If there are fewer than m eigenvectors associated with an eigenvalue λ, we say that λ is *deficient*. Obviously, the occurrence of one or more deficient eigenvalues will lead to a shortage of fundamental solutions.

Suppose that λ is an eigenvalue of A with multiplicity 2. Our experience with nth-order scalar equations suggests that we seek solutions of

$$\frac{dx}{dt} = Ax$$

in the form of an exponential function times a polynomial matrix:

$$x(t) = (z_1 t + z_2)e^{\lambda t}. \tag{7.52}$$

Here z_1 and z_2 are constant column matrices. Substituting the proposed solution into the differential equation gives

$$z_1 e^{\lambda t} + \lambda(z_1 t + z_2)e^{\lambda t} = A(z_1 t + z_2)e^{\lambda t}.$$

We may cancel the exponential from both sides of this equation and then match the coefficients (which are column matrices) of like powers of t to get

$$Az_1 = \lambda z_1, \tag{7.53}$$
$$Az_2 = \lambda z_2 + z_1. \tag{7.54}$$

Thus z_1 must be the eigenvector of A associated with λ, and z_2 is the solution of Eq. (7.54). (It can be proved that such a solution exists.)

Example 5

Consider the matrix differential equation

$$\frac{dx}{dt} = \begin{bmatrix} 1 & 1 & 0 & 0 \\ 0 & 1 & 0 & 0 \\ 0 & 0 & 0 & -2 \\ 1 & 0 & 1 & 2 \end{bmatrix} x.$$

We found in Example 2 of Section 7.3 that the eigenvalues of the coefficient matrix A are $1, 1, 1 + i, 1 - i$. Corresponding to the eigenvalue 1 (which has multiplicity 2), we have just one eigenvector,

$$z_1 = \begin{bmatrix} 1 \\ 0 \\ -2 \\ 1 \end{bmatrix}$$

Now we should solve Eq. (7.54) or $(A - I)z_2 = z_1$, which is

$$\begin{bmatrix} 0 & 1 & 0 & 0 \\ 0 & 0 & 0 & 0 \\ 0 & 0 & -1 & -2 \\ 1 & 0 & 1 & 1 \end{bmatrix} z_2 = \begin{bmatrix} 1 \\ 0 \\ -2 \\ 1 \end{bmatrix}, \quad z_2 = \begin{bmatrix} 0 \\ 1 \\ 0 \\ 1 \end{bmatrix}.$$

The solution (given above) is easily found by row reduction. Now the two independent solutions of the differential equation that correspond to the eigenvalue 1 are

$$z_1 e^t \quad \text{and} \quad (z_1 t + z_2)e^t \quad \text{or} \quad \begin{bmatrix} 1 \\ 0 \\ -2 \\ 1 \end{bmatrix} e^t, \quad \begin{bmatrix} t \\ 1 \\ -2t \\ 1 + t \end{bmatrix} e^t.$$

If an eigenvalue λ has multiplicity $m > 2$ as an eigenvalue of A, one looks for m solutions having the form

$$x_k(t) = (z_1 t^{k-1} + z_2 t^{k-2} + \cdots + z_k)e^{\lambda t}$$

for $k = 1, 2, \ldots, m$. After inserting this expression into the differential equation and equating like terms we get these equations:

$$Az_1 = \lambda z_1,$$
$$Az_2 = \lambda z_2 + (k - 1)z_1,$$
$$\vdots$$
$$Az_{k-1} = \lambda z_{k-1} + 2z_{k-2},$$
$$Az_k = \lambda z_k + z_{k-1}.$$

If A has just one eigenvector corresponding to λ, this process will lead to the required number of solutions. Otherwise, there are further complications, which we shall not discuss here.

Exercises

In Exercises 1–10, solve the given initial value problem.

1. $\dfrac{dx}{dx} = \begin{bmatrix} 1 & 2 \\ 2 & 1 \end{bmatrix} x, \; x(0) = \begin{bmatrix} 1 \\ 0 \end{bmatrix}$

2. $\dfrac{dx}{dt} = \begin{bmatrix} 0 & 1 \\ -1 & 0 \end{bmatrix} x, \; x(0) = \begin{bmatrix} 1 \\ 1 \end{bmatrix}$

3. $\dfrac{dx}{dt} = \begin{bmatrix} 2 & 4 \\ -1 & -2 \end{bmatrix} x, \; x(0) = \begin{bmatrix} 1 \\ 0 \end{bmatrix}$

4. $\dfrac{dx}{dt} = \begin{bmatrix} -2 & 2 & 0 \\ 1 & -2 & 1 \\ 0 & 2 & -2 \end{bmatrix} x, \; x(0) = \begin{bmatrix} 1 \\ 0 \\ 1 \end{bmatrix}$

5. $\dfrac{dx}{dt} = \begin{bmatrix} 0 & 1 & 1 \\ 1 & 0 & 1 \\ 1 & 1 & 0 \end{bmatrix} x, \; x(0) = \begin{bmatrix} 1 \\ 1 \\ 1 \end{bmatrix}$

6. $\dfrac{dx}{dt} = \begin{bmatrix} 3 & 4 & 4 \\ -2 & -3 & -5 \\ 1 & 2 & 4 \end{bmatrix} x, \; x(0) = \begin{bmatrix} 1 \\ 1 \\ 1 \end{bmatrix}$

7. $\dfrac{dx}{dt} = \begin{bmatrix} 0 & 1 & 1 & 0 \\ 1 & 0 & 0 & 1 \\ 1 & 0 & 0 & 1 \\ 0 & 1 & 1 & 0 \end{bmatrix} x, \; x(0) = \begin{bmatrix} 1 \\ 0 \\ 1 \\ 0 \end{bmatrix}$

8. $\dfrac{dx}{dt} = \begin{bmatrix} 0 & 1 & 0 & 1 \\ 1 & 0 & 1 & 0 \\ 0 & 1 & 0 & 1 \\ 1 & 0 & 1 & 0 \end{bmatrix} x, \; x(0) = \begin{bmatrix} 1 \\ 0 \\ 0 \\ 1 \end{bmatrix}$

9. $\dfrac{dx}{dt} = \begin{bmatrix} -2 & 1 & 0 & 1 \\ 1 & -2 & 1 & 0 \\ 0 & 1 & -2 & 1 \\ 1 & 0 & 1 & -2 \end{bmatrix} x, \; x(0) = \begin{bmatrix} 0 \\ 1 \\ 1 \\ 0 \end{bmatrix}$

10. $\dfrac{dx}{dt} = \begin{bmatrix} 0 & 1 & 0 & 0 \\ 0 & 0 & 1 & 0 \\ 0 & 0 & 0 & 1 \\ 1 & 0 & 0 & 0 \end{bmatrix} x, \; x(0) = \begin{bmatrix} 1 \\ 1 \\ 1 \\ 1 \end{bmatrix}$

In Exercises 11–20, find a general solution of $dx/dt = Ax$ if A is the matrix given. In some cases the eigenvalues of A are given.

11. $\begin{bmatrix} 1 & 2 \\ 2 & 1 \end{bmatrix}$ 12. $\begin{bmatrix} 0 & 1 \\ 1 & 0 \end{bmatrix}$ 13. $\begin{bmatrix} 0 & 1 \\ -1 & 0 \end{bmatrix}$

14. $\begin{bmatrix} 2 & 0 & 0 \\ 0 & 2 & 0 \\ 0 & 0 & 2 \end{bmatrix}$ 15. $\begin{bmatrix} 0 & 1 & 1 \\ 1 & 0 & 1 \\ 1 & 1 & 0 \end{bmatrix}$ 16. $\begin{bmatrix} -2 & 2 & 0 \\ 1 & -2 & 1 \\ 0 & 2 & -2 \end{bmatrix}$

17. $\begin{bmatrix} 5 & 2 & 2 \\ 3 & 2 & 1 \\ -9 & -4 & -3 \end{bmatrix}$ 18. $\begin{bmatrix} 5 & 2 & 2 \\ 2 & 2 & 1 \\ -8 & -4 & -3 \end{bmatrix}$

$\qquad$ (1, 1, 2) $\qquad\qquad\qquad\qquad\qquad$ (1, 1, 2)

19. $\begin{bmatrix} 0 & 1 & 1 & 0 \\ 1 & 0 & 0 & 1 \\ 1 & 0 & 0 & 1 \\ 0 & 1 & 1 & 0 \end{bmatrix}$ 20. $\begin{bmatrix} 0 & 1 & 0 & 0 \\ -1 & 0 & 1 & 0 \\ 0 & 0 & 0 & 1 \\ 1 & 0 & -1 & 0 \end{bmatrix}$

$\qquad$ (−2, 0, 0, 2) $\qquad\qquad\qquad$ (0, 0, ±$i\sqrt{2}$)

21. Using the matrix A of Example 4, find the solution of

$$\frac{dx}{dt} = Ax, \qquad x(0) = \begin{bmatrix} 1 \\ 0 \\ 0 \end{bmatrix}.$$

22. Let A, B be $m \times m$ matrices. Show that the $2m \times 2m$ matrix

$$m = \begin{bmatrix} A & B \\ -B & A \end{bmatrix}$$

has eigenvectors (and eigenvalues)

$$\begin{bmatrix} u \\ iu \end{bmatrix}(\gamma), \quad \begin{bmatrix} v \\ -iv \end{bmatrix}(\mu),$$

where $(A + iB)u = \gamma u$, $(A - iB)v = \mu v$.

7.5
Theory of Linear Systems

The general theory dealing with the solutions of systems of n first-order linear equations lies very close to the corresponding theory of nth-order linear equations but is somewhat simpler. The model equation to which we refer is

$$\frac{dx}{dt} = A(t)x + g(t). \tag{7.55}$$

This is the matrix form of Eq. (7.11) of Section 7.1. It is equally correct to call this a first-order matrix differential equation instead of a system. A *solution* of Eq. (7.55) is a differentiable matrix function $x(t)$ for which the equation is identically satisfied.

A system of linear equations often appears in the context of an initial value problem. The following theorem states easily met conditions that guarantee the existence and uniqueness of solutions.

Theorem 7.2

If $A(t)$ and $g(t)$ are continuous functions throughout some interval $\alpha < t < \beta$ containing t_0, then there is one and only one solution of the initial value problem

$$\frac{dx}{dt} = A(t)x + g(t), \qquad x(t_0) = q, \tag{7.56}$$

on that interval.

As in the case of scalar differential equations, the solution of a nonhomogeneous matrix equation can be found once we have solved the corresponding homogeneous equation

$$\frac{dx}{dt} = A(t)x. \tag{7.57}$$

For this reason we concentrate on the homogeneous equation for the rest of this section. Of course, Theorem 7.2 is applicable when $g(t)$ is identically 0.

The fundamental fact about solutions of the homogeneous equation (7.57) is the superposition principle, as expressed in Theorem 7.3.

Theorem 7.3

If $x_1(t), x_2(t), \ldots, x_k(t)$ are solutions of the homogeneous equation

$$\frac{dx}{dt} = A(t)x,$$

then every linear combination of the form

$$x(t) = c_1 x_1(t) + c_2 x_2(t) + \cdots + c_k x_k(t)$$

(where $c_1, c_2, \ldots, c_k$ are constants) is also a solution.

The proof depends on two elementary facts: that the derivative of a linear combination is the same linear combination of the derivatives and that matrix multiplication is distributive. We have seen many applications of the principle of superposition in our work on systems with constant coefficients.

In dealing with nth-order scalar equations it is necessary to introduce the concept of linearly independent functions. One of the advantages of using first-order systems is that independence of solutions as functions coincides with independence as vectors.

Definition 7.2

Suppose $x_1(t), x_2(t), \ldots, x_k(t)$ are solutions of Eq. (7.57) in some interval $\alpha < t < \beta$. They are *independent* if the only linear combination (with constant coefficients) that satisfies

$$c_1 x_1(t) + c_2 x_2(t) + \cdots + c_k x_k(t) = 0, \qquad \alpha < t < \beta, \tag{7.58}$$

is the one with $c_1 = c_2 = \cdots = c_k = 0$.

Note that only *constant* coefficients are allowed in the linear combination and that the equality holds throughout an interval. The most interesting case is the one in which the number of solutions is equal to the number of equations in the original system (that is, the order of the matrix $A(t)$). Let us suppose that $A(t)$ is an $n \times n$ matrix and that $x_1, x_2, \ldots, x_n$ are n solutions of

$$\frac{dx}{dt} = A(t)x$$

in an interval $\alpha < t < \beta$. In order to test for independence we must decide whether nonzero constants $c_1, c_2, \ldots, c_n$ exist for which

$$c_1 x_1(t) + c_2 x_2(t) + \cdots + c_n x_n(t) = 0, \qquad \alpha < t < \beta.$$

Now by the definition of matrix multiplication the equation above is just the same as the matrix equation

$$(x_1(t), x_2(t), \ldots, x_n(t)) \begin{bmatrix} c_1 \\ c_2 \\ \vdots \\ c_n \end{bmatrix} = 0, \qquad \alpha < t < \beta.$$

(The $n \times n$ coefficient matrix in this equation is built up by putting the x's side by side.) We know that the existence of nonzero solutions of an algebraic system such as the one above hinges on the determinant of the coefficient matrix. Thus we have the following definition and theorem.

Definition 7.3

If $x_1(t), x_2(t), \ldots, x_n(t)$ are solutions of a homogeneous $n \times n$ matrix differential equation, then their *Wronskian* is

$$W(x_1, x_2, \ldots, x_n) = \det[x_1(t), x_2(t), \ldots, x_n(t)].$$

Theorem 7.4

Let $A(t)$ be an $n \times n$ matrix, continuous for $\alpha < t < \beta$, and let $x_1(t), x_2(t), \ldots, x_n(t)$ be solutions of the homogeneous system

$$\frac{dx}{dt} = A(t)x$$

throughout the interval $\alpha < t < \beta$. Then
    1. $x_1(t), x_2(t), \ldots, x_n(t)$ are independent if and only if their Wronskian is nonzero;
    2. their Wronskian is either identically zero or never zero throughout the interval $\alpha < t < \beta$.

Example 1

For the 3×3 system below it is easy to verify that $x_1(t), x_2(t)$, and $x_3(t)$ are three solutions:

$$\frac{dx}{dt} = \begin{bmatrix} 0 & 1 & 0 \\ 0 & 0 & 1 \\ 2 & -1 & 2 \end{bmatrix} x;$$

$$x_1(t) = \begin{bmatrix} \cos t \\ -\sin t \\ -\cos t \end{bmatrix}, \qquad x_2(t) = \begin{bmatrix} \sin t \\ \cos t \\ -\sin t \end{bmatrix}, \qquad x_3(t) = \begin{bmatrix} e^{2t} \\ 2e^{2t} \\ 4e^{2t} \end{bmatrix}.$$

The coefficient matrix is constant, so the interval referred to is $-\infty < t < \infty$. According to Theorem 7.4, we can test for the independence of the solutions by computing their Wronskian:

$$W(x_1, x_2, x_3) = \det \begin{bmatrix} \cos t & \sin t & e^{2t} \\ -\sin t & \cos t & 2e^{2t} \\ -\cos t & -\sin t & 4e^{2t} \end{bmatrix} = 5e^{2t}.$$

Since $5e^{2t}$ is never zero, we conclude that $x_1(t), x_2(t), x_2(t)$ are independent solutions of the 3×3 matrix equation, $-\infty < x < \infty$.

The second part of Theorem 7.4 deserves some amplification. It turns out that the Wronskian is always or never zero because it satisfies a simple differential equation itself.

Theorem 7.5

If $x_1(t), x_2(t), \ldots, x_n(t)$ are solutions of the homogeneous $n \times n$ matrix differential equation

$$\frac{dx}{dt} = A(t)x,$$

then their Wronskian W is a solution of the differential equation

$$\frac{dW}{dt} = (\text{tr } A)W.$$

This theorem justifies our practice, in the case A constant, of checking only the independence of solutions in a fundamental set at $t = 0$. In that case, if the solutions are independent at $t = 0$, they are independent for $-\infty < t < \infty$. For instance, in Example 7.12 we found that the Wronskian W was $5e^{2t}$. But W satisfies $dW/dt = 2W$ (since tr $A = 2$) and is always nonzero if it is ever nonzero. Thus checking at $t = 0$ is enough.

A set of n independent solutions of an $n \times n$ system is called a *fundamental set* of solutions. In effect, such a set provides the *general solution*, in the sense that every solution of the homogeneous system is a linear combination (with constant coefficients) of the solutions in a fundamental set.

Theorem 7.6

Let $A(t)$ be continuous in some interval $\alpha < t < \beta$.
1. There is a fundamental set of solutions of

$$\frac{dx}{dt} = A(t)x \tag{7.59}$$

on the interval $\alpha < t < \beta$.

2. Every solution of the system (7.59) is a linear combination, with constant coefficients, of the solutions in a fundamental set.

Proof: (1) Choose a number t_0 in the interval $\alpha < t < \beta$. Theorem 7.2 guarantees the existence of a solution of system (7.59) that satisfies any given condition at t_0. We need only set up n initial value problems consisting of Eq. (7.59) and initial conditions that will make the independence of the solutions at t_0 obvious. One way to do this is to choose initial conditions

$$x_1(t_0) = \begin{bmatrix} 1 \\ 0 \\ 0 \\ \vdots \\ 0 \end{bmatrix}, \quad x_2(t_0) = \begin{bmatrix} 0 \\ 1 \\ 0 \\ \vdots \\ 0 \end{bmatrix}, \quad \ldots, \quad x_n(t_0) = \begin{bmatrix} 0 \\ 0 \\ 0 \\ \vdots \\ 1 \end{bmatrix}$$

The n matrices that satisfy Eq. (7.59) and these conditions are a fundamental set.

(2) Now if $x(t)$ is any solution of the system (7.59), let us designate $x(t_0)$ as q. Thus $x(t)$ solves the initial value problem

$$\frac{dx}{dt} = A(t)x, \qquad x(t_0) = q. \tag{7.60}$$

On the other hand, here is a linear combination of the solutions in the fundamental set that satisfies exactly the same initial value problem:

$$q_1 x_1(t) + q_2 x_2(t) + \cdots + q_n x_n(t) = (x_1(t), x_2(t), \ldots, x_n(t))q.$$

Theorem 7.2 guarantees that there is just one solution for the initial value problem (7.59), so $x(t)$ must be identical with the solution we have found as a linear combination of the solutions in the fundamental set.

The matrix whose columns are $x_1(t), x_2(t), \ldots, x_n(t)$ appears over and over in this section. For this reason it merits a separate definition.

Definition 7.4

Let $x_1(t), x_2(t), \ldots, x_n(t)$ be a fundamental set of solutions of the $n \times n$ differential equation

$$\frac{dx}{dt} = A(t)x,$$

and let $x(t)$ be the $n \times n$ matrix

$$X(t) = (x_1(t), x_2(t), \ldots, x_n(t)).$$

Then $X(t)$ is called a *fundamental matrix* of the $n \times n$ differential equation.

The major properties of a fundamental matrix are just restatements of theorems and observations of this section. Here are some of them.

1. The Wronskian of $x_1(t), x_2(t), \ldots, x_n(t)$ is $W = \det X(t)$.
2. $X(t)$ is nonsingular as long as $A(t)$ is continuous because W is nonzero as long as $A(t)$ is continuous.
3. $X(t)$ satisfies the matrix differential equation

$$\frac{dX}{dt} = A(t)X$$

because each column of X satisfies $dx/dt = Ax$.

4. If $Y(t)$ is any other fundamental matrix, then $X(t)$ and $Y(t)$ are related by

$$Y(t) = X(t)C,$$

where C is a square, nonsingular, constant matrix. Each column of Y satisfies $dy/dt = Ay$ and thus is a linear combination of columns of X.

5. The general solution of the homogeneous system

$$\frac{dx}{dt} = A(t)x$$

is $x(t) = X(t)c$ where c is an arbitrary constant column. This is a general linear combination of columns of X.

6. The solution of the initial value problem

$$\frac{dx}{dt} = A(t)x, \qquad x(t_0) = q$$

is $x(t) = X(t)X^{-1}(t_0)q$.

Example 3

Below are given a 3×3 system and three solutions:

$$\frac{dx}{dt} = \begin{bmatrix} 2 & -2 & -3 \\ 2 & -3 & -6 \\ -1 & 2 & 4 \end{bmatrix} x;$$

$$x_1(t) = \begin{bmatrix} 3 \\ 0 \\ 1 \end{bmatrix} e^t, \qquad x_2 = \begin{bmatrix} 1 \\ -1 \\ 1 \end{bmatrix} e^t, \qquad x_3 = \begin{bmatrix} -2t \\ 1 - 4t \\ 2t \end{bmatrix} e^t.$$

We test these for independence by computing the Wronskian:

$$W(x_1, x_2, x_3) = \det \begin{bmatrix} 3e^t & e^t & -2te^t \\ 0 & -e^t & (1 - 4t)e^t \\ e^t & e^t & 2te^t \end{bmatrix} = -2e^{3t}.$$

Since the Wronskian is nonzero, the three solutions form a fundamental set, and the corresponding fundamental matrix is

$$X(t) = \begin{bmatrix} 3e^t & e^t & -2te^t \\ 0 & -e^t & (1 - 4t)e^t \\ e^t & e^t & 2te^t \end{bmatrix}.$$

In terms of this fundamental matrix the general solution of the given system is $x(t) = X(t)c$. Now suppose that we are to satisfy also the initial condition

$$x(0) = \begin{bmatrix} 1 \\ 0 \\ 0 \end{bmatrix} = e_1.$$

Then we must choose the column c of constants so that $X(0)c = e_1$, or

$$\begin{bmatrix} 3 & 1 & 0 \\ 0 & -1 & 1 \\ 1 & 1 & 0 \end{bmatrix} \begin{bmatrix} c_1 \\ c_2 \\ c_3 \end{bmatrix} = \begin{bmatrix} 1 \\ 0 \\ 0 \end{bmatrix}.$$

This system is easily solved to find $c_1 = -c_2 = -c_3 = \frac{1}{2}$. Thus the solution of the given initial value problem is

$$x(t) = \begin{bmatrix} 3e^t & e^t & -2te^t \\ 0 & -e^t & (1-4t)e^t \\ e^t & e^t & 2te^t \end{bmatrix} \begin{bmatrix} \frac{1}{2} \\ -\frac{1}{2} \\ -\frac{1}{2} \end{bmatrix}.$$

Exercises

In Exercises 1–4 a constant matrix A is given. Find a fundamental set of solutions and test them for independence.

1. $\begin{bmatrix} 2 & 1 \\ 1 & 2 \end{bmatrix}$ 2. $\begin{bmatrix} 2 & 1 \\ -2 & 0 \end{bmatrix}$ 3. $\begin{bmatrix} 0 & 1 \\ -1 & 2 \end{bmatrix}$ 4. $\begin{bmatrix} 1 & 2 \\ 4 & -1 \end{bmatrix}$

In Exercises 5–8, verify that the given x's are solutions of the differential equation and calculate their Wronskian.

5. $\dfrac{dx}{dt} = \begin{bmatrix} 1+t & 1-2t-t^2 \\ 1 & -1-t \end{bmatrix} x$; $x_1(t) = \begin{bmatrix} t \\ 1 \end{bmatrix} e^{-t}$, $x_2(t) = \begin{bmatrix} 2+t \\ 1 \end{bmatrix} e^t$

6. $\dfrac{dx}{dt} = \begin{bmatrix} \cos 2t & 1-\sin 2t \\ -1-\sin 2t & -\cos 2t \end{bmatrix} x$; $x_1(t) = \begin{bmatrix} \cos t \\ -\sin t \end{bmatrix} e^t$,

$x_2(t) = \begin{bmatrix} \sin t \\ \cos t \end{bmatrix} e^{-t}$

7. $\dfrac{dx}{dt} = \begin{bmatrix} 1 & 1 & 0 \\ 0 & 1 & 1 \\ 0 & 0 & 1 \end{bmatrix} x$; $x_1(t) = \begin{bmatrix} 1 \\ 0 \\ 0 \end{bmatrix} e^t$, $x_2(t) = \begin{bmatrix} t \\ 1 \\ 0 \end{bmatrix} e^t$, $x_3(t) = \begin{bmatrix} t^2 \\ 2t \\ 2 \end{bmatrix} e^t$

8. $\dfrac{dx}{dt} = \begin{bmatrix} -2 & 2 & 0 \\ 1 & -2 & 1 \\ 0 & 2 & -2 \end{bmatrix} x$; $x_1(t) = \begin{bmatrix} 1 \\ 1 \\ 1 \end{bmatrix}$, $x_2(t) = \begin{bmatrix} 1 \\ 0 \\ -1 \end{bmatrix} e^{-2t}$,

$x_3(t) = \begin{bmatrix} 1 \\ -1 \\ 1 \end{bmatrix} e^{-4t}$

9. Verify that $x_1(t)$ and $x_2(t)$ are solutions of $\dot{x} = A(t)x$:

$$A(t) = \begin{bmatrix} \dfrac{1}{t+1} & \dfrac{1}{t-1} \\ \dfrac{1}{t-1} & \dfrac{1}{t+1} \end{bmatrix}; \qquad x_1(t) = (t^2 - 1)\begin{bmatrix} 1 \\ 1 \end{bmatrix},$$

$$x_2(t) = \dfrac{t+1}{t-1}\begin{bmatrix} 1 \\ -1 \end{bmatrix}.$$

(a) Find the Wronskian of x_1 and x_2.
(b) Confirm the conclusion of Theorem 7.5.
(c) Find the longest interval containing $t_0 = 0$ on which x_1 and x_2 form a fundamental set.
(d) Find the solution of $\dot{X} = AX$, $X(0) = I$.

10. Verify that $x_1(t)$ and $x_2(t)$ are solutions of $\dot{x} = A(t)x$:

$$A(t) = \begin{bmatrix} 0 & 1 \\ -2 & 2 \\ t^2 & t \end{bmatrix}; \qquad x_1(t) = \begin{bmatrix} t \\ 1 \end{bmatrix}, \qquad x_2(t) = \begin{bmatrix} t^2 \\ 2t \end{bmatrix}.$$

(a) Find the Wronskian of x_1 and x_2.
(b) Where is the Wronskian zero?
(c) Is Theorem 7.4 violated by the fact that the Wronskian is 0 at some point? Explain.
(d) Show that $x_1(t)$ and $x_2(t)$ are independent for all t.
(e) Show that the condition $c_1 x_1(0) + c_2 x_2(0) = [1, 0]^T$ cannot be met by any choice of c's.

In Exercises 11–16, find the solution of the matrix initial value problem

$$\dfrac{dX}{dt} = AX, \qquad X(0) = I.$$

11. $\begin{bmatrix} 0 & 1 \\ -1 & 0 \end{bmatrix}$

12. $\begin{bmatrix} 0 & 1 \\ 1 & 0 \end{bmatrix}$

13. $\begin{bmatrix} 0 & 1 & 1 \\ 1 & 0 & 1 \\ 1 & 1 & 0 \end{bmatrix}$

14. $\begin{bmatrix} -2 & 2 & 0 \\ 1 & -2 & 1 \\ 0 & 2 & -2 \end{bmatrix}$

15. $\begin{bmatrix} 0 & 0 & 2 & 1 \\ 0 & 0 & 1 & 2 \\ 2 & 1 & 0 & 0 \\ 1 & 2 & 0 & 0 \end{bmatrix}$

16. $\begin{bmatrix} 0 & 1 & 0 & 0 \\ 0 & 0 & 1 & 0 \\ 0 & 0 & 0 & 1 \\ 1 & 0 & 0 & 0 \end{bmatrix}$

7.6

Nonhomogeneous Systems

In this section we obtain some properties of the nonhomogeneous equation

$$\frac{dx}{dt} = A(t)x + g(t) \qquad\qquad (7.61)$$

and develop methods for finding solutions. Recall that we have this theorem from Section 7.5.

Theorem 7.2

If $A(t)$ and $g(t)$ are continuous functions throughout some interval $\alpha < t < \beta$ containing t_0, then there is only one and only one solution of the initial value problem

$$\frac{dx}{dt} = A(t)x + g(t), \qquad x(t_0) = q,$$

on that interval.

Also, in Section 7.5 we postponed the study of the nonhomogeneous equation (7.61) on the grounds that solutions could be found from the solutions of the corresponding homogeneous equation

$$\frac{dx}{dt} = A(t)x. \qquad\qquad (7.62)$$

One of the links between the solutions of Eqs. (7.61) and (7.62) is shown by Theorem 7.7.

Theorem 7.7

A general solution of the nonhomogeneous equation

$$\frac{dx}{dt} = A(t)x + g(t)$$

is the sum of any particular solution and the general solution of the corresponding homogeneous equation

$$\frac{dx}{dt} = A(t)x.$$

This should sound familiar. It is virtually the same as a theorem stated for nth-order scalar equations.

Let us recall now some facts about the solution of the homogeneous system (Eq. 7.62)

$$\frac{dx}{dt} = A(t)x.$$

We assume that $A(t)$ is an $n \times n$ matrix, continuous on some interval $\alpha < t < \beta$ containing t_0, so that Theorem 7.2 applies. We know, then, that a fundamental set of solutions exists, and from any such set $x_1(t), x_2(t), \ldots, x_n(t)$, we can form the fundamental matrix

$$X(t) = (x_1(t), x_2(t), \ldots, x_n(t)). \tag{7.63}$$

For our purposes the two most important properties of the fundamental matrix are the following.

1. $X(t)$ is a solution of the matrix differential equation

$$\frac{dX}{dt} = A(t)X. \tag{7.64}$$

2. $X(t)$ is nonsingular as long as $A(t)$ is continuous.

From these flow the other significant properties of $X(t)$. For instance, we know that the general solution of the system (7.62) is

$$x(t) = X(t)c,$$

where c is a constant $n \times 1$ matrix.

A fundamental matrix is not unique; any other matrix of the form

$$Y(t) = X(t)C, \tag{7.65}$$

where C is nonsingular, is also a fundamental matrix. In the context of an initial value problem, however, there is an especially convenient choice of fundamental matrix. If t_0 is an initial point and $X(t)$ a fundamental matrix, we say that

$$\Phi(t; t_0) = X(t)X^{-1}(t_0) \tag{7.66}$$

is the *Green's function* or *state transition matrix*. It satisfies the initial value problem

$$\frac{d\Phi}{dt} = A(t)\Phi, \qquad \Phi(t_0) = I. \tag{7.67}$$

Evidently, the unique solution of the initial value problem

$$\frac{dx}{dt} = A(t)x, \qquad x(t_0) = q, \tag{7.68}$$

can be given in terms of the state transition matrix as

$$x(t) = \Phi(t; t_0)q. \tag{7.69}$$

Variation of Parameters

The method of variation of parameters, as developed for the nth-order nonhomogeneous equation (Section 3.5), involves some lengthy calculations and unnatural assumptions. The method becomes much neater and simpler for $n \times n$ systems.

In order to solve the nonhomogeneous system

$$\frac{dx}{dt} = A(t)x + g(t) \tag{7.70}$$

we assume that $x(t)$ has the form

$$x(t) = X(t)v(t), \tag{7.71}$$

where $X(t)$ is a fundamental matrix for the associated homogeneous system and $v(t)$ is a variable column matrix. Inserting $x(t)$ in this form into the nonhomogeneous equation (7.70) gives

$$\frac{dX}{dt}v + X\frac{dv}{dt} = A(t)Xv + g(t) \tag{7.72}$$

But X is a solution of the matrix differential equation (7.64). Thus two terms may be cancelled from Eq. (7.72), leaving

$$X(t)\frac{dv}{dt} = g(t).$$

However, $X(t)$ is nonsingular, being a fundamental matrix. Therefore we can solve for the derivative of $v(t)$ and then for $v(t)$:

$$\frac{dv}{dt} = X^{-1}(t)g(t)$$

$$v(t) = \int_{t_0}^{t} X^{-1}(t')g(t')\,dt'. \tag{7.73}$$

(In the latter equation, t' is just a dummy variable.) What we want is x, not v. It is necessary only to multiply on the left by the fundamental matrix $X(t)$:

$$x_p(t) = X(t)v(t)$$

$$= X(t)\int_{t_0}^{t} X^{-1}(t')g(t')\,dt'$$

$$= \int_{t_0}^{t} X(t)X^{-1}(t')g(t')\,dt'. \tag{7.74}$$

As we noted above, the product $X(t)X^{-1}(t')$ is actually $\Phi(t; t')$. Thus we have the following theorem.

Theorem 7.8

Let $A(t)$ and $g(t)$ be continuous for all t in some interval $\alpha < t < \beta$ containing t_0. Then the general solution of

$$\frac{dx}{dt} = A(t)x + g(t)$$

is given by

$$x(t) = X(t)c + \int_{t_0}^{t} \Phi(t; t')g(t')\,dt'. \tag{7.75}$$

The solution of the initial value problem

$$\frac{dx}{dt} = A(t)x + g(t), \qquad x(t_0) = q,$$

is given by

$$x(t) = \Phi(t; t_0)q + \int_{t_0}^{t} \Phi(t; t')g(t')\,dt'. \tag{7.76}$$

The purpose of this theorem is to provide a way of *symbolizing* the solution of an initial value problem. Usually, it is not a convenient way to calculate the solution.

Example 1

Find the general solution of the nonhomogeneous system

$$\frac{dx}{dt} = \begin{bmatrix} 0 & 1 \\ -1 & 0 \end{bmatrix} x + \begin{bmatrix} \cos t \\ 0 \end{bmatrix}.$$

It is easy to construct the fundamental matrix for the homogeneous system:

$$X(t) = \begin{bmatrix} \cos t & \sin t \\ -\sin t & \cos t \end{bmatrix}.$$

In order to find a particular solution of the given nonhomogeneous system we compute

$$v(t) = \int_{0}^{t} X^{-1}(t')g(t')\,dt'$$

$$= \int_{0}^{t} \begin{bmatrix} \cos t' & -\sin t' \\ \sin t' & \cos t' \end{bmatrix} \begin{bmatrix} \cos t' \\ 0 \end{bmatrix} dt'$$

$$= \int_0^t \begin{bmatrix} \cos^2 t' \\ \sin t' \cos t' \end{bmatrix} dt'$$

$$= \frac{1}{2} \begin{bmatrix} t + \sin t \cos t \\ \sin^2 t \end{bmatrix}.$$

Then the particular solution is

$$x_p(t) = X(t)v(t)$$

$$= \frac{1}{2} \begin{bmatrix} \cos t & \sin t \\ -\sin t & \cos t \end{bmatrix} \begin{bmatrix} t + \sin t \cos t \\ \sin^2 t \end{bmatrix}$$

$$= \frac{1}{2} \begin{bmatrix} \sin t + t \cos t \\ -t \sin t \end{bmatrix}.$$

The general solution of the system may now be written as

$$x(t) = x_p(t) + X(t)c,$$

where c is an arbitrary constant 2×1 matrix.

Example 2

Find a particular solution of the nonhomogeneous system with nonconstant coefficients:

$$\frac{dx}{dt} = \begin{bmatrix} -\sin 2t & \cos 2t \\ \cos 2t & \sin 2t \end{bmatrix} x + \begin{bmatrix} \cos t \\ \sin t \end{bmatrix}.$$

This system is unusual in that we can write down a fundamental matrix for the corresponding homogeneous system:

$$X(t) = \begin{bmatrix} \cos t & 2t \cos t - \sin t \\ \sin t & 2t \sin t + \cos t \end{bmatrix}.$$

In order to obtain a particular solution of the nonhomogeneous system we need to find

$$X^{-1}(t) = \begin{bmatrix} 2t \sin t + \cos t & -2t \cos t + \sin t \\ -\sin t & \cos t \end{bmatrix}$$

and then to compute

$$v(t) = \int_0^t X^{-1}(t')g(t') \, dt' = \int_0^t \begin{bmatrix} 1 \\ 0 \end{bmatrix} dt' = \begin{bmatrix} t \\ 0 \end{bmatrix}.$$

Now we determine the particular solution

$$x_p(t) = X(t)v(t) = \begin{bmatrix} t \cos t \\ t \sin t \end{bmatrix}.$$

Undetermined Coefficients

As a practical matter, it is farily difficult to apply the formulas of Theorem 7.8. There are frequent instances of nonhomogeneous systems that can be solved by more elementary means—essentially the same as the method of undetermined coefficients used previously. While simple, the method should be applied only to a nonhomogeneous system

$$\frac{dx}{dt} = Ax + g(t) \tag{7.77}$$

in which the coefficient matrix is *constant*.

As a first simplification, let us observe that we can decompose inhomogeneities, in this sense. Let x_1 and x_2 be solutions of two nonhomogeneous systems

$$\frac{dx_1}{dt} = Ax_1 + g_1(t), \qquad \frac{dx_2}{dt} = Ax_2 + g_2(t).$$

Then the sum $x_3(t) = x_1(t) + x_2(t)$ is a solution of

$$\frac{dx_3}{dt} = Ax_3 + g_1(t) + g_2(t).$$

According to the preceding observation, we are free to break up an inhomogeneity into pieces, find particular solutions for each, and add them together.

As in the scalar case, the method of undetermined coefficients works only for certain simple, but common, forms of inhomogeneities.

Case 1. $g(t) = b$, constant

With a constant inhomogeneity the natural guess for a particular solution is also constant;

$$x_p(t) = k.$$

The requirement that this be a solution of the system

$$\frac{dx}{dt} = Ax + b$$

amounts to the condition that

$$0 = Ak + b \qquad \text{or} \qquad Ak = -b$$

When A is a nonsingular matrix, this equation can certainly be solved as

$$k = -A^{-1}b. \tag{7.78}$$

When A is singular, there may or may not be a solution. If not, one might try

$$x_p = k_1 + k_2t \tag{7.79}$$

with constant columns k_1 and k_2.

Example 3

Find a particular solution of

$$\frac{dx}{dt} = \begin{bmatrix} -2 & 1 & 0 \\ 1 & -2 & 1 \\ 0 & 1 & -2 \end{bmatrix} x + \begin{bmatrix} 1 \\ 0 \\ 1 \end{bmatrix}.$$

The inhomogeneity is constant; the particular solution found by the method of undetermined coefficients is $x_p = k$, where k is the solution of

$$\begin{bmatrix} -2 & 1 & 0 \\ 1 & -2 & 1 \\ 0 & 1 & -2 \end{bmatrix} k + \begin{bmatrix} 1 \\ 0 \\ 1 \end{bmatrix} = 0.$$

It is not necessary to calculate the inverse of the coefficient matrix, as Eq. (7.78) might suggest. Rather, we may just solve the algebraic system above to get

$$x_p = k = \begin{bmatrix} 1 \\ 1 \\ 1 \end{bmatrix}.$$

Case 2. $g(t) = be^{\alpha t}$, b constant

The natural guess in this case is again an exponential

$$x_p(t) = ke^{\alpha t}$$

with k a constant column matrix to be determined. When we put the proposed form for x into the system

$$\frac{dx}{dt} = Ax + be^{\alpha t},$$

we find that k must satisfy the algebraic system

$$\alpha k = Ak + b \qquad \text{or} \qquad (\alpha I - A)k = b.$$

Thus if $\alpha I - A$ is nonsingular—that is, if α is not an eigenvalue of A—then we can find

$$k = (\alpha I - A)^{-1}b. \qquad\qquad\qquad (7.80)$$

Example 4

Find a particular solution of

$$\frac{dx}{dt} = \begin{bmatrix} -1 & 1 \\ -1 & -1 \end{bmatrix} x + \begin{bmatrix} e^{-3t} \\ 0 \end{bmatrix}.$$

The particular solution is guessed in the form

$$x_p = ke^{-3t},$$

and then k must satisfy the algebraic condition

$$-3k = \begin{bmatrix} -1 & 1 \\ -1 & -1 \end{bmatrix} k + \begin{bmatrix} 1 \\ 0 \end{bmatrix} \quad \text{or} \quad \begin{bmatrix} 2 & 1 \\ -1 & 2 \end{bmatrix} k = \begin{bmatrix} -1 \\ 0 \end{bmatrix}.$$

When we solve this system, we obtain the particular solution

$$x_p(t) = \begin{bmatrix} -\frac{2}{5} \\ -\frac{1}{5} \end{bmatrix} e^{-3t}.$$

Case 3. $g(t) = b \cos \beta t + d \sin \beta t$, b and d constant

Again we would guess a particular solution to have the same form:

$$x_p(t) = h \cos \beta t + k \sin \beta t.$$

After substituting x_p in this form into the system

$$\frac{dx}{dt} = Ax + b \cos \beta t + d \sin \beta t$$

we find these algebraic conditions for h and k:

$$\begin{aligned} Ah - \beta k &= -b, \\ \beta h + Ak &= -d. \end{aligned} \tag{7.81}$$

Example 5

Find a particular solution of

$$\frac{dx}{dt} = \begin{bmatrix} 0 & 1 \\ 1 & 0 \end{bmatrix} x + \begin{bmatrix} 1 \\ 0 \end{bmatrix} \cos t.$$

In this equation the inhomogeneity contains the cosine function. Even though the sine is absent, we must seek a particular solution in the form

$$x_p(t) = h \cos t + k \sin t.$$

After substituting this into the given differential equation we get

$$-h \sin t + k \cos t = Ah \cos t + Ak \sin t + \begin{bmatrix} 1 \\ 0 \end{bmatrix} \cos t.$$

On equating the coefficients of like functions we get these equations:

$$\left. \begin{aligned} k &= Ah + \begin{bmatrix} 1 \\ 0 \end{bmatrix} \\ -h &= Ak \end{aligned} \right\} \quad \text{or} \quad \left. \begin{aligned} Ah - k &= -\begin{bmatrix} 1 \\ 0 \end{bmatrix} \\ h + Ak &= 0 \end{aligned} \right\}.$$

These are four equations for the four unknown components of h and k.

The matrix form of the algebraic system above is

$$
\begin{bmatrix}
0 & 1 & -1 & 0 \\
1 & 0 & 0 & -1 \\
1 & 0 & 0 & 1 \\
0 & 1 & 1 & 0
\end{bmatrix}
\begin{bmatrix}
h_1 \\ h_2 \\ k_1 \\ k_2
\end{bmatrix}
=
\begin{bmatrix}
-1 \\ 0 \\ 0 \\ 0
\end{bmatrix}.
$$

Its solution is easily found by elimination:

$$
h = \begin{bmatrix} 0 \\ -\frac{1}{2} \end{bmatrix}, \qquad
k = \begin{bmatrix} \frac{1}{2} \\ 0 \end{bmatrix}.
$$

Finally, the particular solution is

$$
x_p(t) = \begin{bmatrix} 0 \\ -\frac{1}{2} \end{bmatrix} \cos t + \begin{bmatrix} \frac{1}{2} \\ 0 \end{bmatrix} \sin t = \frac{1}{2} \begin{bmatrix} \sin t \\ -\cos t \end{bmatrix}.
$$

Exercises

In Exercises 1–7 you are to find the solution of the nonhomogeneous system

$$
\frac{dx}{dt} = Ax + g(t), \qquad x(0) = 0,
$$

by means of variation of parameters. In each exercise this information is given: A, $X(t)$ (a fundamental matrix), and $g(t)$.

1. $A = \begin{bmatrix} 0 & 1 \\ -1 & 0 \end{bmatrix}$, $X(t) = \begin{bmatrix} \cos t & \sin t \\ -\sin t & \cos t \end{bmatrix}$, $g(t) = \begin{bmatrix} 1 \\ 1 \end{bmatrix}$

2. $A = \begin{bmatrix} 0 & 1 \\ 1 & 0 \end{bmatrix}$, $X(t) = \begin{bmatrix} e^{-t} & e^{t} \\ -e^{-t} & e^{t} \end{bmatrix}$, $g(t) = \begin{bmatrix} 1 \\ 1 \end{bmatrix}$

3. $A = \begin{bmatrix} 0 & 1 \\ -1 & -2 \end{bmatrix}$, $X(t) = \begin{bmatrix} 1 & t \\ -1 & 1-t \end{bmatrix} e^{-t}$, $g(t) = \begin{bmatrix} 0 \\ 1 \end{bmatrix} e^{-t}$

4. $A = \begin{bmatrix} 1 & 1 \\ -1 & -1 \end{bmatrix}$, $X(t) = \begin{bmatrix} 1 & t \\ -1 & 1-t \end{bmatrix}$, $g(t) = \begin{bmatrix} 1 \\ -1 \end{bmatrix}$

5. $A = \begin{bmatrix} -2 & 1 & 0 \\ 1 & -2 & 1 \\ 0 & 1 & -2 \end{bmatrix}$,

$$
X(t) = \begin{bmatrix} 1 & 1 & 1 \\ 0 & \sqrt{2} & -\sqrt{2} \\ -1 & 1 & 1 \end{bmatrix} \begin{bmatrix} e^{-2t} & 0 & 0 \\ 0 & e^{(-2+\sqrt{2})t} & 0 \\ 0 & 0 & e^{(-2-\sqrt{2})t} \end{bmatrix},
$$

$$
g(t) = \begin{bmatrix} 1 \\ 0 \\ -1 \end{bmatrix}
$$

6. $A = \begin{bmatrix} 1 & 1 & 0 \\ 1 & 0 & 1 \\ 0 & 1 & 1 \end{bmatrix}$, $X(t) = \begin{bmatrix} 1 & 1 & 1 \\ 1 & 0 & -2 \\ 1 & -1 & 1 \end{bmatrix} \begin{bmatrix} e^{2t} & 0 & 0 \\ 0 & e^t & 0 \\ 0 & 0 & e^{-t} \end{bmatrix}$,

$g(t) = \begin{bmatrix} 1 \\ 1 \\ 1 \end{bmatrix}$

7. $A = \begin{bmatrix} -1 & 1 & 0 \\ 1 & -2 & 1 \\ 0 & 1 & -1 \end{bmatrix}$, $X(t) = \begin{bmatrix} 1 & 1 & 1 \\ 1 & 0 & -2 \\ 1 & -1 & 1 \end{bmatrix} \begin{bmatrix} 1 & 0 & 0 \\ 0 & e^{-t} & 0 \\ 0 & 0 & e^{-3t} \end{bmatrix}$,

$g(t) = \begin{bmatrix} 1 \\ 0 \\ 0 \end{bmatrix}$

In Exercises 8–15, find a particular solution of the system $dx/dt = Ax + g(t)$ by any means. In each case, A and $g(t)$ are given.

8. $A = \begin{bmatrix} 0 & 1 \\ -1 & 0 \end{bmatrix}$, $g(t) = \begin{bmatrix} 1 \\ 1 \end{bmatrix}$

9. $A = \begin{bmatrix} -1 & 1 \\ -1 & -1 \end{bmatrix}$, $g(t) = \begin{bmatrix} 1 \\ 0 \end{bmatrix} e^{-t}$

10. $A = \begin{bmatrix} 0 & 1 \\ 1 & 0 \end{bmatrix}$, $g(t) = \begin{bmatrix} 1 \\ 0 \end{bmatrix}$

11. $A = \begin{bmatrix} 0 & 1 \\ 1 & 0 \end{bmatrix}$, $g(t) = \begin{bmatrix} \cos t \\ 0 \end{bmatrix}$

12. $A = \begin{bmatrix} 0 & 1 & 0 \\ 1 & 0 & 1 \\ 0 & 1 & 0 \end{bmatrix}$, $g(t) = \begin{bmatrix} 1 \\ 0 \\ 1 \end{bmatrix} (1 - e^{-t})$

13. $A = \begin{bmatrix} -2 & 1 & 0 \\ 1 & -2 & 1 \\ 0 & 1 & -2 \end{bmatrix}$, $g(t) = \begin{bmatrix} 0 \\ 1 \\ 0 \end{bmatrix} t$

14. $A = \begin{bmatrix} 0 & 1 & 0 \\ 0 & 0 & 1 \\ 1 & 0 & 0 \end{bmatrix}$, $g(t) = \begin{bmatrix} 1 \\ 1 \\ 1 \end{bmatrix} \cos t$

15. $A = \begin{bmatrix} 0 & 1 & 0 & 0 \\ 0 & 0 & 1 & 0 \\ 0 & 0 & 0 & 1 \\ 1 & 0 & 0 & 0 \end{bmatrix}$, $g(t) = \begin{bmatrix} 1 \\ 1 \\ 1 \\ 1 \end{bmatrix} \cos t$

16. Eliminate k from Eqs. (7.81) to show that h satisfies

$$(A^2 + \beta^2 I)h = -(Ab + \beta d).$$

17. Find a constant column matrix b for which the system

$$\frac{dx}{dt} = \begin{bmatrix} 0 & 1 \\ 0 & 1 \end{bmatrix} x + be^t$$

has a particular solution of the form $x_p = ke^t$, where k is a constant column matrix.

7.7

Qualitative Behavior

One major reason for studying systems of differential equations with the tools of matrix theory is to gain new insight into the character of solutions. Our purpose in this section is to state some important theorems on the qualitative behavior of solutions of linear systems. In particular, we would like to predict what will happen to solutions as time increases, without actually finding the solutions. To simplify our language, we will say that a matrix function of t *vanishes* if every element has limit 0. For example, of the two matrix functions

$$x_1(t) = \begin{bmatrix} 1/t \\ t^2 e^{-t} \end{bmatrix}, \qquad x_2(t) = \begin{bmatrix} \sin t \\ 1/t \end{bmatrix}$$

the first vanishes as $t \to \infty$, while the second does not.

Many conclusions about nonlinear, nonconstant, or nonhomogeneous systems depend directly on knowledge about a linear, constant homogeneous system such as

$$\frac{dx}{dt} = Ax \tag{7.82}$$

in which A is a constant $n \times n$ matrix. Our previous work (Section 7.4) leads us to believe—and it is indeed true—that every solution of Eq. (7.82) is a linear combination of special solutions having the form

$$x(t) = p(t)e^{\lambda t}, \tag{7.83}$$

where λ is an eigenvalue of A and $p(t)$ is an $n \times 1$ matrix with polynomial elements. (Usually, $p(t)$ is just a constant column matrix: an eigenvector. But it may have degree up to $m - 1$, if m is the multiplicity of λ.) The growth or decay of the components of each such special solution is largely determined by the exponential factor and ultimately by the eigenvalue λ.

If λ is real and negative, the exponential in Eq. (7.83) decays faster than any polynomial can grow; thus the special solution vanishes. When λ is complex, $\lambda = \alpha + i\beta$, the exponential is

$$e^{\lambda t} = e^{\alpha t}(\cos \beta t + i \sin \beta t).$$

If α (the real part of λ) is negative, the exponential makes the special solution disappear. On the other hand, if λ or its real part is 0 or positive, the special solution, Eq. (7.83), will not vanish. These observations are summarized in Theorem 7.9.

Theorem 7.9

Let A be a constant matrix. If every eigenvalue of A has negative real part, then every solution of

$$\frac{dx}{dt} = Ax$$

vanishes as $t \to \infty$.

Example 1

Below are three matrices, each with the general solution of

$$\frac{dx}{dt} = Ax.$$

$$A = \begin{bmatrix} -1 & 1 \\ -1 & -1 \end{bmatrix}, \quad x(t) = c_1 e^{-t} \begin{bmatrix} \cos t \\ -\sin t \end{bmatrix} + c_2 e^{-t} \begin{bmatrix} \sin t \\ \cos t \end{bmatrix},$$

$$A = \begin{bmatrix} -2 & 1 \\ 1 & -2 \end{bmatrix}, \quad x(t) = c_1 e^{-t} \begin{bmatrix} 1 \\ 1 \end{bmatrix} + c_2 e^{-3t} \begin{bmatrix} -1 \\ 1 \end{bmatrix},$$

$$A = \begin{bmatrix} 0 & 1 \\ -1 & 0 \end{bmatrix}, \quad x(t) = c_1 \begin{bmatrix} \cos t \\ -\sin t \end{bmatrix} + c_2 \begin{bmatrix} \sin t \\ \cos t \end{bmatrix}.$$

The first matrix has eigenvalues $-1 \pm i$ and thus satisfies the hypotheses; therefore every solution vanishes. The situation is similar for the second matrix, which has eigenvalues -1 and -3. The third matrix, however, has eigenvalues $\pm i$, with zero real parts; no nontrivial solution vanishes (although this is not promised by the theorem).

Definition 7.5

A constant matrix whose eigenvalues all have negative real parts is called a *stable matrix*.

(The reader should be cautious of the word "stable"; different authors use it for many different properties. The definition above is widely accepted, however.)

Theorem 7.9 might be stated in this shorter way: if A is a stable matrix, every solution of Eq. (7.82) vanishes. A typical application of this Theorem 7.9 is to nonhomogeneous problems, as given in Theorem 7.10.

Theorem 7.10

If A is a stable matrix, the difference between any two solutions of

$$\frac{dx}{dt} = Ax + g(t)$$

vanishes as $t \to \infty$.

By way of proof it is necessary only to note that the difference between two solutions of the nonhomogeneous system is a solution of the corresponding homogeneous system. Since A is stable, this difference must vanish. The point of the theorem is that one solution is much like another, so any one will show the behavior of all of them.

Example 2

The nonhomogeneous system below has a periodic inhomogeneity:

$$\frac{dx}{dt} = \begin{bmatrix} -1 & 1 \\ -1 & -1 \end{bmatrix} x + \begin{bmatrix} \cos t \\ 0 \end{bmatrix}.$$

Since the coefficient matrix was found to be stable in Example 1, Theorem 7.11 assures us that as $t \to \infty$, all solutions approach

$$x_p(t) = \frac{1}{5} \begin{bmatrix} 3\cos t + \sin t \\ -\cos t - 2\sin t \end{bmatrix}.$$

This is the solution found by the easiest method, undetermined coefficients.

Another application of Theorem 7.9 is to the study of solutions of systems with nonconstant coefficients.

Theorem 7.11

Let $A(t)$ and $g(t)$ be continuous for $t \geq t_0$. Suppose that $A(t)$ has a limit as $t \to \infty$,

$$\lim_{t \to \infty} A(t) = A_0,$$

and that A_0 is a stable matrix. Then

1. all solutions of the homogeneous system

$$\frac{dx}{dt} = A(t)x$$

vanish as $t \to \infty$; and

2. the difference between any two solutions of the nonhomogeneous system

$$\frac{dx}{dt} = A(t)x + g(t)$$

vanishes as $t \to \infty$.

Example 3

Consider this system with nonconstant coefficients,

$$\frac{dx}{dt} = \begin{bmatrix} 0 & 1 \\ \dfrac{-2 - t}{t} & \dfrac{-2 - 2t}{t} \end{bmatrix} x.$$

The coefficient matrix, $A(t)$, is continuous for $t \geq 1$ and has the limit

$$A_0 = \begin{bmatrix} 0 & 1 \\ -1 & -2 \end{bmatrix},$$

which is stable. (A_0 has -1 as a double eigenvalue.) The first conclusion of the theorem can be verified by inspection of the general solution,

$$x(t) = c_1 e^{-t} \begin{bmatrix} 1 \\ -1 \end{bmatrix} + c_2 \frac{e^{-t}}{t^2} \begin{bmatrix} t \\ -(1 + t) \end{bmatrix}.$$

In addition, all solutions of the nonhomogeneous equation

$$\frac{dx}{dt} = \begin{bmatrix} 0 & 1 \\ \dfrac{-2 - t}{t} & \dfrac{-2 - 2t}{t} \end{bmatrix} x + \begin{bmatrix} 0 \\ 3 + t \end{bmatrix}$$

approach the particular solution

$$x_p = \begin{bmatrix} t - 1 \\ 1 \end{bmatrix}.$$

A difficulty in applying the theorems above is deciding whether a given constant matrix is stable. Finding the eigenvalues of a matrix is a big job. Several tests have been developed that give sufficient conditions for a matrix to be stable. One of the easiest to apply is this one, a consequence of a theorem of Gershgorin.

Theorem 7.12

A real, constant $n \times n$ matrix $A = [a_{ij}]$ is stable if all n of these inequalities are fulfilled:

$$a_{ii} < - \sum_{j \neq i} |a_{ij}|, \qquad i = 1, 2, \ldots, n.$$

The meaning of the inequalities is just this: in each row of the matrix A, the number on the diagonal must be less than the negative of the sum of the absolute values of the other numbers in that row. Of course, all the numbers on the diagonal of A have to be negative.

Example 4

In order to apply Theorem 7.12 to the matrix

$$A = \begin{bmatrix} -3 & 1 & -1 \\ 2 & -4 & 1 \\ -3 & 1 & -7 \end{bmatrix}$$

we consider the three inequalities

$$-3 < -(|1| + |-1|) = -2,$$
$$-4 < -(|2| + |1|) = -3,$$
$$-7 < -(|-3| + |1|) = -4,$$

coming from the first, second, and third rows, respectively. Since they are all true, we may conclude that A is stable. In fact, the eigenvalues of A are -4, $-5 - \sqrt{10}$, $-5 + \sqrt{10}$, the largest of which is the last, approximately -1.84.

Exercises

In Exercises 1–4, determine the behavior as $t \to \infty$ of the solutions, without finding the general solution (if possible).

1. $\dfrac{dx}{dt} = \begin{bmatrix} -1 & 1 \\ -1 & -1 \end{bmatrix} x$

2. $\dfrac{dx}{dt} = \begin{bmatrix} -2 & 1 \\ 1 & -2 \end{bmatrix} x$

3. $\dfrac{dx}{dt} = \begin{bmatrix} -3 & 1 \\ 1 & -1 \end{bmatrix} x + \begin{bmatrix} 1 \\ 0 \end{bmatrix}$

4. $\dfrac{dx}{dt} = \begin{bmatrix} 1 & -1 \\ 0 & -2 \end{bmatrix} x + \begin{bmatrix} 0 \\ t \end{bmatrix}$

In Exercises 5–7 a particular solution of the system is given. Determine the behavior of the general solution as $t \to \infty$.

5. $\dfrac{dx}{dt} = \begin{bmatrix} te^{-t} - 1 & -e^{-t} \\ 1 & -2 \end{bmatrix} x + \begin{bmatrix} 1 \\ 2t \end{bmatrix}$, $x_p = \begin{bmatrix} 1 \\ t \end{bmatrix}$

6. $\dfrac{dx}{dt} = (1 - e^{-t}) \begin{bmatrix} -2 & -1 \\ 1 & 0 \end{bmatrix} x + \begin{bmatrix} 1 - 2e^{-2t} \\ -1 + e^{-2t} \end{bmatrix}$, $x_p = \begin{bmatrix} 1 + e^{-t} \\ -1 \end{bmatrix}$

7. $\dfrac{dx}{dt} = \dfrac{1}{t} \begin{bmatrix} 1 - t & 1 \\ -1 & 1 - t \end{bmatrix} x - \dfrac{1}{t} \begin{bmatrix} 1 \\ t^2 + 1 \end{bmatrix}$, $x_p = \begin{bmatrix} 1 \\ t \end{bmatrix}$

In Exercises 8–11, test the given matrix for stability by using Theorem 7.13.

8. $\begin{bmatrix} -2 & 1 \\ 1 & -2 \end{bmatrix}$

9. $\begin{bmatrix} -2 & 3 \\ -1 & -4 \end{bmatrix}$

10. $\begin{bmatrix} -2 & 1 & 0 \\ 1 & -2 & 1 \\ 0 & 1 & -2 \end{bmatrix}$

11. $\begin{bmatrix} -4 & 2 & 1 \\ 1 & -4 & 1 \\ 1 & 2 & -4 \end{bmatrix}$

12. Verify that the x_p of Example 3 is *not* a solution of

$$\frac{dx}{dt} = A_0 x + g.$$

13. Find a particular solution of the system (with constant coefficients) stated in Exercise 12.

14. Prove the converse of Theorem 7.9: if every solution of $dx/dt = Ax$ vanishes, then every eigenvalue of A has negative real part.

Notes and References

Matrices are now widely used in applications requiring systems of differential equations or equations of high order. Luenberger's book (1979) contains both mathematics and applications to dynamic systems. Its style is lively, and its applications are interesting.

We have not developed the complete theory of matrices as applicable to differential equations. In particular we have not given full details on deficient eigenvalues. A complete treatment is given by Cullen (1979).

The material in Section 7.7 is drawn from Bellman's book (1953), which contains a vast amount of information about differential equations and systems.

Miscellaneous Exercises

In Exercises 1–11 a square matrix A and a column matrix q are given. (a) Find the eigenvalues and eigenvectors of A. (b) Find the general solution of $dx/dt = Ax$. (c) Find the solution that satisfies the initial condition $x(0) = q$. (Exercises 5 and 6 are difficult.)

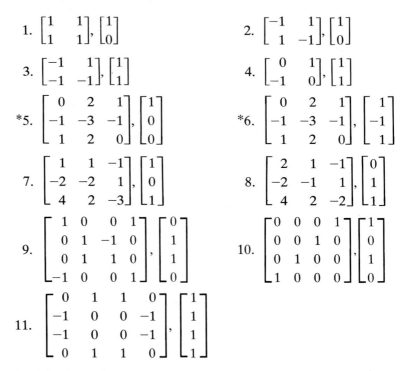

1. $\begin{bmatrix} 1 & 1 \\ 1 & 1 \end{bmatrix}, \begin{bmatrix} 1 \\ 0 \end{bmatrix}$

2. $\begin{bmatrix} -1 & 1 \\ 1 & -1 \end{bmatrix}, \begin{bmatrix} 1 \\ 0 \end{bmatrix}$

3. $\begin{bmatrix} -1 & 1 \\ -1 & -1 \end{bmatrix}, \begin{bmatrix} 1 \\ 1 \end{bmatrix}$

4. $\begin{bmatrix} 0 & 1 \\ -1 & 0 \end{bmatrix}, \begin{bmatrix} 1 \\ 1 \end{bmatrix}$

*5. $\begin{bmatrix} 0 & 2 & 1 \\ -1 & -3 & -1 \\ 1 & 2 & 0 \end{bmatrix}, \begin{bmatrix} 1 \\ 0 \\ 0 \end{bmatrix}$

*6. $\begin{bmatrix} 0 & 2 & 1 \\ -1 & -3 & -1 \\ 1 & 2 & 0 \end{bmatrix}, \begin{bmatrix} 1 \\ -1 \\ 1 \end{bmatrix}$

7. $\begin{bmatrix} 1 & 1 & -1 \\ -2 & -2 & 1 \\ 4 & 2 & -3 \end{bmatrix}, \begin{bmatrix} 1 \\ 0 \\ 1 \end{bmatrix}$

8. $\begin{bmatrix} 2 & 1 & -1 \\ -2 & -1 & 1 \\ 4 & 2 & -2 \end{bmatrix}, \begin{bmatrix} 0 \\ 1 \\ 1 \end{bmatrix}$

9. $\begin{bmatrix} 1 & 0 & 0 & 1 \\ 0 & 1 & -1 & 0 \\ 0 & 1 & 1 & 0 \\ -1 & 0 & 0 & 1 \end{bmatrix}, \begin{bmatrix} 0 \\ 1 \\ 1 \\ 0 \end{bmatrix}$

10. $\begin{bmatrix} 0 & 0 & 0 & 1 \\ 0 & 0 & 1 & 0 \\ 0 & 1 & 0 & 0 \\ 1 & 0 & 0 & 0 \end{bmatrix}, \begin{bmatrix} 1 \\ 0 \\ 1 \\ 0 \end{bmatrix}$

11. $\begin{bmatrix} 0 & 1 & 1 & 0 \\ -1 & 0 & 0 & -1 \\ -1 & 0 & 0 & -1 \\ 0 & 1 & 1 & 0 \end{bmatrix}, \begin{bmatrix} 1 \\ 1 \\ 1 \\ 1 \end{bmatrix}$

In Exercises 12–17 a 2×2 matrix A is given. (a) Find the general solution of $dx/dt = Ax$. (b) If the components of x are interpreted as the coordinates of a point in the x_1x_2-plane, what kind of curve is being traced? These six examples exhibit fundamentally different kinds of behavior, according to relations between eigenvalues: real, distinct, nonzero; real, distinct, one 0; double; double and deficient; imaginary; complex.

12. $\begin{bmatrix} 0 & 1 \\ 1 & 0 \end{bmatrix}$

13. $\begin{bmatrix} 0 & 1 \\ 0 & 1 \end{bmatrix}$

14. $\begin{bmatrix} 1 & 0 \\ 0 & 1 \end{bmatrix}$

15. $\begin{bmatrix} -1 & 0 \\ 1 & -1 \end{bmatrix}$

16. $\begin{bmatrix} 0 & 1 \\ -1 & 0 \end{bmatrix}$

17. $\begin{bmatrix} -1 & 1 \\ -1 & -1 \end{bmatrix}$

Exercises 18–23 refer to Lanchester's model of warfare. It is assumed that each nation's army is worn down at a rate proportional to the size of the opposing army. If $x(t)$ and $y(t)$ are the sizes of the opposing armies, then

$$\frac{dx}{dt} = -ay,$$

$$\frac{dy}{dt} = -bx.$$

(L)

The war is over when one of the two armies is destroyed. Naturally, negative values for x and y are of no interest.

18. Solve system (L) by elimination.

19. Find the relationship between x and y by finding an equation for dy/dx from (L) and solving it.

20. Express system (L) in matrix form and solve it by the eigenvector method.

21. For system (L), show that if parameters and initial conditions are such that $\dot{x}/x = \dot{y}/y$, then $x(t)$ and $y(t)$ both approach 0, and the ratio $y(t)/x(t)$ is constant.

22. What relationship between the initial sizes $x(0)$ and $y(0)$ guarantees that y wins (that is, that $x(t) = 0$ in finite time)?

23. If conditions are such that y will win, how long will it take? (This is a difficult problem.)

24. Richardson's model of an arms race (see Section 7.1) employs the system

$$\frac{dx}{dt} = Ax + g; \qquad A = \begin{bmatrix} -c_1 & d_1 \\ d_2 & -c_2 \end{bmatrix}, \quad g = \begin{bmatrix} g_1 \\ g_2 \end{bmatrix},$$

where the c's, d's, and g's are positive constants. Find a constant particular solution and show that it is positive if and only if $c_1c_2 - d_1d_2$ is positive.

25. Assuming that $c_1c_2 - d_1d_2$ is positive, show that the eigenvalues of A are real and negative.

26. Equation (7.9) of Section 7.1 governs the temperatures in a laminated object. Assume that $C_1 = C_2 = C_3 = C$ (these are heat capacities) and that $k_{01} = k_{34} = k$, $k_{12} = k_{23} = 2k$ (these are conductivities). Find the matrix form for Eq. (7.9) of Section 7.1.

27. Solve the system found in Exercise 26.

28. Let u_1 be the voltage across the capacitor in Fig. 7.4 and let u_2 be the current in the circuit. Then $C\,du_1/dt = u_2$ expresses the voltage–current relation for a capacitor. The Kirchhoff voltage law is $L\,du_2/dt + Ru_2 + u_1 = v(t)$. Combine these two equations to obtain a system $du/dt = Au + g$, and identify A and g.

Figure 7.4

An *RLC* Circuit

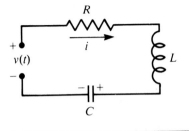

29. Show that the eigenvalues of the matrix A in Exercise 28 are negative or have negative real parts. Under what conditions are the eigenvalues complex?

30. Using u_1 as the current i_1 in Fig. 7.5 and u_2 as the voltage across the

Figure 7.5

A Circuit with Three
Loops

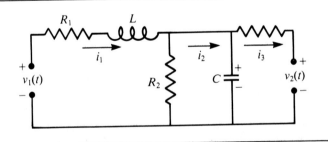

capacitor, these four equations completely specify u_1, u_2, and the other two currents:

$$C\frac{du_2}{dt} = i_2 - i_3,$$

$$L\frac{du_1}{dt} + R_1u_1 + R_2(u_1 - i_2) = v_1(t),$$

$$R_2(u_1 - i_2) = u_2,$$

$$u_2 + R_3i_3 = v_2(t).$$

These are, respectively, the capacitor law and Kirchhoff's voltage law for the three loops. Use the last two equations to eliminate i_2 and i_3 from the first two, leaving two differential equations for u_1 and u_2.

31. Write the equations found for u_1 and u_2 in the form $du/dt = Au + g(t)$.

32. Letting u_1 be the voltage across the capacitor and u_2 the current i_2 in

Figure 7.6

A Circuit with Two
Loops

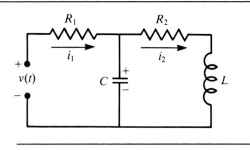

Fig. 7.6, derive this system of equations:

$$\frac{du}{dt} = \begin{bmatrix} \dfrac{-1}{R_1C} & \dfrac{-1}{C} \\[2mm] \dfrac{1}{L} & \dfrac{-R_2}{L} \end{bmatrix} u + \begin{bmatrix} \dfrac{V}{R_1C} \\[2mm] 0 \end{bmatrix}.$$

33. Take $R_1 = 10^3$, $R_2 = 2 \times 10^3$, $C = 10^{-5}$, and $L = 1$ in the circuit of Exercise 32. Find the general solution of the homogeneous system obtained by setting $v(t) \equiv 0$.

34. In Exercises 4 and 5 of Section 7.1 the following system was developed as a model for a national economy:

$$\begin{bmatrix} 1 & 0 \\ -\lambda & 1 \end{bmatrix} \frac{du}{dt} = \begin{bmatrix} -\mu & -\mu r \\ 0 & -\lambda s \end{bmatrix} u + \begin{bmatrix} 0 \\ \lambda A \end{bmatrix}.$$

Transform this system into the usual form and show that the coefficient matrix is stable.

8 Nonlinear Second-Order Equations

8.1

Introduction

In this chapter we study some basic facts about nonlinear second-order differential equations. The emphasis will be on finding qualitative features of the solutions, since, for the most part, exact solutions are impossible to obtain. As an introduction, we derive a few well-known nonlinear equations.

1. In Chapter 2 we derived the equation of motion of a mass–spring–damper system (see Figs. 8.1 and 8.2) under the assumption that the spring and damper acted linearly. However, some dampers act in a decidedly nonlinear fashion. Instead of being proportional to velocity, damping force may be proportional to the square of the velocity, but of the same sign as the velocity:

$$F_d = p\,|v|\,v = p\left|\frac{du}{dt}\right|\frac{du}{dt}. \tag{8.1}$$

Likewise, metal springs that make a fairly small deflection do follow Hooke's law, which says that the spring force is proportional to deflection. Usually, a large deflection leads to a restoring force of a less than proportional magnitude, as shown in Fig. 8.3(b). A spring made of rubber might have a similar characteristic. These are called "soft springs." On the other hand, a "hard spring" requires more than proportional force for larger deflections, as shown in Fig. 8.3c. Several linear springs can be arranged to give a nonlinear hard effect (Fig. 8.4).

A mass–spring–damper system with these nonlinear devices might have as its governing differential equation

$$M\frac{d^2u}{dt^2} + F_d\!\left(\frac{du}{dt}\right) + F_s(u) = 0. \tag{8.2}$$

2. Another simple example of a nonlinear mechanism is the pendulum consisting of a mass M held at distance l from a pivot by a weightless rod. (See Fig. 8.5.) The reference direction is vertical, opposite to the direction of action of gravity. The centripetal force C must balance

Figure 8.1

A Mass–Spring–
Damper System

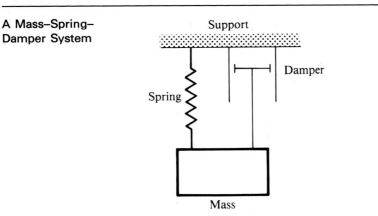

Figure 8.2

Forces on the System

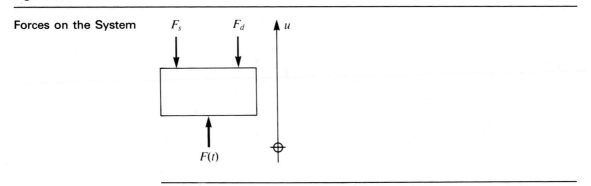

the oppositely directed component of the gravitational force, leaving a force $Mg \sin \theta$ acting at the end of a lever arm of length l to give a restoring torque. Newton's law for rotational systems gives

$$Ml^2 \frac{d^2\theta}{dt^2} + Mgl \sin \theta = 0$$

or

$$\frac{d^2\theta}{dt^2} + \frac{g}{l} \sin \theta = 0 \tag{8.3}$$

as the equation of motion, provided that no other torques come into play.

Figure 8.3

Force Deflection
Curves for (a) Linear
or Hookean Spring;
(b) Soft Spring; (c)
Hard Spring

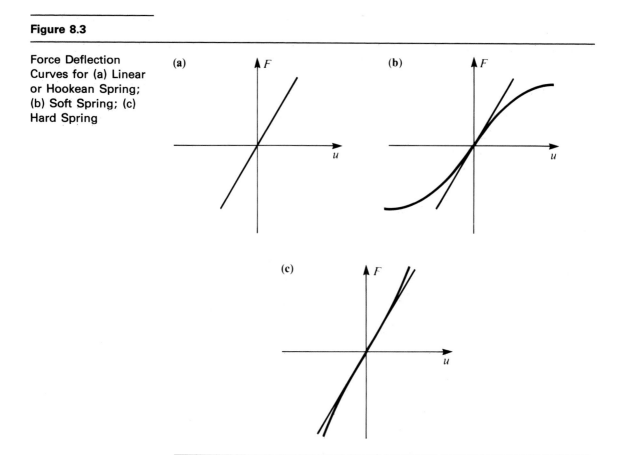

Figure 8.4

Several Linear
Springs Arranged to
Give a "Hard" Effect

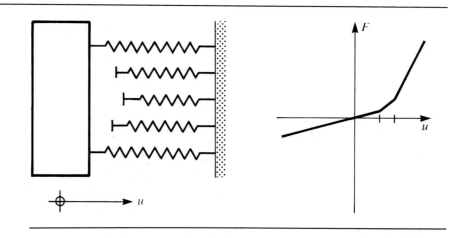

Figure 8.5

Pendulum

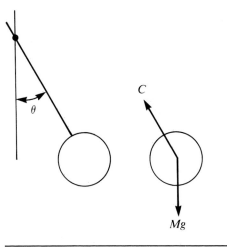

3. The current I in the inner circuit of Fig. 8.6 satisfies the integro-differential equation

$$L\frac{dI}{dt} + RI + \frac{1}{C}\int I\,dt = M\frac{dJ}{dt},\tag{8.4}$$

where L, R, C are inductance, resistance, and capacitance, M is mutual inductance, and J is the current in the outer circuit. The vacuum tube in

Figure 8.6

A Nonlinear Circuit

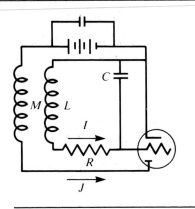

the lower right of Fig. 8.6 has the characteristic

$$J = \sigma\left(E_c - \frac{E_c^3}{3E_s^2}\right),\tag{8.5}$$

where E_s is a saturation voltage and E_c is the voltage drop across the capacitor,

$$E_c = \frac{1}{C}\int I\,dt.\tag{8.6}$$

By various algebraic transformations the equation (8.4) can be simplified to

$$\frac{d^2u}{dt^2} + \varepsilon(u^2 - 1)\frac{du}{dt} + u = 0,\tag{8.7}$$

where ε is a positive parameter and u is proportional to E_c/E_s. Equation (8.7) is called the van der Pol equation.

4. It often happens that the natural way of expressing a relationship is by means of a system of two first-order equations instead of a single second-order equation. A familiar example is the Volterra-Lotka system, already derived in Section 1.5:

$$\frac{dx}{dt} = -mx + axy,$$

$$\frac{du}{dt} = by - kxy.\tag{8.8}$$

In these equations, x and y represent population sizes of predators and prey, respectively, and a, b, m, k are positive constants. Either y or x can be eliminated from the system, leaving one of these nonlinear second-order equations:

$$x\frac{d^2x}{dt^2} - \left(\frac{dx}{dt}\right)^2 = \left(\frac{dx}{dt} + mx\right)x(b - kx),\tag{8.9}$$

$$y\frac{d^2y}{dt^2} - \left(\frac{dy}{dt}\right)^2 = \left(\frac{dy}{dt} - by\right)y(ay - m).\tag{8.10}$$

The second-order equations mentioned above all have one characteristic in common: the independent variable t nowhere appears explicitly. Such an equation is called *autonomous* and is the simplest kind of nonlinear equation. A linear, autonomous equation would have constant coefficients and a constant inhomogeneity, if any. For autonomous equations we have a choice of representations. First is the obvious general representation of a second-order nonlinear equation,

$$\frac{d^2u}{dt^2} = f\left(u, \frac{du}{dt}\right),$$ (8.11)

where f is a given function of two variables. Second, we may convert this second-order equation to a system composed of two first-order equations. One way of doing this is by designating du/dt to be a new variable, v. Then the system

$$\frac{du}{dt} = v,$$

$$\frac{dv}{dt} = f(u, v)$$ (8.12)

is exactly the same as the original second-order equation.

The choice $du/dt = v$ is not the only one that can be made. Furthermore, the system (8.12) is not sufficiently general to include even some simple examples. Therefore we should admit the more general autonomous system

$$\frac{dx}{dt} = g(x, y),$$

$$\frac{dy}{dt} = h(x, y)$$ (8.13)

as a model equation. We have used x and y as the dependent variables to facilitate the geometric interpretations we want to make. If we view $x(t)$ and $y(t)$ in the system (8.13) as the coordinates of a point moving in the xy-plane, the curve traced out, directed by increasing t, is called a *trajectory* of the system. In this interpretation the xy-plane is called the *phase plane*. According to the chain rule, the slope of a trajectory is

$$\frac{dy}{dx} = \frac{dx/dt}{dy/dt} = \frac{h(x, y)}{g(x, y)}.$$ (8.14)

A solution of this first-order differential equation, which shows only the relation between x and y, is called a *first integral* of the system (8.13). A solution curve of (8.14) will be the same as a trajectory, except that it lacks orientation. (If g and h have a common factor, some trajectories may not be given by first integrals.)

Let us recall the fundamental existence theorem from Section 1.7 of Chapter 1. It says that if $h(x, y)/g(x, y)$ is continuous in a region containing a point (x_0, y_0), then there is one and only one solution of

$$\frac{dy}{dx} = \frac{h(x, y)}{g(x, y)} \tag{8.15}$$

through that point. Such a point is called an *ordinary point* of the system (8.13) or of the differential equation (8.15). The theorem says that exactly one trajectory of (8.13) passes through an ordinary point.

Example 1

The second-order differential equation

$$\frac{d^2u}{dt^2} + \omega^2 u = 0,$$

with the definitions $x = u$, $y = du/dt$, becomes the system

$$\frac{dx}{dt} = y, \qquad \frac{dy}{dt} = -\omega^2 x,$$

with solutions

$$x = c_1 \cos \omega t + c_2 \sin \omega t,$$

$$y = -c_1 \omega \sin \omega t + c_2 \omega \cos \omega t.$$

The associated first-order equation is

$$\frac{dy}{dx} = -\omega^2 \frac{x}{y},$$

which has as its solution the family of ellipses

$$y^2 + \omega^2 x^2 = c$$

for $c \geq 0$. Figure 8.7 shows one solution of the system ($c_1 = \frac{1}{2}$, $c_2 = 1$) for $\omega = 2$ and also the corresponding trajectory. Note that many solutions of the system have the same trajectory in the phase plane.

One extremely simple, but important, kind of solution of the autonomous system (8.15) is an *equilibrium solution*

$$x = X, \qquad y = Y,$$

where X and Y are constants. In order for this to be a solution we must have

$$\frac{dx}{dt} = 0 = g(X, Y), \qquad \frac{dy}{dt} = 0 = h(X, Y).$$

Figure 8.7

A Solution and the
Corresponding
Trajectory

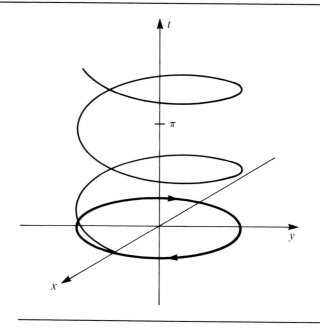

Thus the point in the phase plane corresponding to the constant solution
must be a *critical point* of Eq. (8.15): a point where $dy/dx = 0/0$ is
undefined. More than one trajectory may pass through a critical point. In
the next few sections we shall be concerned with equilibrium solutions
and critical points.

Example 2

The critical points of the Volterra-Lotka system, Eq. (8.8), are the
solutions of the simultaneous equations

$$g(X, Y) = 0: \qquad X(-m + aY) = 0,$$

$$h(X, Y) = 0: \qquad Y(b - kX) = 0.$$

The points are $(0, 0)$ and $(b/k, m/a)$.

Exercises

In Exercises 1–5, turn the second-order differential equation into a
system and find a first integral of the system (that is, solve the corresponding first-order equation).

1. $\dfrac{d^2\theta}{dt^2} + \omega^2 \sin\theta = 0$ \qquad (Pendulum)

2. $\dfrac{d^2u}{dt^2} + u - u^3 = 0$ (Duffing's equation)

3. $\dfrac{d^2u}{dt^2} + \dfrac{u^3 - u}{u^2 + 1} = 0$ (Toggled spring)

4. $\dfrac{d^2u}{dt^2} - \dfrac{2u}{1 - u^2} = 0$ (Steel ball between magnets?)

5. $\dfrac{d^2u}{dt^2} + \left|\dfrac{du}{dt}\right|\dfrac{du}{dt} + u = 0$ (Nonlinear damping)

(Hint: with $x = u$ the system is $dx/dt = y$, $dy/dt = -x \mp y^2$. The equation analogous to Eq. (8.14) is linear in y^2.)

6. If a second-order equation has the form

$$\dfrac{d^2u}{dt^2} + b(u)\dfrac{du}{dt} + c(u) = 0,$$

another way to convert it to a system is to define $x = u$ and $y = (du/dt) + B(u)$, where

$$B(x) = \int_0^x b(z)\, dz.$$

What is the general form of the system that results?

7. Apply the idea of Exercise 6 to van der Pol's equation.

8. In the system below, $g(x, y)$ and $h(x, y)$ have a common factor:

$$\dfrac{dx}{dt} = x + y,$$

$$\dfrac{dy}{dt} = x + y.$$

(a) Find the first integral, the general solution of $dy/dx = 1$, which is analogous to Eq. (8.14). (b) Show that $x(t) = c_1 e^{2t} + c_2$, $y(t) = c_1 e^{2t} - c_2$ is the general solution of the system. (c) Show that the trajectory $y = -x$ (corresponding to $c_1 = 0$, $c_2 \neq 0$) is made up of critical points of the system.

In Exercises 9–12, carefully draw a direction field for the given system. Concentrate on the neighborhoods of critical points.

9. $\dfrac{dx}{dt} = y$

$\dfrac{dy}{dt} = -x + x^2$

10. $\dfrac{dx}{dt} = y$

$\dfrac{dy}{dt} = x$

11. $\dfrac{dx}{dt} = y - y^3$

$\dfrac{dy}{dt} = x$

12. $\dfrac{dx}{dt} = x + y$

$\dfrac{dy}{dt} = y$

8.2
Critical Points of Linear Systems

To set the stage for the study of nonlinear autonomous systems, we first study the linear system with constant coefficients

$$\frac{dx}{dt} = ax + by,$$

$$\frac{dy}{dt} = cx + dy \tag{8.16}$$

and the associated first-order differential equation

$$\frac{dy}{dx} = \frac{cx + dy}{ax + by}. \tag{8.17}$$

The only restriction we make on the coefficients is that $ad - bc$ should be nonzero. If it were 0, Eq. (8.17) would reduce to $dy/dx = \text{const.}$

We are particularly interested in the behavior of solutions near the critical point $(0, 0)$ of the phase plane, corresponding to the trivial solution $x = 0, y = 0$ of the homogeneous system (8.16).

We seek solutions of the system (8.16) that have the special form

$$x = we^{\lambda t}, \qquad y = ze^{\lambda t}, \tag{8.18}$$

where w and z are constants. (This is an extension of the method of Chapter 2 for second-order equations.) Substitution of these functions in to the system (8.16) leads to the equations

$$w\lambda e^{\lambda t} = awe^{\lambda t} + bze^{\lambda t},$$

$$z\lambda e^{\lambda t} = cwe^{\lambda t} + dze^{\lambda t}.$$

After cancellation of the common exponential factor and rearrangement we are left with the two simultaneous homogeneous equations

$$(a - \lambda)w + bz = 0,$$

$$cw + (d - \lambda)z = 0, \tag{8.19}$$

which have a nonzero solution if and only if the determinant is 0.

Equating the determinant to 0 gives the quadratic polynomial equation

$$(a - \lambda)(d - \lambda) - bc = 0$$

or

$$\lambda^2 - (a + d)\lambda + ad - bc = 0. \qquad \textbf{(8.20)}$$

This is called the *characteristic equation* of the system (8.16), and its roots are *characteristic roots*. The most important features of the solution of Eq. (8.16) are dictated by the two characteristic roots. Note that 0 is not a root, since we have required $ad - bc \neq 0$.

Example 1

We stated in Section 8.1 that a second-order autonomous equation could be converted to an autonomous system or to a first-order equation. Let us see how this goes for the equation

$$\frac{d^2x}{dt^2} + \beta\frac{dx}{dt} + \gamma x = 0. \qquad \textbf{(8.21)}$$

First, define dx/dt to be y; then the differential equation becomes

$$\frac{dy}{dt} + \beta y + \gamma x = 0.$$

Thus the system equivalent to Eq. (8.21) is

$$\frac{dx}{dt} = y,$$

$$\qquad \textbf{(8.22)}$$

$$\frac{dy}{dt} = -\gamma x - \beta y,$$

and the associated first-order equation is

$$\frac{dy}{dx} = -\frac{\gamma x + \beta y}{y}. \qquad \textbf{(8.23)}$$

The characteristic equation of this system is

$$\lambda^2 + \beta\lambda + \gamma = 0, \qquad \textbf{(8.24)}$$

which is exactly what we called the characteristic equation of the original second-order equation.

From our experience with second-order equations it is obvious that the solution (8.18) may be increasing or decreasing, oscillatory or not, depending on the positions in the complex plane of the roots of the characteristic equation (8.20). In the rest of this section we study the various possibilities for these roots and the nature of the solutions of the system (8.16) and Eq. (8.17).

Case 1: Real Distinct Roots of Same Sign

Let us call the roots λ_1 and λ_2, with $\lambda_1 > \lambda_2$. The solution of the system (8.16) is

$$x = w_1 e^{\lambda_1 t} + w_2 e^{\lambda_2 t},$$
$$y = z_1 e^{\lambda_1 t} + z_2 e^{\lambda_2 t}. \qquad (8.25)$$

The ratios of w_1 to z_1 and w_2 to z_2 are fixed by the requirement that they must satisfy Eq. (8.19) with λ_1 or λ_2 in place of λ. The actual magnitudes are two free parameters.

If λ_1 and λ_2 are both negative, all solutions of the system approach 0 as t increases. Since λ_1 is the "less negative" of the roots, $e^{\lambda_1 t}$ will be much larger than $e^{\lambda_2 t}$ as t gets large. Thus the ratio y/x will approach z_1/w_1 if either w_1 or z_1 is nonzero; otherwise, y/x will approach z_2/w_2.

In the phase plane we will see all trajectories except two approaching the origin along the line with slope z_1/w_1. The two exceptional ones are halves of the straight line with slope z_2/w_2. The critical point in this case is called a *node*.

Example 2

The system

$$\frac{dx}{dt} = -3x + 4y,$$

$$\frac{dy}{dt} = x - 3y$$

has characteristic equation

$$\lambda^2 + 6\lambda + 5 = 0$$

with roots $\lambda_1 = -1$, $\lambda_2 = -5$. The solutions are

$$x = 2c_1 e^{-t} - 2c_2 e^{-5t},$$
$$y = c_1 e^{-t} + c_2 e^{-5t}.$$

(Comparing with Eq. (8.25), we have $w_1 = 2c_1$, $z_1 = c_1$, etc. The ratio $z_1/w_1 = \frac{1}{2}$, but the magnitudes of w_1 and z_1 are proportional to the arbitrary parameter c_1.) It is clear that as $t \to \infty$, e^{-t} is much greater than

e^{-5t}. Thus the ratio

$$\frac{y}{x} = \frac{c_1 e^{-t} + c_2 e^{-5t}}{2c_1 e^{-t} - 2c_2 e^{-5t}}$$

approaches $\frac{1}{2}$ if $c_1 \neq 0$. If $c_1 = 0$, it approaches $-\frac{1}{2}$. (See Fig. 8.8.)

Figure 8.8

Node, $\lambda_1, \lambda_2 < 0$

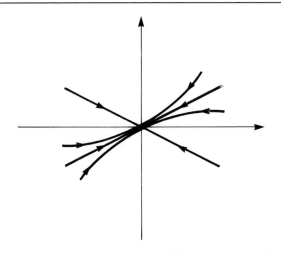

If λ_1 and λ_2 are both positive, then every nontrivial solution of the system (8.16) will approach ∞ as $t \to \infty$. Any solution in the vicinity of the origin will leave it. The geometry of the curves near the origin of the phase plane is just like that for the case where λ_1 and λ_2 are both negative, but with directions on the arrows reversed. The critical point is again called a *node*.

Example 3

The system

$$\frac{dx}{dt} = 3x + y,$$

$$\frac{dy}{dt} = 4x + 3y$$

has characteristic roots $\lambda = 5$, $\lambda_2 = 1$ and the general solution

$$x = c_1 e^{5t} + c_2 e^t,$$
$$y = 2c_1 e^{5t} - 2c_2 e^t.$$

Evidently, as $t \to \infty$, all nontrivial solutions tend to ∞. The geometry near the origin is shown in Fig. 8.9.

Figure 8.9

Node, $\lambda_1, \lambda_2 > 0$

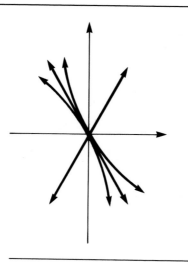

Case 2: Real Roots of Opposite Sign

If λ_1 and λ_2 are of opposite signs ($\lambda_1 > 0 > \lambda_2$), almost all solutions of the system (8.16) approach infinity as $t \to \infty$. The exceptional cases have $w_1 = z_1 = 0$: these approach 0 as $t \to \infty$ and fulfill the relation

$$\frac{y}{x} = \frac{z_2}{w_2}. \tag{8.26}$$

In the phase plane, trajectories coming close to the origin approach near the line (8.26) and leave near the line

$$\frac{y}{x} = \frac{z_1}{w_1}. \tag{8.27}$$

both of these lines are themselves special trajectories. The origin in this case is called a *saddle point* or col.

Example 4

The system

$$\frac{dx}{dt} = y,$$

$$\frac{dy}{dt} = x$$

has characteristic roots ± 1 and the general solution

$$x = c_1 e^t + c_2 e^{-t},$$
$$y = c_1 e^t - c_2 e^{-t}.$$

The phase plane near the origin is shown in Fig. 8.10.

Figure 8.10

Saddle, $\lambda_2 < 0 < \lambda$

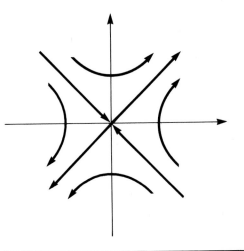

Case 3: Real, Equal Roots

A very special case occurs when $b = c = 0$ and $a = d = \lambda$. Then the system is

$$\frac{dx}{dt} = \lambda x, \qquad \frac{dy}{dt} = \lambda y \tag{8.28}$$

having the general solution

$$x = c_1 e^{\lambda t}, \qquad y = c_2 e^{\lambda t}. \tag{8.29}$$

The two functions x and y are essentially unrelated, and the ratio

$$\frac{y}{x} = \frac{c_2}{c_1} \tag{8.30}$$

is an arbitrary constant. In the phase plane the trajectories are all the straight lines through the origin, which we may call a *star*. If $\lambda > 0$, the trajectories leave the origin; if $\lambda < 0$, the trajectories enter it. (The case $\lambda < 0$ is shown in Fig. 8.11.)

Figure 8.11

Star

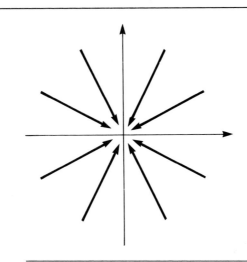

The more common case occurs when either b or c or both are nonzero. In this case the general solution of the system (8.16) is

$$x = (w_1 + w_2t)e^{\lambda t},$$
$$y = (z_1 + z_2t)e^{\lambda t}. \tag{8.31}$$

As t increases, all nontrivial solutions tend to $\pm\infty$ if $\lambda > 0$ or to 0 if $\lambda < 0$. In either case the ratio of y/x has the limit z_2/w_2.

In the phase plane, all trajectories enter the origin (if $\lambda < 0$, or leave it if $\lambda > 0$) along the line

$$\frac{y}{x} = \frac{z_2}{w_2} = \frac{d - a}{2b}. \tag{8.32}$$

The origin is usually called a degenerate node. However, *half-node* would be a better name, since trajectories enter the origin from only two quadrants instead of four, as happens in Case 1.

Example 5

A system that has -2 as a double root is

$$\frac{dx}{dt} = -x + y,$$

$$\frac{dy}{dt} = -x - 3y.$$

The general solution of this system is

$$x(t) = (c_1 + c_2 t)e^{-2t}, \qquad y(t) = (c_2 - c_1 - c_2 t)e^{-2t},$$

where c_1 and c_2 are arbitrary constants. In Fig. 8.12 we see a sketch of some trajectories near the origin. The direction of approach is along the line $y = -x$.

Figure 8.12

Half-node

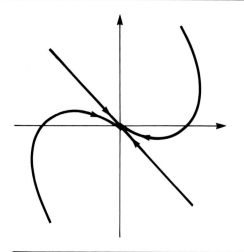

Case 4: Complex Roots

Let us suppose now that our system has two complex conjugate characteristic roots, $\lambda = \alpha \pm i\beta$, $\beta > 0$. The general solution of the system will have the form

$$\begin{aligned} x &= (u_1 \cos \beta t + u_2 \sin \beta t)e^{\alpha t}, \\ y &= (v_1 \cos \beta t + v_2 \sin \beta t)e^{\alpha t}. \end{aligned} \qquad (8.33)$$

We assume for the time being that $\alpha \neq 0$. Then it is clear that all nontrivial solutions approach 0 if $\alpha < 0$ or get far from 0 if $\alpha > 0$.

In the phase plane, trajectories swirl around the origin, which is called a *focus* or spiral point.

Example 6

The system

$$\frac{dx}{dt} = -x + 4y,$$

$$\frac{dy}{dt} = -9x - y$$

has $-1 \pm 6i$ for its characteristic roots and the functions

$$x = (2A \cos 6t + 2B \sin 6t)e^{-t},$$

$$y = (3B \cos 6t - 3A \sin 6t)e^{-t}$$

as the general solution. The trajectories in the phase plane spiral into the origin clockwise. (See Fig. 8.13.)

Figure 8.13

Spiral

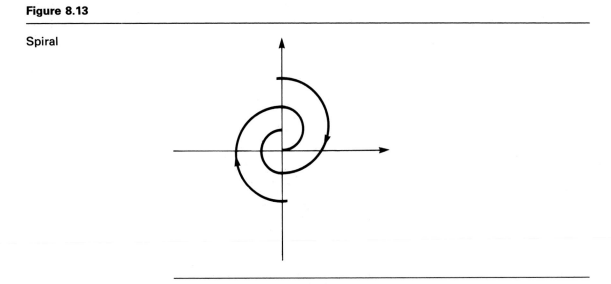

Case 5: Imaginary Roots

When the characteristic roots of the system (8.16) are pure imaginary, $\lambda = \pm i\beta$ ($\beta > 0$), the general solution of the system has the form

$$x = u_1 \cos \beta t + u_2 \sin \beta t,$$

$$y = v_1 \cos \beta t + v_2 \sin \beta t, \tag{8.34}$$

where the u's and v's are (real) constants. These solutions are periodic in time; no solution but the trivial one has any limit, finite or infinite, as t increases.

In the phase plane the trajectories are the ellipses

$$cx^2 - 2axy - by^2 = \text{const} \tag{8.35}$$

surrounding the origin. (Note that c and b have opposite signs and $d = -a$.)

Their orientation is clockwise if b is positive and counterclockwise if b is negative. The origin in this case is called a *center*.

Example 7

The system

$$\frac{dx}{dt} = x - 5y,$$

$$\frac{dy}{dt} = x - y$$

has characteristic roots $\pm 2i$. The general solution of the system is

$$x = 5A \cos 2t + 5B \sin 2t,$$

$$y = (A - 2B) \cos 2t + (2A + B) \sin 2t.$$

In the phase plane the trajectories are the ellipses

$$x^2 - 2xy + 5y^2 = \text{const}$$

with counterclockwise orientation. (See Fig. 8.14.)

Figure 8.14

Center

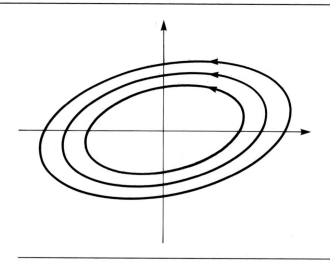

Summary

System: $\dfrac{dx}{dt} = ax + by$

$$ad - bc \neq 0$$

$$\frac{dy}{dt} = cx + dy$$

Associated first-order equation: $\dfrac{dy}{dx} = \dfrac{cx + dy}{ax + by}$

Characteristic equation: $\quad \lambda^2 - (a + d)\lambda + ad - bc$

roots λ_1, λ_2

	Case	Condition on Roots	Name	Figure
Nature of	1	λ_1, λ_2 real, $\lambda_1\lambda_2 > 0$	Node	8.8, 8.9
Trajectories	2	λ_1, λ_2 real, $\lambda_1\lambda_2 < 0$	Saddle	8.10
	3a	$\lambda_1 = \lambda_2$ and $b = c = 0$	Star	8.11
	3b	$\lambda_1 = \lambda_2$ and b or $c \neq 0$	Half-node	8.12
	4	λ_1, λ_2 complex	Focus	8.13
	5	λ_1, λ_2 pure imaginary	Center	8.14

Exercises

In Exercises 1–10 a linear system is given. Find its characteristic roots and classify the origin of the phase plane as node, saddle, etc.

1. $\dfrac{dx}{dt} = 2x + y$

$\dfrac{dy}{dt} = -x + y$

2. $\dfrac{dx}{dt} = x + 2y$

$\dfrac{dy}{dt} = x - y$

3. $\dfrac{dx}{dt} = -3x + y$

$\dfrac{dy}{dt} = -x - y$

4. $\dfrac{dx}{dt} = -3x + y$

$\dfrac{dy}{dt} = x - 3y$

5. $\dfrac{dx}{dt} = -x + 2y$

$\dfrac{dy}{dt} = -5x + y$

6. $\dfrac{dx}{dt} = x + 2y$

$\dfrac{dy}{dt} = 2y + x$

7. $\dfrac{dx}{dt} = y$

$\dfrac{dy}{dt} = x - y$

8. $\dfrac{dx}{dt} = 2x$

$\dfrac{dy}{dt} = 2y$

9. $\dfrac{dx}{dt} = -2x + y$

$\dfrac{dy}{dt} = x - 2y$

10. $\dfrac{dx}{dt} = -2x - y$

$\dfrac{dy}{dt} = x - 2y$

In Exercises 11–18 a linear system is given. (a) Find the characteristic roots; (b) classify the origin of the phase plane; (c) sketch some trajectories near the origin (direction fields may help); (d) solve the associated first-order equation. These are about the simplest examples of the various cases.

11. $\dfrac{dx}{dt} = -x$

$\dfrac{dy}{dt} = -2y$

12. $\dfrac{dx}{dt} = x$

$\dfrac{dy}{dt} = -y$

13. $\dfrac{dx}{dt} = -x + y$

$\dfrac{dy}{dt} = -x - y$

14. $\dfrac{dx}{dt} = y$

$\dfrac{dy}{dt} = -x$

15. $\dfrac{dx}{dt} = -x$

$\dfrac{dy}{dt} = x - y$

16. $\dfrac{dx}{dt} = y$

$\dfrac{dy}{dt} = x$

17. $\dfrac{dx}{dt} = x$

$\dfrac{dy}{dt} = y$

18. $\dfrac{dx}{dt} = x$

$\dfrac{dy}{dt} = \mu y \; (\mu > 0)$

19. Suppose the system (8.16) has characteristic roots λ_1, λ_2. Show that Eq. (8.17) has two straight-line solutions ($y = mx$) if λ_1 and λ_2 are real and distinct; one if $\lambda_1 = \lambda_2$; none if λ_1 and λ_2 are complex.

20. The origin is a node for the system

$$\frac{dx}{dt} = -x + y,$$

$$\frac{dy}{dt} = \varepsilon x - y$$

if $1 > \varepsilon > 0$, a half-node if $\varepsilon = 0$, or a spiral if $\varepsilon < 0$. Check this statement by finding the characteristic roots in the three cases and by sketching trajectories near the origin for $\varepsilon = 0, 0.5, -0.5$. (Direction fields and Exercise 19 will help.)

21. Investigate the case $ad - bc = 0$, using the system

$$\frac{dx}{dt} = -x + y,$$

$$\frac{dy}{dt} = 2x - 2y.$$

(a) Find and solve the equation for dy/dx. (b) Substitute $y(x)$ into the first equation in the system and solve for $x(t)$. (c) Now obtain $y(t)$. (d) Show that $x(t)$ and $y(t)$ both have a limit as $t \to \infty$. (e)

Sketch the phase plane using the information above. (f) Does the equation found in part (a) have a critical point?

8.3

Stability by Linear Comparison

Many systems of physical importance are governed by two nonlinear autonomous equations:

$$\frac{dx}{dt} = g(x, y),$$

$$\frac{dy}{dt} = h(x, y) \tag{8.36}$$

that have an equilibrium solution. Recall that this is a constant solution, $x(t) = X$, $y(t) = Y$, where the constants X and Y necessarily satisfy

$$g(X, Y) = 0, \qquad h(X, Y) = 0. \tag{8.37}$$

In the phase plane, X and Y are coordinates of a critical point, which we assume to be *isolated*. That is, there is some circle centered at (X, Y) that contains no other critical point. The question of interest—and the one we consider in this section and the next—is the stability of the equilibrium solution. Roughly stated, an equilibrium solution is stable if any solution that starts close enough to it stays close to it. In precise terms, we have the following definition.

Definition 8.1

An equilibrium solution $x(t) = X$, $y(t) = Y$ of the system (8.36) is *stable* if, given $\varepsilon > 0$, there is a $\delta > 0$ such that every solution satisfying

$$\sqrt{(x(0) - X)^2 + (y(0) - Y)^2} < \delta$$

also satisfies the inequality

$$\sqrt{(x(t) - X)^2 + (y(t) - Y)^2} < \varepsilon$$

for all $t > 0$. If an equilibrium solution is stable, we say that the corresponding critical point is stable also.

The definition is easiest to understand in the phase plane: the left-hand sides of the inequalities are distances from a trajectory to (X, Y). At a stable critical point, one may specify a tolerance ε. Then any trajectory that starts "close enough" to the critical point stays within the given tolerance forever. (See Figures 8.15(a) and 8.15(b).)

Figure 8.15

(a) Circles of Radii ε (larger) and δ about an Equilibrium Solution; (b) Stability Requires Every Solution Starting within the δ-Circle to Stay in the ε-Circle; (c) Asymptotic Stability Requires Every Solution Starting within the δ- Circle Approach Equilibrium

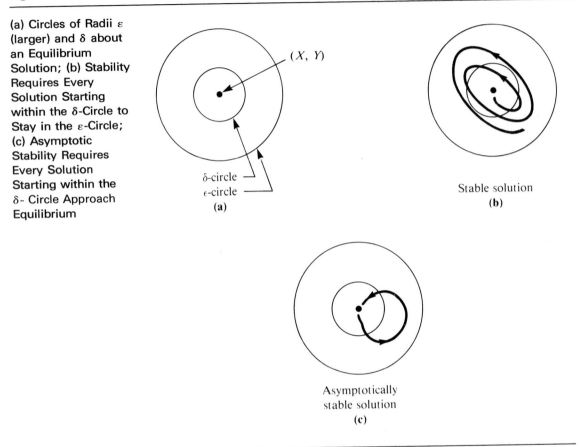

In physical terms, stability means that a sufficiently small deviation from equilibrium will not grow out of bounds. While this is usually a desirable situation, it is even more desirable if every sufficiently small deviation from equilibrium actually returns to equilibrium. (See Fig. 8.15(c).)

Definition 8.2

An equilibrium solution X, Y of the system (8.36) is *asymptotically stable* if there is a $\delta > 0$ such that every solution satisfying

$$\sqrt{(x(0) - X)^2 + (y(0) - Y)^2} < \delta$$

also satisfies the limits:

$$\lim_{t\to\infty} x(t) = X, \qquad \lim_{t\to\infty} y(t) = Y.$$

We also say that the critical point (X, Y) is asymptotically stable.

Example 1

In Section 8.2 we investigated the behavior as $t \to \infty$ of the system

$$\frac{dx}{dt} = ax + by,$$

$$\frac{dy}{dt} = cx + dy$$

$(ad - bc \neq 0)$, which has an isolated equilibrium solution $x = 0$, $y = 0$. When the characteristic roots λ_1, λ_2 are real and distinct, the general solution is

$$x(t) = w_1 e^{\lambda_1 t} + w_2 e^{\lambda_2 t},$$

$$y(t) = z_1 e^{\lambda_1 t} + z_2 e^{\lambda_2 t}.$$

If λ_1 and λ_2 are both negative, all solutions approach 0 as $t \to \infty$. Thus $(0, 0)$ is asymptotically stable (the origin is an asymptotically stable node in the phase plane). If one or both of the characteristic roots are positive, the origin is not stable (saddle or unstable node). Similarly, if the characteristic equation has a double root λ, the origin is asymptotically stable if $\lambda < 0$ and unstable if $\lambda > 0$.

For complex roots $\lambda = \alpha \pm i\beta$, $(0, 0)$ is asymptotically stable if $\alpha = \text{Re}(\lambda) < 0$ (stable spiral point) and unstable if $\alpha > 0$. In the very special case $\alpha = 0$, the roots are imaginary and the origin is stable, but not asymptotically stable.

The situation for a linear homogeneous system

$$\frac{dx}{dt} = ax + by,$$

$$\frac{dy}{dt} = cx + dy$$

can be summarized this way. The origin is asymptotically stable if $ad - bc > 0$ and $a + d < 0$, stable if $ad - bc > 0$ and $a + d = 0$, and unstable otherwise. (See Fig. 8.16.)

For the nonlinear system (8.36) with equilibrium solution $x(t) = X$, $y(t) = Y$ we must consider small deviations from equilibrium. Let us then

Figure 8.16

Stability and Type of
Critical Point at the
Origin for a Linear
Homogeneous
System as Functions
of $t = a + d$ and $\delta = ad - bc$

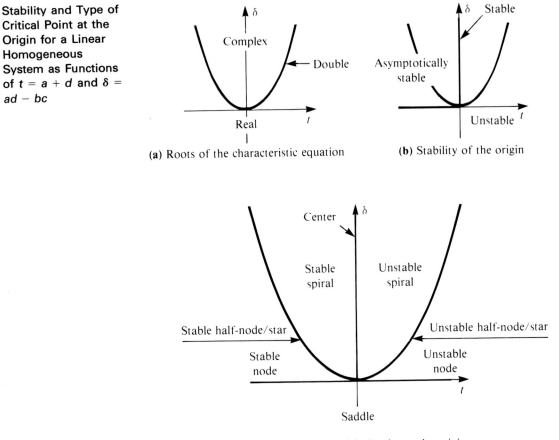

(a) Roots of the characteristic equation

(b) Stability of the origin

(c) Nature of the critical point at the origin

change variables from x and y to the *perturbations*

$$x^* = x - X, \qquad y^* = y - Y. \tag{8.38}$$

In terms of these new variables the system (8.36) becomes

$$\frac{dx^*}{dt} = g(x^* + X, y^* + Y),$$

$$\frac{dy^*}{dt} = h(x^* + X, y^* + Y).$$

Thus the critical point has moved to the origin of the x^*y^*-plane: when

x^* and y^* are 0, both of the right-hand sides in the system are 0 also.

We may determine the dominant behavior near the critical point by expanding g and h in two-variable Taylor polynomials:

$$g(x^* + Y, y^* + Y)$$
$$= g(X, Y) + \frac{\partial g}{\partial x}(X, Y)x^* + \frac{\partial g}{\partial y}(X, Y)y^* + P(x^*, y^*), \quad \textbf{(8.39)}$$

$$h(x^* + X, y^* + Y)$$
$$= h(X, Y) + \frac{\partial h}{\partial x}(X, Y)x^* + \frac{\partial h}{\partial y}(X, Y)y^* + Q(x^*, y^*). \quad \textbf{(8.40)}$$

Here P and Q are functions that contain higher-order terms in x^* and y^*.

By definition, $g(X, Y)$ and $h(X, Y)$ are both 0. Also note that the partial derivatives are evaluated at X, Y and are therefore constants. We shall name them

$$\frac{\partial g}{\partial x}(X, Y) = a, \qquad \frac{\partial g}{\partial y}(X, Y) = b, \qquad \textbf{(8.41)}$$

$$\frac{\partial h}{\partial x}(X, Y) = c, \qquad \frac{\partial h}{\partial y}(X, Y) = d.$$

The system in x^* and y^*, which is fully equivalent to the original system (8.36), has become

$$\frac{dx^*}{dt} = ax^* + by^* + P(x^*, y^*),$$

$$\frac{dy^*}{dt} = cx^* + by^* + Q(x^*, y^*). \qquad \textbf{(8.42)}$$

Now if x^* and y^* are small, their products, squares, and higher-order combinations will be much smaller. Thus it is reasonable to expect that the behavior of the system (8.42) for small x^* and y^* will be essentially the same as that of the *linear comparison system*

$$\frac{dx}{dt} = ax + by,$$

$$\frac{dy}{dt} = cx + dy \qquad \textbf{(8.43)}$$

with a, b, c, and d given by Eq. (8.41). The expectation is justified, as Theorem 8.1 states.

Theorem 8.1

Suppose that the second partial derivatives of $g(x, y)$ and $h(x, y)$ are continuous near the critical point (X, Y) of Eq. (8.36). Then the critical point is (1) asymptotically stable if the origin is an asymptotically stable critical point of the linear comparison system, (2) unstable if the origin is an unstable critical point of the linear comparison system. No conclusion can be drawn if $ad - bc = 0$ or if the origin is merely a stable critical point of the linear comparison system.

Example 2

The van der Pol equation (see Section 8.1),

$$\frac{d^2u}{dt^2} - \varepsilon(1 - u^2)\frac{du}{dt} + u = 0,$$

can be converted to the nonlinear system

$$\frac{dx}{dt} = y,$$

$$\frac{dy}{dt} = -x + \varepsilon(1 - x^2)y$$

($x = u$, $y = du/dt$). There is just one critical point, $X = 0$, $Y = 0$. In this case the perturbations x^*, y^* are the same as the original variables, so Eq. (8.42) are

$$\frac{dx}{dt} = y,$$

$$\frac{dy}{dt} = -x + \varepsilon y - \varepsilon x^2 y.$$

That is, $P(x, y) = 0$, $Q(x, y) = -\varepsilon x^2 y$. The linear comparison system is

$$\frac{dx}{dt} = y,$$

$$\frac{dy}{dt} = -x + \varepsilon y,$$

which has complex characteristic roots $\lambda = \frac{1}{2}\varepsilon \pm i\sqrt{1 - (\frac{1}{2}\varepsilon)^2}$. Thus the origin is an unstable critical point for both the linear comparison system and the original, nonlinear system.

Example 3

The nonlinear pendulum equation (see Section 8.1)

$$\frac{d^2\theta}{dt^2} + \omega^2 \sin\theta = 0$$

is equivalent to the nonlinear system

$$\frac{dx}{dt} = y,$$

$$\frac{dy}{dt} = -\omega^2 \sin x.$$

There are infinitely many critical points: $Y = 0$ and $X = 0, \pm\pi, \pm2\pi, \ldots$.

First, at the critical point $X = 0$, $Y = 0$ the linear comparison system has coefficients

$$a = 0, \qquad b = 1,$$

$$c = \frac{d}{dx}(-\omega^2 \sin x)\Big|_{x=0} = -\omega^2, \qquad d = 0.$$

That is, the comparison system is

$$\frac{dx}{dt} = y,$$

$$\frac{dy}{dt} = -\omega^2 x.$$

The characteristic roots of this system are pure imaginary, $\lambda = \pm\omega i$. Theorem 8.1 provides no information about the critical point $(0, 0)$ of the nonlinear system.

Second, at the critical point $X = \pi$, $Y = 0$ the linear comparison system has coefficients

$$a = 0, \qquad b = 1,$$

$$c = \frac{d}{dx}(-\omega^2 \sin x)\Big|_{x=\pi} = \omega^2, \qquad d = 0.$$

Now the linear comparison system

$$\frac{dx}{dt} = y,$$

$$\frac{dy}{dt} = \omega^2 x$$

has characteristic roots $\lambda = \pm\omega$. The origin is an unstable critical point of the linear comparison system, and $(\pi, 0)$ is likewise an unstable critical point of the original nonlinear system. The remaining critical points are similar to one or the other of these two.

The fact is that even more is true than what Theorem 8.1 says. Except in borderline cases, the *geometry* of the trajectories of the system (8.36) is the same as that of the linear comparison system.

Theorem 8.2

Under the same hypotheses as Theorem 8.1 the nature of the critical point X, Y of the nonlinear is determined by the nature of the critical point $(0, 0)$ of the linear comparison system as follows:

Linear Comparison System	*Nonlinear Systems*
Spiral	Spiral
Center	Spiral or center
Half-node or star	Half-node, star, node, or spiral
Node	Node
Saddle	Saddle

Example 4

We apply Theorem 8.2 to the systems and critical points of Examples 2 and 3.
(a) Van der Pol equation:

$$\frac{dx}{dt} = y,$$

$$\frac{dy}{dt} = -x + \varepsilon(1 - x^2)y. \qquad \text{(NL)}$$

Critical point $(0, 0)$; linear comparison system:

$$\frac{dx}{dt} = y,$$

$$\frac{dy}{dt} = -x + \varepsilon y. \qquad \text{(LCS)}$$

Characteristic roots: $\lambda = \frac{1}{2}\varepsilon \pm i\sqrt{1 - (\frac{1}{2}\varepsilon)^2}$. The critical point of (LCS) is an unstable spiral point. The same is true of (NL).
(b) Pendulum equation:

$$\frac{dx}{dt} = y,$$

$$\frac{dy}{dt} = -\omega^2 \sin x. \qquad \text{(NL)}$$

(i) Critical point $(0, 0)$; linear comparision system:

$$\frac{dx}{dt} = y,$$

$$\frac{dy}{dt} = -\omega^2 x. \tag{LCS}$$

The origin is a center for (LCS). The origin is either a center or a spiral point for (NL). (In fact, it turns out to be a center.)
(ii) Critical point $(0, \pi)$; linear comparison system:

$$\frac{dx}{dt} = y,$$

$$\frac{dy}{dt} = \omega^2 x. \tag{LCS}$$

The origin is a saddle point for (LCS); therefore $(0, \pi)$ is a saddle point for (NL).

Exercises

In Exercises 1–5, identify and classify all the critical points of the given system

1. $\dfrac{dx}{dy} = y$

    $\dfrac{dy}{dt} = -x + x^3$

2. $\dfrac{dx}{dt} = y + 2x + x^2$

    $\dfrac{dy}{dt} = y - 2x - x^2$

3. $\dfrac{dx}{dt} = -x + xy$

    $\dfrac{dy}{dt} = 2y - y^2 - xy$

4. $\dfrac{dx}{dt} = -y$

    $\dfrac{dy}{dt} = \dfrac{x - x^3}{1 + x^2}$

5. $\dfrac{dx}{dt} = x + y - x^3 + xy^2$

    $\dfrac{dy}{dt} = -x + y - x^2 y - y^3$

8.4
Stability by the Direct Method

The linear comparison method of Section 8.3 has two drawbacks. First, it can provide information only about trajectories very close to a critical point—close enough that the nonlinear terms are negligible. Second, it can detect only variation that is of exponential type in time.

To illustrate these points, we set up some examples by means of the polar coordinates r, θ, in terms of which

$$x = r \cos \theta, \qquad y = r \sin \theta. \tag{8.44}$$

Supposing r and θ to be functions of t, we may use the chain rule to find

$$\frac{dx}{dt} = \frac{dr}{dt} \cos \theta - r \frac{d\theta}{dt} \sin \theta,$$

$$\frac{dy}{dt} = \frac{dr}{dt} \sin \theta + r \frac{d\theta}{dt} \cos \theta. \tag{8.45}$$

The sines and cosines can be eliminated by using Eq. (8.44), to give the relations

$$\frac{dx}{dt} = \frac{1}{r} \frac{dr}{dt} x - \frac{d\theta}{dt} y,$$

$$\frac{dy}{dt} = \frac{d\theta}{dt} x + \frac{1}{r} \frac{dr}{dt} y. \tag{8.46}$$

As a first example, suppose that r and θ satisfy the differential equations

$$\frac{dr}{dt} = \alpha r, \qquad \frac{d\theta}{dt} = \beta. \tag{8.47}$$

By substituting these into Eq. (8.46) we get a linear system with constant coefficients

$$\frac{dx}{dt} = \alpha x - \beta y,$$

$$\frac{dy}{dt} = \beta x + \alpha y. \tag{8.48}$$

From Eq. (8.47) or Eq. (8.48) it is clear that the origin of the phase plane is a spiral point ($\beta \neq 0$), asymptotically stable if $\alpha < 0$ and unstable if $\alpha > 0$.

As a second example, suppose that r and θ satisfy these differential equations:

$$\frac{dr}{dt} = r(1 - r^2), \qquad \frac{d\theta}{dt} = 1. \tag{8.49}$$

The corresponding equations for x and y are

$$\frac{dx}{dt} = x - y - x^3 - xy^2,$$

$$\frac{dy}{dt} = x + y - x^2y - y^3. \tag{8.50}$$

The sole critical point is the origin, an unstable spiral point. In confirmation, Eq. (8.49) shows r increasing for small values. Nevertheless, r cannot increase beyond 1, and if r starts larger than 1, it will decrease to 1. Thus although the origin is unstable, all solutions are bounded—they approach the circle $r = 1$. (See Fig. 8.17. This phenomenon is called a limit cycle and can occur only with nonlinear equations.)

Figure 8.17

All Solutions
Approach the Circle
$r = 1$

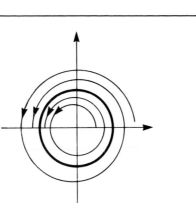

As a final example, let us consider the case in which r and θ satisfy

$$\frac{dr}{dt} = \alpha r^3, \qquad \frac{d\theta}{dt} = \beta, \tag{8.51}$$

where α and β are parameters. The differential equations (8.46) for x and y become

$$\frac{dx}{dt} = -\beta y + \alpha x^3 + \alpha xy^2,$$

$$\frac{dy}{dt} = \beta x + \alpha x^2 y + \alpha y^3. \tag{8.52}$$

The linear comparision system for these equations has imaginary characteristic roots, precluding any conclusion about the nonlinear equations
(8.52), and rightly so. The parameter α does not appear in the linear
comparison system, yet it determines whether r approaches 0 slowly
($\alpha < 0$) or approaches ∞ in finite time ($\alpha > 0$)!

These illustrations show that the polar coordinate r can sometimes
reveal more about the stability of the origin than x and y can. The "direct
method of Lyapunov" establishes the stability of the origin by the
decrease of a function $V(x, y)$, somewhat similar to the radial distance
$r = \sqrt{x^2 + y^2}$, but tailored to a given system.

Definition 8.3

A function $V(x, y)$ is called *positive definite* in a region R containing the
origin if it is continuous and has continuous partial derivatives in R and
    1. $V(0, 0) = 0$,
    2. $V(x, y) > 0$ for all other points in R.

Example 1

The radial distance $r = \sqrt{x^2 + y^2}$ does not have continuous partial derivatives at the origin, but its square $r^2 = x^2 + y^2$ is a positive definite
function in the entire plane. Other examples are

$$V(x, y) = x^2 + xy + y^2 \qquad \text{(all } x, y\text{)}, \tag{8.53}$$

$$V(x, y) = x^2 + y^4 \qquad \text{(all } x, y\text{)}, \tag{8.54}$$

$$V(x, y) = x - \ln|1 + x| + y^2 \qquad (x > -1). \tag{8.55}$$

However, $x^2 + 2xy + y^2 = (x + y)^2$ is not positive definite because it is
0 along the line $y = -x$.

Now the equation $r^2 = c$ represents a family of circles centered on
the origin (for $c > 0$). Similarly, if V is positive definite, the equation
$V(x, y) = c$ represents a family of closed curves surrounding the origin,
for $c > 0$ and small enough. For instance, for the function V of Eq.
(8.53) the curves $V(x, y) = c$ are ellipses centered on the origin and tilted
$45°$ (Fig. 8.18). For Eq. (8.54) they are rounded oblongs (Fig. 8.19), and
for Eq. (8.55) they are ovals stretched out to the right (Fig. 8.20). In
general, the smaller c is, the closer the curve $V(x, y) = c$ is to the origin.
Thus a positive definite function $V(x, y)$ acts like a measure of distance
from the origin.

To prove that the origin is a stable critical point of a given system of
differential equations, we construct a positive definite function $V(x, y)$
and then show that the value of V is a decreasing function of t when x and

Figure 8.18

$x^2 + xy + y^2 = c$

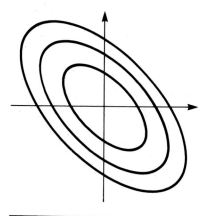

Figure 8.19

$x^2 + y^4 = c$

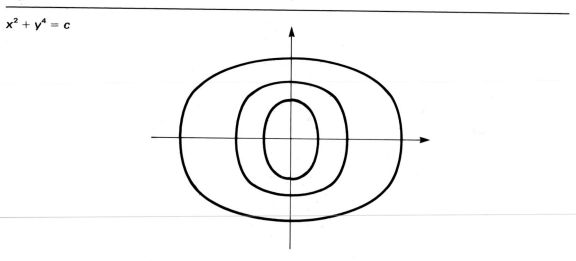

y are solutions of the system. Now if $x(t)$ and $y(t)$ are solutions of an autonomous system

$$\frac{dx}{dt} = g(x, y),$$

$$\frac{dy}{dt} = h(x, y),$$

(8.56)

Figure 8.20

$x - \ln(1 + x) + y^2 = c$

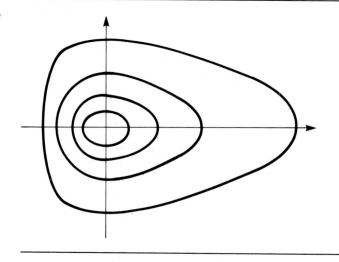

then the value of V at the point $(x(t), y(t))$ is $V(x(t), y(t))$, a function of time. It is decreasing if its derivative with respect to t is negative, $dV/dt < 0$. According to the chain rule,

$$\frac{dV}{dt} = \frac{\partial V}{\partial x}\frac{dx}{dt} + \frac{\partial V}{\partial y}\frac{dy}{dt}. \tag{8.57}$$

But since $x(t)$ and $y(t)$ are to be solutions of Eqs. (8.56), we may write

$$\frac{dV}{dt} = \frac{\partial V}{\partial x} g(x, y) + \frac{\partial V}{\partial y} h(x, y) \tag{8.58}$$

for the rate of change of V along a trajectory. Note, however, that dV/dt is expressed as a function of x and y. Thus it is not necessary to know $x(t)$ and $y(t)$ to determine dV/dt.

Theorem 8.3

Suppose $(0, 0)$ is an isolated equilibrium solution of the system (8.56) and $V(x, y)$ is positive definite in a region R including $(0, 0)$. Then the equilibrium solution is (i) stable if

$$\frac{dV}{dt}(x, y) \leq 0 \text{ in } R;$$

(ii) asymptotically stable if

$$\frac{dV}{dt}(x, y) < 0$$

in R except at $(0, 0)$.

Example 2

In the phase plane the system

$$\frac{dx}{dt} = 2xy - x,$$

$$\frac{dy}{dt} = -x^2 - y^3$$

has a critical point at $(0, 0)$ that cannot be investigated by the linear comparison method because $b = c = d = 0$. However, the function

$$V(x, y) = x^2 + 2y^2$$

is positive definite, and its derivative along the trajectories of the system is

$$\frac{dV}{dt} = 2x\frac{dx}{dt} + 4y\frac{dy}{dt}$$

$$= 2x(2xy - x) + 4y(-x^2 - y^3)$$

$$= -2x^2 - 4y^4.$$

Since this is strictly negative (except at the origin), the equilibrium solution is asymptotically stable.

Example 3

Again the origin is a critical point of the system

$$\frac{dx}{dt} = -x + y + xy,$$

$$\frac{dy}{dt} = x - y - x^2$$

that cannot be analyzed by the linear comparison method. However, the function $V = x^2 + y^2$ has the derivative

$$\frac{dV}{dt} = 2x\frac{dx}{dt} + 2y\frac{dy}{dt}$$

$$= 2x(-x + y + xy) + 2y(x - y - x^2)$$
$$= -2x^2 + 4xy - 2y^2 = -2(x - y)^2$$

along the trajectories of the system. Since dV/dt is negative, but not strictly negative, we may conclude that the origin is stable.

There is an easy geometric interpretation of Theorem 8.3. At any point on a level curve $V(x, y) = c$, the vector

$$N = \frac{\partial V}{\partial x} \hat{\imath} + \frac{\partial V}{\partial y} \hat{\jmath} \tag{8.59}$$

points in the direction of most rapid increase of V. If the curve is closed and V is positive definite, N points away from the interior of the curve. Now the expression (Eq. 8.57)

$$\frac{\partial V}{\partial x} \frac{dx}{dt} + \frac{\partial V}{\partial y} \frac{dy}{dt}$$

is the dot product of the vector N with the tangent vector along a trajectory, $T = \left(\frac{dx}{dt}, \frac{dy}{dt}\right)$. If it is negative, the tangent vector is aimed toward the inside of the curve $V(x, y) = c$, where values of V will be smaller. Figure 8.21 shows the level curve $V(x, y) = x^2 + y^2 = 1$ and

Figure 8.21

Tangents to
Trajectories of
Example 3 and the
Circle $x^2 + y^2 = 1$

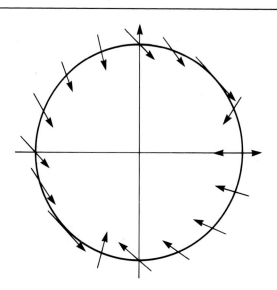

some vectors tangent to trajectories of the system of Example 3. Note that the tangent vectors just graze the level curve along the line $x = y$; this corresponds to the fact that $dV/dt = 0$ there.

We have seen that the direct method can be used near a critical point in cases where linear comparison cannot. It has the additional advantage of being applicable beyond a hard-to-specify neighborhood of a critical point. We can sharpen Theorem 8.3 as follows.

Corollary 8.1

Under the same hypotheses as Theorem 8.3, if R is a bounded region and

$$\frac{dV}{dt}(x, y) < 0$$

in R, except at $(0, 0)$, then every trajectory that enters R approaches the origin.

Example 4

The origin is a critical point of the system

$$\frac{dx}{dt} = -x - y + x^3 + xy^2,$$

$$\frac{dy}{dt} = x - y - x^2 y + y^3.$$

The function $V(x, y) = x^2 + y^2$ is positive definite, and its derivative,

$$\frac{dV}{dt} = 2x\frac{dx}{dt} + 2y\frac{dy}{dt}$$

$$= 2x(-x - y + x^3 + xy^2) + 2y(x - y - x^2 y + y^3)$$

$$= 2(-x^2 - y^2 + x^4 + y^4),$$

is strictly negative on the inside of a region R shown in Fig. 8.22. Therefore any trajectory entering R must approach the origin.

The direct method can also be applied to the question of asymptotic stability *in the large*—that is, in the entire xy-plane.

Theorem 8.4

Suppose $(0, 0)$ is the only critical point of the system (8.56) and $V(x, y)$ is positive definite in the entire xy-plane. If

$$\frac{dV}{dt}(x, y) < 0$$

for all x, y different from $(0, 0)$ and if $V(x, y) \to \infty$ as $x^2 + y^2 \to \infty$, then all solutions of the system (8.58) tend to 0 as $t \to \infty$.

Figure 8.22

Region R where $x^4 + y^4 - x^2 - y^3 < 0$

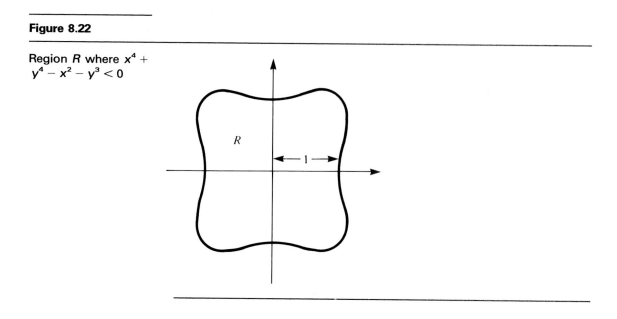

In Example 2 all conditions of Theorem 8.4 are met, so the origin is asymptotically stable in the large for that system. In Example 3, however, the function V that was chosen does not have a negative derivative; asymptotic stability in the large cannot be guaranteed.

The greater power of the direct method has its price: there is no systematic say to construct the function $V(x, y)$ for a given system. There are some guides to choosing V, however.

1. In many physical problems, $V(x, y)$ is proportional to the energy in the system.

2. If there are linear terms in the functions g and h in Eq. (8.56), these do dominate the behavior near the critical point. Choose $V(x, y)$ to exploit them. (See Example 4.)

3. A quadratic function is often a good starting place. The function

$$V(x, y) = Ax^2 + Bxy + Cy^2$$

is positive definite if A and C are positive and

$$B^2 - 4AC < 0.$$

Exercises

Exercises 1–6 are concerned with mass and spring systems having a nonlinear spring. The governing differential equation is supposed to be $d^2x/dt^2 + f(x) = 0$, leading to the system

$$\frac{dx}{dt} = y,$$

$$\frac{dy}{dt} = -f(x).$$

In each exercise, $f(x)$ is given. Find all critical points and investigate their stability. Sketch $f(x)$ versus x.

1. $f(x) = x + ax^3$, $a > 0$ (hard spring)
2. $f(x) = x^3$
3. $f(x) = x/\sqrt{1 + x^2}$ (Soft spring)
4. $f(x) = x/(1 + x^2)$
5. $f(x) = x(x^2 - 1)/(x^2 + 1)$ (Toggled spring)
6. $f(x) = \sin x$ (Pendulum)

In Exercises 7–12, use the direct method to prove stability or asymptotic stability of the origin for the linear system. (Hint: try a quadratic function for $V(x, y)$.)

7. $\dfrac{dx}{dt} = -x + y$

    $\dfrac{dy}{dt} = -x - y$

8. $\dfrac{dx}{dt} = -2x + y$

    $\dfrac{dy}{dt} = x - 2y$

9. $\dfrac{dx}{dt} = -y$

    $\dfrac{dy}{dt} = x - 2y$

10. $\dfrac{dx}{dt} = 3x - 5y$

    $\dfrac{dy}{dt} = 4x - 4y$

11. $\dfrac{dx}{dt} = -x - 2y$

    $\dfrac{dy}{dt} = -y$

12. $\dfrac{dx}{dt} = -x + 2y$

    $\dfrac{dy}{dt} = -3x - 2y$

13. The origin is a spiral point for the system

$$\frac{dx}{dt} = y + xy,$$

$$\frac{dy}{dt} = -x - y - xy - y^2.$$

Prove its asymptotic stability by the direct method. In what region can asymptotic stability be guaranteed?

14. Prove that all solutions of this system spiral into the origin:

$$\frac{dx}{dt} = (x^2 + y^2)(y - x),$$

$$\frac{dy}{dt} = (x^2 + y^2)(-y - x).$$

8.5

Limit Cycles

Let us recall from Section 8.4 the nonlinear system of equations

$$\frac{dx}{dt} = x - y - x^3 - xy^2,$$

$$\frac{dy}{dt} = x + y - x^2y - y^3,$$

      (8.60)

which were created by transforming the following equations in polar coordinates:

$$\frac{dr}{dt} = r(1 - r^2), \qquad \frac{d\theta}{dt} = 1.$$

      (8.61)

Every trajectory that starts inside the circle $r = 1$ spirals outward toward it, and every trajectory that starts outside spirals inward. (See Fig. 8.23.) A simple (nonintersecting) closed curve that trajectories spiral toward or away from is called a *limit cycle*. The circle $r = 1$ is a limit cycle of the system (8.60).

Figure 8.23

Limit Cycle

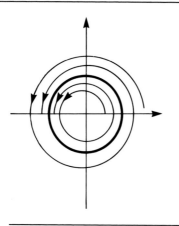

In general, if a system of differential equations

$$\frac{dx}{dt} = g(x, y),$$

$$\frac{dy}{dt} = h(x, y)$$

(8.62)

has a trajectory that is a simple closed curve in the xy-plane, that trajectory corresponds to a periodic solution of the system. That is to say, there is a positive number T, called the period, for which

$$x(t + T) = x(t), \qquad y(t + T) = y(t)$$

(8.63)

for all t. In fact, a simple closed curve corresponds to an infinite number of periodic solutions that differ only by a shift in the time axis (that is, they differ in phase).

If a linear homogeneous system has periodic solutions, the corresponding trajectories form a *family* of ellipses centered on the origin. The period is fixed by the system; the amplitude and phase are fixed by initial conditions. Thus a limit cycle is an inherently nonlinear phenomenon. The period and amplitude of the corresponding periodic solutions* are fixed by the system; the phase is fixed by initial conditions.

Example 1

The general solution of the system (8.60) above is

$$x(t) = \frac{\cos (t + \phi)}{\sqrt{1 + ce^{-2t}}}, \qquad y(t) = \frac{\sin (t + \phi)}{\sqrt{1 + ce^{-2t}}},$$

where ϕ is an arbitrary phase angle and c an arbitrary constant. Both ϕ and c are determined by initial conditions. If c is not 0, the solutions above are *not* periodic. The limit cycle is the trajectory $r = 1$ corresponding to the periodic solutions

$$x(t) = \cos (t + \phi), \qquad y(t) = \sin (t + \phi),$$

with period 2π.

In contrast to this nonlinear example the linear system

$$\frac{dx}{dt} = -y, \qquad \frac{dy}{dt} = x$$

has the general solution

$$x(t) = A \cos (t + \phi), \qquad y(t) = A \sin (t + \phi),$$

* Names: self-excited or entrained oscillations.

where A and ϕ are arbitrary. All solutions are periodic with period 2π, and the corresponding trajectories are circles of radius $|A|$ centered at the origin. There is no spiraling.

In this section we study the nature and existence of limit cycles and the corresponding periodic solutions. Our analysis depends heavily on two-dimensionality: the topology and geometry of the xy-plane. In three or more dimensions the situation is totally different. As an example, we have Theorem 8.5.

Theorem 8.5

(Poincaré-Bendixson)   If a trajectory remains bounded as $t \to \infty$, either
    1. it contains or approaches a critical point, or
    2. it is a closed cycle, or
    3. it approaches a limit cycle.

Example 2

All trajectories of the nonlinear system

$$\frac{dx}{dt} = -y,$$

$$\frac{dy}{dt} = x + 2y - y^3$$

are bounded as $t \to \infty$. This can be seen from the fact that no trajectory can leave the oddly shaped region shown in Fig. 8.24. On the other hand, no trajectory can approach a critical point. The only critical point is the origin, which is an unstable half-node. (The linear comparison system has 1 as a double characteristic root.) Thus trajectories starting far from the origin must come inward, while trajectories starting near the origin must move away from it. In fact, the system does have a limit cycle, shown in Fig. 8.25.

Now we wish to develop methods that will help us to detect and locate limit cycles. A knowledge of the critical points of the system can be of help, as shown by the following theorem.

Theorem 8.6

A trajectory that is a simple closed curve must enclose at least one critical point.

Figure 8.24

Region Whose Boundaries Are: $x = 5$; $x = -3y + y^3$ (an Isocline); $x = 0$; $y = 2$; $x = -4 + y$; $y = 0$; and the Reflections of Those Curves in the Origin

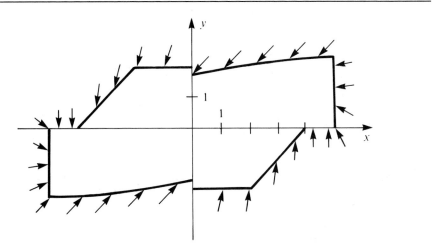

Figure 8.25

A Limit Cycle

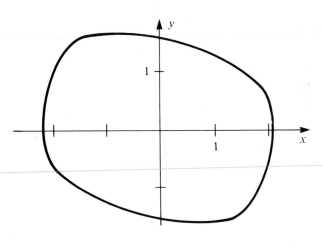

Example 3

A mass–spring system with a toggle (see Fig. 8.26) might give rise to a system of equations like

$$\frac{dx}{dt} = y, \qquad \frac{dy}{dt} = x\frac{1 - x^2}{1 + x^2},$$

Figure 8.26

Toggled Spring–Mass
System

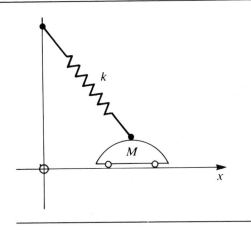

which has three critical points: a saddle point at $(0, 0)$ and centers at $(\pm 1, 0)$. The function

$$Q(x, y) = x^2 + y^2 - 2 \ln (1 + x^2)$$

is a first integral of the system. The existence of periodic solutions is guaranteed by the fact that the trajectories $Q(x, y) = c$ (shown in Fig. 8.27) are simple closed curves—except for $Q(x, y) = 0$, which is a figure eight and therefore not simple.

The trajectories fall into three families: two families of ovals surrounding the centers at $(\pm 1, 0)$ and a family of dumbbell-shaped curves enclosing all three critical points.

We now know that a closed trajectory of the system (8.62) must enclose at least one critical point. Another theorem that can help us to locate them is the following.

Theorem 8.7

(Bendixson) If R is a region in which $\Delta(x, y) = (\partial g/\partial x) + (\partial h/\partial y)$ is always positive or always negative, then no limit cycle or closed trajectory of the system (8.62) lies entirely in R.

Proof: Along a closed curve C that is a closed trajectory of the system (8.62), the equation

Figure 8.27

Trajectories for
Toggled System

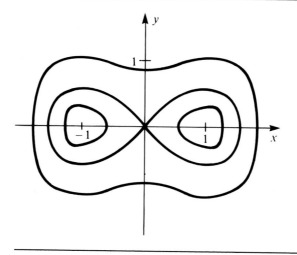

$$g\,dy - h\,dx = 0 \tag{8.64}$$

is satisfied identically. Therefore also this line integral is zero:

$$\int_C g\,dy - h\,dx = 0. \tag{8.65}$$

By Green's theorem in the plane, then,

$$\iint_S \left(\frac{\partial g}{\partial x} + \frac{\partial h}{\partial y} \right) dA = 0, \tag{8.66}$$

where S is the region enclosed by C. It must happen, then, that the integrand $\Delta(x, y)$ is identically zero in S or else it takes both positive and negative values in S. Thus C could not lie entirely in a region where $\Delta(x, y)$ is always positive or always negative.

Theorem 8.7 provides easily used guidance in locating closed trajectories. To apply it, find the locus of $\Delta(x, y) = 0$.

1. If there is no locus—that is, if $\Delta(x, y)$ has the same sign throughout the xy-plane—there can be no closed trajectories.

2. If $\Delta(x, y) \equiv 0$ in the whole plane, a limit cycle may exist, but we gain no information about it.

3. If $\Delta(x, y) = 0$ defines one or more curves in the xy-plane, they divide the plane into regions where $\Delta(x, y)$ is positive or negative. Any closed trajectory must enclose parts of regions corresponding to both signs.

Example 4

In Example 2, we used Theorem 8.5 to prove the existence of a limit cycle for the system

$$\frac{dx}{dt} = -y,$$

$$\frac{dy}{dt} = x + 2y - y^3.$$

By Theorem 8.6 that limit cycle must encircle the origin, which is the only critical point.

Now we apply Theorem 8.7. Since $g(x, y) = -y$ and $h(x, y) = x + 2y - y^3$, we can easily calculate

$$\Delta(x, y) = \frac{\partial g}{\partial x} + \frac{\partial h}{\partial y} = 2 - 3y^2.$$

The locus $\Delta(x, y) = 0$ is the pair of horizontal lines, $y = \pm\sqrt{\frac{2}{3}}$. Figure 8.28 shows the regions where $\Delta(x, y)$ has one sign, along with the limit cycle.

Figure 8.28

Regions Where $\Delta > 0$ and $\Delta < 0$ and a Limit Cycle

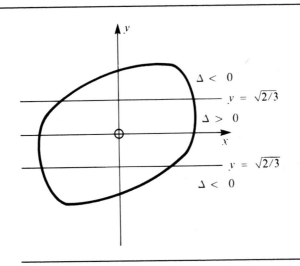

Finally, we cite a theorem concerning the existence of limit cycles for a system derived from a second-order nonlinear differential equation of the form

$$\frac{d^2u}{dt^2} + b(u)\frac{du}{dt} + c(u) = 0. \tag{8.67}$$

This might be thought of as the equation of motion of a mass–spring–damper system having a nonlinear spring (restoring force proportional to $c(u)$) and nonlinear damping (damping force proportional to $b(u)\,du/dt$). This equation can be turned into a system in the usual way. However, it is convenient to use the *Lienard transformation*:

$$x = u, \qquad y = \frac{du}{dt} + B(u), \tag{8.68}$$

in which the function B is defined by

$$B(u) = \int_0^u b(z)\,dz. \tag{8.69}$$

The resulting system in x and y is

$$\frac{dx}{dt} = y - B(x),$$

$$\frac{dy}{dt} = -c(x). \tag{8.70}$$

In the hypotheses of the theorem it is necessary to refer to the function

$$C(x) = \int_0^x c(z)\,dz,$$

which is proportional to the energy stored in the spring, in a mass–spring–damper system.

Theorem 8.8

(Lienard, Levinson, Smith) Suppose that the functions $b(x)$, $c(x)$, and $c'(x)$ are continuous for all x. Furthermore, suppose that
    1. $b(-x) = b(x)$; there is an $x_0 > 0$ such that $B(x) < 0$, $0 < x < x_0$; $B(x) > 0$, $x_0 < x$, and $b(x) > 0$, $x_0 < x$; and $B(x) \to \infty$ as $x \to \infty$;
    2. $c(-x) = -c(x)$; $c(x) > 0$ for $x > 0$; $C(x) \to \infty$ as $x \to \infty$.
Then the system (8.70) has a unique limit cycle.

The hypotheses seem overwhelming, at first. Figure 8.29 shows two functions b and c (along with B and C) that would satisfy the hypotheses.

Figure 8.29

Typical Functions
Satisfying Hypotheses
of Theorem 8.8

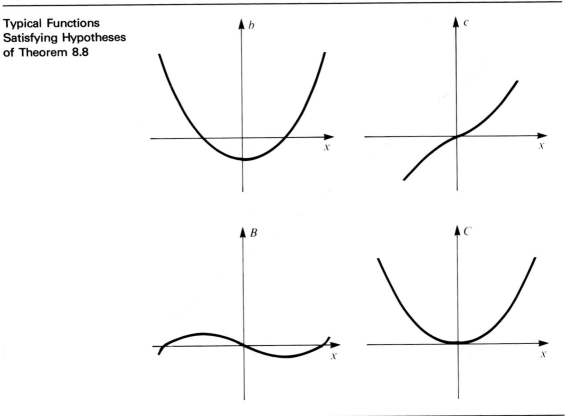

The most important features of the functions are the signs of b, B, and c, rather than the specific shapes of the curves.

Example 5

The van der Pol equation, derived in Section 8.1, is

$$\frac{d^2u}{dt^2} + \varepsilon(u^2 - 1)\frac{du}{dt} + u = 0.$$

We identify the following functions:

$$b(x) = \varepsilon(x^2 - 1), \qquad c(x) = x,$$

$$B(x) = \int_0^x \varepsilon(z^2 - 1)\,dz, \qquad C(x) = \int_0^x z\,dz,$$

$$= \varepsilon\left(\frac{x^3}{3} - x\right), \qquad = \frac{x^2}{2}.$$

It is easy to check that all the hypotheses of Theorem 8.8 are satisfied. In particular, the x_0 mentioned in hypothesis 1 is $\sqrt{3}$. We conclude that the corresponding system

$$\frac{dx}{dt} = y - \varepsilon\left(\frac{x^3}{3} - x\right),$$

$$\frac{dy}{dt} = -x$$

has a unique limit cycle.

Exercises

In Exercises 1–4, decide whether or not the given system has one or more periodic solutions. If so, try to locate and describe the closed trajectory (or trajectories). Besides the theorems of this section, critical point analysis, Lyapunov stability, and graphical methods may be of use.

1. $\dfrac{dx}{dt} = x + y - x^3$

   $\dfrac{dy}{dt} = -x + y - y^3$

2. $\dfrac{dx}{dt} = y$

   $\dfrac{dy}{dt} = (1 - x^2)y - x$

3. $\dfrac{dx}{dt} = 2x - y + x^3 - x^2 y$

   $\dfrac{dy}{dt} = x - y + xy^2$

4. $\dfrac{dx}{dt} = y$

   $\dfrac{dy}{dt} = -x + x^3$

In Exercises 5–7, convert the given second-order equation to a system and study the possibility of periodic solutions.

5. $\dfrac{d^2u}{dt^2} + \left(\dfrac{du}{dt}\right)^3 - \dfrac{du}{dt} + u = 0$ (Rayleigh's equation)

6. $\dfrac{d^2u}{dt^2} + (u^2 - 1)\dfrac{du}{dt} + u = 0$ (van der Pol's equation)

7. $\dfrac{d^2u}{dt^2} + u - u^3 = 0$ (Duffing's equation)

8. Given a differential equation of the form

   $$\frac{d^2u}{dt^2} + f\left(\frac{du}{dt}\right) + u = 0,$$

   differentiate and substitute $v = du/dt$ to obtain an equation of the form required by Theorem 8.8.

9. Assuming that $f'(x)$ is continuous and $f(0) = 0$, what other condi-

tions on f will guarantee the existence of a limit cycle for the equation in Exercise 8?

10. Apply the results of Exercises 8 and 9 to the equation of Exercise 5.

11. Show that a system may have several limit cycles by studying the behavior of the trajectories of the system

$$\frac{dr}{dt} = r(1 - r)(2 - r), \qquad \frac{d\theta}{dt} = 1.$$

8.6

Exact Solution of Nonlinear Equations

In Section 8.1 we observed that the general autonomous second-order equation

$$\frac{d^2u}{dt^2} = f\left(u, \frac{du}{dt}\right),$$ (8.71)

in which the variable t is not explicitly mentioned, is related to a first-order equation. The relation can be made via the equivalent system of first-order equations or by means of the substitution

$$v = \frac{du}{dt}.$$ (8.72)

From this it follows that

$$\frac{d^2u}{dt^2} = \frac{dv}{dt} = \frac{dv}{du}\frac{du}{dt} = \frac{dv}{du} \cdot v.$$ (8.73)

This the original equation (8.71) is converted into the new equation

$$v\frac{dv}{du} = f(u, v).$$ (8.74)

In many important cases, Eq. (8.74) can be solved for v, giving what is called a *first integral* of Eq. (8.71). If it is possible to find $v = du/dt$ explicitly from the first integral, then Eq. (8.71) has been reduced to a first-order equation.

Example 1

A mass–spring system with a soft spring has a differential equation of the form

$$\frac{d^2u}{dt^2} + au - bu^3 = 0, \qquad a, b > 0.$$

Find its first integral.

Using the substitution $v = du/dt$ and its consequences from Eq. (8.73), we obtain

$$v \frac{dv}{du} + au - bu^3 = 0.$$

This is a separable equation. Its separated form and solution are

$$v \, dv = -(au - bu^3) \, du,$$

$$\tfrac{1}{2}v^2 = -(\tfrac{1}{2}au^2 - \tfrac{1}{4}bu^4) + c$$

or

$$v^2 + au^2 - \tfrac{1}{2}bu^4 = C.$$

This is the first integral of the original equation, which can be solved for $v = du/dt$ to give

$$\frac{du}{dt} = \pm\sqrt{C - au^2 + \tfrac{1}{2}bu^4}$$

as a first-order equation for u.

When an autonomous second-order equation can be reduced to a first-order equation via a first integral, the resulting equation often involves square roots and hence inverse functions, which must be interpreted with some care.

Example 2

Use the first integral to solve the linear autonomous equation

$$\frac{d^2u}{dt^2} + au = 0, \qquad a > 0.$$

The substitution $v = du/dt$ reduces the given equation to

$$v \frac{dv}{du} + au = 0.$$

This separable equation can be integrated at once to give the first integral

$$\tfrac{1}{2}v^2 + \tfrac{1}{2}au^2 = C.$$

The constant C can be determined from initial conditions. It can also be identified this way. Let A be the maximum value of u. This value is assumed by u when $v = du/dt$ is 0. Thus

$$C = \tfrac{1}{2}aA^2.$$

Now we may solve the first integral for $v = du/dt$ to obtain this separable first-order equation

$$\frac{du}{dt} = \pm\sqrt{aA^2 - au^2}.$$

The sign in this equation is chosen to match that of du/dt at $t = 0+$.

In separated form the equation above is

$$\frac{du}{\sqrt{A^2 - u^2}} = \pm\sqrt{a}\, dt.$$

If we substitute $u = Aw$, the equation becomes

$$\frac{A\, dw}{\sqrt{A^2 - A^2 w^2}} = \frac{dw}{\sqrt{1 - w^2}} = \pm\sqrt{a}\, dt.$$

This is an elementary form that may be integrated to obtain

$$\sin^{-1} w = \pm\sqrt{a}\, t + c_1.$$

Now we apply the sine function to both sides of this equation and revert to the original variable:

$$w = \sin(\pm\sqrt{a}\, t + c_1),$$

$$u = A\sin(\pm\sqrt{a}\, t + c_1).$$

This last expression is valid for all t, although the expression containing the inverse sine is valid only for a short time after $t = 0$: specifically, until $w^2 = 1$.

The solution obtained is the general solution only if it can meet two arbitrary initial conditions. To do this, two arbitrary parameters must be available. In the expression for u these are the multiplier $\pm A$ and c_1, which appears as a phase angle.

Certain important equations lead to consideration of the function

$$\int_0^\phi \frac{d\theta}{\sqrt{1 - k^2\sin^2\theta}} = F(k, \phi), \qquad 0 < k < 1, \tag{8.75}$$

called the (incomplete) *elliptic integral of the first kind*. Note that it depends on both the parameter k, called the *modulus*, and the variable ϕ. If $\phi = \pi/2$, the result is called the *complete elliptic integral of the first kind* and written

$$K(k) = F\left(k, \frac{\pi}{2}\right) = \int_0^{\pi/2} \frac{d\theta}{\sqrt{1 - k^2\sin^2\theta}}. \tag{8.76}$$

Equation (8.75) defines the integral $F(k, \phi)$ as a function of ϕ. It is often convenient to invert this relationship and think of ϕ as a function of F. This function is called the *amplitude*, written

$$\phi = \text{am} (F).$$

Thus we may write $F = \text{am}^{-1} \phi$ or

$$\int_0^\phi \frac{d\theta}{\sqrt{1 - 2\sin^2\theta}} = \text{am}^{-1}(\phi). \tag{8.77}$$

Of course, the amplitude depends on both ϕ and k, but k is usually suppressed if no confusion will result. Figure 8.30 shows graphs of am and F.

The (Jacobian) *elliptic functions* are defined through the amplitude function. The principal functions are

$$\text{sn } t = \sin (\text{am } t) \tag{8.78}$$

$$\text{cn } t = \cos (\text{am } t) = \pm\sqrt{1 - \text{sn}^2 t} \tag{8.79}$$

$$\text{dn } t = \sqrt{1 - k^2 \text{sn}^2 t}. \tag{8.80}$$

The function sn t is read "s-n of t" or "elliptic sine of t," and the others are read similarly. All three are periodic, with period $4K(k)$. The dn function is always positive. Figure 8.31 shows the sn and cn functions for several values of k.

The inverses of the elliptic functions are related to integrals that contain square roots. They are somewhat analogous to the integrals for the inverse sine and cosine:

$$\int_0^x \frac{dz}{\sqrt{1 - z^2}\,\sqrt{1 - k^2 z^2}} = \text{sn}^{-1}(x), \tag{8.81}$$

$$\int_x^1 \frac{dz}{\sqrt{1 - z^2}\,\sqrt{1 - k^2 + k^2 z^2}} = \text{cn}^{-1}(x). \tag{8.82}$$

From here we can deduce the indefinite integrals

$$\int \frac{dx}{\sqrt{1 - x^2}\,\sqrt{1 - k^2 x^2}} = \text{sn}^{-1}(x) + C, \tag{8.83}$$

$$\int \frac{dx}{\sqrt{1 - x^2}\,\sqrt{1 - k^2 + k^2 x^2}} = -\text{cn}^{-1}(x) + C. \tag{8.84}$$

In practice, changes of variable are often necessary to convert a given integral into the forms shown in Eq. (8.83) or Eq. (8.84). The guiding principle is that the denominator is a biquadratic polynomial with four

Figure 8.30

The Functions (a)
$F(k, x)$ and (b) am x
for $k = 0.2, 0.5, 0.8,$
0.9

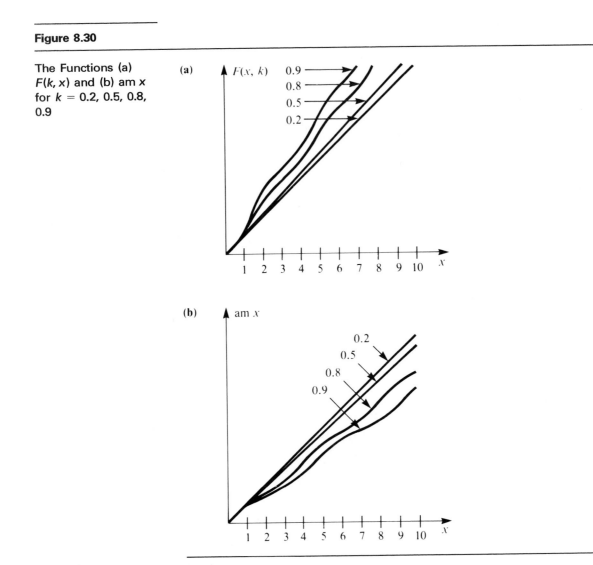

distinct roots. If the roots are all real, we aim for the form in Eq. (8.83).
If two are real and two imaginary, Eq. (8.84) is the desired form.

Example 3

Solve the equation for a mass–spring system with a soft spring:

$$\frac{d^2u}{dt^2} + au - bu^3 = 0, \qquad a, b > 0.$$

Figure 8.31

The Functions sn t and cn t for (a) $k = 0.2$, (b) $k = 0.5$, and (c) $k = 0.8$

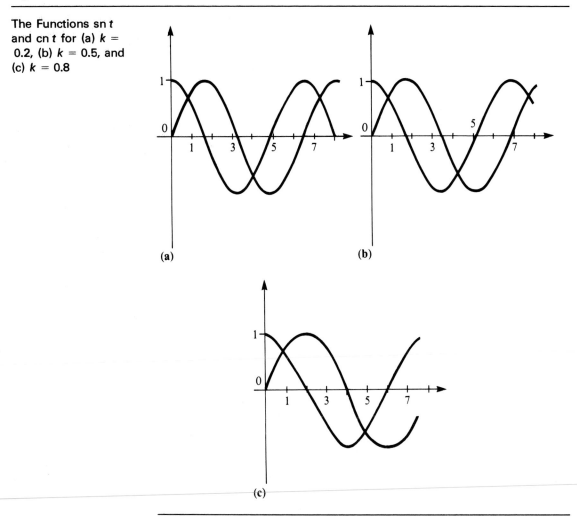

(a)

(b)

(c)

In Example 1, we found the first integral for the equation above:

$$v^2 + au^2 - \tfrac{1}{2}bu^4 = C.$$

The constant C can be expressed in terms of the maximum displacement A. Since u assumes this value when $du/dt = v$ is 0, we find that

$$C = aA^2 - \tfrac{1}{2}bA^4.$$

Now the first integral takes the form

$$v^2 + au^2 - \tfrac{1}{2}bu^4 = aA^2 - \tfrac{1}{2}bA^4$$

or

$$v^2 = a(A^2 - u^2) - \tfrac{1}{2}b(A^4 - u^4). \tag{8.85}$$

When this is solved for $v = du/dt$, we will have a first-order separable equation. Its solution will involve the square root of the right-hand side of Eq. (8.85) in the denominator. To obtain a form like Eq. (8.83) or Eq. (8.84), first factor Eq. (8.85):

$$v^2 = (A^2 - u^2)(a - \tfrac{1}{2}b(A^2 + u^2))$$

$$= A^2\left(1 - \frac{u^2}{A^2}\right)(a - \tfrac{1}{2}bA^2)\left(1 - \frac{b}{2a - bA^2}u^2\right).$$

Next make the change of variable $w = u/A$ and $v = du/dt = A\,dw/dt$:

$$A^2\left(\frac{dw}{dt}\right)^2 = A^2(a - \tfrac{1}{2}bA^2)(1 - w^2)(1 - k^2w^2). \tag{8.86}$$

In the last equation the parameter k is given by

$$k^2 = \frac{bA^2}{2a - bA^2}. \tag{8.87}$$

Since $0 < k < 1$, we must assume that $A^2 < a/b$.

Finally, Eq. (8.86) can be solved as a separable equation, using Eq. (8.83):

$$\frac{dw}{dt} = \pm\sqrt{a - \tfrac{1}{2}bA^2}\,\sqrt{(1 - w^2)(1 - k^2w^2)}$$

$$\frac{dw}{\sqrt{(1 - w^2)(1 - k^2w^2)}} = \pm\sqrt{a - \tfrac{1}{2}bA^2}\,dt$$

$$\operatorname{sn}^{-1}(w) = \pm\sqrt{a - \tfrac{1}{2}bA^2}\,t + c_1$$

$$w = u/A = \operatorname{sn}(\pm\sqrt{a - \tfrac{1}{2}bA^2}\,t + c_1)$$

$$u = \pm A\operatorname{sn}(\sqrt{a - \tfrac{1}{2}bA^2}\,t + c_1). \tag{8.88}$$

Thus the solution is expressed in terms of the elliptic sine function (with modulus k given by Eq. (8.87)) and the two constants A and c_1.

Example 4

Solve the differential equation for the simple pendulum

$$\frac{d^2u}{dt^2} + \omega^2 \sin u = 0.$$

Using the routine substitution of Eq. (8.72), we obtain a first integral:

$$v \frac{dv}{du} + \omega^2 \sin u = 0,$$

$$\tfrac{1}{2}v^2 - \omega^2 \cos u = C.$$

The constant C can be found from initial conditions. Suppose that the pendulum makes a maximum excursion to angle α, assumed to be less than $\pi/2$. Since α is the value assumed by u when $v = 0$, we have

$$\tfrac{1}{2}v^2 - \omega^2 \cos u = -\omega^2 \cos \alpha$$

or

$$v^2 = 2\omega^2(\cos u - \cos \alpha)$$

or

$$v^2 = 4\omega^2(\sin^2 \tfrac{1}{2}\alpha - \sin^2 \tfrac{1}{2}u) \tag{8.89}$$

(the last by means of a trigonometric identity). The separated first-order equation for u,

$$\frac{du}{\sqrt{(\sin^2 \tfrac{1}{2}\alpha - \sin^2 \tfrac{1}{2}u)}} = \pm 2\omega \, dt, \tag{8.90}$$

can be integrated by means of the less than obvious substitution

$$\sin \tfrac{1}{2}u = kz \tag{8.91}$$

with $k = \sin \tfrac{1}{2}\alpha$. The differential of u is then found from

$$\cos (\tfrac{1}{2}u)\tfrac{1}{2}du = k \, dz$$

or

$$du = \frac{2k \, dz}{\sqrt{1 - k^2 z^2}}.$$

Applying the substitution in Eq. (8.91) leads us to

$$\frac{2k \, dz}{\sqrt{1 - k^2 z^2} \sqrt{k^2 - k^2 z^2}} = \pm 2\omega \, dt$$

or

$$\frac{dz}{\sqrt{1 - k^2 z^2} \sqrt{1 - z^2}} = \pm \omega \, dt.$$

Now we integrate, using Eq. (8.83), to get

$$\text{sn}^{-1}(z) = \pm \omega t + c_1,$$

$$z = \text{sn}(\pm \omega t + c_1).$$

In terms of the original variables we have

$$\sin \tfrac{1}{2}u = k \, \text{sn} \, (\pm \omega t + c_1)$$

or

$$u = 2 \sin^{-1} (k \, \text{sn} \, (\pm \omega t + c_1)).$$

Note that the argument of the inverse sine lies within the interval from -1 to 1, so that the inverse sine is single-valued.

Exercises

1. Solve the second-order equation for a mass–spring system with a hard spring:

$$\frac{d^2u}{dt^2} + au + bu^3 = 0, \qquad a, b > 0.$$

2. Find a first integral of this soft spring equation:

$$\frac{d^2u}{dt^2} + \frac{au}{1 + bu^2} = 0, \qquad a, b > 0.$$

3. Solve this soft spring equation (see Examples 1 and 3):

$$\frac{d^2u}{dt^2} + au - bu^3 = 0$$

when A, the value of u at $du/dt = 0$, is equal to $\sqrt{a/b}$.

4. Prove that the solution of the equation in Exercise 3 is not periodic when $A > \sqrt{a/b}$. (In fact, it is not stable.)

5. Set $\phi = \text{am} \, t$—that is,

$$t = \int_0^\phi \frac{d\theta}{\sqrt{1 - k^2 \sin^2 \theta}}$$

—and show that $d\phi/dt = \sqrt{1 - k^2 \sin^2 \phi}$, or $(\text{am} \, t)' = \text{dn} \, t$.

6. Using the results of Exercise 5 and the definitions of the elliptic functions, prove these derivative formulas:

$$\frac{d}{dt} \text{sn} \, t = \text{cn} \, t \, \text{dn} \, t,$$

$$\frac{d}{dt} \text{cn} \, t = -\text{sn} \, t \, \text{dn} \, t,$$

$$\frac{d}{dt} \text{dn} \, t = -k^2 \text{cn} \, t \, \text{sn} \, t.$$

7. Using the results of Exercise 6, find first-order differential equations for $\text{sn} \, t$, $\text{cn} \, t$, and $\text{dn} \, t$.

8. If a thin "beam" is compressed by an axial load (for example, a thin plastic ruler pressed between your index fingers), its shape is described by the equation

$$EI\kappa + Pu = 0,$$

where E is Young's modulus (a material constant), I is the second moment of the cross-section, P is the force at each end, u is the displacement of the centerline from the line between ends, and κ is the curvature. Taking arc length s as the independent variable, we have

$$\kappa = \frac{d\theta}{ds}, \qquad u = \int_0^s \sin \theta(s')\, ds',$$

where θ is the angle between the tangent to the curve and the x-axis. Show that an equation for $\theta(s)$ is

$$\frac{d^2\theta}{ds^2} + \lambda^2 \sin \theta = 0$$

with $\lambda^2 = P/EI$.

9. Suppose that the thin beam of Exercise 8 is such that $\theta(0) = \alpha$, the maximum value of θ. Find an expression for $\theta(s)$. (See Example 4.)

10. The shape of a meandering river is supposed to be described by the same equation as that in Exercise 8, with the same meanings for θ and s. The first integral is

$$\left(\frac{d\theta}{ds}\right)^2 = 2\lambda^2(\cos\theta - \cos\alpha),$$

where α is the maximum value of θ. Sketch $\cos\theta - \cos\alpha$ as a function of θ, $0 \le \theta \le \alpha$, and also sketch the approximation

$$\cos\theta - \cos\alpha \cong (1 - \cos\alpha)\left(1 - \left(\frac{\theta}{\alpha}\right)^2\right).$$

11. Use the approximation mentioned in Exercise 10 to find an approximate solution for the river meander problem.

Notes and References

Nonlinear differential equations have been receiving a great deal of attention lately. Many important physical phenomena cannot be modeled by linear differential equations: their behavior is innately nonlinear. We saw in Section 8.5 that a limit cycle, and the corresponding periodic solution, is a purely nonlinear phenomenon.

There is a close connection between first-order differential equations and nonlinear autonomous second-order differential equations. This fact is apparent in Section 8.6, where the Jacobian elliptic functions are defined. These functions rank with Bessel functions in importance to mathematical applications.

An extensive classification of second-order nonlinear equations was undertaken at the end of the 19th century. Equations whose solutions satisfy certain conditions as functions of a complex variable were classified into 50 canonical types, of which all but six can be solved in terms of known functions. The six remaining equations, of which the first is $y'' = 6y^2 + \lambda x$, define new functions called Painlevé transcendents.

Information about nonlinear equations of all sorts and in all their aspects can be found in the excellent book by Harold T. Davis (1960). Another interesting book is the text of Davies and James (1966). Abraham and Shaw (1983) have produced a spectacular book of drawings about nonlinear mechanics.

Miscellaneous Exercises

1. Let h be the draft of a floating object measured from the waterline downwards. If $V(h)$ is the volume of water displaced by the object, then the equation of vertical motion of the object is

   $$m\frac{d^2h}{dt^2} = mg - \rho g V(h),$$

   where m is the mass of the object and ρ is the density of water. If h_0 is the draft at equilibrium, then $mg = \rho g V(h_0)$, and

   $$\frac{d^2h}{dt^2} = \frac{\rho g}{m}(V(h_0) - V(h)).$$

   Find $V(h)$ if the floating object is a vertical cylinder of cross-section A that (a) cannot or (b) can be completely submerged.

2. What is the natural frequency of oscillation of the floating object in Exercise 1?

3. If the object in Exercise 1 has the shape of a triangular prism (point down), then $V(h) = \alpha h^2$. Study the differential equation

   $$\frac{d^2u}{dt^2} + u^2 + 2u = 0$$

   that arises from the equation of Exercise 1 in the case (after a suitable change of variables). Restrict your attention to $u > -1$.

4. Martini's equations* model the effects of a disease (for example, measles) with the characteristics that (a) all and only sick individuals are infective and (b) the sick either die or become immune on recovery. Let u be the fraction of the population currently sick and v the fraction of the population not infectible. With the assumption that the rate of infection is proportional to the product of the number sick with the number infectible, the equations are

$$\frac{du}{dt} = \alpha u(1 - v) - qu,$$

$$\frac{dv}{dt} = \alpha u(1 - v) - mv,$$

where α is the "infectivity," m is the rate of loss of immunity, and q is the rate at which the sick stop being sick (by recovery or death). (a) Find the critical points of the system. (b) Since u and v cannot be negative, what relation among the parameters must be fulfilled in order for the critical points to be significant?

5. Determine the nature of the critical points of the system in Exercise 4.

6. Find the singular points of the system

$$\frac{dx}{dt} = -y + x^2 + y^2, \qquad \frac{dy}{dt} = x - 2xy.$$

7. Use linear comparison to determine (if possible) the nature of the critical points of the system in Exercise 6.

8. Find a first integral of the system in Exercise 6. An exact equation can be found or the change of variables to $u = x + y$ and $v = x - y$ gives a separable equation.

9. Show that the system of Exercise 6 has two families of periodic solutions whose trajectories surround two of the singular points.

10. Does the system of Exercise 6 have limit cycles?

11. Find the critical points of the system

$$\frac{dx}{dt} = x + y - x^3 + xy^2,$$

$$\frac{dy}{dt} = -x + y - x^2y - y^3$$

and describe them by means of the linear comparison system.

---

* Alfred J. Lotka, *Elements of Mathematical Biology* (New York: Dover, 1956), pp. 79–81.

12. Under certain circumstances the height of the free surface of a water wave in a channel is a function of the form $u(x, t) = y(x - ct)$, where x measures distance along the channel and c is the wave speed. Then $y(x)$ gives the shape of the wave. After a difficult derivation, Lamb* gives this equation for y:

$$\frac{1}{y^2} - \frac{2}{3} y \left(\frac{1}{y}\right)'' + y^2 \left(\left(\frac{1}{y}\right)'\right)^2 = \frac{1}{h^2} - 2g \frac{(y - h)}{c^2 h^2},$$

where g is the acceleration of gravity and h is the depth of the water far from the crest of the wave. Obtain a first-order equation for y by multiplying the given equation by y' and intergrating. (Hint: first carry out the indicated differentiation of the terms in $1/y$ and combine them.)

13. Evaluate the constant of integration by requiring $y' = 0$ when $y = h$.

14. By algebraic manipulation, obtain this equation for y:

$$(y')^2 = 3 \frac{(y - h)^2}{h^2} \left(1 - \frac{g}{c^2} y\right).$$

Note that $y' = 0$ when $y = c^2/g$. This is the maximum value that y has. Make the substitution $z = y - h$ to find this equation:

$$\left(\frac{dz}{dx}\right)^2 = \frac{z^2}{b^2} \left(1 - \frac{z}{a}\right),$$

where $a = (c^2/g) - h$ and $b^2 = h^2 c^2/3ga$.

15. Integrate the equation for z. Determine the constant by requiring that z has its maximum when $x = 0$. Also find $y(x)$. The traveling wave of this shape is called a solitary wave or *soliton*.

* H. Lamb, *Hydrodynamics* (New York: Dover, 1945), p. 425.

9 Boundary Value Problems

9.1
Two-Point Boundary Value Problems

Up to this point, we have been concerned mainly with initial value problems: differential equations with auxiliary conditions at one point. However, many problems in physics and engineering are naturally modeled by a differential equation together with auxiliary conditions at two different points. This combination is called a *boundary value problem*. In this chapter we see the derivation of such problems, methods for their solution, and some of the theory surrounding them.

Hanging Cable

First we consider the problem of finding the shape of a cable, fastened at each end, and carrying a uniformly distributed load. The cables of a suspension bridge provide an important example. Let $u(x)$ denote the position of the centerline of the cable above the x-axis, which we assume to be horizontal. (See Fig. 9.1.) Our objective is to determine this function u.

The shape of the cable is determined by the forces on it. In our analysis we consider the forces that hold a small segment of the cable in place. (See Fig. 9.2.) The key assumption is that the cable is perfectly flexible. This means that it can transmit force only by means of a tension, which is directed at every point tangent to the centerline of the cable. We suppose that the cable is not moving. Thus the horizontal and vertical components of the forces acting on the segment each add up to 0. That is,

$$T(x + \Delta x) \cos \phi(x + \Delta x) - T(x) \cos \phi(x) = 0 \quad \text{(Horizontal)}, \quad \textbf{(9.1)}$$

$$T(x + \Delta x) \sin \phi(x + \Delta x) - T(x) \sin \phi(x) - w\Delta x = 0 \quad \text{(Vertical)}. \quad \textbf{(9.2)}$$

In the second equation, w is the intensity of the distributed load,

Figure 9.1

Hanging Cable

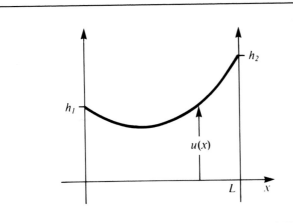

Figure 9.2

Slice Cut Out

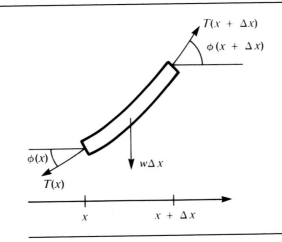

measured in force per unit of horizontal length, $w\Delta x$ is the force due to the loading that is borne by this small segment.

From Eq. (9.1) we see that the horizontal component of the tension is the same at both ends of the segment. In fact, the horizontal component of tension has the same value—call it T—at every point, including the endpoints where the cable is attached to solid supports. By simple algebra we can now find the tension in the cable at the ends of our segment,

$$T(x + \Delta x) = \frac{T}{\cos \phi(x + \Delta x)}, \qquad T(x) = \frac{T}{\cos \phi(x)}, \qquad \textbf{(9.3)}$$

and substitute these into Eq. (9.2), which becomes

$$\frac{T}{\cos \phi(x + \Delta x)} \sin \phi(x + \Delta x) - \frac{T}{\cos \phi(x)} \sin \phi(x) - w\Delta x = 0 \qquad \textbf{(9.4)}$$

or

$$T(\tan \phi(x + \Delta x) - \tan \phi(x)) - w\Delta x = 0. \qquad \textbf{(9.5)}$$

Before going further we should note (Fig. 9.2) that $\phi(x)$ measures the angle between the tangent to the centerline of the cable and the horizontal. Since the position of the centerline is given by $u(x)$, $\tan \phi(x)$ is just the slope of the cable at x. By elementary calculus we have

$$\tan \phi(x) = \frac{du}{dx}(x). \qquad \textbf{(9.6)}$$

Substituting the derivative for the slope in Eq. (9.5) and making some algebraic adjustments, we get

$$\frac{1}{\Delta x} \left(\frac{du}{dx}(x + \Delta x) - \frac{du}{dx}(x) \right) = \frac{w}{T}. \qquad \textbf{(9.7)}$$

Now we may eliminate the auxiliary quantity Δx by taking the limit of Eq. (9.7) as Δx approaches 0. The algebraic adjustments just made were intended to leave the left-hand side of Eq. (9.7) in the form of a difference quotient. In the limit it becomes the derivative of du/dx:

$$\lim_{\Delta x \to 0} \frac{1}{\Delta x} \left(\frac{du}{dx}(x + \Delta x) - \frac{du}{dx}(x) \right) = \frac{w}{T}$$

or

$$\frac{d}{dx} \left(\frac{du}{dx} \right) = \frac{w}{T}. \qquad \textbf{(9.8)}$$

As we well know, the differential equation alone cannot determine $u(x)$ uniquely. We must also specify auxiliary conditions, which in this case relay the information that the ends of the cable are fixed:

$$u(0) = h_1, \qquad u(L) = h_2. \qquad \textbf{(9.9)}$$

To make our problem complete, we need only add to Eq. (9.8) the inequality $0 < x < L$. The differential equation is relevant only over this interval, at whose endpoints the auxiliary conditions are given. Thus we expect that the shape of the cable will be determined by the boundary value problem

$$\frac{d^2 u}{dx^2} = \frac{w}{T}, \qquad 0 < x < L, \tag{9.10}$$

$$u(0) = h_1, \qquad u(L) = h_2. \tag{9.11}$$

We may solve this problem with the same routine used on initial value problems: find the general solution of the differential equation (9.10) and satisfy the two boundary conditions by appropriate choice of the arbitrary constants in the general solution. In the case at hand, since w and T do not vary with x, the general solution of Eq. (9.10) is

$$u(x) = \frac{w}{T} \frac{x^2}{2} + c_1 x + c_2. \tag{9.12}$$

The boundary conditions (9.11) require that

$$u(0) = h_1: \qquad c_2 = h_1,$$
$$u(L) = h_2: \quad \frac{wL^2}{2T} + c_1 L + c_2 = h_2. \tag{9.13}$$

These equations can be solved easily for c_1 and c_2:

$$c_1 = \frac{h_2 - h_1}{L} - \frac{wL}{2T}, \qquad c_2 = h_1. \tag{9.14}$$

From here we can get $u(x)$ in any of several forms. Perhaps the most informative one is

$$u(x) = h_1 \frac{L - x}{L} + h_2 \frac{x}{L} - \frac{w}{2T} x(L - x). \tag{9.15}$$

The first two terms represent a straight line between the endpoints, while the third represents the sag due to the load borne. It is clear that the shape assumed by the cable is parabolic.

In a more general situation we might suppose that the cable carries a distributed load that is not uniform. Indeed, the load may depend on x, u, and even u'. If $F(x, u, u')$ is the intensity of the load (units of force/length, positive up), then the boundary value problem describing the shape of the cable is

$$\frac{d^2 u}{dx^2} = -\frac{F(x, u, u')}{T}, \qquad 0 < x < L, \tag{9.16}$$

$$u(0) = h_1, \qquad u(L) = h_2. \tag{9.17}$$

Adapting this quite general equation to particular circumstances amounts to determining F.

Example 1

Find the boundary value problem for the shape of a uniform cable loaded by its own weight.

Let us suppose that the cable weighs ρ units of weight per unit length. A short segment (as shown in Fig. 9.2) has approximate length

$$\sqrt{1 + \left(\frac{du}{dx}\right)^2}\, \Delta x.$$

(This is the standard "element of arc length.") Then the load on the typical segment is

$$F(x, u, u')\Delta x = -\rho\sqrt{1 + \left(\frac{du}{dx}\right)^2}\, \Delta x.$$

From here we identify F and then obtain the boundary value problem

$$\frac{d^2u}{dx^2} = \frac{\rho}{T}\sqrt{1 + \left(\frac{du}{dx}\right)^2}, \qquad 0 < x < L, \tag{9.18}$$

$$u(0) = h_1, \qquad u(L) = h_2. \tag{9.19}$$

The solution of this problem is left to the exercises.

Heat Flow

Problems involving flows are frequently important to engineers. In most cases, fluid flow problems are too complex for this book, but certain problems in heat flow are well within our range. Let us consider the flow of heat by conduction along a rod of uniform cross-section (not necessarily circular). As shown in Fig. 9.3, the rod lies parallel to the x-axis, between the points $x = 0$ and $x = L$. We analyze a slice of the rod between points x and $x + \Delta x$, under the general assumption that all quantities are constant in time (steady heat flow).

Figure 9.3

Rod

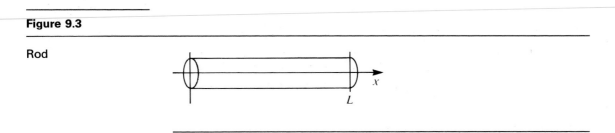

It is convenient to think in terms of a heat flux, which we denote by $q(x)$. This is the rate at which heat flows through a unit of cross-sectional surface (measured in calories/(meter)^2 · second or similar units) and is positive when heat flows to the right. (See Fig. 9.4.)

Figure 9.4

Slice Cut Out

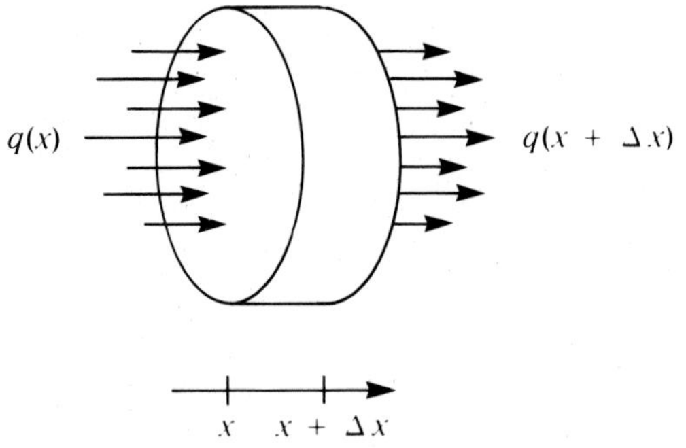

We must take into account the possibility that heat enters or leaves this section of rod by means other than conduction in the axial direction. Two of the mechanisms to be considered are

1. heat conduction through the cylindrical surface of the rod—perhaps by convection or radiation;

2. conversion of other forms of energy to heat within the slice.

Let us suppose that the rate at which heat enters the slice of rod by these means—or by any means other than axial conduction—is given by the "heat generation rate," $H \cdot \Delta x$. Of course, H would be measured in calories/meter · second or similar units.

Now an elementary application of the law of conservation of energy says that the rate at which energy enters our slice must be balanced by the rate at which it leaves. If the q's and H are positive, we then have

$$q(x)A + H\Delta x = q(x + \Delta x)A. \tag{9.20}$$

(This equation remains valid regardless of the signs of the q's and H.) By algebraic manipulation we can convert Eq. (9.20) into

$$\frac{q(x + \Delta x) - q(x)}{\Delta x} = \frac{H}{A}, \tag{9.21}$$

and on taking the limit as Δx approaches 0 we find

$$\frac{dq}{dx} = \frac{H}{A}. \tag{9.22}$$

The last equation simply expresses the law of conservation of energy for the situation we have set up. Unfortunately, it involves the heat flux, which is difficult to observe and measure. However, the heat flux q and the temperature u are related by *Fourier's law*,

$$q = -\kappa \frac{du}{dx}. \tag{9.23}$$

In words, the heat flux is negatively proportional to the temperature gradient. The constant of proportionality κ is called the conductivity. Fourier's law is a specific and precise version of the principle that heat flows from hotter to cooler regions.

We may substitute Fourier's law into the heat balance, Eq. (9.22), to get

$$\frac{d}{dx}\left(-\kappa \frac{du}{dx}\right) = \frac{H}{A} \tag{9.24}$$

or, assuming that κ does not vary with x,

$$\frac{d^2 u}{dx^2} = -\frac{H}{\kappa A}, \qquad 0 < x < L. \tag{9.25}$$

Of course, this equation tells only part of the story. In any particular instance we must specify H and add conditions at $x = 0$ and $x = L$ that describe the situation at the ends of the rod. There are many types of boundary conditions, of which the following are the most important.

1. The temperature is controlled:

$$u(0) = U_0 \qquad \text{or} \qquad u(L) = U_1. \tag{9.26}$$

2. The heat flux is controlled. Since heat flux is proportional to the derivative of temperature, this kind of condition becomes

$$\frac{du}{dx}(0) = S_0 \qquad \text{or} \qquad \frac{du}{dx}(L) = S_1. \tag{9.27}$$

In case one end of the rod is insulated, the heat flux there is 0, so the boundary condition would specify that u have slope 0 at that point.

Example 2

Find the boundary value problem for the temperature in an insulated electrical conductor whose ends are held at temperature U_0.

Assume that conditions are steady and that the conductor has the shape of a rod as analyzed above. Then Eq. (9.25) applies; we need only determine the appropriate form for H. If the electrical insulation is decent thermal insulation, we may suppose that there is no heat transfer across the cylindrical surface of the conductor. Thus H should represent only the rate at which electrical energy is converted to heat in a unit length of the rod; that is,

$$H\Delta x = I^2 R \Delta x, \qquad\qquad\qquad\qquad\qquad (9.28)$$

where I is the current in the rod and R is its resistance per unit length.
Finally, the boundary value problem that we seek is

$$\frac{d^2 u}{dx^2} = -\frac{I^2 R}{\kappa A}, \qquad 0 < x < L, \qquad\qquad (9.29)$$

$$u(0) = U_0, \qquad u(L) = U_0. \qquad\qquad\qquad (9.30)$$

The solution of this problem is left as an exercise.

Example 3

A cooling fin with uniform cross-section is attached to a hot object on its left end and is exposed to cool air along its length and at the right end. Find $u(x)$, the temperature in the fin, which is the solution of this boundary value problem:

$$\frac{d^2 u}{dx^2} = -0.09(20 - u), \qquad 0 < x < 10,$$

$$u(0) = 100, \qquad u(10) = 20.$$

In standard form the differential equation is

$$\frac{d^2 u}{dx^2} - 0.09u = -1.8.$$

Its particular solution is $u_p(x) = 20$, and the general solution is

$$u(x) = c_1 \cosh(0.3x) + c_2 \sinh(0.3x) + 20.$$

The boundary condition at the left end is $u(0) = 100$: $c_1 + 20 = 100$. At the right end the condition is

$$u(10) = 20: \quad c_1 \cosh(3) + c_2 \sinh(3) + 20 = 20.$$

From the two conditions we may determine

$$c_1 = 80, \qquad c_2 = -80 \frac{\cosh(3)}{\sinh(3)} = -80.4.$$

The complete solution is

$$u(x) = 80 \cosh(0.3x) - 80.4 \sinh(0.3x) + 20.$$

Exercises

1. Describe and sketch $u(x)$ as given by Eq. (9.15) when $h_1 = h_2$.

2. If w in Eq. (9.10) is positive, the differential equation tells us that the cable sags down (u cannot have a relative maximum). How?

3. Taking $h_1 = h_2$ in Eq. (9.15), find the total sag, $h_1 - u_{min}$, as a function of w, L, and T.

4. What inequalities must h_1 and h_2 satisfy in order to guarantee that $u(x)$ in Eq. (9.15) has a relative minimum in the interval $0 < x < L$?

5. Solve the boundary value problem in Eqs. (9.18) and (9.19). Note that the differential equation, although nonlinear, is first order in du/dx.

6. If the hanging cable is weightless and carries a load W concentrated at some point x_0, show that the equation describing the shape of the cable is

$$\frac{d^2u}{dx^2} = 0, \qquad 0 < x < x_0, \qquad x_0 < x < L,$$

$$\frac{du}{dx}(x_0+) - \frac{du}{dx}(x_0-) = \frac{W}{T}.$$

(By $f(x_0+)$ we mean $\lim f(x)$ as x approaches x_0 through values $x > x_0$.)

7. Solve the boundary value problem composed of the equations in Exercise 6 and the boundary conditions

$$u(0) = h_0, \qquad u(L) = h_1.$$

8. Sketch your solution to Exercise 7. Can you obtain the solution by another, simpler method (vector addition)?

9. (Poiseuille flow) A viscous fluid is flowing steadily between two parallel walls. (See Fig. 9.5.) The velocity u of the fluid at point (x, y) is supposed to be a function of y only. It may be shown that the

Figure 9.5

Flow

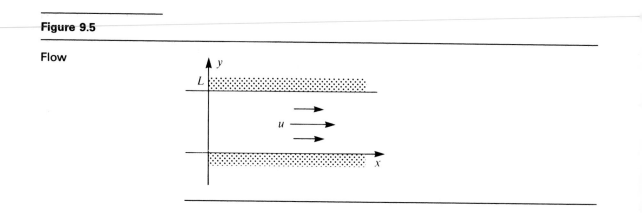

velocity satisfies the equation

$$\frac{d^2u}{dy^2} = \frac{1}{\mu} g, \qquad 0 < y < L$$

where μ is the viscosity and g is the pressure gradient (g must be negative if the fluid flows to the right). Solve the differential equation subject to the "no-slip" boundary conditions that $u(0) = 0$, $u(L) = 0$.

10. If u is the solution to Exercise 9, find the volumetric flow rate,

$$Q = \int_0^L u(y)\, dy.$$

11. Solve the boundary value problem formed by Eqs. (9.29) and (9.30).

12. Sketch the solution found in Exercise 11. Does $u(x)$ have a maximum or a minimum?

13. Solve this problem for the temperature in a cooling fin:

$$\frac{d^2u}{dx^2} = -\frac{hC}{\kappa A}(U_0 - u), \qquad 0 < x < L,$$

$$u(0) = U_1, \qquad \frac{du}{dx}(L) = 0.$$

14. Give a physical interpretation of the boundary value problem in Exercise 13.

15. The temperature in an electrically heated wire (see Fig. 9.6) satisfies

---

Figure 9.6

Hot Wire

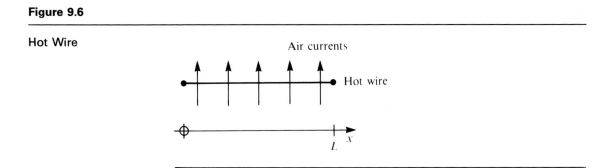

---

the boundary value problem

$$\frac{d^2u}{dx^2} = -\frac{I^2 R}{\kappa A} - \frac{hC}{\kappa A}(U - u), \qquad 0 < x < L,$$

$$u(0) = U_1, \qquad u(L) = U_1.$$

Solve for $u(x)$.

16. In Exercise 15, take $hC/\kappa A = 0.09$, $I^2 R/\kappa A = 50$, $U = U_1 = 30°C$, and $L = 50$ cm. Find maximum value of $u(x)$.

17. In porous soil between two parallel trenches a distance L apart, the depth $u(x)$ to the water table is believed to satisfy this boundary value problem.

$$\frac{d}{dx}\left(u\frac{du}{dx}\right) - e = 0, \qquad 0 < x < L,$$

$$u(0) = h_0, \qquad u(L) = h_1.$$

The boundary conditions say that water is pumped from the trenches to keep the levels as given. The parameter e measures the rate at which water filters down from the surface. Solve this problem. (Note that the differential equation is linear in u^2.)

9.2
Eigenvalue Problems

The boundary value problems that we saw in Section 9.1 did not differ greatly from initial value problems in the solution method. Some differences show up sharply, however, when we consider *homogeneous* boundary value problems—problems in which both the differential equation and boundary conditions are satisfied by the constant 0. As we shall see, it is sometimes possible that nonzero solutions also exist.

Buckling of a Column

First we consider a column bearing a load P directed along its length as shown in Fig. 9.7. The x-axis coincides with the vertical, and $u(x)$ is the distance of the centerline of the column from the x-axis. For simplicity we assume that the column is hinged at both the top and bottom, so it could bend there but not move out of line. Thus $u(x)$ is usually 0 for all x. However, if $u(x)$ is nonzero, the section of the column between the bottom and x experiences vertical forces as shown in Fig. 9.8. Although the forces balance, the fact that $u(x)$ is not 0 means that they create a couple, $-Pu$, which must also be balanced by an internal bending moment M in the beam.

In elasticity theory this expression is derived for the internal bending moment in a beam:

$$M = EI\frac{d^2u}{dx^2}. \tag{9.31}$$

Figure 9.7

Column Bearing Axial
Load

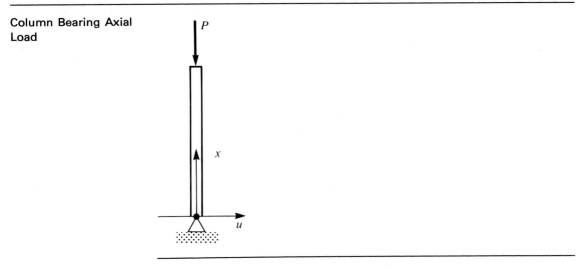

Figure 9.8

Forces and Moment
in a Column

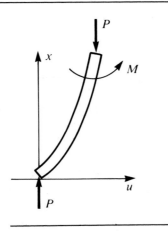

Here E is Young's modulus (a material property) and I is the moment of the cross-section of the beam ($I = ab^3/12$ for a rectangular cross-section of dimensions $a \times b$). Equating the couple to the internal bending moment gives the differential equation

$$EI\frac{d^2u}{dx^2} = -Pu, \qquad 0 < x < L, \tag{9.32}$$

or

$$\frac{d^2u}{dx^2} + \lambda^2 u = 0, \qquad 0 < x < L, \tag{9.33}$$

with $\lambda^2 = P/EI$. The conditions at the top and base of the column give these boundary conditions:

$$u(0) = 0, \qquad u(L) = 0. \tag{9.34}$$

Now it is easy to verify that $u(x) \equiv 0$ is a solution of the boundary value problem composed of Eqs. (9.33) and (9.34). However, let us look carefully for others. First, the general solution of the differential equation (9.33) is

$$u(x) = c_1 \cos \lambda x + c_2 \sin \lambda x, \tag{9.35}$$

with c_1 and c_2 arbitrary. The first boundary condition is $u(0) = 0$, which requires $c_1 = 0$, since $\cos 0 = 1$ and $\sin 0 = 0$. The second condition says $u(L) = 0$, from which we see that

$$c_2 \sin \lambda L = 0. \tag{9.36}$$

Certainly it is true that this condition is satisfied by choosing $c_2 = 0$, which thus gives $u(x) \equiv 0$. Indeed, if $\sin \lambda L \neq 0$, there is no other possibility. However, if the parameters of the problem are such that $\sin \lambda L = 0$, then

$$u(x) = c_2 \sin \lambda x$$

satisfies both differential equation (9.33) and boundary conditions (9.34) for any value of c_2!

What is the physical interpretation of this curious mathematical result? In the usual case, $u(x) \equiv 0$ is the solution. It means that the column is straight and transfers its load to the base. In the exceptional case, however, the column buckles, assuming a succession of shapes until it snaps or collapses. This buckling phenomenon is easily observed by pressing together the ends of a plastic ruler between the forefingers.

The critical combination of parameters for which buckling occurs must be such that

$$\sin \lambda L = 0 \tag{9.37}$$

or $\lambda L = \pi, 2\pi, 3\pi, \ldots$. By using the definition of λ we see that the first of these equalities, $\lambda = \pi/L$, is equivalent to

$$P = EI(\pi/L)^2, \tag{9.38}$$

which is called the Euler buckling load for the column.

The boundary conditions in Eq. (9.34) that correspond to hinged ends on the column are not the only ones possible. One of the ends could be "built in," or clamped, so that the boundary condition would require du/dx equal to 0 at that point, instead of u equal to 0.

The buckling problem introduced above is an example of a special class of boundary value problems, called *eigenvalue problems*. It is convenient to choose the following form for the problem:*

$$\frac{d}{dx}\left(p(x)\frac{du}{dx}\right) - q(x)u + \lambda^2 w(x)u = 0, \qquad \alpha < x < \beta, \tag{9.39}$$

$$au(\alpha) - a'u'(\alpha) = 0, \tag{9.40}$$

$$bu(\beta) + b'u'(\beta) = 0. \tag{9.41}$$

The buckling column problem fits into the form above with $p(x) \equiv 1$, $w(x) \equiv 1$, $q(x) \equiv 0$, $a = b = 1$, $a' = b' = 0$. Although the differential equation seems to have a strange form, it is one that often occurs naturally. Furthermore any second-order linear equation can be put into this form. It is assumed that the coefficient functions p, q and w are given, as well as the interval $\alpha < x < \beta$ and the numerical coefficients a, a', b, b' of the boundary conditions.

It is clear that $u(x) \equiv 0$ is always a solution of the problem in Eqs. (9.39), (9.40), and (9.41). The objective is this: find the numbers λ^2 for which a nonzero solution of the problem exists. Any such special numbers are called the *eigenvalues* of the problem consisting of Eqs. (9.39), (9.40), and (9.41). The nonzero solutions are called the *eigenfunctions* of the problem. Usually, it is desired to find both the eigenvalues and the eigenfunctions. In the general case any nonzero multiple of an eigenfunction is still an eigenfunction. For this reason it is customary to choose one eigenfunction in a simple or convenient form as a representative of all its multiples.

Example 1

We saw that the problem

$$\frac{d^2u}{dx^2} + \lambda^2 u = 0, \qquad 0 < x < L,$$

$$u(0) = 0, \qquad u(L) = 0,$$

has nonzero solutions if λ is one of the values π/L, $2\pi/L$, $3\pi/L,\ldots$. Usually, we write

$$\lambda_n^2 = (n\pi/L)^2, \qquad n = 1, 2, 3, \ldots$$

---

* The prime on a' and b' does not denote differentiation; it reminds us that these coefficients go with the derivatives of u.

to specify the eigenvalues conveniently. Then, "the" eigenfunction corresponding to the eigenvalue λ_n^2 is

$$u_n(x) = \sin \lambda_n x.$$

Let us see how we might go about solving an eigenvalue problem. Suppose first that we are able to find two independent solutions of Eq. (9)—call them $u_1(x)$ and $u_2(x)$. (Note that Eq. (9.39) contains the parameter λ^2, so u_1 and u_2 will also be functions of λ.) Then we may write the general solution

$$u(x) = c_1 u_1(x) + c_2 u_2(x), \tag{9.42}$$

where c_1 and c_2 are arbitrary, for the moment.

Now we apply the boundary conditions to $u(x)$ as given in Eq. (9.42). The first condition is developed as follows:

$$au(\alpha) - a'u'(\alpha) = 0, \tag{9.43}$$

or

$$a(c_1 u_1(\alpha) + c_2 u_2(\alpha)) - a'(c_1 u_1(\alpha) + c_2 u_2(\alpha)) = 0, \tag{9.44}$$

or

$$(au_1(\alpha) - a'u_1'(\alpha))c_1 + (au_2(\alpha) - a'u_2'(\alpha))c_2 = 0. \tag{9.45}$$

Similarly, the second boundary condition reduces to the equation

$$(bu_1(\beta) + b'u_1'(\beta))c_1 + (bu_2(\beta) + b'u_2'(\beta))c_2 = 0. \tag{9.46}$$

The question is: for what values of λ^2 can a nonzero solution of the differential equation (9.39) exist? Evidently, we may just as well ask: for what values of λ^2 are Eqs. (9.45) and (9.46) satisfied by c_1 and c_2 that are not both 0? Since Eqs. (9.45) and (9.46) are simultaneous linear equations in c_1 and c_2, the answer to both questions is: any value of λ^2 that makes the determinant of Eqs. (9.45) and (9.46) equal to 0.

Example 9.5

Solve the eigenvalue problem:

$$\frac{d^2 u}{dx^2} + \lambda^2 u = 0, \qquad 0 < x < L,$$

$$u(0) - u'(0) = 0, \qquad u(L) = 0.$$

The general solution of the differential equation is

$$u(x) = c_2 \cos \lambda x + c_2 \sin \lambda x. \tag{9.47}$$

Now the boundary conditions become these two equations in c_1 and c_2:

$$c_1 - c_2\lambda = 0, \tag{9.45'}$$

$$c_1 \cos \lambda L + c_2 \sin \lambda L = 0. \tag{9.46'}$$

Setting the determinant of this system to 0 gives the equation

$$\sin \lambda L + \lambda \cos \lambda L = 0 \tag{9.48}$$

or

$$\tan \lambda L = -\lambda. \tag{9.49}$$

Figure 9.9 shows the graphs of $y = \tan \lambda L$ and $y = -\lambda$ (for $L = 1$).

Figure 9.9

Graphs of $y = \tan \lambda$
and $y = -\lambda$

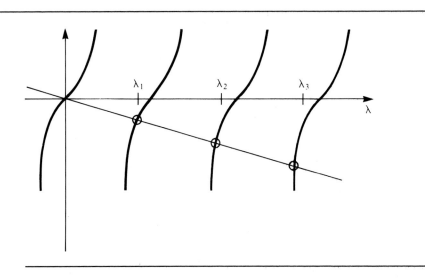

The abcissas of the intersections of these graphs are solutions of Eq. (9.49). It can be seen that there are infinitely many solutions. The first three are

$$\lambda_1 \cong 2.029, \qquad \lambda_2 \cong 4.913, \qquad \lambda_3 \cong 7.979,$$

and the first three eigenvalues are the squares of these numbers.

There is a general theory that covers eigenvalue problems, called Sturm-Liouville theory. First a definition.

Definition 9.1

The eigenvalue problem

$$\frac{d}{dx}\left(p(x)\frac{du}{dx}\right) - q(x)u + \lambda^2 w(x)u = 0, \qquad \alpha < x < \beta,$$

$$au(\alpha) - a'u'(\alpha) = 0,$$

$$bu(\beta) + b'u'(\beta) = 0,$$

is called a *regular Sturm-Liouville problem* if:
   1. α and β are finite,
   2. $p(x)$, $p'(x)$, $q(x)$, and $w(x)$ are continuous, $\alpha \le x \le \beta$;
   3. $p(x) > 0$ and $w(x) > 0$, $\alpha \le x \le \beta$;
   4. at least one of the coefficients a, a' and at least one of b, b' are nonzero;
   5. the parameter λ occurs only where shown; that is, not in the boundary conditions or in the coefficient functions.

Clearly, the two examples we have considered are regular Sturm-Liouville problems. The most common eigenvalue problems that do *not* fit into the definition are those for which the interval $\alpha < x < \beta$ is infinite or for which $p(x)$ is 0 at $x = \alpha$ or β.

The main theorem about eigenvalues is this.

Theorem 9.1

   1. A regular Sturm-Liouville problem has an infinite number of distinct eigenvalues: $\lambda_1^2 < \lambda_2^2 < \lambda_3^2 < \cdots$;
   2. $\lambda_n^2 \to \infty$ as $n \to \infty$;
   3. if $q(x) \ge 0$, $\alpha < x < \beta$, and the coefficients a, a', b, b' are nonnegative, then all eigenvalues are nonnegative.

Example 3

This eigenvalue problem

$$\frac{d}{dx}\left(e^{2x}\frac{du}{dx}\right) + \lambda^2 e^{2x}u = 0, \qquad 0 < x < L,$$

$$u(0) = 0, \qquad u(L) = 0,$$

is a regular Sturm-Liouville problem. According to the theorem, it has an infinite number of distinct eigenvalues, the nth of which grows large as n does. Furthermore, since $q(x) \equiv 0$ and $a = b = 1$, the eigenvalues are all nonnegative.

There are many important physical applications and interpretations of eigenvalues. For instance, the natural frequencies of vibration of strings, beams, structures, fluids, etc. are all intimately related to specific eigenvalue problems.

Exercises

1. Find the Euler buckling load for a wooden yardstick whose cross-section is a rectangle of dimensions $1\frac{1}{2}'' \times \frac{1}{4}''$. Use $E = 2 \times 10^6$ lb/in^2.

2. For a column whose cross-section is an $a \times b$ rectangle, there are two different values for I. Which should be used in computing the buckling load?

3. Suppose that the upper end of a column is constrained to be straight but allowed to move side to side, while the lower end is hinged. The deflection u then satisfies

$$\frac{d^2u}{dx^2} + \lambda^2 u = 0, \qquad 0 < x < L,$$

$$u(0) = 0, \qquad \frac{du}{dx}(L) = 0.$$

where $\lambda^2 = P/EI$. Solve this eigenvalue problem. (Find eigenvalues and eigenfunctions.)

4. How does the critical load in Exercise 3 compare with that found for a column hinged at both ends?

5. Find the eigenvalues and eigenfunctions for this problem:

$$\frac{d^2u}{dx^2} + \lambda^2 u = 0, \qquad 0 < x < L,$$

$$\frac{du}{dx}(0) = 0, \qquad \frac{du}{dx}(L) = 0.$$

(Note: study the possibility that $\lambda = 0$ is an eigenvalue as a special case.)

6. Show that 0 is an eigenvalue of this problem:

$$\frac{d}{dx}\left(p(x)\frac{du}{dx}\right) + \lambda^2 w(x)u = 0, \qquad \alpha < x < \beta,$$

$$\frac{du}{dx}(\alpha) = 0, \qquad \frac{du}{dx}(\beta) = 0.$$

Notice that $q(x)$ is absent from the differential equation. (Hint: set $\lambda = 0$ and solve the resulting differential equation in terms of an integral.)

7. Given a linear second-order differential equation

$$\frac{d^2u}{dx^2} + a_1(x)\frac{du}{dx} + a_2(x)u + \lambda^2 a_3(x)u = 0,$$

show that multiplying through by

$$p(x) = \exp\left(\int a_1(x)\,dx\right)$$

puts the equation into the form required for Sturm-Liouville theory, Eq. (9.49). Identify $q(x)$ and $w(x)$.

8. Only one of the problems below is a regular Sturm-Liouville problem. Which one is it, and where do the others fail?

   (a) $\dfrac{d}{dx}\left(x\dfrac{du}{dx}\right) + \lambda^2 xu = 0, \quad 0 < x < L,$

   $$u'(0) = 0, \quad u(L) = 0$$

   (b) $\dfrac{d^2u}{dx^2} + \lambda^2 u = 0, \quad 0 < x < L,$

   $$u(0) = 0, \quad u'(L) - \lambda^2 u(L) = 0$$

   (c) $\dfrac{d}{dx}\left(x\dfrac{du}{dx}\right) + \dfrac{\lambda^2}{x}u = 0, \quad 0 < x < L,$

   $$u'(0) = 0, \quad u(L) = 0$$

   (d) $\dfrac{d}{dx}\left(x\dfrac{du}{dx}\right) + \lambda^2 u = 0, \quad 1 < x < L,$

   $$u(1) = 0, \quad u(L) = 0$$

9. Solve this eigenvalue problem:

$$\frac{d}{dx}\left(e^{2x}\frac{du}{dx}\right) + \lambda^2 e^{2x}u = 0, \qquad 0 < x < L,$$

$$u(0) = 0, \qquad u(L) = 0.$$

(Hint: the differential equation really has constant coefficients.)

10. Solve the problem of Exercise 8(b). (Leave a transcendental equation to be solved for λ.)

11. Solve this eigenvalue problem:

$$\frac{d}{dx}\left(x\frac{du}{dx}\right) + \frac{\lambda^2}{x}u = 0, \qquad 1 < x < L,$$

$$u(1) = 0, \qquad u(L) = 0.$$

9.3

Singular Problems

Many boundary value problems of practical interest fail to be regular. Usually, such problems are singular (not regular) for one of two reasons.

1. The interval is infinite.

2. The interval is finite, but one endpoint (or both) is a singular point of the differential equation.*

The key to solving singular boundary value problems is to require that the magnitude of the solution—and its derivative, if necessary—have a finite upper bound throughout the interval of interest.

Example 1

A very long cooling fin is exposed all along its length to a medium at temperature U, while its (left) end is in contact with a body at temperature $U_0 > U$. Find the temperature $u(x)$ in the fin.

As in Exercise 13 of Section 9.1, we take the differential equation describing the temperature to be

$$\frac{d^2u}{dx^2} = -\frac{hC}{\kappa A}(U - u), \qquad 0 < x, \tag{9.50}$$

subject to the boundary condition $u(0) = U_0$. By writing the interval as $0 < x$ we announce that there is no right-hand endpoint: we are treating the long fin as if it were infinite. (Sometimes an interval with one finite endpoint is called semi-infinite.)

Designating $hC/\kappa A = \gamma^2$, we may find the general solution of the differential equation (1) to be

$$u(x) = c_1 e^{\gamma x} + c_2 e^{-\gamma x} + U, \tag{9.51}$$

where the arbitrary constants c_1 and c_2 are to be determined by the auxiliary conditions. The boundary condition at $x = 0$ (the only finite boundary) is not enough to determine both constants. We must also require that $|u(x)|$ have a finite upper bound throughout the interval $0 < x < \infty$. This can happen only if $c_1 = 0$; otherwise, $|u(x)|$ will grow exceedingly large as x increases.

With $c_1 = 0$ our solution reduces to

$$u(x) = c_2 e^{-\gamma x} + U, \tag{9.52}$$

and the boundary condition at $x = 0$ requires that

$$c_2 + U = U_0.$$

* See Chapter 4 for a discussion of singular points.

Hence we find that the solution of our singular boundary value problem is

$$u(x) = (U_0 - U)e^{-\gamma x} + U, \qquad 0 < x. \tag{9.53}$$

In the exercises at the end of this section we will check the validity of this result.

The reader might have noticed that we used sinh and cosh to solve the equations $u'' - \gamma^2 u = 0$ in Section 9.1 but used exponentials here. The solution can be expressed either way, of course; however, one form or the other is more convenient algebraically. A rule of thumb is to use sinh and cosh on finite intervals and exponentials on infinite intervals.

Many singular problems of the second type arise because of special coordinate systems. For instance, in polar or cylindrical coordinates we find that

$$\frac{1}{r}\frac{r}{dr}\left(r\frac{du}{dr}\right) \tag{9.54}$$

often appears where one would expect just d^2u/dx^2 in rectangular coordinates. Similarly, in spherical coordinates,

$$\frac{1}{\rho^2}\frac{d}{d\rho}\left(\rho^2\frac{du}{d\rho}\right) \tag{9.55}$$

often plays the role of second derivative. (Each of these is one term in the Laplacian, $\nabla^2 u$, expressed in the nonrectangular coordinate system.) In both cases the center of the coordinate system ($r = 0$ or $\rho = 0$) may be an endpoint of the interval of interest and, if so, is a singular point of the differential equation because some coefficient (in both cases the coefficient of the first derivative) becomes infinite at that point.

Example 2

(Poiseuille flow)   A viscous fluid flows through a tube of cylindrical cross-section. The axial component of velocity at any point is a function of the distance to the axis of the tube. In cylindrical coordinates the differential equation for this component of velocity is

$$\frac{1}{r}\frac{d}{dr}\left(r\frac{du}{dr}\right) = \frac{g}{\mu}, \qquad 0 < r < R, \tag{9.56}$$

where g is the pressure gradient (negative if flow is in the positive axial direction) and μ is the viscosity. The physical boundary is the inner surface of the tube, at $r = R$, where the velocity is 0:

$$u(R) = 0. \tag{9.57}$$

The differential equation (9.56) can be solved easily with just two integrations. We assume g and μ to be constants. Below is the development of the solution:

$$\frac{d}{dr}\left(r\frac{du}{dr}\right) = \frac{g}{\mu}r$$

$$r\frac{du}{dr} = \frac{g}{\mu}\frac{r^2}{2} + c_1 \qquad \text{(Integrated)}$$

$$\frac{du}{dr} = \frac{g}{\mu}\frac{r}{2} + \frac{c_1}{r} \qquad \text{(Divided by } r\text{)}$$

$$u(r) = \frac{g}{\mu}\frac{r^2}{4} + c_1 \ln r + c_2 \qquad \text{(Integrated).} \qquad\qquad (9.58)$$

Of course, c_1 and c_2 are arbitrary constants, to be determined by the auxiliary conditions: first that $|u(r)|$ have a finite upper bound in the interval $0 < r < R$; second, that the boundary condition (9.57) be satisfied.

The function $\ln r$ approaches $-\infty$ as r approaches 0. In order to satisfy the boundedness condition on $u(r)$, then, c_1 must be 0 in Eq. (9.58). Next the boundary condition (9.57) determines

$$c_2 = -\frac{gR^2}{4\mu}.$$

Finally, we have $u(r)$ in the form

$$u(r) = -\frac{g}{4\mu}(R^2 - r^2), \qquad 0 < r < R. \qquad\qquad (9.59)$$

Singular eigenvalue problems are often of interest too. In the case of singularities of the first kind, we frequently find that every positive number is an eigenvalue, so interpretation may be difficult. In singular problems of the other kind, the eigenvalues usually retain their discrete nature.

Example 3

The neutron flux u in a nuclear reactor measures the intensity of the reaction. If the reacting material is a uniform sphere of radius R, the flux satisfies the differential equation

$$\frac{1}{\rho^2}\frac{d}{d\rho}\left(\rho^2\frac{du}{d\rho}\right) + \frac{3(k-1)A}{\delta}u = 0, \qquad 0 < \rho < R, \qquad\qquad (9.60)$$

ture U. A balance of heat flux at the surface gives

$$-\kappa \frac{du}{dr}(R) = h(u(R) - U).$$

Using this boundary condition, find $u(r)$.

8. In a cylindrical bar carrying an alternating current the current density (amp/cm^2) varies linearly with the distance to the axis of the bar. We may assume then that the equation governing the temperature $u(r)$ is

$$\frac{1}{r}\frac{d}{dr}\left(r\frac{du}{dr}\right) = -Ar^2, \qquad 0 < r < R,$$

where A is a constant and r^2 appears on the right because the heating depends on the square of the current density. Find $u(r)$ if the boundary condition is $u(R) = U$.

9. Follow the directions for Exercise 8, but the boundary condition is

$$-\kappa \frac{du}{dr}(R) = h(u(R) - U),$$

indicating heat transfer by convection to a medium at temperature U.

10. Show that, in the limit as $h \to \infty$, the solution of Exercise 9 approaches the solution of Exercise 8. (Large values of h can be achieved by forcing the surrounding medium to flow rapidly.)

11. A spherical body in which heat is generated loses heat from its surface by radiation to the surroundings at temperature U. The internal temperature $u(\rho)$ satisfies the boundary value problem

$$\frac{1}{\rho^2}\frac{d}{d\rho}\left(\rho^2\frac{du}{d\rho}\right) = -\frac{H}{\kappa}, \qquad 0 < \rho < R,$$

$$-\kappa u'(R) = \sigma(u^4(R) - U^4).$$

Find $u(\rho)$ if $U = 0$. (Note: $u(\rho)$ and U must be measured in absolute temperature scales; σ is called the Stefan-Boltzman constant.)

12. Solve the problem in Exercise 11 with U nonzero. How does the value of u at $\rho = 0$ depend on U?

9.4

Green's Functions

In this section we study some aspects of the theory of the boundary value problem

$$\frac{d}{dx}\left(p(x)\frac{du}{dx}\right) - q(x)u = f(x), \qquad \alpha < x < \beta, \tag{9.64}$$

$$au(\alpha) - a'u'(\alpha) = A, \tag{9.65}$$

$$bu(\beta) + b'u'(\beta) = B. \tag{9.66}$$

Our primary objectives are to find conditions under which the problem has a solution and to find a compact representation of the solution, when there is one.

Let us start by writing the general solution of the differential equation (9.64) in the form

$$u(x) = c_1 u_1(x) + c_2 u_2(x) + u_p(x). \tag{9.67}$$

In this expression, u_1 and u_2 are supposed to be independent solutions of the homogeneous equation

$$\frac{d}{dx}\left(p(x)\frac{du}{dx}\right) - q(x)u = 0, \qquad \alpha < x < \beta. \tag{9.68}$$

It will simplify the algebra later if we impose some further requirements on these two solutions. We shall require that u_1 and u_2 satisfy these conditions:

$$au_1(\alpha) - a'u_1'(\alpha) = 0, \tag{9.69}$$

$$bu_2(\beta) + b'u_2'(\beta) = 0. \tag{9.70}$$

The last term in Eq. (9.67) is $u_p(x)$, a particular solution of the differential equation (9.64), which can be rewritten as

$$\frac{d^2u}{dx^2} + \frac{p'(x)}{p(x)}\frac{du}{dx} - \frac{q(x)}{p(x)}u = \frac{f(x)}{p(x)}. \tag{9.71}$$

In Section 3.5 of Chapter 3 we used variation of parameters to develop a formula that gives a particular solution of a second-order linear equation such as Eq. (9.71) in terms of the inhomogeneity. The formula is

$$u_p(x) = \int_\alpha^x \frac{u_2(x)u_1(t) - u_1(x)u_2(t)}{W(t)} \frac{f(t)}{p(t)} dt. \tag{9.72}$$

In the denominator of the integrand is the Wronskian

$$W(t) = \begin{vmatrix} u_1(t) & u_2(t) \\ u_1'(t) & u_2'(t) \end{vmatrix} = u_1(t)u_2'(t) - u_2(t)u_1'(t). \tag{9.73}$$

Later we will need to know something about the derivative of $u_p(x)$. The derivative of $u_p(x)$ as given in Eq. (9.72) can be found using Leibniz's rule (see the Appendix):

$$u_p'(x) = \int_\alpha^x \frac{u_2'(x)u_1(t) - u_1'(x)u_2(t)}{W(t)} \frac{f(t)}{p(t)} dt. \tag{9.74}$$

Incidentally, if we set $x = \alpha$ in Eq. (9.72) or Eq. (9.74), the interval of integration has length 0, and thus we find

$$u_p(\alpha) = 0, \qquad u_p'(\alpha) = 0. \tag{9.75}$$

Now having identified the terms in our expression for the general solution of the differential equation (9.64) or (9.71), let us apply the boundary conditions. The first one, Eq. (9.65), is just

$$ac_2 u_2(\alpha) - a'c_2 u_2'(\alpha) = A.$$

This expression is so simple because of the condition placed on u_1 in Eq. (9.69) and because of Eq. (9.75). Its solution is

$$c_2 = \frac{A}{au_2(\alpha) - a'u_2'(\alpha)}. \tag{9.76}$$

The second boundary condition also simplifies because of the condition (9.70) on u_2. The condition and the value of c_1 that follows from it are

$$c_1(bu_1(\beta) + b'u_1'(\beta)) + bu_p(\beta) + b'u_p'(\beta) = B.$$

$$c_1 = \frac{B}{bu_1(\beta) + b'u_1'(\beta)} - \frac{bu_p(\beta) + b'u_p'(\beta)}{bu_1(\beta) + b'u_1'(\beta)}. \tag{9.77}$$

With c_1 and c_2 determined, the solution, Eq. (9.67), becomes

$$u(x) = \frac{A}{au_2(\alpha) - a'u_2'(\alpha)} u_2(x) + \frac{B}{bu_1(\beta) + b'u_1'(\beta)} u_1(x)$$

$$- \frac{bu_p(\beta) + b'u_p'(\beta)}{bu_1(\beta) + b'u_1'(\beta)} u_1(x) + u_p(x). \tag{9.78}$$

The coefficient of the next-to-last term in Eq. (9.78) involves u_p and its derivative at β. If these values are supplied from Eqs. (9.72) and (9.74) and the condition (9.70) on u_2 is applied, there is a vast simplification. That whole term becomes just

$$u_1(x) \int_\alpha^\beta \frac{u_2(t)}{W(t)} \frac{f(t)}{p(t)} dt. \tag{9.79}$$

Now the last two terms of Eq. (9.78) (the one above and $u_p(x)$) both involve integrals. Together they are

$$\int_\alpha^\beta \frac{u_1(x)u_2(t)}{W(t)} \frac{f(t)}{p(t)} dt + \int_\alpha^x \frac{u_2(x)u_1(t) - u_1(x)u_2(t)}{W(t)} \frac{f(t)}{p(t)} dt.$$

The integrands have a great deal in common, but the intervals of integration are different. For t between α and x the integrands may be added, and there is some cancellation. For t between x and β, only the first integrand need be considered. Thus we can combine the two integrals into one, provided that we define the integrand as

$$
\begin{cases}
\dfrac{u_2(x)u_1(t)}{W(t)}\dfrac{f(t)}{p(t)}, & \alpha \le t \le x, \\[3mm]
\dfrac{u_1(x)u_2(t)}{W(t)}\dfrac{f(t)}{p(t)}, & x \le t \le \beta.
\end{cases}
\tag{9.80}
$$

Our solution will be further simplified by defining the *Green's function* for the problem in Eqs. (9.64), (9.65), and (9.66) as

$$
G(x, t) =
\begin{cases}
\dfrac{u_2(x)u_1(t)}{W(t)p(t)}, & \alpha \le t \le x, \\[3mm]
\dfrac{u_1(x)u_2(t)}{W(t)p(t)}, & x \le t \le \beta.
\end{cases}
\tag{9.81}
$$

Then the last two terms of our solution become

$$
\int_\alpha^\beta G(x, t)f(t)\, dt,
\tag{9.82}
$$

and the whole solution of the boundary value problem, Eqs. (9.64), (9.65), and (9.66) is

$$
u(x) = \frac{A}{au_2(\alpha) - a'u_2'(\alpha)}\, u_2(x) + \frac{B}{bu_1(\beta) + b'u_1'(\beta)}\, u_1(x)
$$

$$
+ \int_\alpha^\beta G(x, t)f(t)\, dt.
\tag{9.83}
$$

It is worthwhile going over the properties of each of the three terms in Eq. (9.83). Recall that $u_1(x)$ and $u_2(x)$ are solutions of the homogeneous differential equation (9.68). In addition,

(i) $\qquad \dfrac{A}{au_2(\alpha) - a'u_2'(\alpha)}\, u_2(x) \qquad\qquad$ (9.84)

satisfies the left-hand boundary condition, Eq. (9.65);

(ii) $\qquad \dfrac{B}{bu_1(\beta) + b'u_1'(\beta)}\, u_1(x) \qquad\qquad$ (9.85)

satisfies the right-hand boundary condition, Eq. (9.66); and

(iii) $\qquad \displaystyle\int_\alpha^\beta G(x, t)f(t)\, dt \qquad\qquad$ (9.86)

satisfies the nonhomogeneous differential equation (9.64) and *homogeneous* boundary conditions *at both ends*.

Example 1

Construct the Green's function for, and solve, the boundary value problem

$$\frac{d^2u}{dx^2} = f(x), \qquad 0 < x < L,$$

$$u(0) = A, \qquad u(L) + 2Lu'(L) = B.$$

The first order of business is to determine $u_1(x)$ and $u_2(x)$. Each of these is to be a solution of the homogeneous differential equation $u'' = 0$. Thus each must have the form $c_1 + c_2x$—with different constants for each, we expect. In addition, u_1 and u_2 have to satisfy

$$u_1(0) = 0, \qquad u_2(L) + 2Lu_2'(L) = 0.$$

Suitable choices are

$$u_1(x) = x, \qquad u_2(x) = 3L - x.$$

The Wronskian of these two functions is

$$W(x) = \begin{vmatrix} x & 3L - x \\ 1 & -1 \end{vmatrix} = -3L.$$

The Green's function can now be constructed, since we have all the data for Eq. (9.81). We obtain

$$G(x, t) = \begin{cases} \dfrac{(3L - x)t}{-3L}, & 0 \le t \le x, \\[2mm] \dfrac{x(3L - t)}{-3L}, & x \le t \le L. \end{cases}$$

In writing out the solution, as symbolized in Eq. (9.83), it is convenient to break the interval of integration at x. The result is

$$u(x) = \frac{A}{3L}(3L - x) + \frac{B}{3L}x + \int_0^x \frac{3L - x}{-3L} tf(t)\, dt + \int_x^L \frac{x}{-3L}(3L - t)f(t)\, dt.$$

Example 2

Same tasks as above, but for the boundary value problem

$$\frac{d^2u}{dx^2} - \gamma^2 u = f(x), \qquad 0 < x < L,$$

$$u'(0) = A, \qquad u(L) = B.$$

The two solutions of $u'' - \gamma^2 u$ are supposed to satisfy

$$u_1'(0) = 0, \qquad u_2(L) = 0,$$

and each must be a linear combination of $\cosh \gamma x$ and $\sinh \gamma x$. Two suitable choices are

$$u_1(x) = \cosh \gamma x, \qquad u_2(x) = \sinh \gamma(L - x),$$

which have for their Wronskian

$$W(x) = \begin{vmatrix} \cosh \gamma x & \sinh \gamma(L - x) \\ \gamma \sinh \gamma x & -\gamma \cosh \gamma(L - x) \end{vmatrix} = -\gamma \cosh \gamma L.$$

(It is necessary to use some hyperbolic identities to obtain this simplified result.)

The Green's function for this problem is

$$G(x, t) = \begin{cases} \dfrac{\sinh \gamma(L - x) \cosh \gamma t}{-\gamma \cosh \gamma L}, & 0 \leq x \leq t, \\[3mm] \dfrac{\cosh \gamma x \sinh \gamma(L - t)}{-\gamma \cosh \gamma L}, & t \leq x \leq L. \end{cases}$$

From here we can follow Eq. (9.83) to obtain this representation for the solution:

$$\begin{aligned} u(x) = {} & \frac{A}{-\gamma \cosh \gamma L} \sinh \gamma(L - x) + \frac{B}{\cosh \gamma L} \cosh \gamma x \\[2mm] & + \int_0^x \frac{\sinh \gamma(L - x) \cosh \gamma t}{-\gamma \cosh \gamma L} f(t) \, dt \\[2mm] & + \int_x^L \frac{\cosh \gamma x \sinh \gamma(L - t)}{-\gamma \cosh \gamma L} f(t) \, dt. \end{aligned}$$

The following theorem summarizes the important properties of the Green's function, although Eq. (9.81) is the most convenient way to construct it. Since the theorem treats the Green's function as a function of x, we repeat Eq. (9.81) with the two lines reversed:

$$G(x, t) = \begin{cases} \dfrac{u_1(x)u_2(t)}{W(t)p(t)}, & \alpha \leq x \leq t, \\[3mm] \dfrac{u_2(x)u_1(t)}{W(t)p(t)}, & t \leq x \leq \beta. \end{cases} \tag{9.87}$$

Theorem 9.2   The Green's function for the problem

$$\frac{d}{dx}\left(p(x)\frac{du}{dx}\right) - q(x)u = f(x), \qquad \alpha < x < \beta, \tag{i}$$

$$au(\alpha) - a'u'(\alpha) = 0, \tag{ii}$$

$$bu(\beta) + b'u'(\beta) = 0 \tag{iii}$$

is a function $G(x, t)$ characterized by these properties.

1. $G(x, t)$ satisfies the homogeneous differential equation

$$\frac{d}{dx}\left(p(x)\frac{dG}{dx}\right) - q(x)G = 0 \tag{iv}$$

for $\alpha < x < t$ and $t < x < \beta$.

2. $G(x, t)$ satisfies the homogeneous boundary condition (ii) at $x = \alpha$ and (iii) at $x = \beta$.

3. $G(x, t)$ is continuous throughout the interval $\alpha < x < \beta$.

4. The derivative of G with respect to x is discontinuous at $x = t$, where it satisfies

$$\frac{d}{dx}G(t+, t) - \frac{d}{dx}G(t-, t) = \frac{1}{p(t)}. \tag{9.88}$$

If the Green's function exists, the solution of the boundary value problem composed of (i), (ii), and (iii) is

$$u(x) = \int_\alpha^\beta G(x, t)f(t)\,dt. \tag{9.89}$$

Proof: The construction of $G(x, t)$ from the properties given is easy. Here is a sketch of the process. First, properties 1 and 2 assure us that

$$G(x, t) = \begin{cases} u_1(x)v_1(t), & \alpha \le x \le t, \\ u_2(x)v_2(t), & t \le x \le \beta, \end{cases}$$

where u_1 and u_2 are solutions of (iv) that satisfy boundary conditions (ii) and (iii), respectively. Next the continuity condition 3 implies that

$$v_2(t) = u_1(t)\phi(t), \qquad v_1(t) = u_2(t)\phi(t),$$

where ϕ is some function. In fact, it is specified by condition 4, which says

$$u_2'(t)u_1(t)\phi(t) - u_1'(t)u_2(t)\phi(t) = \frac{1}{p(t)}. \tag{9.90}$$

The resulting formulas for $G(x, t)$ agree exactly with Eq. (9.87).

The last statement of the theorem (the most important one for applications) can be verified directly.

We must now see whether the solution process carried out at the beginning of this section might fail. There were several tacit assumptions in our work: (1) that solutions of the homogeneous differential equation (9.68) exist; (2) that u_1 and u_2—solutions of Eq. (9.68) that satisfy boundary conditions (9.69) and (9.70)—are independent; (3) that the quantities

$$au_2(\alpha) - a'u_2'(\alpha) \qquad \text{and} \qquad bu_1(\beta) + b'u_1'(\beta) \qquad \textbf{(9.91)}$$

are both nonzero. (See Eqs. (9.76), (9.77), and (9.78).)

The first assumption is made valid by assuming that $p(x)$, $p'(x)$ and $q(x)$ are continuous and that $p(x)$ is never 0 for $\alpha \le x \le \beta$. Then the existence of two independent solutions of Eq. (5) is assured.

The second and third assumptions are linked and are not so easy to dismiss. In fact, if either of the quantities in Eq. (9.91) is 0, then the other one is also, and $u_1(x)$ and $u_2(x)$ are multiples of each other. The whole solution process that we have used then fails, since there would be several divisions by zero. Even under such circumstances, it is not impossible for the given boundary value problem to have a solution; it is impossible for the solution to be unique, as Theorem 9.3 states.

Theorem 9.3

The boundary value problem

$$\frac{d}{dx}\left(p(x)\frac{du}{dx}\right) - q(x)u = f(x), \qquad \alpha < x < \beta,$$

$$au(\alpha) - a'u'(\alpha) = A,$$

$$bu(\beta) + b'u'(\beta) = B$$

has a unique solution if and only if the homogeneous problem

$$\frac{d}{dx}\left(p(x)\frac{du}{dx}\right) - q(x)u = 0,$$

$$au(\alpha) - a'u'(\alpha) = 0,$$

$$bu(\beta) + b'u'(\beta) = 0$$

has no solution other than $u(x) \equiv 0$.

Example 3

Let us solve the simple boundary value problem

$$\frac{d^2u}{dx^2} + \gamma^2 u = f(x), \qquad 0 < x < L,$$

$$u(0) = A, \qquad u(L) = B.$$

The functions u_1 and u_2 that we need in our solution process satisfy $u'' + \gamma^2 u = 0$ and also

$$u_1(0) = 0, \qquad u_2(L) = 0.$$

Evidently, we may choose $u_1(x) = \sin \gamma x$ and $u_2(x) = \sin \gamma(L - x)$. The Wronskian of these two is

$$W(x) = \begin{vmatrix} \sin \gamma x & \sin \gamma(L - x) \\ \gamma \cos \gamma x & -\cos \gamma(L - x) \end{vmatrix} = -\gamma \sin \gamma L.$$

The Green's function for this problem would be (following Eq. (9.81))

$$G(x, t) = \begin{cases} \dfrac{\sin \gamma(L - x) \sin \gamma t}{-\gamma \sin \gamma L}, & 0 \le t \le x, \\[3mm] \dfrac{\sin \gamma x \sin \gamma(L - t)}{-\gamma \sin \gamma L}, & x \le t \le L. \end{cases}$$

Finally, the solution of the problem originally stated can be given as

$$u(x) = \frac{A}{\sin \gamma L} \sin \gamma x + \frac{B}{\sin \gamma L} \sin \gamma(L - x) + \int_0^L G(x, t)f(t) \, dt.$$

Notice that the denominators in both the first and second terms as well as in the Green's function are all 0 if $\sin \gamma L = 0$. If this equation is fulfilled, then $u(x) = \sin \gamma x$ is a nontrivial solution of

$$\frac{d^2 u}{dx^2} + \gamma u = 0, \quad 0 < x < L,$$

$$u(0) = 0, \quad u(L) = 0,$$

just as promised by Theorem 9.3.

Exercises

1. Find the derivative with respect to x of $u_p(x)$ given in Eq. (9.72). (Hint: use Leibniz's rule.)

2. Verify that the next to last term in Eq. (9.78) simplifies to the expression given in Eq. (9.79).

3. Find the first and second derivatives of

$$u(x) = \int_\alpha^\beta G(x, t)f(t) \, dt$$

where $G(x, t)$ is the Green's function for a problem as stated in Theorem 9.2. It is convenient to break the interval of integration at $t = x$.

4. Use the results of Exercise 3 to verify the last statement of Theorem 9.2: $u(x)$, as given in Exercise 3 or Eq. (9.89), is the solution of (i), (ii), and (iii).

In Exercises 5–15, construct the Green's function for the problem as stated.

5. $\dfrac{d^2u}{dx^2} = f(x),\ 0 < x < L,$

$u(0) = 0,\ u(L) = 0.$

6. $\dfrac{d^2u}{dx^2} = f(x),\ 0 < x < L,$

$u(0) = 0,\ u'(L) = 0.$

7. $\dfrac{d^2u}{dx^2} - \gamma^2 u = f(x),\ 0 < x < L,$

$u'(0) = 0,\ u'(L) = 0$

8. $\dfrac{d^2u}{dx^2} - \gamma^2 u = f(x),\ 0 < x < L,$

$u(0) = 0,\ u(L) = 0$

9. $\dfrac{d^2u}{dx^2} + \gamma^2 u = f(x),\ 0 < x < L,$

$u(0) = 0,\ u'(L) = 0$

10. $\dfrac{d^2u}{dx^2} + \gamma^2 u = f(x),\ 0 < x < L,$

$u'(0) = 0,\ u'(L) = 0$

11. $\dfrac{d}{dr}\left(r\dfrac{du}{dr}\right) = rf(r),\ 1 < r < e,$

$u(1) = 0,\ u(e) = 0$

12. $\dfrac{d}{dr}\left(r\dfrac{du}{dr}\right) - \dfrac{\gamma^2}{r}u = rf(r),\ 1 < r < e,$

$u(1) = 0,\ u(e) = 0$

13. $\dfrac{d}{d\rho}\left(\rho^2\dfrac{du}{d\rho}\right) = \rho^2 f(\rho),\ \alpha < \rho < \beta,$

$u(\alpha) = 0,\ u(\beta) = 0\ (\alpha > 0)$

14. $\dfrac{d}{d\rho}\left(\rho^2\dfrac{du}{d\rho}\right) - \gamma^2 u = \rho^2 f(\rho),\ \alpha < \rho < \beta,$

$u'(\alpha) = 0,\ u(\beta) = 0\ (\alpha > 0)$

15. $\dfrac{d}{dx}\left(e^x\dfrac{du}{dx}\right) = e^x f(x),\ 0 < x < 1,$

$u(0) = 0,\ u(1) = 0$

16. For what values of γ does the problem in Exercise 9 fail to have a unique solution (if it has any)?

17. Follow the directions in Exercise 16, but for Exercise 10.

18. With $f(x) = x$, solve the problem stated in Exercise 5, first using the Green's function and then by a more direct method.

19. Solve the problem in Exercise 9, using the Green's function and the direct method. Take $f(x) = \sin \gamma x$.

20. Confirm the statement in the paragraph preceding Theorem 8.3: show that if u_1 and u_2 both satisfy the homogeneous boundary condition Eq. (9.69), then the Wronskian $W(\alpha) = 0$, and therefore $W(t) \equiv 0$, $\alpha \le t \le \beta$.

9.5

Eigenfunction Series: Two Examples

In Section 9.2 we defined and found the eigenvalues and eigenfunctions of the problem

$$\frac{d^2u}{dx^2} + \lambda^2 u = 0, \qquad 0 < x < L, \tag{9.92}$$

$$u(0) = 0, \qquad u(L) = 0. \tag{9.93}$$

If we think of the second derivative as an operator S, then the differential equation (9.92) may be rewritten as

$$S(u) = -\lambda^2 u, \tag{9.94}$$

and we can give a simple verbal description of an eigenfunction: a function, among those satisfying the boundary conditions, on which the effect of the linear operator $S = d^2/dx^2$ is the same as multiplication by a number.

We can use this property of eigenfunction to describe the effect of the second derivative on a linear combination of eigenfunctions of the problem in Eqs. (9.92) and (9.93). Recall that the eigenfunctions and eigenvalues of the problem in Eqs. (9.92) and (9.93) are

$$u_n(x) = \sin \lambda_n x, \qquad \lambda_n^2 = (n\pi/L)^2, \qquad n = 1, 2, 3, \ldots. \tag{9.95}$$

Then a finite linear combination of eigenfunctions is a function of the form

$$f(x) = b_1 \sin \lambda_1 x + b_2 \sin \lambda_2 x + \cdots + b_k \sin \lambda_k x.$$

The second derivative of such a function is

$$\frac{d^2f}{dx^2} = \frac{d^2}{dx^2}(b_1 \sin \lambda_1 x + b_2 \sin \lambda_2 x + \cdots + b_k \sin \lambda_k x)$$

$$= -b_1\lambda_1^2 \sin \lambda_1 x - b_2\lambda_2^2 \sin \lambda_2 x - \cdots - b_k\lambda_k^2 \sin \lambda_k x$$

because of the linearity of the second-derivative operator and the defining property, Eq. (9.94) of its eigenfunctions.

Since there is an infinite family of eigenfunctions of Eqs. (9.92) and (9.93), we should naturally consider infinite linear combinations of them, with the form of series,

$$f(x) = b_1 \sin \lambda_1 x + b_2 \sin \lambda_2 x + \cdots + b_n \sin \lambda_n x + \cdots$$

$$= \sum_{n=1}^{\infty} b_n \sin \lambda_n x.$$

We must ask and answer certain questions about these infinite series.

1. Given a function $f(x)$, how do we find the coefficients $b_1, b_2, \ldots$?
2. What kind of function $f(x)$ can be represented in this form?
3. What kinds of operations can be carried out on such a series?
4. How do these questions and answers generalize to other eigen-value problems?

In this section we will rough out the answers to the first two questions without worrying about mathematical niceties.

As a first step in answering question 1, we might suppose that some function f can be expressed as a series of eigenfunctions,

$$f(x) = \sum_{n=1}^{\infty} b_n \sin \lambda_n x, \qquad 0 < x < L, \tag{9.96}$$

and then try to find the relation between the coefficients b_n and the function f. The solution of this problem depends on a simple fact about definite integrals of sines:

$$\int_0^L \sin \frac{n\pi x}{L} \sin \frac{m\pi x}{L} \, dx = 0 \qquad (n \neq m)$$

or, in view of Eq. (9.95),

$$\int_0^L \sin \lambda_n x \sin \lambda_m x \, dx = 0 \qquad (n \neq m). \tag{9.97}$$

This last equation is called the *orthogonality relation* for the eigenfunctions.

Now if $f(x)$ is indeed equal to the series, as shown in Eq. (9.96), then the same operation applied to both sides should give the same result. In particular, we choose to multiply through Eq. (9.96) by $\sin \lambda_1 x$ and then integrate from 0 to L:

$$\int_0^L f(x) \sin \lambda_1 x \, dx = \int_0^L \sum_{n=1}^{\infty} b_n \sin \lambda_n x \sin \lambda_1 x \, dx$$

$$= \sum_{n=1}^{\infty} b_n \int_0^L \sin \lambda_n x \sin \lambda_1 x \, dx. \tag{9.98}$$

Because of the relation (9.97), all of the integrals in Eq. (9.98) are 0, except the first one, in which $n = 1$:

$$\int_0^L \sin \lambda_1 x \sin \lambda_1 x \, dx = \int_0^L \sin^2 \lambda_1 x \, dx = \frac{L}{2}. \tag{9.99}$$

Thus Eq. (9.98) reduces to

$$\int_0^L f(x) \sin \lambda_1 x \, dx = b_1 \frac{L}{2}, \tag{9.100}$$

from which we can easily find b_1.

This same idea works for other coefficients. We can multiply through Eq. (9.96) by $\sin \lambda_m x$ (where m is any fixed integer) and then integrate from 0 to L to get

$$\int_0^L f(x) \sin \lambda_m x \, dx = \int_0^L \sum_{n=1}^\infty b_n \sin \lambda_n x \sin \lambda_m x \, dx$$

$$= \sum_{n=1}^\infty b_n \int_0^L \sin \lambda_n x \sin \lambda_m x \, dx. \qquad \text{(9.101)}$$

All of the integrals in the series above are 0, according to Eq. (9.97), except the one in which $n = m$. That one is

$$\int_0^L \sin^2 \lambda_m x \, dx = \frac{L}{2}. \qquad \text{(9.102)}$$

Thus Eq. (9.101) reduces to

$$\int_0^L f(x) \sin \lambda_m x \, dx = b_m \frac{L}{2}, \qquad \text{(9.103)}$$

an equation that is valid for any $m = 1, 2, 3, \ldots$. We easily solve Eq. (9.103) for b_m and replace m on both sides by n, to get

$$b_n = \frac{2}{L} \int_0^L f(x) \sin \lambda_n x \, dx \qquad \text{(9.104)}$$

as the equation for the coefficients of the series (9.96). These coefficients are called the *Fourier sine coefficients* of the function f, and the series is called the *Fourier sine series* of f.

Example 1

Find the Fourier sine coefficients and the Fourier sine series of

$$f(x) = x(L - x), \qquad 0 < x < L. \qquad \text{(9.105)}$$

(Assume that the function is equal to its series.)

We use the formula (9.104) to find the coefficients of the series. The development follows:

$$b_n = \frac{2}{L} \int_0^L x(L - x) \sin \lambda_n x \, dx$$

$$= \frac{2}{L} \left(L \int_0^L x \sin \lambda_n x \, dx - \int_0^L x^2 \sin \lambda_n x \, dx \right)$$

$$= 2 \left[\frac{\sin \lambda_n x}{\lambda_n^2} - \frac{x \cos \lambda_n x}{\lambda_n} \right]_0^L - \frac{2}{L} \left[\frac{2x}{\lambda_n^2} \sin \lambda_n x + \left(\frac{2}{\lambda_n^3} - \frac{x^2}{\lambda_n} \right) \cos \lambda_n x \right]_0^L$$

$$= 2 \left[\frac{-L \cos \lambda_n L}{\lambda_n} \right] - \frac{2}{L} \left[\left(\frac{2}{\lambda_n^3} - \frac{L^2}{\lambda_n} \right) \cos \lambda_n L - \left(\frac{2}{\lambda_n^3} \right) \right]$$

$$= \frac{4}{L \lambda_n^3} (1 - \cos \lambda_n L). \qquad \text{(9.106)}$$

It is still more convenient to substitute $n\pi/L = \lambda_n$; then we get this series:

$$f(x) = \frac{4L^2}{\pi^3} \sum_{n=1}^{\infty} \frac{1 - \cos n\pi}{n^3} \sin \lambda_n x, \qquad 0 < x < L, \qquad \textbf{(9.107)}$$

$$f(x) = \frac{4L^2}{\pi^3} \left(2 \sin \frac{\pi x}{L} + \frac{2}{27} \sin \frac{3\pi x}{L} + \frac{2}{125} \sin \frac{5\pi x}{L} + \cdots \right), \qquad 0 < x < L. \qquad \textbf{(9.108)}$$

Note that $1 - \cos n\pi$ is 0 or 2, depending on the parity of n.

Our work above indicates that if a function $f(x)$ is to be equal to a series, as in Eq. (9.96), then the coefficients must be chosen according to Eq. (9.104). Now we must take up the second question about the kind of function that has such a series. The complete answer is not known, but there is a partial answer that is suitable for most applications.

Definition 9.2

A function f is *sectionally smooth* on an interval if that interval can be partitioned into a finite number of subintervals on each of which f is bounded and continuous and has a bounded, continuous derivative.

The fundamental idea in sectional smoothness is that $f(x)$ is pieced together out of continuous, differentiable functions. At the boundaries between the pieces, either f or its derivative may have a jump.

Example 2

Below are listed some functions that are (or are not) sectionally smooth on the interval $0 < x < L$.

(a) $\quad f(x) = \begin{cases} x, & 0 < x < L/2, \\ L - x, & L/2 \le x < L. \end{cases}$

This function is sectionally smooth, $0 < x < L$. It is continuous and has a continuous derivative on each of the subintervals $0 < x < L/2$ and $L/2 < x < L$. At the point $x = L/2$ the derivative jumps from 1 to -1, as indicated by the corner there. (See Fig. 9.10.)

(b) $\quad f(x) = \begin{cases} -1, & 0 < x < L/2, \\ +1, & L/2 \le x < L. \end{cases}$

This function is sectionally smooth, $0 < x < L$. It has a jump at $x = L/2$;

Figure 9.10

A Continuous and
Sectionally Smooth
Function

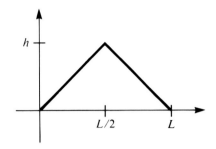

Figure 9.11

A Discontinuous and
Sectionally Smooth
Function

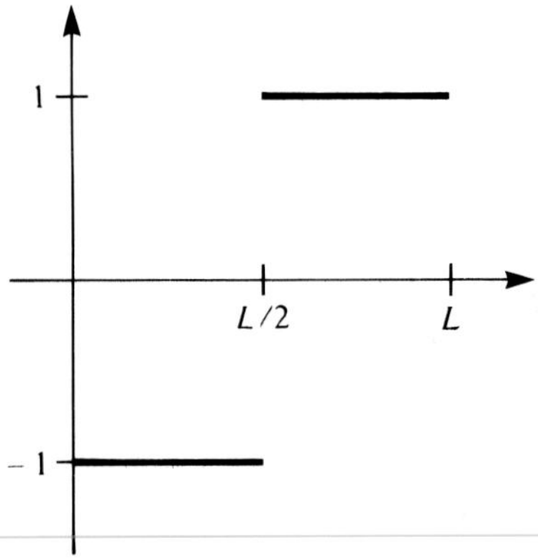

elsewhere, however, it is continuous and has derivative 0. (See Fig. 9.11.)

(c) $f(x) = \sqrt{x}, \quad 0 < x < L.$

This function *fails* to be sectionally smooth on the interval given because its derivative is not bounded (becomes infinite) as x approaches 0. (See Fig. 9.12.)

Figure 9.12

A Continuous
Function that is not
Sectionally Smooth

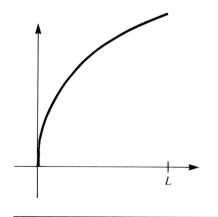

L

The family of sectionally smooth functions includes most of the functions that model physical phenomena. A discontinuity in a function offers a convenient way to represent an abrupt change.

A sectionally smooth function can be represented by means of a series such as Eq. (9.96). A precise statement of the facts is Theorem 9.4.

Theorem 9.4

Let $f(x)$ be sectionally smooth, $0 < x < L$, and let

$$b_n = \frac{2}{L} \int_0^L f(x) \sin \lambda_n x \, dx,$$

with $\lambda_n = n\pi/L$. Then the series

$$\sum_{n=1}^{\infty} b_n \sin \lambda_n x$$

converges at every x, $0 \leq x \leq L$. Its sum is 0 at $x = 0$ and $x = L$, and elsewhere is

$$\lim_{h \to 0} \frac{f(x + h) + f(x - h)}{2}, \qquad 0 < x < L. \tag{9.109}$$

The last expression seems strange, but it amounts to just $f(x)$ at any x where f is continuous. There can be at most a finite number of points in the interval where f is not continuous, and at these points the limit is easily calculated.

Example 3

Find the Fourier sine series for the function

$$f(x) = \begin{cases} -1, & 0 < x < L/2, \\ 1, & L/2 \le x < L, \end{cases}$$

and find the sum of that series.

The coefficients for the series (9.96) are given by Eq. (9.104). In order to carry out the integration it is necessary to break the interval of integration at $L/2$, so that on each subinterval $f(x)$ has a simple formula:

$$\begin{aligned} b_n &= \frac{2}{L} \int_0^L f(x) \sin \lambda_n x \, dx \\ &= \frac{2}{L} \left(\int_0^{L/2} (-1) \sin \lambda_n x \, dx + \int_{L/2}^L (1) \sin \lambda_n x \, dx \right) \\ &= \frac{2}{L} \left(\frac{\cos \lambda_n x}{\lambda_n} \Big|_0^{L/2} + \frac{-\cos \lambda_n x}{\lambda_n} \Big|_{L/2}^L \right) \\ &= 2 \frac{2 \cos (\lambda_n L/2) - \cos (\lambda_n L) - 1}{L \lambda_n} \\ &= 2 \frac{2 \cos (n\pi/2) - \cos (n\pi) - 1}{n\pi}. \end{aligned}$$

The function f is sectionally smooth, as we saw in Example 2. According to Theorem 9.4, then, the series converges at every point of the interval. The only point of discontinuity is at $x = L/2$. Just to the right of $L/2$, $f(x) = 1$; just to the left, $f(x) = -1$. Thus

$$\lim_{h \to 0} \frac{f\left(\frac{L}{2} + h\right) + f\left(\frac{L}{2} - h\right)}{2} = 0.$$

The series and its sum are given below:

$$\frac{2}{\pi} \sum_{n=1}^{\infty} \frac{2 \cos (n\pi/2) - \cos (n\pi) - 1}{n} \sin \frac{n\pi x}{L} = \begin{cases} 0, & x = 0, \\ -1, & 0 < x < L/2, \\ 0, & x = L/2, \\ 1, & L/2 < x < L, \\ 0, & x = L. \end{cases}$$

Usually, if a function f satisfies the hypotheses of the theorem, we would say that it is equal to its series, even though the equality might fail at a finite number of points in the interval. For instance, the sum of the series in Example 3 is different from $f(x)$ at $x = L/2$, where $f(x) = 1$, and at $x = 0$ and L, where $f(x)$ had not been defined.

The eigenvalue problem in Eqs. (9.92) and (9.93) and the corresponding series probably constitute the most important example of eigenfunction series. In second place is the problem

$$\frac{d^2u}{dx^2} + \lambda^2 u = 0, \qquad 0 < x < L, \tag{9.110}$$

$$\frac{du}{dx}(0) = 0, \qquad \frac{du}{dx}(L) = 0, \tag{9.111}$$

which has the solutions (see Exercise 5 of Section 9.2)

$$\begin{aligned} \lambda_0^2 &= 0, \qquad u_0 = 1, \\ \lambda_n^2 &= (n\pi/L)^2, \qquad u_n = \cos \lambda_n x, \qquad n = 1, 2, 3, \ldots. \end{aligned} \tag{9.112}$$

These functions also satisfy an orthogonality relation, similar to Eq. (9.97):

$$\int_0^L u_m(x)u_n(x) \, dx = 0 \qquad (n \neq m). \tag{9.113}$$

It is the orthogonality relation, used in the same fashion as for the Fourier sine series, that allows us to find the coefficients of the *Fourier cosine series*

$$f(x) = a_0 + \sum_{n=1}^{\infty} a_n \cos \lambda_n x, \qquad 0 < x < L. \tag{9.114}$$

The a's, called the *Fourier cosine coefficients*, are then determined to be

$$a_0 = \frac{1}{L} \int_0^L f(x) \, dx, \qquad a_n = \frac{2}{L} \int_0^L f(x) \cos \lambda_n x \, dx. \tag{9.115}$$

Example 4

Find the Fourier cosine series of the function

$$f(x) = x, \qquad 0 < x < L.$$

The formulas (9.115) are to be used to find the coefficients. First, we calculate

$$a_0 = \frac{1}{L} \int_0^L x \, dx = \frac{1}{L} \frac{x^2}{2} \Big|_0^L = \frac{L}{2}.$$

It may be verified that $L/2$ is indeed the average value of $f(x)$, as its formula requires. Next,

$$\begin{aligned} a_n &= \frac{2}{L} \int_0^L x \cos \lambda_n x \, dx = \frac{2}{L} \left(\frac{\cos \lambda_n x}{\lambda_n^2} + \frac{x \sin \lambda_n x}{\lambda_n} \right) \Big|_0^L \\ &= \frac{2}{L} \frac{\cos n\pi - 1}{\lambda_n^2} = -\frac{2L(1 - \cos n\pi)}{n^2 \pi^2}. \end{aligned}$$

Therefore the series required is

$$\frac{L}{2} - \frac{2L}{\pi^2} \sum_{n=1}^{\infty} \frac{1 - \cos n\pi}{n^2} \cos \lambda_n x.$$

The facts about the convergence of Fourier cosine series are similar to those for the sine series.

Theorem 9.5

Let $f(x)$ be sectionally smooth, $0 < x < L$, and let

$$a_0 = \frac{1}{L} \int_0^L f(x) \, dx, \qquad a_n = \frac{2}{L} \int_0^L f(x) \cos \lambda_n x \, dx,$$

with $\lambda_n = n\pi/L$. Then the series

$$a_0 + \sum_{n=1}^{\infty} a_n \cos \lambda_n x$$

converges at every x, $0 \le x \le L$. Its sum is $f(0+)$ at $x = 0$, $f(L-)$ at $x = L$, and elsewhere

$$\lim_{h \to 0} \frac{f(x + h) + f(x - h)}{2}, \qquad 0 < x < L.$$

Exercises

1. Prove the orthogonality relation, Eq. (9.97), by carrying out the integration in detail. Note the importance of the fact that $\lambda_n = n\pi/L$.

2. Carry out the integration to confirm Eq. (9.99).

3. Determine whether or not the given function is sectionally smooth on the interval indicated. Sketch.

   (a) $f(x) = \ln x, 0 < x < 1$ (b) $f(x) = \ln x, 1 < x < 2$

   (c) $f(x) = x \ln x, 0 < x < 1$ (d) $f(x) = \sqrt{1 - x^2}, 0 < x < 1$

   (e) $f(x) = x^{3/2}, 0 < x < 1$ (f) $f(x) = |x - 1|, 0 < x < 2$

   (g) $f(x) = \begin{cases} 0, & 0 < x \le 1 \\ 1, & 1 < x < 2 \end{cases}$

4. Several computer languages supply the functions

   $$\text{abs}(x) = |x| = \begin{cases} x, & \text{if } x \ge 0, \\ -x, & \text{if } x < 0, \end{cases} \qquad \text{sgn}(x) = \begin{cases} 1, & \text{if } x > 0, \\ 0, & \text{if } x = 0, \\ -1, & \text{if } x < 0. \end{cases}$$

   (a) Sketch the graphs of these functions.

(b) Show that sgn (x) is the derivative of abs (x), except at $x = 0$.

(c) Express the functions of Examples 2(a) and 2(b) in terms of abs and sgn.

(d) Express the Heaviside step function (see Section 5.4) in terms of sgn (x).

In Exercises 5–12 a function is given on an interval. (a) Sketch the function, (b) Check whether it is sectionally smooth. (c) Find the coefficients b_n of its Fourier sine series (Eq. 9.96). (d) Find the sum of the series at every point of the interval.

5. $f(x) = \begin{cases} 1, & 0 < x < L/2, \\ 0, & L/2 < x < L \end{cases}$ 6. $f(x) = 1,\ 0 < x < L$

7. $f(x) = x,\ 0 < x < L$ 8. $f(x) = L - x,\ 0 < x < L$

9. $f(x) = \begin{cases} x, & 0 < x < L/2, \\ L - x, & L/2 \le x < L \end{cases}$ 10. $f(x) = \begin{cases} x, & 0 < x < L/2, \\ 0, & L/2 \le x < L \end{cases}$

11. $f(x) = L - 2x,\ 0 < x < L$ 12. $f(x) = \exp(x/L),\ 0 < x < L$

13. Find the Fourier sine series for these functions by using trigonometric identities instead of integration.

(a) $\sin^3 x,\ 0 < x < \pi$

(b) $\sin(3x/2)\cos(x/2),\ 0 < x < \pi$

14. Suppose two functions f and g defined in the interval $0 < x < L$ both have eigenfunction series

$$f(x) = \sum b_n \sin \lambda_n x, \qquad g(x) = \sum B_n \sin \lambda_n x.$$

If f and g are related by the equation $g(x) = f(L - x)$, prove that the coefficients are related by

$$B_n = (-1)^{n+1} b_n.$$

Test this conclusion on the coefficients of Exercises 7 and 8.

15. In Example 1, we found the coefficients of the series

$$f(x) = \sum_{n=1}^{\infty} b_n \sin \lambda_n x, \qquad 0 < x < L,$$

for the function $f(x) = x(L - x),\ 0 < x < L$. Find the series for $f''(x) = -2,\ 0 < x < L$, and compare it to the result of differentiating twice the series for f term by term.

In Exercises 16–21 a function is given on an interval. (a) Sketch the function. (b) Check whether it is sectionally smooth. (c) Find the coefficients of its Fourier cosine series. (d) Find the sum of the series at every point of the interval.

16. $f(x) = \begin{cases} 1, & 0 < x < L/2, \\ -1, & L/2 < x < L \end{cases}$ 17. $f(x) = L - x, \; 0 < x < L$

18. $f(x) = L - 2x, \; 0 < x < L$ 19. $f(x) = \sin x, \; 0 < x < \pi$

20. $f(x) = \sin(\pi x/L), \; 0 < x < L$ 21. $f(x) = \begin{cases} x, & 0 < x < L/2, \\ L - x, & L/2 < x < L \end{cases}$

9.6

Eigenfunction Series: Sturm-Liouville Problems

The results of Section 9.5 can be generalized to allow for other boundary conditions and other differential equations. A convenient starting point is the regular Sturm-Liouville problem introduced in Section 9.2:

$$\frac{d}{dx}\left(p(x)\frac{du}{dx}\right) - q(x)u + \lambda^2 w(x)u = 0, \qquad \alpha < x < \beta, \tag{9.116}$$

$$au(\alpha) - a'u'(\alpha) = 0, \tag{9.117}$$

$$bu(\beta) + b'u'(\beta) = 0. \tag{9.118}$$

There we cited a theorem to the effect that there are infinitely many eigenvalues, $\lambda_1^2, \lambda_2^2, \lambda_3^2, \ldots$, and an eigenfunction $u_1, u_2, u_3, \ldots$ corresponding to each one. To set the stage for series of these eigenfunctions, we prove the *orthogonality relation*.

Lemma 9.1

If u_m and u_n are eigenfunctions of the regular Sturm-Liouville problem in Eqs. (9.116), (9.117), and (9.118) corresponding to different eigenvalues λ_m^2, λ_n^2, then

$$\int_\alpha^\beta u_n(x)u_m(x)w(x)\,dx = 0 \qquad (\lambda_m^2 \neq \lambda_n^2). \tag{9.119}$$

Proof: The proof requires simple manipulations and calculus. First take the equation satisfied by u_n and multiply through it by u_m, and vice versa:

$$(p(x)u_n')'u_m - q(x)u_nu_m = -\lambda_n^2 w(x)u_nu_m,$$

$$(p(x)u_m')'u_n - q(x)u_mu_n = -\lambda_m^2 w(x)u_mu_n.$$

Next subtract the two equations. The terms containing $q(x)$ cancel, leaving

$$(p(x)u_n')'u_m - (p(x)u_m')'u_n = (\lambda_m^2 - \lambda_n^2)w(x)u_nu_m. \tag{9.120}$$

Now integrate both sides over the range $\alpha < x < \beta$. On the left, some integration by parts can be done:

$$\int_\alpha^\beta (p(x)u_n')'u_m\,dx = p(x)u_n'u_m\big|_\alpha^\beta - \int_\alpha^\beta p(x)u_n'u_m'\,dx. \tag{9.121}$$

With an interchange of subscripts the same equation holds for the second term. Thus when they are subtracted, the integral disappears from the left-hand side, and the integrated equation (9.120) becomes

$$\left[p(x)(u_n'(x)u_m(x) - u_m'(x)u_n(x)) \right]\Big|_\alpha^\beta = (\lambda_m^2 - \lambda_n^2) \int_\alpha^\beta w(x)u_n(x)u_m(d) \, dx.$$

(9.122)

Now both u_m and u_n satisfy the boundary condition (9.117) at α:

$$au_m(\alpha) - a'u_m'(\alpha) = 0,$$

$$au_n(\alpha) - a'u_n'(\alpha) = 0.$$

These are two simultaneous, homogeneous equations in a and a'. Since not both of those numbers can be 0 (if they were, there would be no boundary condition), the determinant of the equations must be 0. That is,

$$u_m(\alpha)u_n'(\alpha) - u_n(\alpha)u_m'(\alpha) = 0.$$

The same argument shows also that

$$u_m(\beta)u_n'(\beta) - u_n(\beta)u_m'(\beta) = 0.$$

But these are exactly the expressions that occur in the evaluations on the left-hand side of Eq. (9.122). Thus the left-hand side is just 0, and we conclude that

$$0 = (\lambda_m^2 - \lambda_n^2) \int_\alpha^\beta w(x)u_n(x)u_m(x) \, dx. \qquad (9.123)$$

Finally, Eq. (9.119) follows, since $\lambda_m^2 \neq \lambda_n^2$.

Example 1

Consider this regular Sturm-Liouville problem

$$\frac{d^2u}{dx^2} + \lambda^2 u = 0, \qquad 0 < x < L,$$

$$u'(0) = 0, \qquad u'(L) = 0.$$

The eigenvalues are $\lambda_0 = 0$ and $\lambda_n = n\pi/L$, $n = 1, 2, 3, \ldots$, with corresponding eigenfunctions

$$u_0 = 1 \qquad \text{and} \qquad u_n = \cos \lambda_n x.$$

The orthogonality relation, Eq. (9.119), takes the forms

$$\int_0^L \cos \lambda_n x \, dx = 0, \qquad \int_0^L \cos \lambda_m x \cos \lambda_n x \, dx = 0 \qquad (\lambda_m^2 \neq \lambda_n^2).$$

(Note: when 0 is an eigenvalue, it is common to designate it as λ_0.)

The importance of the orthogonality relation, Eq. (9.119), is that—as in the preceding section—it allows us to find the coefficients in a proposed series of eigenfunctions

$$f(x) = \sum_{n=1}^{\infty} c_n u_n(x), \qquad \alpha < x < \beta. \tag{9.124}$$

Assuming that this equation is valid, we may multiply through it by $w(x)$ and $u_m(x)$, where m is some fixed integer. Then integrate both sides from α to β to obtain

$$\int_{\alpha}^{\beta} f(x)u_m(x)w(x)\,dx = \int_{\alpha}^{\beta} \sum_{n=1}^{\infty} c_n u_n(x)u_m(x)w(x)\,dx$$

$$= \sum_{n=1}^{\infty} c_n \int_{\alpha}^{\beta} u_n(x)u_m(x)w(x)\,dx. \tag{9.125}$$

Now the orthogonality relation (9.119) tells us that the integrals in the last series all are 0, except the one in which $n = m$. Thus

$$\int_{\alpha}^{\beta} f(x)u_m(x)w(x)\,dx = c_m \int_{\alpha}^{\beta} u_m^2(x)w(x)\,dx.$$

Since $w(x)$ (called the *weight* function) is positive in a regular Sturm-Liouville problem, the integral on the right-hand side is positive, and the equation can be solved for c_m (or c_n if we replace m by n in both members of the equation):

$$c_n = \frac{\int_{\alpha}^{\beta} f(x)u_n(x)w(x)\,dx}{\int_{\alpha}^{\beta} u_n^2(x)w(x)\,dx}. \tag{9.126}$$

The salient facts about series of eigenfunctions are given in Theorem 9.6.

Theorem 9.6

Let $\lambda_1^2, \lambda_2^2, \dots$ and $u_1(x), u_2(x), \dots$ be the eigenvalues and corresponding eigenfunctions of the regular Sturm-Liouville problem

$$\frac{d}{dx}\left(p(x)\frac{du}{dx}\right) - q(x)u + \lambda^2 w(x)u = 0, \qquad \alpha < x < \beta,$$

$$au(\alpha) - a'u'(\alpha) = 0,$$

$$bu(\beta) + b'u'(\beta) = 0.$$

If $f(x)$ is sectionally smooth in the interval $\alpha < x < \beta$ and if

$$c_n = \frac{\int_{\alpha}^{\beta} f(x)u_n(x)w(x)\,dx}{\int_{\alpha}^{\beta} u_n^2(x)w(x)\,dx},$$

then the series

$$\sum_{n=1}^{\infty} c_n u_n(x)$$

converges at every x, $\alpha < x < \beta$, to

$$\lim_{h \to 0} \frac{f(x + h) + f(x - h)}{2}.$$

Example 2

Find the representation of the function $f(x) = x$, $0 < x < L$, in terms of the eigenfunctions of the problem

$$\frac{d^2u}{dx^2} + \lambda^2 u = 0, \qquad 0 < x < L,$$

$$u(0) = 0, \qquad \frac{du}{dx}(L) = 0.$$

This is a regular Sturm-Liouville problem with the solution

$$\lambda_n = \frac{(2n - 1)\pi}{2L}, \qquad u_n(x) = \sin \lambda_n x, \qquad n = 1, 2, 3, \ldots.$$

Since $f(x)$ is sectionally smooth, the results of Theorem 9.6 apply. To find the coefficients in the eigenfunction series, we need to know

$$\int_0^L u_n^2(x) \, dx = \int_0^L \sin^2 \lambda_n x \, dx = \frac{L}{2}.$$

Then the coefficients are given by

$$c_n = \frac{2}{L} \int_0^L f(x) u_n(x) \, dx$$

$$= \frac{2}{L} \int_0^L x \sin \lambda_n x \, dx = \frac{2}{L} \frac{\sin \lambda_n L}{\lambda_n^2}.$$

Because the function f has no discontinuities in the given interval, we may write the equality

$$f(x) = \sum_{n=1}^{\infty} \frac{2 \sin \lambda_n L}{\lambda_n^2 L} \sin \lambda_n x, \qquad 0 < x < L.$$

One of the reasons for studying eigenfunctions is to employ them in the solution of boundary value problems. Let us suppose that we want to solve the problem

$$\frac{d}{dx}\left(p(x)\frac{du}{dx}\right) - q(x)u = f(x), \qquad \alpha < x < \beta, \tag{9.127}$$

$$au(\alpha) - a'u'(\alpha) = 0, \tag{9.128}$$

$$bu(\beta) + b'u'(\beta) = 0. \tag{9.129}$$

Now suppose that both the given function $f(x)$ and the unknown function $u(x)$ have series

$$f(x) = \sum_{n=1}^{\infty} c_n u_n(x), \qquad u(x) = \sum_{n=1}^{\infty} C_n u_n(x). \tag{9.130}$$

In these equations the u_n's are supposed to be the eigenfunctions of the problem stated in Theorem 9.6, in which the differential equation is related in an obvious way to Eq. (9.127) and the boundary conditions are the same as Eqs. (9.128) and (9.129).

Since $f(x)$ is a given function, we would expect the c_n's to be known, and the C_n's are to be found. The defining property of eigenfunction is that

$$L(u_n) = \frac{d}{dx}\left(p(x)\frac{du_n}{dx}\right) - q(x)u_n = -\lambda_n^2 u_n. \tag{9.131}$$

Thus we expect that

$$L(u) = \frac{d}{dx}\left(p(x)\frac{du}{dx}\right) - q(x)u = \sum_{n=1}^{\infty} -\lambda_n^2 C_n u_n(x). \tag{9.132}$$

When we equate this to the series for $f(x)$, as in Eq. (9.130), we get

$$\sum_{n=1}^{\infty} -\lambda_n^2 C_n u_n(x) = \sum_{n=1}^{\infty} c_n u_n(x), \tag{9.133}$$

From here, by using orthogonality to justify matching coefficients, we conclude that

$$C_n = -c_n/\lambda_n^2. \tag{9.134}$$

Example 3

Consider the problem

$$\frac{d^2u}{dx^2} = x(L - x), \qquad 0 < x < L,$$

$$u(0) = 0, \qquad u(L) = 0.$$

We have already determined the eigenfunctions of the problem

$$\frac{d^2u}{dx^2} = -\lambda^2 u, \qquad 0 < x < L,$$

$$u(0) = 0, \qquad u(L) = 0,$$

to be $u_n = \sin \lambda_n x$, $\lambda_n^2 = (n\pi/L)^2$, and in Example 1 of Section 9.5 we found the series

$$x(L - x) = \sum_{n=1}^{\infty} \frac{4(1 - \cos \lambda_n L)}{L\lambda_n^3} \sin \lambda_n x, \qquad 0 < x < L.$$

Assuming, as above, that the solution of our problem has a series development

$$u(x) = \sum_{n=1}^{\infty} C_n \sin \lambda_n x, \qquad 0 < x < L,$$

we equate the series for $u''(x)$ to the series for $x(L - x)$:

$$\sum_{n=1}^{\infty} - C_n \lambda_n^2 \sin \lambda_n x = \sum_{n=1}^{\infty} \frac{4(1 - \cos \lambda_n L)}{L\lambda_n^3} \sin \lambda_n x.$$

Matching coefficients of like terms gives this expression for the C_n's:

$$C_n = -\frac{4(1 - \cos \lambda_n L)}{L\lambda_n^5}, \tag{9.135}$$

and hence this series for $u(x)$:

$$u(x) = \sum_{n=1}^{\infty} \frac{-4(1 - \cos \lambda_n L)}{L\lambda_n^5} \sin \lambda_n x, \qquad 0 < x < L.$$

The procedure suggested above is fully justified, as Theorem 9.7 states.

Theorem 9.7

Let $u(x)$ and $u'(x)$ be continuous and $u''(x)$ sectionally smooth, $\alpha < x < \beta$, and suppose that $u(x)$ satisfies the homogeneous boundary conditions stated in Theorem 9.6. If

$$u(x) = \sum_{n=1}^{\infty} C_n u_n(x), \qquad \alpha < x < \beta,$$

is the series for $u(x)$ in eigenfunctions of the problem in Theorem 9.6, then

$$\frac{1}{w(x)} \left[\frac{d}{dx} \left(p(x) \frac{du}{dx} \right) - q(x)u \right] = \sum_{n=1}^{\infty} -\lambda_n^2 C_n u_n(x), \qquad \alpha < x < \beta.$$

This theorem says that the effect of the linear operator L,

$$L(u) = \frac{1}{w(x)}\left[\frac{d}{dx}\left(p(x)\frac{du}{dx}\right) - q(x)u\right],$$

can be calculated term by term for the series of a function that satisfies the hypotheses.

The question that sticks in the mind is this: where does the weight function $w(x)$ come from? The answer is that often it is dictated in a natural way by the problem. For instance, in solving a problem that arises as

$$\frac{1}{x}\frac{d}{dx}\left(x\frac{du}{dx}\right) = f(x),$$

one would naturally set up the eigenvalue problem as

$$\frac{1}{x}\frac{d}{dx}\left(x\frac{du}{dx}\right) = -\lambda^2 u$$

or

$$\frac{d}{dx}\left(x\frac{du}{dx}\right) + \lambda^2 xu = 0.$$

(This is Bessel's equation. See Chapter 4, Section 4.4.) Sometimes, however, $w(x)$ can be chosen to make a problem tractable.

Example 4

Use the eigenfunction method to solve the problem

$$\frac{d}{dx}\left(x\frac{du}{dx}\right) = f(x), \qquad 1 < x < e,$$

$$u(1) = 0, \qquad u(e) = 0.$$

The "natural" choice of $w(x) = 1$ leads to an eigenvalue problem with the differential equation

$$\frac{d}{dx}\left(x\frac{du}{dx}\right) + \lambda^2 u = 0,$$

whose solution involves Bessel functions of $\sqrt{x}$. However, the choice $w(x) = 1/x$ leads to the eigenvalue problem

$$\frac{d}{dx}\left(x\frac{du}{dx}\right) + \frac{\lambda^2}{x}u = 0, \qquad 1 < x < e,$$

$$u(1) = 0, \qquad u(e) = 0,$$

which is easy to solve. (See Exercise 7 below.)

In Exercises 1–8, (a) find the eigenvalues and eigenfunctions for the problem; (b) write the orthogonality relation (see Eq. (9.119)) for the eigenfunctions, and confirm it by doing the integration; (c) find the coefficients of the eigenfunction series for $f(x) = x$.

1. $\dfrac{d^2u}{dx^2} + \lambda^2 u = 0, \; 0 < x < L$

    $u(0) = 0, \; u'(L) = 0$

2. $\dfrac{d^2u}{dx^2} + \lambda^2 u = 0, \; 0 < x < L,$

    $u'(0) = 0, \; u'(L) = 0$

3. $\dfrac{d^2u}{dx^2} + \lambda^2 u = 0, \; 0 < x < L,$

    $u'(0) = 0, \; u(L) = 0$

4. $\dfrac{d^2u}{dx^2} + \lambda^2 u = 0, \; \alpha < x < \beta,$

    $u(\alpha) = 0, \; u(\beta) = 0$

5. $\dfrac{d^2u}{dx^2} + \lambda^2 u = 0, \; -L/2 < x < L/2$

    $u(-L/2) = 0, \; u(L/2) = 0$

6. $\dfrac{d^2u}{dx^2} + \lambda^2 u = 0, \; 0 < x < L$

    $u(0) = 0, \; u(L) + Lu'(L) = 0$

    (The eigenvalues cannot be given in closed form.)

7. $\dfrac{d}{dx}\left(x\dfrac{du}{dx}\right) + \dfrac{\lambda^2}{x} u = 0, \; 1 < x < L,$

    $u(1) = 0, \; u(L) = 0$

    (Hint: this is a Cauchy-Euler equation.)

8. $\dfrac{d}{dx}\left(x^2\dfrac{du}{dx}\right) + \lambda^2 u = 0, \; 1 < x < L,$

    $u(1) = 0, \; u(L) = 0$

9. Let $G(x, t)$ be the Green's function for the problem

    $$\dfrac{d^2u}{dx^2} = f(x), \qquad 0 < x < L,$$

    $$u(0) = 0, \qquad u(L) = 0.$$

    Find a series for $G(x, t)$ (as a function of x) in the eigenfunctions of

    $$\dfrac{d^2u}{dx^2} + \lambda^2 u = 0,$$

    $$u(0) = 0, \qquad u(L) = 0.$$

10. Follow the directions in Exercise 9, but for the problem

    $$\dfrac{d^2u}{dx^2} - \gamma^2 u = f(x), \qquad 0 < x < L,$$

    $$u(0) = 0, \qquad u(L) = 0.$$

11. Solve the problem given in Example 3 by the methods of Section 9.1. Then expand the solution in a series of sines and compare to the series found in the example.

9.7
Other Eigenfunction Series

Many eigenvalue and boundary value problems in applied mathematics do not fit the Sturm-Liouville theory, usually because of a singular point of the differential equation or an infinite range. The general theories covering these cases are too complicated to present here, but the applications are too significant to ignore. Therefore several of the most important cases are assembled here. Many of the solutions were found in Chapter 4 by means of power series.

Periodic Conditions and Fourier Series

Perhaps the most important of all the cases that are not regular Sturm-Liouville problems is this:

$$\frac{d^2u}{dx^2} + \lambda^2 u = 0, \qquad -L < x < L, \tag{F1}$$

$$u(-L) = u(L), \tag{F2}$$

$$u'(-L) = u'(L). \tag{F3}$$

The deviation from regularity is that each boundary condition involves both endpoints of the interval of interest. These are called *periodic boundary conditions* because they say, in essence, that $x = -L$ is the same as $x = L$. Often this kind of problem comes up in polar coordinates where the independent variable is the angle θ and the interval is $-\pi < \theta \le \pi$ (or some other of length 2π). In this setting, the ends of the interval, $\theta = \pi$ and $\theta = -\pi$, are indeed the same.

The peculiarity of the solutions of the eigenvalue problem in Eqs. (F1)–(F3) is that each positive eigenvalue has *two* independent eigenfunctions instead of one. The complete solution of the problem is

$$\lambda_0^2 = 0, \qquad u_0(x) \equiv 1,$$

$$\lambda_n^2 = (n\pi/L)^2, \qquad u_n(x) = \begin{cases} \sin \lambda_n x \\ \cos \lambda_n x \end{cases} \qquad n = 1, 2, 3, \ldots. \tag{F4}$$

It is easy to see that all of the eigenfunctions really are periodic, with period $2L$. That is, $u(x + 2L) = u(x)$ for all x.

The orthogonality relation for the eigenfunctions of this problem takes several forms:

$$\int_{-L}^{L} \sin \lambda_n x \sin \lambda_m x \, dx = 0 \qquad (\lambda_m \neq \lambda_n),$$

$$\int_{-L}^{L} \cos \lambda_n x \cos \lambda_m x \, dx = 0 \qquad (\lambda_m \neq \lambda_n),$$

$$\int_{-L}^{L} \sin \lambda_n x \, dx = 0, \qquad \int_{-L}^{L} \cos \lambda_n x \, dx = 0,$$

$$\int_{-L}^{L} \sin \lambda_n x \cos \lambda_m x \, dx = 0.$$

The second pair of equations expresses the orthogonality of $u_0(x) \equiv 1$ to the eigenfunctions of other eigenvalues. In the last equation the case $\lambda_n = \lambda_m$ is a convenient consequence of choosing sine and cosine as the eigenfunctions.

The expansion of a function in terms of the eigenfunctions of the problem under consideration is called its *Fourier series*. Because of the importance of such series, in applications and in history, we state separately a convergence theorem.

Theorem 9.8

Let $f(x)$ be sectionally smooth, $-L < x < L$, and periodic with period $2L$, and let

$$a_0 = \frac{1}{2L} \int_{-L}^{L} f(x) \, dx, \qquad a_n = \frac{1}{L} \int_{-L}^{L} f(x) \cos \lambda_n x \, dx, \tag{F5}$$

$$b_n = \frac{1}{L} \int_{-L}^{L} f(x) \sin \lambda_n x \, dx. \tag{F6}$$

Then the Fourier series of f converges at every x, to $\lim_{h \to 0} (f(x + h) + f(x - h))/2$:

$$a_0 + \sum_{n=1}^{\infty} a_n \cos \lambda_n x + b_n \sin \lambda_n x = \lim_{h \to 0} \frac{f(x + h) + f(x - h)}{2}. \tag{F7}$$

This theorem can be applied to a function $g(x)$ that has been defined only on the original interval, $-L < x < L$, by creating a periodic function $f(x)$ that is identical with $g(x)$ for $-L < x < L$ and is defined elsewhere through the periodicity condition $f(x + 2L) = f(x)$. (See Fig. 9.13.) The function f is called the *periodic extension* of g. Since f and g coincide for $-L < x < L$, either one can be used in computing the Fourier coefficients—the a's and b's given in Eqs. (F5) and (F6). Furthermore, the function g can be used instead of its periodic extension f in Eq. (F7), provided that the inequality $-L < x < L$ is added, to restrict attention to the interval where g has been defined.

Figure 9.13

Periodic Extension of $g(x)$

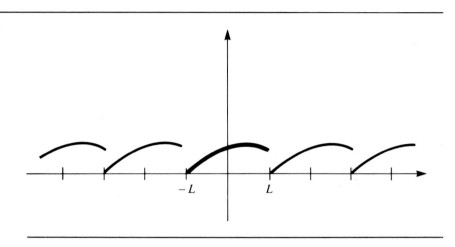

Example

Find the periodic extension and Fourier series of the function

$$g(x) = \begin{cases} L, & -L < x < 0, \\ L - x, & 0 \le x < L. \end{cases}$$

This function's periodic extension is sectionally smooth (see Fig. 9.14). The Fourier coefficients of g and f are

$$a_0 = \frac{1}{2L} \int_{-L}^{L} g(x)\, dx = \frac{1}{2L} \left[\int_{-L}^{0} L\, dx + \int_{0}^{L} (L - x)\, dx \right] = \frac{3L}{4};$$

$$a_n = \frac{1}{L} \int_{-L}^{L} g(x) \cos \lambda_n x\, dx = \frac{1}{L} \left[\int_{-L}^{0} L \cos \lambda_n x\, dx + \int_{0}^{L} (L - x) \cos \lambda_n x\, dx \right]$$

$$= \frac{L}{\pi^2} \frac{1 - \cos n\pi}{n^2};$$

$$b_n = \frac{1}{L} \int_{-L}^{L} g(x) \sin \lambda_n x\, dx = \frac{1}{L} \left[\int_{-L}^{0} L \sin \lambda_n x\, dx + \int_{0}^{L} (L - x) \sin \lambda_n x\, dx \right]$$

$$= -\frac{L}{\pi} \frac{\cos n\pi}{n}.$$

(The details of the integrations have been left out.) Theorem 9.8 states

Figure 9.14

Periodic Extension of
$$g(x) = \begin{cases} L, & -L < x < 0. \\ L - x, & 0 \le x < L \end{cases}$$

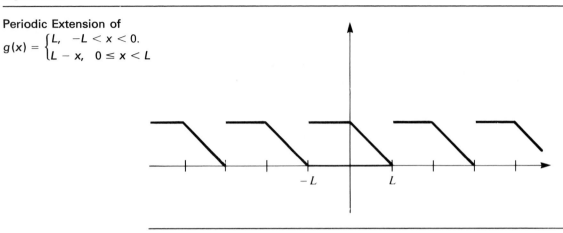

that the Fourier series of g converges to $g(x)$, $-L < x < L$ (because g is continuous in the interval):

$$g(x) = \frac{3L}{4} + L\sum_{n=1}^{\infty} \left(\frac{(1 - \cos n\pi)}{n^2\pi^2} \cos \lambda_n x - \frac{\cos n\pi}{n\pi} \sin \lambda_n x \right), \quad -L < x < L.$$

The series converges to $L/2$ at $x = \pm L$. This value can be determined by examining the appropriate limit of the periodic extension of g:

$$\lim_{h \to 0} \frac{f(L + h) + f(L - h)}{2} = \lim_{h \to 0} \frac{g(-L + h) + g(L - h)}{2}$$

$$= \lim_{h \to 0} \frac{L + (L - (L - h))}{2} = \frac{L}{2}.$$

Bessel functions

This eigenvalue problem often appears when polar or cylindrical coordinate systems are used:

$$\frac{1}{r}\frac{d}{dr}\left(r\frac{du}{dr}\right) - \frac{k^2}{r^2}u + \lambda^2 u = 0, \qquad 0 < r < L, \tag{B1}$$

$$u(r), u'(r) \qquad \text{bounded at } r = 0, \tag{B2}$$

$$bu(L) + b'u'(L) = 0. \tag{B3}$$

(The parameter k is usually an integer.) The differential equation, called

Bessel's equation, has two independent solutions, called Bessel functions of order k of the first and second kinds and denoted by $J_k(\lambda r)$ and $Y_k(\lambda r)$. The first of these is r^k times a power series with center $r = 0$, while the second contains negative powers or logarithms of r. (See Chapter 4, Sections 4.4 and 4.6, for more details.) Thus the bounded solution of the differential equation is the Bessel function of the first kind, $J_k(\lambda r)$.

The eigenvalues are determined by the boundary condition at $r = L$, which becomes

$$bJ_k(\lambda L) + b'\lambda J_k'(\lambda L) = 0. \tag{B4}$$

The cases $b' = 0$ or $b = 0$ are the most frequently encountered. The first few solutions of

$$J_k(\beta) = 0 \qquad \text{and} \qquad J_k'(\beta) = 0$$

are listed in Table 9.1 for $k = 0$ and 1. Many more values for these and other cases are known. (See Abramowitz and Stegun, 1964.) Note that when $b = 0$ in the boundary condition (B3), $\lambda_0 = 0$ is an eigenvalue, and $u_0(r) \equiv 1$ is the corresponding eigenfunction.

Table 9.1	n	$J_0(\beta) = 0$	$J_0'(\beta) = 0$ or $J_1(\beta) = 0$	$J_1'(\beta) = 0$
	1	2.4048	3.8317	1.8412
	2	5.5201	7.0156	5.3314
	3	8.6537	10.1735	8.5363
	4	11.7915	13.3237	11.7060
	5	14.9309	16.4706	14.8636

It is easy to see that the proof of orthogonality given for the eigenfunctions of the regular Sturm-Liouville problem goes through in this case, where $p(r) = r$ is 0 at the left end of the interval. The orthogonality relation is

$$\int_0^L J_k(\lambda_n r)J_k(\lambda_m r) r\, dr = 0 \qquad (\lambda_m^2 \neq \lambda_n^2). \tag{B5}$$

From this relation, one obtains the coefficients of the series (called a *Fourier-Bessel series*)

$$f(r) = \sum_{n=1}^{\infty} c_n J_k(\lambda_n r), \qquad 0 < r < L. \tag{B6}$$

$$c_n = \frac{\int_0^L f(r)J_k(\lambda_n r) r\, dr}{\int_0^L J_k^2(\lambda_n r)\, dr}. \tag{B7}$$

If $f(r)$ is sectionally smooth, $0 < r < L$, the series (B6) converges to

$$\lim_{h \to 0} \frac{f(r + h) + f(r - h)}{2}$$

at all r, $0 < r < L$, just as in the regular cases. The denominator of Eq. (B7) is known in closed form:

$$\int_0^L J_k^2(\lambda_n r) r \, dr = \frac{(\lambda_n L J_k'(\lambda_n L))^2 + k^2 J_k^2(0) + (\lambda_n^2 L^2 - k^2) J_k^2(\lambda_n L)}{2\lambda_n^2}. \tag{B8}$$

The expression can be simplified by using the boundary conditions.

We should note that the nonhomogeneous problem with which Eq. (B1) is associated contains the differential equation

$$\frac{1}{r} \frac{d}{dr} \left(r \frac{du}{dr} \right) - \frac{k^2}{r^2} u = f(r), \qquad 0 < r < L. \tag{B9}$$

This equation is of Cauchy-Euler type and can be solved explicitly.

Spherical Bessel Functions

Problems in spherical (ρ, θ, ϕ) coordinates sometimes lead to the eigenvalue problem

$$\frac{1}{\rho^2} \frac{d}{d\rho} \left(\rho^2 \frac{du}{d\rho} \right) - \frac{n(n + 1)}{\rho^2} u + \lambda^2 u = 0, \qquad 0 < \rho < L, \tag{S1}$$

$$u(\rho), u'(\rho) \qquad \text{bounded at } \rho = 0, \tag{S2}$$

$$bu(L) + b'u'(L) = 0. \tag{S3}$$

The change of variable from $u(\rho)$ to $v(\rho) = \sqrt{\rho} u(\rho)$ transforms Eq. (S1) into

$$\frac{d^2 v}{d\rho^2} + \frac{1}{\rho} \frac{dv}{d\rho} - \frac{(n + \frac{1}{2})^2}{\rho^2} v + \lambda^2 v = 0, \qquad 0 < \rho < L, \tag{S4}$$

which is a form of Bessel's equation. The solution of Eq. (S4) that is bounded at $\rho = 0$ is

$$v(\rho) = J_{n+1/2}(\lambda \rho), \tag{S5}$$

and from here, one can obtain the solution of Eq. (S1) as

$$u(\rho) = \frac{1}{\sqrt{\rho}} J_{n+1/2}(\lambda \rho). \tag{S6}$$

The closely related function

$$j_n(\lambda \rho) = \sqrt{\frac{\pi}{2\lambda \rho}} J_{n+1/2}(\lambda \rho) \tag{S7}$$

is called the *spherical Bessel function* of the first kind and order n. These can be expressed in terms of elementary functions. The first two are

$$j_0(z) = \frac{\sin z}{z}, \qquad j_1(z) = \frac{\sin z}{z^2} - \frac{\cos z}{z}. \tag{S8}$$

In fact, we had already found the solution of Eq. (S1) for $n = 0$ in Section 9.3.

The values of spherical Bessel functions, and their zeros—which are needed to find eigenvalues of the original problems—are tabulated in many books.

It can be seen that the orthogonality principle for the problem in Eqs. (S1)–(S3) takes the form

$$\int_0^L u_n(\rho)u_m(\rho)\rho^2 \, d\rho = 0 \qquad (\lambda_n^2 \neq \lambda_m^2). \tag{S9}$$

Furthermore, if $f(\rho)$ is sectionally smooth on the interval $0 < \rho < L$, then it may be expanded in the series

$$f(\rho) = \sum_{n=1}^{\infty} c_n u_n(\rho), \qquad 0 < \rho < L, \tag{S10}$$

with

$$c_n = \frac{\int_0^L f(\rho)u_n(\rho)\rho^2 \, d\rho}{\int_0^L u_n^2(\rho)\rho^2 \, d\rho}. \tag{S11}$$

This series converges to the usual limit at points of discontinuity of $f(\rho)$.

Legendre Polynomials

Problems in spherical (ρ, θ, ϕ) coordinates sometimes require the solution of

$$\frac{d}{d\phi}\left(\sin \phi \frac{du}{d\phi}\right) + \lambda^2 \sin \phi \, u = 0, \qquad 0 < \phi < \pi. \tag{L1}$$

It can easily be seen that both ends of the interval—corresponding to the north and south poles—are singular points of the differential equation. Therefore the solution is required to satisfy

$$u(\phi), u'(\phi) \qquad \text{bounded at } \phi = 0, \pi. \tag{L2}$$

Changing the independent variable from the co-latitude ϕ to $x = \cos \phi$ leads to this eigenvalue problem:

$$\frac{d}{dx}\left((1 - x^2)\frac{du}{dx}\right) + \lambda^2 u = 0, \qquad -1 < x < 1, \tag{L3}$$

$$u(x), u'(x) \qquad \text{bounded at } x = -1, 1. \tag{L4}$$

In general, the differential equation, called Legendre's equation, has two independent solutions, both unbounded at $x = \pm 1$. However, for

$$\lambda_n^2 = n(n + 1), \qquad n = 0, 1, 2, \ldots, \tag{L5}$$

one solution of Eq. (L3) is a polynomial, which naturally satisfies the boundedness conditions (L4). When normalized by the condition $u_n(1) = 1$, the eigenfunctions are called Legendre polynomials, the first few of which are collected in Table 9.2. (See Chapter 4, Section 4.2.)

Table 9.2

Legendre Polynomials

$P_0(x) = 1$

$P_1(x) = x$

$P_2(x) = \dfrac{3x^2 - 1}{2}$

$P_3(x) = \dfrac{5x^3 - 3x}{2}$

$P_4(x) = \dfrac{35x^4 - 30x^2 + 3}{8}$

$P_5(x) = \dfrac{63x^5 - 70x^3 + 15x}{8}$

Again, the proof of orthogonality used for the regular Sturm-Liouville problem can be modified to show that

$$\int_{-1}^{1} P_n(x)P_m(x)\, dx = 0 \qquad (n \neq m) \tag{L6}$$

Then if $f(x)$ is sectionally smooth, $-1 < x < 1$, it can be expanded in series of Legendre polynomials:

$$f(x) = \sum_{n=0}^{\infty} c_n P_n(x), \qquad -1 < x < 1, \tag{L7}$$

$$c_n = \frac{\int_{-1}^{1} f(x)P_n(x)\, dx}{\int_{-1}^{1} P_n^2(x)\, dx}. \tag{L8}$$

The series converges to $f(x)$ at any point of continuity of f and to the usual limit at any point of discontinuity. The denominator in Eq. (L8) is known in closed form:

$$\int_{-1}^{1} P_n^2(x)\, dx = \frac{2}{2n + 1}. \tag{L9}$$

In terms of the original independent variable ϕ, the solution of Eq. (L1), subject to the boundedness conditions (L2), is

$$\lambda_n^2 = n(n + 1), \qquad u_n(\phi) = P_n(\cos \phi), \qquad n = 0, 1, 2, \dots. \qquad \textbf{(L10)}$$

The orthogonality relation and related series become

$$\int_0^\pi P_n(\cos \phi) P_m(\cos \phi) \sin \phi \, d\phi = 0 \qquad (n \ne m), \qquad \textbf{(L11)}$$

$$g(\phi) = \sum_{n=0}^\infty c_n P_n(\cos \theta), \qquad 0 < \phi < \pi, \qquad \textbf{(L12)}$$

$$c_n = \frac{2n + 1}{2} \int_0^\pi g(\phi) P_n(\cos \phi) \sin \phi \, d\phi. \qquad \textbf{(L13)}$$

Exercises

Fourier Series

F1. A function g defined in a symmetric interval $-L < x < L$ (or $-\infty < x < \infty$) is called *even* if $g(-x) = g(x)$ (x^k, for even k, and $\cos \lambda x$ are examples). Prove that

$$\int_{-L}^L g(x) \, dx = 2 \int_0^L g(x) \, dx.$$

F2. A function h defined in a symmetric interval $-L < x < L$ (or $-\infty < x < \infty$) is called *odd* if $h(-x) = -h(x)$ (x^k, for odd k, and $\sin \lambda x$ are examples). Prove that

$$\int_{-L}^L h(x) \, dx = 0.$$

F3. Suppose that a function f is even and periodic with period $2L$. Show that its Fourier series contains cosines only and that the coefficients are ($\lambda_n = n\pi/L$)

$$a_0 = \frac{1}{L} \int_0^L f(x) \, dx, \qquad a_n = \frac{2}{L} \int_0^L f(x) \cos \lambda_n x \, dx.$$

F4. Suppose $f(x)$ is odd and periodic with period $2L$. Show that its Fourier series contains sines only and that the coefficients are ($\lambda_n = n\pi/L$)

$$b_n = \frac{2}{L} \int_0^L f(x) \sin \lambda_n x \, dx.$$

F5. Show that the integral, over one period, of a periodic function does not depend on the starting point. If f has period $2L$, this means that

$$\int_0^{2L} f(x) \, dx = \int_a^{a+2L} f(x) \, dx \qquad \text{for any } a.$$

In Exercises F6–F12 a periodic function is specified by giving its formula over an interval of length one period. (a) Sketch a graph of the function and confirm that it is sectionally smooth. (b) Find the Fourier series of the function. (c) Determine the sum of the series at every point. It may be convenient to use the ideas of Exercises F1–F5.

F6. $f(x) = x$, $-L < x < L$

F7. $f(x) = \begin{cases} 0, & -L < x < 0, \\ x, & 0 \le x < L \end{cases}$

F8. $f(x) = \begin{cases} 1 + \dfrac{x}{L}, & -L < x < 0, \\[2mm] 1 - \dfrac{x}{L}, & 0 \le x < L \end{cases}$

F9. $f(x) = |\sin x|$ (period π)

F10. $f(x) = \begin{cases} 1, & -\dfrac{L}{3} < x < \dfrac{L}{3}, \\[2mm] 0, & \dfrac{L}{3} \le x < \dfrac{5L}{3} \end{cases}$ (period $2L$)

F11. $f(x) = |x|$, $-L < x < L$

F12. $f(x) = \begin{cases} -1, & -L < x < 0, \\ 1, & 0 < x < L \end{cases}$

F13. The Fourier series of this function is especially important in the theory. Find the series for

$$f_c(x) = \begin{cases} \dfrac{1}{2\varepsilon}, & c - \varepsilon < x < c + \varepsilon, \\[2mm] 0, & c + \varepsilon < x < 2L + c - \varepsilon. \end{cases}$$

You may assume that $-L < c < L$ and that ε is small. A sketch of the function is informative.

F14. Use trigonometric identities, first on the coefficients found in Exercise 13, then on the terms of the Fourier series, to get this result ($\lambda_n = n\pi/L$)

$$f_c(x) = \frac{1}{2L} + \sum_{n=1}^{\infty} \frac{\sin \lambda_n \varepsilon}{n\pi\varepsilon} \cos \lambda_n (x - c).$$

Bessel Functions

B1. Two important facts about the Bessel functions J_0 and J_1 are

$$\frac{d}{dx} J_0(x) = -J_1(x) \qquad \text{and} \qquad \int x J_0(x)\, dx = x J_1(x) + c.$$

Start the proof of the first of these by showing that $J_0'(x)$ satisfies Bessel's equation (B1) with $k = 1$, $\lambda = 1$ (and $x = r$).

B2. Prove the second formula by integrating Eq. (B1) in the form

$$xJ_0 = -(xJ_0')'$$

and then using the first formula.

B3. Find the Fourier-Bessel series, Eq. (B6), for the function $f(r) = 1$, in terms of the eigenfunctions of

$$\frac{1}{r}\frac{d}{dr}\left(r\frac{du}{dr}\right) + \lambda^2 u = 0, \qquad 0 < r < L,$$

$$u(L) = 0.$$

Use the formula in Exercise B1 to do the integration.

B4. Find the eigenfunctions and first few eigenvalues of the problem

$$\frac{1}{r}\frac{d}{dr}\left(r\frac{du}{dr}\right) + \lambda^2 u = 0, \qquad 0 < r < L,$$

$$u'(L) = 0.$$

B5. Find the Fourier-Bessel series of the function $f(r) = 1$, $0 < r < L$, in terms of the eigenfunctions found in Exercise B4.

B6. The temperature $U(r)$ in a disk exposed to a medium at temperature T is governed by the boundary value problem

$$\frac{1}{r}\frac{d}{dr}\left(r\frac{dU}{dr}\right) = \gamma^2(U - T), \qquad 0 < r < L,$$

$$U(L) = T_1.$$

Change the unknown function to

$$u(r) = \frac{U(r) - T_1}{T - T_1}$$

and show that $u(r)$ satisfies

$$\frac{1}{r}\frac{d}{dr}\left(r\frac{du}{dr}\right) = \gamma^2(u - 1), \qquad 0 < r < L,$$

$$u(L) = 0.$$

B7. Solve the problem in Exercise B6 by assuming that u has the form

$$u(r) = \sum_{n=1}^{\infty} C_n u_n(r)$$

where $u_n(r) = J_0(\lambda_n r)$ is the nth eigenfunction of the problem

$$\frac{1}{r}\frac{d}{dr}\left(r\frac{du}{dr}\right) + \lambda^2 u = 0, \qquad 0 < r < L,$$

$u(L) = 0.$

It will be necessary to use the series obtained in Exercise B3.

Spherical Bessel Functions

S1. Find the expansion of the function $f(\rho) = 1$ in terms of the eigen-functions of

$$\frac{1}{\rho^2}\frac{d}{d\rho}\left(\rho^2\frac{du}{d\rho}\right) + \lambda^2 u = 0, \qquad 0 < \rho < L,$$

$u(L) = 0.$

S2. How does the result of Exercise S1 correspond with the Fourier sine series expansion of $f(\rho)$?

S3. Solve the nonhomogeneous problem

$$\frac{1}{\rho^2}\frac{d}{d\rho}\left(\rho^2\frac{du}{d\rho}\right) = -1, \qquad 0 < \rho < L,$$

$u(L) = 0,$

(a) by direct integration; (b) by assuming that the solution has the form

$$u(\rho) = \sum_{n=1}^{\infty} C_n u_n(\rho),$$

where the u_n are the eigenfunctions of the problem in Exercise S1. Replace the constant 1 on the right-hand side by the series found in Exercise S1.

S4. Confirm this property of spherical Bessel functions

$$j_n(z) = \frac{n-1}{z}j_{n-1}(z) - \frac{d}{dz}j_{n-1}(z)$$

by using j_0 and j_1 from Eq. (S8).

S5. Use the formula in Exercise S4 to find $j_2(z)$.

Legendre Polynomials

L1. Express x^2 and x^3 in terms of Legendre polynomials. No integration is necessary.

L2. Explain why x^4 (k a positive integer) can be expressed as a linear combination of $P_k, P_{k-2}, P_{k-4}, \ldots$.

L3. Show that for k a positive integer,

$$\int_{-1}^{1} x^k P_m(x)\,dx = 0, \qquad m = k+1, k+2, \ldots.$$

L4. Confirm the orthogonality relation (L6) for $n = 3$, $m = 0, 1, 2$ by direct integration.

L5. Confirm Eq. (L9) for P_0, P_1, and P_2.

L6. Expand the polynomial $f(x) = 4x^3 - 3x$ in terms of Legendre polynomials.

L7. Find the coefficients through c_3 in the Legendre series for the function $f(x) = \sin \pi x$, $-1 < x < 1$.

L8. Sketch a graph of the third-degree polynomial found in Exercise L7 along with a graph of $\sin \pi x$. Check values at $x = 0.5, 1$, and locate the maximum of the polynomial.

L9. Find the coefficients through c_4 in the Legendre series for the function

$$f(x) = \begin{cases} 1, & 0 < x < 1, \\ 0, & -1 < x \le 0. \end{cases}$$

L10. In spherical coordinates the equation for an ellipsoid of revolution (whose axis is the north–south, or z, axis) is

$$\frac{\rho^2}{a^2} \sin^2 \phi + \frac{\rho^2}{b^2} \cos^2 \phi = 1.$$

Obtain this expression for ρ:

$$\rho = a(1 + kx^2)^{-1/2}, \tag{L14}$$

where $k = (a/b)^2 - 1$ and $x = \cos \phi$.

F11. Expand the function ρ in Exercise L10 in powers of x through x^4. (That is, find the first terms of its Taylor series.) Next replace powers of x by their equivalents in Legendre polynomials to get an approximate expansion for ρ in terms of P_0, P_2, and P_4. The coefficients are functions of k. (This technique is often used in geodesy. The parameter k is usually about 0.01.)

F12. Find the coefficient c_0 of the series for ρ as a function of x by integration.

9.8
Numerical Methods

Linear boundary value problems can be solved numerically by replacing the differential equation and boundary conditions with a system of simultaneous linear algebraic equations. This method is based on the *central difference approximations* for the first and second derivatives,

$$\frac{du}{dx}(x) \cong \frac{u(x + h) - u(x - h)}{2h}, \tag{9.136}$$

$$\frac{d^2u}{dx^2}(x) \cong \frac{u(x+h) - 2u(x) + u(x-h)}{h^2}. \tag{9.137}$$

The right-hand side of Eq. (9.136) resembles the difference quotient in the definition of derivative. The approximation in Eq. (9.137) is obtained by compounding the first one (see Exercise 2 at the end of this section). In order for the indicated approximations to be good, h must be small. In fact, it can be shown that the errors in the approximations are proportional to h^2 (if u has continuous derivatives to order 4 at least).

In numerical methods, instead of obtaining the solution of a problem as a function of x, we are willing to settle for a table of values—or rather approximate values—of the solution at some points $x_0, x_1, x_2, \ldots, x_n$. Usually, the independent variable is adjusted to make the interval $0 < x < 1$, and the *meshpoints* are equally spaced:

$$x_0 = 0, \qquad x_1 = h, \qquad x_2 = 2h, \qquad \ldots, \qquad x_n = nh = 1. \tag{9.138}$$

Thus the spacing is $h = 1/n$; this is the master parameter of the numerical solution.

Now, the central idea is that the values of the solution at the meshpoints—that is, the table entries—satisfy approximately the equations made by replacing the derivatives at the meshpoints with central-difference approximations. As an example, consider the boundary value problem

$$\frac{d^2u}{dx^2} + 2\frac{du}{dx} = -4x, \qquad 0 < x < 1, \tag{9.139}$$

$$u(0) = 0, \qquad u(1) = 1. \tag{9.140}$$

At each meshpoint $x_1, x_2, \ldots, x_{n-1}$, the differential equation is satisfied exactly by the solution function:

$$\frac{d^2u}{dx^2}(x_i) + 2\frac{du}{dx}(x_i) = -4x_i, \qquad i = 1, 2, \ldots, n-1.$$

At each meshpoint, the following relation holds also

$$\frac{u(x_i+h) - 2u(x_i) + u(x_i-h)}{h^2} + 2\frac{u(x_i+h) - u(x_i-h)}{2h} \cong -4x_i$$

or $\tag{9.141}$

$$\frac{u(x_{i+1}) - 2u(x_i) + u(x_{i-1})}{h^2} + 2\frac{u(x_{i+1}) - u(x_{i-1})}{2h} \cong -4x_i \tag{9.142}$$

These are approximate equations because the difference quotients are not exactly equal to the derivatives. Table 9.3 shows values of the solution,

$$u(x) = x - x^2 + (1 - e^{-2x})/(1 - e^{-2}), \tag{9.143}$$

together with values of the approximations for the first and second derivatives, and also the difference ε_i between the left- and right-hand sides of Eq. (9.142).

Table 9.3*	i	x_i	$u(x_i)$	$\delta u(x_i)$	$\delta^2 u(x_i)$	ε_i
	0	0	0			
	1	0.1	0.2996	2.706	−6.642	0.0127
	2	0.2	0.5413	2.161	−5.800	0.0104
	3	0.3	0.7318	1.678	−5.111	0.0085
	4	0.4	0.8769	1.246	−4.547	0.0070
	5	0.5	0.9811	0.857	−4.086	0.0057
	6	0.6	1.0482	0.501	−3.708	0.0047
	7	0.7	1.0813	0.174	−3.398	0.0038
	8	0.8	1.0830	−0.130	−2.937	0.0031
	9	0.9	1.0553	−0.415	−2.767	0.0026
	10	1.0	1.0000			

* $\delta u(x_i)$ and $\delta^2 u(x_i)$ are the central difference approximations to the first and second derivatives, respectively.

To solve the problem in Eqs. (9.139) and (9.140) numerically, we would require numbers

$$u_0, u_1, \ldots, u_n$$

to satisfy exactly an equation corresponding to Eq. (9.142):

$$\frac{u_{i+1} - 2u_i + u_{i-1}}{h^2} + 2\frac{u_{i+1} - u_{i-1}}{2h} = -4x_i, \quad i = 1, 2, \ldots, n - 1.$$

(9.144)

We expect that each u_i will be an approximation to $u(x_i)$. Consequently, we also require that the boundary conditions be satisfied:

$$u_0 = 0, \quad u_n = 1.$$

(9.145)

Equations (9.144) and (9.145) are termed the *replacement equations* for the original problem, Eqs. (9.139) and (9.140).

When h (or n) is chosen, Eqs. (9.144) and (9.145) become an algebraic system to be solved. For instance, if $n = 5$ and $h = \frac{1}{5}$, we get

$$25(u_2 - 2u_1 + u_0) + 5(u_2 - u_0) = -0.8 \quad [i = 1],$$
$$25(u_3 - 2u_2 + u_1) + 5(u_3 - u_1) = -1.6 \quad [i = 2],$$
$$25(u_4 - 2u_3 + u_2) + 5(u_4 - u_2) = -2.4 \quad [i = 3],$$
$$25(u_5 - 2u_4 + u_3) + 5(u_5 - u_3) = -3.2 \quad [i = 4],$$
$$u_0 = 0, \quad u_5 = 1.$$

There are six equations in the six "unknowns" $u_0, u_1, u_2, u_3, u_4, u_5,$

although two of the equations are trivial. The system of equations that results when $h = \frac{1}{10}$ was solved (by microcomputer). The solutions u_i are tabulated in Table 9.4 next to the values of the solution function $u(x)$ at the meshpoints and the difference between them. It can be seen that the approximation is quite good.

Table 9.4

i	x_i	u_i	$u(x_i)$	$u_i - u(x_i)$
0	0.0	0	0	0
1	0.1	0.3001	0.2996	0.00042
2	0.2	0.5419	0.5413	0.00064
3	0.3	0.7325	0.7318	0.00073
4	0.4	0.8776	0.8769	0.00073
5	0.5	0.9817	0.9811	0.00066
6	0.6	1.0487	1.0482	0.00055
7	0.7	1.0817	1.0813	0.00042
8	0.8	1.0833	1.0830	0.00028
9	0.9	1.0555	1.0553	0.00014
10	1.0	1.0000	1.0000	0

In the more general case of a boundary value problem such as

$$\frac{d^2u}{dx^2} + b(x)\frac{du}{dx} + c(x)u = f(x), \qquad 0 < x < 1, \tag{9.146}$$

$$u(0) = A, \qquad u(1) = B, \tag{9.147}$$

the replacement equations become

$$\frac{u_{i+1} - 2u_i + u_{i-1}}{h^2} + b(x_i)\frac{u_{i+1} - u_{i-1}}{2h} + c(x_i)u_i = f(x_i),$$
$$i = 1, 2, \ldots, n - 1 \tag{9.148}$$

together with the boundary conditions

$$u_0 = A, \qquad u_n = B. \tag{9.149}$$

Once n has been chosen, these become a system of simultaneous algebraic equations, which may be solved by Gaussian elimination.

Example 1

Obtain a numerical solution of the boundary value problem

$$\frac{1}{r}\frac{d}{dr}\left(r\frac{du}{dr}\right) - 2ru = 4 - r^2, \qquad 1 < r < 3,$$

$$u(1) = 1, \qquad u(3) = 0.$$

The first step is to change variables from r to $x = (r - 1)/2$. In terms of x the problem becomes

$$\frac{d^2u}{dx^2} + \frac{2}{2x + 1}\frac{du}{dx} - 8(2x + 1)u = 12 - 16x - 16x^2, \qquad 0 < x < 1,$$

$$u(0) = 1, \qquad u(1) = 0.$$

The replacement equations for this problem are

$$\frac{u_{i+1} - 2u_i + u_{i-1}}{h^2} + \frac{2}{2x_i - 1}\frac{u_{i+1} - u_{i-1}}{2h} - 8(2x_i + 1)u_i$$

$$= 12 - 16x_i - 16x_i^2, \qquad i = 1, 2, \ldots, n - 1,$$

$$u_0 = 1, \qquad u_n = 0.$$

The value of n is not specified. Let us choose $n = 4$ and thus $h = \frac{1}{4}$ to get a system that can be written down easily:

$$16(u_2 - 2u_1 + u_0) + \tfrac{8}{3}(u_2 - u_0) - 12u_1 = 7,$$

$$16(u_3 - 2u_2 + u_1) + 2(u_3 - u_1) - 16u_2 = 0,$$

$$16(u_4 - 2u_3 + u_2) + \tfrac{8}{5}(u_4 - u_2) - 20u_3 = -9,$$

$$u_0 = 1, \qquad u_4 = 0.$$

When terms are collected and the boundary values are substituted, these equations result:

$$-44u_1 + \tfrac{56}{3}u_2 \qquad\qquad = -\tfrac{19}{3},$$

$$14u_1 - 48u_2 + 18u_3 = 0,$$

$$\tfrac{72}{5}u_2 - 52u_3 = -9.$$

The solution of this system is given in Table 9.5, along with the corresponding u's obtained with $n = 16$. The error is about 0.02.

Table 9.5	i	x_i	r_i	u_i	$u(x_i)^*$	Error
	0	0	1.0	1	1	0
	1	0.25	1.5	0.2026	0.2014	0.0012
	2	0.50	2.0	0.1384	0.1459	−0.0075
	3	0.75	2.5	0.2114	0.2285	−0.0171
	4	1	3.0	0	0	0

* As approximated by using $n = 16$.

Other boundary conditions can also be treated numerically. For instance, consider the problem

$$\frac{d^2u}{dx^2} + 2\frac{du}{dx} = -4x, \qquad 0 < x < 1, \tag{9.150}$$

$$\frac{du}{dx}(0) = 2, \qquad u(1) = 1. \tag{9.151}$$

The replacement equations for the differential equation have the same form as before (see Eq. (9.144)). However, the replacement for the left boundary condition should be

$$\frac{u_1 - u_{-1}}{2h} = 2, \tag{9.152}$$

in imitation of the central difference approximation to the true condition. Thus a new quantity, u_{-1}, is introduced, and the total number of unknowns is $n + 2$:

$$u_{-1}, u_0, u_1, \ldots, u_{n-1}, u_n.$$

The $n + 2$ equations needed are the two boundary conditions and the replacement equations for $i = 0, 1, 2, \ldots, n - 1$.

Taking $n = 4$, we obtain these six equations in six unknowns:

$$2(u_1 - u_{-1}) = 2,$$
$$16(u_1 - 2u_0 + u_{-1}) + 4(u_1 - u_{-1}) = 0 \qquad [i = 0],$$
$$16(u_2 - 2u_1 + u_0) + 4(u_2 - u_0) = -1 \qquad [i = 1],$$
$$16(u_3 - 2u_2 + u_1) + 4(u_3 - u_1) = -2 \qquad [i = 2],$$
$$16(u_4 - 2u_3 + u_2) + 4(u_4 - u_2) = -3 \qquad [i = 3],$$
$$u_4 = 1.$$

The first equation is the boundary condition Eq. (9.152), the last is the boundary condition at $1 = x_4$, and the other four are Eq. (9.144), the replacement for the differential equation (9.150), for four values of i, and $h = \frac{1}{4}$, of course. Since u_{-1} is not wanted, the usual procedure is to solve the first equation for $u_{-1} = u_1 - 1$ and substitute into the second, which becomes

$$16(2u_1 - 2u_0) = 12.$$

When u_4 is substituted and coefficients are collected, the system to solve is

$$
\begin{aligned}
-32u_0 + 32u_1 \qquad\qquad\qquad &= 12, \\
12u_0 - 32u_1 + 20u_2 \qquad\qquad &= -1, \\
12u_1 - 32u_2 + 20u_3 &= -2, \\
12u_2 - 32u_3 &= -23.
\end{aligned}
$$

The solution of this system, along with values of the exact solution of the problem, is shown in Table 9.6.

Table 9.6	i	x_i	u_i	$u(x_i)$
	0	0.0	0.592	0.5677
	1	0.25	0.967	0.9519
	2	0.5	1.142	1.1337
	3	0.75	1.147	1.1436
	4	1.0	1.0	1.0

In solving the general linear boundary value problem

$$\frac{d^2u}{dx^2} + b(x)\frac{du}{dx} + c(x)u = f(x), \qquad 0 < x < 1, \tag{9.153}$$

$$au(0) - a'u'(0) = A, \tag{9.154}$$

$$bu(1) + b'u'(1) = B, \tag{9.155}$$

the replacement for the differential equation is set up as before. (See Eq. (9.148).) If the derivative is not mentioned in a boundary condition—that is, if $a' = 0$ in Eq. (9.154) or $b' = 0$ in Eq. (9.155)—then that boundary condition is handled as before. (See Eq. (9.149).) If a' or b' is nonzero, then the derivative in the boundary condition is replaced by a central difference. For instance, Eqs. (9.154) and (9.155) would become

$$au_0 - a'\,\frac{u_1 - u_{-1}}{2h} = A,$$

$$bu_n + b'\,\frac{u_{n+1} - u_{n-1}}{2h} = B.$$

Each of these introduces an extraneous unknown, u_{-1} or u_{n+1}, for which an extra version of Eq. (9.154) must be employed, using $i = 0$ or $i = n$, respectively.

Example 2

Set up the numerical solution of the boundary value problem

$$\frac{d^2u}{dx^2} - xu = -1, \qquad 0 < x < 1,$$

$$u(0) - u'(0) = 2, \qquad u(1) + u'(1) = 1.$$

Substituting difference expressions in the boundary conditions gives

$$u_0 - \frac{u_1 - u_{-1}}{2h} = 2, \qquad u_n + \frac{u_{n+1} - u_{n-1}}{2h} = 1.$$

The replacement equations for the differential equation are

$$\frac{u_{i+1} - 2u_i + u_{i-1}}{h^2} - x_i u_i = -1, \qquad i = 0, 1, \ldots, n.$$

Since the derivative appears in both boundary conditions, the replacement for the differential equation is needed for $i = 0$ and n as well as the intermediate values of i.

Now let us take $h = \frac{1}{4}$, $n = 4$. The boundary condition replacements are solved for u_{-1} and $u_{n+1} = u_5$ to get

$$u_{-1} = 1 - \tfrac{1}{2}u_0 + u_1, \qquad u_5 = \tfrac{1}{2} - \tfrac{1}{2}u_4 + u_3.$$

When these are substituted into the replacement equations, we are left with a system of five equations in five unknowns:

$$16(1 - \tfrac{1}{2}u_0 + u_1 - 2u_0 + u_1) = -1,$$
$$16(u_2 - 2u_1 + u_0) - \tfrac{1}{4}u_1 = -1,$$
$$16(u_3 - 2u_2 + u_1) - \tfrac{1}{2}u_2 = -1,$$
$$16(u_4 - 2u_3 + u_2) - \tfrac{3}{4}u_3 = -1,$$
$$16(u_3 - 2u_4 + \tfrac{1}{2} - \tfrac{1}{2}u_4 + u_3) - u_4 = -1.$$

The solution of this system is tabulated in Table 9.7, together with a more precise solution obtained by using $n = 128$.

Table 9.7

i	x_i	$u_i[n = 4]$	$u_i[n = 128]$
0	0.0	1.8237	1.8167
1	0.25	1.7483	1.7443
2	0.5	1.6378	1.6363
3	0.75	1.5160	1.5167
4	1.0	1.4027	1.4054

Exercises

1. Use the Taylor series expansions

$$u(x \pm h) = u(x) \pm hu'(x) + \frac{h^2}{2}u''(x) \pm \frac{h^3}{6}u'''(x) + \cdots$$

to obtain this expression for the central difference:

$$\frac{u(x + h) - u(x - h)}{2h} = u'(x) + \frac{h^2}{3}u'''(x) + \cdots.$$

This confirms the statement that the error in the central difference approximation is proportional to h^2.

2. Combine these central difference formulas to obtain Eq. (9.137):

$$u'\left(x + \frac{h}{2}\right) \cong \frac{u(x + h) - u(x)}{h}, \qquad u'\left(x - \frac{h}{2}\right) \cong \frac{u(x) - u(x - h)}{h},$$

$$u''(x) \cong \frac{u'\left(x + \frac{h}{2}\right) - u'\left(x - \frac{h}{2}\right)}{h}.$$

3. Follow the directions in Exercise 1, for the central difference approximation to the second derivative. Show that

$$\frac{u(x + h) - 2u(x) + u(x - h)}{h^2} = u''(x) + \frac{h^2}{12} u^{iv}(x) + \cdots.$$

4. When we write the matrix of coefficients of the system of equations that replaces a boundary value problem, the nonzero entries are on the diagonal or immediately above or below it. (This is called a tridiagonal matrix.) Explain why this is so.

5. If the error in a numerical solution—that is, $u_i - u(x_i)$—is proportional to h^2, how should the error change when n is doubled?

6. Below are given the exact and approximate solutions of the problem

$$\frac{d^2u}{dx^2} - 2u = 1 - x^2, \qquad 0 < x < 1,$$

$$u(0) = 0, \qquad u(1) = 1.$$

Compare the entries for different values of n and confirm that the errors appear to be proportional to h^2.

x	$u(x)$	$u_i[n = 4]$	$u_i[n = 8]$	$u_i[n = 16]$
0.25	0.1245195	0.1247831	0.1245861	0.1245361
0.5	0.3233196	0.3237578	0.3234304	0.3233472
0.75	0.6096689	0.6100772	0.6097721	0.6096946

7. If $U(x, h)$ represents the numerical solution corresponding to point x, calculated by using step size h, find a way to estimate (extrapolate) the true solution $u(x)$ from $U(x, h)$ and $U(x, h/2)$. Assume that

$$U(x, h) = u(x) + \varepsilon h^2.$$

8. Use the solution of Exercise 7 to extrapolate from adjacent columns of the table in Exercise 6. How do the results compare with the exact solution?

In Exercises 9–19, set up replacement equations for the problem, using the first value of n, and solve them. If a computer is available, set up and solve using the second value n also.

9. $\dfrac{d^2u}{dx^2} - 8u = -4x, \ 0 < x < 1,$

$u(0) = 0, \ u(1) = 0; \ n = 4, \ 8$

10. $\dfrac{d^2u}{dx^2} - 8u = -4x, \ 0 < x < 1,$

$u'(0) = 0, \ u(1) = 0; \ n = 4, \ 8$

11. $\dfrac{d^2u}{dx^2} - 8u = -4x, \ 0 < x < 1,$

$u'(0) = 0, \ u'(1) = 0; \ n = 4, \ 8$

12. $\dfrac{d^2u}{dx^2} - 8u = -4x, \ 0 < x < 1,$

$u(0) = 1, \ u(1) = 0; \ n = 4, \ 8$

13. $\dfrac{d^2u}{dx^2} - 8u = -4x, \ 0 < x < 1,$

$u'(0) = 0, \ u(1) = 1; \ n = 4, \ 8$

14. $\dfrac{d^2u}{dx^2} - 8u = -4x, \ 0 < x < 1,$

$u'(0) = 1, \ u(1) = 0; \ n = 4, \ 8$

15. $\dfrac{d^2u}{dx^2} + 10xu = 0, \ 0 < x < 1,$

$u(0) = 0, \ u(1) = 1; \ n = 5, \ 10$

16. $\dfrac{d^2u}{dx^2} - 10xu = 0, \ 0 < x < 1,$

$u(0) = 0, \ u(1) = 1; \ n = 5, \ 10$

17. $\dfrac{d^2u}{dx^2} + 2x\dfrac{du}{dx} - 8u = -4, \ 0 < x < 1,$

$u(0) = 1, \ u(1) = 0; \ n = 4, \ 8$

18. $\dfrac{d^2u}{dx^2} + 2x\dfrac{du}{dx} - 8u = -4, \ 0 < x < 1,$

$u(0) = 1, \ u'(1) = 0; \ n = 4, \ 8$

19. $\dfrac{d^2u}{dx^2} + \dfrac{10}{(1 + x)^4}u = 0, \ 0 < x < 1,$

$u(0) = 1, \ u(1) = 0; \ n = 5, \ 10$

20. In a problem concerning heat transfer from a washer-shaped plate,

this boundary value problem arises:

$$\frac{1}{r}\frac{d}{dr}\left(r\frac{dU}{dr}\right) = \gamma^2(U - T), \qquad \alpha < r < \beta,$$

$$U(\alpha) = T_1, \qquad U(\beta) = T_1.$$

Here $U(r)$ is the temperature in the plate, T is the ambient temperature, and γ^2 is a combination of physical constants. Make the change of variables

$$x = \frac{r - \alpha}{\beta - \alpha}, \qquad u(x) = \frac{U(r) - T}{T_1 - T},$$

to obtain this problem:

$$\frac{d}{dx}\left((x + k)\frac{du}{dx}\right) - b(x + k)u = 0, \qquad 0 < x < 1,$$

$$u(0) = 1, \qquad u(1) = 1,$$

where $b = \gamma^2(\beta - \alpha)^2$ and $k = \alpha/(\beta - \alpha)$.

21. Set up and solve replacement equations for the problem in Exercise 20, using $k = 1$, $b = 5$, and $n = 4$.

22. Investigate the effect of varying α, β, and γ in the problem of Exercise 20. In particular, what happens if $\beta - \alpha$ and γ are fixed and α decreases toward 0?

Notes and References

Two different kinds of linear boundary value problems are treated in this chapter. In the first kind, a problem with an inhomogeneity in the differential equation or boundary condition is to be solved. We found solutions for specific cases in Section 9.1 and formulated some theory in Section 9.4. The Green's function allows us to represent the solution in a relatively closed form.

In the second kind of problem we have a homogeneous differential equation and homogeneous boundary conditions, and we are to find the special values of a parameter that permit a nonzero solution. These eigenvalue problems are often part of a larger problem; in the next chapter we see how the eigenfunction series can be used to solve partial differential equations. Further information on eigenvalue problems can be found in the books of Leighton (1966) and Churchill and Brown (1978).

The Fourier series is an important tool in many areas of applied and pure mathematics and does not always appear in the context of differential equations. Much information will be found in the book of Churchill and Brown (1978). Tolstov's book (1962) contains an excellent presentation of the theory of Fourier series at an undergraduate level.

Miscellaneous Exercises

1. Find the eigenvalues and eigenfunctions for the problem

$$\frac{d}{dx}\left(e^{kx}\frac{du}{dx}\right) + \lambda^2 e^{kx}u = 0, \qquad 0 < x < L,$$

$$u(0) = 0, \qquad u(L) = 0.$$

2. Discuss the effectiveness of the problem in Exercise 1 as an approximation to

$$\frac{d}{dx}\left((1 + kx)\frac{du}{dx}\right) + \lambda^2(1 + kx)u = 0, \qquad 0 < x < L,$$

$$u(0) = 0, \qquad u(L) = 0.$$

3. Change the independent variable to $y = 1 + kx$ and show that the problem in Exercise 2 can be solved in terms of Bessel functions.

4. Solve the boundary value problem

$$e^{-\alpha x}\frac{d}{dx}\left(e^{\alpha x}\frac{du}{dx}\right) = 1 - x^2, \qquad 0 < x < 1,$$

$$u(0) = 0, \qquad u(1) = 1.$$

5. Solve this problem and compare the results with the solution of the problem in Exercise 4, with $\alpha = \ln 3$.

$$\frac{1}{1 + 2x}\frac{d}{dx}\left((1 + 2x)\frac{du}{dx}\right) = 1 - x^2, \qquad 0 < x < 1,$$

$$u(0) = 0, \qquad u(1) = 1.$$

6. Solve this eigenvalue problem, which comes from the theory of beams:

$$\frac{d^2u}{dx^2} + \frac{\lambda^2}{x^4}u = 0, \qquad \alpha < x < \beta,$$

$$u(\alpha) = 0, \qquad u(\beta) = 0.$$

(Hint: look for the solution of the differential equation by setting $u(x) = xv(1/x)$.)

7. Construct the Green's function for the problem

$$\frac{d}{dr}\left(r\frac{du}{dr}\right) = f(r), \qquad 0 < r < L,$$

$$u, u' \text{ bounded at } 0, \qquad u(L) = 0.$$

Treat the boundedness condition as if it were a boundary condition.

8. Construct the Green's function for the problem

$$\frac{d}{d\rho}\left(\rho^2\frac{du}{d\rho}\right) = f(\rho), \qquad 0 < \rho < L,$$

u, u' bounded at 0, $\qquad u(L) = 0.$

9. Let u measure the deviation of the centerline of a rotating shaft from the x-axis. If the shaft has linear density ρ, Young's modulus E, and second moment I, then u satisfies the differential equation

$$EI\frac{d^4u}{dx^4} = \omega^2\rho u, \qquad 0 < x < L.$$

Assuming that the bearings are at $x = 0$ and $x = L$ and act like simple supports, boundary conditions are

$$u(0) = 0, \qquad u''(0) = 0, \qquad u(L) = 0, \qquad u''(L) = 0.$$

Find the values of $\lambda^4 = \omega^2\rho/EI$ for which nontrivial solutions of the boundary value problem exist. These give the angular speeds (ω) at which the shaft may vibrate.

10. The deflection u of the centerline of a uniform beam from the x-axis is related to the distributed load $w(x)$ through the equation

$$EI\frac{d^4u}{dx^4} = w(x), \qquad 0 < x < L.$$

The quantity $EIu''(x)$ is proportional to the bending moment in the beam. At simple supports, no bending moment is supplied; for such a case the boundary conditions at the supports are

$$u(0) = 0, \qquad u''(0) = 0, \qquad u(L) = 0, \qquad u''(L) = 0.$$

Solve for $u(x)$ if $w(x) = w_0$ is uniform.

11. A uniform beam carrying a uniformly distributed load w_0 is built into a wall at one end ($x = 0$) and free at the other ($x = L$). (This is a cantilevered beam.) Find its deflection $u(x)$ by solving

$$EI\frac{d^4u}{dx^4} = w_0, \qquad 0 < x < L,$$

$$u(0) = 0, \qquad u'(0) = 0, \qquad u''(L) = 0, \qquad u'''(L) = 0.$$

12. Find the values of λ for which this problem has a nontrivial solution:

$$\frac{d^2u}{dx^2} + \lambda^2u = 0, \qquad 0 < x < L,$$

$$u(0) = 0, \qquad u'(L) = \lambda^2u(L).$$

13. Let u_n be the nth eigenfunction of the problem in Exercise 12. Show

that the eigenfunctions are *not* orthogonal:

$$\int_0^L u_n(x)u_m(x)\, dx \neq 0.$$

14. Show that the derivative of the function u in Exercise 12 satisfies a regular Sturm-Liouville problem.

15. As in Section 9.7, it is desired to find a representation for the function

$$\rho = \frac{a}{(1 + k \cos^2 \phi)^{1/2}},$$

where k is a small parameter. Expand in a Taylor series (powers of $\cos^2 \phi$) and replace $\cos^{2k} \phi$ by its equivalent in terms of $\cos 2k\phi$, etc. Take terms through k^2.

16. Let the solution, $H(r)$, of

$$(1 - 2xr + r^2)\frac{dH}{dr} - (x - r)H = 0, \qquad H(0) = 1,$$

be expressed as a Taylor series,

$$H(r) = \sum_0^\infty c_n r^n.$$

Show that the coefficients satisfy the relation

$$(n + 1)c_{n+1} - (2n + 1)xc_n + nc_{n-1} = 0$$

and that $c_0 = 1$, $c_1 = x$. This recursion relation also defines the Legendre polynomials, so $c_n = P_n(x)$.

17. Solve the differential equation in Exercise 16. It is called the generating function for the Legendre polynomials, since

$$H(r, x) = \sum_{n=0}^\infty P_n(x)r^n.$$

10   Partial Differential Equations

10.1

Introduction

A *partial differential equation* is a relation among a function and its partial derivatives with respect to at least two independent variables. Such equations are essential for the description of many significant physical phenomena. In this section we derive some partial differential equations of particular importance.

Vibrating String

First consider a string attached to two pegs on the x-axis at $x = 0$ and $x = L$. (See Fig. 10.1.) As in the case of the hanging cable (see Section 9.1 of Chapter 9), we assume that the string is of uniform material and perfectly flexible, so that it can transmit only tension forces. Now, however, the string is not static; each point is assumed to move, but only in the vertical direction. The string's shape is described by a function $u(x, t)$ that measures the height of the string above the x-axis at time t.

We find the equations of motion of the string by applying Newton's second law to a short segment. (See Fig. 10.2.) The components of force in the x-direction must add up to 0, since there is no horizontal motion:

$$T(x + \Delta x) \cos \phi(x + \Delta x) - T(x) \cos \phi(x) = 0. \tag{10.1}$$

In the y-direction the net force is equated to ma. The mass of the segment is approximately $\rho \Delta x$ (where ρ is the *linear* density, measured in mass/length). Since $u(x, t)$ measures displacement, its second time derivative measures acceleration. Thus we have

$$T(x + \Delta x) \sin \phi(x + \Delta x) - T(x) \sin \phi(x) + \Delta x F = \rho \Delta x \frac{\partial^2 u}{\partial t^2}(x, t) \tag{10.2}$$

from Newton's law in the y-direction. Here F represents any distributed load (positive upwards) that might be imposed on the string.

As in the case of the hanging cable, we deduce from Eq. (10.1) that

555

Figure 10.1

Vibrating String

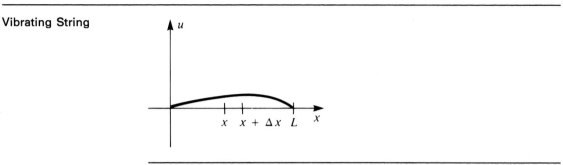

Figure 10.2

Section Cut Out

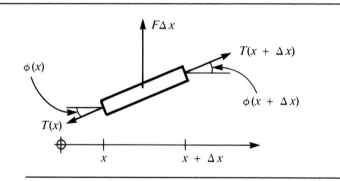

the horizontal component of tension is constant,

$$T(x + \Delta x) \cos \phi (x + \Delta x) = T(x) \cos \phi (x) = T. \tag{10.3}$$

We may now eliminate $T(x)$ and $T(x + \Delta x)$ from Eq. (10.3) to obtain

$$T(\tan \phi (x + \Delta x) - \tan \phi (x)) + \Delta x F = \rho \Delta x \frac{\partial^2 u}{\partial t^2}(x, t). \tag{10.4}$$

Recall that $\tan \phi (x)$ is the slope of the string, equal to the derivative of u with respect to x. Therefore

$$T\left(\frac{\partial u}{\partial x}(x + \Delta x, t) - \frac{\partial u}{\partial x}(x, t)\right) + \Delta x F = \rho \Delta x \frac{\partial^2 u}{\partial t^2}(x, t). \tag{10.5}$$

Finally, we divide through Eq. (10.5) by Δx and take the limit as Δx

approaches 0, obtaining

$$T\frac{\partial^2 u}{\partial x^2} + F = \rho\frac{\partial^2 u}{\partial t^2}, \tag{10.6}$$

which is a partial differential equation. Usually, we isolate the x-derivative and divide through by T to get

$$\frac{\partial^2 u}{\partial x^2} = \frac{1}{c^2}\frac{\partial^2 u}{\partial t^2} - \frac{F}{T} \tag{10.7}$$

with $c^2 = T/\rho$. This equation is valid in the interval $0 < x < L$ and for $t > 0$ (that is, after some arbitrary starting time). It is good practice to write these inequalities next to the equation. Equation (10.7) must be accompanied by the two boundary conditions

$$u(0, t) = 0, \qquad u(L, t) = 0, \qquad 0 < t,$$

which describe the fact that the string is fastened by its ends. Since each particle of the string obeys, in effect, a second-order differential equation in time, each particle's initial position and velocity must be specified by the conditions

$$u(x, 0) = f(x), \qquad \frac{\partial u}{\partial t}(x, 0) = g(x), \qquad 0 < x < L.$$

We now have a complete initial value, boundary value problem for the vibrating string:

$$\frac{\partial^2 u}{\partial x^2} = \frac{1}{c^2}\frac{\partial^2 u}{\partial t^2} - \frac{F}{T}, \qquad 0 < x < L, \qquad 0 < t, \tag{10.8}$$

$$u(0, t) = 0, \qquad u(a, t) = 0, \qquad 0 < t, \tag{10.9}$$

$$u(x, 0) = f(x), \qquad \frac{\partial u}{\partial t}(x, 0) = g(x), \qquad 0 < x < L. \tag{10.10}$$

Heat Conduction

Again we develop the nonsteady version of a problem developed in Section 9.1 of Chapter 9. We study the flow of heat by conduction along a rod of uniform cross-section between $x = 0$ and $x = L$. In the steady case we had identified the rates of heat flow into and out of a short section of the rod (see Fig. 10.3):

$$q(x, t) \cdot A + H\Delta x = \text{rate in},$$

$$q(x + \Delta x, t) \cdot A = \text{rate out}.$$

Figure 10.3

Slice Cut Out of Rod

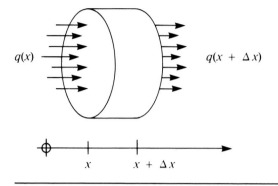

$q(x)$ $q(x + \Delta x)$

x $x + \Delta x$

Here q is the heat flux through a cross-section, measured in cal/sec $\cdot$ cm^2 or similar units, A is the cross-sectional area, and $H\Delta x$ is the rate at which heat enters the section by means other than axial conduction: for example, through the cylindrical surface by convection or by conversion of energy in other forms to heat. To complete a heat balance on the section of rod, we need to know that heat can be stored in the slice by an increase in the temperature, $u(x, t)$. Now the balance has the form

rate in $-$ rate out = storage rate

$$q(x, t) \cdot A + H\Delta x - q(x + \Delta x, t) \cdot A = c\rho A \Delta x \frac{\partial u}{\partial t}(x, t). \tag{10.11}$$

In the last term, c is the specific heat (cal/deg $\cdot$ g), ρ is the density, and $A\Delta x$ the volume of the slice.

After dividing through Eq. (10.11) by Δx and taking the limit as Δx vanishes, we find

$$H - A\frac{\partial q}{\partial x} = c\rho A \frac{\partial u}{\partial t} \tag{10.12}$$

as the representative of the law of conservation of energy.

In order to make a partial differential equation out of Eq. (10.12) we need Fourier's law relating heat flux to temperature gradient,

$$q = -\kappa \frac{\partial u}{\partial x}. \tag{10.13}$$

Substituting this relation into Eq. (10.13) and doing some minor algebra leads to the equation

$$\frac{\partial^2 u}{\partial x^2} = \frac{1}{k}\frac{\partial u}{\partial t} - \frac{H}{\kappa A} \tag{10.14}$$

in which $k = \kappa/c\rho$, called the thermal diffusivity. (See Table 10.1.) We have assumed that κ does not vary with x.

Table 10.1		Water	Concrete	Steel	Silver
Density, ρ(g/(cm)^3)		1	2.3	7.8	10.5
Specific heat, c (cal/g deg C)		1	0.16	0.11	0.056
Conductivity, κ (cal/sec · cm · deg C)		1.4×10^{-3}	4.1×10^{-3}	0.11	1.0
Diffusivity $k = \kappa/c\rho$ (cm^2/sec)		1.4×10^{-3}	1.1×10^{-2}	0.13	1.7

As in the case of the vibrating string, we must add to Eq. (10.14):

1. inequalities showing where and when it is valid;
2. boundary conditions describing the temperature u or the heat flux (proportional to $\partial u/\partial x$) at $x = 0$ and $x = L$;
3. an initial condition showing the temperature at each point at the starting time. Of course, H must also be identified for a particular problem.

A complete initial value, boundary value problem for the temperature in a rod would be

$$\frac{\partial^2 u}{\partial x^2} = \frac{1}{k}\frac{\partial u}{\partial t}, \qquad 0 < x < L, \qquad 0 < t, \tag{10.15}$$

$$u(0, t) = T_0, \qquad \frac{\partial u}{\partial x}(L, t) = 0, \qquad 0 < t, \tag{10.16}$$

$$u(x, 0) = f(x), \qquad 0 < x < L. \tag{10.17}$$

The corresponding physical problem features a cylindrical rod in which there is no energy conversion or loss from the surface ($H \equiv 0$ in Eq. (10.14)); the temperature is held constant, T_0, at the left end; the heat flow is zero at the right end (insulation there); and the initial temperature distribution is the function $f(x)$, yet to be specified.

Part of the vocabulary of partial differential equations is the same as that of ordinary differential equations. Below we list a few important concepts.

Order. The order of a partial differential equation is the order of the highest partial derivative in it. Sometimes one identifies the highest orders of each of the independent variables. Thus the simple *heat equation*

$$\frac{\partial^2 u}{\partial x^2} = \frac{1}{k}\frac{\partial u}{\partial t} \tag{10.18}$$

is second order in x and first in t, therefore second order overall.

Linear/Nonlinear. A partial differential equation is linear if the unknown function and its derivatives do not appear in powers of products and are not arguments of other functions. The heat equation (10.18) is linear. Likewise, the simple *wave equation*

$$\frac{\partial^2 u}{\partial x^2} = \frac{1}{c^2}\frac{\partial^2 u}{\partial t^2} \tag{10.19}$$

is linear (and second order), but the equation

$$\frac{\partial^2 u}{\partial x^2} = \frac{1}{c^2}\frac{\partial^2 u}{\partial t^2}\sqrt{1 + \left(\frac{\partial u}{\partial x}\right)^2} \tag{10.20}$$

is nonlinear. We shall restrict ourselves to linear equations.

Homogeneous/Nonhomogeneous. A linear equation is homogeneous if it contains no term that is free of the unknown function and its derivatives; otherwise, it is nonhomogeneous. For example, the heat equation Eq. (10.18) and the wave equation (10.19) are both homogeneous, while the equation

$$\frac{\partial^2 u}{\partial x^2} = \frac{1}{k}\frac{\partial u}{\partial t} - \frac{I^2 R}{\kappa A}, \tag{10.21}$$

which describes temperature in a current-carrying rod, is nonhomogeneous.

An easy test is this: a linear equation is homogeneous if and only if $u \equiv 0$ satisfies the equation.

Solution. A solution of a partial differential equation is a function, having partial derivatives of orders up to and including those in the equation, that satisfies the equation identically.

Example 1

These functions are both solutions of the simple heat equation (10.18):

$$u_1(x, t) = x^2 + 2kt,$$
$$u_2(x, t) = (\sin \lambda x)e^{-\lambda^2 kt}.$$

(Here λ is an arbitrary parameter). To verify the claim that u_2 is a solution, calculate

$$\frac{\partial^2 u_2}{\partial x^2} = -\lambda^2 (\sin \lambda x) e^{-\lambda^2 kt},$$

$$\frac{\partial u_2}{\partial t} = -\lambda^2 k (\sin \lambda x) e^{-\lambda^2 kt}.$$

Then it is clear that $\partial^2 u_2/\partial x^2 = (1/k)(\partial u_2/\partial t)$.

Example 2

The following functions are all solutions of the wave equation (10.19):

$$u_1(x, t) = (\sin \lambda x)(\cos \lambda ct),$$
$$u_2(x, t) = (\sin \lambda x)(\sin \lambda ct),$$
$$u_3(x, t) = (x - ct)^2.$$

For the first function we have

$$\frac{\partial^2 u_1}{\partial x^2} = -\lambda^2 (\sin \lambda x)(\cos \lambda ct)$$

$$\frac{\partial^2 u_1}{\partial t^2} = -\lambda^2 c^2 (\sin \lambda x)(\cos \lambda ct).$$

It follows that $\partial^2 u/\partial x^2 = (1/c^2)(\partial^2 u/\partial t^2)$, as required. (Again, λ is an arbitrary parameter in u_1 and u_2.)

When it comes to solving an initial value, boundary value problem such as Eqs. (10.8)–(10.10) or Eqs. (10.15)–(10.17), it will be convenient to think always of the partial differential equation and the boundary conditions together. Thus if *both* partial differential equation and boundary conditions are homogeneous, we will speak of a *homogeneous problem*. If either contains an inhomogeneity, we speak of a *nonhomogeneous problem*.

With this agreement we can state two theorems that will sound familiar. The first is called the *principle of superposition*.

Theorem 10.1

Any linear combination of solutions of a homogeneous problem is also a solution.

Example 3

The following problem is homogeneous because both partial differential equation and boundary conditions are homogeneous:

$$\frac{\partial^2 u}{\partial x^2} = \frac{1}{c^2} \frac{\partial^2 u}{\partial t^2}, \qquad 0 < x < L, \qquad 0 < t, \tag{10.22}$$

$$u(0, t) = 0, \qquad u(L, t) = 0, \qquad 0 < t. \tag{10.23}$$

In Example 2 we confirmed that the following two functions are solutions of the partial differential equation above:

$$u_1(x, t) = \sin(\pi x/L) \cos(\pi ct/L),$$

$$u_2(x, t) = \sin(2\pi x/L) \cos(2\pi ct/L).$$

(These correspond to two specific choices for the parameter λ in Example 2.) It is also clear that both functions satisfy the boundary conditions at $x = 0$ and $x = L$, since

$$\sin 0 = \sin \pi = \sin 2\pi = 0.$$

According to Theorem 10.1, then, the linear combination

$$u(x, t) = \sin(\pi x/L) \cos(\pi ct/L) + \tfrac{1}{2} \sin(2\pi x/L) \cos(2\pi ct/L)$$

is also a solution of the problem consisting of the partial differential equation (10.22) and boundary conditions (10.23).

Theorem 10.2

The general solution of a nonhomogeneous problem has the form $u_c + u_p$, where u_p is a particular solution of the nonhomogeneous problem and u_c is the general solution of the corresponding homogeneous problem.

Example 4

The problem

$$\frac{\partial^2 u}{\partial x^2} = \frac{1}{k} \frac{\partial u}{\partial t} - C, \qquad 0 < x < L, \qquad 0 < t, \tag{10.24}$$

$$u(x, 0) = T_0, \qquad \frac{\partial u}{\partial x}(L, t) = 0, \qquad 0 < t, \tag{10.25}$$

has inhomogeneities in both the partial differential equation and the left boundary condition.

It is easy to verify that a particular solution of the problem is

$$u_p(x) = T_0 + \tfrac{1}{2} Cx(2L - x) \tag{10.26}$$

(assuming that C is a constant). That is, u_p satisfies the given partial differential equation and the two boundary conditions.

The complement $u_c(x, t)$ is the general solution of the corresponding homogeneous problem

$$\frac{\partial^2 u}{\partial x^2} = \frac{1}{k}\frac{\partial u}{\partial t}, \qquad 0 < x < L, \qquad 0 < t, \tag{10.27}$$

$$u(0, t) = 0, \qquad \frac{\partial u}{\partial x}(L, t) = 0, \qquad 0 < t. \tag{10.28}$$

Exercises

1. In the derivation of the vibrating string problem (between Eqs. (10.1) and (10.2)) we assumed that the mass of a segment was approximately $m = \rho\Delta x$. Show that a more accurate assumption is $m = \rho\Delta s$, where Δs is an element of arc length. With this new assumption, what equation replaces Eq. (10.6)?

2. Show that the assumption $m = \rho\Delta x$ mentioned in Exercise 1 is equivalent to assuming that $\partial u/\partial x$ is small.

3. In the study of heat flow in a rod we assumed that the cross-sectional area A is uniform. If A is a function of x, how are Eqs. (10.11)–(10.14) changed?

4. Follow the directions in Exercise 3 if A is uniform but κ varies with x.

5. Provide a physical interpretation for this initial value, boundary value problem and sketch $f(x)$:

$$\frac{\partial^2 u}{\partial x^2} = \frac{1}{c^2}\frac{\partial^2 u}{\partial t^2} + \frac{g}{c^2}, \qquad 0 < x < L, \qquad 0 < t,$$

$$u(0, t) = 0, \qquad u(L, t) = 0, \qquad 0 < t,$$

$$u(x, 0) = f(x), \qquad \frac{\partial u}{\partial t}(x, 0) = 0, \qquad 0 < x < L,$$

and

$$f(x) = \begin{cases} 2hx/L, & 0 < x < L/2, \\ 2h(L - x)/L, & L/2 \le x < L. \end{cases}$$

6. Provide a physical interpretation for the problem in Example 4.

7. In the heat conduction problem, suppose that heat is lost through the cylindrical surface of the slice being analyzed by convection to a medium at temperature T. The rate of loss by convection is proportional to the exposed area and the temperature difference between the body and the medium. Find H and then the specific version of Eq. (10.14) that describes temperature in the rod.

8. If the rod in Exercise 7 has a circular cross-section, show how the ratio C/A (circumference/cross-sectional area) depends on the diameter of the rod.

9. If an insulated copper rod carries a current, Eq. (10.21) describes its temperature. Suppose $I = 15$ amps, $R = 1$ ohm/cm, $A = 0.5$ cm^2; for copper, $\kappa = 0.92$ and $k = 1.1$ (units of measurement as in Table 10.1). Find all the terms of Eq. (10.21). Be careful to get the units of measurement correct.

10. In Example 3 the solution $u_1(x, t) = \sin(\pi x/L)\cos(\pi ct/L)$ is periodic in time with frequency $\pi c/L$ radians/sec. A violin string about 50 cm long can produce a sound at about 1000π radians/sec. What value does c have if $u_1(x, t)$ is responsible for that sound?

10.2

The Homogeneous Heat Problem

In this section we develop a method for solving homogeneous problems and apply it to the homogeneous heat problem

$$\frac{\partial^2 u}{\partial x^2} = \frac{1}{k}\frac{\partial u}{\partial t}, \qquad 0 < x < L, \qquad 0 < t, \tag{10.29}$$

$$u(0, t) = 0, \qquad u(L, t) = 0, \qquad 0 < t, \tag{10.30}$$

$$u(x, 0) = f(x), \qquad 0 < x < L. \tag{10.31}$$

Recall that we use the words "homogeneous problem" to refer to the fact that the partial differential equation (10.29) and the boundary conditions (10.30) are all homogeneous.

The key to our method is to seek solutions of the homogeneous problem in the form of a product:

$$u(x, t) = X(x)T(t). \tag{10.32}$$

As indicated by the notation, $X(x)$ is assumed to be a function of x alone and $T(t)$ of t alone. Since both the partial differential equation and boundary conditions are satisfied by $u(x, t) \equiv 0$ (the *trivial* solution), we are interested only in *nonzero* (or nontrivial) solutions.

We now substitute the assumed form of u, Eq. (10.32), into the partial differential equation and boundary conditions. The results are

$$X''(x)T(t) = \frac{1}{k}X(x)T'(t), \qquad 0 < x < L, \qquad 0 < t, \tag{10.33}$$

$$X(0)T(t) = 0, \qquad X(L)T(t) = 0, \qquad 0 < t. \tag{10.34}$$

In these equations the prime is used to denote differentiation with respect to the displayed variable. In writing Eq. (10.33) we have relied on the assumption that $X(x)$ depends on x alone and $T(t)$ on t alone.

From Eq. (10.33) we may draw some inferences. We are assuming $X(x)$ and $T(t)$ to be nontrivial, so we may divide through the equation by the product $X(x)T(t)$ to obtain

$$\frac{X''(x)}{X(x)} = \frac{T'(t)}{kT(t)}, \qquad 0 < x < L, \qquad 0 < t. \tag{10.35}$$

The functions on the left-hand side of this equation depend on x alone, while those on the right-hand side depend on t alone. Since x and t are independent variables, both members of the equation must be constant:

$$\frac{X''(x)}{X(x)} = \text{const} = \frac{T'(t)}{kT(t)}. \tag{10.36}$$

We say now that the *variables* (the two independent variables) have been *separated*.

Equation (10.34) also has implications. Since each of the equalities is to hold for all $t > 0$, and since $T(t)$ is not identically 0, it must be the other factor in each equation that is zero:

$$X(0) = 0, \qquad X(L) = 0. \tag{10.37}$$

With this extra information about $X(x)$ we can make one more deduction about Eq. (10.36): the constant cannot be positive. (See Section 3.6 of Chapter 3 and Exercise 1 at the end of this section.) It is convenient to call the constant $-\lambda^2$.

Now, assembling the information known about $X(x)$, we find that the function must be nonzero and satisfy

$$\frac{X''(x)}{X(x)} = -\lambda^2,$$

or

$$X'' + \lambda^2 X = 0, \qquad 0 < x < L, \tag{10.38}$$

and

$$X(0) = 0, \qquad X(L) = 0. \tag{10.39}$$

From our studies of ordinary boundary value problems in Chapter 9, we recognize Eqs. (10.38) and (10.39) as an eigenvalue problem, called the *eigenvalue problem associated with* the homogeneous heat problem, Eqs. (10.29) and (10.30).

The next objective is to solve the eigenvalue problem, Eqs. (10.38) and (10.39); that is, we must find the values of λ^2 (the eigenvalues) for which nontrivial solutions of Eqs. (10.38) and (10.39) exist, and also find those solutions, the eigenfunctions. We have already solved this problem in Chapter 9, Section 9.2. The results are

$$\lambda_n^2 = (n\pi/L)^2, \qquad X_n(x) = \sin \lambda_n x, \qquad n = 1, 2, 3, \ldots . \tag{10.40}$$

Remember that any constant nonzero multiple of an eigenfunction is still an eigenfunction, but such a multiplier is usually not shown.

We have now found all the possible X's that satisfy Eqs. (10.34) and (10.36). Let us find the other factor, $T(t)$. For each $n = 1, 2, 3, \ldots$, we put $X_n(x)$ into Eq. (10.36) and find that the factor that goes with $X_n(x)$ (call it $T_n(t)$) satisfies

$$\frac{T_n'(t)}{kT_n(t)} = \frac{X_n''(x)}{X_n(x)} = -\lambda_n^2$$

or

$$T_n' + \lambda_n^2 k T_n = 0. \tag{10.41}$$

This equation is easily solved to determine

$$T_n(t) = e^{-\lambda_n^2 kt}. \tag{10.42}$$

Again we have left out the multiplicative constant that would express the general solution of Eq. (10.41).

At this point we have found an infinite number of solutions of the homogeneous problem Eqs. (10.29) and (10.30). They are

$$u_n(x, t) = \sin \lambda_n x \, e^{-\lambda_n^2 kt}, \qquad n = 1, 2, \ldots . \tag{10.43}$$

One of the main properties of a linear, homogeneous problem is that a linear combination of solutions is also a solution. In view of the number of solutions that we have obtained, we must consider not just a sum, but rather an infinite series, of constant multiples of solutions. Therefore we may write

$$u(x, t) = \sum_{n=1}^{\infty} b_n e^{-\lambda_n^2 kt} \sin \lambda_n x \tag{10.44}$$

as the most general solution of Eqs. (10.29) and (10.30). The fact is that some restrictions must be imposed on the coefficients $b_1, b_2, \ldots$ in Eq. (10.44) in order to guarantee that the sum of the series exists and has the partial derivatives needed for Eq. (10.29).* However, examination of these conditions would lead us far afield. Instead, let us see how we can satisfy the intial condition,

$$u(x, 0) = f(x), \qquad 0 < x < L,$$

where f is some given function.

---

* The series in Eq. (10.44) converges and is a solution of the heat equation (10.29) if b_n approaches 0 as n increases.

If we set $t = 0$ in $u(x, t)$ as given by Eq. (10.44), then all of the exponentials have value 1, and the series simplifies to

$$u(x, 0) = \sum_{n=1}^{\infty} b_n \sin \lambda_n x. \tag{10.45}$$

It is this series that must equal the given function $f(x)$:

$$\sum_{n=1}^{\infty} b_n \sin \lambda_n x = f(x), \qquad 0 < x < L. \tag{10.46}$$

But this equation asks for the eigenfunction expansion of the function $f(x)$. To satisfy it, we must take

$$b_n = \frac{2}{L} \int_0^L f(x) \sin \lambda_n x \, dx, \tag{10.47}$$

as derived in Chapter 9, Section 9.5. If $f(x)$ is sectionally smooth, then the series in Eq. (10.46) (with coefficients as given in Eq. (10.47)) actually converges to $f(x)$ at every point where f is continuous, and to

$$\lim_{h \to 0} \frac{f(x + h) + f(x - h)}{2}$$

at any point of discontinuity of f. Even if f should have discontinuities, Eq. (10.46) holds at all but a finite number of points, at worst. We may say that the initial condition is satisfied by choosing the b's by Eq. (10.47). Thus we have in Eq. (10.46) a function that satisfies the partial differential equation (10.29), the boundary conditions (10.30), and the intial condition (10.31). That is, we have solved the given homogeneous heat problem.

Example 1

Solve the heat problem below (units of measurement: x in centimeters, t in seconds, u in degrees Celsius):

$$\frac{\partial^2 u}{\partial x^2} = \frac{1}{0.13} \frac{\partial u}{\partial t}, \qquad 0 < x < 15, \qquad 0 < t,$$

$$u(0, t) = 0, \qquad u(15, t) = 0, \qquad 0 < t,$$

$$u(x, 0) = 100, \qquad 0 < x < 15.$$

The partial differential equation and boundary conditions are homogeneous, so we may proceed as outlined above. First, seek solutions in the product form

$$u(x, t) = X(x)T(t).$$

Next substitute u in this form into the partial differential equation and the boundary conditions

$$X''(x)T(t) = \frac{1}{0.13} X(x)T'(t), \qquad 0 < x < 15, \qquad 0 < t,$$

$$X(0)T(t) = 0, \qquad X(15)T(t) = 0, \qquad 0 < t.$$

Eliminate $T(t)$ from the boundary conditions and separate the variables in the partial differential equation:

$$X(0) = 0, \qquad X(15) = 0,$$

$$\frac{X''(x)}{X(x)} = \frac{T'(t)}{0.13T(t)} = -\lambda^2.$$

The eigenvalue problem associated with the given homogeneous problem is

$$X'' + \lambda^2 X = 0, \qquad 0 < x < 15,$$

$$X(0) = 0, \qquad X(15) = 0,$$

and its solution has already been found, as

$$\lambda_n^2 = (n\pi/15)^2, \qquad X_n(x) = \sin \lambda_n x.$$

The second factor in the product solution is now found from the equation

$$\frac{T_n'}{0.13T_n} = -\lambda_n^2 = -\left(\frac{n\pi}{15}\right)^2.$$

It is easily determined that

$$T_n(t) = \exp\left(-0.13n^2\pi^2 t/15^2\right)$$
$$= \exp\left(-0.0058n^2 t\right).$$

A series of constant multiples of product solutions represents the general solution of the homogeneous problem. It is

$$u(x, t) = \sum_{n=1}^{\infty} b_n \sin(n\pi x/15) \exp(-0.0058n^2 t).$$

At $t = 0$ the temperature is to be 100 deg C throughout the interval $0 < x < 15$. Thus the coefficients b_n must be chosen to make

$$u(x, 0) = \sum_{n=1}^{\infty} b_n \sin(n\pi x/15) = 100, \qquad 0 < x < 15.$$

Using the idea of eigenfunction expansion, we determine

$$b_n = \frac{2}{15} \int_0^{15} 100 \cdot \sin(n\pi x/15)\, dx$$

$$= \frac{200}{15}\left(\frac{1 - \cos n\pi}{n\pi/15}\right) = \frac{200}{\pi}\frac{1 - \cos n\pi}{n}.$$

Finally, the complete solution may be assembled in this form:

$$u(x, t) = \frac{200}{\pi}\sum_{n=1}^{\infty}\frac{1 - \cos n\pi}{n}\sin(n\pi x/15)\exp(-0.0058n^2 t).$$

The temperature "profile" ($u(x, t)$ as a function of x) is shown in Fig. 10.4 for various values of t. Note that after $t = 100$ the graph is almost a pure sine arch, and by $t = 400$ the maximum value of u is only about 10% of the original maximum.

Figure 10.4

Temperature Profiles

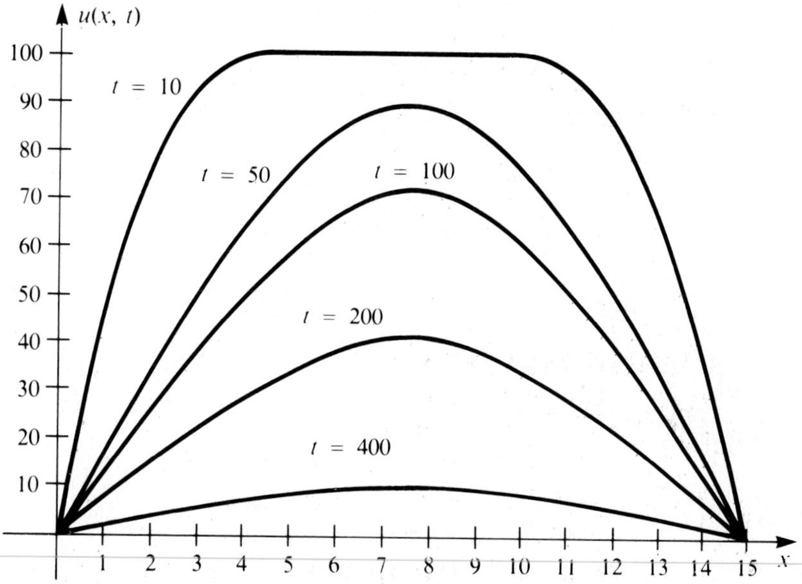

The method we are using is called the *product method, separation of variables,* or the *eigenfunction method.* It applies only to homogeneous problems. We may summarize the procedure as follows:

1. Assume a product solution and substitute into the partial differential equation.
2. Separate the independent variables.
3. Find and solve the associated eigenvalue problem.
4. Find the other factor in the product.

5. Form a series of constant multiples of the product solutions. This is the general solution of the homogeneous problem.

6. Satisfy the initial conditions, using the idea of eigenfunction expansion.

The next section contains further examples of the method.

It is worth noting that the solution of problems—whether in general parameters as in the text or in specific numerical values as in Example 1 is easier if we convert to the dimensionless variables

$$\bar{x} = x/L, \qquad \bar{t} = kt/L^2. \tag{10.48}$$

The first one, $\bar{x}$, measures distance as a proportion of length. Both groupings, x/L and kt/L^2, occur naturally in the problem, since the product solution in Eq. (10.43) is

$$\sin \lambda_n x e^{-\lambda_n^2 kt} = \sin (n\pi x/L) \exp (-n^2\pi^2 kt/L^2).$$

In terms of $\bar{x}$ and $\bar{t}$ the heat equation (10.29) and boundary conditions (10.30) become

$$\frac{\partial^2 u}{\partial \bar{x}^2} = \frac{\partial u}{\partial \bar{t}}, \qquad 0 < \bar{x} < 1, \qquad 0 < \bar{t},$$

$$u(0, \bar{t}) = 0, \qquad u(1, \bar{t}) = 0, \qquad 0 < \bar{t}.$$

Exercises

1. Prove that if X is a nontrivial solution of $X''/X = k^2 > 0$, then $X(x) = 0$ for at most one value of x (k is constant).

2. Prove that if X is a nontrivial solution of $X''/X = k^2 > 0$, then either $X(x) = 0$ for some x or $X'(x) = 0$ for some x, but not both.

In Exercises 3–7, solve the homogeneous heat problem

$$\frac{\partial^2 u}{\partial x^2} = \frac{1}{k}\frac{\partial u}{\partial t}, \qquad 0 < x < L, \qquad 0 < t,$$

$$u(0, t) = 0, \qquad u(L, t) = 0, \qquad 0 < t,$$

$$u(x, 0) = f(x), \qquad 0 < x < L,$$

for the given function $f(x)$. Sketch $f(x)$. It is sectionally smooth?

3. $f(x) = U_0 x/L, \quad 0 < x < L$

4. $F(x) = \begin{cases} 0, & 0 < x < L/2, \\ U_0, & L/2 \le x < L \end{cases}$

5. $f(x) = U_0 \sin (\pi x/L)$

6. $f(x) = U_0(L - x)/L, \quad 0 < x < L$

7. $f(x) = \begin{cases} 2U_0 x/L, & 0 < x < L/2, \\ 2U_0(L - x)/L, & L/2 \le x < L \end{cases}$

8. In Eq. (10.44) it can be seen that each term of the series contains an

exponential, $\exp(-\lambda_n^2 kt)$, which controls the size of the term as a function of t. Sketch graphs of the functions $\exp(-n^2\tau)$ as a function of τ, $0 \leq \tau \leq 1$, for $n = 1, 2, 3$. (Note that $\tau = \pi^2 kt/L^2$ coordinates this function with the exponential above.)

9. A certain heat problem like Eqs. (10.29)–(10.31) has as its solution

$$u(x, t) = \frac{2U_0}{\pi} \sum_{n=1}^{\infty} \frac{1}{n} e^{-n^2\pi^2 t} \sin(n\pi x), \qquad 0 < x < 1, \qquad 0 < t.$$

(In this equation, x and t are dimensionless variables, as in Eq. (10.48).) In order to obtain a graph of $u(x, t)$ versus x for various values of t, it is necessary to add up the first few terms of the series. Suppose we wish to include all terms for which $e^{-n^2\pi^2 t}/n > 0.01$. How many terms are needed at $t = 0.01$? At $t = 0.1$? At $t = 1$?

10. Refer to Example 1. About how long will it take before $u(x, t)$ is everywhere less than 1 degree?

In Exercises 11–14, find and solve the eigenvalue problem associated with the heat problem consisting of

$$\frac{\partial^2 u}{\partial x^2} = \frac{1}{k} \frac{\partial u}{\partial t}, \qquad 0 < x < L, \qquad 0 < t,$$

together with the given boundary conditions.

11. $\dfrac{\partial u}{\partial x}(0, t) = 0,\ u(L, t) = 0$

12. $u(0, t) = 0,\ u(L, t) = 0$

13. $\dfrac{\partial u}{\partial x}(0, t) = 0,\ \dfrac{\partial u}{\partial x}(L, t) = 0$

14. $u(0, t) = 0,\ \dfrac{\partial u}{\partial x}(L, t) = 0$

15. Solve this homogeneous heat problem:

$$\frac{\partial^2 u}{\partial x^2} = \frac{\partial u}{\partial t} + 2u, \qquad 0 < x < 1, \qquad 0 < t,$$

$$u(0, t) = 0, \qquad u(1, t) = 0, \qquad 0 < t,$$

$$u(x, 0) = 100, \qquad 0 < x < 1.$$

(Hint: in separating the variables, keep X''/X alone on one side.)

10.3
Examples

In this section our agenda is to work out some example problems featuring the heat equation. In some cases, however, the physical phenomenon being represented will not be the movement of heat.

Example 1

(a) Solve the heat problem below. (b) Find how long it will take for $u(L, t)$ to drop to 1% of its initial value.

$$\frac{\partial^2 u}{\partial x^2} = \frac{1}{k}\frac{\partial u}{\partial t}, \qquad 0 < x < L, \qquad 0 < t,$$

$$u(0, t) = 0, \qquad \frac{\partial u}{\partial x}(L, t) = 0, \qquad 0 < t,$$

$$u(x, 0) = \frac{U_0 x}{L}, \qquad 0 < x < L.$$

(a) Since the partial differential equation and the boundary conditions are homogeneous, we may start by seeking nontrivial solutions in the product form $u(x, t) = X(x)T(t)$. The partial differential equation and boundary conditions become

$$X''(x)T(t) = \frac{1}{k}X(x)T'(t), \qquad 0 < x < L, \qquad 0 < t,$$

$$X(0)T(t) = 0, \qquad X'(L)T(t) = 0, \qquad 0 < t.$$

The variables are separated by dividing through the partial differential equation by $X(x)T(t)$, with the result

$$\frac{X''(x)}{X(x)} = \frac{T'(t)}{kT(t)}, \qquad 0 < x < L, \qquad 0 < t.$$

Also, $T(t)$ may be eliminated from both boundary conditions, leaving

$$X(0) = 0, \qquad X'(L) = 0.$$

In the variables-separated form of the partial differential equation, both members must be constant. The boundary conditions require that the constant be negative (say, $-\lambda^2$) so that the eigenvalue problem is

$$X'' + \lambda^2 X = 0, \qquad 0 < x < L,$$

$$X(0) = 0, \qquad X'(L) = 0,$$

with solutions

$$\lambda_n^2 = ((2n - 1)\pi/2L)^2, \qquad X_n(x) = \sin \lambda_n x, \qquad n = 1, 2, 3, \ldots .$$

The second factor in the product solution is required to satisfy

$$\frac{T_n'}{kT_n} = \frac{X_n''}{X_n} = -\lambda_n^2,$$

whence $T_n(t) = e^{-\lambda_n^2 kt}$.

Now we may make a general solution by forming a series of

constant multiples of product solutions,

$$u(x, t) = \sum_{n=1}^{\infty} c_n \sin \lambda_n x \, e^{-\lambda_n^2 kt}.$$

To satisfy the initial condition, set $t = 0$ in this expression and equate to $U_0 x/L$:

$$\sum_{n=1}^{\infty} c_n \sin \lambda_n x = \frac{U_0 x}{L}, \qquad 0 < x < L.$$

The coefficients of this eigenfunction series are determined by these integrals:

$$c_n = \frac{2}{L} \int_0^L \frac{U_0 x}{L} \sin \lambda_n x \, dx$$

$$= \frac{2 U_0 \sin \lambda_n L}{(\lambda_n L)^2} = \frac{8 U_0 \sin \lambda_n L}{(2n - 1)^2 \pi^2}.$$

Using these coefficients, we find the complete solution to be

$$u(x, t) = \sum_{n=1}^{\infty} \frac{8 U_0 \sin \lambda_n L}{(2n - 1)^2 \pi^2} \sin \lambda_n x e^{-\lambda_n^2 kt}.$$

(b) To find the value of u at $x = L$, simply substitute in this last expression. Note that $\sin \lambda_n L$ (which is 1 or -1) appears squared, so

$$u(L, t) = \sum_{n=1}^{\infty} \frac{8 U_0}{(2n - 1)^2 \pi^2} e^{-\lambda_n^2 kt}$$

$$= \frac{8 U_0}{\pi^2} \left(e^{-\tau} + \frac{e^{-9\tau}}{9} + \frac{e^{-25\tau}}{25} + \cdots \right),$$

where $\tau = \pi^2 kt/4L^2$. The series converges very rapidly for $\tau > 0$. (That is, only a few terms are needed to obtain a good approximation to the sum.) Supposing that only one term is needed in this approximation, we must solve

$$\tfrac{1}{100} U_0 = \frac{8 U_0}{\pi^2} e^{-\tau}$$

for τ. The solution is

$$\tau = \ln \frac{800}{\pi^2} \cong 4.395,$$

from which we determine

$$t = \frac{4L^2}{\pi^2 k} \tau = \frac{1.78 L^2}{k}.$$

Example 2

A porous rod containing moisture has its left end sealed, has its right end in contact with a dry medium, and loses moisture through its surface to dry air. The concentration of moisture, $u(x, t)$, satisfies the problem

$$\frac{\partial^2 u}{\partial x^2} = \frac{1}{k}\frac{\partial u}{\partial t} + \gamma^2 u, \qquad 0 < x < L, \qquad 0 < t,$$

$$\frac{\partial u}{\partial x}(0, t) = 0, \qquad u(L, t) = 0, \qquad 0 < t,$$

$$u(x, 0) = U_0, \qquad 0 < x < L.$$

Find $u(x, t)$ and determine $u(0, t)$ explicitly.

The partial differential equation and boundary conditions are homogeneous, so we seek nontrivial solutions in the form $u(x, t) = X(x)T(t)$. The partial differential equation becomes

$$X''(x)T(t) = \frac{1}{k}X(x)T'(t) + \gamma^2 X(x)T(t).$$

Separation of variables is accomplished by dividing through by $X(x)T(t)$, leaving

$$\frac{X''(x)}{X(x)} = \frac{T'(t) + k\gamma^2 T(t)}{kT(t)}.$$

Both sides must be constant. The boundary conditions, after dividing out $T(t)$, are

$$X'(0) = 0, \qquad X(L) = 0$$

and require that $X''/X = -\lambda^2$ or

$$X'' + \lambda^2 X = 0, \qquad 0 < x < L.$$

The last two equations make up an eigenvalue problem whose solution is

$$\lambda_n^2 = ((2n - 1)\pi/2L)^2, \qquad X_n(x) = \cos \lambda_n x, \qquad n = 1, 2, 3, \ldots.$$

The second factor in the product solution can now be determined from the variables-separated form of the partial differential equation. Since $X_n''/X_n = -\lambda_n^2$, T_n must satisfy

$$\frac{T_n' + k\gamma^2 T_n}{kT_n} = -\lambda_n^2 \qquad \text{or} \qquad T_n' + k(\gamma^2 + \lambda_n^2)T_n = 0.$$

This equation is easily solved to determine

$$T_n(t) = e^{-k(\gamma^2 + \lambda_n^2)t} = e^{-\gamma^2 kt}e^{-\lambda_n^2 kt}.$$

Now the general solution is assembled as a series of constant multiples of product solutions:

$$u(x, t) = e^{-\gamma^2 kt} \sum_{n=1}^{\infty} c_n \cos \lambda_n x \, e^{-\lambda_n^2 kt}.$$

To satisfy the initial condition, set $t = 0$ and equate to U_0:

$$\sum_{n=1}^{\infty} c_n \cos \lambda_n x = U_0, \qquad 0 < x < L.$$

The coefficients of this eigenfunction series must be

$$c_n = \frac{2}{L} \int_0^L U_0 \cos \lambda_n x \, dx$$

$$= \frac{2U_0 \sin \lambda_n L}{\lambda_n L} = \frac{4U_0 \sin \lambda_n L}{(2n - 1)\pi}.$$

With this choice for c_n the series above for $u(x, t)$ represents the solution of this problem.

The value of $u(0, t)$ can now be found by setting $x = 0$ in our series solution:

$$u(0, t) = e^{-\gamma^2 kt} \frac{4U_0}{\pi} \sum_{n=1}^{\infty} \frac{\sin \lambda_n L}{2n - 1} e^{-\lambda_n^2 kt}$$

$$= \frac{4U_0}{\pi} e^{-\gamma^2 kt} \left(e^{-\tau} - \frac{e^{-9\tau}}{3} + \frac{e^{-25\tau}}{5} - + \cdots \right),$$

where $\tau = \pi^2 kt/4L^2$.

Example 3

Solve the heat/diffusion problem below, which is stated in terms of dimensionless variables:

$$\frac{\partial^2 u}{\partial x^2} = \frac{\partial u}{\partial t}, \qquad 0 < x < 1, \qquad 0 < t,$$

$$\frac{\partial u}{\partial x}(0, t) = 0, \qquad \frac{\partial u}{\partial x}(1, t) = 0, \qquad 0 < t,$$

$$u(x, 0) = 1 + 2x, \qquad 0 < x < 1.$$

The partial differential equation and boundary conditions are homogeneous. After assuming that $u(x, t) = X(x)T(t)$ and preforming the usual manipulations, we obtain the variables-separated form of the partial differential equation,

$$\frac{X''(x)}{X(x)} = \frac{T'(t)}{T(t)}.$$

Both sides of this equation must be constant and cannot be positive. Thus we are led to the eigenvalue problem

$$X'' + \lambda^2 X = 0, \qquad 0 < x < L,$$

$$X'(0) = 0, \qquad X'(L) = 0.$$

The unusual feature of this equation is that its first eigenvalue is 0:

$$\lambda_0^2 = 0, \qquad X_0(x) = 1.$$

The rest of the solution is given by

$$\lambda_n^2 = (n\pi)^2, \qquad X_n(x) = \cos \lambda_n x, \qquad n = 1, 2, 3, \ldots.$$

The second factor in the product solution is found to be

$$T_0(t) = 1, \qquad T_n(t) = e^{-\lambda_n^2 t}, \qquad n = 1, 2, 3, \ldots.$$

We can now put together the infinite series of product solutions that forms the general solution of our problem:

$$u(x, t) = a_0 + \sum_{n=1}^{\infty} a_n \cos \lambda_n x \, e^{-\lambda_n^2 t}.$$

As usual, the initial condition is satisfied by setting $t = 0$ above and equating the series to the initial value of u,

$$a_0 + \sum_{n=1}^{\infty} a_n \cos \lambda_n x = 1 + 2x, \qquad 0 < x < 1.$$

The coefficients are

$$a_0 = \int_0^1 (1 + 2x) \, dx = 2,$$

$$a_n = 2 \int_0^1 (1 + 2x) \cos \lambda_n x \, dx$$

$$= 4 \frac{\cos n\pi - 1}{n^2 \pi^2}.$$

Using these values for the coefficients, we write out the first few terms of our series solution:

$$u(x, t) = 2 - \frac{8}{\pi^2} (\cos \pi x e^{-\pi^2 t} + \tfrac{1}{9} \cos 3\pi x e^{-9\pi^2 t}$$

$$+ \tfrac{1}{25} \cos 5\pi x e^{-25\pi^2 t} + \cdots).$$

Exercises

In Exercises 1–6, solve the given problem by separation of variables.

1. $\dfrac{\partial^2 u}{\partial x^2} = \dfrac{1}{k}\dfrac{\partial u}{\partial t}$, $0 < x < L$, $0 < t$,

$u(0, t) = 0$, $\dfrac{\partial u}{\partial x}(L, t) = 0$, $0 < t$,

$u(x, 0) = U_0$, $0 < x < L$

2. $\dfrac{\partial^2 u}{\partial x^2} = \dfrac{1}{k}\dfrac{\partial u}{\partial t}$, $0 < x < L$, $0 < t$,

$\dfrac{\partial u}{\partial x}(0, t) = 0$, $u(L, t) = 0$, $0 < t$,

$u(x, 0) = \dfrac{U_0 x}{L}$, $0 < x < L$

3. $\dfrac{\partial^2 u}{\partial x^2} = \dfrac{1}{k}\dfrac{\partial u}{\partial t}$, $0 < x < L$, $0 < t$,

$\dfrac{\partial u}{\partial x}(0, t) = 0$, $\dfrac{\partial u}{\partial x}(L, t) = 0$, $0 < t$,

$u(x, 0) = U_0 \sin\dfrac{\pi x}{L}$, $0 < x < L$

4. $\dfrac{\partial^2 u}{\partial x^2} = \dfrac{1}{k}\dfrac{\partial u}{\partial t}$, $0 < x < L$, $0 < t$,

$u(0, t) = 0$, $u(L, t) = 0$, $0 < t$,

$u(x, 0) = \begin{cases} \dfrac{2U_0 x}{L}, & 0 < x < \dfrac{L}{2}, \\[2mm] U_0, & \dfrac{L}{2} \le x < L \end{cases}$

5. $\dfrac{\partial^2 u}{\partial x^2} = \dfrac{1}{k}\dfrac{\partial u}{\partial t}$, $0 < x < L$, $0 < t$,

$\dfrac{\partial u}{\partial x}(0, t) = 0$, $\dfrac{\partial u}{\partial x}(L, t) = 0$, $0 < t$,

$u(x, 0) = \begin{cases} U_0, & 0 < c < \dfrac{L}{2}, \\[2mm] 0, & \dfrac{L}{2} \le x < L \end{cases}$

6. $\dfrac{\partial^2 u}{\partial x^2} = \dfrac{1}{k}\dfrac{\partial u}{\partial t}$, $0 < x < L$, $0 < t$,

$$u(0, t) = 0, \quad \frac{\partial u}{\partial x}(L, t) = 0,$$

$$u(x, 0) = \begin{cases} U_0, & 0 < x < \dfrac{L}{2}, \\ 0, & \dfrac{L}{2} \le x < L \end{cases}$$

7. In Example 1(b) we assumed that one term of the series was adequate to represent u. Confirm that this assumption is reasonable by computing the first term neglected, using the value of τ found.

8. In Example 3, sketch $u(x, t)$ as a function of x at $t = 0, 0.01, 0.1, 1, \infty$. For $t = 0$, use the initial condition. For $t = 0.01$, use series solution through $\cos 3\pi x$; for $t = 0.1$, through $\cos \pi x$. Show that $t = 1$ and $t = \infty$ (that is, the limiting case as $t \to \infty$) are nearly the same.

10.4

The Homogeneous Wave Problem

The problem of determining the shape $u(x, t)$ of a vibrating string was treated in Section 10.1. If deflection due to the weight of the string is negligible (usually the case), then $u(x, t)$ was found to satisfy this homogeneous initial value, boundary value problem:

$$\frac{\partial^2 u}{\partial x^2} = \frac{1}{c^2} \frac{\partial^2 u}{\partial t^2}, \quad 0 < x < L, \quad 0 < t, \tag{10.49}$$

$$u(0, t) = 0, \quad u(L, t) = 0, \quad 0 < t, \tag{10.50}$$

$$u(x, 0) = f(x), \quad \frac{\partial u}{\partial t}(x, 0) = g(x), \quad 0 < x < L. \tag{10.51}$$

The title of this section reflects the fact that Eq. (10.49) is widely known as the (simple) wave equation.

The problem above can be successfully solved by the method of separation of variables. First we seek nontrivial solutions of the homogeneous problem, Eqs. (10.49) and (10.50), in the product form, $u(x, t) = X(x)T(t)$. When the product form for u is substituted into the partial differential equation and boundary conditions, they become

$$X''(x)T(t) = \frac{1}{c^2} X(x)T''(t), \quad 0 < x < L, \quad 0 < t, \tag{10.52}$$

$$X(0)T(t) = 0, \quad X(L)T(t) = 0, \quad 0 < t. \tag{10.53}$$

Neither $X(x)$ nor $T(t)$ may be identically 0; thus we may divide through

Eq. (10.52) by $X(x)T(t)$ to obtain

$$\frac{X''(x)}{X(x)} = \frac{T''(t)}{c^2 T(t)}, \qquad 0 < x < L, \qquad 0 < t. \tag{10.54}$$

As in the case of the heat equation, we argue that the left member cannot vary with t nor the right with x, by assumption. Hence both sides are constant:

$$\frac{X''(x)}{X(x)} = \frac{T''(t)}{c^2 T(t)} = -\lambda^2. \tag{10.55}$$

Again using the fact that $T(t)$ is not identically 0, we may divide that factor from both boundary conditions. Now we find that the factor $X(x)$ must satisfy the eigenvalue problem

$$X'' + \lambda^2 X = 0, \qquad 0 < x < L, \tag{10.56}$$

$$X(0) = 0, \qquad X(L) = 0. \tag{10.57}$$

The solution of this problem is now familiar:

$$X_n(x) = \sin \lambda_n x, \qquad \lambda_n^2 = \left(\frac{n\pi}{L}\right)^2, \qquad n = 1, 2, \ldots. \tag{10.58}$$

The second factor in the product solution must satisfy Eq. (10.55). Knowing what values the constant $-\lambda^2$ can assume, we may write

$$\frac{T_n''}{c^2 T_n} = -\lambda_n^2 \qquad \text{or} \qquad T_n'' + \lambda_n^2 c^2 T_n = 0. \tag{10.59}$$

This is a familiar differential equation, having two independent solutions,

$$\cos \lambda_n ct, \qquad \sin \lambda_n ct. \tag{10.60}$$

Both of these functions (and any linear combination of them) may be the second factor in a product solution of the homogeneous wave problem. Indeed, we have now identified

$$\sin \lambda_n x \cos \lambda_n ct, \quad \sin \lambda_n x \sin \lambda_n ct \tag{10.61}$$

$(n = 1, 2, \ldots)$ as product solutions of that problem. We may construct its general solution be making a series of constant multiples of our product solutions:

$$u(x, t) = \sum_{n=1}^{\infty} (a_n \cos \lambda_n ct + b_n \sin \lambda_n ct) \sin \lambda_n x. \tag{10.62}$$

The next order of business is to satisfy the two initial conditions given in Eq. (10.51). If we set $t = 0$ in the expression for our solution, Eq. (10.62), all the cosines have value 1, while all the sines are 0. Thus the initial condition on u becomes

$$\sum_{n=1}^{\infty} a_n \sin \lambda_n x = f(x), \qquad 0 < x < L. \tag{10.63}$$

This equality can be satisfied only if the coefficients are chosen to be

$$a_n = \frac{2}{L} \int_0^L f(x) \sin \lambda_n x \, dx. \tag{10.64}$$

In addition, we will assume $f(x)$ to be sectionally smooth, so that the theory cited in Section 9.5 guarantees the validity of Eq. (10.63). (It is certainly reasonable to assume that $f(x)$, being the initial shape of a string, is sectionally smooth and continuous, too.)

In order to satisfy the second initial condition, on $\partial u/\partial t$, we will need to differentiate u, as given in Eq. (10.62), with respect to t. This we do by differentiating term by term, finding

$$\frac{\partial u}{\partial t}(x, t) = \sum_{n=1}^{\infty} (-\lambda_n c a_n \sin \lambda_n c t + \lambda_n c b_n \cos \lambda_n c t) \sin \lambda_n x. \tag{10.65}$$

Setting $t = 0$ in this expression and equating it to $g(x)$, as required by the initial condition, we obtain

$$\sum_{n=1}^{\infty} \lambda_n c b_n \sin \lambda_n x = g(x), \qquad 0 < x < L. \tag{10.66}$$

Again we recognize an eigenfunction expansion problem. In order to satisfy Eq. (10.66) the b's must be chosen so that

$$\lambda_n c b_n = \frac{2}{L} \int_0^L g(x) \sin \lambda_n x \, dx$$

or

$$b_n = \frac{2}{\lambda_n c L} \int_0^L g(x) \sin \lambda_n x \, dx. \tag{10.67}$$

Of course, $g(x)$ must also meet some conditions in order for Eq. (10.66) to hold, even when the b's satisfy Eq. (10.67).

If the coefficients are chosen by using Eqs. (10.64) and (10.67), the function $u(x, t)$ in Eq. (10.62) satisfies the wave equation (10.49), the boundary conditions (10.50), and the initial conditions (10.51). That is to say, we have solved the problem originally posed.

Example 1

Solve this vibrating string problem:

$$\frac{\partial^2 u}{\partial x^2} = \frac{1}{c^2} \frac{\partial^2 u}{\partial t^2}, \qquad 0 < x < L, \qquad 0 < t,$$

$$u(0, t) = 0, \qquad u(L, t) = 0, \qquad 0 < t,$$

$$u(x, 0) = f(x), \qquad \frac{\partial u}{\partial t}(x, 0) = 0, \qquad 0 < x < L,$$

$$f(x) = \begin{cases} \dfrac{2hx}{L}, & 0 < x < \dfrac{L}{2}, \\[2mm] \dfrac{2h(L - x)}{L}, & \dfrac{L}{2} \le x < L. \end{cases}$$

The solution by separation of variables has been carried out in general in the text. We need only use Eqs. (10.64) and (10.67) to compute the constants in the expression for $u(x, t)$, Eq. (10.62).

Evidently, all the b's are 0, since $g(x)$ is 0. According to Eq. (10.64), the a's are

$$a_n = \frac{2}{L} \int_0^L f(x) \sin \lambda_n x \, dx$$

$$= \frac{2}{L} \left[\int_0^{L/2} \frac{2hx}{L} \sin \frac{n\pi x}{L} \, dx + \int_{L/2}^L \frac{2h(L - x)}{L} \sin \frac{n\pi x}{L} \, dx \right]$$

$$= \frac{8h}{\pi^2} \frac{\sin (n\pi/2)}{n^2}.$$

The rather lengthy details of the integrations have been left out.

We now are in a position to set down the solution of the problem stated. It is

$$u(x, t) = \sum_{n=1}^{\infty} \frac{8h}{\pi^2} \frac{\sin (n\pi/2)}{n^2} \cos \lambda_n ct \sin \lambda_n x \tag{10.68}$$

together with the information that $\lambda_n = n\pi/L$.

It must be admitted that Eq. (10.68) is not a very satisfying way of expressing the solution of the given problem. Quite a few terms of the series would have to be added up in order to get a reasonable approximation for $u(x, t)$. Fortunately, however, there is a better way to identify the solution.

First, let us simplify the problem and its separation-of-variables solution to what is given in the following.

Theorem 10.3

If f is continuous and sectionally smooth, $0 \le x \le L$ and $f(0) = 0$, $f(L) = 0$, then the solution of

$$\frac{\partial^2 u}{\partial x^2} = \frac{1}{c^2} \frac{\partial^2 u}{\partial t^2}, \qquad 0 < x < L, \qquad 0 < t,$$

$$u(0, t) = 0, \qquad u(L, t) = 0, \qquad 0 < t,$$

$$u(x, 0) = f(x), \qquad \frac{\partial u}{\partial t}(x, 0) = 0, \qquad 0 < x < L,$$

is given by the series

$$u(x, t) = \sum_{n=1}^{\infty} a_n \cos \lambda_n c t \sin \lambda_n x, \tag{10.69}$$

$$a_n = \frac{2}{L} \int_0^L f(x) \sin \lambda_n x \, dx. \tag{10.70}$$

Note that we have taken a zero initial velocity.

Now the product that appears in the series solution, Eq. (10.69) can be transformed by a trigonometric identity

$$\cos \lambda_n c t \sin \lambda_n x = \tfrac{1}{2} \sin \lambda_n (x + ct) + \tfrac{1}{2} \sin \lambda_n (x - ct). \tag{10.71}$$

Thus we may transform the solution into the new form

$$u(x, t) = \frac{1}{2} \sum_{n=1}^{\infty} a_n \sin \lambda_n (x + ct) + \frac{1}{2} \sum_{n=1}^{\infty} a_n \sin \lambda_n (x - ct). \tag{10.72}$$

The task remaining is to identify the sum of each of these series; to do this, we only need to marshall some facts.

First, because of the conditions on f, we have

$$\sum_{n=1}^{\infty} a_n \sin \lambda_n x = f(x), \qquad 0 \le x \le L. \tag{10.73}$$

For values of x that may lie outside this interval we assign a name to the sum of the series:

$$\sum_{n=1}^{\infty} a_n \sin \lambda_n x = \bar{f}_o(x), \qquad -\infty < x < \infty. \tag{10.74}$$

Evidently, $\bar{f}_o(x)$ and $f(x)$ must agree for x between 0 and L:

$$\bar{f}_o(x) = f(x), \qquad 0 \le x \le L. \tag{10.75}$$

Second, it is an elementary fact about the sine function that

$$\sin(-\lambda_n x) = -\sin \lambda_n x.$$

The same property carries over to a linear combination of these sines and thus to $\bar{f}_o$:

$$\bar{f}_o(-x) = -\bar{f}_o(x). \tag{10.76}$$

A function with this property is said to be *odd*; hence the subscript. The graph of an odd function is symmetric in the origin.

Next, because the λ's were determined to be $\lambda_n = n\pi/L$, $\sin \lambda_n$ is *periodic* with period $2L$:

$$\sin \lambda_n(x + 2L) = \sin \frac{n\pi}{L}(x + 2L)$$

$$= \sin \left(\frac{n\pi x}{L} + 2n\pi \right)$$

$$= \sin \frac{n\pi x}{L} = \sin \lambda_n x.$$

This is true for all n. It is easy to see that it must also be true for $\bar{f}_o$:

$$\bar{f}_o(x + 2L) = \bar{f}_o(x). \tag{10.77}$$

The properties expressed by Eqs. (10.75), (10.76, and (10.77) completely determine the function $\bar{f}_0$ for all x. It is called the *odd periodic extension* of f. (Figure 10.5 shows a function and its odd periodic extension.)

Figure 10.5

A Function and its Odd Periodic Extension

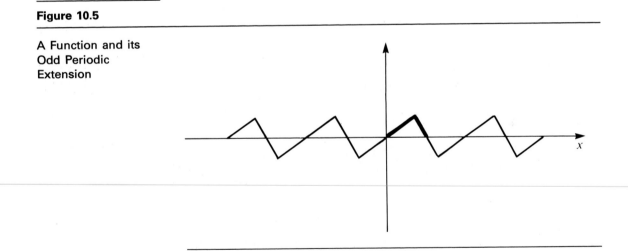

Now that we have identified the sum of the series in Eq. (10.75), we can recognize each of the series in Eq. (10.72). That equation becomes

$$u(x, t) = \tfrac{1}{2}\bar{f}_o(x + ct) + \tfrac{1}{2}\bar{f}_o(x - ct), \tag{10.78}$$

and from here it is a simple matter to actually find values of $u(x, t)$ or to sketch its graph.

We summarize our observations in the form of a theorem analogous to Theorem 10.3.

Theorem 10.4

If f is continuous and sectionally smooth, $0 \le x \le L$ and $f(0) = 0$, $f(L) = 0$, then the solution of

$$\frac{\partial^2 u}{\partial x^2} = \frac{1}{c^2}\frac{\partial^2 u}{\partial t^2}, \qquad 0 < x < L, \qquad 0 < t,$$

$$u(0, t) = 0, \qquad u(L, t) = 0, \qquad 0 < t,$$

$$u(x, 0) = f(x), \qquad \frac{\partial u}{\partial t}(x, 0) = 0, \qquad 0 < x < L,$$

is given by the expression

$$u(x, t) = \tfrac{1}{2}(\bar{f}_o(x + ct) + \bar{f}_o(x - ct))$$

where $\bar{f}_o$ is the odd periodic extension of f.

Example 2

Let $u(x, t)$ be the solution of the problem stated in Example 1. (a) Find the value of $u(x, t)$ at $x/L = \tfrac{1}{8}, \tfrac{1}{4}, \tfrac{1}{2}$, and $t = 3L/4c$. (b) Graph $u(x, t)$ versus x at $t = 3L/4c$.

(a) According to the development in the text, we may express the solution of the initial value, boundary value problem at $t = 3L/4c$ as

$$u(x, t) = \tfrac{1}{2}(\bar{f}_o(x + \tfrac{3}{4}L) + \bar{f}_o(x - \tfrac{3}{4}L)), \tag{10.79}$$

where $\bar{f}_o$ is the odd periodic extension of the given function. Both f and $\bar{f}_o$ are shown in Fig. 10.6. Table 10.2 will help in the calculations. The values of $\bar{f}_o$ were simply read off the graph of $\bar{f}_o(x)$.

(b) In order to graph $u(x, t)$ versus x at $t = 3L/4c$, we again use Eq. (10.79). First, graph $\bar{f}_o(x + \tfrac{3}{4}L)$. This graph has the same shape as that of $\bar{f}_o(x)$ but is shifted $3L/4$ units to the *left*. Then, on the same axis, graph $\bar{f}_o(x - \tfrac{3}{4}L)$, which is the same as the graph of $\bar{f}_o(x)$, shifted $3L/4$ units to the *right*. (See Fig. 10.7.) Finally, make a graphical average (average = half the sum) of the two graphs for x between 0 and L. (See Fig. 10.8.)

The graph confirms the calculations of $u(x, t)$ at $t = 3L/4c$ made above. Note also that the boundary conditions are seen to be satisfied, as they must be.

Figure 10.6

The Function f
(Heavy) and its Odd
Periodic Extension

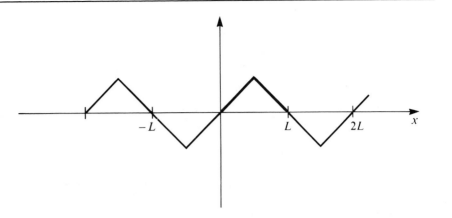

Table 10.2

x	$x + \frac{3}{4}L$	$\bar{f}_o(x + \frac{3}{4}L)$	$x - \frac{3}{4}L$	$\bar{f}_o(x - \frac{3}{4}L)$	$u(x, 3L/4c)$
$\frac{1}{8}L$	$\frac{7}{8}L$	$\frac{1}{4}h$	$-\frac{5}{8}L$	$-\frac{3}{4}h$	$-\frac{1}{4}h$
$\frac{1}{4}L$	L	0	$-\frac{1}{2}L$	$-h$	$-\frac{1}{2}h$
$\frac{1}{2}L$	$\frac{5}{4}L$	$-\frac{1}{2}h$	$-\frac{1}{4}L$	$-\frac{1}{2}h$	$-\frac{1}{2}h$

Figure 10.7

$\bar{f}_o(x + \frac{3}{4}L)$ and
$\bar{f}_o(x - \frac{3}{4}L)$

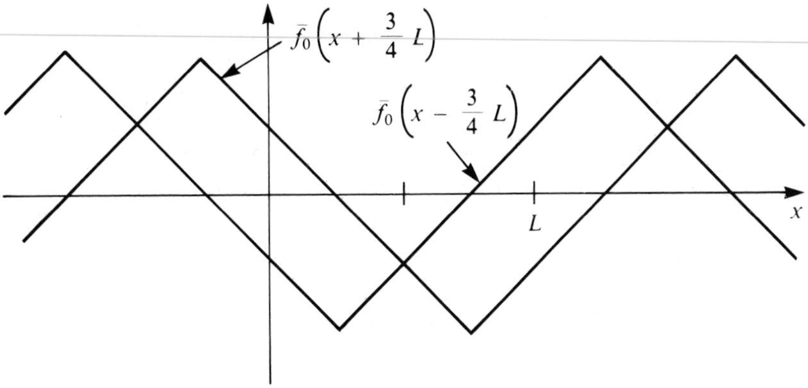

$\bar{f}_0\left(x + \dfrac{3}{4}L\right)$

$\bar{f}_0\left(x - \dfrac{3}{4}L\right)$

Figure 10.8

$u(x, 3L/4c)$ as
Graphical Average of
$\bar{f}_o(x + \frac{3}{4}L)$ and $\bar{f}_o(x - \frac{3}{4}L)$

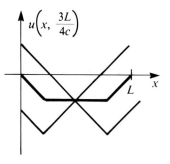

Exercises

In Exercises 1–4 a function f is given. (a) Sketch the odd periodic extension of f. (b) Find the solution, by separation of variables, of the homogeneous wave problem

$$\frac{\partial^2 u}{\partial x^2} = \frac{1}{c^2}\frac{\partial^2 u}{\partial t^2}, \qquad 0 < x < L, \qquad 0 < t,$$

$$u(0, t) = 0, \qquad u(L, t) = 0, \qquad 0 < t,$$

$$u(x, 0) = f(x), \qquad \frac{\partial u}{\partial t}(x, 0) = 0, \qquad 0 < x < L,$$

using the given function for f.

1. $f(x) = h,\ 0 < x < L$

2. $f(x) = \begin{cases} 0,\ 0 \le x \le L/4, \\ h,\ L/4 < x \le 3L/4, \\ 0,\ 3L/4 < x \le L \end{cases}$

3. $f(x) = \begin{cases} 3hx/L,\ 0 \le x < L/3, \\ 3h(L - x)/2L,\ L/3 \le x \le L \end{cases}$

4. $f(x) = \begin{cases} 0,\ 0 \le x < L/4, \\ h(4x - L)/L,\ L/4 < x \le L/2, \\ h(3L - 4x)/L,\ L/2 < x \le 3L/4, \\ 0,\ 3L/4 < x \le L \end{cases}$

5. Which of the functions in Exercises 1–4 satisfy the hypotheses of Theorems 10.3 and 10.4?

For Exercises 6–10, let $u(x, t)$ be the solution of the homogeneous wave problem stated above, where f is a function from Exercises 1–4. Evaluate or sketch $u(x, t)$, using the result of Theorem 10.4, even if f does not satisfy the hypotheses.

6. For f as in Exercise 1, sketch $u(x, t)$ versus x at $t = 0$, $t = L/4c$, $t = L/2c$.

7. For f as in Exercise 3, sketch $u(x, t)$ versus x at $t = L/6c$, $t = L/3c$.

8. For f as in Exercise 4, sketch $u(x, t)$ versus x at $t = L/8c$, $t = L/4c$, $t = L/2c$.

9. For f as in Exercise 1, evaluate $u(L/4, L/2c)$, $u(L/2, L/4c)$, $u(L/2, L/c)$.

10. For f as in Exercise 3, evaluate $u(L/3, L/3c)$, $u(2L/3, L/3c)$, $u(L/3, 2L/3c)$, $u(2L/3, 2L/3c)$.

11. Find the series solution of the homogeneous wave problem in Eqs. (10.49)–(10.51), using $f(x) \equiv 0$ and $g(x) = \alpha c$, $0 < x < L$.

12. The product solutions, given in Eq. (10.61) of the homogeneous wave problem, Eqs. (10.49)–(10.51), are called *standing waves*. Sketch $u(x, t) = \sin(\pi x/L) \cos(\pi ct/L)$ versus x at $t = kL/4c$, $k = 0, 1, 2, 3, 4$.

13. Let $u(x, t)$ be a solution of the problem stated in Eqs. (10.49)–(10.51) with $f(x) \equiv 0$, and let $v = \partial u/\partial t$. Show that v satisfies this problem:

$$\frac{\partial^2 v}{\partial x^2} = \frac{1}{k} \frac{\partial^2 v}{\partial t^2}, \qquad 0 < x < L, \qquad 0 < t,$$

$$v(0, t) = 0, \qquad v(L, t) = 0, \qquad 0 < t,$$

$$v(x, 0) = g(x), \qquad \frac{\partial v}{\partial t}(x, 0) = 0, \qquad 0 < x < L.$$

14. Show that if g has suitable properties, the solution of the problem above is

$$v(x, t) = \tfrac{1}{2}(\bar{g}_o(x + ct) + \bar{g}_o(x - ct)).$$

15. Suppose that G is a periodic function satisfying $G'(x) = \bar{g}_o(x)$. Show that the function u mentioned in Exercise 13 is

$$u(x, t) = \frac{1}{2c}(G(x + ct) - G(x - ct)).$$

(Hint: integrate $v = \partial u/\partial t$ to find u.)

16. The *even periodic extension* of a function $h(x)$ defined on $0 \leq x \leq L$ satisfies (i) $\bar{h}_e(-x) = \bar{h}_e(x)$, (ii) $\bar{h}_e(x + 2L) = \bar{h}_e(x)$. Suppose that $h(x)$ is an integral of $g(x)$: $h'(x) = g(x)$, $0 \leq x \leq L$. Show that $G(x) = \bar{h}_e(x)$ has the properties required in Exercise 15.

17. Sketch the function $G(x)$ needed to find the solution of Exercise 11 for the formula of Exercise 15.

10.5

Nonhomogeneous Problems

In Section 10.1 we derived heat and wave problems that could be nonhomogeneous, depending on specific circumstances. Then we justified restricting ourselves temporarily to homogeneous problems by citing Theorem 10.2.

Theorem 10.2

The general solution of a nonhomogeneous problem has the form $u = u_c + u_p$, where u_p is a particular solution of the nonhomogeneous problem and u_c is the general solution of the corresponding homogeneous problem.

We have studied how to solve homogeneous heat and wave problems. Now we shall look into ways of finding particular solutions of nonhomogeneous problems.

Method 1: Inhomogeneities Independent of t

One can easily find a particular solution of a heat or wave problem if inhomogeneities at the boundaries are constants and any inhomogeneity in the partial differential equation depends on x only. In this case, one can assume that the particular solution also depends on x alone: $u(x, t) = v(x)$. Then the partial differential equation becomes an ordinary one, and the methods of Chapter 9 apply.

Example 1

Find the solution of this problem for the temperature in a current-carrying metal rod.

$$\frac{\partial^2 u}{\partial x^2} = \frac{1}{k}\frac{\partial u}{\partial t} - \frac{I^2 R}{\kappa A}, \qquad 0 < x < L, \qquad 0 < t, \tag{10.80}$$

$$u(0, t) = U_0, \qquad \frac{\partial u}{\partial x}(L, t) = 0, \qquad 0 < t, \tag{10.81}$$

$$u(x, 0) = U_1, \qquad 0 < x < L. \tag{10.82}$$

First we look for a particular solution by Method 1, which is applicable because $I^2 R/\kappa A$ and U_0 are both constants. We may assume that $u_p(x, t) = v(x)$. Substituting into the partial differential equation and boundary conditions, we find that $(C = I^2R/\kappa A)$

$$\frac{d^2v}{dx^2} = -C, \qquad 0 < x < L, \tag{10.83}$$

$$v(0) = U_0, \qquad \frac{dv}{dx}(L) = 0. \tag{10.84}$$

The general solution of the ordinary differential equation for v is

$$v(x) = -C\frac{x^2}{2} + c_1x + c_2 \tag{10.85}$$

with c_1 and c_2 arbitrary. They are determined by the boundary conditions:

$$v(0) = U_0: \quad c_2 = U_0,$$

$$\frac{dv}{dx}(L) = 0: \quad -CL + c_1 = 0.$$

Solving these equations for c_1 and c_2, we find that

$$v(x) = -C\frac{x^2}{2} + CLx + U_0. \tag{10.86}$$

Next we must get the general solution of the corresponding homogeneous problem. This is obtained by discarding the in-homogeneities in the original partial differential equation and boundary conditions. For the case at hand, the homogeneous problem is

$$\frac{\partial^2 u}{\partial x^2} = \frac{1}{k}\frac{\partial u}{\partial t}, \qquad 0 < x < L, \qquad 0 < t, \tag{10.87}$$

$$u(0, t) = 0, \qquad \frac{\partial u}{\partial x}(L, t) = 0, \qquad 0 < t. \tag{10.88}$$

Separation of variables applied to this problem leads to the solution

$$u_c(x, t) = \sum_{n=1}^{\infty} b_n \sin \lambda_n x e^{-\lambda_n^2 kt} \tag{10.89}$$

with

$$\lambda_n^2 = ((2n - 1)\pi/2L)^2, \qquad n = 1, 2, \ldots.$$

Now we may apply the theorem cited at the beginning of this section to assemble the general solution of the original nonhomogeneous problem,

$$u(x, t) = -C\frac{x^2}{2} + CLx + U_0 + \sum_{n=1}^{\infty} b_n \sin \lambda_n x e^{-\lambda_n^2 kt}. \tag{10.90}$$

It remains only to satisfy the initial condition,

$$u(x, t) = U_1, \qquad 0 < x < L,$$

or

$$-C\frac{x^2}{2} + CLx + U_0 + \sum_{n=1}^{\infty} b_n \sin \lambda_n x = U_1, \qquad 0 < x < L.$$

When the series is isolated on the left-hand side, this becomes a simple eigenfunction expansion problem:

$$\sum_{n=1}^{\infty} b_n \sin \lambda_n x = U_1 - U_0 + C\frac{x^2}{2} - CLx, \qquad 0 < x < L. \tag{10.91}$$

According to developments in Chapter 9, Section 9.6, we must choose the coefficients b_n as

$$b_n = \frac{2}{L} \int_0^L \left(U_1 - U_0 + C\frac{x^2}{2} - CLx\right) \sin \lambda_n x \, dx$$

$$= \frac{4(U_1 - U_0) \sin\left(\frac{2n-1}{2}\pi\right)}{(2n-1)\pi} - \frac{16CL^2}{(2n-1)^3\pi^3}. \tag{10.92}$$

Example 2

Find a particular solution of the following problem that satisfies the boundary conditions:

$$\frac{\partial^2 u}{\partial x^2} = \frac{1}{c^2}\frac{\partial^2 u}{\partial t^2} + \frac{g}{c^2}, \qquad 0 < x < L, \qquad 0 < t, \tag{10.93}$$

$$u(0, t) = 0, \qquad u(L, t) = 0, \qquad 0 < t, \tag{10.94}$$

$$u(x, 0) = f(x), \qquad \frac{\partial u}{\partial t}(x, 0) = g(x), \qquad 0 < x < L. \tag{10.95}$$

The unknown function represents the displacement of a string hanging under its own weight.

Using Method 1, we seek a particular solution in the form of a function of x alone: $u_p(x, t) = v(x)$. Then v must satisfy

$$\frac{d^2v}{dx^2} = \frac{g}{c^2}, \qquad 0 < x < L,$$

$$v(0) = 0, \qquad v(L) = 0.$$

The general solution of the differential equation is found to be

$$v(x) = \frac{g}{c^2}\frac{x^2}{2} + c_1 x + c_2 \tag{10.96}$$

with c_1 and c_2 arbitrary. The boundary conditions require

$$v(0) = 0: \quad c_2 = 0,$$

$$v(L) = 0: \quad \frac{g}{c^2}\frac{L^2}{2} + c_1 L + c_2 = 0.$$

From these we find $c_2 = 0$, $c_1 = -L/2c^2$. Then a particular solution that satisfies the boundary conditions is

$$u_p(x, t) = v(x) = \frac{-g(L-x)x}{2c^2}. \tag{10.97}$$

Method 2: Homogeneous Boundary Conditions, Nonhomogeneous Partial Differential Equation

This method is an extension of the method of separation of variables. Instead of finding u_p and u_c separately, we find the whole solution immediately.

First, let us suppose that we have started applying separation of variables to the associated homogeneous problem and that we have found the solution of the associated eigenvalue problem:

$$\lambda_n^2, \quad X_n(x), \quad n = 1, 2, 3, \ldots . \tag{10.98}$$

Next we seek solutions of the nonhomogeneous problem in the form

$$u(x, t) = \sum_{n=1}^{\infty} T_n(t)X_n(x). \tag{10.99}$$

In this equation, $T_n(t)$ is *not* the function we would have found for the homogeneous problem. We also assume that the inhomogeneity in the differential equation is expressed in terms of the X's:

$$f(x, t) = \sum_{n=1}^{\infty} F_n(t)X_n(x). \tag{10.100}$$

Now substitution of u and f, Eqs. (10.99) and (10.100) into the partial differential equation gives rise to an ordinary differential equation for each of the T's. The T's determined by solving these equations will contain some arbitrary constants, which can be used to satisfy the initial conditions.

Example 3

Solve this vibrating string problem. (Imagine a steel string in an oscillating magnetic field.) Dimensionless variables are used.

$$\frac{\partial^2 u}{\partial x^2} = \frac{\partial^2 u}{\partial t^2} + \sin \omega t, \quad 0 < x < 1, \quad 0 < t, \tag{10.101}$$

$$u(0, t) = 0, \qquad u(1, t) = 0, \qquad 0 < t, \tag{10.102}$$

$$u(x, 0) = 0, \qquad \frac{\partial u}{\partial t}(x, 0) = 0, \qquad 0 < x < 1. \tag{10.103}$$

First we write down the associated homogeneous problem. It is

$$\frac{\partial^2 u}{\partial x^2} = \frac{\partial^2 u}{\partial t^2}, \qquad 0 < x < 1, \qquad 0 < t, \tag{10.104}$$

$$u(0, t) = 0, \qquad u(1, t) = 0, \qquad 0 < t. \tag{10.105}$$

This is exactly the problem we treated in Section 10.4. The eigenvalue problem encountered is

$$X'' + \lambda^2 X = 0, \qquad 0 < x < 1,$$

$$X(0) = 0, \qquad X(1) = 0,$$

whose solutions should be recalled as

$$\lambda_n^2 = (n\pi)^2, \qquad X_n(x) = \sin \lambda_n x, \ n = 1, 2, 3, \ldots.$$

According to the outline of the method, we are to assume that the solution has the form

$$u(x, t) = \sum_{n=1}^{\infty} T_n(t) \sin \lambda_n x \tag{10.106}$$

and also are to expand the inhomogeneity in a similar series:

$$\sin \omega t = \sum_{n=1}^{\infty} F_n(t) \sin \lambda_n x. \tag{10.107}$$

From the latter equation we see that

$$F_n(t) = 2 \int_0^1 \sin \omega t \sin \lambda_n x \, dx$$

$$= 2 \sin \omega t \frac{(1 - \cos n\pi)}{n\pi} = B_n \sin \omega t.$$

Next we use the assumed forms for u and the inhomogeneity in the partial differential equation. The result is

$$\sum_{n=1}^{\infty} T_n(t)(-\lambda_n^2) \sin \lambda_n x = \sum_{n=1}^{\infty} T_n''(t) \sin \lambda_n x + \sum_{n=1}^{\infty} F_n(t) \sin \lambda_n x. \tag{10.108}$$

To find the condition that T_n must fulfill, simply match the coefficients of $\sin \lambda_n x$ for each n:

$$-\lambda_n^2 T_n = T_n'' + F_n(t).$$

In a more convenient form this is

$$T_n'' + \lambda_n^2 T_n = -B_n \sin \omega t. \tag{10.109}$$

As long as the impressed frequency ω is different from λ_n for all n, this equation has the solution

$$T_n(t) = a_n \cos \lambda_n t + b_n \sin \lambda_n t - \frac{B_n}{\lambda_n^2 - \omega^2} \sin \omega t, \tag{10.110}$$

where a_n and b_n are arbitrary.

Now we may assemble the solution in the form assumed in Eq. (10.106). For convenience we write the result as a sum of two series

$$u(x, t) = \sum_{n=1}^{\infty} (a_n \cos \lambda_n t + b_n \sin \lambda_n t) \sin \lambda_n x$$

$$+ \sum_{n=1}^{\infty} \frac{-B_n}{\lambda_n^2 - \omega^2} \sin \omega t \sin \lambda_n x. \tag{10.111}$$

We can recognize the first of these as u_c, the general solution of the homogeneous problem, and the second as a particular solution of the nonhomogeneous problem.

It remains to satisfy the initial conditions. Setting $t = 0$ in the expression above for u, we find

$$u(x, 0) = \sum_{n=1}^{\infty} a_n \sin \lambda_n x.$$

Since this is to be 0, according to the condition (10.103), we choose $a_n = 0$.

The second condition is imposed on $\partial u/\partial t$. After differentiating the series (10.111) and setting $t = 0$, we obtain

$$\frac{\partial u}{\partial t}(x, 0) = \sum_{n=1}^{\infty} b_n \lambda_n \sin \lambda_n x - \sum_{n=1}^{\infty} B_n \frac{\omega}{\lambda_n^2 - \omega^2} \sin \lambda_n x. \tag{10.112}$$

Again the initial condition requires that this be 0, so the net coefficient of $\sin \lambda_n x$ must be 0:

$$b_n \lambda_n - B_n \frac{\omega}{\lambda_n^2 - \omega^2} = 0. \tag{10.113}$$

From here we find b_n and finally the complete solution

$$u(x, t) = \sum_{n=1}^{\infty} \frac{B_n \omega}{\lambda_n(\lambda_n^2 - \omega^2)} \sin \lambda_n t \sin \lambda_n x - \sum_{n=1}^{\infty} \frac{B_n}{\lambda_n^2 - \omega^2} \sin \omega t \sin \lambda_n x. \tag{10.114}$$

Moving an Inhomogeneity

To complete the study of nonhomogeneous differential equations, we need to know how to deal with time-varying boundary conditions. If the inhomogeneities in the boundary conditions are sufficiently differentiable, it is possible to "move the inhomogeneity to the differential equation." First, find *any* (twice differentiable) function that satisfies the nonhomogeneous boundary conditions—say $w(x, t)$. Then assume that the solution is

$$u(x, t) = v(x, t) + w(x, t). \tag{10.115}$$

Substitution of u in this form into the partial differential equation and boundary conditions leaves a problem for v with homogeneous boundary conditions and (usually) a nonhomogeneous partial differential equation. Then v can be found by Method 2. Recall that Method 2 will return a general solution—one with enough arbitrary constants to satisfy reasonable initial conditions.

Example 4

Solve this problem for a string with a moving endpoint.

$$\frac{\partial^2 u}{\partial x^2} = \frac{\partial^2 u}{\partial t^2}, \qquad 0 < x < 1, \qquad 0 < t, \tag{10.116}$$

$$u(0, t) = 0, \qquad u(1, t) = \sin \omega t, \qquad 0 < t, \tag{10.117}$$

$$u(x, 0) = 0, \qquad \frac{\partial u}{\partial t}(x, 0) = 0, \qquad 0 < x < 1. \tag{10.118}$$

We must first move the inhomogeneity by finding a function that satisfies the boundary conditions. One simple candidate is

$$w(x, t) = x \sin \omega t,$$

which is just a linear interpolation of the two boundary conditions. Next we assume that $u(x, t) = v(x, t) + w(x, t)$ and substitute into the partial differential equation and boundary conditions Eqs. (10.116) and (10.117). The result is

$$\frac{\partial^2 v}{\partial x^2} + \frac{\partial^2 w}{\partial x^2} = \frac{\partial^2 v}{\partial t^2} + \frac{\partial^2 w}{\partial t^2}, \qquad 0 < x < 1, \qquad 0 < t,$$

$$v(0, t) + w(0, t) = 0, \qquad v(1, t) + w(1, t) = \sin \omega t, \qquad 0 < t.$$

But w is known explicitly, so we may actually write down the required derivatives and boundary values:

$$\frac{\partial^2 v}{\partial x^2} = \frac{\partial^2 v}{\partial t^2} - \omega^2 x \sin \omega t, \qquad 0 < x < 1, \qquad 0 < t, \tag{10.119}$$

$$v(0, t) = 0, \qquad v(1, t) = 0, \qquad 0 < t. \tag{10.120}$$

The general solution of this problem for v can be found by Method 2. (See Exercise 8 at the end of this section.) If $\lambda_n^2 = (n\pi)^2$, we can write v as

$$v(x, t) = \sum_{n=1}^{\infty} (a_n \cos \lambda_n t + b_n \sin \lambda_n t) \sin \lambda_n x$$

$$+ \sum_{n=1}^{\infty} B_n \frac{\omega^2}{\lambda_n^2 - \omega^2} \sin \omega t \sin \lambda_n x, \tag{10.121}$$

where $B_n = -2(\cos n\pi)/n\pi$. (Of course, this is valid only if $\omega \neq \lambda_n$ for $n = 1, 2, \ldots$).

Now the general solution of the original problem is

$$u(x, t) = v(x, t) + x \sin \omega t \tag{10.122}$$

with $v(x, t)$ from Eq. (10.121). It is left as an exercise to satisfy the initial conditions on u. (See Exercise 11.)

Exercises

In Exercises 1–6, find the solution of the problem using Method 1.

1. $\dfrac{\partial^2 u}{\partial x^2} = \dfrac{1}{k} \dfrac{\partial u}{\partial t}, \ 0 < x < L, \ 0 < t,$

   $u(0, t) = U_0, \ u(L, t) = U_1, \ 0 < t,$

   $u(x, 0) = U_0, \ 0 < x < L$

2. $\dfrac{\partial^2 u}{\partial x^2} = \dfrac{1}{k} \dfrac{\partial u}{\partial t} - \gamma^2(T_1 - u), \ 0 < x < L, \ 0 < t,$

   $u(0, t) = U_0, \ u(L, t) = U_0, \ 0 < t,$

   $u(x, 0) = U_0, \ 0 < x < L$

3. $\dfrac{\partial^2 u}{\partial x^2} = \dfrac{1}{k} \dfrac{\partial u}{\partial t} - \gamma^2\left(U_1 \dfrac{x}{L} - u\right), \ 0 < x < L, \ 0 < t,$

   $u(0, t) = 0, \ \dfrac{\partial u}{\partial x}(L, t) = 0, \ 0 < t,$

   $u(x, 0) = 0, \ 0 < x < L$

4. $\dfrac{\partial^2 u}{\partial x^2} = \dfrac{1}{k} \dfrac{\partial u}{\partial t} - C, \ 0 < x < L, \ 0 < t,$

   $u(0, t) = 0, \ u(L, t) = 0, \ 0 < t,$

   $u(x, 0) = 0, \ 0 < x < L$

5. $\dfrac{\partial^2 u}{\partial x^2} = \dfrac{1}{c^2} \dfrac{\partial^2 u}{\partial t^2}, \ 0 < x < L, \ 0 < t,$

$$u(0, t) = 0, \ u(L, t) = h, \ 0 < t,$$

$$u(x, 0) = 0, \ \frac{\partial u}{\partial t}(x, 0) = 0, \ 0 < x < L$$

6. $\dfrac{\partial^2 p}{\partial x^2} = \kappa \rho \dfrac{\partial^2 p}{\partial t^2}, \ 0 < x < L, \ 0 < t,$

$$p(0, t) = p_0, \ \frac{\partial p}{\partial x}(L, t) = 0, \ 0 < t,$$

$$p(x, 0) = p_0 - kx, \ \frac{\partial p}{\partial t}(x, 0) = 0, \ 0 < x < L$$

In this exercise, p represents the pressure in a pipe whose left end is open to a reservoir and right end was closed just at $t = 0$. The parameters are adiabatic compressibility, κ, and density, ρ.

In Exercises 7–10, solve the given problem by Method 2. Dimensionless variables have been used throughout to reduce the number of parameters.

7. $\dfrac{\partial^2 u}{\partial x^2} = \dfrac{\partial u}{\partial t}, \ 0 < x < 1, \ 0 < t,$

$$u(0, t) = f(t), \ u(1, t) = f(t), \ 0 < t,$$
$$u(x, 0) = 0, \ 0 < x < 1,$$
where $f(t) = 1 - e^{-\alpha t}$.

8. $\dfrac{\partial^2 v}{\partial x^2} = \dfrac{\partial^2 v}{\partial t^2} - \omega^2 x \sin \omega t; \ 0 < x < 1, \ 0 < t,$

$$v(0, t) = 0, \ v(1, t) = 0, \ 0 < t.$$
(Assume $\omega^2 \neq \lambda_n^2$ for all n. See Example 4.)

9. $\dfrac{\partial^2 u}{\partial x^2} = \dfrac{\partial u}{\partial t} - 1, 0 < x < 1, 0 < t,$

$$\frac{\partial u}{\partial x}(0, t) = 0, \ \frac{\partial u}{\partial x}(1, t) = 0, \ 0 < t,$$

$$u(x, 0) = 0, \ 0 < x < 1.$$
(Be careful with $T_0(t)$.)

10. $\dfrac{\partial^2 u}{\partial x^2} = \dfrac{\partial u}{\partial t}, \ 0 < x < 1, \ 0 < t,$

$$u(0, t) = 0, \ u(1, t) = \sin \omega t, \ 0 < t,$$
$$u(x, 0) = 0.$$

11. Complete the solution of Example 4 by finding $u(x, t)$, the general solution of Eqs. (10.116) and (10.117), and then satisfying the initial conditions, Eq. (10.118).

12. Find the solution of Example 4 if $\omega = \lambda_1$. In this case, Eq. (10.110) is not valid for $n = 1$.

13. For what values of ω will resonance be observed in the solution of the problem of Example 3? What is the solution in such cases?

14. Attempt the solution of Exercise 9 by Method 1. Why can't it be done?

15. If the pipe of Exercise 6 is an organ pipe, what is the lowest frequency of vibration produced?

16. How does the answer to Exercise 15 change if the pipe is open? That is, the right-hand end condition is $p(L, t) = p_0$.

17. Show that u_0, u_1, u_2, u_3 (sometimes called *heat polynomials*) are solutions of

$$\frac{\partial^2 u}{\partial x^2} = \frac{1}{k} \frac{\partial u}{\partial t}$$

and find a linear combination of them that satisfies

$$u(0, t) = ht, \qquad u(L, t) = ht.$$
$$u_0(x, t) = 1, \qquad u_2(x, t) = x^2 + 2kt,$$
$$u_1(x, t) = x, \qquad u_3(x, t) = x^3 + 6kxt.$$

18. Show that $u(x, t) = e^{-px} \sin(\omega t - px)$, with $p = \sqrt{\omega/2k}$, is a solution of

$$\frac{\partial^2 u}{\partial x^2} = \frac{1}{k} \frac{\partial u}{\partial t}, \qquad u(0, t) = \sin \omega t.$$

10.6
The Potential Equation

In Section 10.1 we derived heat and wave equations for cases in which the unknown function could reasonably be assumed to vary with time and just one spacial variable, x. If there is variation in more than one direction, it almost always happens that the Laplacian operator ∇^2 appears instead of the second partial derivative with respect to x. (Recall that in rectangular coordinates the Laplacian is

$$\nabla^2 u = \frac{\partial^2 u}{\partial x^2} + \frac{\partial^2 u}{\partial y^2} + \frac{\partial^2 u}{\partial z^2}.$$

In vectorial terms, $\nabla^2 u = \bar{\nabla} \cdot \bar{\nabla} u = \text{div} \cdot \text{grad } u$.) Thus the homogeneous multidimensional heat or wave equation would be

$$\nabla^2 u = \frac{1}{k} \frac{\partial u}{\partial t} \qquad \text{or} \qquad \nabla^2 u = \frac{1}{c^2} \frac{\partial^2 u}{\partial t^2}. \qquad \textbf{(10.123)}$$

In this section we are concerned with solving the *potential equation*,

$$\nabla^2 u = 0, \qquad\qquad (\mathbf{10.124})$$

subject to suitable boundary conditions. We can think of this equation as describing the *steady state* temperature in an object: Eq. (10.124) is the multidimensional heat equation with no variation in time. There are, however, many diverse physical phenomena that are described by the potential equation under appropriate conditions. Some of them are summarized in Table 10.3.

Table 10.3	u	$\bar{\nabla}u$
	Temperature	$\bar{\mathbf{q}} = -\kappa \bar{\nabla}u$, heat flux
	Electric potential	$\bar{\mathbf{J}} = -\sigma \bar{\nabla}u$, current
	Velocity potential	$\bar{\mathbf{v}} = -\bar{\nabla}u$, velocity
	Gravitational potential	$\bar{\mathbf{g}} = -\bar{\nabla}u$, acceleration

The boundary conditions on a solution of the potential equation are usually of one of these types: (1) the value of the function u is specified on the boundary (Dirichlet condition); (2) the value of the directional derivative is specified for the outward normal direction on the boundary (Neumann condition). If the boundary consists of several pieces, different pieces may have different types of boundary conditions.

Example 1

A thin rectangular metal plate of dimensions $L \times B$ is sandwiched between sheets of insulation. (See Fig. 10.9.) Because of the insulation

Figure 10.9

Plate between Sheets of Insulation

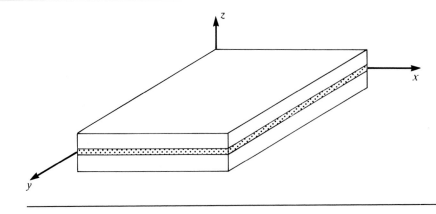

and the thinness of the plate, we may assume temperature does not vary through the plate (that is, in the z-direction). In the steady state the temperature $u(x, y)$ obeys the potential equation

$$\frac{\partial^2 u}{\partial x^2} + \frac{\partial^2 u}{\partial y^2} = 0, \qquad 0 < x < L, \qquad 0 < y < B.$$

Suppose further that the temperature is held at U_0 along the top, U_1 on the bottom, and 0 on the left edge. On the right, suppose that there is insulation; then no heat flows through that side, and the directional derivative in the x-direction is 0. In mathematical form the boundary conditions are

$$u(0, y) = 0, \qquad \frac{\partial u}{\partial x}(L, y) = 0, \qquad 0 < y < B,$$

$$u(x, 0) = U_1, \qquad u(x, B) = U_0, \qquad 0 < x < L.$$

These conditions, along with the potential equation, determine $u(x, y)$ completely. (See Fig. 10.10.)

Figure 10.10

Problem for u

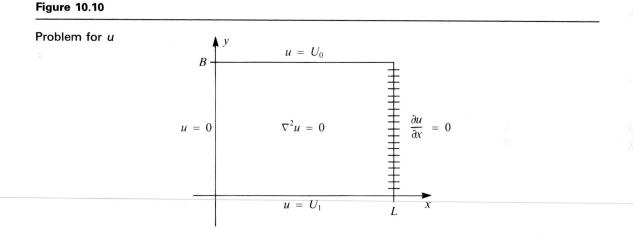

In order to get solutions with the methods already developed, we must restrict the kinds of problems that we consider. To this end, we will assume that
    1. $\nabla^2 u = 0$ holds in a *plane* region;
    2. the region is a rectangle;
    3. there are homogeneous boundary conditions on opposite sides of the rectangle.
A problem conforming to these limitations will be called a homogeneous

problem. The principle of superposition applies to such a problem: a linear combination of solutions is a solution, even in the extended sense of an infinite series of solutions. The coefficients of the linear combination are available to satisfy the nonhomogeneous boundary conditions.

Since the problem stated in Example 1 satisfies the restrictions above, we continue by solving it by separation of variables.

Example 2

Solve this potential problem by separation of variables:

$$\frac{\partial^2 u}{\partial x^2} + \frac{\partial^2 u}{\partial y^2} = 0, \qquad 0 < x < L, \qquad 0 < y < B, \tag{10.125}$$

$$u(0, y) = 0, \qquad \frac{\partial u}{\partial x}(L, y) = 0, \qquad 0 < y < B, \tag{10.126}$$

$$u(x, 0) = U_1, \qquad u(x, B) = U_0, \qquad 0 < x < L. \tag{10.127}$$

First we seek nontrivial solutions in the form $u(x, y) = X(x)Y(y)$, where each of the functions in the product depends only on the variable shown. Next we substitute u in this form into the homogeneous equations: the partial differential equation (10.125) and the homogeneous boundary conditions (10.126). These become

$$X''(x)Y(y) + X(x)Y''(y) = 0, \qquad 0 < x < L, \qquad 0 < y < B,$$

$$X(0)Y(y) = 0, \qquad X'(L)Y(y) = 0, \qquad 0 < y < B.$$

Since we are assuming that X and Y are nontrivial, we may divide the partial differential equation by the product XY and may divide Y out of the boundary conditions. The results are

$$\frac{X''(x)}{X(x)} + \frac{Y''(y)}{Y(y)} = 0, \qquad 0 < x < L, \qquad 0 < y < B, \tag{10.128}$$

$$X(0) = 0, \qquad X'(L) = 0. \tag{10.129}$$

As before, we argue that both ratios X''/X and Y''/Y in the differential equation must be constant—equal in magnitude and opposite in sign. The question of which ratio is negative is easily settled. If X''/X were positive, X could not satisfy the boundary conditions (10.129) without being identically 0. Thus we must have

$$\frac{X''}{X} = -\lambda^2 \qquad \text{or} \qquad X'' + \lambda^2 X = 0, \qquad 0 < x < L, \tag{10.130}$$

$$\frac{Y''}{Y} = \lambda^2 \qquad \text{or} \qquad Y'' - \lambda^2 Y = 0, \qquad 0 < y < B. \tag{10.131}$$

Equation (10.130) and the boundary conditions (10.129) constitute an eigenvalue problem, whose solution is

$$\lambda_n^2 = \left(\frac{(2n-1)\pi}{2L}\right)^2, \qquad X_n(x) = \sin \lambda_n x, \qquad n = 1, 2, \ldots. \qquad \textbf{(10.132)}$$

The equation (10.131) for the second factor, $Y(y)$, is also easily solved. In boundary value problems it is often more convenient to use hyperbolic functions than exponentials, so we choose the solutions as

$$Y_n(y) = \cosh \lambda_n y \qquad \text{or} \qquad \sinh \lambda_n y. \qquad \textbf{(10.133)}$$

Now we may assemble our product solutions. For each $n = 1, 2, 3, \ldots$ we have two solutions,

$$\sin \lambda_n x \cosh \lambda_n y, \qquad \sin \lambda_n x \sinh \lambda_n y.$$

We form the general solution of the homogeneous problem composed of Eqs. (10.125) and (10.126) as an infinite series of constant multiples of solutions:

$$u(x, y) = \sum_{n=1}^{\infty} (a_n \cosh \lambda_n y + b_n \sinh \lambda_n y) \sin \lambda_n x. \qquad \textbf{(10.134)}$$

The coefficients of this series are available to satisfy the two boundary conditions on the sides $y = 0$ and $y = B$. Of course, eigenfunction series will be needed to accomplish this.

At $y = 0$ we have the first condition of Eq. (10.127). In terms of the series of Eq. (10.134) this becomes

$$u(x, 0) = \sum_{n=1}^{\infty} a_n \sin \lambda_n x = U_1, \qquad 0 < x < L, \qquad \textbf{(10.135)}$$

from which we easily determine the a_n to be

$$a_n = \frac{2}{L} \int_0^L U_1 \sin \lambda_n x \, dx = \frac{4U_1}{(2n-1)\pi}. \qquad \textbf{(10.136)}$$

At $y = B$ the second condition of Eq. (10.127) holds. Substituting the series form for u and setting $y = B$, we find

$$\sum_{n=1}^{\infty} (a_n \cosh \lambda_n B + b_n \sinh \lambda_n B) \sin \lambda_n x = U_0, \qquad 0 < x < L. \qquad \textbf{(10.137)}$$

In this case we have to choose b_n so that the quantity in parentheses is the Fourier sine coefficient of the given function of x. That is,

$$a_n \cosh \lambda_n B + b_n \sinh \lambda_n B = c_n \qquad \textbf{(10.138)}$$

where

$$c_n = \frac{2}{L} \int_0^L U_0 \sin \lambda_n x \, dx = \frac{4U_0}{(2n-1)\pi}. \qquad \textbf{(10.139)}$$

We may easily solve Eq. (10.138) for b_n in terms of the explicitly known coefficients a_n and c_n:

$$b_n = \frac{c_n - a_n \cosh \lambda_n B}{\sinh \lambda_n B}.$$ **(10.140)**

Now $u(x, t)$, given in series form in Eq. (10.134), with coefficients as found above, satisfies all of the conditions imposed: the partial differential equation and all four boundary conditions.

The solution for the case $B = 2L$, $U_1 = 10$, $U_0 = 20$ is shown in Fig. 10.11. The curves charted are isotherms, loci of $u(x, y) = T$ for various values of T.

Figure 10.11

Isotherms

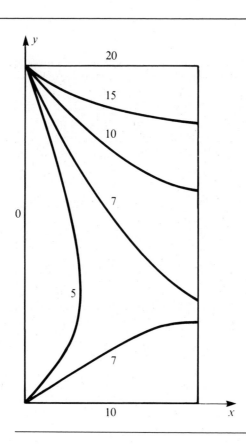

It is evident that we could have solved the problem of the example if the boundary conditions on the top and bottom edges had been any sectionally smooth functions instead of constants. Similarly, we could have dealt with other homogeneous conditions on the left and right edges. The eigenvalue problem would have been different, but the procedure would have been the same. To illustrate these points, we have the following.

Example 3

Obtain a separation-of-variables solution of the problem

$$\frac{\partial^2 u}{\partial x^2} + \frac{\partial^2 u}{\partial y^2} = 0, \qquad 0 < x < L, \qquad 0 < y < B, \tag{10.141}$$

$$u(0, y) = \frac{U_0 y}{B}, \qquad \frac{\partial u}{\partial x}(L, y) = -S, \qquad 0 < y < B, \tag{10.142}$$

$$u(x, 0) = 0, \qquad u(x, B) = 0, \qquad 0 < x < L. \tag{10.143}$$

The problem satisfies all the restrictions stated: the two opposite sides of the rectangle with homogeneous conditions are the bottom ($y = 0$) and top ($y = B$). Thus we seek nontrivial solutions in the product form $u(x, y) = X(x)Y(y)$ and substitute into the partial differential equation (10.141) and the homogeneous conditions (10.143). The results are

$$X''(x)Y(y) + X(x)Y''(y) = 0, \qquad 0 < x < L, \qquad 0 < y < B, \tag{10.144}$$

$$X(x)Y(0) = 0, \qquad X(x)Y(B) = 0, \qquad 0 < x < L. \tag{10.145}$$

After dividing through Eq. (10.144) by $X(x)Y(y)$ and eliminating $X(x)$ from Eq. (10.145) we obtain

$$\frac{X''(x)}{X(x)} + \frac{Y''(y)}{Y(y)} = 0, \qquad 0 < x < L, \qquad 0 < y < B, \tag{10.146}$$

$$Y(0) = 0, \qquad Y(B) = 0. \tag{10.147}$$

The two ratios in Eq. (10.146) must be constants of equal magnitude and opposite sign. The boundary conditions, Eq. (10.147), show that Y''/Y *cannot be positive.* Thus we are lead to the equations

$$\frac{X''}{X} = \lambda^2 \qquad \text{or} \qquad X'' - \lambda^2 X = 0, \qquad 0 < x < L, \tag{10.148}$$

$$\frac{Y''}{Y} = -\lambda^2 \qquad \text{or} \qquad Y'' + \lambda^2 Y = 0, \qquad 0 < y < B. \tag{10.149}$$

Equation (10.149) and boundary conditions Eq. (10.147) form an eigenvalue problem, whose solution is

$$\lambda_n^2 = (n\pi/B)^2, \qquad Y_n(y) = \sin \lambda_n y, \qquad n = 1, 2, 3, \ldots. \tag{10.150}$$

Also, Eq. (10.148) is to be solved for the other factor of the product solutions:

$$X_n(x) = \cosh \lambda_n x \qquad \text{or} \qquad \sinh \lambda_n x.$$

Now we assemble the general solution of our homogeneous problem in the form of a series of constant multiples of product solutions,

$$u(x, y) = \sum_{n=1}^{\infty} (a_n \cosh \lambda_n x + b_n \sinh \lambda_n x) \sin \lambda_n y. \qquad \textbf{(10.151)}$$

The next step is to treat the remaining boundary conditions. First, setting $x = 0$ in the series above we find that the first condition of Eq. (10.142) is

$$\sum_{n=1}^{\infty} a_n \sin \lambda_n y = \frac{U_0 y}{B}, \qquad 0 < y < B.$$

Since this is an eigenfunction series problem, we must choose

$$a_n = \frac{2}{B} \int_0^B \frac{U_0 y}{B} \sin \lambda_n y \, dy = -\frac{2U_0 \cos n\pi}{n\pi}. \qquad \textbf{(10.152)}$$

In order to satisfy the second boundary condition we need $\partial u/\partial x$, obtained by differentiating the series (10.151) term by term.

$$\frac{\partial u}{\partial x}(x, y) = \sum_{n=1}^{\infty} \lambda_n (a_n \sinh \lambda_n x + b_n \cosh \lambda_n x) \sin \lambda_n y.$$

After setting $x = L$ in this series and equating it to the given value of $\partial u/\partial x$ at $x = L$, we have

$$\sum_{n=1}^{\infty} \lambda_n (a_n \sinh \lambda_n L + b_n \cosh \lambda_n L) \sin \lambda_n y = -S, \qquad 0 < y < B.$$

Therefore we must choose b_n so that the coefficient of $\sin \lambda_n y$ will be

$$\lambda_n (a_n \sinh \lambda_n L + b_n \cosh \lambda_n L) = c_n, \qquad \textbf{(10.153)}$$

where

$$c_n = \frac{2}{B} \int_0^B -S \sin \lambda_n y \, dy = \frac{-2S}{\pi} \frac{1 - \cos n\pi}{n}. \qquad \textbf{(10.154)}$$

Finally, we solve Eq. (10.153) to find b_n in terms of the known quantities a_n and c_n:

$$b_n = \frac{\dfrac{c_n}{\lambda_n} - a_n \sinh \lambda_n L}{\cosh \lambda_n L}. \qquad \textbf{(10.155)}$$

The solution is now complete. (See Fig. 10.12.) The function $u(x, y)$

Figure 10.12

Isotherms, with $U_0 = 10$, $S = 1$, $B = 2L$

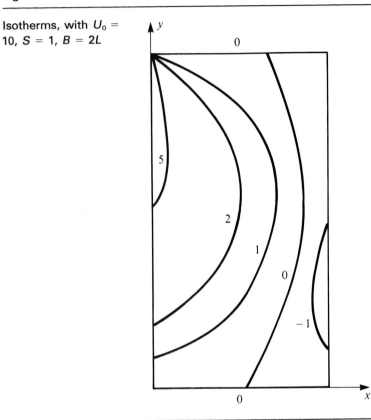

---

given in Eq. (10.151), with coefficients a_n and b_n from Eqs. (10.152) and (10.155), satisfies the potential equation and the boundary conditions assigned in Eqs. (10.142) and (10.143).

Exercises

In Exercises 1–7, solve by separation of variables.

1. $\dfrac{\partial^2 u}{\partial x^2} + \dfrac{\partial^2 u}{\partial y^2} = 0$, $0 < x < 1$, $0 < y < 1$

   $u(0, y) = 0$, $u(1, y) = U_0$, $0 < y < 1$,

   $u(x, 0) = 0$, $u(x, 1) = 0$, $0 < x < 1$

2. $\dfrac{\partial^2 u}{\partial x^2} + \dfrac{\partial^2 u}{\partial y^2} = 0$, $0 < x < 1$, $0 < y < 1$,

$$u(0, y) = U_0, \ u(1, y) = U_0, \ 0 < y < 1,$$
$$u(x, 0) = 0, \ u(x, 1) = 0, \ 0 < x < 1$$

3. $\dfrac{\partial^2 u}{\partial x^2} + \dfrac{\partial^2 u}{\partial y^2} = 0, \ 0 < x < L, \ 0 < y < B,$

$$\dfrac{\partial u}{\partial x}(0, y) = 0, \ \dfrac{\partial u}{\partial x}(L, y) = 0, \ 0 < y < B,$$

$$u(x, 0) = 0, \ u(x, B) = U_0, \ 0 < x < L$$

(Hint: be careful with the solution corresponding to eigenvalue 0.)

4. $\dfrac{\partial^2 u}{\partial x^2} + \dfrac{\partial^2 u}{\partial y^2} = 0, \ 0 < x < L, \ 0 < y < B,$

$$\dfrac{\partial u}{\partial x}(0, y) = 0, \ \dfrac{\partial u}{\partial x}(L, y) = 0, \ 0 < y < B,$$

$$u(x, 0) = U_0, \ u(x, B) = U_0 + Sx, \ 0 < x < L$$

5. $\dfrac{\partial^2 u}{\partial x^2} + \dfrac{\partial^2 u}{\partial y^2} = 0, \ 0 < x < L, \ 0 < y < B,$

$$\dfrac{\partial u}{\partial x}(0, y) = 0, \ u(L, y) = 0, 0 < y < B,$$

$$\dfrac{\partial u}{\partial y}(x, 0) = 0, \ u(x, B) = U_0, \ 0 < x < L$$

6. $\dfrac{\partial^2 u}{\partial x^2} + \dfrac{\partial^2 u}{\partial y^2} = 0, 0 < x < L, 0 < y < B,$

$$u(0, y) = 0, \ u(L, y) = 0, 0 < y < B,$$

$$u(x, 0) = f(x), \ u(x, B) = f(x), 0 < x < L,$$

$$\text{where } f(x) = \begin{cases} \dfrac{2U_0 x}{L}, \ 0 \le x < \dfrac{L}{2}, \\ \dfrac{2U_0(L - x)}{L}, \ \dfrac{L}{2} \le x \le L \end{cases}$$

7. $\dfrac{\partial^2 u}{\partial x^2} + \dfrac{\partial^2 u}{\partial y^2} = 0, \ 0 < x < L, \ 0 < y < B,$

$$\dfrac{\partial u}{\partial x}(0, y) = 0, \ \dfrac{\partial u}{\partial x}(L, y) = 0, \ 0 < y < B,$$

$$\dfrac{\partial u}{\partial x}(x, 0) = 0, \ \dfrac{\partial u}{\partial x}(x, B) = f(x), \ 0 < x < L,$$

$$\text{where } f(x) = \begin{cases} S, \ 0 < x < \dfrac{L}{2}, \\ -S, \ \dfrac{L}{2} < x < L \end{cases}$$

(Note: an additional condition such as $u(0, 0) = 0$ is necessary to determine u completely. In this example, u might be thought of as a velocity potential instead of a temperature. See Table 10.3.)

8. Show that the polynomial $u(x, y) = ax^2 + bxy + cy^2 + dx + ey + f$ ($a, b, \ldots, f$ constants) is a solution of the potential equation if $c + a = 0$.

9. Sketch the level curves of the function $u(x, y) = xy$, which is a solution of the potential equation in the entire xy-plane. What boundary conditions does u satisfy at $x = 0$ and 1, $y = 0$ and 1? (The surface $z = xy$ is called a hyperbolic paraboloid. Roofs are sometimes made in this shape.)

10. Show that the function $u(x, y) = \tan^{-1}(y/x)$ is a solution of the potential equation in any rectangular region not containing the origin.

11. What boundary conditions does the function $u(x, y) = (2/\pi) \tan^{-1}(y/x)$ satisfy at $x = 0$? at $y = 0$? Sketch level curves of u (loci of $u = \text{const}$) in the first quadrant. This function provides a model for discontinuity in boundary conditions at a corner. See Figs. 10.11 and 10.12.

10.7
Other Potential Problems

In Section 10.6 we were able to solve the potential equation on a rectangle, provided that there were homogeneous boundary conditions on opposite sides of the rectangle. Many problems that do not meet this condition can be broken down into two problems that do meet it.

Suppose that u is to satisfy the potential equation in a rectangle. Then assume that $u = u_H + u_V$ where both u_H and u_V are solutions of the potential equation in the rectangle. In addition, u_H satisfies the same boundary conditions as u on the horizontal boundaries but satisfies homogeneous conditions at the left and right—these conditions are formulated by simply zeroing any nonhomogeneity in the corresponding conditions for u. Similarly, u_V satisfies the same conditions as u on the vertical boundaries but homogeneous conditions at top and bottom. Then u_H and u_V may be found by the methods of Section 10.6.

Example 1

Decompose this problem into two problems with homogeneous boundary conditions on opposite sides:

$$\frac{\partial^2 u}{\partial x^2} + \frac{\partial^2 u}{\partial y^2} = 0, \qquad 0 < x < L, \qquad 0 < y < B, \qquad \textbf{(10.156)}$$

$$u(0, y) = U_0 y/B, \qquad \frac{\partial u}{\partial x}(L, y) = -S, \qquad 0 < y < B, \tag{10.157}$$

$$u(x, 0) = U_1, \qquad u(x, B) = U_0, \qquad 0 < x < L. \tag{10.158}$$

According to the text, we are to set $u = u_H + u_V$. The problem statement for u_H is

$$\frac{\partial^2 u_H}{\partial x^2} + \frac{\partial^2 u_H}{\partial y^2} = 0, \qquad 0 < x < L, \qquad 0 < y < B, \tag{10.159}$$

$$u_H(0, y) = 0, \qquad \frac{\partial u_H}{\partial x}(L, y) = 0, \qquad 0 < y < B, \tag{10.160}$$

$$u_H(x, 0) = U_1, \qquad u_H(x, B) = U_0, \qquad 0 < x < L. \tag{10.161}$$

Note that the boundary conditions on the horizontal sides ($y = 0$ and $y = B$, Eq. (10.161)) are identical with those in Eq. (10.158) for u, while the conditions at $x = 0$ and $x = L$, Eq. (10.160), are made by replacing the inhomogeneities in Eq. (10.157) with 0's. The problem statement for u_V is (See Fig. 10.13.)

$$\frac{\partial^2 u_V}{\partial x^2} + \frac{\partial^2 u_V}{\partial y^2} = 0, \qquad 0 < x < L, \qquad 0 < y < B, \tag{10.162}$$

$$u_V(0, y) = \frac{U_0 y}{B}, \qquad \frac{\partial u_V}{\partial x}(L, y) = -S, \qquad 0 < y < B, \tag{10.163}$$

$$u_V(x, 0) = 0, \qquad u_V(x, B) = 0, \qquad 0 < x < L. \tag{10.164}$$

To justify the decomposition above, first observe that $u_H + u_V$ is a solution of the potential equation in the given rectangle. Next, on the line $x = 0$, $u_H + u_V$ adds up to $U_0 y/B$, and on the line $x = L$ the derivative of $u_H + u_V$:

$$\frac{\partial}{\partial x}(u_H + u_V) = \frac{\partial u_H}{\partial x} + \frac{\partial u_V}{\partial x}$$

is just $-S$. These are the conditions on u at $x = 0$ and $x = L$. Similarly, $u_H + u_V$ adds up to U_1 at $y = 0$ and to U_0 at $y = B$. Thus the sum $u_H + u_V$ satisfies all conditions imposed on u.

Incidentally, the problems for u_H and u_V were solved in Examples 2 and 3 of Section 10.6.

A problem that appears to be inhomogeneous, but actually is not, is the potential equation in a circular disk with the function specified all around the circumference. Of course, it is natural to use polar coordinates

Figure 10.13

Original Problem and
Problems into which
It Decomposes

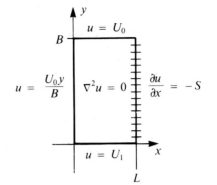

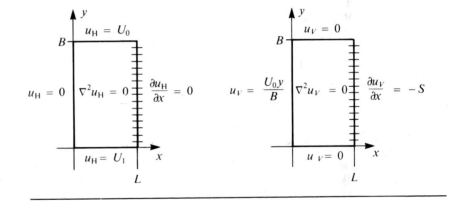

in this problem, so its statement would be

$$\frac{1}{r}\frac{\partial}{\partial r}\left(r\frac{\partial u}{\partial r}\right) + \frac{1}{r^2}\frac{\partial^2 u}{\partial \theta^2} = 0, \qquad 0 \le r < L, \qquad -\pi < \theta \le \pi, \qquad \textbf{(10.165)}$$

$$u(L, \theta) = f(\theta), \qquad -\pi < \theta \le \pi. \qquad \textbf{(10.166)}$$

There are two peculiarities to this problem. One is that the rays $\theta = -\pi$ and $\theta = \pi$ actually coincide. Thus the values of u and of its angular derivative should match there:

$$u(r, -\pi) = u(r, \pi), \qquad \frac{\partial u}{\partial \theta}(r, -\pi) = \frac{\partial u}{\partial \theta}(r, \pi), \qquad 0 \le r < L. \qquad \textbf{(10.167)}$$

The second is that the point $r = 0$ is singular: the coefficient of $\partial^2 u/\partial r^2$ in

Eq. (10.165) is 1, while the coefficients of other terms are $1/r$ and $1/r^2$. We must expect to have to enforce a boundedness condition:

$$u(r, \theta) \text{ bounded as } r \to 0+. \tag{10.168}$$

The problem of potential in a disk can now be treated by separation of variables. In essence, the "periodic conditions" in Eq. (10.167) act like homogeneous conditions on opposite sides of a rectangle.

Example 2

Solve this potential problem on a circular disk:

$$\frac{1}{r}\frac{\partial}{\partial r}\left(r\frac{\partial u}{\partial r}\right) + \frac{1}{r^2}\frac{\partial^2 u}{\partial\theta^2} = 0, \qquad 0 \le r < L, \qquad -\pi < \theta \le \pi, \tag{10.169}$$

$$u(L, \theta) = f(\theta), \qquad -\pi < \theta \le \pi, \tag{10.170}$$

where

$$f(\theta) = \begin{cases} 0, & -\pi < \theta < 0, \\ U_0, & 0 < \theta < \pi. \end{cases} \tag{10.171}$$

We seek nontrivial solutions in the product form $u(r, \theta) = R(r)\Theta(\theta)$. When this expression for u is substituted into the partial differential equation (10.169) and the special continuity conditions (10.167), they become

$$\frac{1}{r}(rR'(r))'\Theta(\theta) + \frac{1}{r^2}R(r)\Theta''(\theta) = 0, \qquad 0 \le r < L, \qquad -\pi < \theta \le \pi, \tag{10.172}$$

$$R(r)\Theta(-\pi) = R(r)\Theta(\pi), \qquad R(r)\Theta'(-\pi) = R(r)\Theta'(\pi), \qquad 0 \le r < L. \tag{10.173}$$

In order to separate the variables we must multiply Eq. (10.172) by r^2 and divide it by $R(r)\Theta(\theta)$. Also, $R(r)$ may be eliminated from Eq. (10.173). The results are

$$\frac{r(rR'(r))'}{R(r)} + \frac{\Theta''(\theta)}{\Theta(\theta)} = 0, \qquad 0 \le r < L, \qquad -\pi < \theta \le \pi, \tag{10.174}$$

$$\Theta(-\pi) = \Theta(\pi), \qquad \Theta'(-\pi) = \Theta'(\pi). \tag{10.175}$$

As usual, each of the terms in Eq. (10.174) must be constant, equal in magnitude and opposite in sign. In order that a nontrivial function satisfy the conditions (10.175) the ratio Θ''/Θ must be negative. That is, the condition (10.175) and the differential equation

$$\frac{\Theta''}{\Theta} = -\lambda^2 \qquad \text{or} \qquad \Theta'' + \lambda^2\Theta = 0$$

form an eigenvalue problem with the solution (see Chapter 9, Section 9.7)

$$\lambda_0^2 = 0, \qquad \Theta_0(\theta) = 1,$$

$$\lambda_n^2 = n^2, \qquad \Theta_n(\theta) = \cos n\theta \quad \text{or} \quad \sin n\theta, \qquad n = 1, 2, 3, \ldots. \quad \textbf{(10.176)}$$

The differential equation for the second factor $R(r)$ is

$$\frac{r(rR_n')'}{R_n} = \lambda_n^2 \quad \text{or} \quad r^2 R_n'' + r R_n' - \lambda_n^2 R_n = 0. \tag{10.177}$$

In the latter form it is readily recognized as an Euler-Cauchy equation, with independent solutions (see Exercise 6 at the end of this section)

$$R_n(r) = r^n \quad \text{and} \quad R_n(r) = r^{-n}. \tag{10.178}$$

The second of these is to be discarded because it does not comply with the boundedness condition (10.168).

We now have the product solutions $u(r, \theta) = 1$ for $n = 0$ and $u(r, \theta) = r^n \cos n\theta$ and $r^n \sin n\theta$ for $n = 1, 2, \ldots$ These are assembled, as a series of constant multiples, into the general solution of the homogeneous problem composed of the potential equation and the periodic boundary conditions (10.167):

$$u(r, \theta) = a_0 + \sum_{n=1}^{\infty} r^n (a_n \cos n\theta + b_n \sin n\theta). \tag{10.179}$$

The nonhomogeneous boundary condition on the physical boundary, $r = L$, may be satisfied by using the coefficients of this series. The condition (10.157) becomes

$$a_0 + \sum_{n=1}^{\infty} L^n (a_n \cos n\theta + b_n \sin n\theta) = f(\theta), \qquad -\pi < \theta \le \pi. \tag{10.180}$$

This is seen to be a Fourier series problem; the coefficients of the functions in the series must be

$$a_0 = \frac{1}{2\pi} \int_{-\pi}^{\pi} f(\theta) \, d\theta, \qquad L^n a_n = \frac{1}{\pi} \int_{-\pi}^{\pi} f(\theta) \cos n\theta \, d\theta,$$

$$L^n b_n = \frac{1}{\pi} \int_{-\pi}^{\pi} f(\theta) \sin n\theta \, d\theta. \tag{10.181}$$

For the function specified in Eq. (10.171) the coefficients are

$$a_0 = \frac{U_0}{2}, \qquad a_n = 0, \qquad b_n = \frac{U_0(1 - \cos n\pi)}{L^n n\pi}. \tag{10.182}$$

The solution of the problem is now complete. With the choice of the coefficients as given in Eq. (10.182) the function in Eq. (10.179) satisfies the problem as originally stated in Eqs. (10.169) and (10.170) and also the continuity conditions in Eqs. (10.167) and (10.168).

Let us generalize from the results of the example. We can solve the potential problem

$$\frac{1}{r}\frac{\partial}{\partial r}\left(r\frac{\partial u}{\partial r}\right) + \frac{1}{r^2}\frac{\partial^2 u}{\partial \theta^2} = 0, \qquad 0 \le r < L, \qquad -\pi < \theta \le \pi, \qquad \textbf{(10.165)}$$

$$u(L, \theta) = f(\theta), \qquad -\pi < \theta \le \pi, \qquad \textbf{(10.166)}$$

subject to the continuity conditions

$$u(r, -\pi) = u(r, \pi), \qquad \frac{\partial u}{\partial \theta}(r, -\pi) = \frac{\partial u}{\partial \theta}(r, \pi), \qquad 0 \le r < L \qquad \textbf{(10.167)}$$

$$u(r, \theta) \text{ bounded as } r \to 0+. \qquad \textbf{(10.168)}$$

The solution can be expressed in this form (slightly different from Eq. (10.179):

$$u(r, \theta) = a_0 + \sum_{n=1}^{\infty} \left(\frac{r}{L}\right)^n (a_n \cos n\theta + b_n \sin n\theta) \qquad \textbf{(10.183)}$$

with these choices for the coefficients (again, slightly different from Eq. (10.181)):

$$a_0 = \frac{1}{2\pi}\int_{-\pi}^{\pi} f(\theta)\, d\theta, \qquad \textbf{(10.184)}$$

$$a_n = \frac{1}{\pi}\int_{-\pi}^{\pi} f(\theta) \cos n\theta\, d\theta, \qquad b_n = \frac{1}{\pi}\int_{-\pi}^{\pi} f(\theta) \sin n\theta\, d\theta. \qquad \textbf{(10.185)}$$

In order to feel confident about the solution we ought to require that $f(\theta)$ be sectionally smooth, so that the series (10.183) will converge to $f(\theta)$, at all points of continuity, at least.

The solution above has deep and important consequences. The first concerns the value of u in the center of the disk. By our notation this is written $u(0, \theta)$, but it is (and must be!) independent of θ. Therefore let us write $u(0)$ for the value of u at the center of the disk.

By setting $r = 0$ in Eq. (10.183) we see that $u(0) = a_0$. This coefficient is identified in Eq. (10.184) as the average value of $f(\theta)$. But $f(\theta)$ is $u(L, \theta)$. Therefore we have

$$u(0) = \frac{1}{2\pi}\int_{-\pi}^{\pi} u(L, \theta)\, d\theta. \qquad \textbf{(10.186)}$$

In words, the value of u at the center of the disk is the average value of u on the circumference of the disk.

We obtained Eq. (10.186) by a series of substitutions. We could also appeal directly to $u(r, \theta)$ as given in Eq. (10.183). Integrate both sides with respect to θ over the range $-\pi$ to π, then divide by 2π. The infinite series on the right disappears, because

$$\int_{-\pi}^{\pi} \cos n\theta \, d\theta = 0, \qquad \int_{-\pi}^{\pi} \sin n\theta \, d\theta = 0.$$

What remains is the equation

$$\frac{1}{2\pi} \int_{-\pi}^{\pi} u(r, \theta) \, d\theta = a_0 = u(0). \tag{10.187}$$

In words, we have: the value of u at the center of the disk is the average value of u on the circumference of *any* concentric disk of radius r, $0 < r \leq L$.

These conclusions appear to be very special in nature, but they are actually quite general. If u satisfies the potential equation in a circular disk with center P—and perhaps in a much larger region—then one could set up a polar coordinate system centered on P and conclude as above. Thus we have the *mean value property*: the value of u at any point P is the average of its values on the circumference of any disk centered at P, provided that $\nabla^2 u = 0$ throughout the disk.

From the mean value property it is an easy step to the *maximum principle*: if $\nabla^2 u = 0$ in a region R, then u cannot have a relative maximum or minimum in the interior of R. If u had a relative maximum at some point P inside R, then $u(P)$ would be greater than u at all nearby points. But then $u(P)$ would be greater than the average value of u on a (small) disk centered at P—a contradiction.

Geometrically, the maximum principle says that the surface $z = u(x, y)$ has no peaks or pits. Equivalently, the level curves of u (loci of $u(x, y) = $ const) do not close in a region where $\nabla^2 u = 0$. (See Figs. 10.11 and 10.12 of Section 10.6.)

Exercises

In Exercises 1–5, solve the potential problem

$$\frac{\partial^2 u}{\partial x^2} + \frac{\partial^2 u}{\partial y^2} = 0, \qquad 0 < x < L, \qquad 0 < y < B,$$

subject to the boundary conditions given.

1. $u(0, y) = 0$, $u(L, y) = 1$, $0 < y < B$,
   $u(x, 0) = 0$, $u(x, B) = 1$, $0 < x < L$
2. $u(0, y) = 0$, $u(L, y) = y/B$, $0 < y < B$,
   $u(x, 0) = 0$, $u(x, B) = x/L$, $0 < x < L$
3. $u(0, y) = 0$, $u(L, y) = 1$, $0 < y < B$,

$$\frac{\partial u}{\partial y}(x, 0) = 0, \frac{\partial u}{\partial y}(x, B) = 1/L, \ 0 < x < L$$

4. $u(0, y) = 1, \ u(L, y) = 1, \ 0 < y < B,$
    $u(x, 0) = -1, \ u(x, B) = -1, \ 0 < x < L$

5. $u(0, y) = 1, \ \dfrac{\partial u}{\partial x}(L, y) = 0, \ 0 < y < B,$

    $u(x, 0) = 0, \ \dfrac{\partial u}{\partial y}(x, B) = 1/L, \ 0 < x < L$

6. We found that $r^n \cos n\theta$ and $r^n \sin n\theta$ are solutions of the potential equation. (a) Show that these are the real and imaginary parts of z^n, if $z = re^{i\theta}$. (b) Setting $z = x + iy$, find the real and imaginary parts of z^2, z^3, z^4. According to part (a), these are solutions of the potential equation.

7. For $u(x, y) = xy$ (known to be a solution of the potential equation) carry out the integration to evaluate

    $$\frac{1}{l} \int_{c_1} u(x, y) \, ds,$$

    where c_1 is a circle of radius $r > 0$ centered at $(\frac{1}{2}, \frac{1}{2})$ traversed counterclockwise and l is the length of this path. Your result should agree with the mean value property.

8. Sketch some level curves of the function $u(x, y) = x^2 - y^2$ in the square $-1 < x < 1, \ -1 < y < 1$. Does your sketch confirm the maximum principle?

In Exercises 9–12, solve the potential problem

$$\frac{1}{r}\frac{\partial}{\partial r}\left(r\frac{\partial u}{\partial r}\right) + \frac{1}{r^2}\frac{\partial^2 u}{\partial \theta^2} = 0, \qquad 0 \le r < L, \qquad -\pi < \theta \le \pi,$$

$$u(L, \theta) = f(\theta), \qquad -\pi < \theta \le \pi,$$

with f as given.

9. $f(\theta) = \theta, \ -\pi < \theta < \pi$

10. $f(\theta) = \begin{cases} 0, & \pi/2 < \theta \le \pi, \\ U_0, & -\pi/2 < \theta \le \pi/2, \\ 0, & -\pi < \theta \le -\pi/2 \end{cases}$

11. $f(\theta) = |\sin \theta|, \ -\pi < \theta \le \pi$

12. $f(\theta) = \begin{cases} \sin \theta, & 0 < \theta \le \pi, \\ 0, & -\pi < \theta \le 0 \end{cases}$

13. Find the value of $u(0)$ for the problems in Exercises 9–12 without solving the problem completely.

14. The functions in Eq. (10.178) are not correct for $n = 0$. Find the two independent solutions of Eq. (10.177) for $\lambda_0 = 0$. Which of these satisfies the boundedness condition at $r = 0$?

15. Find a function of x alone that satisfies the nonhomogeneous equation

$$\frac{\partial^2 u}{\partial x^2} + \frac{\partial^2 u}{\partial y^2} = -C, \qquad 0 < x < L, \qquad 0 < y < B,$$

and the homogeneous boundary conditions

$$u(0, y) = 0, \qquad u(L, y) = 0, \qquad 0 < y < B.$$

This partial differential equation is called *Poisson's equation.*

16. Use the results of Exercise 15 to solve this problem

$$\frac{\partial^2 u}{\partial x^2} + \frac{\partial^2 u}{\partial y^2} = -C, \qquad 0 < x < L, \qquad 0 < y < B,$$

$$u(0, y) = 0, \qquad u(L, y) = 0, \qquad 0 < y < B,$$

$$u(x, 0) = 0, \qquad u(x, B) = 0, \qquad 0 < x < L.$$

17. Find a function of r alone that satisfies Poisson's equation in a circular disk:

$$\frac{1}{r}\frac{\partial}{\partial r}\left(r\frac{\partial u}{\partial r}\right) + \frac{1}{r^2}\frac{\partial^2 u}{\partial \theta^2} = -C, \qquad 0 \le r < L, \qquad -\pi < \theta \le \pi,$$

$$u(L, \theta) = 0, \qquad -\pi < \theta \le \pi.$$

The function must also be bounded at $r = 0$.

10.8
Other Coordinate Systems

Up to this point we have considered problems in very special regions: the potential problem on a rectangle, $0 < x < L$, $0 < y < B$, or a disk, $0 \le r < L$, $-\pi < \theta \le \pi$, and the heat and wave equations for $0 < x < L$, $0 < t$. The important fact is that each spatial variable is restricted by an inequality with constant endpoints. This restriction is essential if separation of variables is to succeed.

In coordinate systems other than the Cartesian it is often possible to solve problems in regions that are physically two- or three-dimensional as if they were one-dimensional. The price for this advantage is that some singularity may be encountered in the differential equation or eigenvalue problems. We already have some experience with such singularity in the solution of the potential problem on the disk.

The most important problems to be considered are the heat, wave, and potential problems in cylindrical and spherical coordinates. All of these will involve the Laplacian operator, which is given in Table 10.4.

Table 10.4

Rectangular	$\nabla^2 u = \dfrac{\partial^2 u}{\partial x^2} + \dfrac{\partial^2 u}{\partial y^2} + \dfrac{\partial^2 u}{\partial z^2}$
Cylindrical	$\nabla^2 u = \dfrac{1}{r}\dfrac{\partial}{\partial r}\left(r\dfrac{\partial u}{\partial r}\right) + \dfrac{1}{r^2}\dfrac{\partial^2 u}{\partial \theta^2} + \dfrac{\partial^2 u}{\partial z^2}$
Spherical	$\nabla^2 u = \dfrac{1}{\rho^2}\left\{\dfrac{\partial}{\partial \rho}\left(\rho^2\dfrac{\partial u}{\partial \rho}\right) + \dfrac{1}{\sin\phi}\dfrac{\partial}{\partial \theta}\left(\sin\phi\dfrac{\partial u}{\partial \phi}\right) + \dfrac{1}{\sin^2\phi}\dfrac{\partial^2 u}{\partial \theta^2}\right\}$

Example 1

The temperature in a long rod of circular section is believed to vary with radial distance. Solve this heat problem which describes the temperature:

$$\frac{1}{r}\frac{\partial}{\partial r}\left(r\frac{\partial u}{\partial r}\right) = \frac{1}{k}\frac{\partial u}{\partial t}, \qquad 0 \le r < L, \qquad 0 < t, \tag{10.188}$$

$$u(L, t) = 0, \qquad 0 < t, \tag{10.189}$$

$$u(r, 0) = f(r), \qquad 0 \le r < L. \tag{10.190}$$

The boundary condition (10.189) is homogeneous. At the other end of the interval for r there is a singular point, so we must require

$$u(r, t) \text{ bounded as } r \to 0+. \tag{10.191}$$

We seek nontrivial solutions in the product form $u(r, t) = R(r)T(t)$. Substitution into the partial differential equation (10.188), boundary condition (10.189), and boundedness condition (10.191) leads to

$$\frac{(rR'(r))'}{rR(r)} = \frac{T'(t)}{kT(t)}, \qquad 0 \le r < L, \qquad 0 < t, \tag{10.192}$$

$$R(L) = 0, \qquad R(r) \text{ bounded as } r \to 0+. \tag{10.193}$$

Evidently both sides of Eq. (10.190) must be the same constant, which we denote by $-\lambda^2$. Then we obtain the eigenvalue problem

$$rR'' + R' + \lambda^2 rR = 0, \qquad 0 \le r < L, \tag{10.194}$$

$$R(L) = 0, \qquad R(r) \text{ bounded as } r \to 0+. \tag{10.195}$$

The differential equation (10.193) is Bessel's equation, solved in Chapter 4, Section 4.4, and Chapter 9, Section 9.7. The solution of the differential equation that is bounded at $r = 0$ is

$$R(r) = J_0(\lambda r),$$

where J_0 is the Bessel function of first kind and order 0. The boundary condition at $r = L$ requires

$$R(L) = J_0(\lambda L) = 0.$$

Therefore the solution of the eigenvalue problem is

$$\lambda_n^2 = (\beta_n/L)^2, \qquad R_n(r) = J_0(\lambda_n r), \tag{10.196}$$

in which β_n is the nth zero of J_0. (See Table 9.1, Section 9.7.)

The second factor in the product solution must satisfy

$$\frac{T_n'}{kT_n} = -\lambda_n^2 \qquad \text{or} \qquad T_n' + \lambda_n^2 kT_n = 0.$$

Thus $T_n(t) = e^{-\lambda_n^2 kt}$, and our most general solution is

$$u(r, t) = \sum_{n=1}^{\infty} c_n e^{-\lambda_n^2 kt} J_0(\lambda_n r). \tag{10.197}$$

The initial condition (10.190) is now to be satisfied. We set $t = 0$ in Eq. (10.196) and equate the resulting series to $f(r)$:

$$\sum_{n=1}^{\infty} c_n J_0(\lambda_n r) = f(r), \qquad 0 \le r < L. \tag{10.198}$$

According to Chapter 9, Section 7, this equality will hold if f is sectionally smooth and the coefficients are chosen to be

$$c_n = \frac{\int_0^L f(r) J_0(\lambda_n r) r \, dr}{\int_0^L J_0^2(\lambda_n r) r \, dr}. \tag{10.199}$$

With this choice for the c's, Eq. (10.96) supplies the solution of the problem originally stated.

Determining a specific solution for a specific function $f(r)$ is left to the Exercises.

The displacement of a flexible circular membrane, such as a drumhead, is determined by a problem such as

$$\frac{1}{r}\frac{\partial}{\partial r}\left(r\frac{\partial u}{\partial r}\right) = \frac{1}{c^2}\frac{\partial^2 u}{\partial t^2}, \qquad 0 \le r < L, \qquad 0 < t, \tag{10.200}$$

$$u(L, t) = 0, \qquad 0 < t, \tag{10.201}$$

$$u(r, 0) = f(r), \qquad \frac{\partial u}{\partial t}(r, 0) = g(r), \qquad 0 \le r < L, \tag{10.202}$$

provided that the displacement does not vary with the angular position.

Of course, a boundedness condition like Eq. (10.191) must be imposed at the origin.

Separation of variables may be applied to the problem above to find a solution in the form

$$u(r, t) = \sum_{n=1}^{\infty} (a_n \cos \lambda_n ct + b_n \sin \lambda_n ct) J_0(\lambda_n r). \tag{10.203}$$

The solution of the vibrating membrane problem, unlike that of the vibrating string problem, cannot be expressed directly in terms of the initial data, f and g. Indeed, although the individual terms of the series in Eq. (10.203)—the product solutions of Eqs. (10.200) and (10.201)—are periodic in time, the sum of the series generally is not periodic, and the original shape of the membrane never reappears. This explains, in part, why the sound of a drum is not as "pleasing" as that of a violin. (See Thomas D. Ressing, "The Physics of Kettledrums," *Scientific American*, Nov. 1982, pp. 172–178.)

Example 2

Solve the potential problem in a sphere:

$$\frac{1}{\rho^2} \left\{ \frac{\partial}{\partial \rho} \left(\rho^2 \frac{\partial u}{\partial \rho} \right) + \frac{1}{\sin \phi} \frac{\partial}{\partial \phi} \left(\sin \phi \frac{\partial u}{\partial \phi} \right) \right\} = 0, \tag{10.204}$$

$$0 \le \rho < L, \qquad 0 \le \phi \le \pi,$$

$$u(L, \phi) = f(\phi), \qquad 0 \le \phi \le \pi. \tag{10.205}$$

(In this problem the boundary condition (10.205) at the surface of the sphere does not vary with the longitude θ, which is therefore absent from the partial differential equation (10.204).)

In the partial differential equation it can be seen that the origin, $\rho = 0$, is a singular point, requiring a boundedness condition,

$$u(\rho, \phi) \text{ bounded as } \rho \to 0+. \tag{10.206}$$

Similarly, since $\sin \phi = 0$ at $\phi = 0$ and π, both of these are singular points, at which we require

$$u(\rho, \phi) \text{ bounded as } \phi \to 0+ \quad \text{and} \quad \phi \to \pi-. \tag{10.207}$$

(The angle ϕ is called the co-latitude, varying from 0 at the north pole to π at the south. Ordinary geographical latitude ranges from $\pi/2$ to $-\pi/2$. See Fig. 10.14.)

We may seek product solutions of Eq. (10.204), subject to the auxiliary conditions in Eqs. (10.206) and (10.207). The assumed form is $u(\rho, \phi) = R(\rho)\Phi(\phi)$, which leads to the equations

$$\frac{(\rho^2 R'(\rho))'}{R'(\rho)} + \frac{(\sin \phi\, \Phi')'}{\sin \phi\, \Phi} = 0, \qquad 0 \le \rho < L, \qquad 0 \le \phi \le \pi. \tag{10.208}$$

Figure 10.14

Spherical Coordinates

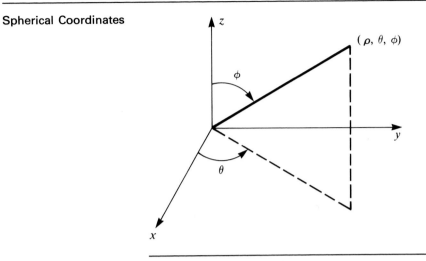

In addition, the individual factors are subject to the conditions

$$R(\rho) \text{ bounded as } \rho \to 0+, \tag{10.209}$$

$$\Phi(\phi) \text{ bounded as } \phi \to 0+ \text{ and } \pi-. \tag{10.210}$$

The usual arguments show that each of the terms in Eq. (10.208) must be constant. It is not immediately obvious which ratio is negative; but the boundary condition (10.205) will require an eigenfunction series, so the second term should be negative:

$$\frac{(\sin \phi \; \Phi')'}{\sin \phi \; \Phi} = -\lambda^2 \quad \text{or} \quad (\sin \phi \; \Phi')' + \lambda^2 \sin \phi \; \Phi = 0, \quad 0 \le \phi \le \pi. \tag{10.211}$$

This differential equation and the condition (10.210) form a singular eigenvalue problem, which was solved in Chapter 9, Section 9.7. The solution is

$$\lambda_n^2 = n(n + 1), \quad \Phi_n(\phi) = P_n(\cos \phi), \quad n = 0, 1, 2, \ldots, \tag{10.212}$$

where P_n is the nth Legendre polynomial.

The second factor in the product solution is now found by solving the differential equation

$$(\rho^2 R_n')' - \lambda_n^2 R_n = 0, \quad 0 \le \rho < L, \tag{10.213}$$

which comes from Eq. (10.208) on equating the first term to λ_n^2. This is an

Euler-Cauchy equation; the solution bounded at $\rho = 0$ is

$$R_n(\rho) = \rho^n. \tag{10.214}$$

The general solution of the potential equation (10.204) subject to the boundedness conditions (10.209) and (10.210) is now seen to be

$$u(\rho, \phi) = \sum_{n=0}^{\infty} c_n \left(\frac{\rho}{L}\right)^n P_n(\cos \phi). \tag{10.215}$$

(The introduction of the constant factor $1/L^n$ in each term will pay off in a moment.)

Finally, the nonhomogeneous boundary condition (10.205) is to be satisfied by appropriate choice of the coefficients in Eq. (10.215). Setting $\rho = L$ in that equation gives this form for the boundary conditions:

$$\sum_{n=0}^{\infty} c_n P_n(\cos \phi) = f(\phi), \qquad -\pi \leq \theta \leq \pi. \tag{10.216}$$

In Section 9.7 of Chapter 9 the coefficients of such a series were shown to be

$$c_n = \frac{2n+1}{2} \int_0^{\pi} f(\phi) P_n(\cos \phi) \sin \phi \, d\phi, \tag{10.217}$$

with convergence of the series (10.216) to $f(\phi)$ guaranteed at points of continuity of f if the function is sectionally smooth.

Again specific problems are left to the Exercises.

These two examples will suffice to show the outlines of the solution process in the presence of singularities. It should be noted that there are many coordinate systems in which the Laplacian operator lends itself to separation of variables. The eigenvalue problems that result give rise to many of the special functions of classical mathematical physics.

Exercises

1. Find the solution of the problem in Eqs. (10.188)–(10.190) if $f(r) = U_0$ (const).
2. Find the solution of the heat problem in a disk, Eq. (10.188), if $u(L, t) = U_0$, $0 < t$, and $u(r, 0) = U_1$, $0 \leq r < L$.
3. Find the form for the solution of the heat equation in a disk, Eq. (10.188), with the boundary condition

$$\frac{\partial u}{\partial r}(L, t) = 0, \qquad 0 < t,$$

and the general initial condition $u(r, 0) = f(r)$, $0 \leq r < L$.

4. Using Eq. (10.203), identify the frequencies of vibration (independent of θ) of a circular membrane in terms of c. If a drum of diameter 1 m has lowest frequency of vibration 120 Hz, find c.

5. Use separation of variables to find the general solution of this problem for spherical waves:

$$\frac{1}{\rho^2}\frac{\partial}{\partial\rho}\left(\rho^2\frac{\partial u}{\partial\rho}\right) = \frac{1}{c^2}\frac{\partial^2 u}{\partial t^2}, \qquad 0 \le \rho < L, \qquad 0 < t.$$

$$u(L, t) = 0, \qquad 0 < t, \qquad u(\rho, t) \text{ bounded at } 0.$$

(Hint: to find R, refer to Chapter 9, Section 9.7, or set $R(\rho) = S(\rho)/\rho$.)

6. Show that $u(\rho, t) = (1/\rho)f(\rho \pm ct)$ is a solution of the partial differential equation in Exercise 5.

7. Solve the potential problem in a sphere (independent of θ), Eqs. (10.204) and (10.205) if

$$f(\phi) = \begin{cases} U_0, & 0 \le \phi < \pi/2, \\ 0, & \pi/2 < \phi \le \pi. \end{cases}$$

Find the first five coefficients of the series (10.216) by integrating.

8. Follow the directions in Exercise 7, but for $f(\phi) = \cos\phi$.

9. Solve the differential equation (10.213) and show that Eq. (10.214) does give the solutions that are bounded at 0.

10. To determine potential (e.g., gravitational) *outside* a sphere, it is necessary to solve Eq. (10.204) for $\rho > L$, subject to the boundedness conditions (10.207) and

$$u(\rho, \phi) \text{ bounded as } \rho \to \infty.$$

Find product solutions for this problem.

11. Waves on a spherical shell ($\rho = L$, const) may satisfy the problem

$$\frac{1}{L^2 \sin\phi}\frac{\partial}{\partial\phi}\left(\sin\phi\frac{\partial u}{\partial\phi}\right) = \frac{1}{c^2}\frac{\partial^2 u}{\partial t^2}, \qquad 0 \le \phi \le \pi, \qquad 0 < t,$$

subject to boundedness at $\phi = 0, \pi$. Find product solutions for this problem.

12. Find the frequencies of vibration for the problem in Exercise 11.

13. Solve this heat problem in a sphere:

$$\frac{1}{\rho^2}\frac{\partial}{\partial\rho}\left(\rho^2\frac{\partial u}{\partial\rho}\right) = \frac{1}{k}\frac{\partial u}{\partial t}, \qquad 0 \le \rho < L, \qquad 0 < t,$$

$$u(L, t) = 0, \qquad 0 < t,$$

$$u(\rho, 0) = U_0, \qquad 0 \le \rho < L.$$

14. Using the first term of the series solution of Exercise 13, approximate $u(0, t)$ as a function of time.

15. Solve this potential problem on a cylinder:

$$\frac{1}{r}\frac{\partial}{\partial r}\left(r\frac{\partial u}{\partial r}\right) + \frac{\partial^2 u}{\partial z^2} = 0, \qquad 0 \le r < L, \qquad 0 < z < B,$$

$$u(L, z) = 0, \qquad 0 < z < B,$$

$$u(r, 0) = 0, \qquad u(r, B) = U_0, \qquad 0 \le r < L.$$

16. Follow the directions in Exercise 15, but the homogeneous boundary condition is

$$\frac{\partial u}{\partial r}(L, z) = 0, \qquad 0 < z < B.$$

10.9

Characteristics and Classification of Equations

Throughout this chapter we have dealt with second-order, linear, partial differential equations. We have solved the heat, wave, and potential equations under certain restrictions. It turns out that these three equations not only are important in mathematical physics, but also provide examples of the three different classes of such equations.

A quite general second-order partial differential equation is

$$A\frac{\partial^2 u}{\partial x^2} + B\frac{\partial^2 u}{\partial x\,\partial y} + C\frac{\partial^2 u}{\partial y^2} + F\left(x, y, u, \frac{\partial u}{\partial x}, \frac{\partial u}{\partial y}\right) = 0. \tag{10.218}$$

This equation is called quasi-linear because the highest order derivatives appear linearly; if the function F depends in a linear way on u, $\partial u/\partial x$, and $\partial u/\partial y$, then the whole equation is linear. The use of x and y for the independent variables is not meant to suggest that they are both space variables. We consider the heat and wave equations to be included in Eq. (10.218). In this section we assume A, B, and C to be *constants*.

After solving the wave equation by separation of variables we found a way to express the solution in terms of an initial value function evaluated at $x + ct$ and $x - ct$. The fact is that if we change variables from x, t to $\xi = x + ct$, $\eta = x - ct$ and set $u(x, t) = v(\xi, \eta)$, the wave equation reduces from

$$\frac{\partial^2 u}{\partial x^2} = \frac{1}{c^2}\frac{\partial^2 u}{\partial t^2} \qquad \text{to} \qquad 4\frac{\partial^2 v}{\partial \xi\,\partial \eta} = 0. \tag{10.219}$$

The latter equation is very easily solved by direct integration. Its general solution is

$$v(\xi, \eta) = \Psi(\xi) + \Phi(\eta), \tag{10.220}$$

where Ψ and Φ are quite arbitrary (except for differentiability requirements). Thus we obtain a general solution of the original wave equation as

$$u(x, t) = \Psi(x + ct) + \Phi(x - ct). \tag{10.221}$$

This is called *D'Alembert's solution*; it is especially convenient for studying waves in infinite media.

Now the question arises: is it possible to use the idea of a change of variables to simplify the general differential equation (10.218)? We propose the change from x, y to

$$\xi = \alpha x + \beta y, \qquad \eta = \gamma x + \delta y \tag{10.222}$$

and set $u(x, y) = v(\xi, \eta)$.* The chain rule is now used to calculate the derivatives of u in terms of those of v:

$$\frac{\partial u}{\partial y} = \frac{\partial v}{\partial \xi} \alpha + \frac{\partial v}{\partial \eta} \gamma, \qquad \frac{\partial u}{\partial y} = \frac{\partial v}{\partial \xi} \beta + \frac{\partial v}{\partial \eta} \delta,$$

$$\frac{\partial^2 u}{\partial x^2} = \frac{\partial^2 v}{\partial \xi^2} \alpha^2 + 2 \frac{\partial^2 v}{\partial \xi \partial \eta} \alpha \gamma + \frac{\partial^2 v}{\partial \eta^2} \gamma^2,$$

$$\frac{\partial^2 u}{\partial x \partial y} = \frac{\partial^2 v}{\partial \xi^2} \alpha \beta + \frac{\partial^2 v}{\partial \xi \partial \eta} (\alpha \delta + \gamma \beta) + \frac{\partial^2 v}{\partial \eta^2} \gamma \delta,$$

$$\frac{\partial^2 u}{\partial y^2} = \frac{\partial^2 v}{\partial \xi^2} \beta^2 + 2 \frac{\partial^2 v}{\partial \xi \partial \eta} \beta \delta + \frac{\partial^2 v}{\partial \eta^2} \delta^2.$$

When these expressions are substituted into the partial differential equation (10.218) and terms are collected, the result is

$$(A\alpha^2 + B\alpha\beta + C\beta^2) \frac{\partial^2 v}{\partial \xi^2} + (2A\alpha\gamma + B(\alpha\delta + \gamma\beta) + 2C\beta\delta) \frac{\partial^2 v}{\partial \xi \partial \eta}$$

$$+ (A\gamma^2 + B\gamma\delta + C\delta^2) \frac{\partial^2 v}{\partial \eta^2} + G\left(\xi, \eta, v, \frac{\partial v}{\partial \xi}, \frac{\partial v}{\partial \eta}\right) = 0. \tag{10.223}$$

Our first goal is to achieve the same reduction that was made for the wave equation: force the coefficients of $\partial^2 u/\partial \xi^2$ and $\partial^2 u/\partial \eta^2$ to be 0. That is, we want

$$A\alpha^2 + B\alpha\beta + C\beta^2 = 0, \tag{10.224}$$

$$A\gamma^2 + B\gamma\delta + C\delta^2 = 0. \tag{10.225}$$

---

* By writing $u(x, y) = v(\xi, \eta)$ we mean: if (x, y) and (ξ, η) designate the same point, then u and v have the same value there. Usually, u and v depend in very different ways on their variables.

Of course, we are assuming here that A or C or both are nonzero (otherwise there is nothing to do). Let us take $A \neq 0$ so that β and δ must be nonzero. Then the equations (10.224) and (10.225) reduce to

$$Az^2 + Bz + C = 0. \tag{10.226}$$

with $z = \alpha/\beta$ or γ/δ. The roots of this quadratic equation are

$$z = \frac{-B \pm \sqrt{B^2 - 4AC}}{2A}, \tag{10.227}$$

and now everything depends on the nature of these roots—more precisely, on the value of $\Delta = B^2 - 4AC$.

Case 1: $\Delta = B^2 - 4AC > 0$. If this is true, we have two real values of z in Eq. (10.227), which we can assign to α/β and γ/δ:

$$\frac{\alpha}{\beta} = \frac{-B + \sqrt{\Delta}}{2A}, \qquad \frac{\gamma}{\delta} = \frac{-B - \sqrt{\Delta}}{2A}.$$

To be specific, let us choose

$$\begin{aligned} \alpha &= -B + \sqrt{\Delta}, & \beta &= 2A, \\ \gamma &= -B - \sqrt{\Delta}, & \delta &= 2A. \end{aligned} \tag{10.228}$$

Then we compute the coefficient of $\partial^2 v/\partial\xi\,\partial\eta$ in Eq. (10.223) and find that the equation becomes

$$-4A\Delta \frac{\partial^2 v}{\partial\xi\,\partial\eta} + G\left(\xi, \eta, v, \frac{\partial v}{\partial\xi}, \frac{\partial v}{\partial\eta}\right) = 0. \tag{10.229}$$

This is the standard form for this case.

Example 1

Find the standard form for the simple wave equation

$$\frac{\partial^2 u}{\partial x^2} - \frac{1}{c^2}\frac{\partial^2 u}{\partial t^2} = 0.$$

First identify y with t; then the wave equation fits the form of Eq. (10.218) with $A = 1$, $C = -1/c^2$, $B = 0$, $F \equiv 0$. The value of Δ is

$$\Delta = B^2 - 4AC = 4/c^2 > 0,$$

so Case 1 applies. The new variables are

$$\xi = \frac{2}{c}x + 2t = \frac{2}{c}(x + ct),$$

$$\eta = -\frac{2}{c}x + 2t = -\frac{2}{c}(x - ct),$$

not the same as, but closely related to, the variables in D'Alembert's solution. From Eq. (10.229) the standard form is

$$-\frac{16}{c^2}\frac{\partial^2 v}{\partial \xi \, \partial \eta} = 0.$$

Case 2: $\Delta = B^2 - 4AC = 0$. Now there is only one value of z that satisfies Eq. (10.226). Since α/β and γ/δ have to be different (otherwise ξ and η are just multiples of each other), we take

$$\alpha = -B, \qquad \beta = 2A \tag{10.230}$$

and leave γ and δ as parameters for a moment. A quick calculation shows that the coefficient of $\partial^2 v/\partial\xi \, \partial\eta$ in Eq. (10.223) vanishes, whatever we choose for γ and δ. Thus we may take $\delta = 0$, $\gamma = 1$ and find that Eq. (10.223) becomes

$$A\frac{\partial^2 v}{\partial \eta^2} + G\left(\xi, \eta, v, \frac{\partial v}{\partial \xi}, \frac{\partial v}{\partial \eta}\right) = 0. \tag{10.231}$$

Example 2

Reduce the equation

$$\frac{\partial^2 u}{\partial x^2} + 2\frac{\partial^2 u}{\partial x \, \partial y} + \frac{\partial^2 u}{\partial y^2} + \frac{\partial u}{\partial x} = 0$$

to the standard form for its type.

First identify $A = C = 1$, $B = 2$ and then find $\Delta = B^2 - 4AC = 0$. The change of variables specified above is from x, y to

$$\xi = -2x + 2y, \qquad \eta = x.$$

The equation given now reduces to

$$\frac{\partial^2 v}{\partial \eta^2} + \left(-2\frac{\partial v}{\partial \xi} + \frac{\partial v}{\partial \eta}\right) = 0,$$

an equation for which product solutions can be found.

Case 3: $\Delta = B^2 - 4AC < 0$. We cannot have real variables and reduce the equation (10.223) to the form in Case 1. But we can do the opposite: make the coefficient of the mixed partial derivative equal to zero and make the other two coefficients equal to one another. This can be done in many ways. In particular, if we choose $\gamma = 1$, $\delta = 0$ and work out the consequences, we find that we must take

$$\alpha = \frac{B}{\sqrt{4AC - B^2}}, \qquad \beta = \frac{-2A}{\sqrt{4AC - B^2}}.$$ (**10.232**)

Then the standard form for the case is

$$A\left(\frac{\partial^2 v}{\partial \xi^2} + \frac{\partial^2 v}{\partial \eta^2}\right) + G\left(\xi, \eta, v, \frac{\partial v}{\partial \xi}, \frac{\partial v}{\partial \eta}\right) = 0.$$ (**10.233**)

Example 3

Obtain the standard form for the equation

$$\frac{\partial^2 u}{\partial x^2} + \frac{\partial^2 u}{\partial x, \partial y} + \frac{\partial^2 u}{\partial y^2} + \frac{\partial u}{\partial y} = 0.$$

We see that A, B, and C are all 1, so $B^2 - 4AC = -3 < 0$, and Eq. (10.233) is the standard form. The new variables are

$$\xi = \frac{x - 2y}{\sqrt{3}}, \qquad \eta = x.$$

With $v(\xi, \eta) = u(x, y)$ the given equation now takes the form

$$\frac{\partial^2 v}{\partial \xi^2} + \frac{\partial^2 v}{\partial \eta^2} - \frac{2}{\sqrt{3}} \frac{\partial v}{\partial \xi} = 0.$$

Because of the similarity between the forms encountered in our calculations and some in geometry, the three different classes of second-order quasilinear equations are given geometric names:

Hyperbolic if $B^2 - 4AC > 0$,

Parabolic if $B^2 - 4AC = 0$,

Elliptic if $B^2 - 4AC < 0$.

Example 4

Classify the heat, wave, and potential equations.
(a) The heat equation,

$$\frac{\partial^2 u}{\partial x^2} - \frac{1}{k} \frac{\partial u}{\partial y} = 0$$

(we have replaced t by y), has $A = 1$, $B = C = 0$. Therefore $B^2 - 4AC = 0$, and the equation is parabolic, already in standard form.
(b) The wave equation,

$$\frac{\partial^2 u}{\partial x^2} - \frac{1}{c^2} \frac{\partial^2 u}{\partial y^2} = 0,$$

has $A = 1$, $B = 0$, $C = -1/c^2$, as we noted in Example 1. It is hyperbolic because $B^2 - 4AC = 4/c^2$ is positive. The standard form, specified in Eq. (10.229) is

$$-\frac{16}{c^2}\frac{\partial^2 v}{\partial\xi\,\partial\eta} = 0.$$

(c) The potential equation,

$$\frac{\partial^2 u}{\partial x^2} + \frac{\partial^2 u}{\partial y^2} = 0,$$

has $A = C = 1$, $B = 0$. Thus $B^2 - 4AC = -4 < 0$, and the equation is elliptic. It is already in standard form.

We see then that our three equations represent the three different possible types. The information about types, variables, and standard forms is summarized in Table 10.5.

Table 10.5	**Type**	**Condition***	**Variables**	**Standard Form**
	Hyperbolic	$\Delta > 0$	$\xi = (-B + \sqrt{\Delta})x + 2Ay$	$-4A\Delta\dfrac{\partial^2 v}{\partial\xi\,\partial\eta} + G = 0$
			$\eta = (-B - \sqrt{\Delta})x + 2Ay$	
	Parabolic	$\Delta = 0$	$\xi = -Bx + 2Ay$	$A\dfrac{\partial^2 v}{\partial\eta^2} + G = 0$
			$\eta = x$	
	Elliptic	$\Delta < 0$	$\xi = \dfrac{Bx - 2Ay}{\sqrt{-\Delta}}$	$A\left(\dfrac{\partial^2 v}{\partial\xi^2} + \dfrac{\partial^2 v}{\partial\eta^2}\right) + G = 0$
			$\eta = x$	

*$\Delta = B^2 - 4AC$.

Let us return for a moment to Eq. (10.223) and our original attempt to make the coefficients of $\partial^2 v/\partial\xi^2$ and $\partial^2 v/\partial\eta^2$ equal 0. If ξ and η are such that Eqs. (10.224) and (10.225) are satisfied, then the lines $\xi = $ const and $\eta = $ const are called *characteristic lines* of the differential equation. Evidently, a hyperbolic equation has two families of characteristic lines, a parabolic equation has one, and an elliptic equation has none.

These lines are of particular importance for hyperbolic equations because a solution may be discontinuous along them. Recall for a moment the vibrating string problem:

$$\frac{\partial^2 u}{\partial x^2} = \frac{1}{c^2}\frac{\partial^2 u}{\partial t^2}, \qquad 0 < x < L, \qquad 0 < t,$$

$$u(0, t) = 0, \qquad u(L, t) = 0, \qquad 0 < t,$$

$$u(x, 0) = f(x), \qquad \frac{\partial u}{\partial t}(x, 0) = 0, \qquad 0 < x < L.$$

We obtained the D'Alembert solution:

$$u(x, t) = \tfrac{1}{2}(\bar{f}_o(x + ct) + \bar{f}_o(x - ct)),$$

where $\bar{f}_o$ is the odd periodic extension of f. This last expression is considered to provide the solution of the vibrating string problem, even if f is only sectionally continuous. Thus if f is discontinuous at x_0, $u(x, t)$ may be discontinuous at any point and time where $x_0 = x + ct$ or $x - ct$, give or take multiples of $2L$.

Exercises

In Exercises 1–4, (a) classify the equation as being hyperbolic, parabolic, or elliptic, (b) find the new variables, (c) transform into standard form.

1. $\dfrac{\partial^2 u}{\partial x^2} + \dfrac{\partial^2 u}{\partial x\,\partial y} + \dfrac{\partial u}{\partial y} = 0$

2. $\dfrac{\partial^2 u}{\partial x^2} - 4\dfrac{\partial u}{\partial x\,\partial y} + 4\dfrac{\partial^2 u}{\partial y^2} - u = 0$

3. $\dfrac{\partial^2 u}{\partial x^2} - 2\dfrac{\partial^2 u}{\partial x\,\partial y} + 2\dfrac{\partial^2 u}{\partial y^2} + u = 0$

4. $\dfrac{\partial^2 u}{\partial x^2} - \dfrac{\partial^2 u}{\partial x\,\partial y} + \dfrac{\partial^2 u}{\partial y^2} = 0$

5. Solve the equation to which the simple wave equation reduces,

$$\frac{\partial^2 v}{\partial \xi\,\partial \eta} = 0.$$

   This is sometimes called *Euler's equation*.

6. Confirm, by differentiating and substituting, that Eq. (10.221) gives a solution of the simple wave equation (10.222).

7. In Example 10.23 we obtained the equation

$$\frac{\partial^2 v}{\partial \eta^2} + \frac{\partial v}{\partial \eta} - 2\frac{\partial v}{\partial \xi} = 0.$$

Show that $w = e^{\eta/2}v$ satisfies the equation

$$\frac{\partial^2 w}{\partial \eta^2} - 2\frac{\partial w}{\partial \xi} - \frac{1}{4}w = 0.$$

8. Show that the locus of points in the $\alpha\beta$-plane that satisfy the equation

$$A\alpha^2 + B\alpha\beta + C\beta^2 = K$$

is a hyperbola if $B^2 - 4AC > 0$ or an ellipse if $B^2 - 4AC < 0$ (for suitable values of K).

9. The result of Case 3 suggests the use of imaginary characteristics. Show that the change of variables from x and y to

$$\xi = x + iy \qquad \text{and} \qquad \eta = x - iy$$

transforms the potential equation to Euler's equation

$$\frac{\partial^2 v}{\partial \xi \, \partial \eta} = 0.$$

10. Linearized perturbation theory for fast flow around a thin body is based on the equation

$$(1 - M^2)\frac{\partial^2 u}{\partial x^2} + \frac{\partial^2 u}{\partial y^2} = 0.$$

Here u is the potential for perturbation velocities and M is the mach number in the free stream flowing in the x-direction. Classify this equation for $M < 1$ (subsonic flow) and for $M > 1$ (supersonic flow).

Notes and References

In this chapter we have studied a method of solution for certain types of partial differential equations that occur frequently in applications. A short list of topics we have not mentioned gives an idea of how vast the subject is: problems in infinite domains; problems in more than two independent variables; solution methods using Laplace, Fourier, and other transforms; similarity solutions; numerical methods; nonlinear equations; higher-order equations; general theory of partial differential equations; and more.

Every item on this list merits a book, but many items are too specialized to interest an undergraduate engineering or science student. Two texts that might be the next step after this one are those by Churchill and Brown (1978) and Powers (1979).

Miscellaneous Exercises

1. The *error function* is defined by

$$\text{erf}(x) = \frac{2}{\sqrt{\pi}} \int_0^x e^{-z^2}\, dz.$$

Show that $f(x) = \text{erf}(x)$ satisfies the initial value problem

$$f'' + 2xf' = 0, \qquad f(0) = 0, \qquad f'(0) = 2/\sqrt{\pi}.$$

2. Show that the function $u(x, t) = \text{erf}(x/\sqrt{4kt})$ satisfies the problem

$$\frac{\partial^2 u}{\partial x^2} = \frac{1}{k}\frac{\partial u}{\partial t}, \qquad 0 < x, \qquad 0 < t,$$

$$u(0, t) = 0, \qquad 0 < t,$$

$$u(x, 0) = 1, \qquad 0 < x.$$

You will need to know that $\text{erf}(x)$ approaches 1 as x tends to $+\infty$.

3. What heat problem is satisfied by $u(x, t) = 1 - \text{erf}(x/\sqrt{4kt})$? The function $1 - \text{erf}(x)$ is called the *complementary error function*.

4. Sketch some level curves of the function $u(x, t)$ of Exercise 2.

5. Solve this nonhomogeneous heat problem:

$$\frac{\partial^2 u}{\partial x^2} = \frac{\partial u}{\partial t} - t, \qquad 0 < x < 1, \qquad 0 < t,$$

$$u(0, t) = 0, \qquad u(1, t) = 0, \qquad 0 < t,$$

$$u(x, 0) = 0, \qquad 0 < x < 1.$$

6. Suppose that $u(x, t)$ and $v(y, t)$ satisfy the equations

$$\frac{\partial^2 u}{\partial x^2} = \frac{1}{k}\frac{\partial u}{\partial t}, \qquad \frac{\partial^2 v}{\partial y^2} = \frac{1}{k}\frac{\partial v}{\partial t}.$$

Show that $w(x, y, t) = u(x, t)v(y, t)$ satisfies the two-dimensional heat equation,

$$\frac{\partial^2 w}{\partial x^2} + \frac{\partial^2 w}{\partial y^2} = \frac{1}{k}\frac{\partial w}{\partial t}.$$

7. Prove that if $X''/X = k^2 > 0$, and X satisfies one of the sets of boundary conditions below, then $X(x) \equiv 0$.
   (a) $X(0) = 0$, $X(L) = 0$; (b) $X(0) = 0$, $X'(L) = 0$;
   (c) $X'(0) = 0$, $X(L) = 0$; (d) $X'(0) = 0$, $X'(L) = 0$.

8. In a uniform electric transmission line the current i and voltage to ground e at the beginning and end of a short segment (see Fig. 10.15)

Figure 10.15

A Short Segment
of a Uniform
Transmission Line

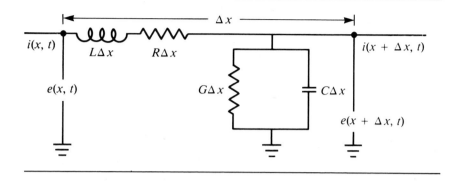

are related by these equations, obtained by applying Kirchhoff's law:

$$i(x, t) = i(x + \Delta x, t) + G\Delta x e(x, t) + C\Delta x \frac{\partial e}{\partial t}(x, t),$$

$$e(x, t) = e(x + \Delta x, t) + R\Delta x i(x, t) + L\Delta x \frac{\partial i}{\partial t}(x, t).$$

The parameters (all per unit length) are resistance R, inductance L, leakage conductance G, and capacitance C. By rearranging terms and taking limits as $\Delta x \to 0$, obtain the *transmission line equations,*

$$\frac{\partial i}{\partial x} + Ge + C\frac{\partial e}{\partial t} = 0,$$

$$\frac{\partial e}{\partial x} + Ri + L\frac{\partial i}{\partial t} = 0.$$

9. In the case of telegraph lines, one may take $G = 0$ and $L = 0$ in the transmission line equations. With this assumption, find a single second-order equation for i and one for e.

10. In the wires connecting computer components, e and i satisfy the transmission line equations with $R = 0$ and $G = 0$. Show that each of them also satisfies the wave equation with $c = 1/\sqrt{LC}$.

11. The displacement $u(x, t)$ of a prismatic beam, simply supported at its ends, satisfies

$$\frac{\partial^4 u}{\partial x^4} + \frac{1}{c^2}\frac{\partial^2 u}{\partial t^2} = 0, \qquad 0 < x < L, \qquad 0 < t,$$

$$u(0, t) = 0, \qquad u(L, t) = 0, \qquad 0 < t,$$

$$\frac{\partial^2 u}{\partial x^2}(0, t) = 0, \qquad \frac{\partial^2 u}{\partial x^2}(L, t) = 0, \qquad 0 < t,$$

plus appropriate initial conditions. Find product solutions of this problem.

12. What are the frequencies of vibration of a beam described in Exercise 11?

13. Compare the sequence of ratios of the nth frequency of vibration to the first, for a beam (Exercise 12), a string (Section 10.4), and a closed pipe (Exercise 15 of Section 10.5).

14. Show that the function $u(x, y) = xy(1 - x - y)$ satisfies the Poisson equation $\nabla^2 u = -2(x + y)$ and boundary condition $u = 0$ on the triangle bounded by $x = 0$, $y = 0$, and $x + y = 1$.

15. Solve the potential problem

$$\frac{1}{r}\frac{\partial}{\partial r}\left(r\frac{\partial u}{\partial r}\right) + \frac{1}{r^2}\frac{\partial^2 u}{\partial \theta^2} = 0, \qquad 0 < r < L, \qquad 0 < \theta < \alpha\pi,$$

$$u(r, 0) = 0, \qquad u(r, \alpha\pi) = 0, \qquad 0 < r < L,$$

$$u(L, \theta) = U_0, \qquad 0 < \theta < \alpha\pi.$$

The region described is a sector of a circle if $\alpha < 2$. If $\alpha = 2$, it is a disk cut along one radius.

16. The potential problem on a hemispherical shell of radius L has the form

$$\frac{1}{L^2 \sin\phi}\frac{\partial}{\partial \phi}\left(\sin\phi\frac{\partial u}{\partial \phi}\right) + \frac{1}{L^2 \sin^2\phi}\frac{\partial^2 u}{\partial \theta^2} = 0, \qquad 0 < \phi < \frac{\pi}{2},$$

$$-\pi < \theta \leq \pi, \qquad u\left(0, \frac{\pi}{2}\right) = f(\theta), \qquad -\pi < \theta \leq \pi.$$

State appropriate continuity and boundedness conditions, set up the separation-of-variables solution, and show that the general solution is

$$u(\theta, \phi) = a_0 + \sum_{n=1}^{\infty}\left(\tan\frac{\phi}{2}\right)^n (a_n \cos n\theta + b_n \sin n\theta).$$

APPENDIX A

Matrix Algebra

A.1

Basic Algebra of Matrices

A *matrix* is an array of numbers or symbols that stand for numbers arranged into rows and columns. The numbers or symbols are called the *elements* or *entries* of the matrix. In dealing with matrices, ordinary numbers are called *scalars*. Usually, we denote a matrix by a single capital letter and its elements by the corresponding lowercase letter with two subscripts. For example,

$$A = \begin{bmatrix} a_{11} & a_{12} & a_{13} \\ a_{21} & a_{22} & a_{23} \end{bmatrix}.$$

The first subscript of an element (row index) tells the row in which it is located, and the second subscript (column index) tells the column. This notation may be further abbreviated to $A = [a_{ij}]$.

The matrix above has two rows and three columns; it is referred to as a 2×3 matrix. The numbers of rows and columns of a matrix are called its *dimensions*. In general, an $m \times n$ matrix has m rows and n columns. *In dimensions and subscripts, rows come first.* When two matrices have the same number of rows and the same number of columns, we say they are of the *same shape.*

The usefulness of matrices lies in the symbolizing of the entire array with a single letter, which can then be manipulated without reference to the individual elements. The manipulations follow rules of algebra as developed below.

Equality

Two matrices of the same shape are *equal* if the elements in corresponding positions are equal. Thus an equality between two $m \times n$ matrices entails equalities between mn pairs of elements.

Example 1

The matrix equality

$$\begin{bmatrix} a_{11} & a_{12} & a_{13} \\ a_{21} & a_{22} & a_{23} \end{bmatrix} = \begin{bmatrix} 1 & 4 & 7 \\ 2 & 3 & 5 \end{bmatrix}$$

implies six equalities: $a_{11} = 1, a_{12} = 4, \ldots, a_{23} = 5$. Incidentally, a_{12} is read "a, one-two."

Addition

Two matrices of the same shape may be added. If $A = [a_{ij}]$ and $B = [b_{ij}]$ are both $m \times n$ matrices, so is their sum $C = A + B$, and the elements of C are given by $c_{ij} = a_{ij} + b_{ij}$.

Example 2

$$\begin{bmatrix} 1 & 4 \\ 0 & 5 \end{bmatrix} + \begin{bmatrix} 2 & -4 \\ 5 & 2 \end{bmatrix} = \begin{bmatrix} 3 & 0 \\ 5 & 7 \end{bmatrix}.$$

Multiplication by a Scalar

The product of a scalar k with an $m \times n$ matrix $A = [a_{ij}]$ is an $m \times n$ matrix denoted by $kA = [ka_{ij}]$. The scalar multiplies each element of a matrix. Usually, we put the scalar multiplier on the left of the matrix, but Ak means the same as kA.

Example 3

$$7\begin{bmatrix} 2 & 0 & 1 \\ 0 & -1 & 4 \end{bmatrix} = \begin{bmatrix} 14 & 0 & 7 \\ 0 & -7 & 28 \end{bmatrix},$$

$$-1\begin{bmatrix} 4 & 0 \\ -1 & 2 \end{bmatrix} = \begin{bmatrix} -4 & 0 \\ 1 & -2 \end{bmatrix}.$$

As a natural extension of the definitions above, we can simplify $A + (-1)B$ to $A - B$. Furthermore, we see that $A - A$ is a matrix with all elements equal to 0. This is called a null or *zero matrix*, designated 0. There is a different zero matrix for every shape, but it is usually not necessary to distinguish them.

Rules

The operations of addition and multiplication by a scalar follow the same laws as in ordinary algebra. The rules of addition are:

A1. $A + B = B + A$ (commutative law),
A2. $(A + B) + C = A + (B + C)$ (associative law),
A3. $A + (-A) = 0$,
A4. $A + 0 = A$.

The rules for multiplication by a scalar are

S1. $(bc)A = b(cA) = c(bA)$,
S2. $1 \cdot A = A$,
S3. $0 \cdot A = 0$.

There are two laws that connect addition with scalar multiplication, called the distributive laws:

S4. $(b + c)A = bA + cA$,
S5. $c(A + B) = cA + cB$.

Example 4

Solve this equation for A:

$$A - B = c(A + B)$$

if A and B are matrices and c is a scalar.

Applying the rules of algebra, we find that

$$A - B = cA + cB,$$

$$A - cA = B + cB,$$

$$(1 - c)A = (1 + c)B,$$

$$A = \frac{1 + c}{1 - c} B \qquad (c \neq 1).$$

In the last step it is necessary to assume $c \neq 1$ in order to divide by $1 - c$. If $c = 1$, the original equation was

$$A - B = A + B,$$

from which it follows that

$$-B = B,$$

$$0 = 2B,$$

and thus no information is obtained about A. However, if $c = 1$, we may conclude that $B = 0$.

Row and Column Matrices

An $n \times 1$ matrix is referred to as a *column matrix* because it consists of just one column. Similarly, a $1 \times n$ matrix is called a *row matrix*. Matrices of these types are often denoted by lowercase letters, and their elements need only one subscript:

$$x = \begin{bmatrix} x_1 \\ x_2 \\ \vdots \\ x_n \end{bmatrix}, \qquad y = [y_1, y_2, \ldots, y_n].$$

In matrix theory a row or column matrix may be called a *vector.*

Geometric Vectors

In elementary physics it is convenient to symbolize certain quantities such as force or displacement by directed line segments. These we shall call *geometric vectors.* They are usually written in terms of the three basis vectors **i**, **j**, and **k**. For instance,

$$\mathbf{a} = 2\mathbf{i} - \mathbf{k}, \qquad \mathbf{v} = x\mathbf{i} + y\mathbf{j} + z\mathbf{k}.$$

The basis vectors function essentially as place holders. It is more convenient to write these geometric vectors as column matrices. For the vectors above, we would write

$$a = \begin{bmatrix} 2 \\ 0 \\ -1 \end{bmatrix}, \qquad v = \begin{bmatrix} x \\ y \\ z \end{bmatrix}.$$

The operations of addition and multiplication by a scalar, as defined for geometric vectors, coincide exactly with the corresponding operations for matrices. Thus

$$\mathbf{v} - 2\mathbf{a} = (x - 4)\mathbf{i} + y\mathbf{j} + (z + 2)\mathbf{k}$$

corresponds to the column matrix

$$\begin{bmatrix} x - 4 \\ y \\ z + 2 \end{bmatrix} = \begin{bmatrix} x \\ y \\ z \end{bmatrix} - 2 \begin{bmatrix} 2 \\ 0 \\ -1 \end{bmatrix} = v - 2a.$$

The correspondence between geometrical vectors and 3×1 matrices allows a geometric interpretation for some matrix concepts and, reciprocally, allows us to deal algebraically with some geometrical problems.

Linear Independence

A sum of scalar multiples of some matrices of the same shape is called a *linear combination* of the matrices. Thus

$$c_1 X_1 + c_2 X_2 + \cdots + c_k X_k$$

symbolizes a linear combination of matrices $X_1, X_2, \ldots, X_k$ with coefficients $c_1, c_2, \ldots, c_k$. For example, the sum $v - 2a$ mentioned above is a linear combination of v and a. Similarly, the expression $\mathbf{a} = 2\mathbf{i} - \mathbf{k}$ exhibits $\mathbf{a}$ as a linear combination of the geometrical vectors $\mathbf{i}$, $\mathbf{j}$, and $\mathbf{k}$. (We may consider $\mathbf{j}$ present in the combination, with a coefficient of 0.)

A question of great importance for solving matrix problems is this. Given matrices $X_1, X_2, \ldots, X_k$, is it possible to find coefficients $c_1, c_2, \ldots, c_k$, *not all* 0, that make a linear combination equal to 0,

$$c_1 X_1 + c_2 X_2 + \cdots + c_k X_k = 0?$$

If such coefficients exist, the matrices $X_1, X_2, \ldots, X_k$ are said to be *linearly dependent*. On the other hand, if the only coefficients that make a linear combination equal 0 is the one with all coefficients equal to 0, the matrices are said to be *linearly independent*.

Sometimes it is easy to decide whether given matrices are dependent or not. For instance, the four matrices

$$X_1 = \begin{bmatrix} 1 & 0 \\ 0 & 0 \end{bmatrix}, \quad X_2 = \begin{bmatrix} 0 & 1 \\ 0 & 0 \end{bmatrix}, \quad X_3 = \begin{bmatrix} 0 & 0 \\ 1 & 0 \end{bmatrix}, \quad X_4 = \begin{bmatrix} 0 & 0 \\ 0 & 1 \end{bmatrix}$$

are linearly independent. To prove this, form a general linear combination of them, which is

$$c_1 X_1 + c_2 X_2 + c_3 X_3 + c_4 X_4 = \begin{bmatrix} c_1 & c_2 \\ c_3 & c_4 \end{bmatrix}.$$

The only way this 2×2 matrix can equal the 2×2 zero matrix is for c_1, c_2, c_3, and c_4 all to be 0.

On the other hand, if the given matrices are

$$X_1 = [1, 3, 2], \quad X_2 = [1, 1, 1], \quad X_3 = [2, 8, 5],$$

it may not be immediately obvious that some linear combination does come to 0. But it is true that

$$3X_1 - X_2 - X_3 = 0,$$

and therefore these three matrices are linearly dependent.

In a later section we will find efficient ways to determine whether given matrices are dependent or independent.

Transposition

There is an operation on matrices that has no analog in operations on real numbers, called transposition. If $A = [a_{ij}]$ is an $m \times n$ matrix, then its *transpose* A^T is an $n \times m$ matrix having a_{ji} in the (i, j) position. For instance,

$$\begin{bmatrix} 1 & 4 \\ 2 & 5 \\ 3 & 6 \end{bmatrix}^T = \begin{bmatrix} 1 & 2 & 3 \\ 4 & 5 & 6 \end{bmatrix}.$$

Another way to describe transposition is this: the elements of the ith row of A become the ith column of A^T. Thus the transpose of a column matrix is a row matrix and vice versa. This fact is often used for typographic convenience: instead of using several lines to write out a column matrix, we designate it as the transpose of a row. For example, $a = [2, 0, -1]^T$ is the 3×1 column matrix

$$a = \begin{bmatrix} 2 \\ 0 \\ -1 \end{bmatrix}.$$

The rules of transposition are very simple:
T1. $(A + B)^T = A^T + B^T$,
T2. $(kA)^T = kA^T$,
T3. $(A^T)^T = A$.

Differentiation and Integration

The elements of a matrix may be functions of some variable. If that is the case, we can define the operations of calculus as follows. Let $A(t) = [a_{ij}(t)]$ be a matrix of functions. Then we define

$$\frac{dA}{dt} = \left[\frac{da_{ij}}{dt} \right],$$

$$\int A(t) \, dt = \left[\int a_{ij}(t) \, dt \right].$$

That is, the indicated operation of differentiation or integration is carried out on each element of the matrix. Partial derivatives and derivatives of higher orders are similarly defined by applying the operation to each element. Because these operations affect each element, we say a matrix is differentiable (or integrable or continuous) if and only if each element has that property.

Example 5

Let $A(t) = [1, \sin t, e^{2t}]$. Then

$$\frac{dA}{dt} = [0, \cos t, 2e^{2t}],$$

$$\int A(t)\, dt = [t, -\cos t, \tfrac{1}{2}e^{2t}] + [c_1, c_2, c_3].$$

The rules for differentiating and integrating matrices are similar to those for ordinary functions, from which they are derived:

C1. $\dfrac{d}{dt}(A + B) = \dfrac{dA}{dt} + \dfrac{dB}{dt}$,

C2. $\dfrac{d}{dt}(kA) = \dfrac{dk}{dt}A + k\dfrac{dA}{dt}$,

C3. $\displaystyle\int (A + B)\, dt = \int A\, dt + \int B\, dt.$

Since the derivative of a constant is 0, a consequence of the second rule is that a constant "comes through" a derivative or an integral. This rule holds whether the constant is a scalar or a matrix.

Exercises

1. Write out the elements of the matrix described.
   (a) A is 3×3, $a_{ij} = 1/(i + j - 1)$
   (b) A is 4×4, $a_{ij} = |i - j|$
   (c) A is 2×3, $a_{ij} = 2 - i - j + ij$

2. Find the matrix result of each expression

$$A = \begin{bmatrix} 1 & 3 \\ 0 & 2 \end{bmatrix}, \quad B = \begin{bmatrix} 1 & -2 \\ 1 & 1 \end{bmatrix}, \quad C = \begin{bmatrix} 0 & 1 \\ 0 & 0 \end{bmatrix}.$$

   (a) $-2A^T$, (b) $5B$, (c) $A^T + B$, (d) $A - 3C$, (e) $B + 2C$, (f) $2B - A + 10C$, (g) $A - B$.

3. Consider the system of three equations in three unknowns:

$$x_1 + 2x_2 + 3x_3 = 0,$$
$$3x_1 + x_2 + 2x_3 = 3,$$
$$2x_1 + 3x_2 + x_3 = -3.$$

   (a) If a_{ij} is the coefficient of x_j in equation i, write out $A = [a_{ij}]$.
   (b) Determine coefficients so that A is a linear combination of I, R, and R^T:

$$I = \begin{bmatrix} 1 & 0 & 0 \\ 0 & 1 & 0 \\ 0 & 0 & 1 \end{bmatrix}, \quad R = \begin{bmatrix} 0 & 1 & 0 \\ 0 & 0 & 1 \\ 1 & 0 & 0 \end{bmatrix}.$$

4. (a) Write the following equations as an equality between two 2×1

matrices:

$$2x + y - z = 1,$$
$$x + y + z = 0.$$

(b) Write the left-hand side as a linear combination of three 2×1 matrices, with x, y, and z as coefficients.

5. Write $X(t)$ as a sum of scalar multiples of constant 2×2 matrices

$$X(t) = \begin{bmatrix} 1+t & 1-t^2 \\ t+t^2 & 1-t \end{bmatrix}.$$

6. Confirm that the matrices satisfy the relation stated.

(a) $X(t) = [e^{2t}, e^{-2t}]; \quad \dfrac{d^2X}{dt^2} = 4X$

(b) $y(t) = \begin{bmatrix} t \\ 2t \end{bmatrix}; \quad t\dfrac{dy}{dt} = y.$

(c) $R_1(t) = [\cos t, \sin t], \; R_2(t) = [-\sin t, \cos t],$

$$\dfrac{dR_1}{dt} = R_2, \qquad \dfrac{dR_2}{dt} = -R_1$$

(d) $A = \begin{bmatrix} x - y \\ 3 \\ e^{x-y} \end{bmatrix}; \quad \dfrac{\partial A}{\partial x} + \dfrac{\partial A}{\partial y} = 0$

7. Write $X(t)$ as a linear combination of constant 1×2 matrices in which e^t and e^{-t} are the coefficients

$$X(t) = [\cosh t, \sinh t].$$

8. Show, by substituting, that x, y, and z, as defined by the equation

$$\begin{bmatrix} x \\ y \\ z \end{bmatrix} = \begin{bmatrix} 3 \\ -1 \\ 0 \end{bmatrix} + t\begin{bmatrix} -3 \\ 2 \\ 1 \end{bmatrix}$$

satisfy this system of two equations in three unknowns, for any value of t:

$$x + 2y - z = 1,$$
$$2x + 3y \quad\;\; = 3.$$

9. Find the missing elements:

$$2\begin{bmatrix} 1 & 1 & -1 \\ 0 & \cdot & \cdot \end{bmatrix} - \begin{bmatrix} 1 & -1 \\ \cdot & 2 \\ \cdot & -1 \end{bmatrix}^T = \begin{bmatrix} \cdot & 3 & 3 \\ \cdot & 4 & -1 \end{bmatrix}.$$

A.2
Matrix Multiplication

One of the fundamental problems of matrix theory is the solution of systems of simultaneous linear equations, such as

$$2x_1 + 3x_2 - 4x_3 = 1,$$
$$3x_1 + 5x_2 + 7x_3 = 2. \tag{A.1}$$

From this system we can form the 3×1 matrix of unknowns

$$x = \begin{bmatrix} x_1 \\ x_2 \\ x_3 \end{bmatrix}, \tag{A.2}$$

the 2×1 matrix of "right-hand sides"

$$b = \begin{bmatrix} 1 \\ 2 \end{bmatrix}, \tag{A.3}$$

and the 3×2 *matrix of coefficients*

$$A = \begin{bmatrix} 2 & 3 & -4 \\ 3 & 5 & 7 \end{bmatrix}. \tag{A.4}$$

Matrix multiplication is defined so that the system above will be written in matrix terms as $Ax = b$.

Matrix Multiplication

The *product AB* of two matrices is defined if the number of columns of A equals the number of rows of B. If A is $m \times n$ and B is $n \times p$, the product $AB = [c_{ij}]$ is an $m \times p$ matrix with elements defined by

$$c_{ij} = a_{i1}b_{1j} + a_{i2}b_{2j} + \cdots + a_{in}b_{nj}.$$

Example 1

The product of the 2×3 matrix A and the 3×1 matrix x defined above is the 2×1 matrix

$$Ax = \begin{bmatrix} 2 & 3 & -4 \\ 3 & 5 & 7 \end{bmatrix} \begin{bmatrix} x_1 \\ x_2 \\ x_3 \end{bmatrix} = \begin{bmatrix} 2x_1 + 3x_2 - 4x_3 \\ 3x_1 + 5x_2 + 7x_3 \end{bmatrix}.$$

Thus the matrix form of the given system (A.1) is $Ax = b$, with A, x, and b as given in Eqs. (A.2), (A.3), and (A.4).

To remember the condition for the existence of the product of two matrices, it is convenient to write their dimensions below the factors. The two dimensions that are adjacent must be equal. The other two dimensions carry over as the dimensions of the product. In Example A.6 we had

$$
\begin{array}{ccc}
A & x & b \\
2 \times 3 & 3 \times 1 & 2 \times 1
\end{array}
$$

And in general the scheme is

$$
\begin{array}{ccc}
A & B & C \\
m \times n & n \times p & m \times p
\end{array}. \tag{A.5}
$$

To remember the rule for calculating the product of two matrices, the two factors can be positioned as shown in Fig. A.1. The i, j element of the product is obtained by adding the term-by-term product of row i of the first factor with column j of the second. This element is located at the point where the ith row of the first matrix and the jth column of the second would intersect.

Figure A.1

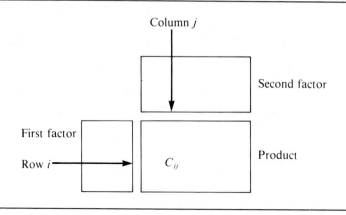

Example 2

Find the product of the matrices A and B:

$$
A = \begin{bmatrix} -3 & 3 \\ 7 & -1 \\ 5 & 1 \end{bmatrix}, \quad B = \begin{bmatrix} 2 & 0 & 4 & -2 \\ -4 & 6 & 8 & -6 \end{bmatrix}.
$$

Since A has two columns and B has two rows, the product is defined and will be a 3×4 matrix. Call $AB = C = [c_{ij}]$. The element in the 1, 1 position of C is

$$c_{11} = -3 \cdot 2 + 3 \cdot (-4) = -18.$$

Other elements of the first column of C are

$$c_{21} = 7 \cdot 2 + (-1) \cdot (-4) = 18,$$

$$c_{31} = 5 \cdot 2 + 1 \cdot (-4) = 6.$$

The rest of the calculation is left as an exercise:

$$\begin{bmatrix} 2 & 0 & 4 & -2 \\ -4 & 6 & 8 & -6 \end{bmatrix}$$

$$\begin{bmatrix} -3 & 3 \\ 7 & -1 \\ 5 & 1 \end{bmatrix} \begin{bmatrix} -18 & \cdot & \cdot & \cdot \\ 18 & \cdot & \cdot & \cdot \\ 6 & \cdot & \cdot & \cdot \end{bmatrix}.$$

Rules

Matrix multiplication follows several rules similar to those for ordinary multiplication. Under the assumption that the indicated operations are possible, the following rules hold.

M1. $A(BC) = (AB)C$ (associative law),

M2. $A(B + C) = AB + AC$ (distributive law),

M3. $(B + C)A = BA + CA$ (distributive law),

M4. $k(AB) = (kA)B = A(kB)$,

M5. $A \cdot 0 = 0,\ 0 \cdot A = 0$.

There is a small surprise in the rule relating transposition and multiplication:

M6. $(AB)^T = B^T A^T$.

By checking dimensions, however, it can easily be seen why this rule must be as it is.

The most remarkable fact about matrix multiplication is that several rules of ordinary algebra are *not* true. In general,

M7. $AB \neq BA$,

M8. $AB = 0$ does not imply $A = 0$ or $B = 0$,

M9. $AB = AC$ does not imply $B = C$.

Note that when AB is defined, BA may be undefined or of a different shape than AB, or just different from AB. For instance, we have these three examples:

$$A = \begin{bmatrix} 1 & -2 \\ 1 & 3 \end{bmatrix}, \quad B = \begin{bmatrix} 4 \\ 1 \end{bmatrix}, \quad AB = \begin{bmatrix} 2 \\ 7 \end{bmatrix}, \quad BA \text{ is not defined.}$$

$$A = \begin{bmatrix} 1 & 3 \end{bmatrix}, \quad B = \begin{bmatrix} 4 \\ 1 \end{bmatrix}, \quad AB = 7, \quad BA = \begin{bmatrix} 4 & 12 \\ 1 & 3 \end{bmatrix}.$$

$$A = \begin{bmatrix} 4 & 2 \\ 2 & 1 \end{bmatrix}, \quad B = \begin{bmatrix} 1 & -2 \\ -3 & 6 \end{bmatrix}, \quad AB = \begin{bmatrix} -2 & 4 \\ -1 & 2 \end{bmatrix}, \quad BA = \begin{bmatrix} 0 & 0 \\ 0 & 0 \end{bmatrix}.$$

In the first case, BA is not defined. In the second case, AB and BA are both defined, but of different shapes. In the third case, AB and BA are both 2×2 matrices but still are not equal. Note also that with

$$A = \begin{bmatrix} 4 & 2 \\ 2 & 1 \end{bmatrix}, \quad B = \begin{bmatrix} 1 & -2 \\ -3 & 6 \end{bmatrix}, \quad C = \begin{bmatrix} 0 & -3 \\ -1 & 8 \end{bmatrix} \tag{A.6}$$

we have $AB = AC$ (see Exercise 10 at the end of this section), but B and C are not equal.

It certainly may happen that $AB = BA$. If this is true, we say that A and B *commute* or are commutative.

Special Kinds of Matrices

If a matrix has dimensions $n \times n$ for some n, we call it a *square* matrix of *order n*. The positions $(1, 1), (2, 2), \ldots, (n, n)$ in a square matrix constitute its (main) *diagonal*, and the elements in these positions are *diagonal elements*. In the matrix

$$A = \begin{bmatrix} 1 & 2 & 3 \\ 4 & 5 & 6 \\ 7 & 8 & 9 \end{bmatrix}$$

the diagonal elements are 1, 5, and 9.

If a matrix has only zeros below (or above) the diagonal, it is called an upper (or lower) *triangular* matrix. For example, the matrix U is an upper triangular matrix, and L is a lower triangular matrix:

$$U = \begin{bmatrix} 1 & 1 & 1 \\ 0 & 3 & -1 \\ 0 & 0 & 5 \end{bmatrix}, \quad L = \begin{bmatrix} 2 & 0 & 0 \\ 1 & 3 & 0 \\ 2 & -1 & 5 \end{bmatrix}.$$

A system of equations with a triangular coefficient matrix is especially easy to solve. Consider the system

$$
\begin{aligned}
x_1 + x_2 + 2x_3 &= 5, \\
3x_2 - x_3 &= 7, \\
5x_3 &= 10,
\end{aligned}
$$

whose coefficient matrix is the upper triangular matrix U above. The last equation is easily solved for $x_3 = 2$. Then the second equation becomes

$$3x_2 - 2 = 7$$

because x_3 is now known; this can be solved for $x_2 = 3$. Finally, with x_2

and x_3 known the first equation becomes

$$x_1 + 3 + 2 \cdot 2 = 5,$$

from which we find $x_1 = -2$.

This process of solving a system with an upper triangular matrix of coefficients from bottom to top is known as *back-substitution*.

If a matrix has zeros both above and below its main diagonal, it is called a *diagonal* matrix. Sometimes a diagonal matrix is denoted by listing the diagonal elements, as in

$$D = \begin{bmatrix} 2 & 0 & 0 \\ 0 & 3 & 0 \\ 0 & 0 & 5 \end{bmatrix} = \text{diag}\,\{2, 3, 5\}.$$

A diagonal matrix with 1's on the diagonal is called an *identity* matrix, denoted by I_n for the $n \times n$ identity, or just by I if the dimensions can be determined from the context. Identity matrices have no effect in multiplication and thus are matrix analogs of the number 1 in ordinary multiplication. For example, one may verify that

$$\begin{bmatrix} 1 & 0 & 0 \\ 0 & 1 & 0 \\ 0 & 0 & 1 \end{bmatrix}\begin{bmatrix} 2 & 1 & 8 \\ 4 & -2 & 7 \\ 3 & -1 & 0 \end{bmatrix} = \begin{bmatrix} 2 & 1 & 8 \\ 4 & -2 & 7 \\ 3 & -1 & 0 \end{bmatrix}\begin{bmatrix} 1 & 0 & 0 \\ 0 & 1 & 0 \\ 0 & 0 & 1 \end{bmatrix}$$

$$= \begin{bmatrix} 2 & 1 & 8 \\ 4 & -2 & 7 \\ 3 & -1 & 0 \end{bmatrix}$$

As suggested by the example above, the $n \times n$ identity commutes with all $n \times n$ matrices: $AI = IA = A$.

Exercises

In Exercises 1–4, identify and give the dimension of A, x, and b so that the given system can be written as $Ax = b$.

1. $\begin{aligned} x_1 + x_2 + x_3 &= 3 \\ x_1 + 2x_2 + 3x_3 &= 6 \end{aligned}$

2. $\begin{aligned} x + y &= 1 \\ y - x &= 3 \end{aligned}$

3. $\begin{aligned} x + 2y &= 1 \\ y + 2z &= 3 \\ z + 2x &= 5 \end{aligned}$

4. $\begin{aligned} c_1 + c_2 &= 1 \\ -c_1 + 2c_2 + c_3 &= 0 \\ c_1 + 4c_2 + 2c_3 &= 0 \end{aligned}$

In Exercises 5–8, write the given matrix equation as a system of scalar equations.

5. $\begin{bmatrix} 1 & 2 & 1 \\ 1 & 0 & 1 \\ 0 & 1 & 0 \end{bmatrix}\begin{bmatrix} x \\ y \\ z \end{bmatrix} = \begin{bmatrix} 4 \\ 2 \\ 1 \end{bmatrix}$

6. $\begin{bmatrix} 1 & 0 & 2 & 2 \\ 3 & 1 & 4 & 7 \end{bmatrix}\begin{bmatrix} x_1 \\ x_2 \\ x_3 \\ x_4 \end{bmatrix} = \begin{bmatrix} 1 \\ 2 \end{bmatrix}$

7. $\begin{bmatrix} 1 & 1 & 1 & 0 \\ -1 & 1 & 0 & 1 \\ 0 & 1 & 1 & 1 \\ 1 & 0 & 1 & 1 \end{bmatrix} \begin{bmatrix} x_1 \\ x_2 \\ x_3 \\ x_4 \end{bmatrix} = \begin{bmatrix} 2 \\ 1 \\ 0 \\ 3 \end{bmatrix}$

8. $\begin{bmatrix} -1 & 1 \\ 0 & 1 \\ 1 & 1 \\ 2 & 1 \end{bmatrix} \begin{bmatrix} c_1 \\ c_2 \end{bmatrix} = \begin{bmatrix} 2 \\ 4 \\ 1 \\ 6 \end{bmatrix}$

9. Complete the matrix multiplication begun in Example 2.

10. Using the matrices in Eq. (A.6), calculate AB and AC and confirm that they are equal. Also find BA and CA.

In Exercises 11 and 12, use A, B, and C as given to calculate all of the expressions listed that can be calculated. If any operation is not defined, explain why.

11. $A = \begin{bmatrix} 1 & -1 \\ 0 & 0 \\ 2 & 1 \end{bmatrix}$, $B = \begin{bmatrix} 0 & 1 & 1 \\ 2 & 0 & 3 \end{bmatrix}$, $C = \begin{bmatrix} 2 \\ 1 \end{bmatrix}$.

   (a) AB, (b) BA, (c) AC, (d) CA, (e) BAC, (f) $(I + BA)C$, (g) $AB - BA$, (h) CB.

12. $A = \begin{bmatrix} 1 & 0 & 2 \\ 0 & 2 & 1 \end{bmatrix}$, $B = \begin{bmatrix} 1 \\ -1 \\ 1 \end{bmatrix}$, $C = \begin{bmatrix} x & y \end{bmatrix}$

   (a) AB, (b) BA, (c) AC, (d) CA, (e) BC, (f) CB, (g) CAB, (h) $I + A$, (i) $AB + C^T$, (j) ABC, (k) C^TA^T, (l) B^TB, (m) CC^T, (n) C^TC.

In Exercises 13–20, perform the indicated multiplication.

13. $\begin{bmatrix} 1 & 0 & 2 & 2 \\ 3 & 1 & 4 & 7 \end{bmatrix} \begin{bmatrix} -2 & -2 \\ -1 & 2 \\ 0 & 1 \\ 1 & 0 \end{bmatrix}$

14. $\begin{bmatrix} 0 & 1 & 0 & 2 \\ 2 & 0 & 1 & 0 \\ 0 & 2 & 0 & 1 \\ 1 & 0 & 2 & 0 \end{bmatrix} \begin{bmatrix} w \\ x \\ y \\ z \end{bmatrix}$

15. $\begin{bmatrix} 1 & -2 & 1 \end{bmatrix} \begin{bmatrix} 1 & 1 \\ 1 & 2 \\ 1 & 3 \end{bmatrix}$

16. $\begin{bmatrix} 1 & 2 & 1 \\ 3 & 0 & 2 \end{bmatrix} \begin{bmatrix} 1 & 1 & 0 \\ 1 & -1 & 1 \\ 2 & 0 & 1 \end{bmatrix}$

17. $\begin{bmatrix} 1 & 2 & 1 \\ 0 & 2 & 1 \\ 0 & 0 & -1 \end{bmatrix} \begin{bmatrix} 1 & 2 & 0 \\ 0 & -1 & 1 \\ 0 & 0 & 3 \end{bmatrix}$

18. $\begin{bmatrix} 2 & 1 \\ 3 & 2 \\ 0 & 1 \end{bmatrix} \begin{bmatrix} 2 & 2 & -1 \\ 0 & 1 & 3 \end{bmatrix}$

19. $\begin{bmatrix} 3 & 0 & 0 \\ 0 & 5 & 0 \\ 0 & 0 & 7 \end{bmatrix} \begin{bmatrix} 2 & 1 \\ -1 & 3 \\ 1 & -1 \end{bmatrix}$

20. $\begin{bmatrix} 2 & -1 & -2 \\ 1 & 3 & 1 \end{bmatrix} \begin{bmatrix} 3 & 0 & 0 \\ 0 & 5 & 0 \\ 0 & 0 & 7 \end{bmatrix}$

In Exercises 21–24, solve the given system by back-substitution.

21. $\begin{bmatrix} 1 & 1 \\ 0 & 2 \end{bmatrix} \begin{bmatrix} x \\ y \end{bmatrix} = \begin{bmatrix} 3 \\ 4 \end{bmatrix}$

22. $\begin{bmatrix} 1 & 2 & 1 \\ 0 & 1 & 1 \\ 0 & 0 & 1 \end{bmatrix} \begin{bmatrix} x \\ y \\ z \end{bmatrix} = \begin{bmatrix} 3 \\ 2 \\ 3 \end{bmatrix}$

23. $\begin{aligned} x + 2y + z &= 3 \\ y - z &= 3 \\ 7z &= -14 \end{aligned}$

24. $\begin{aligned} x - z &= 2 \\ y + 2z &= -1 \\ z &= -2 \end{aligned}$

In Exercises 25 and 26, find the missing elements

25. $\begin{bmatrix} 1 & \cdot \\ \cdot & 0 \end{bmatrix}\begin{bmatrix} \cdot & 1 \\ 0 & 1 \end{bmatrix} = \begin{bmatrix} 2 & 3 \\ 4 & \cdot \end{bmatrix}$

26. $\begin{bmatrix} 1 & 1 & 0 \\ 2 & \cdot & 1 \\ 1 & \cdot & 1 \end{bmatrix}\begin{bmatrix} 0 & 1 & 2 & 1 \\ \cdot & \cdot & -1 & 3 \\ 0 & 1 & \cdot & \cdot \end{bmatrix} = \begin{bmatrix} 2 & 1 & 1 & \cdot \\ \cdot & \cdot & 2 & \cdot \\ 2 & \cdot & 0 & 1 \end{bmatrix}$

27. Find an upper triangular matrix U, having positive elements on its diagonal, that satisfies

$$U^T U = \begin{bmatrix} 1 & -1 & 1 \\ -1 & 5 & 3 \\ 1 & 3 & 6 \end{bmatrix}$$

28. Show that whatever the shape of A, both AA^T and A^TA are defined.

29. (a) Compute the indicated products

$$\begin{bmatrix} d_1 & 0 & 0 \\ 0 & d_2 & 0 \\ 0 & 0 & d_3 \end{bmatrix}\begin{bmatrix} 1 & 2 \\ 3 & -1 \\ 0 & 5 \end{bmatrix}, \qquad \begin{bmatrix} 1 & 2 \\ 3 & -1 \\ 0 & 5 \end{bmatrix}\begin{bmatrix} d_1 & 0 \\ 0 & d_2 \end{bmatrix}$$

(b) Describe in words the effect of multiplying a given matrix A on the left by a diagonal matrix D.

30. Write the following system of differential equations in the form $\dot{x} = Ax + g$ identifying A, x, and g:

$$\frac{dx_1}{dt} = -2x_1 + x_2 + 1,$$

$$\frac{dx_2}{dt} = x_1 - 2x_2 + x_3 + 2,$$

$$\frac{dx_3}{dt} = x_2 - 2x_3 + 1.$$

A.3

Elimination

One of the major problems of matrix theory is solving and studying systems of simultaneous linear equations. A system of m equations in n unknowns would be written as

$$a_{11}x_1 + a_{12}x_2 + \cdots + a_{1n}x_n = b_1,$$
$$a_{21}x_1 + a_{22}x_2 + \cdots + a_{2n}x_n = b_n,$$
$$\vdots$$
$$a_{m1}x_1 + a_{m2}x_2 + \cdots + a_{mn}x_n = b_m,$$

or, in matrix form, $Ax = b$. By *solution* we mean n numbers $x_1, x_2, \ldots, x_n$ that satisfy all of the given equations. If a solution exists, we say that the system is *consistent*; otherwise, it is *inconsistent*. In the rest of this section we deal only with systems having unique solutions.

Solutions by Elimination

In order to solve a system of two equations in two unknowns, such as

$$x + y = 3, \tag{A.7a}$$

$$-x + y = 1 \tag{A.7b}$$

we might add the first equation to the second, obtaining the equation $2y = 4$, which we can readily solve. In other words, we replace the original system by a new system

$$x + y = 3, \tag{A.8a}$$

$$2y = 4, \tag{A.8b}$$

which has two important characteristics: (1) its coefficient matrix is triangular, and (2) it has exactly the same solution as the original system. As evidence for the second statement, note that we can subtract Eq. (A.8a) from Eq. (A.8b) to obtain

$$-x + y = 1,$$

which is just Eq. (A.7a).

The importance of the triangular coefficient matrix is that it allows solution immediately by back-substitution. We easily see that $y = 2$, and then determine from Eq. (A.8a) that $x = 1$.

In order to solve a system of three equations in three unknowns, the first step is again to eliminate the first unknown from all but the first equation. For instance, to solve the system

$$x + 2y + z = 3, \tag{A.9a}$$

$$x + 3y + 2z = 5, \tag{A.9b}$$

$$2x + y + 4z = 15, \tag{A.9c}$$

we would subtract the first equation from the second to get

$$(x + 3y + 2z) - (x + 2y + z) = 5 - 3$$

or

$$y + z = 2.$$

Then subtract twice the first from the third to get

$$(2x + y + 4z) - 2(x + 2y + z) = 15 - 2 \cdot 3$$

or

$-3y + 2z = 9.$

Thus we trade the original system for a new one,

$$x + 2y + z = 3, \tag{A.10a}$$

$$y + z = 2, \tag{A.10b}$$

$$-3y + 2z = 9, \tag{A.10c}$$

which has the same solutions as the original system. The new one does not have a triangular coefficient matrix, but it is on its way. We just treat Eqs. (A.10b) and (A.10c) as a system of two equations in two unknowns as in the previous example. Adding three times Eq. (A.10b) to Eq. (A.10c) gives $5z = 15$, or $z = 3$. Thus the original system has the same solutions as

$$x + 2y + z = 3, \tag{A.11a}$$

$$y + z = 2, \tag{A.11b}$$

$$z = 3. \tag{A.11c}$$

This triangular system can be solved easily by back-substitution.

Before going further it is important to note that the operations of the elimination process are actually carried out on the coefficients and right-hand sides of the equations. The symbols for the unknowns are just place holders. Thus we can record the results of operating on the equations of a system $Ax = b$ by using the *augmented matrix* of the system, $[A, b]$. This is formed by writing the column of right-hand sides next to the matrix of coefficients.

The augmented matrix of the system in Eq. (A.9) is

$$\begin{bmatrix} 1 & 2 & 1 & 3 \\ 1 & 3 & 2 & 5 \\ 2 & 1 & 4 & 15 \end{bmatrix}. \tag{A.12}$$

In the first step of elimination we subtracted Eq. (A.9a) from Eq. (A.9b). We might as well have subtracted the first row of the augmented matrix from the second, leading to

$$\begin{bmatrix} 1 & 2 & 1 & 3 \\ 0 & 1 & 1 & 2 \\ 2 & 1 & 4 & 15 \end{bmatrix}.$$

Next, twice Eq. (A.9a) was subtracted from Eq. (A.9c). Carrying out the analogous operation on the rows of the augmented matrix above gives

$$\begin{bmatrix} 1 & 2 & 1 & 3 \\ 0 & 1 & 1 & 2 \\ 0 & -3 & 2 & 9 \end{bmatrix}.$$

This is just the augmented matrix of the system in Eq. (A.10).

Elementary Operations

The operations we use in the elimination process are called *elementary row operations* because they are carried out on the rows of the augmented matrix of a system. They are:

1. add to some row a multiple of another row;
2. multiply a row by a nonzero scalar;
3. interchange two rows.

Each of these operations on a row of an augmented matrix records the result of the analogous operation on an equation of the corresponding system. Since each of these operations can be undone, any sequence of them can be carried out without affecting the solution of the system. That is, the solution of any system obtained from $Ax = b$ via elementary row operations on the augmented matrix is exactly the solution of $Ax = b$.

Gaussian Elimination

The systematic application of elementary operations to solve a system of equations is called *Gaussian elimination*. The first step is to use the first equation to eliminate the first unknown from all other equations. (Should it happen that the first given equation does not contain the first unknown, simply interchange it with another one that does.) The result will be one equation containing the first unknown and others, together with a subsystem of $n - 1$ equations in the remaining $n - 1$ unknowns. This smaller subsystem is then reduced in the same way, and so on, until we obtain a triangular system, which can then be solved by back-substitution.

Example 1

Solve the system of equations

$$y + 2z = 3,$$
$$x + 2y + z = 1,$$
$$x + y = 0.$$

The augmented matrix of the system is given below.

$$\begin{bmatrix} 0 & 1 & 2 & 3 \\ 1 & 2 & 1 & 1 \\ 1 & 1 & 0 & 0 \end{bmatrix}.$$

Since the first given equation does not contain x, we interchange it with the second equation. That is, since there is a 0 in the (1, 1) position of the augmented matrix, interchange rows 1 and 2 to obtain

$$\begin{bmatrix} 1 & 2 & 1 & 1 \\ 0 & 1 & 2 & 3 \\ 1 & 1 & 0 & 0 \end{bmatrix}.$$

Next subtract row 1 from row 3:

$$\begin{bmatrix} 1 & 2 & 1 & 1 \\ 0 & 1 & 2 & 3 \\ 0 & -1 & -1 & -1 \end{bmatrix}.$$

Now the system contains one equation involving x and a subsystem of two equations in y and z. The latter is easily reduced by adding the second equation to the third. (Add row 2 to row 3.) The result is

$$\begin{bmatrix} 1 & 2 & 1 & 1 \\ 0 & 1 & 2 & 3 \\ 0 & 0 & 1 & 2 \end{bmatrix}.$$

This is the augmented matrix of the triangular system

$$x + 2y + \quad z = 1,$$
$$y + 2z = 3,$$
$$z = 2,$$

which can now be solved by back-substitution to get

$$z = 2, \quad y = -1, \quad x = 1.$$

Example 2

Solve the system of four equations in four unknowns whose augmented matrix is

$$\begin{bmatrix} 1 & 2 & 1 & 1 & 3 \\ 1 & 3 & 1 & -1 & 0 \\ -1 & 0 & 4 & 2 & 0 \\ 2 & 1 & -4 & -1 & 0 \end{bmatrix}.$$

The first steps, as carried out on the augmented matrix, are: subtract row 1 from row 2; add row 1 to row 3; subtract twice row 1 from row 4.

The result is

$$\begin{bmatrix} 1 & 2 & 1 & 1 & 3 \\ 0 & 1 & 0 & -2 & -3 \\ 0 & 2 & 5 & 3 & 3 \\ 0 & -3 & -6 & -3 & -6 \end{bmatrix}.$$

Notice that in using row 1 to modify the other rows, row 1 itself is not changed! Now eliminate the second unknown from equations 3 and 4. That is, add appropriate multiples of row 2 to rows 3 and 4 so as to create 0's in the $(3, 2)$ and $(4, 2)$ positions. The result is

$$\begin{bmatrix} 1 & 2 & 1 & 1 & 3 \\ 0 & 1 & 0 & -2 & -3 \\ 0 & 0 & 5 & 7 & 9 \\ 0 & 0 & -6 & -9 & -15 \end{bmatrix}.$$

Finally, eliminate the third unknown from the fourth equation. It is convenient to divide row 3 by 5. The final triangular system is

$$\begin{bmatrix} 1 & 2 & 1 & 1 & 3 \\ 0 & 1 & 0 & -2 & -3 \\ 0 & 0 & 1 & 1.4 & 1.8 \\ 0 & 0 & 0 & -0.6 & -4.2 \end{bmatrix}.$$

To carry out the back-substitution many people find it convenient to write out the equations, which are in this case

$$x_1 + 2x_2 + x_3 + \quad x_4 = 3,$$
$$x_2 \quad - \quad 2x_4 = -3,$$
$$x_3 + 1.4x_4 = 1.8,$$
$$- 0.6x_4 = -4.2.$$

The solution is $x_1 = -18$, $x_2 = 11$, $x_3 = -8$, $x_4 = 7$.

Checking

The definitive check for any proposed solution of a system of equations is to substitute into all of the original equations and see whether they are all satisfied. When time is limited, such a complete check may be impossible. In this case, one should substitute the proposed solution into the equation that was most operated on in the solution process. In Example 1, the first and second equations survive, unaltered, to the triangular system. It is the third equation, $x + y = 0$, that was most altered.

Similarly, in Example 2, the last equation of the original set,

$$2x_1 + x_2 - 4x_3 - x_4 = 0,$$

was affected in every stage of the elimination process. Thus we would check that

$$2(-18) + 11 - 4(-8) - 7 = 0 \quad \checkmark$$

It is more useful to catch errors as they are made and before they cause a waste of time. In manual calculations it is common to use a *check column*. To the extreme right of each row of the initial augmented matrix, write the sum of the other numbers in that row. As elimination proceeds, carry out operations on the numbers in the check column, as well as on the rest of the row. At every stage, each number in the check column should equal the sum of the other numbers in that row.

Example 3

Solve, using a check column, the system of equations

$$
\begin{aligned}
x_1 + 2x_2 + x_3 &= 3, \\
-2x_1 \quad\;\; + x_3 &= -8, \\
x_1 + x_2 + 2x_3 &= 0.
\end{aligned}
$$

The augmented matrix, with check column, is

$$
\left[\begin{array}{rrr|r}
1 & 2 & 1 & 3 \\
-2 & 0 & 1 & -8 \\
1 & 1 & 2 & 0
\end{array}\right]
\begin{array}{r}
7 \\
-9 \\
4
\end{array}.
$$

The first two operations are: add twice row 1 to row 2; subtract row 1 from row 3. The resulting augmented matrix is

$$
\left[\begin{array}{rrr|r}
1 & 2 & 1 & 3 \\
0 & 4 & 3 & -2 \\
0 & -1 & 1 & -3
\end{array}\right]
\begin{array}{r}
7 \\
5 \\
-3
\end{array}.
$$

Each entry in the check column is still the sum of the numbers to its left.

Computation is easier if we interchange rows 2 and 3 and then change signs in the new row 2:

$$
\left[\begin{array}{rrr|r}
1 & 2 & 1 & 3 \\
0 & 1 & -1 & 3 \\
0 & 4 & 3 & -2
\end{array}\right]
\begin{array}{r}
7 \\
3 \\
5
\end{array}.
$$

Now subtract four times row 2 from row 3 to obtain the triangular system:

$$\left[\begin{array}{cccc|c} 1 & 2 & 1 & 3 & 7 \\ 0 & 1 & -1 & 3 & 3 \\ 0 & 0 & 7 & -14 & -7 \end{array}\right].$$

Again each entry in the check column is the sum of the numbers to its left.

Finally, back-substitution on the triangular system gives

$$x_1 = 3, \qquad x_2 = 1, \qquad x_3 = -2.$$

A quick check is to substitute into the second of the original equations.

Exercises

Solve each of the following systems by Gaussian elimination. Use a check column.

1. $x + y = 1,$
   $x + 2y = 4$

2. $x + y = 1,$
   $x - y = 3$

3. $x + y = a,$
   $x - y = b$

4. $x + y + z = 1,$
   $x + 2y + 2z = 4,$
   $x + 2y + 3z = 7$

5. $x + y + 2z = 5,$
   $x - y + 2z = 3,$
   $x - y + z = 0$

6. $x - y + z = 2,$
   $x + y - z = 0,$
   $-x + y + z = 2$

7. $2x + 3y - z = 1,$
   $x - 2y + 6z = 11,$
   $5x + y - z = 12$

8. $x + y + z = 5,$
   $x \quad - z = -1,$
   $x - y + z = 2$

9. $x + y = 3,$
   $y + z = 1,$
   $x + z = 0$

10. $\left[\begin{array}{ccc} 2 & -1 & 0 \\ -1 & 2 & -1 \\ 0 & -1 & 2 \end{array}\right]\left[\begin{array}{c} x_1 \\ x_2 \\ x_3 \end{array}\right] = \left[\begin{array}{c} 1 \\ 0 \\ 0 \end{array}\right]$

11. $\left[\begin{array}{ccc} 2 & -1 & 0 \\ -1 & 2 & -1 \\ 0 & -1 & 2 \end{array}\right]\left[\begin{array}{c} x_1 \\ x_2 \\ x_3 \end{array}\right] = \left[\begin{array}{c} 0 \\ 1 \\ 0 \end{array}\right]$

12. $\left[\begin{array}{ccc} 2 & -1 & 0 \\ -1 & 2 & -1 \\ 0 & -1 & 2 \end{array}\right]\left[\begin{array}{c} x_1 \\ x_2 \\ x_3 \end{array}\right] = \left[\begin{array}{c} 0 \\ 0 \\ 1 \end{array}\right]$

In Exercises 13–18, the augmented matrix of the system is given.

13. $\left[\begin{array}{ccc|c} -1 & 0 & 1 & 0 \\ 1 & -2 & 1 & 0 \\ 1 & 1 & 1 & 1 \end{array}\right]$

14. $\left[\begin{array}{ccc|c} 1 & 2 & 1 & 3 \\ 1 & 3 & 2 & 5 \\ 2 & 1 & 4 & 15 \end{array}\right]$

15. $\left[\begin{array}{cccc|c} 1 & 0 & 1 & 1 & 2 \\ 1 & 1 & 0 & 1 & 5 \\ 1 & 1 & 1 & 0 & 2 \\ 0 & 1 & 1 & 1 & 0 \end{array}\right]$

16. $\left[\begin{array}{cccc|c} 1 & 1 & 1 & 1 & 3 \\ 1 & 1 & -1 & -1 & 1 \\ 1 & -1 & -1 & 1 & 3 \\ 1 & -1 & 1 & -1 & -3 \end{array}\right]$

17. $\begin{bmatrix} 1 & 1 & 1 & 1 & 1 \\ -1 & 0 & 1 & 2 & 2 \\ 1 & 0 & 1 & 4 & 3 \\ -1 & 0 & 1 & 8 & 4 \end{bmatrix}$
18. $\begin{bmatrix} 2 & -2 & 0 & 0 & 1 \\ -1 & 2 & -1 & 0 & 0 \\ 0 & -1 & 2 & -1 & 0 \\ 0 & 0 & -1 & 2 & 0 \end{bmatrix}$

19. Find a formula for x_n in terms of the b's. (This result was known to the ancient Greeks.)

$$x_1 + x_n = b_1,$$
$$x_2 + x_n = b_2,$$
$$\vdots$$
$$x_{n-1} + x_n = b_{n-1},$$
$$x_1 + x_2 + \cdots + x_{n-1} + x_n = b_n.$$

20. Use the result of Exercise 19 to solve the system

$$x_1 + x_4 = 1,$$
$$x_2 + x_4 = 2,$$
$$x_3 + x_4 = 5,$$
$$x_1 + x_2 + x_3 + x_4 = 4.$$

A.4

Inverse

The *inverse* of an $n \times n$ matrix A is the $n \times n$ matrix X that satisfies the matrix equation $AX = I$. If such a matrix exists, A is called an *invertible* matrix, and the inverse is symbolized as $X = A^{-1}$.

It can be proved that if A is invertible, its *inverse is unique*—there cannot be a different solution of $AX = I$. Furthermore, $A^{-1}A = AA^{-1} = I$; that is, A^{-1} commutes with A.

Example 1

A 3×3 matrix and its inverse are given below. The reader should carry out multiplications to verify that $AA^{-1} = A^{-1}A = I$:

$$A = \begin{bmatrix} 1 & 1 & 1 \\ 1 & 2 & 3 \\ 2 & 3 & 5 \end{bmatrix}, \qquad A^{-1} = \begin{bmatrix} 1 & -2 & 1 \\ 1 & 3 & -2 \\ -1 & -1 & 1 \end{bmatrix}.$$

Rules

The inverse of a matrix is the analog of the reciprocal of a number. The most profound difference is that many matrices have no inverse. For instance, the 2×2 matrix

$$A = \begin{bmatrix} 1 & -2 \\ -2 & 4 \end{bmatrix}$$

can have no inverse. For those matrices that have an inverse, several rules hold.

I1. $(A^{-1})^{-1} = A$,

I2. $(A^T)^{-1} = (A^{-1})^T$,

I3. $(kA)^{-1} = \dfrac{1}{k} A^{-1}$ $(k \neq 0)$,

I4. $(AB)^{-1} = B^{-1}A^{-1}$.

In Rule 4 it is assumed that both A and B are square, invertible matrices. To prove the rule, simply test the defining property of inverses: is it true that

$$(AB)(B^{-1}A^{-1}) = I?$$

The associative law and rules governing I allow us to transform the left-hand side of the equation above as

$$(AB)(B^{-1}A^{-1}) = A(BB^{-1})A^{-1} = AIA^{-1} = AA^{-1} = I.$$

A further consequence of Rule 4 can be stated as follows: any product of invertible matrices is invertible, and the inverse of the product is the product of the inverses in reverse order.

The cancellation laws that hold for nonzero numbers are valid for invertible matrices.

I5. If A is invertible and $AB = 0$ or $BA = 0$, then $B = 0$.

I6. If A is invertible and $AB = AC$ or $BA = CA$, then $B = C$.

The proofs of these rules are simple. Suppose $AB = AC$. Multiply both members of the equation on the left by A^{-1} and apply the associative law:

$$A^{-1}(AB) = A^{-1}(AC),$$
$$(A^{-1}A)B = (A^{-1}A)C.$$

One more step leads to the conclusion that $B = C$.

Special Matrices

The inverse of an invertible 2×2 matrix can be given by the formula:

$$\begin{bmatrix} a_{11} & a_{12} \\ a_{21} & a_{22} \end{bmatrix}^{-1} = \frac{1}{a_{11}a_{22} - a_{12}a_{21}} \begin{bmatrix} a_{22} & -a_{12} \\ -a_{21} & a_{11} \end{bmatrix}.$$

It is necessary to assume that $a_{11}a_{22} - a_{12}a_{21} \neq 0$. For example,

$$\begin{bmatrix} 1 & 2 \\ 3 & 4 \end{bmatrix}^{-1} = \frac{1}{1 \cdot 4 - 2 \cdot 3}\begin{bmatrix} 4 & -2 \\ -3 & 1 \end{bmatrix} = \begin{bmatrix} -2 & 1 \\ 1.5 & -0.5 \end{bmatrix}.$$

The formula for the reverse of a 3×3 matrix is rarely worth the effort to write it down.

The inverse of a diagonal matrix is diagonal, and

$$(\text{diag }\{d_1, d_2, \ldots, d_n\})^{-1} = \text{diag }\left\{\frac{1}{d_1}, \frac{1}{d_2}, \ldots, \frac{1}{d_n}\right\}.$$

Of course, this is valid only if all the d's are nonzero. The inverse of an upper (or lower) triangular matrix is also upper (lower) triangular. The description of the off-diagonal elements is complicated.

Solving Systems

The importance of the matrix inverse lies in its role in the solution of matrix equations, as explained in Theorem A.1.

Theorem A.1

Let A be an invertible matrix. Then the system of equations (or matrix equation) $Ax = b$ has one and only one solution, which is $x = A^{-1}b$.

Proof: $A(A^{-1}b) = (AA^{-1})b = b$, so $x = A^{-1}b$ is indeed a solution. If y is also a solution, $Ay = b$, then $Ax = Ay$, and by the cancellation rule, $x = y$.

Example 2

Use the inverse supplied in Example 1 to solve the system of equations

$$x_1 + x_2 + x_3 = 2,$$
$$x_1 + 2x_2 + 3x_3 = 7,$$
$$2x_1 + 3x_2 + 5x_3 = 11.$$

This system can be written in matrix form as $Ax = b$, and its solution can be written as $x = A^{-1}b$. The inverse of the coefficient matrix A is given in Example 2, so the solution is

$$x = A^{-1}b = \begin{bmatrix} 1 & -2 & 1 \\ 1 & 3 & -2 \\ -1 & -1 & 1 \end{bmatrix}\begin{bmatrix} 2 \\ 7 \\ 11 \end{bmatrix} = \begin{bmatrix} -1 \\ 1 \\ 2 \end{bmatrix} = \begin{bmatrix} x_1 \\ x_2 \\ x_3 \end{bmatrix}.$$

Substitution in the original system confirms that this is indeed a solution.

The theorem allows us to *symbolize* the solution of $Ax = b$ as $x = A^{-1}b$ when A is nonsingular. The symbolism is somewhat misleading: if A and b are numerical matrices, it is usually more efficient to solve the system by elimination, rather than to find A^{-1} and then multiply $A^{-1}b$.

Elimination

We defined the inverse of an $n \times n$ matrix A as the solution of $AX = I$. It is easy to see that the jth column of X satisfies $Ax = e_j$, where e_j is the jth column of I_n. Thus finding $A^{-1} = X$ amounts to solving n systems of equations. But because all of the systems have the same coefficient matrix, they can be solved simultaneously.

The usual procedure is to start with the $n \times 2n$ matrix $[A, I]$, which is the augmented matrix of $AX = I$. Then this matrix is modified by means of row operations until the *left* half becomes an $n \times n$ identity. The right half is then A^{-1}, the solution of $AX = I$.

Example 3

Compute the inverse of the matrix A given in Example 1.

Below are the matrix $[A, I]$ and other matrices into which it is transformed, ending with $[I, A^{-1}]$.

$$\begin{bmatrix} 1 & 1 & 1 & 1 & 0 & 0 \\ 1 & 2 & 3 & 0 & 1 & 0 \\ 2 & 3 & 5 & 0 & 0 & 1 \end{bmatrix}$$ This is $[A, I]$.

Subtract row 1 from row 2; subtract $2 \times$ row 1 from row 3.

$$\begin{bmatrix} 1 & 1 & 1 & 1 & 0 & 0 \\ 0 & 1 & 2 & -1 & 1 & 0 \\ 0 & 1 & 3 & -2 & 0 & 1 \end{bmatrix}$$

Subtract row 2 from row 1; subtract row 2 from row 3.

$$\begin{bmatrix} 1 & 0 & -1 & 2 & -1 & 0 \\ 0 & 1 & 2 & -1 & 1 & 0 \\ 0 & 0 & 1 & -1 & -1 & 1 \end{bmatrix}$$

Add row 3 to row 1; subtract $2 \times$ row 3 from row 2.

$$\begin{bmatrix} 1 & 0 & 0 & 1 & -2 & 1 \\ 0 & 1 & 0 & 1 & 3 & -2 \\ 0 & 0 & 1 & -1 & -1 & 1 \end{bmatrix}$$ This is $[I, A^{-1}]$.

The elimination process used in finding the inverse differs from the one used for solving systems in that we combine each row with rows both below and above it to obtain the I on the left. This version is called *Gauss–Jordan* elimination. Its benefit is that it makes back-substitution unnecessary.

Checking

As with ordinary elimination, it is convenient to use a check column to help catch errors as they are made. The initial check column contains row sums. It is operated on just as the rest of its row. At every stage of elimination, each element in the check column is the sum of the rest of the numbers in that row.

Example 4

Compute the inverse, using a check column, of a 3×3 matrix.

The matrix appears as the left half of the 3×6 matrix $[A, I]$. To the right are the numbers of the check column.

$$\left[\begin{array}{cccccc} 0 & 1 & 0 & 1 & 0 & 0 \\ -1 & 0 & 1 & 0 & 1 & 0 \\ 1 & -2 & 1 & 0 & 0 & 1 \end{array}\right] \begin{array}{c} 2 \\ 1 \\ 1 \end{array}$$

Interchange rows 1 and 2.
Divide (new) row 1 by -1.
Subtract row 1 from row 3.

$$\left[\begin{array}{cccccc} 1 & 0 & -1 & 0 & -1 & 0 \\ 0 & 1 & 0 & 1 & 0 & 0 \\ 0 & -2 & 2 & 0 & 1 & 1 \end{array}\right] \begin{array}{c} -1 \\ 2 \\ 2 \end{array}$$

Add $2 \times$ row 2 to row 3.

$$\left[\begin{array}{cccccc} 1 & 0 & -1 & 0 & -1 & 0 \\ 0 & 1 & 0 & 1 & 0 & 0 \\ 0 & 0 & 2 & 2 & 1 & 1 \end{array}\right] \begin{array}{c} -1 \\ 2 \\ 6 \end{array}$$

Divide row 3 by 2. Add (new) row 3 to row 1.

$$\left[\begin{array}{cccccc} 1 & 0 & 0 & 1 & -\frac{1}{2} & \frac{1}{2} \\ 0 & 1 & 0 & 1 & 0 & 0 \\ 0 & 0 & 1 & 1 & \frac{1}{2} & \frac{1}{2} \end{array}\right] \begin{array}{c} 2 \\ 2 \\ 3 \end{array}$$

The right half of the last matrix is indeed A^{-1}, as direct multiplication confirms. Note that each operation was carried out on an entire row, including number in the check column, and that at every stage, each number in the check column is the sum of the numbers to its left.

Exercises

Find the inverses of the matrices given in Exercises 1–13.

1. $\begin{bmatrix} 1 & 1 \\ 1 & -1 \end{bmatrix}$

2. $\begin{bmatrix} 1 & 1 \\ 1 & 2 \end{bmatrix}$

3. $\begin{bmatrix} 10 & 9 \\ 11 & 10 \end{bmatrix}$

4. $\begin{bmatrix} 1 & 1 & 0 \\ 0 & 1 & 1 \\ 0 & 0 & 1 \end{bmatrix}$

5. $\begin{bmatrix} 1 & 2 & 1 \\ 1 & 3 & 0 \\ 1 & 0 & 4 \end{bmatrix}$

6. $\begin{bmatrix} 20 & 3 & -4 \\ -7 & 0 & 1 \\ -4 & -1 & 1 \end{bmatrix}$

$$7. \begin{bmatrix} -2 & 1 & 3 \\ 3 & -1 & 2 \\ 3 & -1 & 1 \end{bmatrix} \qquad 8. \begin{bmatrix} 1 & 1 & 1 \\ 5 & 4 & 8 \\ 5 & 4 & 7 \end{bmatrix} \qquad 9. \begin{bmatrix} 1 & -4 & 5 \\ 3 & -11 & 13 \\ 0 & 1 & -1 \end{bmatrix}$$

$$10. \begin{bmatrix} 1 & 0 & 1 & 0 \\ 0 & 1 & 0 & 1 \\ 0 & 1 & 0 & -1 \\ 1 & 0 & -1 & 0 \end{bmatrix} \qquad 11. \begin{bmatrix} 1 & 2 & 1 & 1 \\ 1 & 1 & -1 & -1 \\ 1 & -1 & -1 & 1 \\ 1 & -2 & 1 & -1 \end{bmatrix}$$

$$12. \begin{bmatrix} 1 & -1 & 0 & 0 \\ -1 & 2 & -1 & 0 \\ 0 & -1 & 2 & -1 \\ 0 & 0 & -1 & 2 \end{bmatrix} \qquad 13. \begin{bmatrix} 2 & -2 & 0 & 0 \\ -1 & 2 & -1 & 0 \\ 0 & -1 & 2 & -1 \\ 0 & 0 & -1 & 2 \end{bmatrix}$$

14. Let B and C be matrices of the same shape, such that $I + C^T B$ is an invertible matrix. Verify this formula by multiplying out:

$$(I + BC^T)^{-1} = I - B(I + C^T B)^{-1} C^T.$$

15. Use the formula above to find the inverse of $I + ee^T$, where e is an $n \times 1$ matrix of 1's.

16. (a) Verify that the matrices below are inverses of each other:

$$\begin{bmatrix} 2 & -1 & 0 & 0 \\ -1 & 2 & -1 & 0 \\ 0 & -1 & 2 & -1 \\ 0 & 0 & -1 & 2 \end{bmatrix}, \quad \begin{bmatrix} 0.8 & 0.6 & 0.4 & 0.2 \\ 0.6 & 1.2 & 0.8 & 0.4 \\ 0.4 & 0.8 & 1.2 & 0.6 \\ 0.2 & 0.4 & 0.6 & 0.8 \end{bmatrix}.$$

(b) Use the inverse to solve the system of equations:

$$2x_1 - x_2 \qquad = 1,$$
$$-x_1 + 2x_2 - x_3 = 0,$$
$$-x_2 + 2x_3 - x_4 = 1,$$
$$- x_3 + 2x_4 = 2,$$

17. Suppose that A is a square matrix and $I - A$ is invertible. If $B = (I + A)(I - A)^{-1}$, find A in terms of B. What matrix do you have to assume invertible to find A?

A.5

General Systems

A general system of m equations in n unknowns may have no solutions, one solution, or an infinite number of solutions. In the last case we seek a *general solution*, one that includes all particular solutions.

Example 1

The system of two equations in three unknowns

$$x + 3y - z = 2,$$
$$y + z = 1$$

has an infinite number of solutions. The third unknown, z, can be assigned any value, and then x and y can be determined. Direct substitution shows that

$$x = -1 + 4z, \qquad y = 1 - z, \qquad z = z$$

is a solution for any value of z, and clearly every particular solution can be obtained by specifying z.

Row-Echelon Form

The method of solution for a general system is to use elementary row operations to transform the coefficient matrix into a form that reveals any inconsistency and, in the case of a consistent system, allows the general solution to be obtained easily.

We say that a matrix is a *row-echelon* matrix or is in row-echelon form if it meets the following requirements.

1. Any zero rows are below all nonzero rows.

2. In each nonzero row the first nonzero element (reading left to right) is a 1. This is called a *leading* 1.

3. In each nonzero row after the first, the leading 1 is to the right of the leading 1 in the row above it.

Example 2

Each of the matrices below is a row-echelon matrix. A cross denotes an element whose value is immaterial:

$$\begin{bmatrix} 0 & 1 & \times & \times \\ 0 & 0 & 0 & 0 \end{bmatrix} \qquad \begin{bmatrix} 1 & \times & \times & 1 \\ 0 & 0 & 1 & \times \end{bmatrix}$$

$$\begin{bmatrix} 1 & \times & \times \\ 0 & 1 & \times \\ 0 & 0 & 0 \end{bmatrix} \qquad \begin{bmatrix} 1 & \times & \times & \times & \times \\ 0 & 0 & 1 & \times & \times \\ 0 & 0 & 0 & 1 & 0 \end{bmatrix} \qquad \begin{bmatrix} 1 & \times & \times \\ 0 & 1 & \times \\ 0 & 0 & 1 \\ 0 & 0 & 0 \\ 0 & 0 & 0 \end{bmatrix}.$$

The augmented matrix of the system in Example 1 is also a row-echelon matrix.

Any matrix can be transformed by elementary row operations into a row-echelon matrix. The procedure, called *row reduction,* is about like the Gaussian elimination used previously, except that we must move to the right when we encounter a column of 0's.

Example 3

Transform the matrix below into a row-echelon matrix:

$$\begin{bmatrix} 2 & 4 & -2 & -2 & 0 \\ -1 & -2 & 1 & 5 & -4 \\ 2 & 4 & -4 & -4 & -4 \\ 1 & 2 & -3 & 1 & -8 \end{bmatrix}.$$

First, divide row 1 by 2 to create the leading 1 in that row. Add appropriate multiples to rows below to create 0's in the rest of the first column:

$$\begin{bmatrix} 1 & 2 & -1 & -1 & 0 \\ 0 & 0 & 0 & 4 & -4 \\ 0 & 0 & -2 & -2 & -4 \\ 0 & 0 & -2 & 2 & -8 \end{bmatrix}.$$

Clearly, we cannot obtain a leading 1 in the $(2, 2)$ position by any row operations. But by interchanging row 2 with row 3 we obtain

$$\begin{bmatrix} 1 & 2 & -1 & -1 & 0 \\ 0 & 0 & -2 & -2 & -4 \\ 0 & 0 & 0 & 4 & -4 \\ 0 & 0 & -2 & 2 & -8 \end{bmatrix}.$$

Divide row 2 by -2 to get a leading 1 in the $(2, 3)$ position; then add 2 times row 2 to row 4:

$$\begin{bmatrix} 1 & 2 & -1 & -1 & 0 \\ 0 & 0 & 1 & 1 & 2 \\ 0 & 0 & 0 & 4 & -4 \\ 0 & 0 & 0 & 4 & -4 \end{bmatrix}.$$

Now divide row 3 by 4 and subtract 4 times row 3 from row 4 to obtain this row-echelon matrix:

$$\begin{bmatrix} 1 & 2 & -1 & -1 & 0 \\ 0 & 0 & 1 & 1 & 2 \\ 0 & 0 & 0 & 1 & -1 \\ 0 & 0 & 0 & 0 & 0 \end{bmatrix}.$$

For an $m \times n$ row-echelon matrix R, let r be the number of nonzero rows and let $l(1), l(2), \ldots, l(r)$ be the locations of the leading 1's. The last matrix in the example above is a 4×5 row-echelon matrix for which $r = 3$ and $l(1) = 1$, $l(2) = 3$, $l(3) = 4$.

Algorithm

We can write out an algorithm for reducing a given $m \times n$ matrix to a row-echelon matrix. Suppose the matrix is nonzero and, in order to get started, take $k = 1$ and $l(0) = 0$.

1. Consider the part of the matrix from row k to row m and from column $l(k - 1) + 1$ to column n. Find the first column in this block that is nonzero and record its number as $l(k)$. If there is none, decrement k and stop.

2. Search column $l(k)$ from row k to row m for the first nonzero element. Interchange rows, if necessary, to put this element in row k so that there is a nonzero element in the $(k, l(k))$ position.

3. Divide through row k by the element in the $(k, l(k))$ position.

4. If $k = m$, stop. If $k < m$, add (or subtract) multiples of row k to rows $k + 1, \ldots, m$ to create 0's below the $(k, l(k))$ position. If $l(k) = n$, stop.

5. Increment k and return to step 1.

The algorithm above will be stopped by an instruction in step 1 or step 4. When it stops, the matrix is in row-echelon form, the value of k is r (the number of nonzero rows), and the positions of the leading 1's are recorded as $l(1), l(2), \ldots, l(r)$.

Solution of Systems

The method for solving a general system is to reduce the augmented matrix of the system to a row-echelon matrix. Since this is done by row operations, the solution of the original system is also a solution of the system whose augmented matrix is in row-echelon form. Furthermore, the latter system is easily solved.

Example 4

Solve the system of four equations in four unknowns whose augmented matrix is

$$\begin{bmatrix} 2 & -2 & 0 & 2 & 2 \\ 1 & -1 & 2 & 5 & -1 \\ -1 & 1 & 1 & 2 & -3 \\ 2 & -2 & 3 & 9 & -2 \end{bmatrix}.$$

First row-reduce the augmented matrix to the row-echelon matrix

$$\begin{bmatrix} 1 & -1 & 0 & 1 & 1 \\ 0 & 0 & 1 & 2 & -1 \\ 0 & 0 & 0 & 1 & -1 \\ 0 & 0 & 0 & 0 & 0 \end{bmatrix},$$

in which $r = 3$, $l(1) = 1$, $l(2) = 3$, $l(3) = 4$. The first three rows of this matrix give three equations:

$$\begin{aligned} x_1 - x_2 \quad + \quad x_4 &= 1, \\ x_3 + 2x_4 &= -1, \\ x_4 &= -1. \end{aligned}$$

The fourth row translates into the equation $0 = 0$, which contains no information and can be discarded.

The last two equations determine $x_4 = -1$ and $x_3 = 1$. The first equation becomes $x_1 - x_2 = 2$. We may choose x_2 arbitrarily and determine x_1 in terms of x_2. The general solution is thus

$$x_1 = 2 + x_2, \qquad x_2 = x_2, \qquad x_3 = 1, \qquad x_4 = -1.$$

Example 5

Solve the system of three equations in three unknowns whose augmented matrix is

$$\begin{bmatrix} 1 & 1 & 1 & 2 \\ 1 & 2 & 3 & 3 \\ 2 & 5 & 8 & 8 \end{bmatrix}.$$

The given matrix is row-reduced to the row-echelon matrix

$$\begin{bmatrix} 1 & 1 & 1 & 2 \\ 0 & 1 & 2 & 1 \\ 0 & 0 & 0 & 1 \end{bmatrix}.$$

This is the augmented matrix of the following three equations:

$$\begin{aligned} x_1 + x_2 + x_3 &= 2, \\ x_2 + x_3 &= 1, \\ 0 &= 1. \end{aligned}$$

Since the last equation cannot be satisfied by any choice of x's, the original system was inconsistent. There is no solution.

We are now in a position to give a description of the process for solving a system of m equations in n unknowns, $Ax = b$.

1. Form the augmented matrix $[A, b]$ and transform it by row operations to $[R, f]$, where R is a row-echelon matrix having r nonzero rows.

2. If $r < m$, examine $f_{r+1}, f_{r+2}, \ldots, f_m$. If any of these is nonzero, the system is inconsistent, and there is no solution.

3. The r unknowns whose subscripts are $l(1), l(2), \ldots, l(r)$ are found in terms of the other $n - r$ unknowns, which are arbitrary.

Example 6

Find the value of k for which the given system is consistent, and find the general solution for that value:

$$[A, b] = \begin{bmatrix} 1 & 1 & 1 & 2 & 1 \\ 1 & 2 & 3 & 3 & 0 \\ 2 & 5 & 8 & 7 & k \end{bmatrix}.$$

First transform this matrix by row operations until the first four columns are a row-echelon matrix. The result is

$$[R, f] = \begin{bmatrix} 1 & 1 & 1 & 2 & 1 \\ 0 & 1 & 2 & 1 & -1 \\ 0 & 0 & 0 & 0 & k + 1 \end{bmatrix}.$$

The matrix R has $r = 2$ nonzero rows and $l(1) = 1$, $l(2) = 2$.

If $k + 1 \neq 0$, the system is inconsistent. If $k + 1 = 0$ (that is, $k = -1$), the system is consistent. Using this value of k, we have $r = 2$ significant equations, which are

$$x_1 + x_2 + x_3 + 2x_4 = 1,$$
$$x_2 + 2x_3 + x_4 = -1.$$

In solving these, x_1 and x_2 (subscripts $l(1)$ and $l(2)$) are to be found in terms of the $n - r = 2$ others, x_3 and x_4, which may be chosen arbitrarily. The general solution is thus

$$x_1 = 2 + x_3 - x_4, \qquad x_2 = -1 - 2x_3 - x_4, \qquad x_3 = x_3, \qquad x_4 = x_4.$$

Checking

As in the case of systems with unique solutions, the best check on a proposed solution is substitution into the original equations, and the next best is substitution into the most altered equation(s). The idea of the check column is valid any time row operations are being used and is highly recommended.

Exercises

1. Which of the following are row-echelon matrices, and where do the others fail?

(a) $\begin{bmatrix} 0 & 0 \\ 0 & 0 \end{bmatrix}$ (b) $\begin{bmatrix} 0 & 0 & 1 \\ 0 & 0 & 1 \end{bmatrix}$ (c) $\begin{bmatrix} 0 & 2 & 0 \\ 0 & 0 & 1 \end{bmatrix}$ (d) $\begin{bmatrix} 1 & 0 & 1 \\ 0 & 0 & 1 \end{bmatrix}$

(e) $\begin{bmatrix} 1 & 0 & 0 \\ 0 & 0 & 1 \\ 0 & 0 & 0 \end{bmatrix}$ (f) $\begin{bmatrix} 1 & -5 \\ 0 & 1 \\ 0 & 0 \end{bmatrix}$ (g) $\begin{bmatrix} 1 & 0 & 1 \\ 0 & 1 & 0 \\ 1 & 1 & 1 \end{bmatrix}$ (h) $\begin{bmatrix} 0 & 1 & 0 \\ 0 & 0 & 0 \\ 0 & 0 & 1 \end{bmatrix}$

2. Use row operations to reduce to row-echelon form the matrices in Exercise 1 that are not already row-echelon matrices.

In Exercises 3–15 you are given the augmented matrix of a system, with the coefficient matrix in row-echelon form. (a) Determine whether the system is consistent. If it is, (b) state how many of the unknowns are arbitrary, and (c) find the general solution of the system.

3. $\begin{bmatrix} 1 & 0 & 1 & 2 \\ 0 & 1 & -1 & 1 \\ 0 & 0 & 0 & 0 \end{bmatrix}$ 4. $\begin{bmatrix} 1 & 1 & 1 & 2 \\ 0 & 1 & 1 & 0 \\ 0 & 0 & 1 & 2 \end{bmatrix}$ 5. $\begin{bmatrix} 1 & 0 & 1 & 0 \\ 0 & 0 & 1 & 0 \\ 0 & 0 & 0 & 1 \end{bmatrix}$

6. $\begin{bmatrix} 1 & 1 & 1 & 2 \\ 0 & 0 & 1 & 0 \\ 0 & 0 & 0 & 0 \end{bmatrix}$ 7. $\begin{bmatrix} 0 & 1 & 1 & 2 \\ 0 & 0 & 1 & 1 \end{bmatrix}$ 8. $\begin{bmatrix} 1 & 2 & 1 \\ 0 & 1 & 2 \\ 0 & 0 & 0 \end{bmatrix}$

9. $\begin{bmatrix} 1 & 1 & 0 & 1 & 0 \\ 0 & 0 & 1 & -2 & 1 \\ 0 & 0 & 0 & 0 & 0 \end{bmatrix}$ 10. $\begin{bmatrix} 1 & 1 & 0 & 1 & 0 \\ 0 & 0 & 1 & -2 & 1 \\ 0 & 0 & 0 & 1 & 0 \end{bmatrix}$

11. $\begin{bmatrix} 1 & 2 & 1 & 0 \\ 0 & 1 & 1 & 0 \\ 0 & 0 & 0 & 0 \end{bmatrix}$ 12. $\begin{bmatrix} 1 & 4 & 0 & 1 & 3 \\ 0 & 1 & 2 & 1 & 0 \\ 0 & 0 & 1 & 2 & 2 \\ 0 & 0 & 0 & 1 & 1 \\ 0 & 0 & 0 & 0 & 1 \end{bmatrix}$

13. $\begin{bmatrix} 1 & 2 & 0 & -1 & 1 & 0 \\ 0 & 0 & 1 & 0 & 1 & 0 \\ 0 & 0 & 0 & 1 & 2 & -1 \\ 0 & 0 & 0 & 0 & 1 & 0 \\ 0 & 0 & 0 & 0 & 0 & 0 \end{bmatrix}$

14. $\begin{bmatrix} 1 & 1 & 0 & 0 & 1 & 1 & 0 & 1 & 0 \\ 0 & 0 & 0 & 1 & 0 & 0 & 0 & 1 & 0 \\ 0 & 0 & 0 & 0 & 1 & 0 & 0 & 0 & 1 \end{bmatrix}$

15. $\begin{bmatrix} 1 & -1 & 0 & 1 & 0 & 0 & 0 & 1 \\ 0 & 0 & 1 & 0 & 1 & 1 & 0 & 1 \\ 0 & 0 & 0 & 0 & 0 & 0 & 1 & 0 \end{bmatrix}$

In Exercises 16–21, find the value(s) of k for which the system is consistent and find the general solution using that value of k.

16. $\begin{bmatrix} 1 & 2 & 1 & 0 \\ 1 & 2 & 2 & 1 \\ 2 & 4 & 5 & k \end{bmatrix}$
    17. $\begin{bmatrix} 1 & 1 & 1 & 1 \\ 1 & 2 & 2 & 2 \\ 3 & 4 & 4 & k \end{bmatrix}$
    18. $\begin{bmatrix} 1 & 1 & 2 & 0 \\ 2 & 1 & 2 & k \\ 0 & 1 & 2 & 1 \end{bmatrix}$

19. $\begin{bmatrix} 2 & 1 & 1 & k \\ 1 & 2 & 2 & 1 \\ 3 & 3 & 3 & k \end{bmatrix}$
    20. $\begin{bmatrix} 1 & 1 & 2 \\ 2 & 1 & 3 \\ 1 & k & 2 \end{bmatrix}$
    21. $\begin{bmatrix} 1 & 0 & 1 & 1 \\ 0 & 1 & 0 & 2 \\ 2 & 2 & k & 0 \end{bmatrix}$

In Exercises 22–30, use row reduction to find the general solution, if the system is consistent. In each case the augmented matrix of the system is given.

22. $\begin{bmatrix} 1 & 1 & 1 & 1 \\ 1 & 2 & 3 & 0 \\ 4 & 5 & 6 & 3 \end{bmatrix}$
    23. $\begin{bmatrix} 1 & 2 & 3 & 2 \\ 2 & 5 & 8 & 2 \\ 5 & 8 & 11 & 14 \end{bmatrix}$

24. $\begin{bmatrix} 1 & 1 & 1 \\ 2 & 2 & 3 \\ 3 & 3 & 4 \end{bmatrix}$
    25. $\begin{bmatrix} 0 & 1 & 1 & 1 \\ 0 & 2 & 1 & 5 \end{bmatrix}$

26. $\begin{bmatrix} 1 & 2 & -1 & -3 \\ -3 & -6 & 3 & 9 \end{bmatrix}$
    27. $\begin{bmatrix} 2 & 1 & -1 & 3 \\ 0 & 1 & 2 & 1 \\ 4 & 2 & -2 & 7 \end{bmatrix}$

28. $\begin{bmatrix} 1 & 2 & 1 & 3 & 1 \\ 1 & 2 & 1 & 4 & 0 \\ 2 & 4 & 2 & 5 & 3 \\ 1 & 3 & 1 & 2 & 4 \end{bmatrix}$
    29. $\begin{bmatrix} 1 & 0 & 1 & 1 \\ 0 & 1 & 1 & 2 \\ 0 & 1 & 1 & 2 \\ 1 & 0 & 1 & 1 \end{bmatrix}$

30. $\begin{bmatrix} 1 & 1 & 2 \\ 1 & -1 & 0 \\ 1 & 1 & 2 \\ 1 & -1 & 0 \end{bmatrix}$

31. Prove that the system of equations $Ax = 0$ is consistent, whatever matrix A may be.

A.6

Rank

If a given matrix A can be transformed by a sequence of elementary row operations into a row-echelon matrix R, the *rank* of A is defined to be the number r of nonzero rows in R: rank $A = r$. It can be proved that the rank of A does not depend on the particular sequence of operations used.

The concept of rank is closely associated with the solution of systems. In Section A.5 we specified a test to decide whether a system $Rx = f$, with coefficient matrix R in row-echelon form, has a solution or not. In terms of rank we can say: $Rx = f$ is consistent if and only if rank $[R, f]$ = rank R. This translates into the following theorem.

Theorem A.2

The system $Ax = b$ is consistent if and only if rank $[A, b]$ = rank A.

　　Proof: It is necessary only to note that if $[A, b]$ is row-reduced into $[R, f]$, where R is a row-echelon matrix, then rank A = rank R and rank $[A, b]$ = rank $[R, f]$.

Example 1

Determine the value of k that makes the system $Ax = b$ consistent:

$$[A, b] = \begin{bmatrix} 1 & 2 & 1 & 1 & 0 \\ 0 & 1 & 1 & 2 & -1 \\ 1 & 3 & 2 & 4 & 1 \\ 1 & 4 & 3 & 6 & k \end{bmatrix}.$$

The augmented matrix above can be row-reduced to

$$[R, f] = \begin{bmatrix} 1 & 2 & 1 & 1 & 0 \\ 0 & 1 & 1 & 2 & -1 \\ 0 & 0 & 0 & 1 & 2 \\ 0 & 0 & 0 & 0 & k \end{bmatrix}.$$

Now rank A = rank R = 3, and rank $[A, b]$ = rank $[R, f]$ = 3 if $k = 0$. But rank $[A, b]$ = 4 if $k \neq 0$. Thus the system is consistent if and only if $k = 0$.

The rank has a further role, in determining how many solutions a consistent system has.

Theorem A.3

If A is an $m \times n$ matrix and the system $Ax = b$ is consistent, there is exactly one solution if rank $A = n$, or infinitely many solutions if rank $A < n$.

This theorem is also just a translation of our observations of the last section. If rank $A = r < n$, then r of the unknowns are determined in terms of the other $n - r$, which are arbitrary. Sometimes we say that the solution has $n - r$ *degrees of freedom*.

Example 2	Determine how many solutions there are to the system in Example 1 when it is consistent.

Taking $k = 0$ to obtain consistency, we have rank $A = 3$ and $n = 4$, so there is an infinite number of solutions (one degree of freedom).

Homogeneous Systems

A system of the form $Ax = 0$ is said to be *homogeneous*. It is always consistent: $x = 0$ is a solution, called the *trivial* solution. The key question about a homogeneous system is whether it has any nontrivial solutions. The answer again comes in terms of rank.

Theorem A.4	If A is an $m \times n$ matrix, the homogeneous system $Ax = 0$ has a nontrivial solution if and only if rank $A < n$.

Proof: This theorem is just an application of Theorem A.3 to the consistent system $Ax = 0$. If rank $A = n$, the trivial solution is unique. If rank $A < n$, there are infinitely many nontrivial solutions.

Example 3	Determine whether $Ax = 0$ has nontrivial solutions, for

$$A = \begin{bmatrix} 1 & 2 & 1 & 1 \\ 0 & 1 & 1 & 2 \\ 1 & 3 & 2 & 4 \\ 1 & 4 & 3 & 6 \end{bmatrix}.$$

Row reduction of this matrix leads to the row-echelon matrix

$$\begin{bmatrix} 1 & 2 & 1 & 1 \\ 0 & 1 & 1 & 2 \\ 0 & 0 & 0 & 1 \\ 0 & 0 & 0 & 0 \end{bmatrix}.$$

Since rank $A = 3 < n = 4$, there is a nontrivial solution. We may choose x_3 arbitrarily. Then $x_4 = 0$, $x_2 = -x_3$, $x_1 = x_3$ is the general solution.

Linear Independence

Rank can also be used to answer questions about linear independence of column matrices. Given some $m \times 1$ column matrices, $z_1, z_2, \ldots, z_n$, they are independent if the only linear combination equal to 0 has 0 coefficients: that is, if $c_1z_1 + c_2z_2 + \cdots + c_nz_n = 0$ implies $c_1 = c_2 = \cdots = c_n = 0$. A question about linear independence of column matrices turns into a question about nontrivial solutions of a homogeneous system, as this example shows.

Example 4

Determine whether these column matrices are independent or dependent:

$$z_1 = \begin{bmatrix} 1 \\ 2 \\ 1 \end{bmatrix}, \qquad z_2 = \begin{bmatrix} 1 \\ 0 \\ -1 \end{bmatrix}, \qquad z_3 = \begin{bmatrix} 1 \\ -2 \\ 1 \end{bmatrix}.$$

First form a general linear combination and equate it to 0:

$$c_1z_1 + c_2z_2 + c_3z_3 = c_1 \begin{bmatrix} 1 \\ 2 \\ 1 \end{bmatrix} + c_2 \begin{bmatrix} 1 \\ 0 \\ -1 \end{bmatrix} + c_3 \begin{bmatrix} 1 \\ -2 \\ 1 \end{bmatrix} = 0.$$

By the definition of the matrix operations we must have

$$\begin{aligned} c_1 + c_2 + c_3 &= 0, \\ 2c_1 \quad - 2c_3 &= 0, \\ c_1 - c_2 + c_3 &= 0, \end{aligned} \qquad \text{or} \qquad \begin{bmatrix} 1 & 1 & 1 \\ 2 & 0 & -2 \\ 1 & -1 & 1 \end{bmatrix} \begin{bmatrix} c_1 \\ c_2 \\ c_3 \end{bmatrix} = 0.$$

The question is whether all c's are 0: is there a nontrivial solution of this homogeneous system?

Row reduction of the coefficient matrix of the system above reveals that its rank is 3. Since $n = 3$ also, Theorem A.4 tells us that there is no nontrivial solution of the system. The given columns are independent.

We can summarize the observations above as follows. To determine whether the $m \times 1$ column matrices $z_1, z_2, \ldots, z_n$ are independent, form the $m \times n$ matrix Z whose columns are the given columns. Then $c_1z_1 + \cdots + c_nz_n = 0$ if and only if

$$Zc = 0 \qquad (c = [c_1, c_2, \ldots, c_n]^T).$$

Thus a nontrivial linear combination of the z's is 0 if and only if $Zc = 0$ has a nontrivial solution.

Theorem A.5

Let $z_1, z_2, \ldots, z_n$ be given $m \times 1$ column matrices and let Z be the $m \times n$ matrix whose columns are the given columns.

1. The z's are independent if and only if rank $Z = n$.
2. There is a set of $r = $ rank Z of the z's that are independent. Any larger set of z's cannot be independent.

3. If $l(1), l(2), \ldots, l(r)$ are determined as before from a row-echelon matrix R to which Z is row-reduced, then $z_{l(1)}, z_{l(2)}, \ldots, z_{l(r)}$ are r independent columns.

Example 5

Given the five column matrices below, determine whether they are independent or not. If not, determine the largest number of them that are independent and which ones.

$$z_1 = \begin{bmatrix} 1 \\ -1 \\ 2 \\ 1 \end{bmatrix}, \quad z_2 = \begin{bmatrix} 1 \\ 2 \\ 2 \\ -2 \end{bmatrix}, \quad z_3 = \begin{bmatrix} 2 \\ 1 \\ 4 \\ -1 \end{bmatrix},$$

$$z_4 = \begin{bmatrix} 1 \\ -3 \\ 2 \\ 2 \end{bmatrix}, \quad z_5 = \begin{bmatrix} 1 \\ 1 \\ 2 \\ 1 \end{bmatrix}.$$

The matrix Z that has the given matrices as columns is below. Next to it is a row-echelon matrix obtained from Z by row operations.

$$Z = \begin{bmatrix} 1 & 1 & 2 & 1 & 1 \\ -1 & 2 & 1 & -3 & 1 \\ 2 & 2 & 4 & 2 & 2 \\ 1 & -2 & -1 & 2 & 1 \end{bmatrix}, \quad R = \begin{bmatrix} 1 & 1 & 2 & 1 & 1 \\ 0 & 1 & 1 & -1 & 0 \\ 0 & 0 & 0 & 1 & 2 \\ 0 & 0 & 0 & 0 & 0 \end{bmatrix}.$$

Since $r = \operatorname{rank} Z = 3$, but $n = 5$, the given matrices are not independent. Theorem A.5 then tells us that there is a set of 3 z's that are independent.

Since $l(1) = 1$, $l(2) = 2$, $l(3) = 4$, we are assured that columns z_1, z_2, and z_4 are independent. To confirm this, we form the matrix having three z's as columns and verify that its rank is 3:

$$\begin{bmatrix} 1 & 1 & 1 \\ -1 & 2 & -3 \\ 2 & 2 & 2 \\ 1 & -2 & 2 \end{bmatrix} \rightarrow \begin{bmatrix} 1 & 1 & 1 \\ 0 & 1 & -1 \\ 0 & 0 & 1 \\ 0 & 0 & 0 \end{bmatrix}.$$

Exercises

1. The augmented matrices of several systems are given below. For each one, find the rank of the coefficient matrix and the rank of the augmented matrix.

(a) $\begin{bmatrix} 1 & 2 & 1 \\ 0 & 1 & 1 \\ 0 & 0 & 1 \end{bmatrix}$
(b) $\begin{bmatrix} 2 & 1 & 3 \\ 2 & 1 & 4 \\ 0 & 0 & 0 \end{bmatrix}$

(c) $\begin{bmatrix} 1 & 3 & 1 & 2 \\ 4 & 1 & 4 & -3 \\ -1 & 2 & -1 & 3 \end{bmatrix}$
(d) $\begin{bmatrix} 4 & 2 & 3 \\ 7 & 1 & 4 \\ 5 & 3 & 4 \end{bmatrix}$

(e) $\begin{bmatrix} 1 & 1 & 0 & 1 & 2 \\ 2 & 1 & 3 & 1 & 4 \\ -1 & 0 & -3 & 0 & -2 \end{bmatrix}$
(f) $\begin{bmatrix} 1 & 2 & 3 & 4 \\ 5 & 6 & 7 & 8 \\ 0 & 0 & 1 & 0 \end{bmatrix}$

2. Find the number of arbitrary unknowns in the solution of each of the consistent systems of Exercise 1.

3. Find the general solution for each of the consistent systems in Exercise 1.

4. Find the rank of A and of $[A, b]$ and find the general solution of $Ax = b$ when the system is consistent:

$$[A, b] = \begin{bmatrix} 1 & 1 & k \\ 1 & k & 1 \\ k & 1 & 1 \end{bmatrix}.$$

In Exercises 5–11, obtain the general solution of the homogeneous system $Ax = 0$. In each case, A is given

5. $\begin{bmatrix} 0 & 1 & 1 \\ 0 & 0 & 1 \\ 0 & 0 & 0 \end{bmatrix}$
6. $\begin{bmatrix} 1 & 2 & 1 \\ 0 & 1 & 0 \\ 1 & 0 & 1 \end{bmatrix}$

7. $\begin{bmatrix} 1 & 1 & 1 \\ 1 & 2 & 3 \end{bmatrix}$
8. $\begin{bmatrix} 0 & 1 & 0 & 1 \\ 1 & 0 & 1 & 0 \\ 0 & 1 & 0 & 1 \\ 1 & 0 & 1 & 0 \end{bmatrix}$

9. $\begin{bmatrix} -2 & 2 & 0 & 0 \\ 1 & -2 & 1 & 0 \\ 0 & 1 & -2 & 1 \\ 0 & 0 & 2 & -2 \end{bmatrix}$
10. $\begin{bmatrix} 0 & 1 & 0 & 0 \\ -1 & 0 & 1 & 0 \\ 0 & 0 & 0 & 1 \\ 1 & 0 & -1 & 0 \end{bmatrix}$

11. $A = 4I - ee^T$, where $e^T = [1, 1, 1, 1]$.

In Exercises 12–17, (a) determine the largest number, r, of the given matrices that are independent; (b) choose r of them that are independent.

12. $\begin{bmatrix} 1 \\ 1 \\ 1 \\ 1 \end{bmatrix}, \begin{bmatrix} 1 \\ 2 \\ 2 \\ 1 \end{bmatrix}, \begin{bmatrix} -1 \\ 4 \\ 4 \\ -1 \end{bmatrix}, \begin{bmatrix} 0 \\ 1 \\ 1 \\ 0 \end{bmatrix}$
13. $\begin{bmatrix} 1 \\ 2 \\ 1 \\ 3 \end{bmatrix}, \begin{bmatrix} 1 \\ 2 \\ 1 \\ 4 \end{bmatrix}, \begin{bmatrix} 1 \\ 2 \\ 1 \\ 5 \end{bmatrix}, \begin{bmatrix} 2 \\ 4 \\ 2 \\ 6 \end{bmatrix}$

14. $\begin{bmatrix} 1 \\ 0 \\ 0 \end{bmatrix}, \begin{bmatrix} 1 \\ -1 \\ 1 \end{bmatrix}, \begin{bmatrix} 1 \\ 2 \\ 4 \end{bmatrix}$

15. $\begin{bmatrix} 1 \\ 1 \\ 1 \end{bmatrix}, \begin{bmatrix} 1 \\ 2 \\ 4 \end{bmatrix}, \begin{bmatrix} 1 \\ 3 \\ 9 \end{bmatrix}, \begin{bmatrix} 1 \\ -1 \\ 1 \end{bmatrix}$

16. $\begin{bmatrix} 2 \\ 1 \\ 3 \end{bmatrix}, \begin{bmatrix} 0 \\ 1 \\ 4 \end{bmatrix}, \begin{bmatrix} 1 \\ 0 \\ 0 \end{bmatrix}, \begin{bmatrix} 2 \\ 3 \\ 3 \end{bmatrix}$

17. $\begin{bmatrix} 1 \\ 1 \\ 1 \end{bmatrix}, \begin{bmatrix} 2 \\ 0 \\ -1 \end{bmatrix}, \begin{bmatrix} 2 \\ 2 \\ 2 \end{bmatrix}, \begin{bmatrix} 3 \\ 1 \\ 0 \end{bmatrix}$

In Exercises 18–21, express the last vector as a linear combination of the others, if possible.

18. $\begin{bmatrix} 1 \\ 2 \\ 1 \end{bmatrix}, \begin{bmatrix} 1 \\ 0 \\ 1 \end{bmatrix}; \begin{bmatrix} 2 \\ 2 \\ 2 \end{bmatrix}$

19. $\begin{bmatrix} 1 \\ 2 \\ 1 \end{bmatrix}, \begin{bmatrix} -1 \\ 0 \\ 1 \end{bmatrix}, \begin{bmatrix} 2 \\ 1 \\ 1 \end{bmatrix}; \begin{bmatrix} 1 \\ 0 \\ 1 \end{bmatrix}$

20. $\begin{bmatrix} 1 \\ 2 \\ 1 \\ 2 \end{bmatrix}, \begin{bmatrix} 0 \\ 1 \\ 0 \\ 1 \end{bmatrix}; \begin{bmatrix} 2 \\ 1 \\ 2 \\ 1 \end{bmatrix}$

21. $\begin{bmatrix} 1 \\ 1 \\ 1 \\ 1 \end{bmatrix}, \begin{bmatrix} 1 \\ 0 \\ -1 \\ 0 \end{bmatrix}, \begin{bmatrix} 0 \\ 1 \\ 0 \\ -1 \end{bmatrix}, \begin{bmatrix} 1 \\ -1 \\ 1 \\ -1 \end{bmatrix}; \begin{bmatrix} 0 \\ 1 \\ 1 \\ 0 \end{bmatrix}$

A.7

Determinant

In our discussion of the matrix inverse we saw a formula for the inverse of a general 2×2 matrix,

$$\begin{bmatrix} a & b \\ c & d \end{bmatrix}^{-1} = \frac{1}{ad - bc} \begin{bmatrix} d & -b \\ -c & a \end{bmatrix}.$$

The quantity in the denominator, which must be nonzero in order for the matrix to exist, is called the *determinant* of the 2×2 matrix:

$$\det \begin{bmatrix} a & b \\ c & d \end{bmatrix} = \begin{vmatrix} a & b \\ c & d \end{vmatrix} = ad - bc. \tag{A.14}$$

In general we define the *determinant* of an $n \times n$ matrix $A = [a_{ij}]$ as the number obtained by the following process, called *expansion by minors*.

1. Choose a row of A, say row i.
2. Let M_{ij} denote the determinant of the $(n-1) \times (n-1)$ matrix obtained from A by deleting the ith row and jth column. (M_{ij} is called the *minor* of a_{ij}.)
3. $\det A = a_{i1}(-1)^{i+1}M_{i1} + a_{i2}(-1)^{i+2}M_{i2} + \cdots + a_{in}(-1)^{i+n}M_{in}$.

It can be proved that this definition does not depend on which row is chosen.

Example 1　　　Evaluate the determinant of the 4×4 matrix

$$A = \begin{bmatrix} 1 & 2 & 3 & 4 \\ 3 & 0 & 0 & 0 \\ 5 & 2 & 1 & 2 \\ 7 & 0 & 1 & 0 \end{bmatrix}.$$

We are free to choose i. Since the elements of the ith row are coefficients in the expansion, choosing $i = 2$ will give many 0's and little work. By the definition, then,

$$\det A = 3(-1)^{2+1}M_{21} + 0 \cdot (-1)^{2+2}M_{22}$$
$$+ 0 \cdot (-1)^{2+3}M_{23} + 0 \cdot (-1)^{2+4}M_{24}.$$

The only minor that need be evaluated is M_{21}:

$$\det A = -3 \det \begin{bmatrix} 2 & 3 & 4 \\ 2 & 1 & 2 \\ 0 & 1 & 0 \end{bmatrix}.$$

Now expand the 3×3 determinant by its third row to get

$$\det A = -3 \cdot 1(-1)^{3+2} \det \begin{bmatrix} 2 & 4 \\ 2 & 2 \end{bmatrix}.$$

Finally, the 2×2 determinant can be evaluated by using the formula (A.14). Thus

$$\det A = (-3)(-1)(2 \cdot 2 - 4 \cdot 2) = 3(-4) = -12.$$

Vector Algebra

It is worth noting that the "expansion by minors" is used in calculating the cross product of geometrical vectors. If $\mathbf{a} = a_1\mathbf{i} + a_2\mathbf{j} + a_3\mathbf{k}$, and $\mathbf{b} = b_1\mathbf{i} + b_2\mathbf{j} + b_3\mathbf{k}$, their cross product is calculated as

$$\mathbf{a} \times \mathbf{b} = \begin{vmatrix} \mathbf{i} & \mathbf{j} & \mathbf{k} \\ a_1 & a_2 & a_3 \\ b_1 & b_2 & b_3 \end{vmatrix}.$$

(Strictly speaking, this is nonsense because there are vectors in the first row instead of numbers.) Expansion by minors of the first row gives

$$\mathbf{a} \times \mathbf{b} = (a_2b_3 - a_3b_2)\mathbf{i} - (a_1b_3 - a_3b_1)\mathbf{j} + (a_1b_2 - a_2b_1)\mathbf{k}.$$

Properties

The expansion by minors used in the definition is a computational disaster for $n > 4$. Several properties of determinants can be used to develop more efficient techniques.

D1. If A is a triangular matrix, then det $A = a_{11} \cdot a_{22} \cdots a_{nn}$. This property can be proved by expanding by minors of the row having the most 0's.

D2. If two rows of a matrix are interchanged, the sign of the determinant is changed.

D3. If all elements of a row are multiplied by a scalar, the determinant is multiplied by the same scalar.

D4. If a multiple of one row is added to another, the determinant is unchanged.

D5. det A^T = det A. Since the rows of A^T are the columns of A, this property means that properties D1–D4 remain valid if the word "row" is replaced by "column."

Example 2

Evaluate the determinant of the following matrix.

$$X = \begin{bmatrix} 1 & x & e^x & e^{2x} \\ 0 & 1 & e^x & 2e^{2x} \\ 0 & 0 & e^x & 4e^{2x} \\ 0 & 0 & e^x & 8e^{2x} \end{bmatrix}.$$

First use property D4 to factor out e^x from the third column and e^{2x} from the fourth.

$$\det X = e^x \cdot e^{2x} \det \begin{bmatrix} 1 & x & 1 & 1 \\ 0 & 1 & 1 & 2 \\ 0 & 0 & 1 & 4 \\ 0 & 0 & 1 & 8 \end{bmatrix}.$$

Next subtract row 3 from row 4:

$$\det X = e^{3x} \det \begin{bmatrix} 1 & x & 1 & 1 \\ 0 & 1 & 1 & 2 \\ 0 & 0 & 1 & 4 \\ 0 & 0 & 0 & 4 \end{bmatrix}.$$

Finally, use property D1 to evaluate the determinant of the triangular matrix. det $X = 4e^{3x}$.

It should be noted that we may freely alternate row and column operations in manipulating determinants, as was done in the example.

Practical Evaluation

Properties D2, D3, and D4 tell us what effects the row operations have on a determinant. Property D1 tells how to find the determinant of a triangular matrix. Consequently, we can evaluate a determinant by using the same routine as in solving systems, while keeping track of the effects of our operations. This procedure is often called *pivotal condensation*.

Example 3

Evaluate the determinant of the following matrix:

$$A = \begin{bmatrix} 2 & 2 & 2 & 2 \\ 1 & 4 & 1 & 4 \\ 1 & 2 & -1 & -2 \\ 1 & 8 & -1 & -8 \end{bmatrix}.$$

Below, we carry out the same sequence of row operations as in row reduction. The factor in front records any effect of the operations. The property used is noted in parentheses:

$$\det A = 2 \det \begin{bmatrix} 1 & 1 & 1 & 1 \\ 1 & 4 & 1 & 4 \\ 1 & 2 & -1 & -2 \\ 1 & 8 & -1 & -8 \end{bmatrix} \qquad \text{(D3)}$$

$$= 2 \cdot \det \begin{bmatrix} 1 & 1 & 1 & 1 \\ 0 & 3 & 0 & 3 \\ 0 & 1 & -2 & -3 \\ 0 & 7 & -2 & -9 \end{bmatrix} \qquad \text{(D4)}$$

$$= 6 \cdot \det \begin{bmatrix} 1 & 1 & 1 & 1 \\ 0 & 1 & 0 & 1 \\ 0 & 1 & -2 & -3 \\ 0 & 7 & -2 & -9 \end{bmatrix} \qquad \text{(D3)}$$

$$= 6 \det \begin{bmatrix} 1 & 1 & 1 & 1 \\ 0 & 1 & 0 & 1 \\ 0 & 0 & -2 & -4 \\ 0 & 0 & -2 & -16 \end{bmatrix} \qquad \text{(D4)}$$

$$= 6 \det \begin{bmatrix} 1 & 1 & 1 & 1 \\ 0 & 1 & 0 & 1 \\ 0 & 0 & -2 & -4 \\ 0 & 0 & 0 & -12 \end{bmatrix} \qquad \text{(D4)}$$

$$= 6(-2)(-12) = 144 \qquad \text{(D1)}.$$

Systems of Equations

Determinants play an important role in the theory of solutions of a system $Ax = b$, where A is a square matrix. The link to our previous work is through row operations. A sequence of row operations on the augmented matrix of a system does not change the existence or identity of the solution. On the other hand, a sequence of row operations on the coefficient matrix of a system cannot change its determinant from zero to nonzero or vice versa.

Theorem A.6

Let A be an $n \times n$ matrix.
1. $Ax = b$ has a unique solution if and only if $\det A \neq 0$.
2. A^{-1} exists if and only if $\det A \neq 0$.
3. $Ax = 0$ has a nontrivial solution if and only if $\det A = 0$.

A square matrix is said to be *singular* if its determinant is 0 and *nonsingular* otherwise. The theorem says: (1) $Ax = b$ has a unique solution if and only if A is nonsingular; (2) A is invertible if and only if A is nonsingular; (3) $Ax = 0$ has a nontrivial solution if and only if A is singular.

Example 4

For what value of k does the system below have a solution?

$$\begin{bmatrix} 1 & -1 & 2 \\ 2 & -1 & 0 \\ 3 & -2 & k \end{bmatrix} \begin{bmatrix} x_1 \\ x_2 \\ x_3 \end{bmatrix} = \begin{bmatrix} 3 \\ 3 \\ 6 \end{bmatrix}.$$

Expand the determinant of the coefficient matrix by the last column:

$$\det \begin{bmatrix} 1 & -1 & 2 \\ 2 & -1 & 0 \\ 3 & -2 & k \end{bmatrix} = 2(-4 + 3) + k(-1 + 2)$$
$$= -2 + k.$$

Thus if $k \neq 2$, there is a unique solution of the system, which could be symbolized as $x = A^{-1}b$, since the coefficient matrix is invertible if its determinant is not 0.

If $k = 2$, then the coefficient matrix has determinant 0. The theorem then tells us that there cannot be a *unique* solution. In fact, when $k = 2$, the general solution is

$$x_1 = 2x_3, \qquad x_2 = -3 + 4x_3, \qquad x_3 = x_3.$$

Example 5

Find all values of k for which $(A - kI)x = 0$ has nontrivial solutions:

$$A = \begin{bmatrix} 1 & 1 & 0 \\ 1 & 0 & 1 \\ 0 & 1 & 1 \end{bmatrix}.$$

According to Theorem A.6, $(A - kI)x = 0$ has nontrivial solutions if and only if its determinant is 0:

$$\det \begin{bmatrix} 1-k & 1 & 0 \\ 1 & -k & 1 \\ 0 & 1 & 1-k \end{bmatrix} = 0.$$

To simplify expansion, interchange rows 1 and 2, then rows 2 and 3. Two row interchanges leave the determinant unchanged, so

$$\det (A - kI) = \det \begin{bmatrix} 1 & -k & 1 \\ 0 & 1 & 1-k \\ 1-k & 1 & 0 \end{bmatrix}.$$

Now triangularize the matrix by row operations:

$$\det (A - kI) = \det \begin{bmatrix} 1 & -k & 1 \\ 0 & 1 & 1-k \\ 0 & p & k-1 \end{bmatrix}, \qquad p = 1 + k(1 - k),$$

$$= \det \begin{bmatrix} 1 & -k & 1 \\ 0 & 1 & 1-k \\ 0 & 0 & q \end{bmatrix}$$

$$= q = (k - 1) - p(1 - k).$$

When q is written out in terms of k, we obtain

$$\det (A - kI) = (k - 1)(2 + k - k^2).$$

The roots of this polynomial are $k = 1, 2$, and -1. Thus if k has any of these three values, $(A - kI)x = 0$ has a nontrivial solution x.

Exercises

In Exercises 1–9, use pivotal condensation to evaluate the determinant of the given matrix.

1. $\begin{bmatrix} 2 & 1 & 4 \\ 1 & 0 & 2 \\ 3 & 1 & 6 \end{bmatrix}$ 2. $\begin{bmatrix} 0 & 1 & 2 \\ 1 & 2 & 3 \\ 2 & 3 & 4 \end{bmatrix}$ 3. $\begin{bmatrix} 1 & \frac{1}{2} & \frac{1}{3} \\ \frac{1}{2} & \frac{1}{3} & \frac{1}{4} \\ \frac{1}{3} & \frac{1}{4} & \frac{1}{5} \end{bmatrix}$

4. $\begin{bmatrix} 1 & 1 & 2 \\ 2 & 1 & 5 \\ 0 & -1 & 2 \end{bmatrix}$ 5. $\begin{bmatrix} -7 & 4 & -3 \\ 4 & -2 & 1 \\ 2 & -1 & 1 \end{bmatrix}$ 6. $\begin{bmatrix} 1 & 1 & 1 \\ 1 & -1 & 2 \\ 1 & 1 & 4 \end{bmatrix}$

7. $\begin{bmatrix} 2 & -1 & 0 & 0 \\ -1 & 2 & -1 & 0 \\ 0 & -1 & 2 & -1 \\ 0 & 0 & -1 & 2 \end{bmatrix}$ 8. $\begin{bmatrix} 0 & 1 & 0 & 1 \\ 0 & 0 & 1 & 0 \\ 0 & 0 & 0 & 1 \\ 1 & 2 & 1 & 3 \end{bmatrix}$

9. $\begin{bmatrix} 1 & 2 & 0 & 0 \\ 2 & 1 & 0 & 0 \\ 3 & 5 & 2 & 3 \\ 7 & 8 & 3 & 4 \end{bmatrix}$

In Exercises 10–15, find the determinants of $A - xI$, and then find the values of the scalar x for which this determinant is 0. In each case, A is given.

10. $\begin{bmatrix} 4 & 2 \\ 0 & 3 \end{bmatrix}$ 11. $\begin{bmatrix} 0 & 3 \\ 4 & 2 \end{bmatrix}$ 12. $\begin{bmatrix} 1 & 3 \\ 3 & 1 \end{bmatrix}$

13. $\begin{bmatrix} 1 & 0 & 0 \\ 4 & 2 & 1 \\ 1 & 0 & 3 \end{bmatrix}$ 14. $\begin{bmatrix} 2 & -2 & 0 \\ -1 & 2 & -1 \\ 0 & -2 & 2 \end{bmatrix}$ 15. $\begin{bmatrix} 1 & 0 & -1 \\ 1 & -2 & 1 \\ 2 & -3 & 1 \end{bmatrix}$

The Wronskian of given functions $u_1(t), \ldots, u_n(t)$ is an $n \times n$ determinant with the functions $u_1 \cdots u_n$ in the first row, their derivatives in the second row, etc. Find the Wronskian for each set of functions.

16. $u_1 = e^{-t}, u_2 = e^{3t}, W = \det \begin{bmatrix} e^{-t} & e^{3t} \\ -e^{-t} & 3e^{3t} \end{bmatrix}$

17. $u_1 = e^t, u_2 = te^t, W = \det \begin{bmatrix} e^t & te^t \\ e^t & (t + 1)e^t \end{bmatrix}$

18. $u_1 = \cos t, u_2 = \sin t$

19. $u_1 = t, u_2 = t \ln t$

20. $u_1 = t, u_2 = t^2$

21. $u_1 = \cos (\ln t), u_2 = \sin (\ln t)$

22. $u_1 = 1, u_2 = t, u_3 = t^2$

23. $u_1 = e^t, u_2 = \cos t, u_3 = \sin t$

24. $u_1 = e^t, u_2 = e^{-t}, u_3 = \cos t, u_4 = \sin t$

APPENDIX B

Mathematical

References

Trigonometry

$$\sin (A \pm B) = \sin A \cos B \pm \cos A \sin B$$

$$\cos (A \pm B) = \cos A \cos B \mp \sin A \sin B$$

$$\sin A + \sin B = 2 \sin \tfrac{1}{2}(A + B) \cos \tfrac{1}{2}(A - B)$$

$$\sin A - \sin B = 2 \cos \tfrac{1}{2}(A + B) \sin \tfrac{1}{2}(A - B)$$

$$\cos A + \cos B = 2 \cos \tfrac{1}{2}(A + B) \cos \tfrac{1}{2}(A - B)$$

$$\cos A - \cos B = 2 \sin \tfrac{1}{2}(A + B) \sin \tfrac{1}{2}(A - B)$$

$$\sin A \sin B = \tfrac{1}{2}(\cos (A - B) - \cos (A + B))$$

$$\sin A \cos B = \tfrac{1}{2}(\sin (A - B) + \sin (A + B))$$

$$\cos A \cos B = \tfrac{1}{2}(\cos (A - B) + \cos (A + B))$$

Complex Numbers

1. A complex number may always be written as $a + ib$ where a and b are real numbers and i is the *imaginary unit*, having the property that $i^2 = -1$.

a is called the *real part* of $a + ib$: $a = \text{Re} (a + ib)$;

b is called the *imaginary part* of $a + ib$: $b = \text{Im} (a + ib)$.

Note that the imaginary part is a real number.

2. Two complex numbers are equal if and only if their real parts are equal and their imaginary parts are equal: $a + ib = c + id$ if and only if $a = c$ and $b = d$. A complex equality is equivalent to two real equalities.

3. The arithmetic operations are defined by

$$(a + ib) \pm (c + id) = (a \pm c) + i(b \pm d), \tag{B.1}$$

$$(a + ib)(c + id) = (ac - bd) + i(ad + bc), \tag{B.2}$$

$$\frac{a + ib}{c + id} = \frac{a + ib}{c + id} \cdot \frac{c - id}{c - id} = \frac{(ac + bd) + i(bc - ad)}{c^2 + d^2}. \tag{B.3}$$

Example 1

$$\frac{2 - i}{1 + i} = \frac{(2 - i)(1 - i)}{(1 + i)(1 - i)} = \frac{1 - 3i}{2}; \quad \frac{1}{2 + 3i} = \frac{2 - 3i}{(2 + 3i)(2 - 3i)} = \frac{2 - 3i}{13}.$$

4. The usual rules of algebra are valid in complex algebra. An easy way to deal with complex algebra is this: treat i as you would treat the symbol x in manipulating polynomials. At any stage you may simplify any expression using the identity $i^2 = -1$ and its consequences: $i^3 = -i$, $i^4 = 1$.

Example 2

$$(2 + i)^3 = 2^3 + 3 \cdot 2^2 \cdot i + 3 \cdot 2 \cdot i^2 + i^3$$
$$= 8 - 6 + (12 - 1)i = 2 + 11i.$$

5. The number $\bar{z} = a - ib$ is the *complex conjugate* of $z = a + ib$ and vice versa. The real and imaginary parts of z are given by

$$\text{Re } z = \frac{1}{2}(z + \bar{z}), \quad \text{Im } z = \frac{1}{2i}(z - \bar{z}). \tag{B.4}$$

The *absolute value* or *magnitude* of z is

$$|z| = \sqrt{z\bar{z}} = \sqrt{(\text{Re } z)^2 + (\text{Im } z)^2}. \tag{B.5}$$

6. Functions of a complex variable are often defined by their power series (for example, polynomials in the complex variable z). If this is done for the exponential function, one may prove that the usual rules of exponents are valid:

$$e^{z+w} = e^z e^w, \quad e^{-z} = 1/e^z, \quad e^{az} = (e^z)^a.$$

Furthermore, collecting real and imaginary terms of the series gives

$$e^{iy} = \cos y + i \sin y, \quad e^{-iy} = \cos y - i \sin y. \tag{B.6}$$

From these follow the exponential definitions of sine and cosine:

$$\cos y = \frac{1}{2}(e^{iy} + e^{-iy}), \quad \sin y = \frac{1}{2i}(e^{iy} - e^{-iy}). \tag{B.7}$$

7. A complex number $a + ib$ can be written in the form

$$a + ib = \rho e^{i\theta}, \tag{B.8}$$

where $\rho = \sqrt{a^2 + b^2} = |a + ib|$ is a nonnegative real number and θ satisfies

$$\cos \theta = \frac{a}{\rho}, \qquad \sin \theta = \frac{b}{\rho}.$$

(If a and b are thought of as Cartesian coordinates of a point in the plane, then ρ and θ are its polar coordinates.) We call θ the *argument* of the complex number $a + ib$. The *polar form* in Eq. (B.8) is convenient for extracting roots. Since $e^{2\pi i} = 1$, also $e^{i(\theta + 2k\pi)} = e^{i\theta}$ for any integer k. Then

$$(a + ib)^{1/n} = \rho^{1/n} e^{i(\theta + 2k\pi)/n}$$

(where $\rho^{1/n}$ is the positive nth root of ρ) gives n different values for $k = 0, 1, \ldots, n - 1$ and repeats those values for larger k.

Example 3

Find the solutions of $s^4 + 4 = 0$. The equation requires that $s^4 = -4$; thus a solution is a fourth root of $-4 = 4e^{i\pi}$. The four roots are $\sqrt{2}e^{i\pi/4} = 1 + i$, $\sqrt{2}e^{3i\pi/4} = -1 + i$, $\sqrt{2}e^{5i\pi/4} = -1 - i$, $\sqrt{2}e^{7i\pi/4} = 1 - i$.

Polynomials

The typical polynomial can be put in the form

$$p(s) = s^n + a_1 s^{n-1} + \cdots + a_{n-1}s + a_n. \tag{B.9}$$

(Note that the coefficient of the term of highest degree is 1; this is called a *monic* polynomial.) The *degree* of the polynomial is n.

1. A (real or complex) number r is a *root* or *zero* of $p(s)$ if $s - r$ divides $p(s)$. Equivalently, $p(r) = 0$. A root is said to have *multiplicity* m if $(s - r)^m$ divides $p(s)$ but $(s - r)^{m+1}$ does not. Equivalently, $p(r) = 0$, $p'(r) = 0, \ldots, p^{(m-1)}(r) = 0$, but $p^{(m)}(r) \neq 0$.

2. If r is a root of $p(s)$, the remaining roots of $p(s)$ are the roots of $p(s)/(s - r)$. ("Dividing out a root.")

3. A polynomial of degree n has exactly n (real or complex) roots, counting a root of multiplicity m as m roots (fundamental theorem of algebra).

4. If $r_1, r_2, \ldots, r_n$ are the n roots of $p(s)$, then

$$p(s) = (s - r_1)(s - r_2) \cdots (s - r_n) \tag{B.10}$$

(factored form of a polynomial).

5. The coefficients $a_1, a_2, \ldots, a_n$ in the form (B.9) are related to the roots. Two that are easy to find are

$$a_1 = -(r_1 + r_2 + \cdots + r_n) \tag{B.11}$$

and

$$a_n = (-1)^n r_1 r_2 \cdots r_n. \tag{B.12}$$

6. If z is any number (real or complex), there is a root of $p(s)$ within or on the circle in the complex plane having center z and radius $|p(z)|^{1/n}$.

7. The quadratic $p(s) = s^2 + bs + c$ has roots given by

$$r_1, r_2 = \frac{-b \pm \sqrt{b^2 - 4c}}{2}.$$

The roots of a polynomial of the form $p(s) = s^{2k} + bs^k + c$ may be found by substituting $z = s^k$. There are formulas for the roots of polynomials of degrees 3 and 4, but they are rarely used.

8. If the a's are all real, the roots are either real or pairs of complex conjugates. (That is, if $r = \alpha + i\beta$ is a root, so is $\bar{r} = \alpha - i\beta$.)

9. If the a's are all real and n is odd, $p(s)$ has at least one real root.

10. If the a's are all real and if r_1 and r_2 are two consecutive real roots, then $p'(s)$ has a root between r_1 and r_2 (by Rolle's theorem).

11. If the a's are all integers, any integer root must be a factor (positive or negative) of a_n (by Gauss's theorem).

Example 4

$p(s) = s^3 - s^2 - 4s - 6$ has at least one real root, by 9. If the root is an integer, it must be one of the numbers $\pm 1, \pm 2, \pm 3, \pm 6$, by 11. Evaluating $p(s)$ for each of these shows that $s = 3$ is a root, and $s - 3$ divides $p(s)$ by 1. By long division, find that $p(s)/(s - 3) = s^2 + 2s + 2$. The roots of this polynomial are $-1 \pm i$, by 7. According to 2, these are the two remaining roots of $p(s)$.

Determinants

A determinant is a number calculated from a square array of numbers. The order of the determinant is the number of rows and columns in the array.

1. Determinants of order 2:

$$\begin{vmatrix} a & b \\ c & d \end{vmatrix} = ad - bc.$$

2. Determinants of order 3:

$$\begin{vmatrix} a_1 & a_2 & a_3 \\ b_1 & b_2 & b_3 \\ c_1 & c_2 & c_3 \end{vmatrix} = a_1 b_2 c_3 + a_2 b_3 c_1 + a_3 b_1 c_2 - a_1 b_3 c_2 - a_2 b_1 c_3 - a_3 b_2 c_1.$$

The terms in a third-order (only!) determinant can be remembered by this figure:

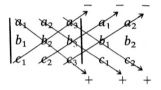

3. Determinants of any order. This rule is called "expansion by minors." To evaluate the $n \times n$ determinant

$$\Delta = \begin{vmatrix} a_{11} & a_{12} & \cdots & a_{1n} \\ a_{21} & a_{22} & \cdots & a_{2n} \\ \vdots & \vdots & & \vdots \\ a_{n1} & a_{n2} & \cdots & a_{nn} \end{vmatrix},$$

choose any row, say row i, and calculate the n numbers $M_{i1}, M_{i2}, \ldots, M_{in}$, called minors. The *minor* M_{ik} is the $(n-1) \times (n-1)$ determinant made from Δ by crossing out the ith row and kth column of Δ. Then

$$\Delta = a_{i1}(-1)^{i+1}M_{i1} + a_{i2}(-1)^{i+2}M_{i2} + \cdots + a_{in}(-1)^{i+n}M_{in}.$$

Example 5

Expand by the second row:

$$\begin{vmatrix} 1 & 3 & 2 \\ t & u & v \\ 0 & 7 & 5 \end{vmatrix} = t(-1)^3 \begin{vmatrix} 3 & 2 \\ 7 & 5 \end{vmatrix} + u(-1)^4 \begin{vmatrix} 1 & 2 \\ 0 & 5 \end{vmatrix}$$

$$+ v(-1)^5 \begin{vmatrix} 1 & 3 \\ 0 & 7 \end{vmatrix} = -t + 5u - 7v.$$

The same kind of expansion can be made using some column instead of a row, with the same result.

The evaluation of determinants can be simplified by using these rules:

1. If k is a common factor of all elements in some row (or column), then k may be factored out of the determinant.

2. The value of a determinant is unchanged if a multiple of some row is added to another row or if a multiple of some column is added to another column.

Example 6

$$\begin{vmatrix} e^{2t} & \cos t & \sin t \\ 2e^{2t} & -\sin t & \cos t \\ 4e^{2t} & -\cos t & -\sin t \end{vmatrix} = e^{2t} \begin{vmatrix} 1 & \cos t & \sin t \\ 2 & -\sin t & \cos t \\ 4 & -\cos t & -\sin t \end{vmatrix}$$

$$= e^{2t} \begin{vmatrix} 1 & \cos t & \sin t \\ 2 & -\sin t & \cos t \\ 5 & 0 & 0 \end{vmatrix} = e^{2t} \cdot 5(\cos^2 t + \sin^2 t).$$

Successive steps employed: factoring out e^{2t} from the first column (rule 1); adding row 1 to row 3 (rule 2); expanding by the third row.

Theorem B.1

The system of linear equations

$$a_{11}x_1 + a_{12}x_2 + \cdots + a_{1n}x_n = b_1,$$
$$a_{21}x_1 + a_{22}x_2 + \cdots + a_{2n}x_n = b_2,$$
$$\vdots$$
$$a_{n1}x_1 + a_{n2}x_2 + \cdots + a_{nn}x_n = b_n$$

has a *unique* solution if and only if the *determinant* of the system

$$\Delta = \begin{vmatrix} a_{11} & a_{12} & \cdots & a_{1n} \\ a_{21} & a_{22} & \cdots & a_{2n} \\ \vdots & & & \\ a_{n1} & a_{n2} & \cdots & a_{nn} \end{vmatrix}$$

is different from zero.

Corollary B.1

The *homogeneous* system of linear equations

$$a_{11}x_1 + a_{12}x_2 + \cdots + a_{1n}x_n = 0,$$
$$a_{21}x_1 + a_{22}x_2 + \cdots + a_{2n}x_n = 0,$$
$$\vdots$$
$$a_{n1}x_1 + a_{n2}x_2 + \cdots + a_{nn}x_n = 0$$

always has solution $x_1 = x_2 = \cdots = x_n = 0$. This is the only ($=$ unique) solution *unless* $\Delta = 0$.

Calculus

1. Higher derivatives of a product:

$$(uv)^{(n)} = u^{(n)}v + \binom{n}{1}u^{(n-1)}v' + \cdots + \binom{n}{n-1}uv^{(n-1)} + uv^{(n)}.$$

In this formula the coefficients are the binomial coefficients:

$$\binom{n}{k} = \frac{n!}{k!\,(n-k)!}.$$

In the special case $n = 2$ the rule above is

$$(uv)'' = u''v + 2u'v' + uv''.$$

2. Integration by parts. The repeated use of integration by parts gives

$$\int v^{(n)}u\,dx = v^{(n-1)}u - v^{(n-2)}u' + v^{(n-3)}u'' - + \cdots + (-1)^n \int vu^{(n)}\,dx.$$

In particular, for $n = 2$ we have

$$\int v''u\,dx = v'u - vu' + \int vu''\,dx.$$

3. Partial derivatives. If $f(x, y)$ is a function of two variables, its partial derivatives are

$$\frac{\partial f}{\partial x}(x, y) = \lim_{h\to 0}\frac{f(x + h, y) - f(x, y)}{h},$$

$$\frac{\partial f}{\partial y}(x, y) = \lim_{k\to 0}\frac{f(x, y + k) - f(x, y)}{k}.$$

The total differential of f is

$$df = \frac{\partial f}{\partial x}\,dx + \frac{\partial f}{\partial y}\,dy.$$

A partial derivative with respect to x is calculated by differentiating in the usual way with respect to x while treating all other variables as constants.

Example 7

Let $f(x, y) = y^2 \sin(x^3 + kxy)$. Then

$$\frac{\partial f}{\partial x} = y^2 \cos(x^3 + kxy) \cdot (3x^2 + ky)$$

$$\frac{\partial f}{\partial y} = 2y \sin(x^3 + kxy) + y^2 \cos(x^3 + kxy) \cdot (kx).$$

Higher partial derivatives are calculated successively:

$$\frac{\partial^2 f}{\partial x^2} = \frac{\partial}{\partial x}\left(\frac{\partial f}{\partial x}\right), \qquad \frac{\partial^2 f}{\partial x\,\partial y} = \frac{\partial}{\partial x}\left(\frac{\partial f}{\partial y}\right), \qquad \text{etc.}$$

Theorem B.2

If $\partial f/\partial x$, $\partial f/\partial y$, and $\partial^2 f/\partial x\,\partial y$ are continuous in some rectangle, then throughout that rectangle

$$\frac{\partial^2 f}{\partial x\,\partial y} = \frac{\partial^2 f}{\partial y\,\partial x}.$$

Because of this theorem, it is not usually necessary to keep partial differentiations in order.

    4. Derivative of an integral.

(a) $\displaystyle \frac{d}{dt}\int_a^t f(x)\,dx = f(t)$ (a is constant),

(b) $\displaystyle \frac{d}{dt}\int_a^b f(x, t)\,dx = \int_a^b \frac{\partial f}{\partial t}(x, t)\,dx$ (a, b are constants),

(c) $\displaystyle \frac{d}{dt}\int_{u(t)}^{v(t)} f(x, t)\,dx = \int_{u(t)}^{v(t)} \frac{\partial f}{\partial t}(x, t)\,dx + f(v(t), t)v'(t) - f(u(t), t)u'(t).$

(This is called Leibniz's rule; it generalizes both a and b.)

Example 8

Use Leibniz's rule to find the derivative of

$$F(t) = \int_0^{t^2} \frac{\sin(t\sqrt{x})}{x}\,dx. \qquad F'(t) = \int_0^{t^2} \frac{\sqrt{x}\cos(t\sqrt{x})}{x}\,dx + \frac{\sin(t^2)}{t^2}\cdot 2t.$$

Series

    1. Taylor polynomial.

Theorem B.3

If $f(x)$ has a derivative of order $n + 1$ at all points of an interval $a \le x \le b$, then

$$f(x) = f(a) + f'(a)(x - a) + \frac{f''(a)}{2!}(x - a)^2 + \cdots$$

$$+ \frac{f^{(n)}}{n!}(x - a)^n + \frac{f^{(n+1)}(\xi)}{(n + 1)!}(x - a)^{n+1}$$

holds for $a \leq x \leq b$; the number ξ depends on x and satisfies $a \leq \xi \leq x$.

2. Taylor series. For many functions it can be shown that the last term (remainder) in the expression in Theorem B.3 vanishes as n increases for all x in an interval around a. Such a function is equal to its *Taylor series with center a:*

$$f(x) = f(a) + f'(a)(x - a) + \cdots + \frac{f^{(n)}}{n!}(x - a)^n + \cdots$$

$$= \sum_{n=0}^{\infty} \frac{f^{(n)}(a)}{n!}(x - a)^n.$$

The equality holds in a certain interval $a - R < x < a + R$. The largest possible value of R is called the *radius of convergence* of the series, and the interval is the *interval of convergence* of the series.

3. Properties. If $f(x)$ has a Taylor series with center a and radius of convergence R, the following operations may be carried out term by term on the series: multiplication by a power of $x - a$; multiplication by a constant; integration; differentiation any number of times. The result is again a Taylor series with center a and radius of convergence R. If $f(x)$ and $g(x)$ both have Taylor series with center a and radius of convergence R or larger, the two series may be added, subtracted, and multiplied to obtain another valid series with radius of convergence at least R.

4. Important series (radius of convergence in parentheses).

$$\frac{1}{1 - x} = 1 + x + x^2 + \cdots = \sum_{n=0}^{\infty} x^n \qquad (R = 1),$$

$$\frac{1}{(1 - x)^2} = 1 + 2x + 3x^2 + \cdots = \sum_{n=0}^{\infty} (n + 1)x^n \qquad (R = 1),$$

$$\ln(1 + x) = x - \frac{x^2}{2} + \frac{x^3}{3} - \cdots = \sum_{n=1}^{\infty} \frac{(-1)^{n+1}}{n} x^n \qquad (R = 1),$$

$$\sqrt{1 + x} = 1 + \tfrac{1}{2}x - \tfrac{1}{8}x^2 + \tfrac{1}{16}x^3 - + \cdots$$

$$= 1 + \sum_{n=1}^{\infty} \frac{(-1)^{n+1}}{n \cdot 2^{2n-1}} \binom{2n - 2}{n - 1} x^n \qquad (R = 1),$$

$$e^x = 1 + x + \frac{x^2}{2} + \frac{x^3}{6} + \cdots = \sum_{n=0}^{\infty} \frac{x^n}{n!} \qquad (R = \infty),$$

$$\sin x = x - \frac{x^3}{6} + \frac{x^5}{120} - + \cdots = \sum_{n=0}^{\infty} \frac{(-1)^n}{(2n + 1)!} x^{2n+1} \qquad (R = \infty),$$

$$\cos x = 1 - \frac{x^2}{2} + \frac{x^4}{24} - + \cdots = \sum_{n=0}^{\infty} \frac{(-1)^n}{(2n)!} x^{2n} \qquad (R = \infty).$$

Bibliography

Abraham, Ralph H. and Christopher D. Shaw (1983), *Dynamics—The Geometry of Behavior, Part 1*, Aerial Press, P.O. Box 1360, Santa Cruz, CA.

Abramowitz, Milton and Irene A. Stegun (1964), *Handbook of Mathematical Functions*, National Bureau of Standards, Washington, D.C.

Bellman, Richard (1953), *Stability Theory of Differential Equations*, McGraw-Hill, N.Y. Reprinted by Dover, N.Y., 1969.

Birkhoff, Garrett and Gian-Carlo Rota (1978), *Ordinary Differential Equations*, Third ed., Wiley, N.Y.

Braun, Martin (1975), *Differential Equations and their Applications*, Springer-Verlag, N.Y.

Burden, Richard L., J. Douglas Faires and Albert C. Reynolds (1978), *Numerical Analysis*, Prindle, Weber & Schmidt, Boston, MA.

Churchill, Ruel V. (1972), *Operational Mathematics*, Third ed., McGraw-Hill, N.Y.

Churchill, Ruel V. and James W. Brown (1978), *Fourier Series and Boundary-Value Problems*, Third ed., McGraw-Hill, N.Y.

Cullen, Charles G. (1979), *Linear Algebra and Differential Equations*, Prindle, Weber & Schmidt, Boston, MA.

Dahlquist, Germund and Ake Björk (1974), *Numerical Methods*, Prentice-Hall, Englewood Cliffs, N.J.

Davies, T.V. and Eleanor M. James (1966), *Nonlinear Differential Equations*, Addison-Wesley, Reading, MA.

Davis, Harold T. (1960), *Introduction to Nonlinear Differential and Integral Equations*, U.S. Atomic Energy Commission, Washington, D.C. Reprinted by Dover, N.Y., 1962.

Hagin, Frank G. (1975), *A First Course in Differential Equations*, Prentice-Hall, Englewood Cliffs, N.J.

Lamb, Sir Horace (1932), *Hydrodynamics*, Cambridge University. Reprinted by Dover, N.Y., 1945.

Leighton, Walter (1967), *Ordinary Differential Equations*, Wadsworth, Belmont, CA.

Luenberger, David G. (1979), *Introduction to Dynamic Systems*, Wiley, N.Y.

Powers, David L. (1979), *Boundary Value Problems*, Second ed., Academic Press, N.Y.

Roberts, G.E. and H. Kaufman (1966), *Table of Laplace Transforms*, Saunders, Philadelphia, PA.

Tolstov, Georgi P. (1962), *Fourier Series*, Prentice-Hall, Englewood Cliffs, N.J. Reprinted by Dover, N.Y., 1976.

Tranter, C.J. (1966), *Integral Transforms in Mathematical Physics*, Third ed., Methuen, London.

Answers to Odd-Numbered Exercises

Chapter 1

Section 1.1

1. $u(t) = 1 + 4t$

3. $u(t) = \dfrac{1 - \cos 2t}{2}$

5. $u(t) = c + \ln(1 + t)$

7. $u(t) = c + \sqrt{1 + t^2}$

9. $u(t) = c + 3\ln(t + 2) - 2\ln(t + 1)$

11. $\dfrac{dV}{dt} = q(t), \; 0 < t; \; V(0) = 0$

13. $V(t) = \begin{cases} 4 - \dfrac{t^2}{20}, & 0 < t \le 40 \\ 80, & 40 < t \end{cases}$

15. Rate in = 0; rate out = $5u(t)/50$; accumulation rate = du/dt. The initial value problem is

$$\frac{du}{dt} = -\frac{u}{10}, \quad 0 < t; \; u(0) = 500$$

17. (a) The rates dV/dt and dR/dt are related by $dV/dt = 4\pi R^2 \, dR/dt$. Hence

$$4\pi R^2 \frac{dR}{dt} = -k4\pi R^2(c_s - c) \quad \text{or} \quad \frac{dR}{dt} = -k(c_s - c).$$

(b) Since k and $c_s - c$ are constants, we find $R(t) = R_0 - k(c_s - c)t$. This is valid until $R(t)$ becomes 0.

Section 1.2

1. $u(t) = ce^{2t}$

3. $u(t) = ce^{0.4t}$

5. $u(t) = ct$; the coefficient $1/t$ is discontinuous at $t = 0$

7. $u(t) = c(\sec t + \tan t)$; discontinuous at $t = \pm\dfrac{\pi}{2}, \pm\dfrac{3\pi}{2}, \ldots$

9. $u(t) = c \exp(e^{-t})$

11. $u(t) = c \exp\left(\dfrac{-t^2}{2}\right)$; $u(t) = 3 \exp\left(\dfrac{-t^2}{2}\right)$

13. $u(t) = c \sin t$; the coefficient $\cot t$ is discontinuous at $t = 0, \pm\pi, \pm2\pi, \ldots$

15. $u(t) = c/t^k$

17. $b = \dfrac{\ln(U_2/U_1)}{t_2 - t_1}$

19. $u(t) = c \exp(-(t + 1)e^{-t})$

21. $u(t) = c \exp(t - \cos t)$

23. We can solve for $u(t) = c \exp\left(-\int_0^t a(x)\, dx\right)$. Then $v(t) = c \exp(\phi(t))$ where

$$\phi(t) = \alpha t - \int_0^t a(x)\, dx,$$

and $\phi(t)$ is periodic: $\phi(t + p)\phi t$. Therefore $v(t)$ is periodic, too.

25. T_1/T_0 is greatest when the cable is at the point of slipping. Then $T_1 = T(\alpha) = T_0 e^{\mu\alpha}$. For $\mu = 0.5$, $\alpha = \pi$, $T_1/T_0 = e^{\pi/2} \cong 4.8$; for $\alpha = 3\pi$, $T_1/T_0 = e^{3\pi/2} \cong 111$.

Section 1.3

1. $u(t) = \dfrac{t}{3} + \dfrac{1}{9} + ce^{3t}$

3. $u(t) = e^{-t} + ce^{-2t}$

5. $u(t) = 1 + c \exp(t^2/2)$

7. $u(t) = \dfrac{t^3}{3} + \dfrac{c}{t}$

9. $u(t) = 1 + c \exp(-\sin t)$; $c = -1$

11. $u(t) = -\cos t + c \sin t$

13. $u(t) = te^{-bt} + ce^{-bt}$

15. $u(t) = e^{-t^2/2}\left(\displaystyle\int_0^t e^{z^2/2} + c\right)$

17. $u(t) = kV + ce^{-Qt/V}$, $c = k_0 - k$

19. $z = y^{1-n}$, $\dfrac{dz}{dx} = (1 - n)y^{-n}\dfrac{dy}{dx}$, so the equation becomes

$$\dfrac{1}{1 - n}\dfrac{dz}{dx} = a(x)z + b(x).$$

21. $z = y^2$; $y(x) = \sqrt{1 + ce^{2x}}$ is discontinuous where $e^{-2x} = -c$ (c must be negative for this equation to have a real root).

23. $z = \sqrt{y}$; $y(x) = (x + 2 + ce^{x/2})^2$; to satisfy the initial condition, $c = -1$.

25. $z = u^{-1}$; $u(t) = \left(\dfrac{m}{b} + ce^{-bt}\right)^{-1}$ is discontinuous when $e^{-bt} = -\dfrac{m}{bc}$ (c must be negative).

Section 1.4

1. By (1.21) the decay rate is proportional to the amount present, so

$$t = \frac{1}{\lambda} \ln \frac{r(t)}{r(0)}$$

gives the time, where $r = |dc/dt|$ is the decay rate ($r(0)$ is found from living material).

3. If T is the half-life for a substance decaying according to Eq. (1.21), then

$$c(T) = c_0 e^{-\lambda T} = \tfrac{1}{2}c_0.$$

Thus $\lambda T = -\ln \tfrac{1}{2} = 0.693$.

5. $q(t) = 10^{-5} \dfrac{\sin 20t - 2\cos 20t + 2e^{-10t}}{5}$, $V_c(t) = 10^6 q(t)$.

7. $i(t) = \dfrac{V_0}{(\omega L)^2 + R^2}(R \sin \omega t - \omega L \cos \omega t + \omega L e^{-Rt/L})$

9. Solve $R\, dq/dt + q/C = 0$, $q(0) = 9C$ for $q(t)$, and then find $V_c(t) = 9e^{-t/RC}$. If T is the time delay, 120 sec, then $V_c(T) = 9e^{-T/RC} = 4$. Hence $T/RC = \ln(9/4)$, giving $RC = 120/\ln(9/4) \cong 148$.

11.

Year	1800	10	20	30	40	50	60
Rate	3.5%	3.6	3.3	3.4	3.3	3.6	3.5

13.

Year	1870	80	90	1900	10	20	30
Rate	1.6%	1.9	1.5	1.5	0.9	0.9	1.2

15.

Rate r	1	3	5	10	20	100
"72 Rule"	72	24	14.4	7.2	3.6	0.72
Correct	69.6	23.4	14.2	7.27	3.8	1

17. The principal is $P(t) = \left(P(0) - \dfrac{M}{r}\right)e^{rT} + \dfrac{M}{r}$. Set $P(T) = 0$ and solve for

$$M = P(0)\frac{re^{rT}}{e^{rT} - 1}.$$

19. Let $u(t)$ be the amount of salt in the reservoir. The rate at which salt enters is 0; the rate at which it leaves is (concentration × fluid flow rate) $(u(t)/10^9 \,\text{m}^3) \times 250 \,\text{m}^3/\text{sec}$ in kilograms per second if $u(t)$ is measured in kilograms. Thus the differential equation for u is $du/dt = -0.25 \times 10^{-6}u$, with t measured in seconds. The initial amount is

$$u(0) = \frac{300\,\text{g}}{\text{liter}} \times \frac{1\,\text{kg}}{1000\,\text{g}} \times \frac{1000\,\text{liters}}{\text{m}^3} \times 4.5 \times 10^6\,\text{m}^3$$

$$= 1.35 \times 10^9\,\text{kg}.$$

21. The concentration is $u(t)/10^9$ or $1.35 \exp(-0.25 \times 10^{-6} t)\,\text{kg/m}^3$. This decreases by 50% when $-0.25 \times 10^6 t = \ln\tfrac{1}{2} = -0.69$: $t = 2.76 \times 10^6$ sec, or about 32 days.

Section 1.5

1. $\frac{1}{2}y^2 = \frac{1}{2}x^2 + c$ or $y = \pm\sqrt{x^2 + 2c}$

3. $\frac{1}{2}y^2 = x - \ln|1 + x| + c$ or $y = \pm\sqrt{2(x - \ln|1 + x| + c)}$

5. $\frac{1}{2}\ln\left|\dfrac{1 + y}{1 - y}\right| = x + C$ or $y = \dfrac{ce^{2x} - 1}{ce^{2x} + 1}$; $c = 1$

7. $\tan^{-1} y = \tan^{-1} x + c$ or $y = \tan(\tan^{-1} x + c)$; $c = \dfrac{\pi}{4}$

9. $2\sqrt{y} = \frac{2}{3}x^{3/2} + c$ or $y = (\frac{1}{3}x^{3/2} + \frac{1}{2}c)^2$

11. (a) $u(t) = e^{-t}$; (b) $u(t) = 1/(t + 1)$; (c) $u(t) = (1 - \frac{1}{2}t)^2$, until $t = 2$. Solution (c) reaches 0 at time $t = 2$. Solutions (a) and (b) approach 0 asymptotically, (a) more rapidly than (b).

13. $\dfrac{1}{b}\ln\left|\dfrac{u}{b - mu}\right| = t + C$ or $u(t) = \dfrac{bce^{bt}}{1 + mce^{bt}}$. To obtain the solution in Eq. (1.45), it is necessary to divide the numerator and denominator by ce^{bt} and rename c.

15. (a) $u(t) \rightarrow b/m$ $(c > 0)$; (b) $u(t) \rightarrow b/m$ $(c < -1/m)$; (c) $u(t)$ becomes infinite in finite time $(0 > c > -1/m)$.

17. The solution is $x + y - \ln xy = 10 - 2\ln 5$. This equation is also satisfied by $x = y = 0.0348$. (A sketch of the function $2x - 2\ln x$ shows that each value greater than 1 comes from two values of x.)

19. $bu - a^2/u = -t + c$ or $u(t) = (c - t \pm \sqrt{(c - t)^2 + 4a^2 b})/2b$. Since $u > 0$ is given, take the solution with the plus sign.

21. This is seen most clearly from the first form of the solution in 19 above: $t \rightarrow \infty$ as $u \rightarrow 0$ or vice versa.

Section 1.6

1. $\exp\left(\tan^{-1}\dfrac{y}{x}\right)\sqrt{x^2 + y^2} = 2$

3. $\sqrt{\left|\dfrac{x + y}{x - y}\right|}\dfrac{1}{x^2 - y^2} = c$ or $(x - y)^3(x + y) = \dfrac{1}{c}$

5. The separated equation is

$$\frac{(1 + v^2)\,dv}{3 - v - v^2 - v^3} = \frac{dx}{x}.$$

The denominator on the left-hand side has $v - 1$ as a factor. Use partial fractions to get

$$\left(-\frac{1}{3}\frac{1}{v - 1} - \frac{2}{3}\frac{v}{v^2 + 2v + 3}\right)dv = \frac{dx}{x}.$$

Solution (in terms of $v = y/x$):

$$-\frac{1}{3}\ln|v-1| - \frac{2}{3}\left[\frac{1}{2}\ln(v^2 + 2v + 3) - \frac{1}{\sqrt{2}}\tan^{-1}\left(\frac{v+1}{\sqrt{2}}\right)\right] = \ln|x| + c$$

7. $\sqrt{\left(\dfrac{y}{x}\right)^2 + 2\dfrac{y}{x} + } = cx;\ c = \sqrt{5}$

9. $\exp(y/x) = \ln x + C$ or $y = x\ln|\ln x + C|$

11. $h = 2,\quad k = 1;\quad \ln\left|\dfrac{1+v}{1-v}\right| - \frac{1}{2}\ln|1 - v^2| = \ln X + C$ or $X + Y = c(X-Y)^3$ or $x + y - 2 = c(x - y - 1)^3$. In the last expression, choose $c = 2$.

13. $h = -2,\ k = 0;\ \exp(X/Y) = cY$. This can be solved for $X = Y\ln(cY)$ or $x + 2 = y\ln cy$

15. $\ln(\ln v) = \ln x + C$ or $y = xe^{cx}$

17. $y + \sqrt{x^2 + y^2} = c$ or $y = \dfrac{c^2 - x^2}{2c}$

19. (a) The determinant of Eq. (1.66) is zero. (b) If the numerator and denominator of the given equation are not proportional, then the substitution gives $u = (z - Ax - C)/B$, and

$$\frac{1}{B}\left(\frac{dz}{dx} - A\right) = \frac{\lambda z + \alpha}{z},$$

where $\lambda = a/A = b/B$ and $\alpha = c - \lambda C$. This equation is separable.

21. The solution is $\exp(\tan^{-1}(y/x)) = c\sqrt{x^2 + y^2}$. In terms of polar coordinates it is $cr = e^\theta$, the equation of a spiral.

23. The equation is linear fractional:

$$\frac{dy}{dx} = \frac{x - 3y + 1}{-3x + y + 5} \quad \text{or} \quad \frac{dY}{dX} = \frac{X - 3Y}{-3X + Y}$$

with $x = X + 2,\ y = Y + 1$. The latter has solution $Y - X = C(Y + X)^2$. The solution curves are parabolas through $X = 0,\ Y = 0$ or $x = 2,\ y = 1$.

25. $x = c\exp(\tan^{-1}v),\ y = cv\exp(\tan^{-1}v)$.

Section 1.7

1. The isoclines are lines through the origin: $y = -x/m$. The solution curves are circles centered at 0.

3. The isoclines are lines through 0: $y = x/(2 - m)$. (All the isoclines corresponding to negative slopes are in the wedge between the x-axis and the line $y = x/2$.) All solutions are asymptotic to the solution $y = x$ as $x \to \infty$ and also as $x \to 0$.

5. Isoclines are $y = (1 + m)x/(1 - m)$. Solutions spiral counterclockwise into the origin.

7. The origin is the only critical point. Isoclines are the parabolas $y = x(x - m)$. All solutions approach the solution $y = x^2/3$, as $x \to \pm\infty$.

9. Isoclines are the circles $x^2 + y^2 = m$ $(m > 0)$. There is no critical point. Solutions all have (different) vertical asymptotes.

11. There are three critical points: $(0, 0)$, $(0, 1)$, and $(\frac{1}{2}, \frac{1}{2})$. Solutions in quadrant I form spirals or closed curves about the last. In quadrants III and IV, solutions look roughly like hyperbolas with asymptotes on the axes. In quadrant II there are two families of solutions; each has one axis as an asymptote, and both have $y = -x/3$ as an asymptote.

13. (a) $y = cx$, infinitely many; (b) $x^2 + y^2 = c$, none; (c) $y^2 - x^2 = c$, two: $y = x$ and $y = -x$; (d) $xy = c$, one: $y = 0$.

15. If the equation is $y' = f(y, x)$, then the isoclines are given by $f(y/x) = m$. But if $f(y/x)$ is constant, so is y/x. Thus the isoclines are $y/x = $ const., straight lines through the origin.

Section 1.7A

1. $\dfrac{dy}{dx} = y$, $y(0) = 1$; $y(x) = e^x$ 3. $\dfrac{dy}{dx} = e^{-y}$, $y(1) = 0$; $y(x) = \ln x$

5. (a) $y(x) = \displaystyle\int_0^x (z - y^2(z))\, dz$;   (b) $\phi_0(x) = 0$, $\phi_1(x) = x^2/2$, $\phi_2(x) =$
$x^2/2 - x^5/20$, $\phi_3(x) = \dfrac{x^2}{2} - \dfrac{x^5}{20} + \dfrac{x^8}{160} - \dfrac{x^{11}}{4400}$

7. (a) $y(x) = 1 + \displaystyle\int_0^x \sqrt{y(z)}\, dz$;   (b) $\phi_0(x) = 1$, $\phi_1(x) = 1 + x$, $\phi_2(x) =$
$\dfrac{(3(1 + x)^{3/2} - 1)}{2}$

9. (a) $y(x) = 1/(1 - x)$; (b) the interval is $-\infty < x < 1$ because y has a vertical asymptote—and is therefore discontinuous—at $x = 1$; (c) $F(x, y)$ is continuous in the entire xy-plane.

Section 1.8

1. Exact; $x + y = c$ 3. Exact; $xy = c$

5. Not exact, homogeneous; $y = \dfrac{cx}{(x - c)}$ or $\dfrac{xy}{(x + y)} = c$

7. Exact; $\dfrac{x^3}{3} + xy + e^y = c$ 9. Exact; $x^2 y + y^2 x = c$

11. Not exact, separable and linear; $y = \dfrac{c}{\sin x}$

13. Exact; $x^2 e^y = c$ 15. Not exact, separable; $\dfrac{y}{x} = c$

17. Not exact, becomes exact on eliminating the common factor $1 + x$; $x^2 - y^2 = c$

19. Exact; $y \ln |x| = c$

21. $\dfrac{\partial M}{\partial y} = \dfrac{\partial N}{\partial x}$ by exactness. Therefore $\dfrac{\partial}{\partial y}[M + f(x)] = \dfrac{\partial M}{\partial y} = \dfrac{\partial N}{\partial x} = \dfrac{\partial}{\partial x}[N + g(y)]$.

23. The solutions of $x\,dx + y\,dy = 0$ are circles centered at the origin; $x^2 + y^2 = c$. The orthogonal trajectories are solutions of $y\,dx - x\,dy = 0$ (separable) or $y/x = c'$. These are lines through the origin.

25. Since $M\,dx + N\,dy = 0$ is exact, $M = \partial Q/\partial x$, $N = \partial Q/\partial y$. In order for $N\,dx - M\,dy$ to be exact, $\partial N/\partial y = -\partial M/\partial x$ or $\partial^2 Q/\partial x^2 + \partial^2 Q/\partial y^2 = 0$ must be fulfilled.

27. $Q(x, y) = \tfrac{1}{2}ax^2 + bxy - \tfrac{1}{2}By^2 + cx - Cy = \text{const.}$

Section 1.9

1. i.f.: $1/x^2$; $\ln x - (y/x) = c$ 3. i.f.: e^{-y}; $x^2 y + xe^{-y} = c$

5. i.f.: $\exp(-x^2/2)$; $ye^{-x^2/2} - \displaystyle\int e^{-x^2/2}\,dx = c$

7. From $\dfrac{\phi'}{\phi} = \dfrac{2}{(4 - x)}$, find $\phi(x) = \dfrac{1}{(4 - x)^2}$. The exact equation has solution $\dfrac{y^2 + 4 - x}{(4 - x)^2} = c$.

9. Assume $\phi(x, y) = \theta(y/x)$. Then Eq. (1.89) becomes

$$((-y/x^2)n - (1/x)m)\theta' = ((1/x)m' + (y/x^2)n')\theta$$

where the prime denotes differentiation with respect to $v = y/x$. Now multiply by x:

$$-(vn + m)\theta' = (vn' + m')\theta.$$

This equation contains only functions of v, showing the consistency of the assumption.

11. Divide by x^2 to get homogeneity: $m = 1 + v^2$, $n = -v$. Then $\phi(x, y) = \theta(v)$ satisfies

$$-(-v^2 + 1 + 2v^2)\theta' = (-v + 4v)\theta.$$

Thus $\theta = (1 + v^2)^{-3/2}$.

Miscellaneous Exercises

1. Separable; $y = (\tfrac{1}{3}x^{3/2} + c)^2$; $c = 1$

3. Separable; $y = \tan(x + c)$; $c = \pi/4$

5. Homogeneous-polar; $x = cv \exp(-1/2v^2)$, $y = vx$

7. Separable; $y = 1/(x^2 + c)$

9. Linear; $y = x^2 - 2x + 2 + ce^{-x}$

11. Exact; $\frac{1}{4}(x^4 + 2x^2y^2 + y^4 - 2x^2 + 2y^2) = c$. These curves are ovals of Cassini.

13. Separable and homogeneous-polar; $y = \dfrac{1}{\dfrac{1}{x} - c} = \dfrac{x}{1 - cx}$

15. Homogeneous-polar; $2x + \sqrt{y^2 + 4x^2} = cx^2y$

17. Separable; $y = \operatorname{sech}(x + c)$ 19. Exact; $x^2 + 3xy + 2y^2 = c$

21. Linear; $r(\theta) = \tan\theta + c\sec\theta$

23. Homogeneous-polar as is and linear when x is the dependent variable; $x(y) = y\ln y + c\ln y$

25. The isoclines are the parabolas $x = (m + 1)y^2 + m$ (a line for $m = -1$). As $x \to \infty$, all solutions in the first quadrant approach $\sqrt{x}$.

27. From Newton's second law,

$$M\frac{dv}{dt} = -(av + bv^2), \qquad v(0) = v_0.$$

29. Solve the equation above for v:

$$v = -\frac{M}{a + bv}\frac{dv}{dt}.$$

Now integrate both sides from $t = 0$ to ∞. Since $v(t) = dx/dt$, the integral of v will be the distance traveled:

$$d = \int_0^\infty v(t)\,dt = -\int_0^\infty \frac{M}{a + bv}\frac{dv}{dt}\,dt.$$

Change variables in the second integral to v: $v(0) = v_0$, $v \to 0$ as $t \to \infty$:

$$d = -\int_{v_0}^0 \frac{M}{a + bv}\,dv = \frac{M}{b}\ln\left(\frac{a + bv_0}{a}\right).$$

31. (a) $y = \left(\dfrac{x}{2} + 1\right)^2$; (b) $y = e^x$; (c) $y = 1/(1 - x)$. Solution (c) becomes infinite at $x = 1$. The other two increase with x, but the exponential increases much more rapidly. Note that the three graphs are tangent at $x = 0$.

33. $\dfrac{dV}{dt} = A\dfrac{dh}{dt}$ and $\dfrac{dV}{dt} = -k\sqrt{h}$,

where A is the cross-sectional area of the tank and $k\sqrt{h}$ is the rate of outflow. Thus $dh/dt = -(k/A)\sqrt{h}$.

35. 500 seconds

37. The form of the solution depends on the sign of y: $y(x) = ce^{-x} + 1$ if $y(x) \geq 0$; $y(x) = Ce^x - 1$ if $y(x) \leq 0$. (To solve, use $|y| = y$ if $y > 0$ and $|y| = -y$ if $y < 0$.) (a) $y(x) = e^{-x} + 1$; (b) $y(x) = -\frac{1}{2}e^{-x} + 1$; (c) $y(x) = \frac{1}{2}e^x - 1$ for $0 \leq x < \ln 2$, and $y(x) = -2e^{-x} + 1$ for $\ln 2 \leq x < \infty$; (d) $y(x) = -e^x - 1$. Note that solution (c) crosses the x-axis at $x = \ln 2$.

Chapter 2

Section 2.1

1. $M\dfrac{d^2u}{dt^2} + p\dfrac{du}{dt} = F(t)$

   $u(0) = u_0,\ \dfrac{du}{dt}(0) = v_0$

3. $M\dfrac{d^2u}{dt^2} + ku = F(t)$

   $u(0) = u_0,\ \dfrac{du}{dt}(0) = v_0$

5. Derive this equation by equating the spring and damper forces where spring and damper meet:

   $ky = p(\dot{u} - \dot{y}).$

   Apply Newton's law to the mass to get

   $M\ddot{u} = p(\dot{y} - \dot{u}).$

   By eliminating one or the other variable between equations, find

   $$M\dfrac{d^2y}{dt^2} + \dfrac{Mk}{p}\dfrac{dy}{dt} + ky = 0$$

   $$M\dfrac{d^3u}{dt^3} + \dfrac{Mk}{p}\dfrac{d^2u}{dt^2} + k\dfrac{du}{dt} = 0$$

7. $R\dfrac{dq}{dt} + \dfrac{1}{C}q = E(t)$

   $q(0) = q_0,\ \dfrac{dq}{dt}(0) = i_0$

9. The units of weight must be kilograms-force. The spring constant is $350/12.5 = 28$ kgf/cm.

11. $c_1 = 1,\ c_2 = -\frac{1}{2}$

13. $c_1 = -6,\ c_2 = 8$

Section 2.2

1. $m^2 + 5m + 6;\ u(t) = c_1e^{-3t} + c_2e^{-2t}$
3. $m^2 + 2m = 0;\ u(t) = c_1 + c_2e^{-2t}$
5. $m^2 + 2m + 2 = 0;\ u(t) = e^{-t}(c_1\cos t + c_2\sin t)$
7. $m^2 + 2m - 3 = 0;\ u(t) = c_1e^{-3t} + c_2e^{t}$
9. $m^2 - 2m + 10 = 0;\ u(t) = e^{t}(c_1\cos 3t + c_2\sin 3t)$
11. $m^2 - 1 = 0;\ u(t) = c_1e^{t} + c_2e^{-t}$
13. $u(t) = c_1\cos 2t + c_2\sin 2t;\ c_1 = 1,\ c_2 = -\frac{1}{2}$
15. $u(t) = (c_1 + c_2t)e^{-2t};\ c_1 = 2,\ c_2 = 4$
17. $u(t) = c_1e^{2t} + c_2e^{-2t};\ c_1 = \frac{5}{4},\ c_2 = -\frac{1}{4}$
19. $u(t) = c_1\cos 10t + c_2\sin 10t;\ c_1 = 0,\ c_2 = \frac{1}{10}$
21. The characteristic polynomial $m^2 + bm + c$ has two negative real roots, or a negative double root, or complex conjugate roots with negative real part. In any case the exponential approach 0 as t increases.
23. (a) Since V is the sum of two squares, it cannot be negative and can be 0 only

if both terms are 0; that is,

$$u = 0 \quad \text{and} \quad \frac{du}{dt} + \frac{b}{2}u = 0.$$

(b) Differentiate $V(t)$ as given in the beginning of the exercise, and make the replacement

$$\frac{d^2u}{dt^2} = -b\frac{du}{dt} - cu$$

(from the differential equation). After algebra,

$$\frac{dV}{dt} = -b\left(\frac{du}{dt}\right)^2 - bcu^2.$$

(c) Since $dV/dt < 0$ if u or du/dt is nonzero, V decreases toward 0 where u and du/dt are both 0.

25. $u(t) = c_1 t + c_2$. The equation says that the curvature of u is 0, since curvature is

$$\kappa = \frac{\ddot{u}}{(1 + \dot{u}^2)^{3/2}}.$$

Of course, $u(t)$ is a straight line and has curvature 0.

27. (b) The ratio of the two functions is not constant.
    (c) $c_1 = u_0$, $c_2 = v_0 + bu_0/2$

29. If $u_2/u_1 = $ const., then $(u_2/u_1)' = 0$. That is,

$$\frac{u_1 u_2' - u_2 u_1'}{u_1^2} = 0.$$

But then the numerator must be 0, and the numerator is just the given determinant.

Section 2.3

1.

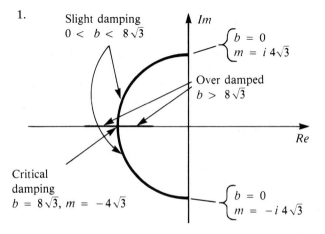

Slight damping
$0 < b < 8\sqrt{3}$

$\begin{cases} b = 0 \\ m = i\,4\sqrt{3} \end{cases}$

Over damped
$b > 8\sqrt{3}$

Im

Re

Critical damping
$b = 8\sqrt{3},\ m = -4\sqrt{3}$

$\begin{cases} b = 0 \\ m = -i\,4\sqrt{3} \end{cases}$

3. $b = 16$: $u(t) = e^{-12t}$; $b = 19$: $u(t) = (e^{-4t} + 2e^{-16t})/3$; $b = 26$: $u(t) = (6e^{-2t} + 5e^{-24t})/11$; $b = 49$: $u(t) = (36e^{-t} + 11e^{-48t})/47$.

5. Example 1: $t = \pi/24\sqrt{3} \cong 0.0766$; Example 2: $t = (\frac{1}{2}\pi - 0.955)/4\sqrt{2} \cong 0.189$; Example 3: $t = 1/(12 - 4\sqrt{3}) \cong 0.197$; Example 4: $t = \frac{1}{2}\ln(3/2) \cong 0.2027$; Exercise 3: $b = 16$; $t = \infty$.

7. (a) Critical; (b) $u(t) = (c_1 + c_2 t)e^{-3t}$; (c) $u(t) = te^{-3t}$

9. (a) Supercritical; (b) $u(t) = c_1 e^{-t} + c_2 e^{-9t}$; (c) $u(t) = 12.5e^{-t} - 2.5e^{-9t}$

11. (a) Supercritical; (b) $u(t) = c_1 e^{-t} + c_2 e^{-t/4}$; (c) $u(t) = \dfrac{-e^{-t} + 4e^{-t/4}}{3}$

13. $\left(\dfrac{du}{dt} = v\right)$ $\dfrac{d^2 v}{dt^2} + b\dfrac{dv}{dt} + \omega^2 v = 0$,

$v(0) = v_0$, $\dfrac{dv}{dt}(0) = -bv_0 - \omega^2 u_0$

15. $A = \sqrt{u_0^2 + \left(v_0 + \dfrac{b}{2}u_0\right)^2 \Big/ \beta^2}$,

17. $c_1 = u_0$, $c_2 = v_0 + \omega u_0$

$\cos\phi = \dfrac{u_0}{A}$, $\sin\phi = \dfrac{v_0 + \dfrac{b}{2}u_0}{\beta A}$

19. $u(0) = c_1 + c_2 = u_0$

$\dfrac{du}{dt}(0) = m_1 c_1 + m_2 c_2 = v_0$

$c_1 = \dfrac{m_2 u_0 - v_0}{m_2 - m_1}$, $c_2 = \dfrac{v_0 - m_1 u_0}{m_2 - m_1}$

21. We must have

$t = \dfrac{\ln(-c_1/c_2)}{m_2 - m_1}$.

Using the c's from Exercise 19 and the m's from Eq. (2.57),

$t = \dfrac{\ln\left(\dfrac{v_0 - m_2 u_0}{v_0 - m_1 u_0}\right)}{2\sqrt{\left(\dfrac{b}{2}\right)^2 - \omega^2}}$.

Of course, v_0 and u_0 must be such that the argument of the logarithm is greater than 1 if t is to be positive.

23. (a) We may suppose that $u(t)$ is given by Eq. (2.35) for as long as it takes until du/dt reaches 0—that is,

$u(t) = \dfrac{v_0}{\omega}\sin\omega t$, $0 \le t \le \pi/2\omega$.

Since $u(\pi/2\omega) = v_0/\omega = 1.5$ m, we have

$\omega^2 = \dfrac{k}{M} = \left(\dfrac{v_0}{1.5}\right)^2$, and $k = M\left(\dfrac{v_0}{1.5}\right)^2 = \dfrac{700 \text{ kg}}{9.8 \text{ m/sec}^2}\left(\dfrac{25 \text{ m/sec}}{1.5 \text{ m}}\right)^2$

$= 19{,}841 \text{ kg/m}$.

(b) To obtain critical damping, the parameter b should be 2ω. The damping constant is $Mb = 2M\omega$:

$$Mb = 2M\omega = 2 \cdot \frac{700 \text{ kg}}{9.8 \text{ m/sec}^2} \left(\frac{25 \text{ m/sec}}{1.5 \text{ m}} \right)$$

$$= 2381 \text{ kg sec/m}.$$

Section 2.4

1. $u_c = c_1 + c_2 e^{-t}$, $u_p = t$

3. $u_c = e^{-2t}(c_1 \cos t + c_2 \sin t)$, $u_p = \frac{1}{5}$

5. $u_c = c_1 e^{m_1 t} + c_2 e^{m_2 t}$, $u_p = \dfrac{1}{b\omega} \sin \omega t$

7. $u_c = c_1 \cos \omega t + c_2 \sin \omega t$, $u_p = \dfrac{t}{2\omega} \sin \omega t$

9. $u_c = c_1 e^{-t} + c_2 e^{-2t}$, $u_p = \dfrac{\cos t + 3 \sin t}{10}$

11. $u_c = c_1 \cos t + c_2 \sin t$, $u_p = A + Bt + Ce^t$

13. $u_c = (c_1 + c_2 t)e^{-t}$, $u_p = A + Bt^2 e^{-t}$

15. $u_c = c_1 + c_2 t$, $u_p = t^2(A + Ct)$

17. $u_c = c_1 e^t + c_2 e^{-t}$, $u_p = t(Ae^t + Be^{-t})$

19. $u_c = c_1 \cos t + c_2 \sin t$,
    $u_p = A + t(B_1 \cos t + B_2 \sin t) + C_1 \cos 2t + C_2 \sin 2t$

21. $u(t) = c_1 + c_2 e^{-t} - t$; $c_1 = 11$, $c_2 = -1$

23. $u(t) = c_1 \cos 2t + c_2 \sin 2t - \frac{1}{4}t \cos 2t$; $c_1 = 0$, $c_2 = \frac{1}{8}$

25. $u(t) = \frac{1}{2}t^2 e^{-t} + (c_1 + c_2 t)e^{-t}$; $c_1 = c_2 = 1$

27. $u(t) = 1 + (c_1 + c_2 t)e^{-t/2}$; $c_1 = c_2 = 0$

29. $u(t) = e^{-2t}(c_1 \cos 2t + c_2 \sin 2t) + \frac{1}{8}t - \frac{1}{16}$; $c_1 = \frac{1}{16}$, $c_2 = 0$

Section 2.5

1. $u(t) = A - Ae^{-bt/2}\left(\cos \beta t + \dfrac{b}{2\beta} \sin \beta t \right)$ where $A = f_0/\omega_0^2$, $\beta = \sqrt{\omega_0^2 - (b/2)^2}$

3. (a) $u(t) = A(1 - \cos \omega_0 t)$
   (b) $u(t) = A - A(1 - \omega_0 t)e^{-\omega_0 t}$
   (c) $u(t) = A - \dfrac{A}{m_2 - m_1}(m_1 e^{m_2 t} - m_2 e^{m_1 t})$
   where $m_1 = (-2 - \sqrt{3})\omega_0$, $m_2 = (-2 + \sqrt{3})\omega_0$ and $A = f_0/\omega_0^2$.

5. $u(t) = \dfrac{k}{\omega_0^2}\left(t - \dfrac{b}{\omega_0^2} \right) + \dfrac{k}{\omega_0^4}e^{-bt/2}\left(b \cos \beta t + \dfrac{b^2 - 2\omega_0^2}{2\beta} \sin \beta t \right)$
   where β is as in the answer to Exercise 1 above.

7. Take $f(t) = 1 - e^{-kt}$, $k > 0$. The persistent part is $1/\omega_0^2$, if $b > 0$.

9. The persistent part of the solution is

$$u_p(t) = \frac{f_0}{\omega_0 b} \sin \omega_0 t.$$

11. The derivative of $\Delta^{-1/2}$ with respect to ω is 0 when

$$b^2 + 2(\omega^2 - \omega_0^2) = 0.$$

13. Assuming $\omega_0 + \omega$ much greater than $|\omega_0 - \omega|$, the time required for u to achieve its maximum is about the time from 0 to the first maximum or minimum of $\sin[(\omega_0 - \omega)t/2]$, which occurs when

$$|\omega_0 - \omega| t/2 = \pi/2.$$

15. $M = \dfrac{2f_0}{|\omega_0^2 - \omega^2|}, \quad T = \dfrac{\pi}{|\omega_0 - \omega|}$

$$M/T = \frac{2f_0}{\pi(\omega_0 + \omega)} \rightarrow \frac{f_0}{\pi\omega_0}$$

Section 2.6

1. The differential equation is

$$10^{-4}q'' + 10q' + 10^6 q = v_0 e^{i\omega t}.$$

The steady state solution is

$$q(t) = \frac{v_0}{(10^6 - 10^{-4}\omega^2) + j10\omega} e^{j\omega t}.$$

3. $q = \rho e^{j(\omega t - \phi)}$, where

$$\rho = \frac{v_0}{\sqrt{(10^6 - 10^{-4}\omega^2)^2 + (10\omega)^2}} = \frac{v_0}{\Delta}$$

and $\cos \phi = (10^6 - 10^{-4}\omega^2)/\Delta$, $\sin \phi = 10\omega/\Delta$. The phase angle ϕ is between 0 and π:

$$i = \rho j\omega e^{j(\omega t - \phi)} = \rho\omega e^{j(\omega t - \phi + \pi/2)}.$$

5. Since $Z = R + j\left(L\omega - \dfrac{1}{C\omega}\right)$, it is minimized when $LC = 1/\omega^2$. Thus $C = 1/L\omega^2 = 1/(10^{-4})(10^8 \cdot 2\pi)^2 \cong 2.5 \times 10^{-6}$ farad.

7. $Ri + q/C = R \, dq/dt + q/C = v(t)$,

$$q = \frac{v_0}{Rj\omega + 1/C} e^{j\omega t}.$$

9. $V_0(t)/V(t) = (j + (1/RC\omega))^{-1}$. This ratio approaches $-j$ if ω is very large but approaches 0 if ω is very small. (This circuit is called a high-pass filter.)

11. $V_0(t)/V(t) = (1 + jRC\omega)^{-1}$. This ratio approaches 1 if ω is very small but approaches 0 if ω is very large. (This is a low-pass filter.)

Miscellaneous Exercises

1. $u(t) = ce^{-5t} + \dfrac{5 \sin t - \cos t}{26}$

3. $u(t) = c_1 e^t + c_2 e^{-6t} + \dfrac{te^t}{7}$

5. $u(t) = c_1 e^{kt} + c_2 e^{-kt} - \dfrac{1 + 3t}{k^2}$

7. $u(t) = c_1 + c_2 e^{-t} - \dfrac{\sin t + \cos t}{2}$

9. $u(t) = c_1 \cos 2t + c_2 \sin 2t - \dfrac{t}{4} \cos 2t$

11. $u = e^{-2t}(c_1 \cos \sqrt{2}t + c_2 \sin \sqrt{2}t);\ c_1 = 0,\ c_2 = \dfrac{1}{\sqrt{2}}$

13. $u = e^{-4t}(c_1 + c_2 t);\ c_1 = -1,\ c_2 = -4$

15. $u = c_1 e^{-t} + c_2 e^{-4t};\ c_1 = c_2 = 0$ 17. $u(t) = te^{-t}$

19. $u = \frac{1}{2}e^{-t} + c_1 \cos t + c_2 \sin t;\ c_1 = -\frac{1}{2},\ c_2 = \frac{1}{2}$

21. $u = 1 + e^{-2t}(c_1 \cos \sqrt{12}t + c_2 \sin \sqrt{12}t);\ c_1 = -1,\ c_2 = \dfrac{-1}{\sqrt{3}}$

23. $u = \frac{1}{2}t \sin t + c_1 \cos t + c_2 \sin t$ 25. $u = \frac{1}{2} \sin t + e^{-t}(c_1 + c_2 t)$

27. $u = t + c_1 \cos t + c_2 \sin t$ 29. $v(t) \to Mg/p$

Chapter 3

Section 3.1

1. $L(c_1 u_1 + c_2 u_2) = a_1(t)L_1(c_1 u_1 + c_2 u_2) + a_2(t)L_2(c_1 u_1 + c_2 u_2)$

$\quad = a_1(t)[c_1 L_1(u_1) + c_2 L_1(u_2)] + a_2(t)[c_1 L_2(u_1) + c_2 L_2(u_2)]$

$\quad = c_1[a_1(t)L_1(u_1) + a_2(t)L_2(u_1)] + c_2[a_1(t)L_1(u_2) + a_2(t)L_2(u_2)]$

$\quad = c_1 L(u_1) + c_2 L(u_2)$

3. In all cases the linearity of the operator follows from (1) the linearity of the integral and (2) the distributive, associative, and commutative laws of algebra.

5. The general solution may be written $c_1 u_1(t) + c_2 u_2(t)$, where u_1 and u_2 are given by these formulas:

$$u_1(t) = \begin{cases} 1, & t < 0 \\ \cos t, & 0 < t \end{cases}, \qquad u_2(t) = \begin{cases} t, & t < 0 \\ \sin t, & 0 < t \end{cases}.$$

7. The first row of the Wronskian of these three functions is all 0's at $t = -1$. Therefore the Wronskian is 0 there and, by Theorem 3.4, everywhere.

9. The Wronskian of the two functions is $W = -\cos^2 t^{-1} - \sin^2 t^{-1} = -1$.

11. The given functions are solutions of this equation:

$$\frac{d^2 u}{dt^2} + (m_1 + m_2)\frac{du}{dt} + m_1 m_2 u = 0.$$

Their Wronskian is $W = \pm(m_2 - m_1)e^{(m_1 + m_2)t}$, which is never zero.

13. (a) The differential equation is

$$(u_1 u_2' - u_2 u_1')u'' - (u_1 u_2'' - u_2 u_1'')u' + (u_1' u_2'' - u_2' u_1'')u = 0.$$

(b) Obviously, both u_1 and u_2 are solutions, and they were assumed to be independent in the statement.

15. $\begin{vmatrix} t & e^t & u \\ 1 & e^t & u' \\ 0 & e^t & u'' \end{vmatrix} = 0$, or $(t-1)u'' - tu' + u = 0$

17. If the Wronskian of some functions is identically 0, some combination of the functions must be identically 0, but the coefficients need not be constants. For example, let

$$u_1(t) = t^2, \qquad u_2(t) = \begin{cases} t^2, & t \geq 0 \\ -t^2, & t < 0 \end{cases}.$$

Then u_1 and u_2 have continuous first derivatives and $W(u_1, u_2) \equiv 0$ (all t). The combination equal to 0 is $u_1(t) - u_2(t)(t \geq 0)$ and $u_1(t) + u_2(t)(t < 0)$. But no linear combination with constant coefficients is identically zero for all t.

Section 3.2

1. $m^4 - k^4 = 0$, $u(t) = c_1 \sin kt + c_2 \cos kt + c_3 e^{kt} + c_4 e^{-kt}$

3. $m^4 + k^4 = 0$, $\quad u(t) = e^{bt}(c_1 \cos bt + c_2 \sin bt) + e^{-bt}(c_3 \cos bt + c_4 \sin bt)$, where $b = k/\sqrt{2}$.

5. $m^3 - k^3 = 0$, $u(t) = c_1 e^{kt} + e^{-kt/2}\left(c_2 \cos \dfrac{\sqrt{3}kt}{2} + c_3 \sin \dfrac{\sqrt{3}kt}{2}\right)$

7. The characteristic polynomial is $m^3 + am^2 + \omega^2 m = 0$, where $a = k/p$, $\omega^2 = k/M$. The general solution of the differential equation is

$$u = c_1 + e^{-at/2}(c_2 \cos \beta t + c_3 \sin \beta t),$$

where $\beta = \sqrt{\omega^2 - (a/2)^2}$ (assuming that $a < 2\omega$).

9. No. the coefficient is

$$\omega_1^2 \omega_2^2 - \lambda^4 = \frac{(k_1 + k)(k_2 + k) - k^2}{M_1 M_2} > 0.$$

11. Since $\omega_1^2 = \omega_2^2 = 2\lambda^2$, the characteristic polynomial is $m^4 + 4\lambda^2 m^2 + 3\lambda^4$, and the general solution is

$$u(t) = c_1 \cos \sqrt{3}\lambda t + c_2 \sin \sqrt{3}\lambda t + c_3 \cos \lambda t + c_4 \sin \lambda t.$$

13. The characteristic polynomial has roots

$$m = \pm\sqrt{-b \pm \sqrt{b^2 - c}}.$$

If $b^2 > c > 0$, then all the roots are simple and imaginary, and no solution grows without bound.

15. $u_p = (\sin kt)/2k^4$

17. $u_p = \dfrac{1}{4k^3} t \cos kt$

19. $u_p = e^{-t}/(1 - k^4) \quad (k \neq 1)$,
$\quad = -te^{-t} \quad (k = 1)$.

21. $u_p = t^2/2$

23. $u_p = 1$

25. $u_p = (A_0 t + A_1)e^{-t}$

27. $u_p = t^3(A_0 t + A_1)e^t$

29. $u_p = t^3(A_0 t + A_1) + Bte^t$

31. The characteristic polynomial is $m^n - 1$, having roots $m_k = \alpha_k + i\beta_k$, where $\alpha_k = \cos(2\pi k/n)$, $\beta_k = \sin(2\pi k/n)$, for $k = 0, 1, \ldots, n - 1$. The solutions:

  (a) If n is even,

$$u_k(t) = e^{\alpha_k t}(c_k \cos \beta_k t + c'_k \sin \beta_k t) \qquad \left(k = 1, 2, \ldots, \frac{n}{2} - 1\right),$$

  and

$$u_0(t) = c_0 e^t, \qquad u_{n/2}(t) = c_{n/2}e^{-t}$$

  (b) If n is odd, $u_k(t)$ has the same formula as above, for $k = 1, 2, \ldots, (n-1)/2$, and for $k = 0$.

33. (a) $u^{(k+1)} = 0$; (b) m^{k+1}

35. (a) $u'' - 2\alpha u' + (\alpha^2 + \beta^2)u = 0$; (b) $(m - \alpha)^2 + \beta^2$

37. (b) $(m - \alpha)^{k+1}$

39. (b) $(m^2 + \beta^2)^{k+1}$

41. (b) $(m^2 - 2\alpha m + \alpha^2 + \beta^2)^{k+1}$

Section 3.3

1. $q(m) = m + 2$, $u = c/t^2$

3. $q(m) = m^2 - 2m$, $u = c_1 + c_2 t^2$

5. $q(m) = m^2 + 1$, $u = c_1 \cos(\ln t) + c_2 \sin(\ln t)$

7. $q(m) = m^2 - m - 2$, $u = c_1 t^2 + c_2/t$

9. $q(m) = m^2 - 3m + 2$, $u = c_1 t^2 + c_2 t$

11. $u(r) = -\dfrac{r^2}{4} + c_1 \ln r + c_2$

13. $u(r) = c_1 \cos(k \ln r) + c_2 \sin(k \ln r)$

15. $u(\rho) = -\dfrac{\rho^2}{6} + \dfrac{c_1}{\rho} + c_2$

17. $u(\rho) = c_1 \rho^k + c_2/\rho^{k+1}$

19. Since $\dot{x} = 1/t$, $\ddot{x} = -1/t^2$, we get

$$\frac{d^2 v}{dx^2}\left(\frac{1}{t}\right)^2 + \frac{dv}{dx}\left(\frac{-1}{t^2}\right) + \frac{b}{t}\frac{dv}{dx}\left(\frac{1}{t}\right) + \frac{c}{t^2}v = 0$$

  or

$$\frac{d^2 v}{dx^2} + (b - 1)\frac{dv}{dx} + cv = 0.$$

Section 3.4

1. $u_2 = 1 + \frac{1}{2}t \ln\left|\dfrac{t - 1}{t + 1}\right|$

3. $u_2(t) = t^3 \ln t$

5. $u_2(t) = te^{-2t}$

7. $u_p(t) = \dfrac{t^2}{2}e^{-t}$

9. $u_p(t) = \dfrac{(\ln t)^2}{2} t$

11. $u_p(t) = -\cos t \ln |\sec t + \tan t|$

13. $W(u_1, u_2) = c_1 e^{-A_1(t)}$

Section 3.5

1. $U(t) = \dfrac{u_1(z)u_2(t) - u_2(z)u_1(t)}{W(z)}$

is a solution of the homogeneous equation because it is a linear combination of $u_1(t)$ and $u_2(t)$. The first "initial condition" is obviously satisfied. The second follows because the derivative of the numerator, evaluated at $t = z$, is exactly the Wronskian of u_1 and u_2.

3. $\dfrac{du_p}{dt} = G(t, t)f(t) + \displaystyle\int_{t_0}^{t} \dfrac{dG}{dt} f(z)\, dz$

$\dfrac{d^2 u_p}{dt^2} = \dfrac{dG}{dt}(t, t)f(t) + \displaystyle\int_{t_0}^{t} \dfrac{d^2 G}{dt^2} f(z)\, dz$

$\dfrac{d^2 u_p}{dt^2} + a_1(t)\dfrac{du_p}{dt} + a_2 u_p = f(t) + \displaystyle\int_{t_0}^{t} \left(\dfrac{d^2 G}{dt^2} + a_1 \dfrac{dG}{dt} + a_2 G \right) f(z)\, dz$

5. $u_p(t) = \dfrac{t \sin \omega t}{2\omega} + \dfrac{\cos \omega t}{4\omega^2}$. Note that the second term is a solution of the homogeneous equation.

7. $u'(t) = \displaystyle\int_0^t f(x)\, dx, \quad u(t) = \int_0^t \int_0^z f(x)\, dx\, dz.$

9. $u(t) = \displaystyle\int_0^t (t - z)f(z)\, dz$

11. $G(t, z) = \dfrac{z}{2k} \left[\left(\dfrac{t}{z}\right)^k - \left(\dfrac{z}{t}\right)^k \right]$

13. $u(t) = \dfrac{1}{2k} \left[\dfrac{t^2}{4 - k^2} - \left(\dfrac{t^k}{2k(2 - k)} - \dfrac{t^{-k}}{2k(2 + k)} \right) \right]$

15. $u_p(t) = \displaystyle\int_0^t \dfrac{\sinh k(t - z)}{k} f(z)\, dz$ and $u_p(t) = (\cosh kt - 1)/k^2.$

17. $v_1' = -(t^3/6)f, \; v_2' = (t^2/2)f, \; v_3' = -(t/2)f, \; v_4' = \tfrac{1}{6}f,$ and

$u_p(t) = \displaystyle\int_0^t \dfrac{(t - z)^3}{6} f(z)\, dz.$

19. With $u_1(t) = e^{-t/4} \cos (t/2)$ and $u_2(t) = e^{-t/4} \sin (t/2)$, find $W(t) = \tfrac{1}{2}e^{t/2}$ and

$v_1 = \displaystyle\int_0^t e^{z/4} \sin (z/2)\sqrt{z}\, dz, \qquad v_2 = \int_0^t e^{z/4} \cos (z/2)\sqrt{z}\, dz.$

The integrals cannot be evaluated in closed form.

Section 3.6

1. General solution $u(t) = c_1 \cos (\omega \ln t) + c_2 \sin (\omega \ln t)$. Theorem 3.12 would require $(\omega^2 + \frac{1}{4})/t^2 \geq k^2$ (some k); this condition is not met. The conditions of Theorem 3.13 are not met either: $a_2 = \omega^2/t^2$ is not $\geq k^2$ for any $k > 0$ and all $t > t_0$.

3. Theorem 3.13 works: $a_2 - \frac{1}{4}a_1^2 - \frac{1}{2}a_1' = 1$, which is positive.

5. Since $\cos kt$ has zeros spaced π/k units apart, it has at least $(T\pi/k) - 1$ zeros in any interval of length T. The function u must have that many or more.

7. $2a_1a_2 + a_2' = 2\dfrac{1}{t} \cdot \dfrac{\omega^2}{t^2} - \dfrac{2\omega^2}{t^3} \equiv 0$.

9. (a) Theorem 3.12 shows the solutions oscillatory without condition on α.
   (b) Theorem 3.13 requires $a_1 = -2\alpha/t \geq 0$ and $a_2' = -2(\alpha^2 + \alpha)/t^3 \leq 0$. Thus $\alpha^2 + \alpha \geq 0$ and $\alpha \leq 0$; together these give $\alpha \leq -1$.

11. Theorem 3.12 requires $4t^2 - (5/4t^2) \geq k^2$ for $t > 1$; this is certainly satisfied by $k = 1$. Exercise 6 requires $2a_1a_2 + a_2' \geq 0$; the expression comes out to be exactly 0.

Miscellaneous Exercises

1. (a) $u''' - 2u'' + u' = 0$; $u = c_1 + c_2 e^t + c_3 t e^t$

3. (a) $u''' - u = 0$; $u = c_1 e^t + e^{-t/2}(c_2 \cos \beta t + c_3 \sin \beta t)$; $\beta = \sqrt{3}/2$

5. (a) $u^{(4)} + 2u''' - 2u' - u = 0$; (b) $u = c_1 e^t + (c_2 + c_3 t + c_4 t^2)e^{-t}$

7. (a) $u''' - 3u'' + 3u' - u = 0$; (b) $W = 2e^{3t}$

9. (a) $u''' = 0$, (b) $W = -2$

11. (a) $u_1(t) = 1$, $u_2(t) = e^t$, $u_3(t) = e^{-t}$; (b) $W = 2$

13. The Wronskian $W(t, t \ln t, t(\ln t)^2) = 2$. The differential equation is $t^3 u''' + tu' - u = 0$.

15. If t is a solution, so is 1; $u'' = 0$.

17. If $\sin t$ is a solution, so is $\cos t$, and $\pm i$ are roots of the characteristic polynomial; similarly, $\pm 2i$ are roots. Therefore $u^{(4)} + 5u'' + 4u = 0$.

19. $t \cosh t = \frac{1}{2}t(e^t + e^{-t})$: $u^{(4)} - 2u'' + u = 0$

21. $u = c_1 e^t + c_2 e^{-t} + c_3 \cos t + c_4 \sin t$; $c_1 = -c_2 = \frac{1}{4}$, $c_3 = 0$, $c_4 = -\frac{1}{2}$

23. $u = (c_1 + c_2 t)e^t + (c_3 + c_4 t)e^{-t} + (c_5 + c_6 t) \cos t + (c_7 + c_8 t) \sin t$

25. $u = c_1 + c_2 \cos (\ln t) + c_3 \sin (\ln t)$

27. $u = t^{-1/2}(c_1 \cos (\frac{1}{2}\sqrt{3} \ln t) + c_2 \sin (\frac{1}{2}\sqrt{3} \ln t))$

29. $u = c_1 t + c_2 t^{-4} + c_3 t^{-1}$

31. $u = cr^2$

33. The homogeneous equation has solution $u(t) = t$. Use reduction of order to find general solution $u(t) = \frac{1}{6}(\ln t)^3 + (c_1 + c_2 \ln t)t$.

35. Assume $y = ve^x$. Then $xv'' - v' = 0$ gives $v = c_1 x^2 + c_2$ and $y = (c_1 x^2 + c_2)e^x$.

37. Assume $u = v \cos t$. Then $v'' \cos t - 2v' \sin t = 1/\cos t$ gives $v = t \tan t + \ln |\cos t| + c_1 \tan t + c_2$ and $u = t \sin t + \cos t \ln |\cos t| + c_1 \sin t + c_2 \cos t$.

39. Use $y_2 = xv$. Then $xv'' + 2(1 + x^2)v' = 0$ gives $v' = (\exp(-x^2))/x^2$, $v = -(\exp(-x^2))/x - 2\int \exp(-x^2)\,dx$.

41. Using $u_1(t) = e^t$, $u_2(t) = te^t$, find

$$G(t, z) = \frac{te^t e^z - e^t t' e^z}{e^{2z}} = (t - z)e^{(t-z)}.$$

Chapter 4

Section 4.1

1. (a) $(n + 1)c_{n+1} - c_{n-1} = 0$; (b) $c_2 = c_0/2$, $c_4 = c_0/8$, $c_6 = c_0/48$; in general, $c_n = 0$ if n is odd, $c_n = c_0/(2 \cdot 4 \cdots n)$, if n is even; (c) $\rho = \infty$; (d) $y(x) = c_0 e^{x^2/2}$.

3. (a) $(n + 1)c_{n+1} + c_{n-1} = 0$, $c_1 = 1$, $c_0 = 0$; (b) $c_3 = -\frac{1}{3}$, $c_5 = 1/(3 \cdot 5)$, $c_7 = -1/(3 \cdot 5 \cdot 7)$; (c) $\rho = \infty$; (d) $y(x) = e^{-x^2/2}\int_0^x e^{t^2/2}\,dt$; the integral is called Dawson's integral.

5. (a) $(n + 2)(n + 1)c_{n+2} + c_n = 0$, $c_0 = 1$, $c_1 = 0$; (b) $c_0 = 1$, $c_2 = -\frac{1}{2}$, $c_4 = \frac{1}{24}$, $c_6 = -\frac{1}{720}$, and in general, $c_n = 0$ (n odd), $= (-1)^{n/2}/n!$ (n even); (c) $\rho = \infty$; (d) $y = \cos x$.

7. (a) $(n + 2)(n + 1)c_{n+2} - (n + 1)(1 + 2n)c_{n+1} + (n^2 - 1)c_n = 0$ or $(n + 2)c_{n+2} - (1 + 2n)c_{n+1} + (n - 1)c_n = 0$; (b) $c_2 = (c_1 + c_0)/2$, $c_3 = c_4 = c_5 = \cdots = c_2$. (c) $\rho = 1$; (d) $y = A(x - 1) + B/(1 - x)$.

9. (a) $(n + 2)(n + 1)c_{n+2} - c_n - c_{n-2} = 0$; (b) $c_2 = c_0/2$, $c_3 = c_1/6$, $c_4 = 3c_0/4!$, $c_5 = 7c_1/5!$; (c) $\rho = \infty$; (d) ?

11. (a) $(n + 2)(n + 1)c_{n+2} + (n + 1)c_n = 0$; (b) $c_2 = -c_0/2$, $c_3 = -c_1/3$, $c_4 = c_0/(2 \cdot 4)$, $c_5 = c_1/(3 \cdot 5)$; (c) $\rho = \infty$; (d) one solution is $e^{-x^2/2}$; a second is Dawson's integral.

13. $-2c_2 = 0$, $2c_0 - 6c_3 = 0$ and $-(n + 2)(n + 1)c_{n+2} + n(n - 1)c_n + 2c_{n-1} + c_{n-2} = 0$; c_0, c_1 are arbitrary; $c_2 = 0$, $c_3 = c_0/3$, $c_4 = (c_0 + 2c_1)/12$, $c_5 = (2c_0 + c_1)/20$, $c_0 = (5c_0 + 6c_1)/90$.

Section 4.2

1. For $k = 2$, $P_2(x) = y_0(x)/y_0(1)$, and $y_0(x) = c_2 x^2 - \frac{1 \cdot 2}{2 \cdot 3}c_2$. Thus $P_2(x) = (x^2 - \frac{1}{3})/(1 - \frac{1}{3}) = \frac{1}{2}(3x^2 - 1)$. Similarly, for $k = 2$, $P_3(x) = y_1(x)/y_1(1)$, and $y_1(x) = c_3 x^3 - \frac{2 \cdot 3}{2 \cdot 5}c_3 x$. Then $p_3(x) = (x^3 - \frac{3}{5}x)/(1 - \frac{3}{5}) = \frac{1}{2}(5x^3 - 3x)$.

3. Assume that the solution has the form $y = vx$. Substitution into Eq. (4.11b) leads to $v''x(1 - x^2) + v'(2(1 - x^2) - 2x^2) = 0$. Then $v' = 1/x^2(1 - x^2)$ and

$$v = \frac{1}{2}\ln\left(\frac{1 + x}{1 - x}\right) - \frac{1}{x}.$$ Finally, the second solution is $y = xv = \frac{1}{2}x\ln\left(\frac{1 + x}{1 - x}\right) - 1$.

5. $P_0(\cos \phi) = 1$; $P_1(\cos \phi) = \cos \phi$; $P_2(\cos \phi) = (1 + 3 \cos 2\phi)/4$; $P_3(\cos \phi) = (3 \cos \phi + 5 \cos 3\phi)/8$

7. Differentiating $k + 1$ times gives

$$(x^2 - 1)F^{(k+2)} + (k + 1)2xF^{(k+1)} + \frac{(k + 1)k}{2} \cdot 2F^{(k)}$$
$$= 2k(xF^{(k+1)} + (k + 1)F^{(k)}).$$

When terms are collected, this is seen to be Legendre's equation with $F^{(k)}$ in place of y.

9. $c_{n+2} = \dfrac{2(n - k)}{(n + 1)(n + 2)} c_n$. All the coefficients with even (odd) indices are multiples of c_0 (of c_1). Because of the $(n - k)$ in the numerator of the relation above, when k is an integer, $c_{k+2} = 0$ as do c_{k+4}, c_{k+6}, etc.

11. For $k = 2$, $y_0 = 1 - 2x^2$; for $k = 3$, $y_1 = x - \frac{2}{3}x^3$.

13. $y_0(0) = 1$, $y_0'(0) = 0$, $y_1(0) = 0$, $y_1'(0) = 1$. Thus $W(y_1, y_2) = 1$ at $x = 0$. Since the coefficients of Legendre's equation in normal form (coefficient of y'' is 1) have zero denominators at $x = \pm 1$, the interval is $-1 < x < 1$.

Section 4.3

1. (a) $r(r - 1) = 0$; (b) $c_1 = \frac{1}{2}c_0$, $c_2 = \frac{1}{12}c_0$, $c_3 = \frac{1}{144}c_0$; (c) $n(n + 1)c_n - c_{n-1} = 0$; (d) $c_n = c_0/n!(n + 1)!$

3. (a) $4r(r - 1) + 1 = 0$; (b) $c_1 = -c_0$, $c_2 = c_0/4$, $c_3 = -c_0/36$; (c) $n^2c_n + c_{n-1} = 0$; (d) $c_n = (-1)^n c_0/(n!)^2$

5. (a) $r^2 = 0$; (b) $c_4 = -c_0/16$, $c_8 = c_0/1024$, $c_{12} = -c_0/1024 \cdot 144$; (c) $n^2c_n + c_{n-4} = 0$; (d) $c_{4m} = (-1)^m c_0/4^{2m}(m!)^2$. Coefficients with indices not divisible by 4 are all 0.

7. Indicial equation: $2r^2 = 0$

$$c_1 = -\frac{M}{2} c_0, \quad c_2 = -\frac{(M - 2)M}{16} c_0, \quad c_3 = -\frac{(M - 6)(M - 2)M}{32 \cdot 9} c_0$$

Relation $c_{n+1} = \dfrac{M - n(n + 1)}{2(n + 1)^2} c_n$

9. $y(x) = c_0(1 - 6x - \frac{15}{2}x^2 - \frac{5}{2}x^3)$

11. (a) $x = \pm 1$, regular; (b) $x = 0, 1$, regular; (c) $x = 0, -2$, regular; $x = 2$, irregular; (d) $x = 0$, regular.

Section 4.4

1. Indicial equation $r^2 - k^2 = 0$. Take $r = k \geq 0$ and then $c_1 = 0$, $c_n = c_{n-2}/n(n + 2k)$, $n \geq 2$. The c's are just as in J_k, except that here, all have the same sign.

3. $\dfrac{dy}{dx} = \lambda J_k'$, $\dfrac{d^2y}{dx^2} = \lambda^2 J_k''$, where the prime denotes differentiation with respect

to the argument. Substituting the expressions above into the equation in the Summary gives

$$x^2\lambda^2 J_k''(\lambda x) + x\lambda J_k'(\lambda x) + (\lambda^2 x^2 - k^2)J_k(\lambda x) = 0,$$

which is just Bessel's equation with λx in place of x.

5. If $y(x)$ is a solution of the equation in Exercise 1 with $k = 0$, then $u(r) = y(\gamma r)$ is a solution of the homogeneous equation in Exercise 4.

7. Reduction of order gives a second independent solution as

$$y(x) = J_0(x) \int \frac{dx}{xJ_0^2(x)} \cong J_0(x) \ln x, \qquad x \ll 1.$$

9. $\dfrac{d^2 v}{dx^2} + \left(1 + \dfrac{1 - 4k^2}{4x^2}\right)v = 0.$

11. With $k = \frac{1}{2}$ the equation above becomes $v'' + v = 0$ with general solution $v = A \cos x + B \sin x$. Then the solution of Bessel's equation with $k = \frac{1}{2}$ is $v/\sqrt{x}$.

Section 4.5

1. After substituting the form for y_2 as given in case 3 of Theorem 4.3, one finds

$$2\beta \sum_1^\infty (n - \tfrac{1}{2})c_{n-1}x^{n-1/2} + \sum_0^\infty ((n - \tfrac{1}{2})^2 - \tfrac{1}{4})b_n x^{n-1/2} + \sum_2^\infty b_{n-2}x^{n-1/2} = 0.$$

Equate net coefficients to zero, to obtain

$$[(0 - \tfrac{1}{2})^2 - \tfrac{1}{4}]b_0 = 0 \qquad (\text{coeff. of } x^{-1/2}),$$

$$2\beta(1 - \tfrac{1}{2})c_0 + [(1 - \tfrac{1}{2})^2 - \tfrac{1}{4}]b_1 = 0 \qquad (\text{coeff. of } x^{1/2}).$$

Since $c_0 \neq 0$, β must be 0.

3. The indicial equation is $r^2 - 1 = 0$, so the roots differ by an integer. Assume $y_2 = \beta y_1 \ln x + \sum b_n x^{n-1}$. After substituting in the equation, one finds that $-b_1 = 0$ (coeff. of x^0); $\beta \cdot 2c_0 + b_0 = 0$ (coeff. of x); and in general $(n^2 - 1)b_{n+1} + b_{n-1} + 2n\beta c_{n-1} = 0$ (coeff. of x^n). If β were 0, b_0 would have to be 0. Thus $\beta \neq 0$, and there is a logarithmic term.

5. From Exercise 1 of Section 4.3, $y_1 = x \sum_0^\infty \dfrac{x^n}{n!(n+1)!}$ and $r_2 = 0$. Then $y_2 = \beta y_1 \ln x + v(x)$ where $v = b_0 + b_1 x + \cdots$. Use the result of Exercise 2: $x^2 v'' - xv + \beta(2xy_1' - y) = 0$. The coefficients of the resulting series are $\beta - b_0 = 0$; $2\beta - \tfrac{1}{2}\beta + 2b_2 - b_1 = 0$; and in general

$$\beta\left(\frac{2}{(n-1)!(n-1)!} - \frac{1}{n!(n-1)!}\right) + n(n-1)b_n - b_{n-1} = 0.$$

From the first equation, β must be nonzero; take $\beta = 1 = b_0$. From the second, b_1 is arbitrary; take $b_1 = 0$ and find $b_2 = -\tfrac{3}{4}$. Then $b_3 = -\tfrac{7}{36}$, $b_4 = -35/1728$.

7. Since $r = 0$ is a double root of the indicial equation, the second solution has

the form

$$y_2(x) = y_1 \ln x + (b_1 x + b_2 x^2 + \cdots).$$

Section 4.6

1. If k is not an integer, $y_2(x)$ can be written in the power-times-series form, with a *negative* power of x as multiplier. If k is an integer other than 0, y_2 has the logarithmic term and also a power-times-series term, starting with a negative power of x. Finally, if $k = 0$, y_2 contains the term $J_0(x) \ln x$, and $|\ln x| \to \infty$ as $x \to 0$.

3. $Y_k = -J_{-k}/\sin k\pi = (-1)^{k+1/2} J_{-k}$ for $k = \frac{1}{2}, \frac{3}{2}$, etc.

5. Substitute c_{2m} and b_{2m} into Eq. (4.98) to get

$$4m \frac{(-1)^m}{2^{2m}(m!)^2} + 4m^2 \frac{(-1)^{m-1} h_m}{2^{2m}(m!)^2} + \frac{(-1)^{m-2} h_{m-1}}{2^{2m-2}((m-1)!)^2} = 0.$$

Multiply through by $(-1)^m 2^{2m}(m!)^2$ to get

$$4m - 4m^2 h_m + 4m^2 h_{m-1} = 0.$$

Then verify that this is true, from the definition of h_m, Eq. (4.99).

Section 4.7

1. According to Example 4.11, the equation is $xy'' + 2y' = 0$, with solution $y = c_0 + c_1/x$.

3. The candidates for singular points are $0, 1, \infty$. Both 0 and 1 are regular points if $c = 0$, $ab = 0$, and $a + b + 1 = 0$; 0 is regular, and 1 is regular singular if $ab = c = 0$, $a + b + 1 \neq 0$; 1 is regular and 0 is regular singular if $ab = 0$, $c = a + b + 1 \neq 0$. The point ∞ is regular if $ab = 0$ and $a + b + 1 = -2$ and regular singular otherwise.

5. Using $y = e^{mx} \sum_0^\infty c_n x^{n+r}$, $c_0 \neq 0$, substituting into the differential equation, and equating net coefficients to 0, we get $(m^2 + 1)c_0 = 0$ from the lowest power of x; $(m^2 + 1)c_1 + (2r + 1)mc_0 = 0$ (next); $(m^2 + 1)c_n + (2(r - n + 1) + 1)mc_{n-1} + (r - n + 2)^2 c_{n-2} = 0$ (general). Since $c_0 \neq 0$, $m^2 + 1 = 0$. Take $m = i$. Then $(2r + 1)ic_0 = 0$, so $r = -\frac{1}{2}$. Then, in general,

$$c_{n-1} = -\frac{(\frac{5}{2} - n)^2}{(4 - 2n)i} c_{n-2}.$$

For instance, $c_1 = -ic_0/8$, $c_2 = -9c_0/128$, etc. The coefficients with odd index are imaginary multiples of c_0, and those with even index are real multiples. Writing $y = e^{ix}(s_0 + is_1)/\sqrt{x} = (\cos x + i \sin x)(s_0 + is_1)/\sqrt{x}$, find

$$y = \frac{c_0}{\sqrt{x}} \left[\cos x \left(1 - \frac{9}{128x^2} + \cdots \right) + \sin x \left(\frac{1}{8x} - \frac{125}{1024x^3} + \cdots \right) \right].$$

7. The parameters are determined by the equations $m^2 + a = 0$, $2mr + b = 0$.

Section 4.8

1. $A = \dfrac{C_0 m_2 + C_1 R}{m_2 - m_1}, \qquad B = \dfrac{C_0 m_1 + C_1 R}{m_1 - m_2}$

3. Since $-M_a R = m_1 m_2 < 0$, the two roots cannot be complex conjugates and must have opposite signs.

5. Since $\ln(1 - x)$ has a series with center 0 and radius of convergence 1, the expected radius of convergence is 1.

Miscellaneous Exercises

1. Using $(1 - x)^{-1} = 1 + x + x^2 + \cdots + x^n + \cdots$, all c's are determined by the relation

   $$(n + 2)(n + 1)c_{n+2} + c_n = 1, \qquad n = 0, 1, 2, \ldots.$$

   The first few are $c_0 = c_1 = 0$, $c_2 = \frac{1}{2}$, $c_3 = \frac{1}{6}$, $c_4 = c_5 = \frac{1}{24}$.

3. The series for y^2 has general term

   $$(c_0 c_n + c_1 c_{n-1} + \cdots + c_{n-1} c_1 + c_n c_0) x^n.$$

   The initial condition gives $c_0 = 1$, and the first few coefficients are $c_1 = c_2 = 1$, $c_3 = \frac{4}{3}$, $c_4 = \frac{7}{6}$, $c_5 = \frac{6}{5}$.

5. $c_0 = 1$, $c_1 = -\frac{1}{30}$, $c_2 = \frac{1}{240}$.

7. Using $y = c_0 + c_1 x + \cdots$ and $(1 - x)^{-1}$ as in the solution to Exercise 1 above, we find that

   $$(n + 2)(n + 1)c_{n+2} + c_n + c_{n-1} + \cdots + c_0 = 0$$

   is valid for $n \geq 0$. Thus $c_2 = -c_0/2$, $c_3 = -(c_1 + c_0)/6$, $c_4 = -(c_0 + 2c_1)/24$. The radius of convergence should be 1.

9. Use $y = c_0 + c_1 x + \cdots$ and $e^x = \sum_0^\infty x^n/n!$. The general relation for the c's is

   $$(n + 2)(n + 1)c_{n+2} = \frac{c_0}{n!} + \frac{c_1}{(n - 1)!} + \cdots + c_{n-1} + c_n.$$

   The first few c's are $c_2 = c_0/2$, $c_3 = (c_0 + c_1)/2$, $c_4 = (c_0 + c_1)/12$, $c_5 = (5c_0 + 4c_1)/120$.

11. $x^2 + 2x + 1 = \frac{2}{3}P_2 + 2P_1 + \frac{4}{3}P_0$

13. Assuming $y = c_0 + c_1 x + \cdots + c_n x^n + \cdots$, we find that the c's are determined by the relation

    $$(n + 1)c_{n+1} + \beta^n c_n = 0.$$

    The general solution is $c_n = c_0 \beta^N/n!$, where $N = n(n - 1)/2$.

15. By the usual process, find $c_0 r(r - 1) = 0$. Since $c_0 \neq 0$, $r(r - 1) = 0$. Take $r = 1$. Then $c_1 = c_2 = c_3 = c_4 = 0$, but $c_5 = 4c_0/35$. In general,

    $$\left(\frac{n}{2} + 3\right)\left(\frac{n}{2} + 2\right)c_{n+4} - c_{n-1} = 0$$

determines the c's. Only those with subscripts divisible by 5 are nonzero: $c_{10} = 4c_0/420$, $c_{15} = 16c_0/17 \cdot 15 \cdot 7 \cdot 5$.

17. Taking $y = \sum c_n x^{n+r}$, $c_0 \neq 0$, find the indicial equation $r^2 = 0$, and then the relation

$$n^2 c_n + c_{n-k} = 0, \qquad n \geq k.$$

(As $c_1 = c_2 = \cdots = c_{k-1} = 0$, the solution is really a series of powers of x^k.) A solution is $c_{mk} = (-1)^m / k^{2m} (m!)^2$.

19. The indicial equation is $r(r - 1) - k(k + 1) = 0$. Taking $r = k + 1$, the relation between coefficients in the series for y is

$$n(n + 2k + 1)c_n + c_{n-2} = 0, \qquad n = 2, 3, \ldots.$$

21. Take $y = \sum c_n x^{n+r}$, $c_0 \neq 0$. Then the indicial equation is $r^2 - \frac{1}{4} = 0$. Use $r = \frac{1}{2}$. The relation between the c's is $n(n + 1)c_n - c_{n-1} = 0$. A solution is $c_n = 1/(n + 1)!n!$.

23. Take $y = \beta y_1 \ln x + \sum b_n x^{n-1/2}$, with y_1 the solution of Exercise 21. The usual manipulations lead to $\beta c_0 - b_0 = 0$ and for $n \geq 1$,

$$\beta(2n + 1)c_n + n(n + 1)b_{n+1} - b_n = 0.$$

If $\beta = 0$, the second solution is not independent.

25. Take $y = \sum c_n x^{r-n}$, $c_0 \neq 0$. The indicial equation is $r(r + 1) - k(k + 1) = 0$. Use $r = k$ and obtain the relation

$$n(n - 2k - 1)c_n + (n - k - 2)(n - k - 1)c_{n-2} = 0.$$

The series converges for $|x| > 1$.

Chapter 5

Section 5.1

1. $F(s) = \dfrac{\omega}{(s - a)^2 + \omega^2}$

3. $F(s) = \dfrac{1}{2}\left(\dfrac{1}{s - a} + \dfrac{1}{s + a}\right) = \dfrac{s}{s^2 + a^2}$

5. $F(s) = \dfrac{1}{(s - m)(s - n)}$

7. $F(s) = \dfrac{1}{(s - k)^2}$

9. $F(s) = \dfrac{a}{(s + b)^2 - a^2}$

11. $\lim_{n \to m} f(t) = te^{mt}$; $\lim F(s) = \dfrac{1}{(s - m)^2} = \mathcal{L}(te^{mt})$

13. $\cos^2 kt = \frac{1}{2}(1 + \cos 2kt)$, so $\mathcal{L}(\cos^2 kt) = \dfrac{1}{2}\left(\dfrac{1}{s} + \dfrac{s}{s^2 + 4k^2}\right) = \dfrac{s^2 + 2k^2}{s(s^2 + 4k^2)}$

15. $\mathcal{L}(\sin kt \cos kt) = \dfrac{k}{s^2 + 4k^2}$

17. If $s > \alpha$, then the integrand is $e^{-st}f(t) = e^{-(s-\alpha)t} \cdot [e^{-\alpha t}f(t)]$. The function in brackets is continuous and tends to 0 as $t \to \infty$. Therefore the improper integral converges, by the comparison test using $e^{-(s-\alpha)t}$.

Section 5.2

1. $U(s) = \dfrac{1}{(s + 1)(s + 2)} + \dfrac{3}{s + 2}$, $u(t) = e^{-t} + 2e^{-2t}$

3. $U(s) = \dfrac{1}{(s + 2)^2}$, $u(t) = te^{-2t}$

5. $U(s) = \dfrac{s + 2}{s^2 + 2s + 4}$, $u(t) = e^{-t}\left(\cos \sqrt{3}t + \dfrac{1}{\sqrt{3}} \sin \sqrt{3}t\right)$

7. $U(s) = \dfrac{1}{(s + 1)^3}$, $u(t) = \dfrac{t^2}{2} e^{-t}$

9. $U(s) = \dfrac{1}{s^2(s + 2)}$, $u(t) = -\dfrac{1}{4} + \dfrac{t}{2} + \dfrac{e^{-2t}}{4}$

11. $U(s) = \dfrac{1}{(s^2 + 4)(s + 2)}$, $u(t) = \frac{1}{8}(e^{-2t} - \cos 2t + \sin 2t)$

13. $U(s) = \dfrac{s}{(s^2 + 4)(s^2 + 1)} + \dfrac{s}{s^2 + 4}$, $u(t) = \dfrac{2 \cos 2t + \cos t}{3}$

15. $U(s) = \dfrac{1}{s^3}$, $u(t) = \dfrac{t^2}{2}$

Section 5.3

1. $u_p = te^{-4t}$

3. $u_p = -\frac{1}{3} \sin 2t$

5. $u_p = \frac{1}{2}e^{-t}$

7. $u_p = \frac{1}{4}(t^2 \sin t + t \cos t)$

9. $u_p = \frac{1}{5} \sin t - \frac{2}{5} \cos t$

11. (a) $u = c_1 \sinh \lambda x + c_2 \sin \lambda x$;

   (b) $u = c_1(\sinh \lambda x - \sin \lambda x) + c_2(\cosh \lambda x - \cos \lambda x)$;

   (c) $u = c_1(\sinh \lambda x + \sin \lambda x) + c_2(\cosh \lambda x + \cos \lambda x)$

13. $\mathrm{Re}\,(\Delta(i\theta)) = (\omega^2 - \theta^2)(5 + \lambda^2 - \theta^2) - \lambda^2\omega^2$, $\mathrm{Im}\,(\Delta(i\theta)) = (\omega^2 - \theta^2)2\theta$. The imaginary part can be 0 if and only if $\theta^2 = \omega^2$, but then the real part is $-\lambda^2\omega^2 \neq 0$.

15. $U(s) = \dfrac{s + 1 - e^{-s}}{s(s - e^{-s})}$

Section 5.4

1. $U(s) = \dfrac{1 - e^{-s}}{s(s + a)}$, $u(t) = \dfrac{1 - e^{-at}}{a} - h(t - 1)\dfrac{1 - e^{-a(t - 1)}}{a}$

3. $U(s) = \dfrac{1 - 2e^{-s} + e^{-2s}}{s(s + a)}$, $u(t) = g(t) - 2h(t - 1)g(t - 1) + h(t - 2)g(t - 2)$

   where $g(t) = (1 - e^{-at})/a$.

5. $U(s) = \dfrac{1 - e^{-s}}{s(s^2 + 1)}$, $u(t) = 1 - \cos t - h(t - 1)(1 - \cos (t - 1))$

7. $U(s) = \dfrac{1 - 2e^{-s\pi} + e^{-2s\pi}}{s(s^2 + 1)}$, $\quad u(t) = h(t) - 2h(t - \pi) + h(t - 2\pi)$

$\qquad + [h(t) + 2h(t - \pi) + h(t - 2\pi)] \cos t$

9. $U(s) = \dfrac{1 - e^{-s}}{s^2(s^2 + 3s + 2)}$, $\quad u(t) = g(t) + h(t - 1)g(t - 1)$, $\quad$ where $\quad g(t) =$

$\dfrac{t}{2} - \dfrac{3}{4} + e^{-t} - \dfrac{1}{4}e^{-2t}$

11. $U(s) = \dfrac{1}{s^2 + 1}$, $\quad u(t) = \sin t$

13. $\mathcal{L}(f) = \displaystyle\sum_{k=0}^{\infty} \dfrac{e^{-ks}}{s} = \dfrac{1}{s(1 - e^{-s})}$

15. The Heaviside function $h(t)$ has no derivative at $t = 0$, and $h'(t) = 0$ for $t \neq 0$. However, if Eq. (5.54) is accepted, $h(t - a) = \int_0^t \delta(\tau - a)\, d\tau$.

17. $U(s) = \dfrac{1}{(s + 1)^2}(1 + e^{-\tau s} + \cdots + e^{-K\tau s})$

$\qquad u(t) = te^{-t} + h(t - \tau)(t - \tau)e^{-(t - \tau)} + \cdots + h(t - K\tau)(t - K\tau)e^{-(t - K\tau)}$

Section 5.5

1. (a) $(t - 1) + e^{-t}$; (b) $\frac{1}{2}(e^{-t} - \cos t + \sin t)$; (c) $\sin t$

3. $u(t) = \displaystyle\int_0^t f(z) \sin \omega(t - z)\, dt/\omega$

5. (a) te^{at}; (b) $\dfrac{1}{a^2 + 1}(e^{-at} + a \sin t - \cos t)$; (c) $(\sin \omega t - \omega t \cos \omega t)/2\omega$

7. $u(t) = \displaystyle\int_0^t \dfrac{e^{(t-z)} - e^{-(t-z)}}{(e^z + e^{-z})}\, dz = \dfrac{e^t}{2}\displaystyle\int_0^t \dfrac{e^{-z}}{e^z + e^{-z}}\, dz - \dfrac{e^{-t}}{2}\displaystyle\int_0^t \dfrac{e^z}{e^z + e^{-z}}\, dz$

$\qquad = -\dfrac{e^t}{4}\ln\left(\dfrac{1 + e^{-2t}}{2}\right) - \dfrac{e^{-t}}{4}\ln\left(\dfrac{1 + e^{2t}}{2}\right)$

The form of the solution can be changed in many ways by using properties of exponentials.

9. $u(t) = \dfrac{1}{\omega}\displaystyle\int_0^t \sin \omega(t - z) \sin \omega z\, dz = \dfrac{\sin \omega t}{2\omega^2} - \dfrac{t \cos \omega t}{2\omega}$

11. $U(s) = n(0)\left(\dfrac{1}{s} - \dfrac{1}{s + \alpha}\right)(s + \alpha)$; $u(t) = n(0)\alpha$

13. $U(s) = n(0)\dfrac{2s^2 + 1}{2s^3 - \frac{8}{3}s^2 + 2s - \frac{4}{3}} = \dfrac{n(0)}{8}\left[\dfrac{9}{s - 1} - \dfrac{s - 2}{s^2 - \frac{1}{3}s + \frac{2}{3}}\right]$

$\qquad u(t) = \dfrac{n(0)}{8}\left(9e^{-t} - e^{t/6}\left(\cos \alpha t - \dfrac{11}{6\alpha}\sin \alpha t\right)\right)$ where $\alpha = \sqrt{23}/6$

Section 5.6

1. The function dx/dt is zero at time 0, and its zeros are spaced π/β apart. The function x has a minimum at time 0 and a maximum at $T = \pi/\beta$.

3. Using β from Eq. (5.79), find

$$\frac{d}{db}(e^{-b\pi/2\beta}) = -e^{-b\pi/2\beta}(\pi c/2\beta^3) < 0,$$

$$\frac{dT}{db} = \pi b/2\beta^3 > 0.$$

5. (a) $M\dfrac{d^2x}{dt^2} + p\dfrac{dx}{dt} = k(u - x)$

(b) $M\dfrac{d^2x}{dt} = k(u - x) + p\left(\dfrac{du}{dt} - \dfrac{dx}{dt}\right)$

7. $x(t) = \alpha t - d + d\cos\sqrt{k}t - \dfrac{\alpha}{\sqrt{k}}\sin\sqrt{k}t$

9. The characteristic polynomial is $m^2 + 2km + k = 0$, with roots $m = -k \pm \sqrt{k^2 - k}$. For any positive k these roots are negative or have negative real part.

11. The equation to solve is

$$J\frac{d^2x}{dt^2} + p\frac{dx}{dt} = k(u - x) + h\frac{du}{dt}$$

With $u = \alpha t$ the persistent part of $x(t)$ is $\alpha t + (h - p)\alpha/k$.

13. The factorized characteristic polynomials are

$$m^3 + 2m^2 + 4m + 3 = (m + 1)(m^2 + m + 3)$$

$$m^3 + 2m^2 + 4m + 8 = (m + 2)(m^2 + 4)$$

$$m^3 + 2m^2 + 4m + 21 = (m + 3)(m^2 - m + 7)$$

15. $x_p(t) = A\cos\omega t + B\sin\omega t$

$A = ((c - \omega^2)a\omega - (a + b)\omega c)/\Delta$

$B = ((a + b)a\omega^2 + (c - \omega^2)c)/\Delta$

$\Delta = (c - \omega^2)^2 + (a + b)^2\omega^2$

17. $x_p(t) = A\cos\omega t + B\sin\omega t$

$A = (c\omega(h - \omega^2b) - h\omega(c - \omega^2))/\Delta$

$B = (c\omega^2(c - \omega^2) + h(h - \omega^2b))/\Delta$

$\Delta = (h - \omega^2b)^2 + \omega^2(c - \omega^2)^2$

Section 5.7

1. $G(s) = \omega(1 - e^{-2\pi s/\omega})/(s^2 + \omega^2)$, $F(s) = \omega/(s^2 + \omega^2)$

3. $G(s) = (1 - 2e^{-as} + e^{-2as})/s^2 = (1 - e^{-as})^2/s^2$, $F(s) = (1 - e^{-as})/s^2(1 + e^{-as})$.
   Note that this function is also the integral of the square wave in Example 5.21.

5. $G(s) = \omega(1 + e^{-s\pi/\omega})/(s^2 + \omega^2)$, $F(s) = G(s)/(1 - e^{-s\pi/\omega})$

7. Both numerator and denominator are zero when $s = -\alpha$. L'Hopital's rule gives the limit as $s \to -\alpha$ as $p/(1 - e^{-\alpha p})$.

9. The pole at $s = -\alpha$ contributes this term to the charge:

$$\frac{E(1 + e^{\alpha \pi})}{R(\alpha^2 + 1)(1 - e^{\alpha \pi})} e^{-\alpha t}.$$

Miscellaneous Exercises

1. $U(s) = \dfrac{1}{(s - 2)^2(s - 3)}$; $u(t) = e^{3t} - (t + 1)e^{2t}$

3. $U(s) = \dfrac{s}{(s^2 + 1)(s^2 + 2s + 10)}$ with zero initial conditions. Discard the term of the partial fractions expansion that corresponds to the factor $(s^2 + 2s + 10)$. A particular solution is $u(t) = (9 \cos t + 2 \sin t)/85$.

5. $U(s) = 1/(s - 1)^4$ with zero initial conditions. $u(t) = t^3 e^t/6$

7. $U(s) = \dfrac{1 - s - e^{-s}}{s^2(s^2 + 1)} = \dfrac{-1}{s} + \dfrac{1}{s^2} + \dfrac{s - 1}{s^2 + 1} - e^{-s}\left(\dfrac{1}{s^2} - \dfrac{1}{s^2 + 1}\right)$;

$u(t) = t - 1 + \cos t - \sin t - h(t - 1)(t - 1 - \sin (t - 1))$.

9. $\dfrac{du}{dt} + u = 0$, $u(0) = 1$

11. $\dfrac{d^2u}{dt^2} + 3\dfrac{du}{dt} + 2u = 0$, $u(0) = 1$, $\dfrac{du}{dt}(0) = -3$

13. $\dfrac{d^2u}{dt^2} + u = -2 \sin 2t$, $u(0) = 0$, $\dfrac{du}{dt}(0) = 1$; or

$\dfrac{d^2u}{dt^2} + 4u = -\sin t$, $u(0) = 0$, $\dfrac{du}{dt}(0) = 1$

15. $\dfrac{dF}{ds} = \displaystyle\int_0^\infty \dfrac{d}{ds}(e^{-st}f(t))\, dt = \int_0^\infty -te^{-st}f(t)\, dt = \mathcal{L}(-tf(t))$

17. $U(s) = \dfrac{1}{(s + 1)^2}\displaystyle\sum_{k=0}^\infty e^{-k\tau s}$

19. $U(s) = \dfrac{1}{(s + 1)^2(1 - e^{-\tau s})}$. The denominator is 0 at $s = -1$ (double root) and at $s = 0$. The terms in the partial fractions expansion are $\dfrac{A}{(s + 1)^2} + \dfrac{B}{s + 1} + \dfrac{C}{s}$ with

$A = \dfrac{1}{1 - e^\tau}$, $B = \dfrac{-e^\tau \tau}{(1 - e^\tau)^2}$, $C = \dfrac{1}{\tau}$.

Appendix to Chapter 5

1. $\dfrac{1}{2}\left(\dfrac{1}{(s - 1)^2} - \dfrac{1}{s - 1} + \dfrac{s}{s^2 + 1}\right)$

3. $-\dfrac{1}{s^2} + \dfrac{3}{s} - \dfrac{2}{(s + 1)^2} - \dfrac{3}{s + 1}$

5. $\dfrac{-\frac{1}{3}}{s-1} + \dfrac{\frac{1}{3}}{s+1} + \dfrac{\frac{5}{12}}{s-2} - \dfrac{\frac{5}{12}}{s+2}$ 7. $\dfrac{-\frac{2}{3}}{s^2+1} + \dfrac{\frac{5}{3}}{s^2+4}$

9. $\dfrac{1}{4}\left(\dfrac{1}{s-1} + \dfrac{1}{(s-1)^2} - \dfrac{1}{s+1} + \dfrac{1}{(s+1)^2}\right)$

11. $\dfrac{s}{s^2+1} - \dfrac{s}{(s^2+1)^2}$

Chapter 6

Section 6.1

1. $u_{n+1} = 1.1u_n,\ u_0 = 1;\ u(0.5) \cong 1.611,\ u(1) \cong 2.594$

3. $u_{n+1} = u_n + 0.1(1 + u_n^2),\ u_0 = 0;\ u(0.5) \cong 0.532,\ u(1) \cong 1.396.$ The keystroke sequence for a simple calculator is shown at left.

```
 +
 1
 ÷
RM
 ×
.1
 =
M+
RM
```

5. $u_{n+1} = 0.8u_n,\ u_0 = 1;\ u(0.5) \cong 0.328,\ u(1) \cong 0.107$

7. (1) $u = e^t,\ u(1) = 2.718\ldots$
   (2) $u = \sinh(t + \sinh^{-1} 1),\ u(1) = 3.20\ldots$
   (3) $u = \tan t,\ u(1) = 1.557\ldots$
   (4) $u - \ln(1 + u) = t + 1 - \ln 2,\ u(1) = 3.583\ldots$
   (5) $u = e^{-2t},\ u(1) = 0.135\ldots$

9. $u(1) \cong 0.2315\ (h = 0.2)$ and $u(1) \cong 0.2319\ (h = 0.1)$

11. By using the modified Euler method and $h = 0.1$, the following approximations for $u(1)$ were found:

$u(0) = 0.2$	0.4	0.6	0.8	1.0	1.2	1.4	1.6	1.8	2.0
$u(1) \cong 0.4046$	0.6444	0.8030	0.9157	1.0	1.0655	1.1180	1.1611	1.1972	1.2279

All solution curves with $u(0)$ between 0 and 1 rise toward 1; those with $u(0)$ between 1 and 2 fall toward 1. This is in agreement with graphical analysis.

13.

h	Euler $u(1) \cong$	Modified Euler $u(1) \cong$
0.2	1.5819	1.48598422
0.1	1.5340	1.48593131
0.05	1.5099	1.48591821
0.025	1.4979	1.48591495

Section 6.2

1.

t	Extrapolated	Exact	Error
0.2	1.4399	1.44	0.0001
0.4	1.9491	1.96	0.0105
0.6	2.5483	2.56	0.0177
0.8	3.2274	3.24	0.0126
1.0	3.9866	4.0	0.0134

3. Equation (6.39) becomes $(16U(h/2) - U(h))/15$. With $h = 1.0$ the extrapolated value is 3.43589409. With $h = 0.5$ it is 3.43653505. The errors are about 6.7×10^{-4} and 2.9×10^{-5}, respectively.

5. The extrapolated values found in Exercise 4 are 0.5571626 and 0.5571618. Extrapolating these with $r = 3$ in Eq. (6.39) gives 0.5571617.

7. By successive differentiations we get

$$u' = -2tu^2$$
$$u'' = -2(u^2 + 2tuu')$$
$$u''' = -4(2uu' + tuu'' + tu'^2)$$

Therefore $u(0.5) \cong 1 + 0 - \frac{2}{2}(\frac{1}{2})^2 + 0 = \frac{3}{4}$. Here is a table of values of derivatives calculated from formulas above:

$t = 0$	$t = 1/2$
$u = 1$	$u = 3/4$
$u' = 0$	$u' = -9/16$
$u'' = -2$	$u'' = -9/32$
$u''' = 0$	$u''' = -405/128$

From the values of the second column, $u(1) \cong 0.3676758$. Using $h = 1$ gives $u(1) \cong 0$. The Taylor series is particularly bad for this case ($u(t) = 1/(1 + t^2)$).

Section 6.3

1.

t	u(t)	t	u(t)
0.1	0.914	0.6	0.762
0.2	0.851	0.7	0.769
0.3	0.808	0.8	0.784
0.4	0.780	0.9	0.805
0.5	0.765	1.0	0.833

3. $u(1) \cong 3.73$

5.

t	u(t)	t	u(t)
0.1	0.0998	0.9	0.783
0.2	0.199	1.0	0.841
0.3	0.296	1.1	0.891
0.4	0.389	1.2	0.932
0.5	0.479	1.3	0.964
0.6	0.565	1.4	0.985
0.7	0.644	1.5	0.997
0.8	0.717	1.6	1.002

Note: the value listed for $u(1.6)$ was obtained by using an absolute value inside the radical.

7.

t	u(t)	t	u(t)
0.1	0.995	0.6	0.800
0.2	0.980	0.7	0.714
0.3	0.954	0.8	0.600
0.4	0.917	0.9	0.436
0.5	0.866	1.0	0.049

After $t = 1$ the behavior of the numerical solution is unpredictable. The exact solution is $u(t) = \sqrt{(1 - t^2)}$, which does not exist for $t > 1$.

9. Approximate results for $h = 0.2$:

t	u(t)	t	u(t)
0.2	0.003	1.0	0.318
0.4	0.021	2.0	1.626
0.6	0.072	3.0	2.814
0.8	0.167	4.0	3.869
		5.0	4.897

For $h = 0.1$, $u(5) \cong 4.8967$ (all digits good). The numerical solution of this problem goes haywire at about $t = 2/h$. Graphical analysis shows that $u(t) \cong t$ for large t.

11. The ratio apparently approaches 0.044.

13. Partial results for $h = 0.5$:

t	u(t)	t	u(t)
0	1	4	−0.655
1	0.541	5	0.281
2	−0.415	6	0.958
3	−0.989	7	0.755

Also $u(7) \cong 0.7463$ ($h = 1.0$), 0.7540 ($h = 0.25$).

15. Results with $h = 0.2$:

t	u(t)	t	u(t)
0.2	0.999	1.2	0.728
0.4	0.989	1.4	0.583
0.6	0.964	1.6	0.405
0.8	0.916	1.8	0.202
1.0	0.839	2.0	−0.015

Also $u(1) \cong 0.83880956$ ($h = 0.2$), 0.83881219 ($h = 0.1$), 0.83881230 ($h = 0.05$).

17. Partial results with $h = 0.25$:

t	u(t)	t	u(t)
0.25	0.247	2.0	1.005
0.5	0.480	3.0	0.363
0.75	0.683	4.0	−0.589
1.0	0.848	5.0	−1.046
1.25	0.966	6.0	−0.678
1.5	1.032	7.0	0.254

19. With $h = 0.25$, one finds the first zero between 3.75 and 4.0 and the second at about 22.25.

21. With $h = 0.2$ the first zero is seen to be between 1.8 and 2.0. With $h = 0.1$ it is between 1.9 and 2.0. With $h = 0.05$, using interpolation between the values $u(1.95) \cong 0.0399$ and $u(2.0) \cong -0.0150$, the zero is approximately 1.987.

Section 6.4

1. $u(1) \cong 0.56109$; the error is about 0.004.

3. The ratios of exact global error $\div h^2$ are approximately 0.028, 0.11, 0.16, 0.19. These seem to be approaching a limit.

5. $u_{n+1} = (4u_{n+1}^c + u_{n+1}^p)/5$

7. $u(1) \cong 0.717743$

9. Since the error in the tabular values is proportional to h^3, we use $u(1) \cong [8U(h/2) - U(h)]/7$, where $U(h)$ is the approximation to $u(1)$ computed with h. The result from the last two lines ($h = 0.025$) is $u(1) \cong 0.71828181$.

11. $u(0.8) \cong 0.6087$, $u(1) \cong 0.4994$

13. With $h = 0.2$ the solution goes haywire before $t = 10$. With $h = 0.1$, $u(10) \cong 9.94961714$; with $h = 0.05$, $u(10) \cong 9.94961726$.

15.

h	$t = 1$	2	10
0.5	0.49970152	0.22598740	0.0099180686
0.2	0.49934285	0.20022490	0.0099011407
0.1	0.49997941	0.20000427	0.0099009844
0.05	0.49999903	0.20000010	0.0099009897

17.

t	$u(t)$	t	$u(t)$
0.1	0.91379	0.6	0.76205
0.2	0.85119	0.7	0.76859
0.3	0.80762	0.8	0.78346
0.4	0.77978	0.9	0.80543
0.5	0.76525	1.0	0.83335

Section 6.5

1. Set $h = T/n$. Then $u_n = (1 + TA/n)^n$ has limit $\exp(TA)$. (This is a well-known limit theorem.)

3. Abbreviate Eq. (6.83) as $u_{n+1} = au_n + bu_{n-1}$ and substitute from Eq. (6.84) to get

$$c_1\rho_1^{n+1} + c_2\rho_2^{n+1} = a(c_1\rho_1^n + c_2\rho_2^n) + b(c_1\rho_1^{n-1} + c_2\rho_2^{n-1}).$$

By rearranging terms and using Eq. (6.85) this equality is confirmed.

5. The relation is $u_{n+1} = u_{n-1} + 2hAu_n$. Substituting $u_k = \rho^k$ and canceling ρ^{n-1} leads to the characteristic polynomial $\rho^2 = 1 + 2hA\rho$, which has roots $\rho = hA \pm \sqrt{(1 + h^2A^2)}$.

7. For large t, $u(t) \cong t$, $\partial f/\partial u = -2u \cong -2t$ becomes large and negative.

9. Since $f^p_{n+1} = f(t_{n+1}, u^p_{n+1})$ and $f_{n+1} = f(t_{n+1}, u_{n+1})$ have to be computed, use

$$\frac{\partial f}{\partial u}(t_{n+1}, u_{n+1}) \cong \frac{f^p_{n+1} - f_{n+1}}{u^p_{n+1} - u_{n+1}}.$$

Miscellaneous Exercises

1.

t	u (mod. Euler)	u (RK4)
0	1	1
0.5	2.2	2.2467
1.0	3.876	3.9928

3.

t	u (mod. Euler)	u (RK4)
0	1	1
0.5	2.225	2.2493
1.0	3.948	3.9988

5.

t	u
0	0
0.5	0.7373
1.0	1.0834
1.5	1.5082
2.0	1.9023

These numbers come from solving $du/dt = \sqrt{|(1 - u^2)(3 - u^2)|}$.

7. Change the independent variable from t to $\tau = t/T$. If $u(t) = u(T\tau) = v(\tau)$, then $dv/d\tau = T\sqrt{(1 - v^2)(3 - v^2)}$, $v(0) = 0$. Now search for T so that $v(1) = 1$. Using RK4 and $h = 0.01$, find $T \cong 0.995$.

9. $u' = tu^2$, $u(0) = 1$, so the second derivative is $u'' = 2tuu' + u^2 = 2tu \cdot tu^2 + u^2 = u^2(2t^2u + 1)$. The Taylor-series approximation is

$$u(t_{k+1}) = u(t_k) + 0.2u'(t_k) + 0.02u''(t_k).$$

Below is a table of u, u', u'' at various t_k.

t	0	0.2	0.4	0.6	0.8	1
u	1	1.020	1.083	1.208	1.438	1.886
u'	0	0.208	0.469	0.876	1.65	
u''	1	1.04	1.58	2.73	5.88	

11. Table of values found with $RK4$:

t	0	0.2	0.4	0.6	0.8	1.0	1.2
u	1	0.980	0.917	0.800	0.600	0.068	−0.844

13. The equation for $v = 1/u$ is $v' = -tz^2 - 1$, $v(0) = 1$. At first try using RK4 with $h = 0.1$ shows t between 0.9 and 1.0. Refining h shows t around 0.93.

15. Numerov's method with $h = \pi/4$, $p = 1$, and $f = 0$ gives the difference equation

$$u_{n+1} = 1.41331u_n - u_{n-1}.$$

n	2	3	4	5	6	7	8
u_n	0.99935	0.70529	−0.0026	−0.7089	−0.99934	−0.7035	0.0051

17. The solution is a strongly damped oscillation. By using RK4 with $h = 0.1$ the zeros are roughly located at 2.55, 3.73, 4.63, 5.37, ... with maxima or minima between them having values about -0.009, $+0.008$, -0.0015, etc.

Chapter 7

Section 7.1

1. With $\omega_1^2 = k_1/M_1$, $\omega_2^2 = k_2/M_2$, $\beta_1^2 = k/M_1$, $\beta_2^2 = k/M_2$,

$$\frac{dx_1}{dt} = x_3$$

$$\frac{dx_2}{dt} = x_4$$

$$\frac{dx_3}{dt} = -(\omega_1^2 + \beta_1^2)x_1 + \beta_1^2 x_2$$

$$\frac{dx_4}{dt} = \beta_2^2 x_2 - (\omega_2^2 + \beta_2^2)x_2$$

3. Since $x_2 = \dfrac{1}{C}\displaystyle\int i_2\, dt$, $i_2 = C\dfrac{dx_2}{dt}$. Use this in the first equation to get

$$\frac{dx_1}{dt} = -\frac{1}{R_1 C}x_1 + \frac{1}{R_1 C}u$$

$$\frac{dx_2}{dt} = -\frac{R_2}{L}x_2 + \frac{1}{L}u$$

5. (a) $\dfrac{dK}{dt} = \mu r Y - \mu K$ (b) $\dfrac{dK}{dt} = -\mu K + \mu r Y$

   $\dfrac{dY}{dt} - \lambda\dfrac{dK}{dt} = \lambda(-sY + A)$ $\dfrac{dY}{dt} = -\lambda\mu K + \lambda(\mu r - s)Y + \lambda A$

Section 7.2

1. $x = c_1 e^{3t} + c_2 e^t$ 3. $x = e^t(c_1 \cos t + c_2 \sin t)$
   $y = c_1 e^{3t} - c_2 e^t$ $y = e^t((c_2 - c_1)\cos t - (c_1 + c_2)\sin t)$

5. $x = c_1 \cos t + c_2 \sin t$ 7. $x = (c_1 t + c_2)e^t$
   $y = c_2 \cos t - c_1 \sin t$ $y = c_1 e^t$

9. $x = c_1 e^{2t} + c_2 e^{-t} + c_3 e^t$ 11. $x = c_1 e^{2t} + c_2 + c_3 e^{-2t}$
   $y = c_1 e^{2t} - 2c_2 e^{-t}$ $y = 2c_1 e^{2t} - 2c_3 e^{-2t}$
   $z = \frac{1}{2}c_1 e^{2t} - c_2 e^{-t} + c_3 e^t$ $z = c_1 e^{2t} - c_2 + c_3 e^{-2t}$

13. $x = c_1 e^t + c_2 e^{-t} + \frac{1}{2}t(e^t - e^{-t})$
    $y = c_1 e^t - c_2 e^{-t} + \frac{1}{2}(t - 1)(e^t + e^{-t})$

15. $\dfrac{d^3 u_1}{dt^3} + 6\dfrac{d^2 u_1}{dt^2} + 10\dfrac{du}{dt} + 4u = 3T_0 + T_4$

17. Add the equations to find that $(d/dt)(x + y) = x + y$. Therefore $x + y = ce^t$. Now substitute for $x + y$ to obtain

$$\frac{dx}{dt} = x + tce^t, \qquad \frac{dy}{dt} = y - tce^t.$$

The solutions are

$$x = ae^t + ct^2 e^t/2, \qquad y = a'e^t - ct^2 e^t/2,$$

but $a + a' = c$. A neat way to express the solutions: set $a = b + c/2$, $a' = -b + c/2$; then b and c are two arbitrary constants.

19. The particular solution is $K_p = vA/s$. $K(t)$ will approach K_p if the roots of the polynomial

$$m^2 + (\mu - \lambda\mu r + s\lambda)m + s\lambda\mu = 0$$

are negative or have negative real part. Since s, λ, and μ are positive, it is sufficient that

$$\mu - \lambda\mu r + s\lambda > 0.$$

The solutions are oscillatory if also

$$(\mu - \lambda\mu r + s\lambda)^2 < 4s\lambda\mu$$

Section 7.3

1. $\begin{bmatrix} 1 \\ 1 \end{bmatrix}(3); \begin{bmatrix} 1 \\ -1 \end{bmatrix}(-1)$

3. $\begin{bmatrix} 1 \\ 1 \end{bmatrix}(a + b); \begin{bmatrix} 1 \\ -1 \end{bmatrix}(a - b)$

5. $\begin{bmatrix} 1 \\ i \end{bmatrix}(a + ib); \begin{bmatrix} 1 \\ -i \end{bmatrix}(a - ib)$

7. $\begin{bmatrix} 1 \\ 0 \end{bmatrix}(a); \begin{bmatrix} 0 \\ 1 \end{bmatrix}(b)$

9. $\begin{bmatrix} 1 \\ 1 \end{bmatrix}(-1); \begin{bmatrix} 0 \\ 1 \end{bmatrix}(-2)$

11. $\begin{bmatrix} 1 \\ 0 \end{bmatrix}(a); \begin{bmatrix} b \\ c - a \end{bmatrix}(c)$

13. $\begin{bmatrix} 1 \\ 1 \\ 1 \end{bmatrix}(a + 2b), \begin{bmatrix} 1 \\ 0 \\ -1 \end{bmatrix}, \begin{bmatrix} 1 \\ -2 \\ 1 \end{bmatrix}(a - b)$

15.   (0)   (−1)   (−3)   (−4)

$$\begin{bmatrix} 1 \\ 1 \\ 1 \\ 1 \end{bmatrix} \quad \begin{bmatrix} 2 \\ 1 \\ -1 \\ -2 \end{bmatrix} \quad \begin{bmatrix} 2 \\ -1 \\ -1 \\ 2 \end{bmatrix} \quad \begin{bmatrix} 1 \\ -1 \\ 1 \\ -1 \end{bmatrix}$$

17.   (4)   (3)   (1)   (−2)

$$\begin{bmatrix} 0 \\ 2 \\ 1 \\ 1 \end{bmatrix} \quad \begin{bmatrix} 0 \\ 1 \\ 0 \\ 0 \end{bmatrix} \quad \begin{bmatrix} 1 \\ -1 \\ 0 \\ 0 \end{bmatrix} \quad \begin{bmatrix} 10 \\ -2 \\ -15 \\ 15 \end{bmatrix}$$

19. (4)   (0)   (−2)   (The eigenvalue 4 is deficient.)

$$\begin{bmatrix} 1 \\ 1 \\ 1 \\ 1 \end{bmatrix} \quad \begin{bmatrix} 0 \\ 1 \\ 0 \\ -1 \end{bmatrix} \quad \begin{bmatrix} -2 \\ 1 \\ -8 \\ 1 \end{bmatrix}$$

21. $A(c_1 z_1 + c_2 z_2) = c_1 A z_1 + c_2 A z_2 = c_1 \lambda z_1 + c_2 \lambda z_2 = \lambda(c_1 z_1 + c_2 z_2)$

23. If $c_1 z_1 + c_2 z_2 = 0$, then also $0 = (A - \lambda_2 I)(c_1 z_1 + c_2 z_2) = c_1(\lambda_1 - \lambda_2)z_1$. Hence $c_1 = 0$, and it follows that $c_2 = 0$.

25. Suppose the equation for y is solved by the eigenvector method, so $y = c_1 z_1 e^{\lambda_1 t} + \cdots + c_n z_n e^{\lambda_n t}$. Then z_i is an eigenvector of A corresponding to eigenvalue λ_i; likewise, z_i is an eigenvector of $aI + A$ corresponding to eigenvalue $a + \lambda$. The same linear combination of special solutions of $dx/dt = Ax$ is

$$x = c_1 z_1 e^{(a+\lambda_1)t} + \cdots + c_n z_n e^{(a+\lambda_n)t} = e^{at} y.$$

This satisfies the initial value problem for x.

27. Use $A = \begin{bmatrix} 0 & 1 \\ 3 & 2 \end{bmatrix}$, $B = \begin{bmatrix} 1 & 0 \\ 0 & 1 \end{bmatrix}$. Then we need these matrices, eigenvalues, and eigenvectors:

$$A + B = \begin{bmatrix} 1 & 1 \\ 3 & 3 \end{bmatrix}; \quad \begin{bmatrix} 1 \\ -1 \end{bmatrix}(0), \quad \begin{bmatrix} 1 \\ 3 \end{bmatrix}(4)$$

$$A - B = \begin{bmatrix} -1 & 1 \\ 3 & 1 \end{bmatrix}; \quad \begin{bmatrix} 1 \\ -1 \end{bmatrix}(-2), \quad \begin{bmatrix} 1 \\ 3 \end{bmatrix}(2)$$

(Why are the eigenvectors the same for $A + B$ and $A - B$?) The eigenvectors (and eigenvalues) of M are

$$\begin{bmatrix} 1 \\ -1 \\ 1 \\ -1 \end{bmatrix}(0), \quad \begin{bmatrix} 1 \\ 3 \\ 1 \\ 3 \end{bmatrix}(4), \quad \begin{bmatrix} 1 \\ -1 \\ -1 \\ 1 \end{bmatrix}(-2), \quad \begin{bmatrix} 1 \\ 3 \\ -1 \\ -3 \end{bmatrix}(2).$$

Section 7.4

1. $x(t) = \dfrac{1}{2}\begin{bmatrix} 1 \\ 1 \end{bmatrix} e^{3t} + \dfrac{1}{2}\begin{bmatrix} 1 \\ -1 \end{bmatrix} e^{-t}$

3. $x(t) = \begin{bmatrix} 1 + 2t \\ -t \end{bmatrix}$ (The eigenvalue 1 is deficient.)

5. $x(t) = \begin{bmatrix} 1 \\ 1 \\ 1 \end{bmatrix} e^{2t}$

7. $x(t) = \dfrac{1}{2}\begin{bmatrix} 1 \\ 1 \\ 1 \\ 1 \end{bmatrix} e^{2t} + \dfrac{1}{2}\begin{bmatrix} 1 \\ -1 \\ 1 \\ -1 \end{bmatrix}$

9. $x(t) = \dfrac{1}{2}\begin{bmatrix}1\\1\\1\\1\end{bmatrix} + \dfrac{1}{2}\begin{bmatrix}-1\\1\\1\\-1\end{bmatrix}e^{-2t}$

11. $x(t) = c_1\begin{bmatrix}1\\1\end{bmatrix}e^{3t} + c_2\begin{bmatrix}-1\\1\end{bmatrix}e^{-t}$

13. $x(t) = c_1\begin{bmatrix}-i\\1\end{bmatrix}e^{it} + c_2\begin{bmatrix}i\\1\end{bmatrix}e^{-it}$ or $x(t) = c_1\begin{bmatrix}\cos t\\-\sin t\end{bmatrix} + c_2\begin{bmatrix}\sin t\\\cos t\end{bmatrix}$

15. $x(t) = c_1\begin{bmatrix}1\\1\\1\end{bmatrix}e^{2t} + c_2\begin{bmatrix}-1\\0\\1\end{bmatrix}e^{-t} + c_3\begin{bmatrix}-1\\1\\0\end{bmatrix}e^{-t}$

17. $X(t) = \begin{bmatrix}0 & -e^t & -2e^{2t}\\-e^t & (2-t)e^t & -3e^{2t}\\e^t & te^t & 6e^{2t}\end{bmatrix}\begin{bmatrix}12 & 6 & 7\\3 & 2 & 2\\-2 & -1 & -1\end{bmatrix}$

19. $X(t) = \dfrac{1}{2}\begin{bmatrix}C+1 & S & S & C-1\\S & C+1 & C-1 & S\\S & C-1 & C+1 & S\\C-1 & S & S & C+1\end{bmatrix}$, $C = \cosh 2t$, $S = \sinh 2t$.

21. $x(t) = \dfrac{1}{7}\begin{bmatrix}4\\2\\1\end{bmatrix} - \dfrac{1}{7}e^{-t}\left(c\begin{bmatrix}-1\\0\\1\end{bmatrix} - s\sqrt{3}\begin{bmatrix}-1\\1\\0\end{bmatrix}\right)$

$- \dfrac{2}{7\sqrt{3}}e^{-t}\left(s\begin{bmatrix}-1\\0\\1\end{bmatrix} + c\sqrt{3}\begin{bmatrix}-1\\1\\0\end{bmatrix}\right),$

where $c = \cos\sqrt{3}t$, $s = \sin\sqrt{3}t$.

Section 7.5

1. $x_1(t) = \begin{bmatrix}1\\1\end{bmatrix}e^{3t}$, $x_2(t) = \begin{bmatrix}-1\\1\end{bmatrix}e^{-t}$ 3. $x_1(t) = \begin{bmatrix}1\\1\end{bmatrix}e^t$, $x_2(t) = \begin{bmatrix}t\\t+1\end{bmatrix}e^t$

5. $W(x_1, x_2) = -2$ 7. $W(x_1, x_2, x_3) = 2e^{3t}$

9. (a) $W(x_1, x_2) = -2(t+1)^2$. (b) $\mathrm{tr}\,(A) = 2/(t+1)$, and $dW/dt = 2W/(t+1)$ is true. (c) The longest interval is $-1 < t < 1$: $x_2(t)$ is discontinuous at $t = 1$, and $x_1(-1) = x_2(-1) = 0$. Also $A(t)$ is discontinuous at $t = -1$ and 1.

(d) $X(t) = \dfrac{1+t}{2(1-t)}\begin{bmatrix}t^2 - 2t + 2 & t^2 - 2t\\t^2 - 2t & t^2 - 2t + 2\end{bmatrix}$

11. $X(t) = \begin{bmatrix}\cos t & \sin t\\-\sin t & \cos t\end{bmatrix}$

13. $X(t) = \dfrac{1}{3}\begin{bmatrix}-1 & -1 & 1\\1 & 0 & 1\\0 & 1 & 1\end{bmatrix}\begin{bmatrix}e^{-t} & & \\ & e^{-t} & \\ & & e^{2t}\end{bmatrix}\begin{bmatrix}-1 & 2 & -1\\-1 & -1 & 2\\1 & 1 & 1\end{bmatrix}$

15. $X(t) = \dfrac{1}{4}\begin{bmatrix} 1 & 1 & 1 & 1 \\ 1 & -1 & -1 & 1 \\ 1 & 1 & -1 & -1 \\ 1 & -1 & 1 & -1 \end{bmatrix}\begin{bmatrix} e^{3t} & & & \\ & e^{t} & & \\ & & e^{-t} & \\ & & & e^{-3t} \end{bmatrix}\begin{bmatrix} 1 & 1 & 1 & 1 \\ 1 & -1 & 1 & -1 \\ 1 & -1 & -1 & 1 \\ 1 & 1 & -1 & -1 \end{bmatrix}$

Section 7.6

1. $x(t) = \displaystyle\int_0^t X(t)X^{-1}(s)g(s)\,ds = \begin{bmatrix} \sin t - \cos t + 1 \\ \sin t + \cos t - 1 \end{bmatrix}$

3. $x(t) = \dfrac{1}{2}\begin{bmatrix} t^2 \\ 2t - t^2 \end{bmatrix}e^{-t}$

5. $x(t) = \begin{bmatrix} 1 \\ 0 \\ -1 \end{bmatrix}\dfrac{1 - e^{-2t}}{2}$

7. $x(t) = \dfrac{t}{3}\begin{bmatrix} 1 \\ 1 \\ 1 \end{bmatrix} + \dfrac{1}{9}\begin{bmatrix} 5 \\ -1 \\ -4 \end{bmatrix} + \dfrac{1}{2}\begin{bmatrix} -1 \\ 0 \\ 1 \end{bmatrix}e^{-t} + \dfrac{1}{18}\begin{bmatrix} -1 \\ 2 \\ -1 \end{bmatrix}e^{-3t}$

9. $x_p = e^{-t}\begin{bmatrix} 0 \\ -1 \end{bmatrix}$

11. $x_p = \dfrac{1}{2}\begin{bmatrix} 0 \\ -1 \end{bmatrix}\cos t + \dfrac{1}{2}\begin{bmatrix} 1 \\ 0 \end{bmatrix}\sin t$

13. $x_p = \dfrac{-1}{2}\begin{bmatrix} 2 \\ 3 \\ 2 \end{bmatrix} + \dfrac{1}{2}\begin{bmatrix} 1 \\ 2 \\ 1 \end{bmatrix}t$

15. $x_p = \tfrac{1}{2}(\sin t - \cos t)\begin{bmatrix} 1 \\ 1 \\ 1 \\ 1 \end{bmatrix}$

17. The desired solution exists if and only if b is chosen so that the system of algebraic equations $(I - A)k = b$, that is,

$$\begin{bmatrix} 1 & -1 \\ -1 & 1 \end{bmatrix}k = b,$$

has a solution. It is necessary and sufficient that $[1, 1] \cdot b = 0$, which means

$$b = \alpha\begin{bmatrix} 1 \\ -1 \end{bmatrix},$$

where α is a scalar.

Section 7.7

1. All solutions vanish.

3. All solutions approach $x_p = \dfrac{1}{2}\begin{bmatrix} 1 \\ 1 \end{bmatrix}$.

5. A_0 is stable, so all solutions approach x_p.

7. A_0 is stable, so all solutions approach x_p.

9. This matrix does not satisfy the conditions of Theorem 7.13, but its transpose does. Since A and A^T have the same eigenvalues, this matrix is stable.

11. Stable, by Theorem 7.13.

13. $x_p = \begin{bmatrix} t + 1 \\ 1 \end{bmatrix}$.

Miscellaneous Exercises

1. (a) $\begin{bmatrix} 1 \\ 1 \end{bmatrix}(2)$, $\begin{bmatrix} -1 \\ 1 \end{bmatrix}(0)$; (b) $x(t) = c_1 e^{2t}\begin{bmatrix} 1 \\ 1 \end{bmatrix} + c_2\begin{bmatrix} -1 \\ 0 \end{bmatrix}$;

   (c) $c_1 = c_2 = \frac{1}{2}$

3. (a) $\begin{bmatrix} -i \\ 1 \end{bmatrix}(-1 + i)$, $\begin{bmatrix} i \\ 1 \end{bmatrix}(-1 - i)$;

   (b) $x(t) = c_1 e^{(-1+i)t}\begin{bmatrix} -i \\ 1 \end{bmatrix} + c_2 e^{(-1-i)t}\begin{bmatrix} i \\ 1 \end{bmatrix}$;

   (c) $x(t) = e^{-t}\begin{bmatrix} \cos t + \sin t \\ \cos t - \sin t \end{bmatrix}$

5. (a) $\lambda = -1$ is the only eigenvalue and is deficient. Two eigenvectors are
   $\begin{bmatrix} -1 \\ 0 \\ 1 \end{bmatrix}$, $\begin{bmatrix} -2 \\ 1 \\ 0 \end{bmatrix}$;

   (b) $x(t) = c_1 e^{-t}\begin{bmatrix} -1 \\ 0 \\ 1 \end{bmatrix} + c_2 e^{-t}\begin{bmatrix} -2 \\ 1 \\ 0 \end{bmatrix} + c_3 e^{-t}\left(\begin{bmatrix} 1 \\ -1 \\ 1 \end{bmatrix}t + \begin{bmatrix} 1 \\ 0 \\ 0 \end{bmatrix}\right)$;

   (c) $c_1 = c_2 = 0$, $c_3 = 1$

7. (a) $\begin{bmatrix} 2 \\ 0 \\ 1 \end{bmatrix}(-1)$, $\begin{bmatrix} -2 \\ 1 \\ 0 \end{bmatrix}(-1)$, $\begin{bmatrix} 1 \\ -1 \\ 2 \end{bmatrix}(-2)$;

   (b) $x(t) = c_1 e^{-t}\begin{bmatrix} 2 \\ 0 \\ 1 \end{bmatrix} + c_2 e^{-t}\begin{bmatrix} -2 \\ 1 \\ 0 \end{bmatrix} + c_3 e^{-2t}\begin{bmatrix} 1 \\ -1 \\ 2 \end{bmatrix}$;

   (c) $c_1 = \frac{3}{4}$, $c_2 = \frac{1}{4}$, $c_3 = 0$

9. (a) $\begin{bmatrix} -i \\ 0 \\ 0 \\ 1 \end{bmatrix}(1 + i)$, $\begin{bmatrix} 0 \\ i \\ 1 \\ 0 \end{bmatrix}(1 + i)$, $\begin{bmatrix} i \\ 0 \\ 0 \\ 1 \end{bmatrix}(1 - i)$, $\begin{bmatrix} 0 \\ -i \\ 1 \\ 0 \end{bmatrix}(1 - i)$;

   (b) $x(t) = e^t\left\{ c_1\begin{bmatrix} \sin t \\ 0 \\ 0 \\ \cos t \end{bmatrix} + c_2\begin{bmatrix} -\cos t \\ 0 \\ 0 \\ \sin t \end{bmatrix} + c_3\begin{bmatrix} 0 \\ -\sin t \\ \cos t \\ 0 \end{bmatrix} + c_4\begin{bmatrix} 0 \\ \cos t \\ \sin t \\ 0 \end{bmatrix} \right\}$

   (c) $x(t) = \begin{bmatrix} 0 \\ \cos t - \sin t \\ \cos t + \sin t \\ 0 \end{bmatrix}$

11. (a) $\begin{bmatrix} -1 \\ 0 \\ 0 \\ 1 \end{bmatrix}(0)$, $\begin{bmatrix} 0 \\ -1 \\ 1 \\ 0 \end{bmatrix}(0)$, $\begin{bmatrix} 1 \\ i \\ i \\ 1 \end{bmatrix}(2i)$, $\begin{bmatrix} 1 \\ -i \\ -i \\ 1 \end{bmatrix}(-2i)$;

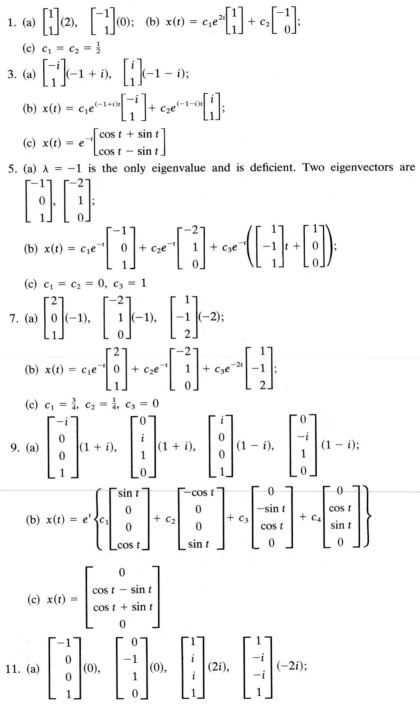

(b) $x(t) = c_1 \begin{bmatrix} -1 \\ 0 \\ 0 \\ 1 \end{bmatrix} + c_2 \begin{bmatrix} 0 \\ -1 \\ 1 \\ 0 \end{bmatrix} + c_3 \begin{bmatrix} \cos 2t \\ -\sin 2t \\ -\sin 2t \\ \cos 2t \end{bmatrix} + c_4 \begin{bmatrix} \sin 2t \\ \cos 2t \\ \cos 2t \\ \sin 2t \end{bmatrix}$

(c) $x(t) = \begin{bmatrix} 1 \\ 1 \\ 1 \\ 1 \end{bmatrix} \cos 2t + \begin{bmatrix} 1 \\ -1 \\ -1 \\ 1 \end{bmatrix} \sin 2t$

13. (a) $x(t) = c_1 e^t \begin{bmatrix} 1 \\ 1 \end{bmatrix} + c_2 \begin{bmatrix} 1 \\ 0 \end{bmatrix}$. (b) The curves are half-lines, starting from the x_1-axis and inclined at 45° to it. The x_1-axis itself is a collection of static solutions: $x_1 = c$, $x_2 = 0$.

15. (a) $x(t) = c_1 e^{-t} \begin{bmatrix} 0 \\ 1 \end{bmatrix} + c_2 e^{-t} \begin{bmatrix} 1 \\ t \end{bmatrix}$. (b) All curves approach the origin tangent to the vertical axis. The curves have the equations $x_2 = x_1(k - \ln x_1)$.

17. (a) $x(t) = c_1 e^{-t} \begin{bmatrix} \cos t \\ -\sin t \end{bmatrix} + c_2 e^{-t} \begin{bmatrix} \sin t \\ \cos t \end{bmatrix}$.

(b) The curves are all clockwise spirals heading into the origin.

19. $\dfrac{dy}{dx} = \dfrac{bx}{ay}$, so $ay^2 = bx^2 + c$. The solution curves are hyperbolas.

21. If $(1/x)(dx/dt) = (1/y)(dy/dt)$, then $-(ay/x) = -(bx/y)$, or $ay^2 = bx^2$. (This is the condition for equal force.) If the condition holds, then $ay\dot{y} = bx\dot{x} = -abxy$. Since a, b, x, and y are all positive, $\dot{x}$ and $\dot{y}$ are negative. Also, $y/x = \sqrt{b/a}$.

23. Using the solution of Exercise 20 in the form

$$\begin{bmatrix} x \\ y \end{bmatrix} = c_1 \begin{bmatrix} \sqrt{a} \\ \sqrt{b} \end{bmatrix} e^{-\sqrt{ab}t} + c_2 \begin{bmatrix} -\sqrt{a} \\ \sqrt{b} \end{bmatrix} e^{\sqrt{ab}t},$$

find that $x(t) = 0$ when $t = (1/2\sqrt{ab}) \ln (c_1/c_2)$. The initial condition applied to the solution above gives

$$\frac{c_1}{c_2} = \frac{y_0\sqrt{a} + x_0\sqrt{b}}{y_0\sqrt{a} - x_0\sqrt{b}}$$

Of course, the denominator must be positive for y to win.

25. The characteristic polynomial of A is $\lambda^2 + (c_1 + c_2)\lambda + (c_1 c_2 - d_1 d_2)$. The roots are $-\frac{1}{2}(c_1 + c_2) \pm \sqrt{\frac{1}{4}(c_1 + c_2)^2 - (c_1 c_2 - d_1 d_2)}$. The radicand is equal to $\frac{1}{4}(c_1 - c_2)^2 + d_1 d_2$, which is always positive; thus the eigenvalues are real. Since $\lambda_1 \lambda_2 = \det A = c_1 c_2 - d_1 d_2 > 0$, both eigenvalues are negative.

27. The system is $du/dt = Au + g$, where

$$A = \beta \begin{bmatrix} -3 & 1 & 0 \\ 1 & -2 & 1 \\ 0 & 1 & -3 \end{bmatrix}, \qquad g = \beta \begin{bmatrix} T_0 \\ 0 \\ T_4 \end{bmatrix}$$

and $\beta = k/C$. The eigenvalues of A are $-\beta, -3\beta, -4\beta$. The general solution is

$$u(t) = c_1 \begin{bmatrix} 1 \\ 2 \\ 1 \end{bmatrix} e^{-\beta t} + c_2 \begin{bmatrix} -1 \\ 0 \\ 1 \end{bmatrix} e^{-3\beta t} + c_3 \begin{bmatrix} 1 \\ -1 \\ 1 \end{bmatrix} e^{-4\beta t} + \frac{1}{12} \begin{bmatrix} 5T_0 + T_4 \\ 3T_0 + 3T_4 \\ T_0 + 5T_4 \end{bmatrix}.$$

29. The matrix is $A = \dfrac{1}{L} \begin{bmatrix} 0 & L/C \\ -1 & -R \end{bmatrix}$. The characteristic polynomial of the matrix LA is $\lambda^2 + R\lambda + L/C$. Thus the eigenvalues of A are $(-\frac{1}{2}R \pm \sqrt{\frac{1}{4}R^2 - L/C})/L$. These are complex if $L/C > \frac{1}{4}R^2$.

31. $\dfrac{du}{dt} = \begin{bmatrix} -R_1/L & -1/L \\ 1/C & 1/R_3 C - 1/R_2 C \end{bmatrix} u + \begin{bmatrix} v_1/L \\ v_2/R_3 C \end{bmatrix}$

33. $A = \begin{bmatrix} -10^2 & -10^5 \\ 1 & -2 \times 10^3 \end{bmatrix}$ has characteristic polynomial $\lambda^2 + 2100\lambda + 3 \times 10^5$, with roots $\lambda = 1946, -154$.

Chapter 8

Section 8.1

1. $(\theta = x)\ \dfrac{dx}{dt} = y, \dfrac{dy}{dt} = -\sin x;\ \frac{1}{2}y^2 + (1 - \cos x) = \text{const}$

3. $(u = x)\ \dfrac{dx}{dt} = y, \dfrac{dy}{dt} = \dfrac{x - x^3}{1 + x^2};\ \frac{1}{2}(x^2 + y^2) - \ln(1 + x^2) = \text{const}$

5. Treat the equation in two parts, accordingly as $y > 0$ or $y < 0$. If $y > 0$,
$$\dfrac{dy}{dx} = \dfrac{-x - y^2}{y}, \text{ giving } \tfrac{1}{2}y^2 = c_1 e^{2x} + \tfrac{1}{2}x + \tfrac{1}{4} \text{ or } y = \sqrt{(2c_1 e^{-2x} + x + \tfrac{1}{2})}. \text{ If }$$
$y < 0,\ \dfrac{dy}{dx} = \dfrac{-x + y^2}{y}$ gives $y = -\sqrt{(2c_2 e^{2x} + x + \tfrac{1}{2})}$. The constants c_1 and c_2 must be matched to join the curves at $y = 0$. The curves are spirals.

7. $b(x) = \varepsilon(x^2 - 1),\ B(x) = \varepsilon(\frac{1}{3}x^3 - x);\ \dfrac{dx}{dt} = y - \varepsilon(\frac{1}{3}x^3 - x);\ \dfrac{dy}{dt} = -x$

9. A first integral is $y^2 + x^2 - 2x^3/3 = c$. Critical points at $(0,0)$ and $(1,0)$.

11. A first integral is $x^2 - y^2 + y^4/2 = c$. Critical points at $(0,0), (0,\pm 1)$.

Section 8.2

1. $(3 \pm i\sqrt{3})/2$; spiral point

3. $-2, -2$; half-node

5. $\pm 3i$; center

7. $(-1 \pm \sqrt{5})/2$; saddle

9. $-1, -3$; node

11. (a) $-1, -2$, (b) node, (d) $y = cx^2$

13. (a) $-1 \pm i$, (b) focus, (d) solving as a homogeneous equation gives $cx\sqrt{1 + v^2} = \exp(\tan^{-1} v)$. In polar coordinates this is $cr = e^\theta$.

15. (a) $-1, -1$, (b) half-node, (d) solved as a linear equation: $y = cx - x \ln|x|$.

17. (a) $1, 1$, (b) star, (d) $y = cx$

19. If $y = mx$ is a solution of Eq. (8.17), m must satisfy

$$m = \frac{c + dm}{a + bm}$$

or $bm^2 + (a - d)m - c = 0$. There are 2, 1, or 0 real values of m as $\Delta = (a - d)^2 - 4bc$ is positive, zero, or negative. But Δ is also the discriminant of the characteristic equation.

21. (a) $dy/dx = -2$; $y(x) = -2x + c$. (b) $dx/dt = -3x + c$; $x(t) = c_2 e^{-3t} + c/3$. (c) $y(t) = -2c_2 e^{-3t} + c/3$. (d) $x(t)$ and $y(t)$ approach $c/3$. (e) All trajectories are half-lines with slope -2 approaching the line $y = x$. (f) The common factor $y - x$ was cancelled, eliminating a lineful of critical points.

Section 8.3

1. $(0, 0)$ is a center; $(\pm 1, 0)$ are saddle points.

3. $(0, 0)$ and $(0, 2)$ are saddle points; $(1, 1)$ is a stable focus.

5. The locus for $dx/dt = 0$ is $(x + y)(1 - x^2 + xy) = 0$: $y = -x$ or $y = (x^2 - 1)/x$. The locus for $dy/dt = 0$ is $r^2 = 1 - \cot \theta$ in polar coordinates. There are five critical points: $(0, 0)$ is an unstable spiral point; $(1, -1)$ and $(-1, 1)$ are stable half-nodes (characteristic roots -2, -2); the remaining two at (x, y) and $(-x, -y)$ for $x = 0.6$, $y = 1.09$ (approximately) are saddle points.

Section 8.4

1. The origin is the only critical point. $V(x, y) = \frac{1}{2}x^2 + \frac{a}{4}x^4 + \frac{1}{2}y^2$ has $dV/dt \equiv 0$. ($V = \text{const}$ is a first integral of the system; V represents energy). The origin is a center.

3. Critical point at $(0, 0)$. $V(x, y) = \sqrt{1 + x^2} + y^2/2 - 1$ has $dV/dt \equiv 0$. (See the answer to Exercise 1.)

5. Critical points at $y = 0$, $x = 0, 1, -1$. $V(x, y) = \frac{1}{2}(x^2 + y^2) - \ln(1 + x^2)$ has $dV/dt \equiv 0$ but is not positive definite. The origin is a saddle. To study the critical point at $(1, 0)$, you can change variables (see Section 8.3) by $x = 1 + x^*$, $y = y^*$. Then $x^* = 0$, $y^* = 0$ is a critical point of

$$\frac{dx^*}{dt} = y^*, \qquad \frac{dy^*}{dt} = \frac{x^*(x^* + 1)(x^* + 2)}{x^{*2} + 2x^* + 2}.$$

The function $V(x^*, y^*) = x^* + \frac{1}{2}(x^{*2} + y^{*2}) - \ln(1 + x^* + \frac{1}{2}x^{*2})$ is positive definite near $x^* = 0$, $y^* = 0$ and has $dV/dt \equiv 0$. This critical point is a center.

7. $V = x^2 + y^2$ (not unique)

9. $V = x^2 + y^2 - xy$ (not unique)

11. $V = x^2 - 2xy + 5y^2$ (not unique)

13. The function $V = x^2 + xy + y^2$ has a negative derivative ($dV/dt < 0$) in the

region defined by $(1 - y)(1 - 2x - 2y) > \frac{1}{4}$, which includes the origin. Note that there are singular points at $(-1, \pm 1)$.

Section 8.5

1. The loci of $\dot{x} = 0$ and $\dot{y} = 0$ intersect only at the origin, which is an unstable spiral. The derivative of $V(x, y) = (x^2 + y^2)/2$ along the trajectories is $x^2 + y^2 - (x^4 + y^4)$, which is negative for $x^2 + y^2 > 2$. By Theorem 8.5, at least one periodic solution exists, and its trajectory must lie inside $x^2 + y^2 \leq 2$; by Theorem 8.6 it surrounds the origin.

3. This system can have no periodic solutions because $\Delta(x, y) = 1 + 3x^2$ is positive in the whole plane.

5. With $x = u$, $y = \dot{u}$ the system is

$$\frac{dx}{dt} = y, \qquad \frac{dy}{dt} = y - x - y^3.$$

The origin is the only critical point, an unstable spiral. One can construct a region (see Example 2), from which no trajectory escapes, using $x = -b$, $x^3 = (1 - m)y - y^3$ (an isocline—use $m < 0$) and $y = c - x$ in the upper half-plane and its negative in the lower half-plane. (See Fig. 8.24.) By Theorem 8.5 a periodic solution of the original equation exists.

7. If this is taken as a nonlinear mass–spring system, the energy, $\frac{1}{2}(du/dt)^2 + \frac{1}{2}u^2 - \frac{1}{4}u^4$, is constant. In the phase plane the function $V = \frac{1}{2}y^2 + \frac{1}{2}x^2 - \frac{1}{4}x^4$ is constant. There are three critical points: a center at $(0, 0)$ and saddle points at $(\pm 1, 0)$. The closed curves $V = \text{const}$ surrounding the origin represent periodic solutions.

9. The function $b(x) = f'(x)$ must satisfy $f'(-x) = f'(x)$. Also, $\int_0^x f'(z)\,dz = f(x)$ must be negative, $0 < x < x_0$, and both $f(x)$ and $f'(x)$ positive, $x > x_0$; finally, $f(x) \to \infty$ as $x \to \infty$.

Section 8.6

1. First integral: $v^2 = a(A^2 - u^2)\dfrac{b}{2}(A^4 - u^4) = (A^2 - u^2)\left(a + \dfrac{b}{2}A^2 + \dfrac{b}{2}u^2\right)$.

Let $w = Au$. Then $A^2\left(\dfrac{dw}{dt}\right)^2 = A^2(1 - w^2)\left(a + \dfrac{b}{2}A^2 + \dfrac{b}{2}A^2w^2\right)$. Since the polynomial on the right has two real and two imaginary roots (see Eqs. (8.83), (8.84) and what follows), set $A^2b/(2a + A^2b) = k^2$ and $(dw/dt)^2 = (1 - w^2)\left(a + \dfrac{b}{2}A^2\right)(1 - k^2 + k^2w^2)$. Now $\text{cn}^{-1}w = \sqrt{a + \dfrac{b}{2}A^2}\,t + c_1$ and

$$u = A\,\text{cn}\left(\sqrt{a + \dfrac{b}{2}A^2}\,t + c_1\right).$$

3. First integral: $v^2 = (b/2)(A^2 - u^2)^2$. Solution: $u = \pm A \tanh(\alpha t + c)$, $\alpha = A\sqrt{b/2}$.

5. By the fundamental theorem of calculus, $dt/d\phi = 1/\sqrt{1 - k^2 \sin^2 \phi}$, so $d\phi/dt = \sqrt{1 - k^2 \sin^2 \phi} = \text{dn } t$.

7. $\left(\dfrac{du}{dt}\right)^2 = (1 - u^2)(1 - k^2 u^2)$, $u = \text{sn } t$;

   $\left(\dfrac{du}{dt}\right)^2 = (1 - u^2)(1 - k^2 + k^2 u^2)$, $u = \text{cn } t$;

   $\left(\dfrac{du}{dt}\right)^2 = (1 - u^2)(k^2 - 1 + u^2)$, $u = \text{dn } t$.

9. $\theta(s) = 2 \sin^{-1}(k \, \text{sn}\,(\pm \lambda s + c_1))$ is the general solution, with $k = \sin \frac{1}{2}\alpha$. Since $\theta(0) = \alpha = 2 \sin^{-1} k$, we must choose c_1 so that $\text{sn } c_1 = 1$ (or $c_1 = K(k)$) and take the minus sign, as θ decreases with s. Thus $\theta = 2 \sin^{-1}(k \, \text{sn}\,(K - \lambda s))$.

11. $\dfrac{d\theta}{\sqrt{1 - (\theta/\alpha)^2}} = \pm \lambda \sqrt{2(1 - \cos \alpha)} \, ds$ gives

    $\alpha \sin^{-1}(\theta/\alpha) = \pm \lambda \sqrt{2(1 - \cos \alpha)}\, s + c$ or $\theta = \alpha \sin(\omega s + c)$,

    $\omega = \lambda \sqrt{2(1 - \cos \alpha)}/\alpha$.

Miscellaneous Exercises

1. (a) $V(h) = Ah$, (b) $V(h) = \begin{cases} Ah, & 0 < h < l, \\ Al, & h > l, \end{cases}$

   where l is the length of the cylinder. Note that $h < 0$ means that the object is out of the water, so $V(h) = 0$ if $h < 0$.

3. The system $\dot{u} = v$, $\dot{v} = -u^2 - 2u$ has a first integral $\frac{1}{3}u^3 + u^2 + \frac{1}{2}v^2 = c$. The critical points are $(0, 0)$, where the linear comparison system has a center, and $(-2, 0)$, where the linear comparison system has a saddle point. The level curve of the first integral that passes through $(-2, 0)$ has $c = \frac{4}{3}$. This loops around the origin, and smaller values of c give closed curves surrounding $(0, 0)$, which is therefore a stable critical point.

5. At $(0, 0)$ the linear comparison system has characteristic roots $\alpha - q$ and $-m$. The second critical point, at $\left(\dfrac{m(\alpha - q)}{\alpha q}, \dfrac{\alpha - q}{\alpha}\right)$ is significant only if $\alpha - q > 0$ (that is, if $(0, 0)$ is unstable). Then the linear comparison system has characteristic polynomial $\lambda^2 + (m\alpha/q)\lambda + m(\alpha - q)$. The roots always have negative real part, so this is stable. Oscillations can occur.

7. $(0, 0)$ and $(0, 1)$ are centers of the linear comparison system, and $(\pm\frac{1}{2}, \frac{1}{2})$ are saddle points.

9. Among the level curves of the first integral are two families of closed curves surrounding $(0, 0)$ and $(0, 1)$. These correspond to the two families of periodic solutions.

11. There are five critical points: an unstable spiral at $(0, 0)$, stable half-nodes at $(1, -1)$ and $(-1, 1)$, and saddle points at $(0.6, -1.09)$ and $(-0.6, 1.09)$ (approximately).

13. The integrated equation is

$$-\frac{1}{y} + \frac{1}{3}\frac{(y')^2}{y} = \frac{y}{h^2} - \frac{g(y - h)^2}{c^2 h^2} + K.$$

Then $y' = 0$ at $y = h$ gives $K = -2/h$ or

$$\frac{1}{h} - \frac{1}{y} + \frac{1}{3}\frac{(y')^2}{y} = \frac{y - h}{h^2} - \frac{g(y - h)^2}{c^2 h^2}.$$

15. $z = a\,\text{sech}^2(x/b)$, and hence $y(x) = h + a\,\text{sech}^2(x/b)$.

Chapter 9

Section 9.1

1. The curve $u(x)$ is a section of a parabola, opening up.

3. Set $du/dx = 0$ to find that x at which u has its minimum ($x = L/2$). Then find u_{min} and $h_1 - u_{min} = wL^2/8T$.

5. Set $v = du/dx$. Then Eq. (9.18) becomes $v'/\sqrt{1 + v^2} = \rho/T$, with solution $v = \sinh(\rho x/T + c_1)$. Another integration gives

$$u(x) = \frac{T}{\rho}\cosh\left(\frac{\rho x}{T} + c_1\right) + c_2.$$

To evaluate c_1 and c_2, it is necessary to solve $\cosh(c_1 + \rho L/T) - \cosh c_1 = (h_2 - h_1)\rho/T$. By using an identity for the difference of two hyperbolic cosines, one may find

$$c_1 = \sinh^{-1}\left(\frac{(h_2 - h_1)\rho}{2T\sinh(\rho L/2T)}\right) - \frac{\rho L}{2T}.$$

7. Since $d^2 u/dx^2 = 0$, u is a straight line: $u(x) = A_0 + B_0 x$, $0 < x < x_0$; and $u(x) = A_1 + B_1 x$, $x_0 < x < L$. The second condition from Exercise 6 requires $B_1 - B_0 = W/T$, and the boundary conditions require $A_0 = h_0$, $A_1 + B_1 L = h_1$. Finally, we must also have $A_0 + B_0 x_0 = A_1 + B_1 x_0$. These four equations in four unknowns may be solved to find

$$A_0 = h_0, \qquad A_1 = h_0 - x_0\omega, \qquad B_1 = (h_1 - h_0 + x_0\omega)/L,$$
$$B_0 = (h_1 - h_0 + (x_0 - L)\omega)/L, \qquad \text{where } \omega = W/T.$$

9. $u(y) = -gy(L - y)/2\mu$

11. $u(x) = U_0 + \dfrac{I^2 R}{\kappa A}\dfrac{x(L - x)}{2}$

13. $u(x) = U_0 + (U_1 - U_0)(\cosh \gamma x - \tanh \gamma L \sinh \gamma x)$
    $= U_0 + (U_1 - U_0)\cosh \gamma(L - x)/\cosh \gamma L$, where $\gamma = \sqrt{hC/\kappa A}$

15. $u(x) = B + c_1 \cosh \gamma x + c_2 \sinh \gamma x$ with $B = U + I^2 R/hC$, $\gamma = \sqrt{hC/\kappa A}$. From the boundary conditions, $c_1 = U_1 - B$, $c_2 = (U_1 - B)(1 - \cosh \gamma L)/\sinh \gamma L$. An easier way to write the solution is $u(x) = B + c_3 \cosh \gamma(x - \frac{1}{2}L)$ and $c_3 = (U_1 - B)/\cosh(\gamma L/2)$.

17. $u(x) = [-ex(L - x) + (h_1^2 - h_0^2)x/L + h_0^2]^{1/2}$

Section 9.2

1. $P = \pi^2 EI/L^2$
   $= \pi^2 \times 2 \times 10^6 \times \frac{1}{12} \times \frac{3}{2} \times (\frac{1}{4})^3/(36)^2$
   $= 29.7\,\text{lb}$

3. Eigenvalues are given by $\lambda_n = (2n - 1)\pi/2L$ and eigenfunctions are $\sin \lambda_n x$, for $n = 1, 2, 3, \ldots$.

5. The first eigenvalue is 0, and its accompanying eigenfunction is $u(x) \equiv 1$. The rest of the eigenvalues and eigenfunctions are

$$\lambda_n^2 = (n\pi)^2 \ (n = 1, 2, \ldots) \qquad \text{and} \qquad u_n(x) = \cos \lambda_n x.$$

7. Note that $p' = a_1 p$. Thus after multiplication the first two terms are $(pu')'$. Then $q(x) = -a_2(x)p(x)$, and $w(x) = a_3(x)p(x)$.

9. $\lambda_n^2 = 1 + (n\pi/L)^2$ and $u_n = e^{-x} \sin \sqrt{\lambda_n^2 - 1} x$

11. $\lambda_n^2 = (n\pi/\ln L)^2$ and $u_n = \sin (\lambda_n \ln x)$.

Section 9.3

1. The limiting form is just $u(x)$ as in Eq. (9.53).

3. $Q = -g\pi R^4/8\mu$

5. $u(\rho) = \dfrac{H}{6\kappa} (R^2 - \rho^2) + U$

7. $u(r) = \dfrac{H}{4\kappa} (R^2 - r^2) + \dfrac{HR}{2h} + U$

9. $u(r) = \dfrac{A}{16} (R^4 - r^4) + \dfrac{\kappa A R^3}{4h} + U$

11. $u(\rho) = \dfrac{H}{6\kappa} (R^2 - \rho^2) + \left(\dfrac{HR}{3\sigma}\right)^{1/4}$

Section 9.4

1. Call the integrand in Eq. (9.72) $h(x, t)$. Then $u_p(x) = \int_\alpha^x h(x, t) \, dt$ and

$$\frac{du_p}{dx} p = h(x, x) + \int_\alpha^x \frac{\partial h}{\partial x} (x, t) \, dt.$$

The first term is identically 0, and the second is just what appears in Eq. (9.74).

3. $u_p(x) = \int_\alpha^x G(x, t)f(t) \, dt + \int_x^\beta G(x, t)f(t) \, dt$. Use Leibniz's rule to find the first derivative:

$$u_p'(x) = G(x, x-)f(x) - G(x, x+)f(x) + \int_\alpha^x + \int_x^\beta \frac{\partial G}{\partial x} (x, t)f(t) \, dt$$

$$= \int_\alpha^\beta \frac{\partial G}{\partial x} (x, t)f(t) \, dt.$$

Similarly for the second derivative:

$$u_p''(x) = \frac{\partial G}{\partial x} (x, x-)f(t) - \frac{\partial G}{\partial x} (x, x+)f(t) + \int_\alpha^x + \int_x^\beta \frac{\partial^2 G}{\partial x^2} (x, t)f(t) \, dt$$

$$= \frac{f(t)}{p(t)} + \int_\alpha^\beta \frac{\partial^2 G}{\partial x^2} (x, t)f(t) \, dt.$$

5. $G(x, t) = \begin{cases} \dfrac{x(L - t)}{-L}, & 0 \le x \le t, \\ \dfrac{t(L - x)}{-L}, & t \le x \le L \end{cases}$

7. $G(x, t) = \begin{cases} \dfrac{\cosh \gamma x \cosh \gamma(L - t)}{-\gamma \cosh \gamma L}, & 0 \le x \le t, \\ \dfrac{\cosh \gamma(L - x) \cosh \gamma t}{-\gamma \cosh \gamma L}, & t \le x \le L \end{cases}$

9. $G(x, t) = \begin{cases} \dfrac{\sin \gamma x \cos \gamma(L - t)}{-\gamma \cos \gamma L}, & 0 \le x \le t, \\ \dfrac{\cos \gamma(L - x) \sin \gamma t}{-\gamma \cos \gamma L}, & t \le x \le L \end{cases}$

11. $G(r, t) = \begin{cases} \dfrac{\ln r(1 - \ln t)}{(-1)}, & 1 \le r \le t, \\ \dfrac{(1 - \ln r) \ln t}{(-1)}, & t \le r \le e \end{cases}$

13. $G(\rho, t) = \begin{cases} \dfrac{1}{\delta}\left(\dfrac{1}{\alpha} - \dfrac{1}{\rho}\right)\left(\dfrac{1}{t} - \dfrac{1}{\beta}\right), & \alpha \le \rho \le t, \\ \dfrac{1}{\delta}\left(\dfrac{1}{\rho} - \dfrac{1}{\beta}\right)\left(\dfrac{1}{\alpha} - \dfrac{1}{t}\right), & t \le \rho \le \beta \end{cases}$

where $\delta = -\left(\dfrac{1}{\alpha} - \dfrac{1}{\beta}\right)$.

15. $G(x, t) = \begin{cases} \dfrac{(1 - e^{-x})(e^{-t} - e^{-1})}{-(1 - e^{-1})}, & 0 \le x \le t, \\ \dfrac{(e^{-x} - e^{-1})(1 - e^{-t})}{-(1 - e^{-1})}, & t \le x \le 1 \end{cases}$

17. The solution fails to be unique (if one exists) when $\gamma^2 = 0, \pi^2, (2\pi)^2, \dots$.

19. $u(x) = \left(\dfrac{1}{2\gamma^2} - \dfrac{L}{2\gamma} \tan \gamma L\right) \sin \gamma x - \dfrac{x}{2\gamma} \cos \gamma x$ provided that $\cos \gamma L \ne 0$.

Section 9.5

1. $\displaystyle\int_0^L \sin \lambda_n x \sin \lambda_m x \, dx = \left[\dfrac{\sin(\lambda_n - \lambda_m)x}{2(\lambda_n - \lambda_m)} - \dfrac{\sin(\lambda_n + \lambda_m)x}{2(\lambda_n + \lambda_m)}\right]_0^L = 0.$
   The fact that $\sin(\lambda_n \pm \lambda_m)L = 0$ comes from the choice of λ_n. The integration above is false if $\lambda_n = \lambda_m$.

3. (a) No; unbounded at 0. (b) Yes. (c) No; derivative unbounded at 0 (vertical tangent there). (d) No; derivative unbounded at $x = 1$. (e) Yes. (f) Yes. (g) Yes.

5. (c) $b_n = 2(1 - \cos(n\pi/2))/n\pi$. (d) Series sums to 0 at $x = 0, L$, to $\frac{1}{2}$ at $x = \dfrac{L}{2}$, to $f(x)$ elsewhere.

7. (c) $b_n = -(2L/n\pi)\cos n\pi$. (d) Series sums to 0 at $x = L$, to $f(x)$ for $0 \le x \le L$.

9. (a) See Fig. 9.10. (c) $b_n = \dfrac{4L}{n^2\pi^2}\sin\dfrac{n\pi}{2}$. (d) Series converges to $f(x)$, $0 \le x \le L$.

11. (c) $b_n = 2L(1 + \cos n\pi)/n\pi$. (d) Series converges to 0 at $x = 0, L$ and to $f(x)$, $0 < x < L$.

13. (a) $\sin^3 x = (\frac{3}{4})\sin x - (\frac{1}{4})\sin 3x$. (b) $\sin(3x/2)\cos(x/2) = \frac{1}{2}(\sin x + \sin 2x)$.

15. Both the series for $f''(x) = -2$, $0 < x < L$, and the twice differentiated series for f are

$$\sum_{n=1}^{\infty} \frac{-4(1 - \cos n\pi)}{n\pi}\sin \lambda_n x.$$

17. (c) $a_0 = L/2$, $a_n = 2L(1 - \cos n\pi)/n^2\pi^2$.
    (d) Series sums to $L - x$, $0 \le x \le L$.

19. (c) $a_0 = 2/\pi$, $a_1 = 0$, $a_n = -2(1 + \cos n\pi)/(n^2 - 1)\pi$, $n = 2, 3, \ldots$.
    (d) Series sums to $\sin x$, $0 \le x \le \pi$.

21. (c) $a_0 = L/4$, $a_n = 2L\left(2\cos\dfrac{n\pi}{2} - \cos n\pi - 1\right)\Big/n^2\pi^2$.

    (d) Series sums to $f(x)$, $0 \le x \le L$.

Section 9.6

1. (a) $u_n(x) = \sin \lambda_n x$, $\lambda_n^2 = ((2n - 1)\pi/L)^2$, $n = 1, 2, 3, \ldots$;

    $\displaystyle\int_0^L \sin \lambda_n x \sin \lambda_m x \, dx = 0$; $n \ne m$, (c) $c_n = 2\sin \lambda_n L/\lambda_n^2 L$.

3. (a) $u_n(x) = \cos \lambda_n x$, λ_n^2 as in 1(a);

    (b) $\displaystyle\int_0^L \cos \lambda_n x \cos \lambda_m x \, dx = 0$; $n \ne m$, (c) $c_n = \dfrac{2}{L}\left[\dfrac{L\sin \lambda_n L}{\lambda_n} - \dfrac{1}{\lambda_n^2}\right]$.

5. (a) $u_n(x) = \begin{cases} \cos \lambda_n x, & n \text{ odd} \\ \sin \lambda_n x, & n \text{ even} \end{cases}$, $\lambda_n^2 = (n\pi/2L)^2$, $n = 1, 2, 3, \ldots$;

    (b) $\displaystyle\int_{-L/2}^{L/2} \cos \lambda_n x \cos \lambda_m x \, dx = 0 \ (m \ne n, \text{ even})$,

    $\displaystyle\int_{-L/2}^{L/2} \sin \lambda_n x \cos \lambda_m x \, dx = 0 \ (n \text{ even}, m \text{ odd})$,

    $\displaystyle\int_{-L/2}^{L/2} \sin \lambda_n x \sin \lambda_m x \, dx = 0 \ (m \ne n, \text{ odd})$;

    (c) $c_n = -2\cos(\lambda_n L/2)/\lambda_n$.

7. (a) $u_n(x) = \sin(\lambda_n \ln x)$, $\lambda_n^2 = (n\pi/\ln L)^2$, $n = 1, 2, 3, \ldots$;

    (b) $\displaystyle\int_1^L \sin(\lambda_n \ln x)\sin(\lambda_m \ln x)\dfrac{dx}{x}$, $n \ne m$;

(c) $c_n = \dfrac{2\lambda_n(1 - L\cos n\pi)}{(\lambda_n^2 + 1)\ln L}$.

9. $G(x, t) = \displaystyle\sum_{n=1}^{\infty} \dfrac{-2\sin\lambda_n t}{L\lambda_n^2}\sin\lambda_n x$

11. $u(x) = \frac{1}{12}(2x^3L - x^4 - xL^3)$

Section 9.7

Fourier Series

F1. $\int_{-L}^{L} g(x)\,dx = \int_{-L}^{0} g(x)\,dx + \int_0^L g(x)\,dx$. Now change variable to $y = -x$ in the integral over $-L < x < 0$ to get

$$\int_L^0 g(-y)(-dy) + \int_0^L g(x)\,dx = \int_0^L g(y)\,dy + \int_0^L g(x)\,dx.$$

F3. If f is even, $f(x)\sin\lambda_n x$ is odd, so by Exercise F2

$$b_n = \frac{1}{L}\int_{-L}^{L} f(x)\sin\lambda_n x\,dx = 0.$$

Also, $f(x)\cos\lambda_n x$ is even, so the formulas for a_0 and a_n follow from Exercise F1.

F5. First assume that $0 < a < 2L$. Then

$$\int_a^{a+2L} f(x)\,dx = \int_a^{2L} f(x)\,dx + \int_{2L}^{2L+a} f(x)\,dx.$$

Now use the periodicity to equate

$$\int_{2L}^{2L+a} f(x)\,dx = \int_0^a f(x)\,dx.$$

F7. $a_0 = L/4$, $a_n = L(\cos n\pi - 1)/n^2\pi^2$, $b_n = -L\cos n\pi/n\pi$. The series converges to $f(x)$ except at $x = \pm L, \pm 3L, \ldots$, where it converges to $L/2$.

F9. $a_0 = 2/\pi$, $a_n = -4/\pi(4n^2 - 1)$, $b_n = 0$. Remember that $\lambda_n = 2n$. The series converges everywhere to $|\sin x|$.

F11. Note that $f(x)$ is even. $a_0 = L/2$, $a_n = 2L(1 - \cos n\pi)/n^2\pi^2$, $b_n = 0$. The series converges everywhere to the periodic function $f(x)$.

F13. $a_0 = 1/2L$, $a_n = (\sin\lambda_n(c + \varepsilon) - \sin\lambda_n(c - \varepsilon))/2n\pi\varepsilon$,
$b_n = (\cos\lambda_n(c - \varepsilon) - \cos\lambda_n(c + \varepsilon))/2n\pi\varepsilon$.

Bessel Functions

B1. Rewrite Eq. (B1) as $\dfrac{1}{x}(xJ_0')' + J_0 = 0$, then differentiate. The result is

$$J_0''' + \frac{1}{x}J_0'' - \frac{1}{x^2}J_0' + J_0' = 0.$$

This proves that $J_0' = c_1 J_1 + c_2 Y_1$, for some c_1 and c_2.

B3. $\displaystyle\int_0^L f(r)J_0(\lambda_n r)r\,dr = \frac{1}{\lambda_n^2}\int_0^{\lambda_n L} J_0(x)x\,dx = \frac{L}{\lambda_n}J_1(\lambda_n L),$

$\displaystyle\int_0^L J_0^2(\lambda_n r)r\,dr = L^2 J_1^2(\lambda_n L)/2.$ Therefore $c_n = 2/J_1(\lambda_n L)\lambda_n L.$

B5. Since $u_0(r) \equiv 1$ is an eigenfunction, the series is just $f(r) = u_0(r)$.

B7. Substitute the series for u and the series for 1 in the differential equation to get

$$\sum_{n=1}^{\infty} -\lambda_n^2 C_n u_n(r) = \sum_{n=1}^{\infty} \gamma^2(C_n - c_n)u_n(r),$$

where c_n are given in the solution to Exercise B3. Now equate coefficients of like terms to get

$$C_n = c_n/(\lambda_n^2 + \gamma^2).$$

Spherical Bessel Functions

S1. The eigenvalues and eigenfunctions are $\lambda_n^2 = (n\pi/L)^2$, $u_n = (\sin\lambda_n\rho)/\rho$, $n = 1, 2, 3, \ldots$. Then

$$c_n = \frac{2}{L}\int_0^L (\sin\lambda_n\rho)\rho\,d\rho = -2L\cos n\pi/n\pi.$$

S3. Inserting $u(r)$ and 1 in the form of series into the differential equation leads to

$$\sum_{n=1}^{\infty}(-\lambda_n^2)C_n u_n(\rho) = \sum_{n=1}^{\infty} -c_n u_n(\rho)$$

and thus $C_n = c_n/\lambda_n^2$, where c_n is given in Exercise S1.

S5. $\displaystyle j_2(z) = \left(\frac{3}{z^3} - \frac{1}{z}\right)\sin z - \frac{3}{z^2}\cos z$

Legendre Polynomials

L1. $x^2 = \frac{2}{3}P_2(x) + \frac{1}{3}P_0(x);\quad x^3 = \frac{2}{5}P_3(x) + \frac{3}{5}P_1(x).$

L3. According to Exercise L2, x^k is a linear combination of P_k, P_{k-2}, etc. The orthogonality of these with P_m then guarantees the result.

L5. $\displaystyle\int_{-1}^{1}P_2^2(x)\,dx = \frac{1}{4}\int_{-1}^{1}(9x^4 - 6x^2 + 1)\,dx = \frac{1}{4}\left(\frac{9\cdot 2}{5} - \frac{6\cdot 2}{3} + 2\right) = \frac{2}{5}$

L7. $c_0 = c_2 = 0$, $c_1 = 3/\pi$, $c_3 = -7(15 - \pi^2)/\pi^3$ $(c_1 \cong 0.955,\ c_3 \cong -1.158)$

L9. $c_0 = \frac{1}{2}$, $c_1 = \frac{3}{4}$, $c_2 = c_4 = 0$, $c_3 = -\frac{7}{16}$

L11. $\rho/a = (1 + kx^2)^{-1/2} = 1 - \frac{1}{2}kx^2 + \frac{3}{8}k^2x^4 - + \cdots$

$$= P_0 - \frac{1}{2}k\left(\frac{2}{3}P_2 + \frac{1}{3}P_0\right) + \frac{3}{8}k^2\left(\frac{8}{35}P_4 + \frac{4}{7}P_2 + \frac{1}{5}P_0\right) + \cdots$$

$$= \left(1 - \frac{k}{3} + \frac{3k^2}{40} + \cdots\right)P_0(x) - \left(\frac{k}{3} - \frac{3k^2}{14} + \cdots\right)P_2(x)$$

$$+ \left(\frac{3k^2}{35} + \cdots\right)P_4(x) + \cdots$$

Section 9.8

1. Substitute the series in the numerator. The terms in $h^2, h^4, \ldots$ cancel out.

3. Same as Exercise 1. It is necessary to take terms of the Taylor series through $h^4 u^{iv}(x)/24$.

5. When n is doubled, $h^2 = 1/n^2$ is quartered. The error should also be quartered.

7. If ε were independent of h (it depends weakly on h), then from

$$U(x, h) = u(x) + \varepsilon h^2$$

and

$$U(x, h/2) = u(x) + \varepsilon h^2/4$$

we would find $u(x) = (4U(x, h/2) - U(x, h))/3$.

For Exercises 9–19 the first, last and general line of the replacement equations are given, along with the solutions for both values of n.

9. $-40u_1 + 16u_2 = -1$
   $16u_1 - 40u_2 + 16u_3 = -2$
   $16u_2 - 40u_3 = -3$

x_i	0	0.25	0.5	0.75	1	n
u_i	0	0.0779	0.1324	0.1279	0	4
u_i	0	0.0791	0.1345	0.1303	0	8

11. $-40u_0 + 32u_1 = 0$
    $16u_{i-1} - 40u_i + 16u_{i+1} = -i, \ i = 1, 2, 3$
    $32u_3 - 40u_4 = -4$

x_i	0	0.25	0.5	0.75	1	n
u_i	0.1471	0.1838	0.25	0.3162	0.3529	4
u_i	0.1544	0.1864	0.25	0.3136	0.3456	8

13. $-40u_0 + 32u_1 = 0$
    $16u_{i-1} - 40u_i + 16u_{i+1} = -i, \ i = 1, 2$
    $16u_3 - 40u_4 = -16$

x_i	0	0.25	0.5	0.75	1	n
u_i	0.2276	0.2845	0.4212	0.6435	1	4
u_i	0.2326	0.2847	0.4193	0.6410	1	8

15. $-48u_i + 25u_2 = 0$
    $25u_{i-1} - (50 - 2i)u_i + 25u_{i+1} = 0, \ i = 1, 2, 3$
    $25u_3 - 42u_4 = -25$

x_i	0	0.2	0.4	0.6	0.8	1	n
u_i	0	0.5778	1.1095	1.4636	1.4664	1	5
u_i	0	0.5772	1.1028	1.4491	1.4510	1	10

17. $-40u_1 + 17u_2 = -19$
    $14u_1 - 40u_2 + 18u_3 = -4$
    $13u_2 - 40u_3 = -4$

x_i	0	0.25	0.5	0.75	1	n
u_i	1	0.6626	0.4415	0.2435	0	4
u_i	1	0.6627	0.4424	0.2447	0	8

19. $-45.17747u_i + 25u_2 = -25$
$25u_{i-1} - (50 - 10/(1 + i/5)^4)u_i + 25u_{i+1} = 0, \; i = 2, 3$
$25u_3 - 49.0474u_4 = 0$

x_i	0	0.2	0.4	0.6	0.8	1	n
u_i	1	1.0297	0.8608	0.6023	0.3070	0	5
u_i	1	1.0401	0.8736	0.6127	0.3127	0	10

21.

x_i	0	0.25	0.5	0.75	1
u_i	1	0.6726	0.5957	0.7025	1

Miscellaneous Exercises

1. $u_n(x) = e^{-kx/2} \sin \mu_n x$, $\mu_n = n\pi/L$, $\lambda_n^2 = \mu_n^2 + (k/2)^2$, $n = 1, 2, \ldots$

3. If $v(y) = u(x)$, the equation for v is $v'' + v'/y + (\lambda/k)^2 v = 0$, which is Bessel's equation of order 0. The general solution is
$$v(y) = AJ_0(\lambda y/k) + BY_0(\lambda y/k).$$

5. $u(x) = \dfrac{13}{48} x + \dfrac{11}{48} x^2 + \dfrac{1}{36} x^3 - \dfrac{1}{16} x^4 + \dfrac{77}{144} \dfrac{\ln(2x + 1)}{\ln 3}.$

The solution of Exercise 4 deviates by about 0.025 from this solution. The parameter $\alpha = \log 3$ is the average value of $1/(1 + 2x)$ (the coefficient of u') in the interval $0 < x < 1$.

7. $G(r, t) = \begin{cases} \ln(r/L), & 0 < t < r, \\ \ln(t/L), & r < t < L \end{cases}$

9. $\lambda_n^4 = (n\pi/L)^4 = \omega^2 \rho/EI$. The critical speeds are $\omega = (n\pi/L)^2 (EI/\rho)^{1/2}$.

11. $u(x) = (w_0/24EI)(x^4 - 4LX^3 + 6L^2 x^2).$

13. $u_n(x) = \sin \lambda_n x$. After using trigonometric identities and the equation $\cos \lambda_n L = \lambda_n \sin \lambda_n L$, which defines the λ's, the given integral reduces to $-\sin \lambda_n L \sin \lambda_m L$.

15. $\rho/a = 1 - \dfrac{k}{2} \cos^2 \phi + \dfrac{3k^2}{8} \cos^4 \phi - \cdots$

$= 1 - \dfrac{k}{4}(1 + \cos 2\phi) + \dfrac{3k^2}{64}(3 + 4\cos 2\phi + \cos 4\phi) + \cdots$

$= \left(1 - \dfrac{k}{4} + \dfrac{9k^2}{64} + \cdots\right) - \left(\dfrac{k}{4} - \dfrac{3k^2}{16}\right)\cos 2\phi$

$\quad + \left(\dfrac{3k^2}{64} + \cdots\right)\cos 4\phi + \cdots$

17. $H(r) = (1 - 2xr + r^2)^{-1/2}.$

Chapter 10

Section 10.1

1. If the string is uniform in size and density, then the length of the string above the interval from x to $x + \Delta x$ is approximately $\Delta s = \Delta x \sqrt{1 + (\partial u/\partial x)^2}$. With this assumption, Eq. (10.6) would be replaced by
$$T\frac{\partial^2 u}{\partial x^2} + F = \rho \sqrt{1 + \left(\frac{\partial u}{\partial x}\right)^2} \frac{\partial^2 u}{\partial t^2}.$$

3. If $A = A(x)$ is variable, Eqs. (10.11) and (10.12) become

$$q(x, t)A(x) + H\Delta x - q(x + \Delta x, t)A(x + \Delta x) = c\rho A(x)\frac{\partial u}{\partial t}(x, t),$$

$$H - \frac{\partial}{\partial x}(Aq) = c\rho A\frac{\partial u}{\partial t}.$$

5. One possible interpretation is the following. The differential equation refers to a string loaded by its own weight: $F = -g\rho$. The string is fixed at both ends. At $t = 0$ the string is given the shape of an isosceles triangle ($f(x)$) and is released (at $t = 0$, $\partial u/\partial t \equiv 0$).

7. $H\Delta x = hC\Delta x(T - u(x, t))$, where h is a constant of proportionality and C is the circumference of the rod. Recall that H is positive if heat is gained by the slice; heat is gained if $T > u(x, t)$. Finally, Eq. (10.14) becomes

$$\frac{\partial^2 u}{\partial x^2} = \frac{1}{k}\frac{\partial u}{\partial t} - \frac{hC}{\kappa A}(T - u).$$

9. $\dfrac{I^2 R}{A\kappa} = \dfrac{225 \text{ watts/cm}}{0.5 \text{ cm}^2 \cdot 0.92 \text{ cal/sec} \cdot \text{cm} \cdot \text{C deg}} \cdot \dfrac{0.239 \text{ cal}}{\text{watt sec}} = 116.9 \text{ C deg/cm}^2$. If u is in degrees Celsius, x in centimeters, and t in seconds, then the dimensionally correct version of Eq. (10.21) is

$$\frac{\partial^2 u}{\partial x^2} = \frac{1}{1.1}\frac{\partial u}{\partial t} - 116.9.$$

Section 10.2

1. The general solution of $X''/X = k^2$ is $X(x) = c_1 e^{kx} + c_2 e^{-kx} = e^{-kx}(c_1 e^{2kx} + c_2)$. Then $X(x) = 0$ if and only if $c_1 e^{2kx} + c_2 = 0$, which can happen for one value of x if $c_1 c_2 < 0$ or for no x if $c_1 c_2 \geq 0$. (Recall that X is nontrivial, so not both c_1 and c_2 can be 0.)

3. The solution is given by Eq. (10.44) with $b_n = -2U_0 \cos n\pi/n\pi$. Sectionally smooth.

5. $u(x, t) = U_0 \sin(\pi x/L) \exp(-\pi^2 kt/L^2)$. (In Eq. (10.44), all b's are 0 except b_1.)

7. Use Eq. (10.44) with $b_n = 8U_0 \sin(n\pi/2)/n^2\pi^2$. Sectionally smooth.

9. $t = 0.01$, n up to 7; $t = 0.1$, n up to 2; $t = 1$, all terms are less than 0.01. See the rough table of values of $e^{-n^2\pi^2 t}$ below

$n =$	1	2	3	4	5	6	7
$t = 0.01$	0.91	0.67	0.41	0.21	0.08	0.02	0.008
$t = 0.1$	0.37	0.02					

11. $X'' + \lambda^2 X = 0$, $0 < x < L$; $X'(0) = 0$, $X(L) = 0$. Solution: $X_n(x) = \cos \lambda_n x$, $\lambda_n^2 = ((2n - 1)\pi/2L)^2$, $n = 1, 2, 3, \ldots$. (That is, $\lambda_1 = \pi/2L$, $\lambda_2 = 3\pi/2L, \ldots$.)

13. $X'' + \lambda^2 X = 0$, $0 < x < L$; $X'(0) = 0$, $X'(L) = 0$. Solution: $X_0(x) = 1$, $\lambda_0^2 = 0$; $X_n(x) = \cos \lambda_n x$, $\lambda_n^2 = (n\pi/L)^2$, $n = 1, 2, 3, \ldots$.

15. Separation of variables: $\dfrac{X''}{X} = \dfrac{T' + 2T}{T}$.

Eigenvalue problem: $X'' + \lambda^2 X = 0$, $X(0) = 0$, $X(1) = 0$

Second factor: $T_n(t) = \exp\left(-(\lambda_n^2 + 2)t\right)$

Solution: $u(x, t) = e^{-2t} \sum b_n \sin \lambda_n x\, e^{-\lambda_n^2 t}$,

$b_n = 200(1 - \cos n\pi)/n\pi$, $\lambda_n^2 = (n\pi)^2$.

Section 10.3

1. $u(x, t) = \displaystyle\sum_{n=1}^{\infty} c_n \sin \lambda_n x\, e^{-\lambda_n^2 kt}$, $\lambda_n^2 = ((2n - 1)\pi/2L)^2$,

$c_n = 2U_0/\lambda_n L$.

3. $u(x, t) = a_0 + \displaystyle\sum_{n=1}^{\infty} a_n \cos \lambda_n x\, e^{-\lambda_n^2 kt}$, $\lambda_n^2 = (n\pi/L)^2$, $a_0 = 2U_0/\pi$,

$a_1 = 0$, $a_n = -4U_0(1 + \cos n\pi)/(n^2 - 1)\pi$.

5. Same form as in 3 above, but $a_0 = U_0/2$, $a_n = 2U_0 \sin \dfrac{n\pi}{2}\Big/n\pi$.

7. The first term neglected is $e^{-9\tau}/9$, compared with $e^{-\tau}$, the last term included. Since $\tau \cong 4.4$, $e^{-9\tau}/9 \cong 7 \times 10^{-19}$.

Section 10.4

1. (a)

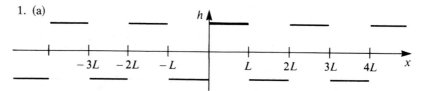

(b) Use Eq. (10.69) with $a_n = 2h(1 - \cos n\pi)/n\pi$.

3. (a) See Fig. 10.5. (b) Use Eq. (10.69) with $a_n = 9h \sin (n\pi/3)/n^2\pi^2$

5. Those of Exercises 3 and 4 satisfy the hypotheses. From Exercise 1: $f(0)$ and $f(L)$ are not 0—or else f is not continuous. From Exercise 2, f is not continuous.

7.

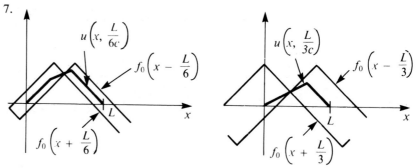

9. $u(L/4, L/2c) = \frac{1}{2}(h - h) = 0$; $u(L/2, L/4c) = \frac{1}{2}(h + h) = h$;
$u(L/2, L/c) = \frac{1}{2}(-h - h) = -h$

11. Use Eq. (10.62) with $a_n \equiv 0$, $b_n = 2\alpha L(1 - \cos n\pi)/n^2\pi^2$.

13. Differentiate Eqs. (10.49) and (10.50) to get

$$\frac{\partial^3 u}{\partial x^2 \partial t} = \frac{1}{c^2}\frac{\partial^3 u}{\partial t^3}, \qquad 0 < x < L, \qquad 0 < t,$$

$$\frac{\partial u}{\partial t}(0, t) = 0, \qquad \frac{\partial u}{\partial t}(L, t) = 0, \qquad 0 < t.$$

Our initial condition is $\partial u/\partial t(x, 0) = g(x)$. This becomes $v(x, 0) = g(x)$. Now

$$\frac{\partial v}{\partial t} = \frac{\partial^2 u}{\partial t^2} = c^2\frac{\partial^2 u}{\partial x^2}$$

Since $u(x, 0) = 0$, so is $\dfrac{\partial^2 u}{\partial x^2}(x, 0) = 0$.

15. Exercise 14 gives $v(x, t)$ which is $\partial u/\partial t$. Now

$$u(x, t) = u(x, 0) + \int_0^t \frac{\partial u}{\partial t}(x, t')\, dt'$$

$$= 0 + \int_0^t \tfrac{1}{2}(\bar{g}_0(x + ct') + \bar{g}_0(x - ct'))\, dt'$$

$$= \frac{1}{2}\left[\frac{1}{c}(G(x + ct) - G(x)) + \frac{1}{-c}(G(x - ct) - G(x))\right].$$

The c and $-c$ in the denominators occur because you change variables and integrate with respect to $x \pm ct'$. The result can also be confirmed by substituting directly into the problem for u.

17.

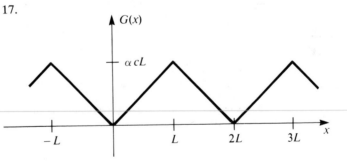

Section 10.5

1. $u_p = T_0 + (U_1 - U_0)x/L$. The general solution is $u = u_p + \sum b_n e^{-\lambda_n^2 kt} \sin \lambda_n x$ with $\lambda_n^2 = (n\pi/L)^2$. To satisfy the initial condition, take

$$b_n = 2(U_1 - U_0)\cos n\pi/n\pi$$

3. $u_p = \dfrac{U_1}{L}\left(x - \dfrac{\sinh \gamma x}{\gamma \cosh \gamma L}\right)$. General solution: $u = u_p + \sum c_n e^{-\lambda_n^2 kt} \sin \lambda_n x$ with

$\lambda_n = (2n - 1)\pi/2L$. To satisfy initial conditions,

$$c_n = -\frac{2}{L} \int_0^L u_p(x) \sin \lambda_n x \, dx = -\frac{2U_1\gamma^2 \sin \lambda_n L}{L^2\lambda_n^2(\lambda_n^2 + \gamma^2)}.$$

5. $u_p = hx/L$. General solution: $u(x, t) = u_p + \sum (a_n \cos \lambda_n ct + b_n \sin \lambda_n ct) \times$
   $\sin \lambda_n x$. To satisfy initial conditions, take $b_n = 0$, $a_n = \dfrac{2h}{L} \dfrac{\cos n\pi}{n\pi}$.

7. $u(x, t) = w(x, t) + v(x, t)$, where $w(x, t) = f(t)$ satisfies the boundary conditions. Then v must satisfy

$$\frac{\partial^2 v}{\partial x^2} = \frac{\partial v}{\partial t} + \frac{\partial w}{\partial t} = \frac{\partial v}{\partial t} + \alpha e^{-\alpha t},$$

$$v(0, t) = 0, \qquad v(1, t) = 0.$$

Now use Method 2 on v, setting $(\lambda_n = n\pi)$ $\alpha e^{-\alpha t} = \sum F_n \sin \lambda_n x$, $v = \sum T_n(t) \sin \lambda_n x$ with $F_n = \alpha e^{-\alpha t} 2(1 - \cos n\pi)/n\pi$. Equation for T_n: $T_n' + \lambda_n^2 T_n = -F_n$. Solution: $T_n = b_n e^{-\lambda_n^2 t} - F_n/(\lambda_n^2 - \alpha)$, provided that the last denominator is not 0 for any n. Finally,

$$u(x, t) = f(t) + \sum_{n=1}^{\infty} b_n e^{-\lambda_n^2 t} \sin \lambda_n x - \sum_{n=1}^{\infty} \frac{F_n(t)}{\lambda_n^2 - \alpha} \sin \lambda_n x.$$

To satisfy the initial conditions, choose $b_n = F_n(0)/(\lambda_n^2 - \alpha)$, since $f(0) = 0$.

9. Assume $u(x, t) = T_0(t) + \sum T_n(t) \cos \lambda_n x$, $\lambda_n = n\pi$ for $n = 1, 2, 3, \ldots$. Recall that $\lambda_0 = 0$ is an eigenvalue, $X_0(x) = 1$. Equations for T's: $T_0' - 1 = 0$; $T_n' = -\lambda_n^2 T_n$, $n = 1, 2, 3, \ldots$. Then

$$u(x, t) = t + a_0 + \sum_{n=1}^{\infty} a_n e^{-\lambda_n^2 t} \cos \lambda_n x$$

is the general solution. To satisfy initial conditions, all a's are 0.

11. The general solution of Eqs. (10.116) and (10.117) is given by Eqs. (10.121) and (10.122). To satisfy the initial conditions, choose $a_n = 0$ and
    $$b_n\lambda_n + B_n\frac{\omega^3}{\lambda_n^2 - \omega^2} + B_n\omega = 0.$$ Note that $x = \sum B_n \sin \lambda_n x$, $0 < x < 1$.

13. Since $B_n = 0$ for n even, resonance is observed if ω equals any λ_n with n odd. If $\omega = \lambda_N$, N odd, then

$$T_N(t) = a_N \cos \lambda_N t + b_N \sin \lambda_N t + \frac{B_N}{\lambda_N} t \cos \omega t$$

in Eqs. (10.110)–(10.112). In Eq. (10.113), for $n = N$: $b_N\lambda_N + B_N/\lambda_N = 0$.

15. Product solutions are $\sin \lambda_n x$ times sine or cosine of $\lambda_n ct$, $c = 1/\sqrt{\kappa\rho}$. Thus the lowest frequency is $\lambda_1 c$ rad/sec or $\lambda_1 c/2\pi$ Hz. Recall that $\lambda_1 = \pi/2L$.

17. The linear combination is $h(u_2 - Lu_1)/2k$.

Section 10.6

1. $u(x, y) = \displaystyle\sum_{n=1}^{\infty} \frac{2U_0(1 - \cos n\pi)}{n\pi} \frac{\sinh \lambda_n x}{\sinh \lambda_n} \sin \lambda_n y$, $\lambda_n = n\pi$

3. The general solution of the homogeneous problem is

$$u(x, y) = a_0 + b_0 y + \sum_{n=1}^{\infty} (a_n \cosh \lambda_n y + b_n \sinh \lambda_n y) \cos \lambda_n x,$$

with $\lambda_n = n\pi/L$. However, all coefficients are 0 except b_0, so $u(x, y) = U_o y/B$.

5. $u(x, y) = \sum_{n=1}^{\infty} (a_n \cosh \lambda_n y + b_n \sinh \lambda_n y) \cos \lambda_n x$, $\lambda_n = (2n - 1)\pi/2L$,

$$a_n = 0, \, b_n = \frac{4U_0 \sin \lambda_n L}{(2n - 1)\pi \sinh \lambda_n B}$$

7. $u(x, y) = a_0 + b_0 y + \sum_{n=1}^{\infty} (a_n \cosh \lambda_n y + b_n \sinh \lambda_n y) \cos \lambda_n x$,

$$\lambda_n^2 = (n\pi/L)^2, \, b_0 = 0, \, b_n = 0, \, a_0 = 0, \, a_n = 4S \sin \frac{n\pi}{2} \Big/ (\lambda_n^2 L \sinh \lambda_n B)$$

9. The level curves are hyperbolas with the x and y axes as asymptotes. $u(0, y) = 0$, $u(1, y) = y$, $u(x, 0) = 0$, $u(x, 1) = x$.

11. $u(x, 0) = 0$, $u(0, y) = 1$. Level curves are rays issuing from the origin.

Section 10.7

1. $u = u_H + u_V$. $u_H = \sum_{n=1}^{\infty} (a_n \cosh \lambda_n y + b_n \sinh \lambda_n y) \sin \lambda_n x$,
$\lambda_n^2 = (n\pi/L)^2$, $a_n = 0$, $b_n = 2(1 - \cos n\pi)/n\pi \sinh \lambda_n B$.
$u_V = \sum_{n=1}^{\infty} (A_n \cosh \mu_n x + B_n \sinh \mu_n x) \sin \mu_n y$, $\mu_n^2 = (n\pi/B)^2$,
$A_n = 0$, $B_n = 2(1 - \cos n\pi)/n\pi \sinh \mu_n L$.

3. $u = u_H + u_V$. $u_H = \sum (a_n \cosh \lambda_n y + b_n \sinh \lambda_n y) \sin \lambda_n x$,
$\lambda_n^2 = (n\pi/L)^2$, $b_n = 0$, $a_n = 2(1 - \cos n\pi)/(n^2\pi^2 \sinh \lambda_n B)$.
$u_V = A_0 + B_0 x + \sum (A_n \cosh \mu_n x + B_n \sinh \mu_n x) \cos \mu_n y$, $\mu_n^2 = (n\pi/B)^2$,
$A_0 = 0$, $A_n = 0$, $B_0 = 1/L$, $B_n = 0$. That is, $u_V = x/L$.

5. $u = u_H + u_V$. $u_H = \sum (a_n \cosh \lambda_n y + b_n \sinh \lambda_n y) \sin \lambda_n x$,
$\lambda_n^2 = ((2n - 1)\pi/2L)^2$, $a_n = 0$, $b_n = 2/(\lambda_n L)^2 \cosh \lambda_n B$.
$u_V = \sum (A_n \cosh \mu_n x + B_n \sinh \mu_n x) \sin \mu_n y$,
$\mu_n^2 = ((2n - 1)\pi/2B)^2$, $A_n = 2/\mu_n B$, $B_n = -A_n \coth \mu_n L$.

7. Parameterize the path by setting $x = \frac{1}{2} + r \cos t$, $y = \frac{1}{2} + r \sin t$, $0 \le t < 2\pi$. The integral then becomes

$$\frac{1}{2\pi r} \int_0^{2\pi} (\tfrac{1}{2} + r \cos t)(\tfrac{1}{2} + r \sin t) \cdot r \, dt$$

$$= \frac{1}{2\pi} \int_0^{2\pi} \left(\frac{1}{4} + \frac{r}{2} (\cos t + \sin t) + r^2 \cos t \sin t \right) dt$$

$$= \frac{1}{2\pi} \cdot \frac{1}{4} \, 2\pi = \frac{1}{4} = u\left(\frac{1}{2}, \frac{1}{2}\right).$$

9. Use Eq. (10.183). $a_0 = 0$, $a_n = 0$, $b_n = -2 \cos n\pi/n$.

11. Use Eq. (10.183). $a_0 = 2/\pi$, $a_n = -2(1 + \cos n\pi)/(n^2 - 1)\pi$, $b_n = 0$.

13. Exercise 9, $u(0) = 0$; Exercise 10, $u(0) = U_0/2$; Exercise 11, $u(0) = 2/\pi$;

Exercise 12, $u(0) = 1/\pi$. In each case, $u(0) = \dfrac{1}{2\pi}\displaystyle\int_{-\pi}^{\pi} f(\theta)\,d\theta$.

15. $u(x, y) = Cx(L - x)/2$.

17. $u(r, \theta) = C(L^2 - r^2)/4$.

Section 10.8

1. The solution is Eq. (10.197), with $c_n = 2U_0 J_1(\beta_n)/\beta_n$. See Exercise 3 (Bessel functions) of Section 9.7.

3. $u(r, t) = \sum_{n=0}^{\infty} c_n J_1(\lambda_n r)e^{-\lambda_n^2 kt}$, where $\lambda_0^2 = 0$, $\lambda_n L = \beta_n$ is the nth zero of the Bessel function J_1. The coefficients are

$$c_n = \frac{\int_0^L f(r)J_1(\lambda_n r)r\,dr}{\int_0^L J_1^2(\lambda_n r)r\,dr}.$$

5. $u(r, t) = \displaystyle\sum_{n=1}^{\infty} (a_n \cos \lambda_n ct + b_n \sin \lambda_n ct)\dfrac{\sin \lambda_n \rho}{\rho}$, $\lambda_n^2 = (n\pi/L)^2$.

The a's and b's are available to satisfy an initial condition.

7. The solution is given in Eq. (10.215). The coefficients can be evaluated by direct integration:

$$c_n = \frac{2n + 1}{2}\int_0^{\pi/2} U_0 P_n(\cos\phi)\sin\phi\,d\phi = \frac{2n + 1}{2}U_0\int_0^1 P_n(x)\,dx.$$

Thus $c_0 = U_0/2$, $c_1 = 3U_0/4$, $c_3 = -7U_0/16$, $c_2 = c_4 = 0$.

9. Equation (10.213) is an Euler–Cauchy equation with solutions ρ^n and $\rho^{-(n+1)}$, since $\lambda_n^2 = n(n + 1)$.

11. Product solutions are $a_0 + b_0 t$ and $(a_n \cos(\lambda_n ct/L) + b_n \sin(\lambda_n ct/L)) \cdot P_n(\cos\phi)$, where P_n is the nth Legendre polynomial, $n = 1, 2, \ldots$.

13. General solution: $u(\rho, t) = \sum_{n=1}^{\infty} c_n e^{-\lambda_n^2 kt}(\sin \lambda_n\rho)/\rho$, $\lambda_n^2 = (n\pi/L)^2$. For the given initial condition, $c_n = -2U_0 L(\cos n\pi)/n\pi$.

15. General solution: $u(r, z) = \sum_{n=1}^{\infty} (a_n \cosh \lambda_n z + b_n \sinh \lambda_n z)J_0(\lambda_n r)$, $\lambda_n^2 = (\beta_n/L)^2$, where β_n is the nth zero of J_0. For the given conditions at $z = 0$ and B, $a_n = 0$, $b_n = 2U_0 J_1(\beta_n)/(\beta_n \sinh \lambda_n B)$. See Exercise 1.

Section 10.9

1. (a) Hyperbolic; (b) $\xi = 2y$, $\eta = -2x + 2y$;

(c) $-4\dfrac{\partial^2 v}{\partial\xi\,\partial\eta} + 2\left(\dfrac{\partial v}{\partial\xi} + \dfrac{\partial v}{\partial\eta}\right) = 0$.

3. (a) Elliptic; (b) $\xi = -x - y$, $\eta = x$; (c) $\dfrac{\partial^2 v}{\partial\xi^2} + \dfrac{\partial^2 v}{\partial\eta^2} + v = 0$.

5. Integrate once with respect to ξ: $\partial v/\partial\eta = \theta(\eta)$. Now integrate with respect to η: $v = \int \theta(\eta)\,d\eta + \psi(\xi)$. The functions θ and ψ are playing the role of integration "constants" and are arbitrary. Rename $\int \theta(\eta)\,d\eta = \phi$ to get $v(\xi, \eta) = \phi(\eta) + \psi(\xi)$.

7. $v = e^{-\eta/2}w$; $\partial v/\partial \eta = e^{-\eta/2}(\partial w/\partial \eta - \frac{1}{2}w)$;
$\partial^2 v/\partial \eta^2 = e^{-\eta/2}(\partial^2 w/\partial \eta^2 - \partial w/\partial \eta + \frac{1}{4}w)$;
$\partial v/\partial \xi = e^{-\eta/2}\, \partial w/\partial \xi$. Now substitute.

9. Substitute formally into Eq. (10.223) with $A = C = 1$, $B = 0$, $\alpha = \gamma = 1$, $\beta = -\delta = i$.

Miscellaneous Exercises

1. $f'(x) = (2/\sqrt{\pi})e^{-x^2}$, $f''(x) = (2/\sqrt{\pi})(-2x)e^{-x^2}$.

3. $u(x, t)$ satisfies the heat equation for $0 < x$, and $u(0, t) = 1$, $u(x, 0) = 0$.

5. Assume the form $u(x, t) = \sum_{n=1}^{\infty} T_n(t) \sin \lambda_n x$, $\lambda_n^2 = (n\pi)^2$. Expand $t = \sum_{n=1}^{\infty} F_n(t) \sin \lambda_n x$, with $F_n(t) = 2t(1 - \cos n\pi)/n\pi$. Then $-\lambda_n^2 T_n = T_n' - F_n(t)$
or $\qquad T_n' + \lambda_n^2 T_n = F_n(t)$, $\qquad$ and $\qquad T_n(0) = 0$. $\qquad$ Solution: $\qquad T_n = $
$\left(\dfrac{t}{\lambda_n^2} - \dfrac{1 - e^{-\lambda_n^2 t}}{\lambda_n^4}\right)\dfrac{2(1 - \cos n\pi)}{n\pi}$.

7. $X(x) = c_1 \cosh kx + c_2 \sinh kx$ is the general solution of $X''/X = k^2$. In (a) and (b), $X(0) = 0$ means $c_1 = 0$. Then $c_2 \sinh kL = 0$ in (a) or $kc_2 \cosh kL = 0$ in (b) requires $c_2 = 0$. In (c) and (d), $X'(0) = 0$ means $c_2 = 0$. Then $c_1 \cosh kL = 0$ in (c) or $c_1 k \sinh kL = 0$ in (d) requires $c_1 = 0$.

9. The equations are $\partial i/\partial x = -C\, \partial e/\partial t$ and $\partial e/\partial x = -Ri$. Solve the second for i and substitute in the first to get the heat equation for e. Or solve the first for $\partial e/\partial t$ and differentiate with respect to x; differentiate the second with respect to t; then eliminate $\partial^2 e/\partial x\, \partial t$ between equations to get a heat equation for i. In both cases, $k = 1/RC$.

11. Separation of variables leads to $X^{(4)}/X = -T''/c^2 T = \text{const} = +\lambda^4$, $X(0) = 0$, $X''(0) = 0$, $X(L) = 0$, $X''(L) = 0$. The eigenfunctions are $\sin \lambda_n x$, $\lambda_n^4 = (n\pi/L)^4$. Product solutions are $\sin \lambda_n x \cos \lambda_n^2 ct$ and $\sin \lambda_n x \sin \lambda_n^2 ct$.

13. For the beam the ratios are $\lambda_n^2 c/\lambda_1^2 c = n^2$ or $1, 4, 9, \ldots$. For a string they are $\lambda_n c/\lambda_1 c$ with $\lambda_n = n\pi/L$, giving $1, 2, 3, \ldots$. For a closed pipe they are $\lambda_n c/\lambda_1 c$ with $\lambda_n = (2n - 1)\pi/2L$, giving $1, 3, 5, \ldots$.

15. No special continuity condition is required for θ, but boundedness at $r = 0$ is required. Product solutions are $r^{\lambda_n} \sin \lambda_n \theta$, with $\lambda_n^2 = (n\pi/\alpha\pi)^2 = (n/\alpha)^2$, $n = 1, 2, 3, \ldots$. The full solution is

$$u(r, \theta) = \sum_{n=1}^{\infty} b_n (r/L)^{\lambda_n} \sin \lambda_n \theta,$$

$$b_n = 2U_0(1 - \cos n\pi)/n\pi.$$

Appendix A   Section A.1

1. (a) $A = \begin{bmatrix} 1 & \frac{1}{2} & \frac{1}{3} \\ \frac{1}{2} & \frac{1}{3} & \frac{1}{4} \\ \frac{1}{3} & \frac{1}{4} & \frac{1}{5} \end{bmatrix}$ (b) $A = \begin{bmatrix} 0 & 1 & 2 & 3 \\ 1 & 0 & 1 & 2 \\ 2 & 1 & 0 & 1 \\ 3 & 2 & 1 & 0 \end{bmatrix}$

   (c) $A = \begin{bmatrix} 1 & 1 & 1 \\ 1 & 2 & 3 \end{bmatrix}$

3. $A = \begin{bmatrix} 1 & 2 & 3 \\ 3 & 1 & 2 \\ 2 & 3 & 1 \end{bmatrix} = I + 2R + 3R^{\mathsf{T}}$

5. $X(t) = \begin{bmatrix} 1 & 1 \\ 0 & 1 \end{bmatrix} + t\begin{bmatrix} 1 & 0 \\ 1 & -1 \end{bmatrix} + t^2\begin{bmatrix} 0 & -1 \\ 1 & 0 \end{bmatrix}$

7. $X(t) = e^t[\frac{1}{2}, \frac{1}{2}] + e^{-t}[\frac{1}{2}, -\frac{1}{2}]$

9. $2\begin{bmatrix} 1 & 1 & -1 \\ 0 & 3 & -1 \end{bmatrix} - \begin{bmatrix} 1 & -1 & -5 \\ -1 & 2 & -1 \end{bmatrix} = \begin{bmatrix} 1 & 3 & 3 \\ 1 & 4 & -1 \end{bmatrix}$

Section A.2

1. $A = \begin{bmatrix} 1 & 1 & 1 \\ 1 & 2 & 3 \end{bmatrix}$, 2×3; $x = \begin{bmatrix} x_1 \\ x_2 \\ x_3 \end{bmatrix}$, 3×1; $b = \begin{bmatrix} 3 \\ 6 \end{bmatrix}$, 2×1

3. $A = \begin{bmatrix} 1 & 2 & 0 \\ 0 & 1 & 2 \\ 2 & 0 & 1 \end{bmatrix}$, 3×3; $x = \begin{bmatrix} x \\ y \\ z \end{bmatrix}$, 3×1; $b = \begin{bmatrix} 1 \\ 3 \\ 5 \end{bmatrix}$, 3×1

5. $\begin{aligned} x + 2y + z &= 4, \\ x \quad\quad + z &= 2, \\ y \quad\quad &= 1 \end{aligned}$ 7. $\begin{aligned} x_1 + x_2 + x_3 \quad\quad &= 2, \\ -x_1 + x_2 \quad\quad + x_4 &= 1, \\ x_2 + x_3 + x_4 &= 0, \\ x_1 \quad\quad + x_3 + x_4 &= 3 \end{aligned}$

9. The product is $\begin{bmatrix} -18 & 18 & 12 & -12 \\ 18 & -6 & 20 & -8 \\ 6 & 6 & 28 & -16 \end{bmatrix}$.

11. (a) $AB = \begin{bmatrix} -2 & 1 & -2 \\ 0 & 0 & 0 \\ 2 & 2 & 5 \end{bmatrix}$;   (b) $BA = \begin{bmatrix} 2 & 1 \\ 8 & 1 \end{bmatrix}$;   (c) $AC = \begin{bmatrix} 1 \\ 0 \\ 5 \end{bmatrix}$;

   (d) $(2 \times 1) \cdot (3 \times 2)$ not   defined;   (e) $BAC = \begin{bmatrix} 5 \\ 17 \end{bmatrix}$;   (f) $(I + BA)C = \begin{bmatrix} 7 \\ 18 \end{bmatrix}$;   (g) $(3 \times 3) - (2 \times 2)$ not defined;   (h) $(2 \times 1) \cdot (2 \times 3)$ not defined.

13. $\begin{bmatrix} 0 & 0 \\ 0 & 0 \end{bmatrix}$ 15. $[0, 0]$ 17. $\begin{bmatrix} 1 & 0 & 5 \\ 0 & -2 & 5 \\ 0 & 0 & -3 \end{bmatrix}$

19. $\begin{bmatrix} 6 & 3 \\ -5 & 15 \\ 7 & -7 \end{bmatrix}$

21. $x = 1$, $y = 2$

23. $x = 3$, $y = 1$, $z = -2$

25. $\begin{bmatrix} 1 & 2 \\ 2 & 0 \end{bmatrix}\begin{bmatrix} 2 & 1 \\ 0 & 1 \end{bmatrix} = \begin{bmatrix} 2 & 3 \\ 4 & 2 \end{bmatrix}$

27. $U = \begin{bmatrix} 1 & -1 & 1 \\ 0 & 2 & 2 \\ 0 & 0 & 1 \end{bmatrix}$

29. (b) If $D = \text{diag}\ \{d_1, d_2, \ldots\}$, then the ith row of DA is the ith row of A multiplied through by d_i.

Section A.3

1. $x = -2$, $y = 3$

3. $x = (a + b)/2$, $y = (a - b)/2$

5. $x = -2$, $y = 1$, $z = 3$

7. $x = 3$, $y = -1$, $z = 2$

9. $x = 1$, $y = 2$, $z = -1$

11. $x = [\frac{1}{4}, \frac{1}{2}, \frac{3}{4}]^T$

13. $x_1 = x_2 = x_3 = \frac{1}{3}$

15. $x_1 = 3$, $x_2 = 1$, $x_3 = -2$, $x_4 = 1$

17. $x_1 = \frac{1}{6}$, $x_2 = -1$, $x_3 = \frac{3}{2}$, $x_4 = \frac{1}{3}$

19. $x_n = (b_1 + b_2 + \cdots + b_{n-1} - b_n)/(n - 2)$

Section A.4

1. $\frac{1}{2}\begin{bmatrix} 1 & 1 \\ 1 & -1 \end{bmatrix}$

3. $\begin{bmatrix} 10 & -9 \\ -11 & 10 \end{bmatrix}$

5. $\begin{bmatrix} 12 & -8 & -3 \\ -4 & 3 & 1 \\ -3 & 2 & 1 \end{bmatrix}$

7. $\begin{bmatrix} 1 & -4 & 5 \\ 3 & -11 & 13 \\ 0 & 1 & -1 \end{bmatrix}$

9. $\begin{bmatrix} -2 & 1 & 3 \\ 3 & -1 & 2 \\ 3 & -1 & 1 \end{bmatrix}$

11. $\frac{1}{12}\begin{bmatrix} 3 & 3 & 3 & 3 \\ 2 & 2 & -2 & -2 \\ 3 & -3 & -3 & 3 \\ 2 & -4 & 4 & -2 \end{bmatrix}$

13. $\begin{bmatrix} 2 & 3 & 2 & 1 \\ 1.5 & 3 & 2 & 1 \\ 1 & 2 & 2 & 1 \\ 0.5 & 1 & 1 & 1 \end{bmatrix}$

15. $(I + ee^T)^{-1} = I - ee^T/(n + 1)$

17. $A = (I + B)^{-1}(B - I)$ if $I + B$ is invertible

Section A.5

1. a, d, e, f are row-echelon matrices; (b) leading 1 in row 2 is not to the right of the leading 1 in row 1; (c) leading element in row 1 is not a 1; (g) leading 1 in row 3 is not to the right of the leading 1 in row 1; (h) a zero row is above a nonzero row.

3. (a) consistent; (b) 1; (c) $x_1 = 2 - x_3$, $x_2 = 1 + x_3$, $x_3 = x_3$

5. (a) inconsistent

7. (a) consistent; (b) 1; (c) $x_1 = x_1$, $x_2 = 1$, $x_3 = 1$

9. (a) consistent; (b) 2; (c) $x_1 = -x_2 - x_4$, $x_2 = x_2$, $x_3 = 1 + 2x_4$, $x_4 = x_4$

11. (a) consistent; (b) 1; (c) $x_1 = x_3$, $x_2 = -x_3$, $x_3 = x_3$

13. (a) consistent; (b) 1; (c) $x_1 = 1 - 2x_2$, $x_2 = x_2$, $x_3 = 0$, $x_4 = -1$, $x_5 = 0$

15. (a) consistent; (b) 4; (c) $x_1 = 1 + x_2 - x_4$, $x_2 = x_2$, $x_3 = 1 - x_5 - x_6$, $x_4 = x_4$, $x_5 = x_5$, $x_6 = x_6$, $x_7 = 0$

17. $k = 4$; $x_1 = 0$, $x_2 = 1 - x_3$, $x_3 = x_3$

19. Not consistent for any value of k

21. Any $k \neq 2$; $x_1 = 1 + \dfrac{6}{k - 2}$, $x_2 = 2$, $x_3 = \dfrac{-6}{k - 2}$

23. $x_1 = 6 + x_3$, $x_2 = -2 - 2x_3$, $x_3 = x_3$

25. $x_1 = x_1$, $x_2 = 4$, $x_3 = -3$

27. Inconsistent

29. $x_1 = 1 - x_3$, $x_2 = 2 - x_3$, $x_3 = x_3$

31. $x = 0$ is a solution

Section A.6

1. The numbers following each letter are rank A and rank $[A, b]$. (a) 2, 3 (b) 1, 2 (c) 2, 2 (d) 2, 2 (e) 2, 2 (f) 3, 3

3. (c) $x_1 = -1 - x_3$, $x_2 = 1$, $x_3 = x_3$, (d) $x_1 = x_2 = \frac{1}{2}$. (e) $x_1 = 2 - 3x_3$, $x_2 = -x_4 + 3x_3$, $x_3 = x_3$, $x_4 = x_4$, (f) $x_1 = -2$, $x_2 = 3$, $x_3 = 0$

5. $x = x_1[1, 0, 0]^T$ 7. $x = x_3[1, -2, 1]^T$ 9. $x = x_4[1, 1, 1, 1]^T$

11. $x = x_4[1, 1, 1, 1]^T$ 13. $r = 2$; any two except z_1 and z_4

15. $r = 3$; any three   17. $r = 2$; any two except z_1 and z_3

19. $\begin{bmatrix} 1 \\ 0 \\ 1 \end{bmatrix} = -\frac{1}{2}x_1 + \frac{1}{2}x_2 + x_3$

21. $\begin{bmatrix} 0 \\ 1 \\ 1 \\ 0 \end{bmatrix} = \frac{1}{2}x_1 - \frac{1}{2}x_2 + \frac{1}{2}x_3$

Section A.7

1. 0   3. 1/2160   5. -1   7. 5   9. 3

11. $x^2 - 2x - 12$; $x = 1 \pm \sqrt{13}$

13. $(1 - x)(2 - x)(3 - x)$; $x = 1, 2, 3$

15. $-x^3 - 2x$; $x = 0, \pm\sqrt{2}i$ 17. $W = 2t$

19. $W = \det \begin{bmatrix} t & t \ln t \\ 1 & 1 + \ln t \end{bmatrix} = t$

21. $W = \det \begin{bmatrix} \cos(\ln t) & \sin(\ln t) \\ -\dfrac{1}{t}\sin(\ln t) & \dfrac{1}{t}\cos(\ln t) \end{bmatrix} = \dfrac{1}{t}$

23. $W = \det \begin{bmatrix} e^t & \cos t & \sin t \\ e^t & -\sin t & \cos t \\ e^t & -\cos t & -\sin t \end{bmatrix} = 2e^t$

Index